Lectures on
Functional Analysis
and Applications

Lectures on
Functional Analysis
and Applications

V. S. PUGACHEV & I. N. SINITSYN

Russian Academy of Sciences, Moscow

Translated by

I. V. Sinitsyna

World Scientific
Singapore • New Jersey • London • Hong Kong

Published by

World Scientific Publishing Co. Pte. Ltd.

P O Box 128, Farrer Road, Singapore 912805

USA office: Suite 1B, 1060 Main Street, River Edge, NJ 07661

UK office: 57 Shelton Street, Covent Garden, London WC2H 9HE

Library of Congress Cataloging-in-Publication Data
Pugachev, V. S. (Vladimir Semenovich)
    Lectures on functional analysis and applications / V.S.
Pugachev & I.N. Sinitsyn.
        p.   cm.
    Includes bibliographical references and index.
    ISBN 9810237227   ISBN 9810237235 (pbk)
    1. Functional analysis.   I. Sinitsyn, I. N. (Igor' Nikolaevich)
II. Title.   III. Title: Functional analysis and applications
    QA320 .P844   1999
    515'.7--dc21   99-26757
                                                        CIP

British Library Cataloguing-in-Publication Data
A catalogue record for this book is available from the British Library.

Printed in Singapore.

PREFACE

The book is written on the basis of lectures which the authors presented during many years to the undergraduates and researchers of Russian Technical Universities, to engineers and other specialists working in various areas of applications of mathematics. So, contrary to known books on functional analysis in the world literature, this book is addressed not to mathematicians–professionals but to large circles of specialists in various branches of knowledge having only a moderate mathematical background who need to increase their mathematical knowledge for successful work in their professional areas and reading the respective special literature. It may be useful also for mathematicians who deal with applications of functional analysis.

The experience of the authors of many years working in various fields of applications of mathematics enabled them to outline in a single book practically all the material on functional analysis necessary for applications. The main destination of the book determined the method and style of presentation of the material. The book is written simply and is easy for understanding. To master the material outlined in the book the reader must have only the usual knowledge of linear algebra, mathematical analysis and differential equations in the frames of the usual mathematical education of engineers.

In accordance with the destination of the book it contains the general foundations of the functional analysis necessary for those working in applications without mathematical details. In particular, some questions of the theory of functions of a real variable which are traditional in the courses of functional analysis are left apart from this book as well as other questions interesting only for professional mathematicans.

Owing to the destination of the book the choice of material involved, which includes practically all the information on functional analysis that may be necessary for those working in various areas of applications of mathematics, simplicity and availability of exposition the book has no analogues in the world literature.

In Chapter 1 the subject of functional analysis is determined and the general notions of a function and a space are given. After a brief survey of the general notions of the sets theory and studying the properties of images and inverse images the choice axiom is formulated and the equivalent theorems of Zermelo and Hausdorff and Zorn lemma are proved. The general propositions of the theory of metric spaces are given: the notions of a metric, convergence of sequences, continuity of functions,

completeness and separability of metric spaces. Then the general theory
of linear (vector) spaces is exposed, the notions of a norm and a scalar
product are introduced and their properties are studied. Normed and
Euclidean spaces are determined. Then Banach and Hilbert spaces are
introduced (B–spaces and H–spaces). In Section 1.4 the general definition of a linear function is given and linear functionals are studied
in detail including the existence and the extension of linear functionals.
General extension Hahn – Banach theorem is proved.

Chapter 2 is devoted to the elements of measure theory. Classes
of sets are determined: semi-algebras, algebras, sigma-algebras
(σ-algebras) monotone classes. Their general properties are studied.
The general definitions of a set function and a measure are given.
The measure is defined here as an additive or a countably additive
(σ-additive) set function with values in a normed space. General properties of such measures are studied. Numerical measures, in particular, nonnegative measures are studied as special cases. This enables one
to cover by the general notion of a measure operator-valued and function space-valued measures encountered in the sequel. After studying
the general properties of measures, in particular, nonnegative ones, the
general theorem is proved on extension of a numerical measure.

Chapter 3 contains the theory of integrals. The notion of a measurable functions is introduced and general properties of measurable functions are studied. The main modes of convergence of sequences of functions are considered, i.e. almost everywhere convergence, convergence
in measure, almost uniform convergence. The general definition of an
integral of a function with values in a separable B-space (a Bochner
integral) is given and its general properties are studied. The abstract
Lebesgue integral, the Lebesgue integral by the Lebesgue measure and
the Lebesgue – Stieltjes integral are considered as special cases. The relation between the Lebesgue integral by the Lebesgue measure and the
Riemann integral as well as between the Lebesgue – Stieltjes integral and
the Riemann – Stieltjes integral is established. The Riemann integral of
a function with values in a separable B-space is defined and its relation
to the Bochner integral is established. General theorems on passing to
the limit under the integral sign are proved. The notions of absolute continuity and singularity of measures are introduced. The definition of the
Radon – Nikodym derivative is given. Then the definition of Lebesgue
spaces is given which constitute one of the most important classes of
spaces. After that Sobolev spaces are defined. In the last Section 3.8
some methods of constructing and studying measures in product spaces

are outlined and the theory of multiple and iterated integrals is given.

Chapter 4 contains the elements of topology and theory of topological spaces. General definitions are given of a topology and topological space, of basis and subbasis of a topology. Separability and countability axioms are introduced. The general method for assigning a topology in a space is considered. The notions of compact sets and spaces are defined and their general properties are studied. Theorems on compact sets in metric spaces are proved. Then the notion of a topological linear space is introduced and the general method for assigning a topology in a linear space is given. The notion of a weak topology in a topological linear space is introduced and its relation to the strong topology is considered.

Chapter 5 is devoted to the spaces of operators and the spaces of functionals. The notions of spaces of linear operators and spaces of bounded linear operators are introduced. The topologies in the space of bounded linear operators are defined. The definition is given of the dual space for a given topological linear space and the properties of dual spaces are studied. Section 5.2 contains the theory of weak integrals (Pettis integrals) which play an important role in probability theory. Section 5.3 is devoted to the theory of generalized functions. The notions of the space of test functions and the space of generalized functions are introduced, operations on generalized functions are defined and their main properties important for applications are studied. In the last Section 5.4 dual spaces are studied for the main types of functions spaces, i.e. for the space of bounded functions, space of continuous functions, spaces of differentiable functions and Lebesgue spaces.

Chapter 6 contains the general theory of linear operators. The definition of closed operators is given and their main properties are studied. The definition of an adjoint operator is introduced and the conditions for its existence are established. The main theorems concerning bounded operators are proved. The definitions of isometric and unitary operators are given and the properties of isometric, unitary and unitarly equivalent operators are studied. Section 6.2 stands in somewhat apart from the basic topic of Chapter 6, namely from the theory of linear operators. In Section 6.2 operator equations are considered linear and nonlinear as well. The method of contraction images is presented as one of the main methods for proving the existence and uniqueness of solutions of various equations. This method is necessary, in particular, for further exposition of the theory of linear operators. The last Section 6.3 is devoted to the notion of spectrum of an operator. After introducing the main definitions concerning the spectra of operators, the operator equations connected with the notion of spectrum are considered. The general properties of the spectra of linear operators are studied, in particular, the spectra of unitary operators are studied.

Chapter 7 contains the theory of linear operators in H-spaces. The definitions of orthogonal subspaces, their orthogonal sums and orthogonal complements are given. The problem of finding the projection of a vector on a subspace which does not contain this vector is solved. The general form of a continuous linear functional on a H-space is established. Bounded and unbounded linear operators and unitary operators are studied. As a special case of an unitary operator Fourier – Plancherel operator is considered. Symmetric and self-adjoint operators and their spectra are studied. The general theory of orthogonal projection operators (orthoprojectors) is outlined. Orthogonal, orthonormal, biorthogonal and biorthonormal systems of vectors are studied. The notion of a basis is introduced and the condition of existence of a basis in a H-space is established. Then the expansions of any vectors in terms of a basis are derived. As special cases the expansions of functions are studied in terms of systems of orthonormal functions forming bases in the respective function spaces representing H-spaces. In the last Section 7.6 the general theory of compact linear operators is outlined and spectra of compact operators on H-spaces are studied. Hilbert – Schmidt operators and trace type operators are considered as special cases. Fredholm linear integral equations are also considered and the general theorems of the theory of such equations are formulated following from the general theorems of the theory of compact linear operators.

Chapter 8 contains the spectral theory of linear operators in H-spaces. The theorem of existence of eigenvalues is proved for self-adjoint compact operators. The spectral decomposition of such an operator is derived. The application of this theory is given to Fredholm integral equations with symmetric kernels. The method for solving integral equations of a sufficiently large class is outlined. The definition of the operator-valued measure is given whose values represent orthoprojectors. Such a measure represents the decomposition of the identity. Then the theory of integrals of numerical functions by operator-valued measures is outlined and the properties of operators are studied determined by the decomposition of the identity. Also the properties of spectra of such operators are studied. A wide class of functions of an unitary and self-adjoint operators is considered. This theory implies the expression for the decomposition of the identity, i.e. the spectral measure of a given self-adjoint operator. Special Section 8.3 is devoted to the functions of operators. In Section 8.4 using this spectral measure the representation of the self-adjoint operator in the form of the integral by its spectral measure is derived (the spectral decomposition of a self-adjoint operator). Hence the spectral decompositions of an unitary operator, a group of unitary operators and a normal operator are derived.

Chapter 9 is devoted to the theory of nonlinear operators and functionals. The strong differential and derivative (Frechét differential and derivative) and weak differential and derivative (Gâteaux differential and derivative) of an operator or a functional are defined. The relations between these two kinds of differentials and derivatives are studied. The strong and weak differentials and derivatives of higher orders are defined. The rule for the differentiation of a composite function, the finite increments formula, the Taylor formula and Taylor series are derived for operators and functionals. The general necessary and sufficient conditions of the extremum of a functional are established. In the last Section 9.3 the elements of the general theory of linear and nonlinear differential equations in B-spaces are outlined.

In the last Chapter 10 the elements of the basic approximate methods of numerical analysis in abstract spaces are exposed. One of the most efficient methods for approximate solving operator equations, namely Newton method is presented as well as modified Newton method. Some basic methods of numerical analysis in abstract spaces, namely Rayleigh – Ritz method, Galerkin methods and the finite elements method are outlined. Special attention is paid to the numerical solution of real time problems. The last Section is devoted to the methods for solving improper problems based on Tychonoff regularization method.

Appendices contain auxiliary materials (tables of basic characteristics of some typical linear control systems with the lumped and distributed parameters, tables of operators of two–dimensional signal and image processing systems, some definite integrals and special functions, tables of the continuous and discrete Laplace and the Fourier, Cosine and Sine transforms) necessary for solving examples and problems.

The list of references contains only the books which enable a reader to learn more profoundly the parts of the course in which he is interested and get additional information about applications considered in examples given in the book. The authors do not pretend in any way to provide a complete list of literature references. Only those sources are given in the list which are cited in the text. The Harvard system of references is used in the book.

To facilitate the mastering of the theory presented and to provide the successful work of a reader without assistance about 300 examples and over 500 problems are given in the book which illustrate the theory outlined and its main applications in applied mathematics, probability theory and statistics, mathematical physics and mechanics, control systems analysis and design, signal and image processing. Best Russian books of problems on functional analysis are used as sources for examples and problems collection. Some problems are equipped with instructions varying from a hint to the detailed instructions containing the sketch of the solution. Several especially difficult problems are given with complete solutions. Some of the problems contain much additional information about possible generalizations and development of the methods presented in the book.

Every subsection contains short bibliographical remarks and recommendations for more detailed study. For information about the notions and theorems from various parts of mathematics used in the book we advise (Korn and Korn 1968). For recalling linear algebra a reader may use (Lancaster 1969, Noble and Daniel 1977, Wilkinson 1965). For recalling mathematical analysis refer to (Burkill and Burkill 1970, Il'in and Posnajk 1982, Rudin 1964).

Every chapter of the book has its own numeration of theorems, formulae and examples. Problems are given practically in every section and are enumerated within the section. The formulations of all basic theorems and statements are given in italics. The beginning and the end of the evaluations, proofs and discussions which lead to certain results are indicated by triangular indices ▷ and ◁.

The authors are very much obliged to Mrs I.V.Sinitsyna for her excellent translation of the textbook into English. The authors also owe thanks to V.A.Il'in, M.U.Khafizov, G.I.Marchuk, E.I.Moiseev, O.S.Ogneva, A.R.Pankov, V.A.Sadovnichii, A.V.Safonov, A.L.Skybachevskii for their valuable remarks, to A.P.Nosov, V.I.Shin, V.I.Sinitsyn for assistance and E.N.Fedotova and I.V.Makarenkova for typesetting the camera–ready manuscript.

CONTENTS

CHAPTER 1
SETS. SPACES. FUNCTIONS

In this Chapter the subject of functional analysis is determined and the general notions of a function and a space are given. In Section 1.1 after a brief survey of the general notions of the sets theory and studying the properties of images and inverse images the choice axiom is formulated and the equivalent theorems of Zermelo and Hausdorff and Zorn lemma are proved. The general propositions of the theory of metric spaces: the notions of a metric, convergence of sequences, continuity of functions, completeness and separability of metric spaces are given in Section 1.2. Then in Section 1.3 the general theory of linear (vector) spaces is exposed, the notions of a norm and a scalar product are introduced and their properties are studied. Normed and Euclidean spaces are determined. Then Banach and Hilbert spaces (B-spaces and H-spaces) are introduced. In Section 1.4 the general definition of a linear function is given. The linear functionals are studied in detail including existence and extension of linear functionals. General extension Hahn–Banach theorem is proved.

1.1. Functions and Mappings

1.1.1. What is Functional Analysis

Functional analysis represents one of the most important branches of mathematical sciences. Together with abstract algebra and mathematical logics it serves as a foundation of many other branches of mathematics. Functional analysis is, in particular, widely used in probability theory and random functions theory and their numerous applications. Functional analysis serves also as a powerful tool in modern control and information sciences.

The main subject of mathematical analysis represent scalar and finite–dimensional vector functions of scalar or finite–dimensional vector variables. Functional analysis is studying more general functions whose arguments and values may be the elements of any sets. While studying functions in mathematical analysis and linear algebra geometrical presentations are widely used; a function is considered as the mapping of one finite–dimensional space into another finite–dimensional space. For instance, the scalar function of one scalar variable represents the mapping of the real axis R into the real axis R. The scalar function of two (three) scalar variables represents the mapping of the plane R^2 (the three–dimensional space R^3 respectively) into R. While studying more general functions whose arguments and values may be the elements of any sets wonderful analogies appear between many properties of functions and the visual geometric properties of more simple functions. You meet such analogies in linear algebra where the spaces of any finite dimensions are considered (the n–dimensional spaces R^n at any finite n). In particular, the properties of linear functions in R^n are absolutely identical with the properties of linear functions in one–, two– and three–dimensional spaces. These properties of functions caused the generalization of the notion of a space and wide application of intuitive geometrical presentations and geometrical terminology while studying any functions.

Functional analysis was born in the works of Italian mathematician *Vito Volterra* (Volterra 1913, Volterra and Pérés 1935). He was the first who considered functions as the points of some space. The spaces whose points are functions are called *function spaces*.

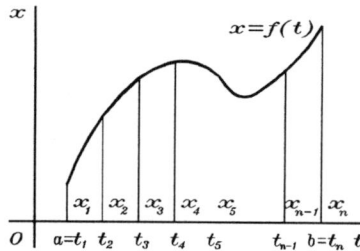

Fig. 1.1

E x a m p l e 1.1. Let us consider the real function $y = x(t)$ on the interval $[a, b]$. We partition the interval $[a, b]$ into $n - 1$ equal parts and denote the points of division $t_1 = a, t_2, \ldots, t_{n-1}, t_n = b$, $t_k = a + (k - 1)(b - a)/(n - 1)$ and the values of $x(t)$ at these points by $x_1, \ldots, x_n$, $x_k = x(t_k)$ respectively (Fig.1.1). These n values of the function $x(t)$ form the n–dimensional vector. Thus any such function may be approximately represented by the n–dimensional vector, i.e. by the point of the n–dimensional space R^n. It is clear that such representation of a function is realized by the tabular assignment of a function: any table assigns a function by its values at a finite number of points. In the case of a continuous function its representation as the n–dimensional vector may determine it with any degree of accuracy at sufficiently large n. The larger is n the higher is the accuracy of such a representation. In the limit as n indefinitely increases the function $x(t)$ will be exactly represented by the vector (the point) of the infinite dimensional space.

Volterra defined also a real function whose argument represents the set of all the values of a continuous function in the interval $[a, b]$. Such a function he called a *functional*. This was the reason to call the branch of mathematics studying functionals a *functional analysis*.

It is worth while to recall that long before Volterra some functionals where considered by great Euler who created calculus of variations, though he did not use the term "functional".

Primarily functional served as the main object of study in functional analysis. In further development the notion of a function was essentially generalized. Respectively the range of interests of functional analysis was considerably extended. So, the object of functional analysis represents now the study of functions whose arguments and values may be the elements of any sets which are usually called *spaces*.

1.1.2. Sets

We often meet sets in mathematics and its applications.

E x a m p l e 1.2. Examples of sets are: the set of all real numbers (points of the real axis $R^1 = R$), the set of all rational numbers, the set of all the points of the Euclidean plane R^2, the set of points of some region in the plane or space, the set of functions of some variable possessing an assigned property, the set of all plane curves, the set of all possible input signals to a control system etc.

The set may consist, in particular, of a single element. It is expedient also to consider the *empty set* for generality, containing no elements. The empty set is denoted $\emptyset$.

The infinite set is called a *denumerable set* if all its elements may be put in one–to–one correspondence with natural numbers so as a single natural number corresponds to each element of the set and vice versa, a single element of the set corresponds to each natural number.

A set is called *countable* if it is either a finite set or a denumerable set. In other words the countable set is the set whose elements may be enumerated. The set whose elements cannot be enumerated is called an *uncountable set*.

E x a m p l e 1.3. As examples of a countable set may serve: the set of all the integers, the set of all the rational numbers, the set of all the points of a plane or of a finite–dimensional space with rational coordinates.

1.1.3. Spaces

Considering the set of *all* the elements of a set possessing some common property one usually calls this set a *space*. For instance, the set of all real numbers (the real axis R) is usually called an *one–dimensional space*; the set of all the points of a plane R^2 is called a *two–dimensional space*; the set of all the points of the usual space R^3 studied in elementary geometry is called a *three–dimensional space*. Generalizing these definitions the set of all ordered sequences of n real numbers is called an *n–dimensional space* R^n. The set of all continuous scalar functions is called a *space of continuous functions*.

Nevertheless it is expedient in some cases to consider a given space as a *subspace* of some "wider" space. For instance, the two–dimensional space R^2 (the plane) may be considered as a subspace of the three–dimensional space R^3. The space of all continuous scalar functions in some region may be considered as a subspace of the space of all the scalar functions in this region.

We shall denote spaces preferably by capital letters from the end of latin alphabet and their elements by respective small letters. For instance, denoting spaces by X, Y, Z, U, ... we shall denote their elements by x, y, z, u, ... The sets of points of spaces we shall denote by capital letters from the beginning of Latin alphabet, for instance, A, B, C, D, ... If the element (the point) x belongs to the set A we write $x \in A$; if x does not belong to A we write $x \notin A$ or $x \bar{\in} A$. If all the elements of the set B belong also to the set A, we call B a *subset* of the set A and write $B \subset A$. Obviously A and B coincide if $A \subset B$, $B \subset A$ what is written in the form of equality $A = B$. It is also clear that $B \subset A$, $C \subset B$ imply $C \subset A$.

1.1.4. Sets Operations

The set of all the elements of a space X not belonging to the set A is called the *complement of the set A* and is denoted $\bar{A}$ in the sequel. It is evident that the empty set $\emptyset$ is the complement of the whole space X and vice versa, and A is the complement of $\bar{A}$.

The *union* $A \bigcup B$ of sets A and B is the smallest set of elements each of which belongs to one of the sets A, B (Fig.1.2).

The *intersection* $A \bigcap B = AB$ of sets A and B is the largest set of elements belonging to both sets A and B (Fig.1.3).

Let $\{A_\alpha\}$ be any set of sets either countable or uncountable.

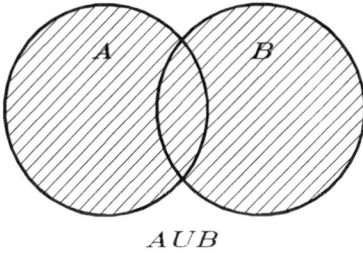

$A \cup B$

Fig. 1.2

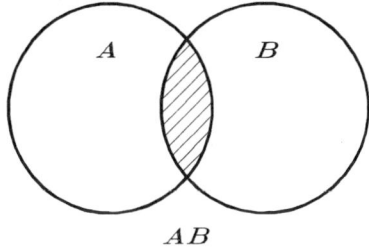

AB

Fig. 1.3

The *union* $\bigcup_{\alpha} A_\alpha$ *of the sets* A_α is the set whose each element belongs to at least one of the sets A_α.

The *intersection* $\bigcap_{\alpha} A_\alpha$ *of the sets* A_α is the set whose each element belongs to all the sets A_α.

The operations of union and intersection of sets possess the following obvious properties:

(i) $A \bigcup B = B \bigcup A$, $AB = BA$ (commutativity),

(ii) $A \bigcup (B \bigcup C) = (A \bigcup B) \bigcup C$, $A(BC) = (AB)C$ (associativity),

(iii) $(A \bigcup B)C = AC \bigcup BC$ (distributivity),

(iv) $(AB) \bigcup C = (A \bigcup C) \bigcap (B \bigcup C)$ (distributivity).

The *difference* $A \backslash B$ *of sets* A and B is the set of all the elements of A which do not belong to B (Fig.1.4). Obviously $A \backslash B = A\bar{B}$ and $X \backslash A = X\bar{A} = \bar{A}$.

In some cases it is expedient to consider the *symmetric difference* of sets A and B defined by $A \triangle B = B \triangle A = (A \backslash B) \bigcup (B \backslash A)$ (Fig.1.5).

From the above definitions directly follow the following relations:

$$\overline{A \bigcup B} = \bar{A} \bigcap \bar{B}, \quad \overline{A \bigcap B} = \bar{A} \bigcup \bar{B}, \tag{1.1}$$

$$\overline{\bigcup_{\alpha} A_\alpha} = \bigcap_{\alpha} \bar{A}_\alpha, \quad \overline{\bigcap_{\alpha} A_\alpha} = \bigcup_{\alpha} \bar{A}_\alpha. \tag{1.2}$$

These formulae represent the so called *duality principle* playing an important role in sets theory. By this principle from any relation follows the dual relation obtained by replacing all the sets by their complements, unions by intersections and intersections by unions.

Let $\{A_\alpha\}$ be some set of subsets of some set C. We may consider C as a space. Then the complements of sets A_α in C will be equal to the differences $C \backslash A_\alpha$ and duality principle (1.2) will imply

$$C \backslash \bigcup_{\alpha} A_\alpha = \bigcap_{\alpha} (C \backslash A_\alpha), \quad C \backslash \bigcap_{\alpha} A_\alpha = \bigcup_{\alpha} (C \backslash A_\alpha). \tag{1.3}$$

These relations will often be used in the sequel.

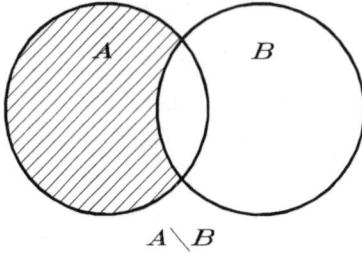

$A \setminus B$

Fig. 1.4

$A \triangle B$

Fig. 1.5

It is easy to see that *the union of any countable set of countable sets is also a countable set*. Really, denote the elements of the set A_1 by a_{11}, a_{12}, a_{13}, ..., the elements of the set A_2 not belonging to A_1 by a_{21}, a_{22}, a_{23}, ... and so on, the elements of A_n not belonging to $\bigcup\limits_{k=1}^{n-1} A_k$ by a_{n1}, a_{n2}, a_{n3}, ... and dispose them in the form of the Table 1.1

Table 1.1

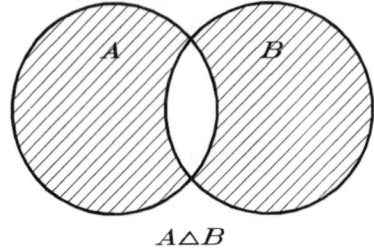

Then we may enumerate them in the order shown by arrows: $x_1 = a_{11}$, $x_2 = a_{21}$, $x_3 = a_{12}$, $x_4 = a_{31}$, $x_5 = a_{22}$, $x_6 = a_{13}$, $x_7 = a_{41}$, $x_8 = a_{32}$, $x_9 = a_{23}$, ...

From this general proposition follows that the set of all rational numbers is countable. Really, the set of all the fractions with the denominator n is countable and the set of all rational numbers represents the union of the sets of fractions with the denominators $1, 2, 3, \ldots, n, \ldots$.

The set of all real numbers (points of the real axis R) is uncountable. It is sufficient to prove the uncountability of the set of points of the interval $[0,1]$. Assuming that this set is countable enumerate all the numbers of the interval $[0,1]$. As a result we obtain a numerical sequence $\{a_n\}$. Represent each number of this sequence as an infinite decimal fraction

$$a_n = 0, a_n^{(1)} a_n^{(2)} \ldots a_n^{(k)} \ldots$$

Determine the number

$$a = 0, a^{(1)} a^{(2)} \ldots a^{(k)} \ldots$$

in such a way that the first its decimal digit $a^{(1)}$ does not coincide with that of the number a_1, $a^{(1)} \neq a_1^{(1)}$, the second one $a^{(2)}$ does not coincide with that of the number a_2, $a^{(2)} \neq a_2^{(2)}$, and so on, the k-th decimal digit $a^{(k)}$ does not coincide with that of the number a_k, $a^{(k)} \neq a_k^{(k)}$. It is clear that the number a cannot coincide with any of the numbers a_n (it differs from every a_n by the n-th decimal digit). Consequently, no countable set can coincide with the set of all real numbers.

1.1.5. General Definition of a Function

If some points of a space X are put in correspondence with some points of another space Y in such a way that one and only one point $y \in Y$ corresponds to a given point x of a certain subset of X then this correspondence is called a *function* and is denoted $y = f(x)$ exactly in the same way as in elementary mathematical analysis[a] . One says in this case that the function $y = f(x)$ acts from X into Y.

The set of points of the space X at which the function $y = f(x)$ is defined is called the *domain of the function* f and is denoted D_f. The set of points of the space Y which corresponds to the points $x \in D_f$ is called the *range of the function* f and is denoted R_f.

Any function defines the mapping of D_f on R_f. If $R_f = Y$ we say that the function $y = f(x)$ is mapping X *onto* Y. If $R_f \neq Y$, $R_f \subset Y$ we say that the function $y = f(x)$ is mapping X *into* Y. Accordingly a function is also called a *mapping* what is written in the form $f : X \to Y$. This record is equivalent to $y = f(x)$. If the function is the one–to–one

---

[a] This definition means that only univalent functions are considered in functional analysis. If a countable set of points $y \in Y$ corresponds to a single point $x \in X$ then the function is multivalent. Each branch of this multivalent function is considered as a function in functional analysis. For instance, $y = \sqrt{x}$ and $y = -\sqrt{x}$ are considered as two different functions.

mapping of D_f onto R_f then there exists the inverse function $x = f^{-1}(y)$ with the domain R_f and the range D_f.

The function mapping a whole space X onto a space Y is called a *surjective mapping* or a *surjection*. The function representing the one–to–one mapping of a whole space X into a space Y is called an *injective mapping* or an *injection*. The injection represents thus the function mapping a whole space X into Y which has the inverse function. The function representing the one–to–one mapping of a whole space X onto Y is called a *bijective mapping* or a *bijection*.

The function with values on the real axis R or on the complex plane C (in the *field of scalars*) is called a *functional*. Functionals of the variable x are often denoted by small letters fx, gx, ... without enclosing the argument x in brackets.

If the space Y is not the real axis R or the complex plane C then the function $f(x)$ is called an *operator*. Operators of the variable x are often denoted by capital letters Ax, Tx, ... without enclosing the argument x in brackets.

We shall now introduce the notation which will be useful in the sequel. Namely, we denote the set of values of a variable x satisfying the condition P by $\{x : P\}$.

The set of ordered pairs $\{\{x, y\} : x \in D_f, y = f(x) \in R_f\}$ is called a *plot of the function* $y = f(x)$ and is denoted $\mathrm{Gr}(f)$.

The function equal to unity at all the points x of a set A and zero outside A is called the *indicator of the set* A and denoted $\mathbf{1}_A(x)$:

$$\mathbf{1}_A(x) = \begin{cases} 1 & \text{at } x \in A, \\ 0 & \text{at } x \notin A. \end{cases} \tag{1.4}$$

E x a m p l e 1.4. The area y of the region of a plane under the plot of the scalar function of a scalar variable $x = x(t) > 0$ on the interval $[a, b]$, $y = f(x)$ (Fig.1.6) presents a functional. The domain D_f of this functional is the set of all the curves for which this area is defined. The range R_f represents the set of all positive numbers, $R_f = (0, \infty)$.

E x a m p l e 1.5. Under the conditions of Example 1.4 the area y of the region of a plane under the plot of the scalar variable $x = x(t) > 0$ on the interval $[a, s]$, s being a variable, $s \in [b, c]$, $b > a$ presents an operator. The domain. D_f of this operator is the set of all curves for which this area is defined. The range R_f is the set of all positive numbers $R_f = (0, \infty)$.

E x a m p l e 1.6. The area $y = f(x)$ of the plane region bounded by a closed loop (Fig 1.7) presents a functional. The domain D_f of this functional is the set of all closed loops for which the areas bounded by them are defined. The range R_f is the set of all positive numbers, $R_f = (0, \infty)$.

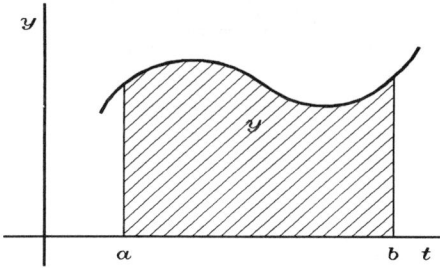 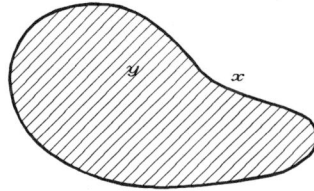

Fig. 1.6   Fig. 1.7

E x a m p l e 1.7. Some feature y of a plane curve x, for instance, the number of self–intersections, crossings of a level, extreme values (or generally any feature y of some image x) $y = f(x)$ presents a functional. In this case the domain D_f is the set of all the curves for which the notions of self–intersection, crossing of a level, extreme values are defined and the range R_f is the set of all nonnegative integers.

E x a m p l e 1.8. Consider one–dimensional (Fig.1.8) or multi–dimensional (Fig.1.9) control system with lumped parameters (D'Azzo and Houpis 1975, Zadeh and Desoer 1963). The relation between the input signal (input x) and the output signal (response or output y) determines the operator. In this case the domain D_f is the set of all inputs and the range R_f is the set of all outputs. Basic characteristics of some linear control systems with lumped parameters determined by the ordinary linear differential equations are given in Appendix 1 (Table A.1.1).

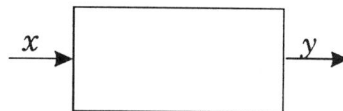

Fig. 1.8

E x a m p l e 1.9. Let us consider control system (Fig.1.8 and Fig.1.9) whose input signals are continuosly distributed along some line on the surface or in the space. Such systems may be considered as the systems with the continuous set of inputs (Butkovskiy 1983, Curtain and Pritchard 1978). Analogously the systems with the continuous outputs are considered. By the definition the systems whose inputs

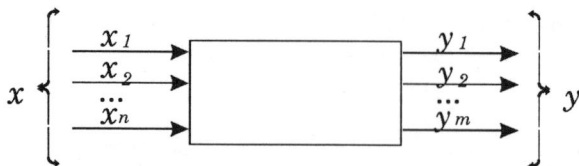

Fig. 1.9

and outputs are continuously distributed in some regions are called control systems with *distributed parameters*.

It should be noticed that input signals are introduced into the system almost in finite number of definite points and the outputs are also measured at finite number of the points. Therefore at practice a system with distributed parameters may be considered as a system with a finite number of the inputs and the outputs (Fig.1.9).

Basic characteristics of some linear systems with distributed parameters defined by linear equations in partial derivatives are given in Appendix 1 (Table A.1.2).

E x a m p l e 1.10. Consider at first an optical system relation between two–dimensional continuous image $x(\xi, \eta, t)$ and the light source spectral sensitivity $s(\lambda)$ given by formula (Pratt 1978)

$$x(\xi, \eta, t) = \int\limits_0^\infty \mathcal{C}(\xi, \eta, t, \lambda) s(\lambda) d\lambda .$$

Here ξ, η are coordinates of space point, t is time variable, the kernel $\mathcal{C} = \mathcal{C}(\xi, \eta, t, \lambda)$ is called distribution function of the light source. For the single colour optical system the signal x characterizes the distribution of the image brightness. In two–dimensional optical processing systems the input image $x = x(\xi, \eta, t)$ and the output image $y = y(\xi, \eta, t)$ are usually determined by some space–time operator A, $y = Ax$ (Appendix 1, Table A.1.3). For this system the domain D_f is the set of all input images and the range R_f is the set of all output images.

1.1.6. Images and Inverse Images of Sets. Inverse Mappings

The value of a function $f(x)$ at a point $x \in X$ is called the *image of the point x*.

The set of all the points $x \in X$ to which corresponds the same value y of the function $f(x)$ is called the *inverse image of the point y* and is denoted $f^{-1}(y)$, $f^{-1}(y) = \{x : f(x) = y\}$.

E x a m p l e 1.11. The set of points $\{x_1, x_2, x_3\}$ (Fig.1.10) represents the inverse image of the point y while y serves as the image of each of three points x_1, x_2, x_3.

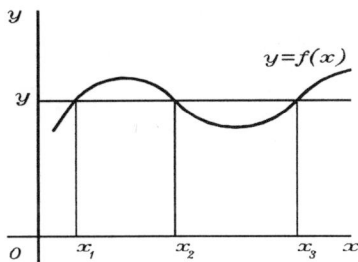

Fig. 1.10

The notion of the inverse image should not be confused with the notion of the inverse function. The inverse function $f^{-1}(y)$ of a function $f(x)$ exists if $f(x)$ represents the one–to–one mapping, whereas the inverse image of a function exists always and is denoted also by $f^{-1}(y)$. The reason for such a generalization of the notation $f^{-1}(y)$ is that in the case of an one–to–one mapping $f(x)$ the inverse image coincides with the inverse function.

If the function $y = f(x)$ maps a set A onto the set B then B is called the *image of the set A* and is denoted $B = f(A) = \{y : y = f(x), x \in A\}$. The set A of all the points x for which $y = f(x) \in B$ is called the *inverse image of the set B* and is denoted $A = f^{-1}(B) = \{x : f(x) \in B\}$. The function $f^{-1}(B)$ putting in correspondence to each set $B \subset Y$ its inverse image is called an *inverse mapping*.

In exactly the same way the images and inverse images of the classes of sets are defined. The set of images $\mathcal{B}$ of all the sets of a class of sets $\mathcal{A}$ is called the *image of the class of sets $\mathcal{A}$* and is denoted $\mathcal{B} = f(\mathcal{A}) = \{B : B = f(A), A \in \mathcal{A}\}$. The set $\mathcal{A}$ of inverse images of all the sets of a class of sets $\mathcal{B}$ is called the *inverse image of the class of sets $\mathcal{B}$* and is denoted $\mathcal{A} = f^{-1}(\mathcal{B}) = \{A : A = = f^{-1}(B), B \in \mathcal{B}\}$.

E x a m p l e 1.12. In the case of the function whose plot is presented on Fig.1.11 the interval (c, d) serves as the image of the interval (a, b) and the interval (a, b) represents the inverse image of the interval (c, d). In the case of the function whose plot is presented on Fig.1.12 the interval B represents the image of each of the intervals A_1, A_2, A_3 and any their unions, $B = f(A_1) = f(A_2) = f(A_3) = f(A_1 \bigcup A_2) = f(A_1 \bigcup A_3) = f(A_2 \bigcup A_3) = f(A_1 \bigcup A_2 \bigcup A_3)$, whereas the union $A_1 \bigcup A_2 \bigcup A_3$ represents the inverse image of the interval B. The last example shows that if B is the image of A then A may be not the inverse image of B but only a subset of the set of all the points at which $f(x) \in B$.

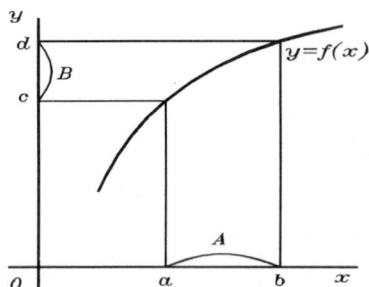

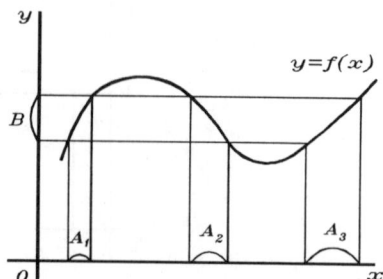

Fig. 1.11   Fig. 1.12

1.1.7. Properties of Inverse Mappings

Now we shall study the properties of inverse mappings.

1. If $B_1 \subset B_2$ then $f^{-1}(B_1) \subset f^{-1}(B_2)$. Really, if $y \in B_1$ then $y \in B_2$ and therefore $f^{-1}(B_1) \subset f^{-1}(B_2)$.

2. The inverse image of the union of sets is the union of the inverse images of these sets:

$$f^{-1}\left(\bigcup_\alpha B_\alpha\right) = \bigcup_\alpha f^{-1}(B_\alpha). \qquad (1.5)$$

Really, if $y \in B_\alpha$ at some α then $f^{-1}(y) \subset f^{-1}(B_\alpha) \subset \bigcup_\alpha f^{-1}(B_\alpha)$.

3. The inverse image of the intersection of sets is the intersection of the inverse images of these sets:

$$f^{-1}\left(\bigcap_\alpha B_\alpha\right) = \bigcap_\alpha f^{-1}(B_\alpha). \qquad (1.6)$$

Really, if $y \in B_\alpha$ at all α then $f^{-1}(y) \subset f^{-1}(B_\alpha)$ at all α and therefore $f^{-1}(y) \subset \bigcap_\alpha f^{-1}(B_\alpha)$.

4. The inverse image of the complement $\bar{B}$ of a set B represents the complement of its inverse image in the domain D_f of the function f:

$$f^{-1}(\bar{B}) = D_f \backslash f^{-1}(B) = \overline{f^{-1}(B)} D_f. \qquad (1.7)$$

5. If, in particular, the domain of the function $f(x)$ represents the whole space X, $D_f = X$ then the inverse image of the complement $\bar{B}$ of the set B represents the complement of its inverse image:

$$f^{-1}(\bar{B}) = \overline{f^{-1}(B)}. \tag{1.8}$$

6. The inverse image of the empty set $\varnothing$ is the complement of the domain of the function f, $f^{-1}(\varnothing) = \bar{D}_f = X \backslash D_f$. If the function f is defined on the whole space X, $D_f = X$, then the inverse image of the empty set represents the empty set: $f^{-1}(\varnothing) = \varnothing$.

The results obtained may be formulated in the form of a theorem.

Theorem 1.1. *Inverse mappings leave invariant all the relations among sets.*

E x a m p l e 1.13. It is easily seen that straight–forward mappings do not possess this property. To make this sure it is sufficient to consider the function $f(x)$ whose plot is presented on Fig.1.13. The sets A_1 and A_2 are nonoverlapping while their images have the nonempty intersection. The inverse image of the intersection of the sets $B_1 = f(A_1)$ and $B_2 = f(A_2)$ is the interval $A_3 \neq A_1 A_2$.

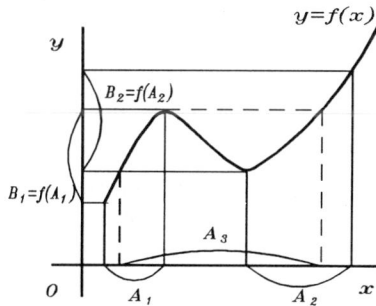

Fig. 1.13

1.1.8. Relation of Order

Any set of pairs $\{a, b\}$, a being the element of a set A and b the element of a set B is called a *relation*, more precisely a *binary relation*.

The relation is generally written as aRb what means that the pair $\{a, b\}$ belongs to the relation R. The set B may in particular coincide with the set A. In this case R represents a *relation on the set* A. The examples of relations are the equality $a = b$ (the set of pairs of a set A whose second elements coincide with the first ones), the inequality $a < b$ (the set of pairs of real numbers the first of which is smaller than the second), the inclusion $A \subset B$ (the set of pairs of sets, the first of which represents a subset of the second).

It is clear that any function represents a relation which does not contain two pairs $\{x, y\}$ with the same x.

The relation for some of those pairs of elements is defined which of them precedes the other is called a *relation of order*. If the element a precedes the element b then it is written $a \leq b$ in the same way as for real numbers or $a \prec b$. This relation means that a precedes b or coincides with b. If a precedes b but does not coincide with b then one says that a strictly precedes b and writes $a < b$.

The set is called a *partially ordered* set if it is equipped with the relation of order possessing the properties

(i) $a \leq b$, $b \leq c$ imply $a \leq c$;

(ii) $a \leq a$ for any element a;

(iii) $a \leq b$, $b \leq a$ imply $a = b$.

E x a m p l e 1.14. As an example of a partially ordered set the set of all subsets of some set A may be indicated. The relation of order represents in this case the inclusion $\subset$. As another example of a partially ordered set we mention the space R^n in which the inequality between the vectors a and b is defined as the set of similar inequalities between all their components. In other words, $a \leq b$ if and only if $a_1 \leq b_1$, $\ldots$, $a_n \leq b_n$.

The elements a, b of a partially ordered set are called *incomparable* if neither of the relations $a \leq b$, $b \leq a$ holds. In particular, in the set of all subsets of some set A partially ordered by inclusion any two sets, neither of which is a subset of the other, are incomparable.

The partially ordered set is called simply *ordered* or *linearly ordered* if any two its elements are comparable.

E x a m p l e 1.15. As examples of ordered sets are the set of all natural numbers and the set of all real numbers (the real axis R).

The element s of a partially ordered set S is called a *majorant of the subset* $A \subset S$ if $x \leq s$ at any $x \in A$.

The element a of a partially ordered set S is called an *upper bound* of the subset $A \subset S$ if $x \leq a \ \forall x \in A$ and $a \leq s$ for any majorant s of A.

The element a of the set A is called the *maximal element of the set A* if A does not contain elements following a. Similarly a *minorant* and *lower bound* of the subset of a partially ordered set and the *minimal element* of a set are defined.

Any ordered subset of a partially ordered set is called a *chain*.

An ordered set is called a *well ordered set* if each its subset has the minimal element.

E x a m p l e 1.16. As an example of well ordered set we mention the set of all natural numbers. As an example of ordered but not well ordered set the interval $[a, b]$ in R may serve. This interval has the minimal element a but its subsets (α, β), $(\alpha, \beta]$, $a \leq \alpha < \beta \leq b$ have no minimal element.

The *segment* of an ordered set A *determined by an element* $a \in A$ is defined as the set of all the elements of A strictly preceding a. This definition implies that the element of an ordered set does not belong to the segment determined by this element.

1.1.9. Choice Axiom

The choice axiom underlies many constructions of modern mathematics. In particular, it is of the utmost importance for functional analysis. The *choice axiom* is formulated as follows: *for any collection of sets* $\{X_t\}$, $t \in T$, *the function* $x(t)$, $t \in T$, *may be determined by choosing at any t an arbitrary element* $x_t \in X_t$ *and putting* $x(t) = x_t$.

This axiom is quite natural. If all the spaces X_t represent "samples" of the same space X corresponding to various values of $t \in T$ then the choice axiom asserts the existence of a function $x(t)$ with values in X whose value at each t may be arbitrarily assigned. The existence of such a function appears quite obvious especially in the case where T and X represent the real axis R.

The choice axiom implies some theorems often used in mathematics, in particular, in functional analysis.

Theorem 1.2. *Any set may be well ordered* (Zermelo theorem).

▷ Let S be any set. Determine the function φ by choosing arbitrarily an element from each nonempty subset of S. This is possible by virtue of the choice axiom.

Consider the class $\mathcal{C}$ of all ordered subsets of the set S possessing the following property: each element x of the subset $A \in \mathcal{C}$ is determined as the value of the function φ at the complement in S of the set B_x of all the elements of A preceding x:

$$x = \varphi(S \backslash B_x). \tag{1.9}$$

Obviously C is not empty. It contains in particutar all the finite subsets of the set S with the first element $x_1 = \varphi(S \backslash \varnothing) = \varphi(S)$ and subsequent elements determined by formula (1.9): $x_n = \varphi(S \backslash \{x_1, \ldots, x_{n-1}\})$.

At first we prove that all the sets of the class C are well ordered. To do this we notice that any subset of the set A containing some segment of A has the minimal element $\varphi(S)$ which is the common first element of all the sets of the class C. For each subset B of the set A which contains no a segment of A there exists the maximal segment C of the set A non–intersecting with B. It represents the union of all the segments of A non–intersecting with B. The element $\varphi(A \backslash C)$ of the set A determining the segment C is the minimal element of the subset $B \subset A$. Thus every subset of the set $A \in C$ has the minimal element, i.e. the set A is well ordered.

Now we prove that any two of sets of the class C either coincide or one of them is the segment of the other. Really, any two sets $A, B \in C$ have common segments since $\varphi(S)$ is the first element of both. Let C be the maximal common segment of A, B. The following element of both sets A, B is $\varphi(S \backslash C)$ by formula (1.9). Hence C can be the maximal common segment of the sets A, B only if $C = A \subset B$ or $C = B \subset A$ or $C = A = B$.

Consider now the union L of all the sets of the class C. The set L is ordered. Really, any two elements $x, y \in L$ belong to some sets of the class C, say $x \in A$, $y \in B$, $A, B \in C$. By proved above x, y belong to the maximal of the sets A, B and therefore are comparable. Any element $x \in L$ belongs to some set $A \in C$ and consequently is determined by (1.9). It remains to notice that A is a segment of L. Thus L belongs to the class C, and consequently, is well ordered. But L is the maximal element of the class C by definition and therefore must coincide with S. Assuming contrary, $L \neq S$, we might determine the following element $\varphi(S \backslash L)$ by formula (1.9) in contradiction with the maximality of L. ◁

Corollary *Any set may be well ordered in the infinite variety of ways depending on the choice of the functions φ.*

Theorem 1.3. *Each chain of a partially ordered set is contained in some maximal chain* (Hausdorff theorem).

▷ Let C be any chain in a partially ordered set S, $B = S \backslash C$. By Theorem 1.2 the set B may be well ordered. Certainly this relation of order will not coincide with the relation of order determined on S. Let us partition all the elements of the well ordered set B into two classes in the following way. The first (minimal) element of B we refer to the first class if it is comparable in S with all the elements of the chain C, and to the second class in the opposite case. Each following element of the well ordered set B we refer to the first class if it is comparable in S with all the elements of the chain C and with all preceding elements of B belonging to the first class, and to the second class in the opposite case. Adding to C all the elements of B of the first class we obtain the chain $C_1 \supset C$. This chain is maximal since any element of $S \backslash C_1$ is incomparable with at least one element of the chain C_1. ◁

Theorem 1.4. *If any chain of a partially ordered set S has a majorant then S has the maximal element* (Zorn lemma).

▷ Let C be any maximal chain in S, c its majorant. If S contains an element s following c, $c \leq s$, then adding this element to C we obtain a wider chain. This contradiction makes sure that S does not contain elements following c, i.e. c is the maximal element of the set S. ◁

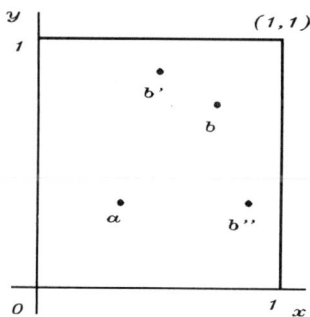   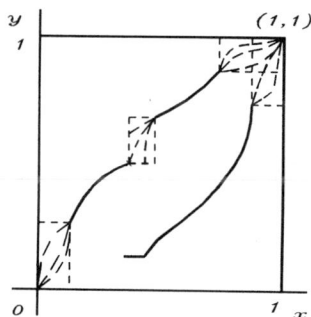

Fig. 1.14   Fig. 1.15

E x a m p l e 1.17. Fig.1.14 and Fig.1.15 illustrate Theorems 1.3 and 1.4. The closed square S with corners at the points $(0,0), (0,1), (1,0), (1,1)$ represents a partially ordered set in which the point a precedes the point b if all the coordinates of the point a do not exceed the corresponding coordinates of the point b. The points b, b', b'' on Fig.1.14 follow the point a and the points b, b', b'' are incomparable with one another. Solid lines on Fig.1.15 represent chains and by dotted lines their possible extensions to maximal ones are shown. It is clear that the point (1.1) is the end point of all maximal chains and at the same time serves as the common majorant of all maximal chains and is the maximal element of the square S.

It appears that each of Zermelo and Hausdorff theorems as well as Zorn lemma are equivalent to the choice axiom, and consequently, each of them may be taken as an axiom instead of the choice axiom. To make this statement sure it is sufficient to prove that the choice axiom is a consequece of Zorn lemma.

▷ Let $\{X_t\}$, $t \in T$ be any system of sets. Denote by $\mathcal{S}$ the set of all the subsystems of sets $\{X_t\}$, $t \in S \subset T$, for which the choice axiom is valid. Such subsystems exist. For instance such are subsystems $\{X_t\}$, $t \in S$, corresponding to all finite subsets S of the set T. Determine the function $x_S(t)$ on each subsystem $\{X_t\}$, $t \in S$, of the set $\mathcal{S}$ in such a way that $x_{S_1}(t) = x_{S_2}(t)$ at $t \in S_1 S_2$ for all S_1, S_2 corresponding to subsystems of the set $\mathcal{S}$. The set $\mathcal{S}$ is partially ordered by inclusion of the corresponding sets S. Let $\mathcal{C}$ be any chain of subsystems of the set $\mathcal{S}$, consisting of subsystems $\{X_t\}$, $t \in S_\alpha$ of the system $\{X_t\}$, $t \in T$. Consider the subsystem $\{X_t\}$, $t \in S$, where $S = \bigcup S_\alpha$. Determine the function $x_S(t)$ coinciding with $x_{S_\alpha}(t)$ on S_α. Then the subsystem $\{X_t\}$, $t \in S$, will be the maximal element of the chain $\mathcal{C}$. Thus any chain of the set $\mathcal{S}$ has the maximal element. By Zorn lemma the set $\mathcal{S}$ has the maximal element $\{X_t\}$, $t \in S_0$. This maximal element cannot differ from $\{X_t\}$, $t \in T$. Really, if $S_0 \neq T$ then taking any element $t_0 \in T \backslash S_0$ and determining the value $x_{S_0}(t_0)$ of the function $x_{S_0}(t)$ we extend the function $x_{S_0}(t)$ to the set $S_1 = S_0 \bigcup \{t_0\}$ what contradicts to maximality of the subsystem $\{X_t\}$, $t \in S_0$. Therefore $S_0 = T$ and putting $x(t) = x_{S_0}(t)$, $t \in T$, we make sure that the system $\{X_t\}$, $t \in T$ belongs to the set $\mathcal{S}$ for which the choice axiom is valid. ◁

For further study of the sets theory see (Halmos 1960, Suppes 1960).

Problems

1.1.1. Let X be any nonempty set and $\mathcal{F}$ be the class of all subsets of X. Prove that $\mathcal{F}$ is Boolean algebra.

1.1.2. Let $X = R$ be the real axis. Define the set by the following condition: $\mathcal{I} = \{$all intervals of the form $(-\infty, +\infty)$, $(-\infty, a]$, $(a, +\infty)$, $(a, b]$ where $a, b \in R\}$. Prove that the system $\mathcal{F} = \{A \ : \ A \subset R, A = \bigcap_{i=1}^{n} A_i, A_i \in \mathcal{I}, A_i \cap A_j \neq \emptyset$ at $i \neq j$ for some integer $n\}$ is Boolean algebra.

1.1.3. Find the domains D_f and ranges R_f of functions, functionals and operators for control systems given in Tables A.1.1 and A.1.2 (Appendix 1).

1.1.4. In probability theory the results x of some statistical experiments form the sample space X. Let $\mathcal{F}$ be some Boolean algebra of subsets of sample space X. Establish the correspondence between probability theory and sets theory given in Table 1.2.

Table 1.2

	Probability Theory	Sets Theory
1.	Event A	$A \in \mathcal{F}$
2.	Event A implies event B	$A \subset B$
3.	Event A does not occur	$X - A$
4.	At least one of events A and B occurs	$A \cup B$
5.	Both events A and B occur	AB
6.	Event that always occurs	X
7.	Impossible event	$\varnothing$
8.	Events that cannot occur simultaneously	$A \cap B \neq \varnothing$

1.2. Metric Spaces

1.2.1. Main Properties of Spaces

Nominating a set a space one usually equips this set with one or several of the properties of usual spaces studied in elementary geometry. Such properties are:

(i) the distance between any two points is defined;

(ii) it is possible to pass from anyone of the points to another point continuously, and each point may be enclosed into some arbitrarily small "vicinity" of this point;

(iii) the concept of a vector is defined in the space together with operations of addition of vectors and multiplication of a vector by a number.

Having equipped an abstract space with any one or any two or with all three of these properties we obtain various types of spaces studied in functional analysis.

1.2.2. Definition of a Metric Space

Based on known properties of the distance between points of a plane and of usual three–dimensional space the distance between two points may be defined in any space.

At first we introduce the concept of a product space. We call the *product A × B* (*direct product* or *Cartesian product*) *of sets A and B* the set of all ordered pairs $\{\{a, b\} : a \in A, b \in B\}$.

Let X be any space (set). The distance between any two points of the space X is defined as the numerical function of two points $d(x, y)$, $d : X \times X \to R$ possessing the properties:

(i) $d(x, y) \geq 0$ and $d(x, y) = 0$ if and only if $x = y$ (nonnegativity),

(ii) $d(x, y) = d(y, x)$ (symmetry),

(iii) $d(x, z) \leq d(x, y) + d(y, z)$ (triangle inequality).

The function possessing these properties is called a *metric*.

The space equipped with a metric is called a *metric space*. Properties (i), (ii), (iii) of a metric are called *metric axioms*.

The metric space, i.e. the space X with the metric d is usually denoted (X, d). If the same space is equipped with two different metrics d_1 and d_2, two different metric spaces are obtained (X, d_1) and (X, d_2).

E x a m p l e 1.18. The usual Euclidean metric on the plane with a rectangular Cartesian system of coordinates is determined by

$$d(x, y) = \sqrt{(x_1 - y_1)^2 + (x_2 - y_2)^2}.$$

E x a m p l e 1.19. The usual Euclidean metric in the n–dimensional space R^n: the distance between the points $x = (x_1, \ldots, x_n)$ and $y = (y_1, \ldots, y_n)$ is determined by

$$d(x, y) = \sqrt{\sum_{k=1}^{n} (x_k - y_k)^2}.$$

E x a m p l e 1.20. The non–Euclidean metric in the n–dimensional space R^n may be determined by

$$d(x, y) = \max_{1 \leq k \leq n} \{ |x_k - y_k| \}.$$

E x a m p l e 1.21. The Poincaré model of the Lobachevski plane: in theoretical physics is the upper halfplane with the metric

$$d(x, y) = \inf \int_x^y \frac{\sqrt{du_1^2 + du_2^2}}{u_2} = \inf \int_{x_1}^{y_1} \sqrt{1 + [f'(u)]^2} \, \frac{du}{f(u)},$$

the infinum being taken over the set of all the curves $u_2 = f(u_1)$ passing through the points x and y. The straight lines in this model are simulated by arches of circumferences whose centers lie on the absciss axis (Fig.1.16). The infinitely distant

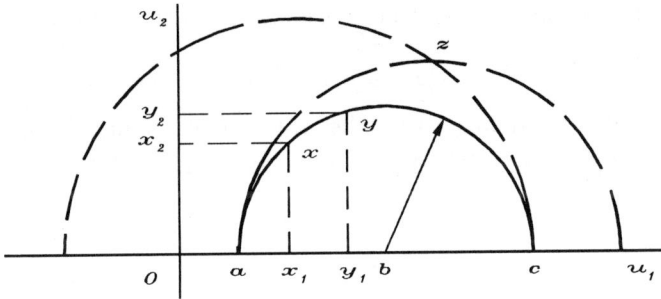

Fig. 1.16

straight line is simulated by the absciss axis. The infinitely many "straight lines" non–intersecting with the "straight line" $axyc$ (parallel to $axyc$) are passing through any point z disposed outside this "straight line". The set of these "straight lines" parallel to $axyc$ is bounded on Fig.1.16 by dotted "straight lines".

E x a m p l e 1.22. The metric in the space $C(T)$ of continuous functions with domain T is generally determined by

$$d(x,y) = \sup_{T} \mid x(t) - y(t) \mid .$$

E x a m p l e 1.23. The metric in the space $C_2(T)$ of continuous functions with the domain T is determined by

$$d(x,y) = \left\{ \int_{T} \mid x(t) - y(t) \mid^2 dt \right\}^{1/2} .$$

It is easy to prove that the metrics determined in previous examples satisfy the metric axioms.

1.2.3. Open and Closed Sets. Vicinities of Points

Starting from the definitions of a circle and a ball in elementary geometry the concept of a ball may be extended to all metric spaces.

The *open ball* $S_r(a)$ of radius r with the center at the point a in a metric space (X, d) is defined as the set of all the points $x \in X$ whose distance from the point a is smaller then r:

$$S_r(a) = \big\{ x : d(x,a) < r \big\}.$$

The *closed ball* of radius r with the center at the point a is defined as the set of points $x \in X$ whose distance from the point a does not exceed r:

$$\{x \ : \ d(x, a) \le r\}.$$

The *sphere* of radius r with the center at the point a is defined as the set of points $y \in X$ whose distance from the point a is equal to r:

$$\{x \ : \ d(x, a) = r\}.$$

Obviously the closed ball represents the union of the open ball and the sphere of the same radius with the center at the same point.

Any open ball with the center at the point x is called a *ball vicinity* of the point x. The vicinity $S_\varepsilon(x)$ is shortly called an *ε-vicinity* of the point x.

To find an approach to the general definition of an open set in a metric space we first establish one important property of the open balls.

Let $S_r(a) = \{x \ : \ d(x, a) < r\}$ be an open ball. For any point $x \in S_r(a)$ $d(x, a) < r$, the number $\delta > 0$ exists (certainly depending on x, $\delta = \delta_x$) such that $d(x, a) < r - \delta$ at this x. The ball vicinity $S_\delta(x)$ of the point x is entirely contained in the ball $S_r(a)$ since

$$d(y, a) \le d(x, a) + d(x, y) < r - \delta + \delta = r$$

for any point $y \in S_\delta(x)$.

Thus any open ball contains any its point together with some its ball vicinity. In view of this fact the *open set* in a metric space is defined as the set containing any its point together with some its ball vicinity.

The following evident theorems are valid.

Theorem 1.5. *Any union of open sets is an open set.*

Theorem 1.6. *Any nonempty intersection of a finite number of open sets is an open set.*

R e m a r k. Pay attention to the fact that an *infinite* intersection of open sets may be not an open set, while *any* (even uncountable) union of open sets is always an open set. For instance, the intersection of open balls

$$\bigcap_{n=1}^{\infty} S_{r+\varepsilon_n}(a) = \{x \ : \ d(x, a) \le r\}$$

where $\varepsilon_n > 0$, $\varepsilon_n \to 0$ as $n \to \infty$ represents a closed set since any ball $S_{r+\varepsilon_n}(a)$ contains all the points of the sphere $\{x \ : \ d(x, a) = r\}$. At the same time the union of all such balls is the open ball $S_{r+\varepsilon_n}(a)$.

To define closed sets we define at first the notion of the *boundary* of a set. The point x is called a *boundary point* of the set A in a metric space if any ball vicinity of x contains some points of A and some points of its complement $\bar{A}$ as well. The set of all the boundary points of the set A is called a *boundary* of the set A. Obviously the boundary of the set A serves at the same time as the boundary of its complement $\bar{A}$.

An open set contains none of its boundary points.

The set of a metric space X is called a *closed set* if it contains its boundary (i.e. all its boundary points). This definition together with the definition of an open set imply that the complement of any open set is a closed set and the complement of any closed set is an open set.

The whole metric space X contains any its point together with all its vicinities. Consequently, the whole space must be considered as an open set. But the empty set $\varnothing$ will in this case be a closed set. On the other hand, the empty set $\varnothing$ is open as the intersection of any two nonintersecting open sets. This contradiction may be removed only if we assume that the whole space X and the empty set are both open and closed sets $\varnothing$ simultaneously.

The concept of an open set enables one to extend the notion of the vicinity of a point. So, any open set containing the point x is called a *vicinity* of the point x.

1.2.4. Convergence in a Metric Space

The sequence of points $\{x_n\}$ of a metric space is called *convergent* to the point x if any its ε–vicinity $S_\varepsilon(x)$ contains all the points x_n corresponding to $n \geq N = N_\varepsilon$. In other words, the sequence $\{x_n\}$ converges to x if any $\varepsilon > 0$ corresponds such natural $N = N_\varepsilon$ (depending on ε) that $d(x_n, x) < \varepsilon$ at any $n \geq N$. We write $x_n \to x$ or $x = \lim_{n \to \infty} x_n$ or shortly $x = \lim x_n$ if the sequence of points $\{x_n\}$ converges to x.

The sequence of the points $\{x_n\}$ of a metric space is called *fundamental* or a *Cauchy sequence* if Cauchy condition holds $d(x_n, x_m) \to 0$ as any $n, m \to \infty$.

Theorem 1.7. *Any convergent sequence is a fundamental sequence.*

▷ From the triangle inequality follows

$$d(x_n, x_m) \leq d(x_n, x) + d(x_m, x).$$

Take any $\varepsilon > 0$ and choose N in such a way that $d(x_n, x) < \varepsilon/2$, $d(x_m, x) < \varepsilon/2$ at $n, m > N$. This is possible due to the convergence of $\{x_n\}$ to x. So, we get $d(x_n, x_m) < \varepsilon$ whence $d(x_n, x_m) \to 0$ as n, $m \to \infty$. ◁

R e m a r k. On the real axis R and in the finite–dimensional space R^n the Cauchy condition is not only necessary but also sufficient for the convergence of a sequence. But it is not so in the general case.

1.2.5. Complete Metric Spaces

A metric space is called *complete* if any fundamental sequence converges to some point of this space.

In a *complete* metric space *Cauchy condition is necessary and sufficient for the convergence of a sequence.*

E x a m p l e 1.24. As examples of complete metric spaces we mention all the finite–dimensional spaces R^n and the space $C([a,b])$ of continuous functions on $[a,b]$ with the sup as a norm (supnorm) of Example 1.22.

E x a m p l e 1.25. As an example of the incomplete metric space the space $C_2([-2,2])$ of continuous functions on the interval $[-2,2]$ of Example 1.23 may serve. It is easy to make sure that the sequence of continuous functions (Fig.1.17)

$$f_n(t) = \begin{cases} 0 & \text{at} \quad |t| > 1 + 1/2n, \\ n(1- |t|) + 1/2 & \text{at} \quad 1 - 1/2n \le |t| \le 1 + 1/2n, \\ 1 & \text{at} \quad |t| < 1 - 1/2n \quad (n = 1, 2 \ldots) \end{cases}$$

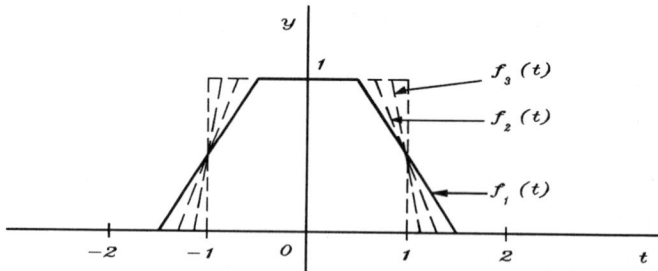

Fig. 1.17

is a fundamental sequence in the metric of the space $C_2([-2,2])$. But it is convergent (also in the metric of the space $C_2([-2,2])$) to the indicator of the set

$[-1, 1]$, $\mathbf{1}_{[-1,1]}(t)$, which is evidently a discontinuous function. We have thus an example where the same space of continuous functions is complete in the metric of the space $C([-2, 2])$ of Example 1.22 and is incomplete in the quadratic metric of the space $C_2([-2, 2])$ of Example 1.23.

1.2.6. Separable Metric Spaces

The set A of points in a metric space X is called *dense in* X if any vicinity of any point $x \in X$ contains points of the set A, i.e. if for any $\varepsilon > 0$ and $x \in X$ there are such points a of the set A that $d(x, a) < \varepsilon$.

A metric space is called *separable* if it contains a countable dense set. In other words, the metric space X is separable if there exists such a countable set of points $\{x_n\} \subset X$ that for any $\varepsilon > 0$ and any point $x \in X$ $d(x, x_n) < \varepsilon$ for some of the points x_n.

E x a m p l e 1.26. Examples of separable spaces are the real axis R and all the finite–dimensional spaces R^n, since the countable set of rational numbers (points with rational coordinates) is dense in the set of all real numbers (all points of R^n).

1.2.7. Completion of a Metric Space

Complete spaces are exceedingly important in functional analysis. Therefore the question naturally arises about the possibility of extension of an incomplete metric space to a complete one. Examples 1.24 and 1.25 suggest a natural approach which is efficient in many cases, i.e. to include in the space the limits of all fundamental sequences of this space belonging to some more wide space with the same metric[a] . However, it is not clear either the space obtained in this way is complete or not, and what are the cases in which this approach solves the problem. Answers to these questions are given by two basic theorems about completing spaces.

The metric space $\hat{X}$ is called a *completion of the metric space* X (certainly with the same metric) if $X \subset \hat{X}$, X is dense in $\hat{X}$ and $\hat{X}$ is complete.

Theorem 1.8. *If* $X \subset \hat{X}$, X *is dense in* $\hat{X}$ *and any fundamental sequence of points* X *has a limit in* $\hat{X}$ *then* $\hat{X}$ *is a completion of* X.

▷ Since X and $\hat{X}$ satisfy the first two conditions of the definition of a completion it is sufficient to prove that $\hat{X}$ is complete. Let $\{\hat{x}_n\}$

---

[a] To follow this way it is necessary that the given space X be a subspace of some other space $\hat{X}$ with the same metric.

be a fundamental sequence of points of $\hat{X}$. Since X is dense in $\hat{X}$ there exists for any $\delta > 0$ such a sequence $\{x_n\}$ of points of the space X that $d(x_n, \hat{x}_n) < \delta/2^n$. This sequence is a fundamental sequence since

$$d(x_n, x_m) \leq d(x_n, \hat{x}_n) + d(\hat{x}_n, \hat{x}_m) + d(x_m, \hat{x}_m) \to 0 \quad \text{as} \quad n, m \to \infty.$$

By the assumption of the theorem the sequence $\{x_n\}$ has a limit $\hat{x} \in \hat{X}$. This limit is at the same time the limit of the sequence $\{\hat{x}_n\}$. Really,

$$d(\hat{x}_n, \hat{x}) \leq d(x_n, \hat{x}_n) + d(x_n, \hat{x}) \to 0 \quad \text{as} \quad n \to \infty. \quad \triangleleft$$

Theorem 1.9. *Any metric space admits a completion.*

▷ Let X be a metric space. It is clear that in the general theory dealing with abstract spaces we have no "material" for its completion. Consequently, it is necessary to replace the space X by another space which may be identified with X and at the same time to get "material" for completing X. The classical way to reach this goal consists in replacing each element $x \in X$ by the set of all the sequences converging to x. This set contains always the *stationary sequence* whose all the members are equal to x. This is the reason for such a construction of the new space X. To assign the metric in the new space X identical with the metric of the initial space X we define the distance between two sequences. Let $\{x_n\}$ and $\{y_n\}$ be any two fundamental sequences in the initial space X. We shall call the *distance between the sequences* $\{x_n\}$ and $\{y_n\}$ the quantity

$$d(\{x_n\}, \{y_n\}) = \lim_{n \to \infty} d(x_n, y_n). \tag{1.10}$$

The limit exists here since $\{x_n\}$ and $\{y_n\}$ are the fundamental sequences. Really,

$$| d(x_n, y_n) - d(x_m, y_m) | \leq d(x_n, x_m) + d(y_n, y_m) \to 0 \quad \text{as} \quad n, m \to \infty.$$

Thus the numerical sequence $\{d(x_n, y_n)\}$ is a fundamental sequence and therefore is convergent.

The distance (1.10) between sequences possesses the following obvious properties:

(i) if two sequences $\{x_n\}$ and $\{x'_n\}$ converge to the same point x then $d(\{x_n\}, \{x'_n\}) = 0$;

(ii) if $x_n \to x$, $y_n \to y$ then $d(\{x_n\}, \{y_n\}) = d(x, y)$.

The first property serves as a reason to assume as *equivalent* any sequences the distance between which is equal to zero. So, the new space X represents a *space of classes of equivalent sequences.*

The second property shows that the distance between classes of equivalent convergent sequences is independent of the choice of specific representatives from these classes and is equal to the distance between their limits. Thus we have replaced the initial space X by the new space X whose elements represent classes of equivalent convergent sequences. At the same time we have got the "material" for completing X, namely all the classes of equivalent fundamental sequences which have no limit in X. Extending X by adding to it all such classes we obtain the space $\hat{X}$ representing the space of all the classes of equivalent fundamental sequences of the initial space X. The distance between the elements $\hat{x}$, $\hat{y}$ of the space $\hat{X}$ is defined by (1.10)

$$d(\hat{x}, \hat{y}) = \lim_{n \to \infty} d(x_n, y_n). \tag{1.11}$$

It is clear that the spaces X and $\hat{X}$ constructed in such a way satisfy the condition $X \subset \hat{X}$. And each fundamental sequence of points $\{x_n\}$ of the space X has the limit in $\hat{X}$ representing the class of all fundamental sequences equivalent to $\{x_n\}$. It remains to prove that X is dense in $\hat{X}$ and to apply Theorem 1.8. For this purpose we take any element $\hat{x} \in \hat{X}$ and some fundamental sequence of points $\{x_n\}$ of the initial space X forming $\hat{x}$. Since $\{x_n\}$ is a fundamental sequence $d(x_n, x_m) < \varepsilon$ for any $\varepsilon > 0$ and all sufficiently large m, n. Consequently,

$$\lim_{m \to \infty} d(x_n, x_m) \leq \varepsilon$$

at any sufficiently large n. But the left–hand side of this inequality represents by virtue of (1.11) the distance between the points $\hat{x} \in \hat{X}$ and $\hat{x}_n \in X$, $\hat{x}_n$ being the class of equivalent sequences containing the stationary sequence whose all the members are equal to x_n. Consequently, there is a point $\hat{x}_n$ of the space X in any vicinity of any point $\hat{x}$ of the space $\hat{X}$, i.e. X is dense in $\hat{X}$. Thus the space $\hat{X}$ satisfies all the conditions of Theorem 1.8. Consequently, $\hat{X}$ is a completion of X. ◁

The method of completing the space used in Theorem 1.9 is by no means a constructive one. It is based on replacing the initial space X by some more complicated space which may be identified with X in some sense and on extending this new space by adding new elements to it.

In specific situations often occurs that all the fundamental sequences of a given space X have the limits but some of these limits do not belong to X and represent the elements of some wider space $\hat{X}$ with the same metric. Theorem 1.8 enables one in such cases to realize the completion of the space X by simple adding to X the limits of all the fundamental sequences of X. We shall meet such a situation in Subsections 3.7.7 and 3.7.8.

1.2.8. Continuous Functions in Metric Spaces

The function $y = f(x)$ mapping a metric space (X, d) into another metric space (Y, r) is called *continuous at the point* x if for any $\varepsilon > 0$ there exists such $\delta = \delta_\varepsilon > 0$ that $r(f(x), f(x')) < \varepsilon$ at any x', $d(x, x') < \delta$. Thus the function $y = f(x)$ is continuous in a metric space if $d(x, x') \to 0$ implies $r(f(x), f(x')) \to 0$.

The function $y = f(x)$ is called *continuous on some set* if it is continuous at all the points of this set. If the function $y = f(x)$ is continuous on its whole domain it is called simply *continuous*.

The number $\delta = \delta_\varepsilon$ depends in the general case not only on ε but on the point x as well. The function is called *uniformly continuous* on the set A if for any $\varepsilon > 0$ such a number $\delta = \delta_\varepsilon > 0$ exists depending only on ε and independent of x that $r(f(x), f(x')) < \varepsilon$ at all x, $x' \in A$, $d(x, x') < \delta$.

The functions of the set $\{f_\alpha(x)\}$ are called *equicontinuous* on the set A if for any $\varepsilon > 0$ exists such $\delta = \delta_\varepsilon > 0$ that $r(f_\alpha(x), f_\alpha(x')) < \varepsilon$ at all x', $d(x, x') < \delta$ *for all* the functions of the set $\{f_\alpha(x)\}$, $x \in A$.

For further study of the metric spaces basic notions refer to (Dunford and Schwarz 1958, Halmos 1957).

P r o b l e m s

1.2.1. Picture the ball of radius r with the center at the point x with the coordinates x_1, x_2 on the plane with the metric of Example 1.20.

1.2.2. What represents the ball of radius r with the center at the point x in the space R^n with the metric of Example 1.20?

1.2.3. The same question for the space $C([a, b])$ of continuous functions of a scalar variable with the metric of Example 1.22. Picture this ball on the plane with the Cartesian coordinates t, x.

1.2.4. The same question for the space $C([a, b])$ of Example 1.22 of continuous two–dimensional vector functions with values on the plane with Euclidean metric of Example 1.18. Picture this ball in the three–dimensional space with Cartesian coordinates t, x_1, x_2.

1.2.5. Let $d(x, y)$ be the metric in a space X, $\varphi(u)$ the strictly increasing positive function satisfying the conditions $\varphi(u_1 + u_2) \leq \varphi(u_1) + \varphi(u_2)$ $\forall u_1, u_2$, $\varphi(0) = 0$. Prove that the function $\delta(x, y) = \varphi(d(x, y))$ represents the metric in a space X.

1.2.6. Prove that the space with the metric $d(x, y) = |x^3 - y^3|$ is complete.

1.2.7. Are the spaces with the metrics

1) $d(x, y) = |\arctg x - \arctg y|$;

2) $d(x, y) = |e^x - e^y|$

incomplete? Find the corresponding completions.

1.2.8. Prove that the set of polynomials is dense in the space of continuous functions $C([a, b])$ of Example 1.22.

1.2.9. Prove that the space $C([a, b])$ of Example 1.22 is separable.

1.2.10. Prove that the space $C_2([a, b])$ of Example 1.23 is separable.

1.2.11. Prove that any open set in a separable metric space represents the countable union of all ball vicinities of the points of some dense set with rational radii contained in this open set.

1.2.12. Let $y = f(x)$ be the continuous function mapping a metric space X into the real axis R. Prove that the set $\{x : f(x) < c\} \subset X$ is open at any $c \in R$.

1.3. Linear Spaces

1.3.1. Definition of a Linear Space

An important property of finite–dimensional spaces is the possibility to define *vectors* in them with the operations of addition and multiplication by numbers yielding other vectors. In this case to any point x corresponds the vector whose beginning is at the origin and the end at this point x. This vector is called the *radius–vector of the point x*. This correspondence is evidently one–to–one. This is the reason to consider any finite–dimensional space as the space whose points represent vectors.

Generalizing the properties of the operations of addition of vectors and multiplication of a vector by a number one may define these operations in any space.

A space X is called a *linear* or a *vector space* if

(i) for any two elements $x, y \in X$ their sum $x + y \in X$ is uniquely defined;

(ii) for any element $x \in X$ and number α the product $\alpha x \in X$ so as $1 \cdot x = x$ is uniquely determined;

(iii) there exists the unique zero element $0 \in X$ such that $0 \cdot x = 0^a$ for any $x \in X$ (the existence of zero);

(iv) the operations of addition of vectors and multiplication of a vector by a number possess usual properties:

$$x + y = y + x \qquad \text{(commutativity)},$$
$$(x + y) + z = x + (y + z), \ \alpha(\beta x) = (\alpha\beta)x \quad \text{(associativity)},$$
$$(\alpha + \beta)x = \alpha x + \beta x, \ \alpha(x + y) = \alpha x + \alpha y \quad \text{(distributivity)}.$$

In other words, a linear space is the space in which the operations of addition of vectors and multiplication of a vector by a number are defined possessing the usual properties of commutativity, associativity and distributivity and the inique element exists adopted as zero (zero element corresponding to the origin in geometry). Conditions (i)–(iv) represent the *linear space axioms*. The elements (the points) of a linear space are called *vectors*.

If the operation of multiplication of a vector by a number is defined only for real numbers then the linear space is called a *real linear space* (a *linear space over the field of real numbers R*). If the operation of multiplication of a vector by a number is defined for all complex numbers then the linear space is called a *complex linear space* (a *linear space over the field of complex numbers C*). We shall deal in the sequel chiefly with complex linear spaces.

Real and complex numbers represent scalars. This is the reason to use the term *linear space over the field of scalars K* assuming $K = R$ in the case of a real linear space and $K = C$ in the case of a complex linear space[b] .

Since the operation of multiplication of a vector by the number in a complex linear space is defined for real numbers any complex linear space is at the same time a real linear space. As for a real linear space it cannot be considered as a complex linear space.

E x a m p l e 1.27. As examples of linear spaces may serve all the finite–dimensional spaces and the spaces $C(T)$ and $C_2([a, b])$ of Examples 1.22 and 1.23.

Axioms (i)–(iv) imply directly the following properties of linear spaces.

---

[a] Here $0 \cdot x$ is the product of the element $x \in X$ by the number 0 and 0 on the right–hand side represents zero element of the space X.

[b] Many statements of the theory of linear spaces are valid also in more general case of an arbitrary (abstract) field K which is also called a *field of scalars*.

(i) $x + 0 = x$ at any x since $x + 0 = x + 0 \cdot x = (1 + 0)x = x$.

(ii) Each element x has the unique *opposite element* $-x$ such as $x + (-x) = 0$. Really, $x + (-1)x = (1 + (-1))x = 0 \cdot x = 0$. Hence $-x = (-1)x$. Assuming that there is another element x' for which $x + x' = 0$ we get

$$x' = x' + 0 = x' + [x + (-x)] = (x + x') + (-x) = -x.$$

(iii) For any x and $\alpha \in K$

$$-(\alpha x) = (-1)\alpha x = (-\alpha x) = \alpha(-1)x = \alpha(-x).$$

The vector $z = x + (-y)$ is called a *difference of vectors* x, y and is denoted $x - y$.

(iv) If $z = x - y$ then $x = y + z$ since $x = x + 0 = x + y + (-y) = y + x - y = y + z$.

The concept of a difference of two vectors enables one to transfer items from one part of an equality to the other part with changed sign as in usual algebraic equalities.

(v) $x - y = 0$ if and only if $y = x$.

If $y = x$ then $x - y = x + (-y) = x + (-x) = 0$. If $x - y = 0$ then

$$y = y + x + (-1)x = x + (y - x) = x\,.$$

(vi) $\alpha x = 0$ if and only if either $\alpha = 0$ or $x = 0$.

If $\alpha = 0$ then $\alpha x = 0$ at any x. If $x = 0$ then $\alpha x = \alpha \cdot 0 = \alpha(0 \cdot x) = (\alpha \cdot 0)x = 0 \cdot x = 0$ at any α. If $\alpha x = 0$ and $\alpha \neq 0$ then

$$x = (\alpha \frac{1}{\alpha})x = \frac{1}{\alpha}(\alpha x) = \frac{1}{\alpha} \cdot 0 = 0.$$

If $\alpha x = 0$ and $x \neq 0$ then $\alpha = 0$ since $\alpha \neq 0$ implies $x = 0$ as was just proved.

1.3.2. Linear Dependence and Independence of Vectors

Vectors $x_1, \ldots, x_n$ are called *linearly independent* if the equality

$$\alpha_1 x_1 + \cdots + \alpha_n x_n = 0 \tag{1.12}$$

is possible only at $\alpha_1 = \cdots = \alpha_n = 0$. If (1.12) is valid at $\alpha_1, \ldots, \alpha_n$ some of which are different from zero then the vectors $x_1, \ldots, x_n$ are called *linearly dependent*.

If a linear space X contains no more than a finite number n of linearly independent vectors then the space X is called a n-*dimensional space* and the number n is called the *dimension* of the space X. Let prove that in this case any vector $x \in X$ is expressed in terms of linearly independent vectors $x_1, \ldots, x_n$ by

$$x = c_1 x_1 + \cdots + c_n x_n.$$

▷ From linear dependence of vectors $x, x_1, \ldots, x_n$ follows that

$$\alpha x + \alpha_1 x_1 + \cdots + \alpha_n x_n = 0,$$

where $\alpha \neq 0$ owing to the linear independence of $x_1, \ldots, x_n$. It follows from this equality

$$x = \frac{1}{\alpha}(\alpha x) = -\frac{\alpha_1}{\alpha} x_1 - \cdots - \frac{\alpha_n}{\alpha} x_n. \quad ◁$$

If n linearly independent vectors exist at any n then the space is called *infinite-dimensional*.

E x a m p l e 1.28. The space of all continuous scalar functions of the variable t with the domain $[a, b]$ is an infinite dimensional space. Really, the functions $t, t^2, \ldots, t^n$ are linearly independent at any n since

$$c_1 t + c_2 t^2 + \cdots + c_n t^n = 0 \quad \forall t \in [a, b]$$

if and only if $c_1 = \cdots = c_n = 0$ (no power t^n can be expressed in terms of $t, \cdots, t^{n-1}$).

The infinite set of vectors $\{x_\alpha\}$ (in an infinite-dimensional space) is called *linearly independent* if the vectors $x_{\alpha_1}, \ldots, x_{\alpha_n}$ are linearly independent at any $n, \alpha_1, \ldots, \alpha_n$.

1.3.3. Subspaces. Linear Spans

A subset Y of a linear space X is called a *subspace* of X if it is a linear space itself with the same operations of addition of vectors and multiplication of a vector by a number.

The intersection of any set of subspaces is also a subspace. Really, $x, y \in \in \bigcap_\alpha Y_\alpha$ implies that x, y belong to all Y_α, and consequently, $x + y$ and cx belong to all Y_α, i.e. to their intersection $\bigcap_\alpha Y_\alpha$.

The smallest subspace of X containing a set of vectors $A \subset X$ is called a *linear span* of the set of vectors A or a *subspace formed by the set of vectors A* and is denoted $L(A)$.

Obviously $L(A)$ represents the intersection of all the subspaces containing the set of vectors A and therefore exists always (since at least one subspace, i.e. X, contains A).

Theorem 1.10. *The linear span $L(A)$ of the set of vectors A represents the set L_A of all finite linear combinations of vectors belonging to the set A.*

▷ Since $L(A)$ is a subspace it contains all finite linear combinations of its vectors, in particular, of vectors of $A \subset L(A)$. Consequently, $L(A) \supset L_A$. On the other hand, L_A represents evidently a linear space containing A. Hence L_A is a subspace containing A. But $L(A)$ is the smallest subspace containing A. Hence $L(A) \subset L_A$. Two opposite inclusions obtained yield $L_A = L(A)$. ◁

Corollary. *Any vector $x \in L(\{x_\alpha\})$ where $\{x_\alpha\}$ is a linearly independent set of vectors is expressed by*

$$x = c_1 x_{\alpha_1} + \cdots + c_n x_{\alpha_n} \tag{1.13}$$

at some n, $\alpha_1, \ldots, \alpha_n$, $c_1, \ldots, c_n$.

▷ The uniqueness of the representation of a vector $x \in L(\{x_\alpha\})$ by formula (1.13) follows immediately from the linear independence of the vectors $x_{\alpha_1}, \ldots \ldots, x_{\alpha_n}$. ◁

If the linear span of a linearly independent set of vectors $\{x_\alpha\}$ coincides with the whole space X then the set of vectors $\{x_\alpha\}$ is called *Hammel basis* of the space X. In this case any vector of the space X represents a finite linear combination of vectors of the set $\{x_\alpha\}$.

1.3.4. Quotient Spaces

Let L be the subspace of a linear space X. It is necessary in some cases to reckon vectors $x_1, x_2 \in X$ equivalent if their difference $x_1 - x_2$ belongs to L. So, the relation of equivalence will be determined in X. Certainly any vector $x \in X$ is self–equivalent since any subspace contains zero element. As a result the space X will represent the union of various *classes of equivalence* (*quotients* or *cosets*).

If the quotients of the space X are considered as elements of the space Y then Y will be a linear space whose zero element will be L, i.e. the class of all the elements of X equivalent to zero. Really, defining the sum of two quotients $\bar{x}$, $\bar{y}$ as the set of sums $x + y$, $x \in \bar{x}$, y

$\in \bar{y}$ and the product of the quotient $\bar{x}$ by the number α as the set of products αx, $x \in \bar{x}$ we shall have for any $x_1, x_2 \in \bar{x}$, $y_1, y_2 \in \bar{y}$ by definition $x_1 + y_1 - (x_2 + y_2) = (x_1 - x_2) + (y_1 - y_2) \in L$ and $\alpha x_1 - \alpha x_2 = \alpha(x_1 - x_2) \in L$. Thus $\bar{x} + \bar{y}$ and $\alpha\bar{x}$ represent quotients, i.e. belong to the space Y.

The space Y of quotients of the space X is called the *quotient space* (*factor* or *coset space*) *of the space* X *by the subspace* L. The subspace L represents generally the subspace whose elements possess some common property which enables one to reckon them equivalent.

1.3.5. Norm of a Vector

Extending the notion of the vector modulus in a finite–dimensional space the notion of the norm of a vector may be defined in any linear space, naturally preserving the main properties of the vector moduli.

The *norm of a vector* x is defined as the function $\| x \|$ of this vector possessing the following properties:

(i) $\| x \| \geq 0$, $\| x \| = 0$ only at $x = 0$,
(ii) $\| \alpha x \| = | \alpha | \| x \|$ for any number α,
(iii) $\| x + y \| \leq \| x \| + \| y \|$ (triangle inequality).

These properties of the norm are called the *norm axioms*.

The norm of a vector in the space X represents a function mapping X into the real axis R, $\| x \| : X \to R$.

The norm of a vector x is sometimes denoted $| x |$ as the modulus of the vector x in a finite–dimensional space.

1.3.6. Normed Linear Spaces

The linear space X equipped with the norm is called a *normed linear space*. This definition implies that the normed space is always a linear space. Any normed linear space X will be a metric space if we define the metric in it induced by the following norm:

$$d(x, y) = \| x - y \| .$$

This metric satisfies all the metric axioms (Subsection 1.2.2).

E x a m p l e 1.29. The finite–dimensional space R^n in which the norm is defined as the modulus of a vector

$$\| x \| = \sqrt{\sum_{k=1}^{n} x_k^2} .$$

E x a m p l e 1.30. The linear space $C(T)$ of continuous functions, $T \subset R^n$, with the norm

$$\| x \| = \sup_{t \in T} | x(t) |.$$

This norm is usually called *supnorm*.

E x a m p l e 1.31. The linear space $C_2(T)$ of continuous functions, $T \subset R^n$, with the norm

$$\| x \| = \left\{ \int_T |x(t)|^2 \, dt \right\}^{1/2}.$$

The metric is induced by the norm in the spaces of Examples 1.18, 1.19, 1.22 and 1.23 defined in Examples 1.29, 1.30 and 1.31.

1.3.7. Banach Spaces

The normed space complete in the metric induced by the norm is called a *Banach space* or shortly a *B–space* by the name of one of the founders of functional analysis the Polish mathematician Banach.

E x a m p l e 1.32. As an example of a B–space we mention the space of continuous functions $C(T)$ of Example 1.30.

E x a m p l e 1.33. The space $C_2(T)$ of Example 1.31 is not a B–space since it is incomplete (Subsection 1.2.5).

1.3.8. Scalar Product

In a normed linear space, in particular, in a B–space, the distances between points and the norms of vectors are defined but such notions as angles between vectors and orthogonality of vectors are absent. To introduce the notion of orthogonality of vectors in any linear space the scalar product of vectors is usually defined.

The *scalar product of vectors* x, y of a linear space X is defined as the function of these vectors (x, y) possessing the following properties:

(i) $(x, y) = \overline{(y, x)}$,

(ii) $(\alpha x, y) = \alpha(x, y)$,

(iii) $(x_1 + x_2, y) = (x_1, y) + (x_2, y)$,

(iv) $(x, x) \geq 0$, $(x, x) = 0$ only at $x = 0$.

The scalar product is thus mapping the product space $X \times X$ into the field of scalars K, $(x, y) : X \times X \to K$.

Accordingly the scalar product in a real linear space is always real and property 1 represents the commutativity of factors in the scalar product: $(y, x) = (x, y)$.

E x a m p l e 1.34. In the Euclidean space R^n with rectangular Cartesian coordinates the scalar product of the vectors $x = \{x_1, \ldots, x_n\}$ and $y = \{y_1, \ldots, y_n\}$ is determined by

$$(x, y) = \sum_{k=1}^{n} x_k y_k .$$

In the complex n–dimensional linear space C^n the scalar product is determined by

$$(x, y) = \sum_{k=1}^{n} x_k \bar{y}_k .$$

This definition of a scalar product is extended also to the spaces of sequences R^∞ and C^∞ under the condition of convergence of the series determining (x, x) and (y, y).

E x a m p l e 1.35. In the linear space $C_2(T)$, $T \subset R^n$ of continuous functions the scalar product of the vectors $x = x(t)$ and $y = y(t)$ is determined by

$$(x, y) = \int_T x(t)\overline{y(t)} \, dt.$$

E x a m p l e 1.36. If $K(t, s)$ is a continuous function of $t, s \in [a, b]$ with the properties $K(s, t) = \overline{K(t, s)}$ and

$$\int_a^b \int_a^b K(t, s)\varphi(t)\overline{\varphi(s)} \, dt \, ds > 0$$

for any continuous function $\varphi(t) \not\equiv 0$, then the scalar product in the space of continuous functions on $[a, b]$ may be determined by

$$(x, y) = \int_a^b \int_a^b K(t, s)x(t)\overline{y(s)} \, dt \, ds.$$

Let $x_1, \ldots, x_n$ be any n vectors of a space X with the scalar product. The matrix

$$\Gamma = \begin{bmatrix} (x_1, x_1) & (x_1, x_2) & \cdots & (x_1, x_n) \\ (x_2, x_1) & (x_2, x_2) & \cdots & (x_2, x_n) \\ \vdots & \vdots & \ddots & \vdots \\ (x_n, x_1) & (x_n, x_2) & \cdots & (x_n, x_n) \end{bmatrix} \tag{1.14}$$

is called the *Gram matrix* of the vectors $x_1, \ldots, x_n$.

Theorem 1.11.   *The Gram matrix is a Hermite nonnegatively definite matrix.*

▷ Scalar product axioms imply $\mathbf{\Gamma}^* = \mathbf{\Gamma}$ and

$$\left(\sum_{k=1}^{n} \alpha_k x_k, \sum_{k=1}^{n} \alpha_k x_k \right) \geq 0 \quad \forall \alpha_1, \ldots, \alpha_n, \; x_1, \ldots, x_n \;^a \qquad (1.15)$$

or

$$\sum_{k,l=1}^{n} \alpha_k \bar{\alpha}_l (x_k, x_l) \geq 0 \quad \forall \alpha_1, \ldots, \alpha_n, \; x_1, \ldots, x_n. \quad ◁$$

Theorem 1.12.   *The determinant of the Gram matrix is non-negative and is equal to zero if and only if the vectors $x_1, \ldots, x_n$ are linearly dependent.*

▷ Since the matrix $\mathbf{\Gamma}$ is a nonnegatively definite Hermite matrix it has n nonnegative eigenvalues $\lambda_1, \ldots, \lambda_n$ and $|\mathbf{\Gamma}| = \lambda_1 \ldots \lambda_n$.

If m of the eigenvalues $\lambda_1, \ldots, \lambda_n$ are equal to zero then $|\mathbf{\Gamma}| = 0$ and the equation

$$\mathbf{\Gamma} x = 0$$

has m linearly independent solutions $\varphi_1, \ldots, \varphi_m$ representing the eigenvectors of the matrix $\mathbf{\Gamma}$ corresponding to the m–tiple zero eigenvalue:

$$\mathbf{\Gamma} \varphi_p = 0 \quad (p = 1, \ldots, m).$$

Multiplying this equality from the left by $\varphi_p^* = \bar{\varphi}_p^T$ we get

$$\varphi_p^* \mathbf{\Gamma} \varphi_p = \sum_{k,l=1}^{n} (x_k, x_l) \bar{\varphi}_{pk} \varphi_{pl} = \sum_{k,l=1}^{n} (\bar{\varphi}_{pk} x_k, \bar{\varphi}_{pl} x_l) =$$

$$= \left(\sum_{k=1}^{n} \bar{\varphi}_{pk} x_k, \sum_{k=1}^{n} \bar{\varphi}_{pk} x_k \right) = 0 \quad (p = 1, \ldots, m).$$

Hence

$$\sum_{k=1}^{n} \bar{\varphi}_{pk} x_k = 0 \quad (p = 1, \ldots, m). \qquad (1.16)$$

---

^{a} The asterisk means the transposition of a matrix with replacing all its complex elements by the respective conjugate numbers.

Thus if some m of the eigenvalues of the matrix $\boldsymbol{\Gamma}$ are equal to zero then $|\boldsymbol{\Gamma}| = 0$ and the vectors $x_1, \ldots, x_n$ are connected by m linear relations (1.16).

If the vectors $x_1, \ldots, x_n$ are connected by m linear relations then such numbers $c_{p1}, \ldots, c_{pn}$ exist that

$$c_{p1}x_1 + \cdots + c_{pn}x_n = 0 \quad (p = 1, \ldots, m).$$

Multiplying this equlity scalary from the left by $x_1, \ldots, x_n$ we get

$$\bar{c}_{p1}(x_k, x_1) + \cdots + \bar{c}_{pn}(x_k, x_n) = 0 \quad (k = 1, \ldots, n; \quad p = 1, \ldots, m)$$

or in the matrix form

$$\boldsymbol{\Gamma} c_p = 0 \quad (p = 1, \ldots, m),$$

where $c_p = [\bar{c}_{p1} \ldots \bar{c}_{pn}]^T$ are linearly independent vectors (matrices–columns). These vectors are obviously the eigenvectors of the Gram matrix $\boldsymbol{\Gamma}$ corresponding to the m–tiple zero eigenvalue. Consequently, $|\boldsymbol{\Gamma}| = \lambda_1 \ldots \lambda_n = 0.$ ◁

The nonnegativity of the Gram determinant implies in particular

$$\begin{vmatrix} (x, x) & (x, y) \\ (y, x) & (y, y) \end{vmatrix} \geq 0,$$

for any $x, y \in X$, whence

$$|(x, y)|^2 \leq (x, x)(y, y), \tag{1.17}$$

The equality sign here occurs if and only if $y = \alpha x$. Inequality (1.17) is called a *Cauchy–Schwarz inequality*.

E x a m p l e 1.37. For the scalar product of Example 1.34 the inequality (1.17) takes the form

$$\left| \sum_k x_k \bar{y}_k \right|^2 \leq \sum_k |x_k|^2 \cdot \sum_k |y_k|^2.$$

This inequality holds for finite sums and for convergent series as well. In the latter case the convergence of the series on the right implies the convergence of the series on the left.

E x a m p l e 1.38. For the scalar product of Example 1.35 inequality (1.17) takes the form

$$\left| \int_a^b x(t)\overline{y(t)}\,dt \right|^2 \le \int_a^b |x(t)|^2\,dt \cdot \int_a^b |y(t)|^2\,dt.$$

E x a m p l e 1.39. For the scalar product of Example 1.36 inequality (1.17) takes the form

$$\left| \int_a^b\int_a^b K(t,s)x(t)\overline{y(s)}\,dt\,ds \right|^2 \le$$

$$\le \int_a^b\int_a^b K(t,s)x(t)\overline{x(s)}\,dt\,ds \cdot \int_a^b\int_a^b K(t,s)y(t)\overline{y(s)}\,dt\,ds.$$

R e m a r k. These examples show what a powerful tool for research is functional analysis. The direct derivation of the inequalities of above examples is very tedious while they are simple consequences of general formula (1.17) which was derived rather simply.

1.3.9. Hilbert Spaces

If the scalar product is defined in a linear space X then this space may be normed by putting

$$\| x \| = \sqrt{(x,x)}. \tag{1.18}$$

Really, (1.18) and the scalar product axioms imply $\| x \| \ge 0$, $\| x \| = 0$ only at $x = 0$ and $\| \alpha x \| = \sqrt{(\alpha x, \alpha x)} = |\alpha|\,\| x \|$. Furthermore

$$\| x + y \|^2 = \| x \|^2 + \| y \|^2 + (x,y) + (y,x) \le$$

$$\le \| x \|^2 + \| y \|^2 + 2\| x \| \cdot \| y \|,$$

whence

$$\| x + y \| \le \| x \| + \| y \|.$$

Thus the function $\| x \| = \sqrt{(x,x)}$ possesses all the properties of the norm.

Norm (1.18) is called the *norm induced by the scalar product*.

The complete normed space with the norm induced by the scalar product is called a *Hilbert space* or shortly a *H–space* by the name of the famous German mathematician Hilbert who was the first who studied such spaces. In other words, the H–space is the B–space with the norm induced by the scalar product.

R e m a r k. Any linear spaces, B–spaces and H–spaces may be either real or complex. In accordance with the general definition of Subsection 1.3.8 the scalar product is the real function of two elements in a real H–space and the complex function of two elements in a complex H–space.

The concept of a scalar product enables one to speak about the orthogonality of vectors. In the same way as in the case of the finite–dimensional Euclidean spaces the vectors x, y of an H–space X are called *orthogonal* if their scalar product is equal to zero, $(x, y) = 0$.

In a real H–space the notion of the angle between two vectors may also be introduced exactly as in the finite–dimensional Euclidean spaces. Namely, the angle between the vectors x, y of a real H–space is defined by formula

$$\cos \theta = \frac{(x, y)}{\| x \| \cdot \| y \|} \quad (x \neq 0, \ y \neq 0).$$

Inequality (1.17) implies that the angle θ may assume any value in the interval $[0, \pi]$ and is equal to $\pi/2$ if and only if the vectors x, y are orthogonal, is equal to 0 or π if and only if the vectors x, y are collinear, i.e. if and only if $y = \alpha x$ ($\alpha > 0$ at $\theta = 0$ and $\alpha < 0$ at $\theta = \pi$).

From the scalar product axioms (Subsection 1.3.8) immediately follows the *parallelogram identity*:

$$\| x + y \|^2 + \| x - y \|^2 = 2\| x \|^2 + 2\| y \|^2 , \qquad (1.19)$$

well known in the elementary geometry. This identity represents the specific property of the norm induced by the scalar product. If the norm does not satisfy (1.19) then it cannot be induced by some scalar product. At the same time any norm satisfying (1.19) at any x, y determines uniquely the inducing it scalar product.

1.3.10. Algebras

The above examples of linear spaces show that the linear spaces exist in which the product of two elements is defined except of usual operations of addition of vectors and multiplication of a vector by a number.

The linear space with the operation of multiplication of two elements such that for any two elements x, y there exists the unique element z

$= xy$, is called an *algebra*. If the operation of multiplication of the elements is commutative then the algebra is called a *commutative algebra*.

E x a m p l e 1.40. The space $C(T)$ of Example 1.30 may serve as an example of the commutative algebra.

If an algebra contains the element e for which $ex = xe = x$ at any x, then the algebra is called an *algebra with the identity*.

E x a m p l e 1.41. The algebra with the identity of the same space $C(T)$ of Example 1.30 may be mentioned. Here the function $e(t) \equiv 1$ represents the unit element e.

An algebra with the identity representing a normed space whose norm satisfies the additional condition $\| e \| = 1$ is called a *normed algebra*. If a normed algebra represents a B–space then it is called a *Banach algebra*. Banach algebra theory plays important role in functional analysis (Rudin 1973).

1.3.11. Spaces of Continuous Functions and Spaces of Bounded Functions

The space $C(T)$ represents the space of bounded continuous scalar functions with the domain $T \subset R^n$ and the norm

$$\| x \| = \sup_{t \in T} | x(t) |. \tag{1.20}$$

The space $B(T)$ represents the space of bounded scalar functions with the domain $T \subset R^n$ and the same norm (1.20). It is clear that the space $C(T)$ is a subspace of the space $B(T)$.

The convergence in $C(T)$ and $B(T)$ represents the uniform convergence of the sequence of functions $\{x_n(t)\}$ from $B(T)$ or $C(t)$ since

$$\| x_n - x \| = \sup_{t \in T} | x_n(t) - x(t) | < \varepsilon$$

implies $| x_n(t) - x(t) | < \varepsilon$ at all $t \in T$.

All the spaces of bounded continuous functions $C(T)$ and the spaces of bounded functions $B(T)$ represent normed spaces. Moreover they are B–spaces as we shall prove.

Theorem 1.13. *The spaces $C(T)$ and $B(T)$ are complete.*

▷ If $\{x_n(t)\}$ is a fundamental sequence of functions from $C(T)$ or $B(T)$ then for any $\varepsilon > 0$ $\| x_n - x_m \| < \varepsilon$ implies $| x_n(t) - x_m(t) | < \varepsilon \, \forall t$

$\in T$. Hence the numerical sequence $\{x_n(t)\}$ is a fundamental sequence at any $t \in T$. Therefore the sequence $\{x_n(t)\}$ converges to some function $x(t)$ at any t. Taking any $\varepsilon > 0$ we shall have

$$|x_n(t) - x_m(t)| < \varepsilon/2 \quad \forall t \in T$$

at all n, m greater than some N_ε since $\{x_n(t)\}$ is a fundamental sequence. Taking now some $t \in T$ and sufficiently large $m > N_\varepsilon$ we shall have

$$|x_m(t) - x(t)| < \varepsilon/2$$

at these values of t and m by virtue of convergence of $\{x_n(t)\}$ to $x(t)$. Two inequalities obtained yield

$$|x_n(t) - x(t)| \leq |x_n(t) - x_m(t)| + |x_m(t) - x(t)| < \varepsilon \quad \forall n > N_\varepsilon.$$

Since $t \in T$ was taken arbitrarily this inequality is valid at all $t \in T$ and $n > N_\varepsilon$ (the result is independent of the choice of the function $x_m(t)$ at each t). Thus

$$\| x_n - x \| = \sup_{t \in T} |x_n(t) - x(t)| < \varepsilon \quad \forall n > N_\varepsilon.$$

This inequality proves the uniform convergence of the sequence $\{x_n(t)\}$ to $x(t)$. The boundedness of the limit function $x(t)$ follows from the boundedness of any fundamental sequence, i.e. from the uniform boundedness of the sequence $\{x_n(t)\}$.

Thus we proved that any fundamental sequence of functions from $C(T)$ or $B(T)$ converges to some bounded function, i.e. to an element of the space $B(T)$. This proves that $B(T)$ is complete.

To prove that $C(T)$ is complete it is necessary to prove that $x(t)$ is continuous in the case of the fundamental sequence $\{x_n(t)\}$ in $C(T)$. But this is an immediate consequence of the continuity of the functions $x_n(t)$. Taking any $\varepsilon > 0$ we have $|x_n(t) - x(t)| < \varepsilon/3 \ \forall t \in T$ for all sufficiently large n. By continuity of $x_n(t)$ we have $|x_n(t') - x_n(t)| < \varepsilon/3$ at all t' sufficiently near to t. These two inequalities yield

$$|x(t') - x(t)| \leq |x(t') - x_n(t')| + |x_n(t') - x_n(t)| + |x_n(t) - x(t)| < \varepsilon$$

at any $t \in T$ and at all t' sufficiently near to t. ◁

R e m a r k. Theorem 1.13 may be extended without any changes in proof to spaces of functions mapping any metric space T into any complete metric space.

1.3.12. Spaces of Differentiable Functions

Now we consider the *space* $C^n(T)$ of continuous functions in the finite interval $T = [a, b]$ with continuous derivatives of orders up to n. The norm of the element $x(t) \in C^n(T)$ is determined by

$$\| x \| = \sum_{p=0}^{n} \sup_{t \in T} | x^{(p)}(t) |. \qquad (1.21)$$

This formula shows that the convergence of the sequence of functions of the space $C^n(T)$ represents the uniform convergence of this sequence of functions and the sequences of all their derivatives of orders up to n.

For further study of the linear spaces basic notions refer to (Balakrishnan 1976, Dunford and Schwarz 1958).

P r o b l e m s

1.3.1. Prove the following inequalities:
1) $| \, \| x \| - \| y \| \, | \leq \| x - y \|$;
2) $\| x \| \leq \max \{ \| x + y \|, \| x - y \| \}$;
3) $\| x - y \| \leq \frac{q}{1-q} \| y \|$ at $\| x - y \| \leq q \| x \|, q < 1$,

$x, y \in X$, X being some linear normed space.

1.3.2. Is the set of all integer nonnegative powers of the variable t linearly independent in the space of all scalar functions of t on the interval $[a, b]$? Find the linear span of this set.

1.3.3. Is the set of trigonometric functions $\sin nt$, $\cos nt$ $(n = 0, 1, 2, \ldots)$ linearly independent in the space of continuous functions on the interval $[-\pi, \pi]$? Find the subspace formed by the set of trigonometric functions.

1.3.4. Prove the existence of Hamel basis in any linear space.

I n s t r u c t i o n. The set of all possible linear systems of vectors $\{x_\alpha\}$ and the subspaces formed by them represent the partially ordered set. Prove that any chain (ordered subset) has in this set a majorant and use Zorn lemma.

1.3.5. Characterize the following functions in the space $C([-1, 1])$ (which of them are functionals or operators?):
1) $y = a_1 x(t_1) + a_2 x(t_2) + \cdots + a_n x(t_n)$, $t_1, \ldots, t_n$ being fixed values of the argument of the function $x(t)$;
2) $y = [x(t) + x(-t)]$;
3) $y = \int_a^b \varphi(t) x(t) \, dt$;
4) $y = \int_a^b K(s, t) x(t) \, dt$;

5) $y = \int\limits_a^b \varphi(t)x(t)\, dt + \alpha x(a) + \beta x(b);$

6) $y = \int\limits_a^b K(s,t)x^2(t)\, dt.$

1.3.6. May the norm in the space of continuous scalar functions on the interval $[a, b]$ be determined by

$$\int\limits_a^b |x(t)|\, dt\,?$$

May the norm in R^n be determined by

$$\max_{1 \le k \le n} |x_k|\,?$$

1.3.7. Let K be a strictly positive definite $(n \times n)$–matrix. Prove that the bilinear form $x^T K y$ may serve as the scalar product in the space R^n. Show that the norm induced by this scalar product satisfies the parallelogram identity.

1.3.8. Prove the completeness and separability of the following spaces:

1) the space m of bounded numerical sequences with the norm

$$\| x \| = \sup_k |x_k|;\qquad (1.22)$$

2) the space c of convergent numerical sequences with the same norm (1.22);

3) the space c_0 of sequences convergent to zero with the same norm (1.22);

4) the space of sequences l_p with the norm

$$\| x \|^p = \sum_{k=1}^\infty |x_k|^p,\quad p \ge 1;\qquad (1.23)$$

5) the space of sequences s $(x = \{\xi_n\}_{n=1}^\infty \in S,\ \sup_n |\xi_n| < \infty)$ with the norm

$$\| x \| = \sum_{n=1}^\infty 2^{-n} \frac{|\xi_n|}{a+|\xi_n|},\quad a > 1.\qquad (1.24)$$

1.3.9. Prove that the operations of addition of vectors and multiplication of a vector by a number in a normed space are continuous.

1.3.10. The closed unit ball in the normed space X represents the complete metric space with the metric induced by the norm of the space X. Show that X is a B–space.

1.3.11. Prove that if the norm satisfies the parallelogram identity (1.19) then there exists the unique scalar product inducing this norm.

I n s t r u c t i o n. Supposing that such a norm is induced by the scalar product, $\| x \|^2 = (x, x)$, show that for any x, y

$$(x, y) = (\| x + y \|^2 - \| x - y \|^2)/4 \tag{1.25}$$

in the case of a real normed space and

$$(x, y) = (\| x + y \|^2 - \| x - y \|^2 + i \| x + iy \|^2 - i \| x - iy \|^2)/4 \tag{1.26}$$

in the case of a complex normed space. Show that the expressions on the right–hand sides of (1.25) and (1.26) possess all the properties of the scalar product.

1.3.12. Prove the following inequality:

$$\| a_1 - a_3 \| \cdot \| a_2 - a_3 \| \leq \| a_1 - a_2 \| \cdot \| a_3 - a_4 \| + \| a_2 - a_3 \| \cdot \| a_1 - a_4 \|,$$

where a_1, a_2, a_4, $a_4 \in H$.

1.4. Linear Functions and Functionals

1.4.1. Sets Operations in a Linear Space

Besides the ordinary operations on sets it is expedient to introduce the operations of the sets addition and the multiplication of the set by a number in a linear space.

A sum of the vector x and the set B in a linear space is called a set of all the vectors of the form $x + y$ where y is any vector which belongs to the set B, $x + B = \{z \; : \; z = x + y, \; y \in B\}$.

A sum of the sets A and B in a linear space is called a set of all the vectors of the form $x + y$ where x is any vector from the set A and y is any vector from the set B, $A + B = \{z \; : \; z = x + y, \; x \in A, \; y \in B\}$.

A product of the set A by the number c in a linear space is called a set of all the vectors of the form cx where x is any vector of the set A, $cA = \{z \; : \; z = cx, \; x \in A\}$.

It is easy to understand that the sum of the subspaces L_1 and L_2 of the linear space X also represents the subspace, namely the minimal subspace which contains the subspaces L_1 and L_2.

1.4.2. Convex Sets

The set A of a linear space is called *convex* if it contains along with any two vectors x, y their linear combination $\alpha x + (1 - \alpha)y$ at any $\alpha \in (0, 1)$. In other words, the set A is convex if it contains rectlines which connect its any points.

It is evident that the sum of any convex sets and the product of any convex set by any number are convex sets.

1.4.3 Linear Functions

The function $y = f(x)$ mapping the linear space X into the linear space Y is called *linear* if at any $n, x_1, \ldots, x_n \in D_f$ and any numbers $\alpha_1, \ldots, \alpha_n$

$$f\left(\sum_{k=1}^{n} \alpha_k x_k\right) = \sum_{k=1}^{n} \alpha_k f(x_k). \tag{1.27}$$

It is natural to consider the subspace of the space X as the domain D_f of the linear function f as if it is determined at $x = x_1, \ldots, x_n$ then formula (1.27) also determines it for all linear combination of the vectors $x_1, \ldots, x_n$.

It is easy to see that the range R_f of the linear function f represents a subspace of the space Y as for any $n, x_1, \ldots, x_n \in D_f, \alpha_1, \ldots, \alpha_n$ from

$$y_1 = f(x_1) \in R_f, \ldots, y_n = f(x_n) \in R_f$$

follows

$$\sum_{k=1}^{n} c_k y_k = \sum_{k=1}^{n} c_k f(x_k) = f\left(\sum_{k=1}^{n} c_k x_k\right) \in R_f.$$

It shows that R_f is a linear space and consequently, a subspace of the space Y of the values of the function f.

Theorem 1.14. *For the linearity of the function f it is necessary and sufficient that for any $x, x_1, x_2 \in D_f$ and any number α the following conditions should be fulfilled:*
(i) $f(\alpha x) = \alpha f(x)$,
(ii) $f(x_1 + x_2) = f(x_1) + f(x_2)$.

▷ The necessity of these conditions follows directly from the definition of a linear function. For proving the sufficiency we notice that from the inequalities

$$f(\alpha_1 x_1 + \alpha_2 x_2) = f(\alpha_1 x_1) + f(\alpha_2 x_2) = \alpha_1 f(x_1) + \alpha_2 f(x_2),$$

$$f\left(\sum_{k=1}^{n}\alpha_k x_k\right) = f\left(\sum_{k=1}^{n-1}\alpha_k x_k + \alpha_n x_n\right) = f\left(\sum_{k=1}^{n-1}\alpha_k x_k\right) + \alpha_n f(x_n),$$

by the induction of n follows the validity of property (1.27) at all n, $x_1, \ldots, x_n \in D_f, \alpha_1, \ldots, \alpha_n$. ◁

If the linear function $y = f(x)$ maps X into the real axis R or the complex plane C then $y = f(x)$ represents *a linear functional* $f(x) = fx$.

The linear function $f(x)$ mapping the linear space X into another linear space Y different from the set of the numbers represents *a linear operator* $f(x) = Ax$.

It is clear that in a complex linear space the linear functions possess property (1.27) for any complex numbers $\alpha_1, \ldots, \alpha_n$ and in the real space–only for real $\alpha_1, \ldots, \alpha_n$. In the real linear space X we usually restrict ourselves to the real linear functionals mapping this space into the real axis R.

Besides the linear functions in the complex linear space we may determine the analogous functions which possess the symmetric properties.

The function $y = f(x)$ mapping the linear space X into the linear space Y is called *conjugate linear function* if for any $n, x_1, \ldots, x_n \in D_f$ and any numbers $\alpha_1, \ldots, \alpha_n$ we have

$$f\left(\sum_{k=1}^{n}\alpha_k x_k\right) = \sum_{k=1}^{n}\bar{\alpha}_k f(x_k). \tag{1.28}$$

1.4.4. Linear Functionals

Theorem 1.15. *In any linear space there exists a linear functional which takes given values on any finite set of linearly independent vectors.*

▷ Let $x_1, \ldots, x_n$ be linearly independent vectors of the linear space X, $y_1, \ldots, y_n$ be arbitrary numbers (real in the case of the real space X and complex numbers in the case of the complex space X). We denote by L the subspace generated by the vectors $x_1, \ldots, x_n$ (i.e. the linear span of the vectors $x_1, \ldots, x_n$, see Subsection 1.3.3). Any vector x of the subspace L represents a linear combination of the vectors $x_1, \ldots, x_n$:

$$x = \alpha_1 x_1 + \ldots + x_n. \tag{1.29}$$

Denote the functional on L

$$f(x) = \alpha_1 y_1, + \ldots + \alpha_n y_n. \tag{1.30}$$

It is evident that this functional is linear as formula (1.30) establishes the correspondence between any linear combination of the vectors (1.29) and the same linear combination[a] of the values of the functional $f(x)$. Further $f(x_k) = y_k$ $(k = 1, \ldots, n)$ as $x = x_k$ if and only if where $\alpha_k = 1$, $\alpha_l = 0$ at $l \neq k$. ◁

1.4.5. Extension of a Linear Functional

Theorem 1.15 establishes the existence in any linear space a linear functional which takes the given values in any finite set of the points. This functional is determined on the correspondent finite–dimensional subspace of the given linear space X. But for constructing the practically useful theory the linear functionals determined on the whole space X are needed. Therefore the problem of extension of a linear functional determined on some subspace on the whole space X arises.

Theorem 1.16. *Any linear functional determined on the subspace of a linear space may be extended on more wide subspace.*

▷ Let $f_0(x)$ be a linear functional determined on the subspace L_0 of the linear space X, z_1 be any vector which does not belong to the subspace L_0, L_1 be the subspace formed by the subspace L_0 and the vector z_1:

$$L_1 = \{x \: : \: x = y + \alpha z_1, \quad y \in L_0, \quad \alpha \in K\}.$$

We determine on L_1 a linear functional

$$f_1(x) = f_1(y + \alpha z_1) = f_0(y) + \alpha f_1(z_1), \tag{1.31}$$

where $f_1(z_1)$ is an arbitrary given value of the functional f_1 in the point z_1. It is evident that $f_1(x) = f_0(x)$ for any $x \in L_0$. Really, $x = y + \alpha z_1 \in L_0$ at $y \in L_0$ iff $\alpha = 0$, $x = y$. Consequently, formula (1.31) determines the extension of the functional f_0 from the subspace L_0 on the subspace $L_1 \supset L_0$. ◁

The proved theorem establishes the principal possibility of non–restricted expansion of the domain of a linear functional. If remains to solve the problem: is it possible to extend the domain of a functional till the whole space? It is expedient to restrict, in particular, the possibilities of the choice of the functional values at each new point z_1.

---

[a] It means a linear combination with the same coefficients.

1.4.6. Convex Functionals

The nonnegative functional $p(x)$ determined on the whole linear space X is called convex if it satisfies the following conditions:

(i) $p(x + y) \leq p(x) + p(y) \quad \forall x, y$;

(ii) $p(\alpha x) = |\alpha| \, p(x) \quad \forall x, \alpha \in K$.

From the definition of vector norm in the linear normed space (Subsection 1.3.6) follows that a norm represents a convex functional. It is evident that the product of the vector norm of any positive number also represents a convex functional. Therefore a convex functional is often called *a semi-norm*.

1.4.7. Hahn–Banach Theorem on Extension of a Linear Functional

Let us consider one of the fundamental theorems of functional analysis which establishes the possibility of the extension of the linear functional on the whole space.

Theorem 1.17. *Any linear functional $f_0(x)$ determined on the subspace L_0 of the linear space X and satisfying the condition*

$$|f_0(x)| \leq p(x) \quad \forall x \in L_0, \tag{1.32}$$

where $p(x)$ is some convex functional, may be extended to the whole space X with retention of this condition (Hahn–Banach theorem).

At first we shall prove this theorem for the real linear space X and correspondingly for the real linear functional $f_0(x)$ and afterwards extend it to the case of the complex linear space X and the complex linear functional $f_0(x)$.

▷ We shall show that in Theorem 1.16 the number $f_1(z_1)$, i.e. the value of the extension of the functional $f_0(x)$ at the point $z_1 \notin L_0$ may be always chosen in such a way that the functional $f_1(x)$ would safisfy condition (1.32) on the subspace L_1. It is evident that condition (1.32) will be satisfied on L_1 if at all $x \in L_1$ $f_1(x) \leq p(x)$. Really, then at all $x \in L_1$

$$f_1(x) = -f_1(-x) \geq -p(-x) = -p(x).$$

Thus the problem is reduced to such a choice of $f_1(z_1)$ that at all $y \in L_0$ and α

$$f_1(x) = f_1(y + \alpha z_1) \leq p(y + \alpha z_1)$$

or

$$f_0(y) + \alpha f_1(z_1) \le |\,\alpha\,|\, p\!\left(\frac{y}{\alpha} + z_1\right).$$

Hence it follows

$$f_0\!\left(\frac{y}{\alpha}\right) + f_1(z_1) \le p\!\left(\frac{y}{\alpha} + z_1\right) \quad \text{at} \quad \alpha > 0,$$

$$f_0\!\left(\frac{y}{\alpha}\right) + f_1(z_1) \ge -p\!\left(\frac{y}{\alpha} + z_1\right) \quad \text{at} \quad \alpha < 0$$

or as y/α is an arbitrary vector from L_0,

$$f_0(u) + f_1(z_1) \le p(u + z_1), \quad f_0(v) + f_1(z_1) \ge -p(v + z_1)$$

at any $u,\ v \in L_0$. We shall prove that $f_1(z_1)$ may be always chosen in such a way that these inequalities will be satisfied at any $u,\ v \in L_0$. For this purpose we estimate the difference

$$[-f_0(u) + p(u + z_1)] - [-f_0(v) - p(v + z_1)]$$

$$= p(u + z_1) + p(v + z_1) - f_0(u - v).$$

On the basis of the determination of a convex functional we have

$$p(u + z_1) + p(v + z_1) = p(u + z_1) + p(-v - z_1) \ge p(u - v).$$

Consequently, by virtue of condition (1.32)

$$[-f_0(u) + p(u + z_1)] - [-f_0(v) - p(v + z_1)]$$

$$\ge p(u - v) - f_0(u - v) \ge 0 \quad \forall u,\ v \in L_0.$$

Hence it follows that

$$\inf_{u \in L_0} [-f_0(u) + p(u + z_1)] \ge \sup_{v \in L_0} [-f_0(v) - p(v + z_1)].$$

Thus at any value of $f_1(z_1)$ satisfying the inequalities

$$\sup_{v \in L_0} [-f_0(v) - p(v + z_1)] \le f_1(z_1) \le \inf_{u \in L_0} [-f_0(u) + p(u + z_1)],$$

formula (1.31) determines the extension of the functional f_0 on the subspace L_0 with the retention of the conformity condition (1.32). It is evident that this extension in general case will be not unique.

If in the space X there exists a countable set of the vectors $\{z_n\}$ generating together with the subspace L_0 the whole space X then continuing the process of the extension of a linear functional domain which was described at the proof of the Theorem 1.16 we convince ourselves that this functional may be extended to the whole space X. If there is no such a countable set of the vectors then for accomplishing the proof we have to apply Zorn lemma (Subsection 1.1.9). We shall call the extensions of the functional f_0 on two different spaces *conformed* if they coincide on the intersection of these two subspaces, i.e. if the extensions f_1 and f_2 of the functional f_0 on the subspaces L_1 and L_2 satisfy the condition $f_1(x) = f_2(x)$ at $x \in L_1$, L_2. Let consider the set $\mathcal{M}$ of all conformed extensions of the functional f_0 on the different subspaces with the retention of the conformity condition (1.32). On the basis of the proved part of the theorem the set $\mathcal{M}$ is not empty. It is partially ordered by the sign of the including the functional domain extensions. Let $\mathcal{M}_0$ be an ordered subset (a chain) of the set $\mathcal{M}$, $\{L_\alpha\}$ be a set of domains of all the functional f_0 extensions from $\mathcal{M}_0$, $L_{\alpha_1} \subset L_{\alpha_2}$ at any $\alpha_1 < \alpha_2$, f_α be the functional f_0 extension from L_0 to L_α. It is clear that the functional $f_{\mathcal{M}_0}$ determined on the union of all the subspaces L_α

$$L_{\mathcal{M}_0} = \bigcup_\alpha L_\alpha$$

by formula

$$f_{\mathcal{M}_0}(x) = f_\alpha(x) \quad \text{at} \quad x \in L_\alpha \, ,$$

represents the extension of the functional f_0 from L_0 to $L_{\mathcal{M}_0}$ with the retention of condition (1.32), i.e. belongs to the set $\mathcal{M}$. It is also clear that this functional $f_{\mathcal{M}_0}$ represents an upper bound of the set $\mathcal{M}_I$. Thus every chain of a partially ordered set $\mathcal{M}$ has an upper bound. According to Zorn lemma there exists in the set $\mathcal{M}$ the maximal element, i.e. the extension f of the functional f_0 with the maximal domain. This maximal domain coincides with the whole space X as otherwise it may be extended by virtue of the proved first part of theorem, i.e. it may not be the maximal. ◁

We pass on to the proof Hahn–Banach theorem in the case of the complex linear space X.

▷ Let X be a complex linear space, L_0 be its subspace, $f_0(x)$ be a linear functional determined on L_0 and satisfying condition (1.32). Let X_R and L_R be the space X and the subspace L_0 which are considered as real linear spaces, $f_R^0(x)$ and $f_I^0(x)$ be a real and imaginary parts of the complex functional $f_0(x)$. So,

$$f_R^0(x) = \frac{f_0(x) + \overline{f_0(x)}}{2}, \quad f_I^0(x) = \frac{f_0(x) - \overline{f_0(x)}}{2i}, \tag{1.33}$$

$$f_0(x) = f_R^0(x) - if_R^0(ix), \quad f_I^0(x) = -f_R^0(ix). \tag{1.34}$$

From formulae (1.32) and (1.33) follows that $f_R^0(x)$ represents a linear functional on L_R (but not on L_0!!) satisfying condition (1.32) $|f_R^0(x)| \leq p(x)$. According to Hahn–Banach theorem for a real linear space the functional f_R^0 may be extended to the whole space X_R with the retention of condition (1.32). As a result we obtain a linear functional $f_R(x)$ determined on the whole space X_R and satisfying the condition $|f_R(x)| \leq p(x)$. In accordance with formula (1.34) this functional determines a linear functional

$$f(x) = f_R(x) - if_R(ix) \tag{1.35}$$

on the whole space X. Really, for any x_1, $x_2 \in X$ and $\alpha = \alpha_R + i\alpha_I$, $\beta = \beta_R + i\beta_I$ we have

$$f(\alpha x_1 + \beta x_2) = \alpha_R f_R(x_1) + \alpha_I f_R(ix_1) + \beta_R f_R(x_2) + \beta_I f_R(ix_2)$$

$$-i\alpha_R f_R(ix_1) + i\alpha_I f_R(x_1) - i\beta_R f_R(ix_2) + i\beta_I f_R(x_2)$$

$$= \alpha f_R(x_1) - i\alpha f_R(ix_1) + \beta f_R(x_2) - i\beta f_R(ix_2) = \alpha f(x_1) + \beta f(x_2),$$

$$|f(x)| \leq p(x) \quad \forall x \in X.$$

Supposing that there exists the vector x_0 for which $|f(x_0)| = \rho > p(x_0)$ we put $f(x_0) = \rho e^{i\varphi}$, $y_0 = x_0 e^{-i\varphi}$. Then we shall have $f(y_0) = f(x_0)e^{-i\varphi} = \rho > p(x_0)$, $p(y_0) = p(x_0)$ and $f_R(y_0) = \rho > p(y_0)$, i.e. a real linear functional f_R does not satisfy condition (1.32) at the point y_0. The obtained contradiction proves that the linear functional $f(x)$ determined by formula (1.35) represents an extension of the functional $f_0(x)$ on the whole space X satisfying condition (1.32). ◁

Corollary. *For any two different points x_1, x_2 of the linear space X there exists a linear functional determined on the whole space X which takes in these points different values $f(x_1) \neq f(x_2)$.*

1.4.8. Kernel of a Linear Functional

A set of all the vectors of the linear space X for which the value of the linear functional f is equal to zero is called *a kernel* of the functional f and is denoted by $\ker f$:

$$\ker f = \{x \; : \; f(x) = 0\}. \tag{1.36}$$

It is evident that a kernel of a linear functional represents a subspace as from $f(x_1) = \ldots = f(x_n) = 0$ follows

$$f(\alpha_1 x_1 + \ldots + \alpha_n x_n) = \sum_{k=1}^{n} \alpha_k f(x_k) = 0.$$

Theorem 1.18. *A subspace formed by a kernel of a linear functional and a vector which does not belong to it coincides with the whole space X.*

▷ Let x_0 be any vector which does not belong to a kernel of the linear functional f, $f(x_0) \neq 0$. Without loss of generality we may suppose that $f(x_0) = 1$ as if $f(x_0) \neq 1$ then instead of x_0 we may take the vector $x_0' = x_0/f(x_0)$ and receive $f(x_0') = 1$. We take any vector $x \in X$ and form a vector $y = x - x_0 f(x)$. This vector belongs to the kernel of the functional f, $y \in \ker f$ as $f(y) = f(x) - f(x_0) f(x) = 0$. Thus any vector of the space X may be represented in the form $x = y + \alpha x_0$, $\quad y \in \ker f$. Consequently, we get $\{x \; : \; x = y + \alpha x_0, y \in \ker f\} = X$. ◁

Corollary 1. *The quotient space $X/\ker f$ is one–dimensional for any linear functional f.*

▷ The set of the vectors for which the functional f has one and the same value $y = \{x \; : \; f(x) = k\}$ serve as the elements of the quotient space $X/\ker f$. It is clear that $y = ky_0$ where $y_0 = \{x \; : \; f(x) = 1\}$. ◁

Corollary 2. *For any subspace L which does not coincide with the whole space X there exists a linear functional whose kernel contains L.*

▷ Let $x_0 \notin L$. We determine on the subspace $L_1 = \{x \; : \; x = y + \alpha x_0, y \in L\}$ a linear functional $f(x) = \alpha f(x_0)$. From $f(x) = f(y) + \alpha f(x_0)$ follows that $f(y) = 0$ for any vector $y \in L$. According to Theorem 1.17 the functional f may be extended on the whole space X. Hence we shall have $L \subset \ker f$. ◁

Theory of linear functions and functionals will be continued in Chapters 5–8.

P r o b l e m s

1.4.1. Prove that any linear combination $c_1 A_1 + \ldots + c_n A_n$ of the convex sets $A_1 + \ldots + A_n$ is convex.

1.4.2. Prove that any intersection of the convex sets is convex.

1.4.3. Whether the union of any convex sets is convex one?

1.4.4. Prove that the direct product $A \times B$ $(A_1 \times \ldots \times A_n)$ of the convex sets A and B $(A_1, \ldots, A_n)$ is the convex set.

1.4.5. Prove that for any points $x_1, \ldots, x_n$ of the convex set A and any $\alpha_1, \ldots, \alpha_n \in (0, 1)$, $\alpha_1 + \ldots + \alpha_n = 1$, $\alpha_1 x_1 + \ldots + \alpha_n x_n \in A$.

1.4.6. Find linear functional $f_0(x)$ in the space of continuous functions $C([0, 1])$ taking values $1/k$ on t^{k-1} $(k = 1, \ldots, n)$. What is the domain L_0 for this funcitional? Prove that the functional $f_0(x)$ yields on L_0 the following condition $| f_0(x) | \leq \| x \| = \sup | x(t) |$. After supplying L_0 by function t^n (not belonging to L_0) find the extension of the functional $f_0(x)$ on $L_1 = \{x(t) : y(t) + \alpha t^n, y(t) \in L_0\}$ with the retention of condition $| f_0(x) | \leq \| x \|$. Find the extension of the functional $f_0(x)$ on the whole space $C([0, 1])$ with the retention of condition $| f_0(x) | \leq \| x \|$.

1.4.7. Let X be the H–space, $x_1, \ldots, x_n \in X$ the vectors satisfying the condition $(x_k, x_l) = \delta_{kl}$. Find linear functional $f_0(x)$ taking at the points $x_1, \ldots, x_n$ the values $y_1, \ldots, y_n$. Show that the functional yields the condition $|f_0(x)| \leq \sqrt{y_1^2 + \cdots + y_n^2} \| x \|$. Find the domain of $f_0(x)$. Represent $f_0(x)$ in the form of scalar $f \in L_0$. Give the explicit expression of $f_0(x)$ extension of the whole space X with the retention of inequality mentioned above.

I n s t r u c t i o n. Use the following formula $x = (x, x_1) x_1 + \ldots + (x, x_n) x_n$ valid for any vector $x \in L_0$.

1.4.8. Prove that the functional $| g(x) |$ is convex for any linear functional g on the whole space X. Prove Hahn–Banach Theorem 1.17 (Subsection 1.4.7) for $p(x) = | g(x) |$.

1.4.9. Let take n linear functionals $f_1, \ldots, f_n$ on the space X with different kernels $N_1 = \ker f_1, \ldots, N_n = \ker f_n$. Prove that the quotient space X/N where $N = N_1 \ldots N_n$ is n–dimensional.

1.4.10. Prove Theorem 1.14 for conjugate linear function satisfying condition (1.28).

CHAPTER 2
MEASURE THEORY

This Chapter is devoted to measure theory. In Section 2.1 classes of sets are determined: algebra, semi–algebra, σ-algebra, monotone classes. Their general properties are studied. The general definitions of a set function and a measure are given in Section 2.2, the measure is defined here as an additive or a countably additive (σ-additive) set function with values in a normed space. General properties of such measures are studied. Numerical measures, in particular, nonnegative measures are studied as special cases. This enables one to cover by the general notion of a measure operator-valued and function space-valued measures encountered in the sequel. In Section 2.3 after studying the general properties of measures, in particular, nonnegative ones, the general theorem is proved on extension of a numerical measure.

2.1. Classes of Sets

2.1.1. Set Functions

In geometry and physics we always meet with set functions, i.e. the functions whose sets serve as arguments. So, for instance, the areas of the geometric figures depend on the points set of the figure; to each points set of the plane (to each geometric figure) corresponds some number, i.e. the area of the figure. The mass or electric charge distributed in the space represent set functions; to each points set (space region) corresponds the number, i.e. the mass or the charge which is in the given domain.

According to the general definition of a function *a set function* represents the correspondence between the sets of some space X and the elements of some other space Y. Some class $\mathcal{C}$ (a set) of sets serves as a domain of a set function. The set function $y = \varphi(A)$ maps the sets class $\mathcal{C}$ into the space Y, $\varphi : \mathcal{C} \to Y$. In this connection the necessity of learning different sets classes arises.

2.1.2. Semi–Algebras of Sets

The sets class $\mathcal{C}$ is called *a semi–algebra* if it contains the empty set $\emptyset$, the whole space X, the finite intersections of the sets which form this space and the complement of any set which represents a finite union of pairwise nonintersecting sets of this space. In other words, C is a semi–algebra, if $\emptyset$, $X \in C$, $A,B \in C \Rightarrow AB \in C$ and $A \in C \Rightarrow \bar{A} = \bigcup\limits_{k=1}^{n} A_k$, $A_1, \ldots, A_n \in C$, $A_k A_h = \emptyset$, at $h \neq k$.

The importance of the notion of a semi–algebra is illustrated by the following examples.

E x a m p l e 2.1. The set of all intervals of the real axis R represents a semi–algebra (the sets $(-\infty, \infty)$, $(-\infty, a)$, (b, ∞), $(-\infty, a]$, $[b, \infty)$, $(a, a) = \emptyset$ are also considered as the intervals).

E x a m p l e 2.2. The set of all rectangles of the finite–dimensional space R^n (in particular, of the plane) represents a semi–algebra (the rectangles may have infinite or empty sides).

Theorem 2.1. *Any countable sets union of the semi–algebra C may be presented in the form of the countable union of pairwise non-intersecting sets of the semi–algebra C.*

▷ Let $\{C_n\}$ be any sets sequence of the semi–algebra C. It is evident that

$$\bigcup_{n=1}^{\infty} C_n = C_1 \cup C_2 \bar{C}_1 \cup \ldots \cup C_n \bar{C}_1 \ldots \bar{C}_{n-1} \cup \ldots = \bigcup_{n=1}^{\infty} \left(C_n \bigcap_{k=1}^{n-1} \bar{C}_k \right).$$

According to the definition of a semi–algebra

$$\bar{C}_k = \bigcup_{l=1}^{N_k} C_{kl}, \quad C_{kl} \in C, \quad C_{kl} C_{kh} = \emptyset \quad \text{at} \quad h \neq l.$$

Consequently,

$$\bigcap_{k=1}^{n-1} \bar{C}_k = \bigcup_{l_1=1}^{N_1} \cdots \bigcup_{l_{n-1}=1}^{N_{n-1}} C_{1l_1} \ldots C_{n-1,l_{n-1}},$$

$$\bigcup_{n=1}^{\infty} C_n = \bigcup_{n=1}^{\infty} \bigcup_{l_1=1}^{N_1} \cdots \bigcup_{l_{n-1}=1}^{N_{n-1}} C_n C_{1l_1} \ldots C_{n-1,l_{n-1}}. \qquad (2.1)$$

As according to the definition of a semi–algebra $C_n C_{1l_1} \ldots C_{n-1,l_{n-1}} \in C$ and the sets $C_n C_{1l_1} \ldots C_{n-1,l_{n-1}}$ and $C_n C_{1h_1} \ldots C_{n-1,h_{n-1}}$ do not intersect even if one of the indexes $h_1, \ldots, h_{n-1}$ does not coincide with the correspondent index $l_1, \ldots, l_{n-1}$ then equality (2.1) proves the theorem. ◁

2.1.3. Algebras of Sets

Algebra of sets (of some space X) is such a class of sets which as well as with any set A also contains its complement $\bar{A}$ and as well as with any two sets A and B also contains their union $A \bigcup B$.

Directly from this definition follow the properties of algebra of sets.

1. The whole space X and the empty set $\emptyset$ belong to the sets algebra. Really, if the set A belongs to the sets algebra $\mathcal{A}$ then according to the definition also $\bar{A} \in \mathcal{A}$, and consequently, $X = A \bigcup \bar{A} \in \mathcal{A}$. But $\emptyset = \bar{X}$, and consequently, from $X \in \mathcal{A}$ follow $\emptyset \in \mathcal{A}$.

2. The union of any finite number of sets $A_1, \ldots, A_n \in \mathcal{A}$ belongs to algebra $\mathcal{A}$. It follows from the definition according to the induction as

$$\bigcup_{k=1}^{n} A_k = \left(\bigcup_{k=1}^{n-1} A_k \right) \bigcup A_n .$$

3. From the duality principle follows that the algebra $\mathcal{A}$ also contains all finite intersections of sets which include it as from $A_1, \ldots, A_n \in \mathcal{A}$ according to the definition follows $\bar{A}_1, \ldots, \bar{A}_n \in \mathcal{A}$. Consequently,

$$\bigcup_{k=1}^{n} \bar{A}_k \in \mathcal{A}, \quad \bigcap_{k=1}^{n} A_k = \overline{\bigcup_{k=1}^{n} \bar{A}_k} \in \mathcal{A} .$$

4. Algebra $\mathcal{A}$ contains together with any two sets A and B the differences $A \setminus B$, $B \setminus A$ and $A \triangle B$, as from $A, B \in \mathcal{A}$ follows $\bar{A}, \bar{B} \in \mathcal{A}$, and consequently, $A \setminus B = A\bar{B} \in \mathcal{A}$ and $B \setminus A = \bar{A}B \in \mathcal{A}$, $A \triangle B = (A \setminus B) \bigcup (B \setminus A) \in \mathcal{A}$. Thus the sets algebra is *closed* relative to the operations of complement, difference, finite unions and intersections.

The algebra of sets $\mathcal{A}$ is called a σ-*algebra* of the sets if it contains all countable unions of the sets which are included in it, i.e. if from $A_k \in \mathcal{A}$ $(k = 1, 2, \ldots)$ follows $\bigcup_{k=1}^{\infty} A_k \in \mathcal{A}$ [a] .

Directly from this definition and the duality principle follows that a σ-algebra contains all countable intersections of the sets which are included in it. Thus a σ-algebra is *closed* relative to the operations of complement, countable unions and intersections.

The space X with a given σ-algebra of its sets $\mathcal{A}$ is called *a countable space* and is denoted $(X, \mathcal{A})$ and the sets belonging to σ-algebra $\mathcal{A}$ are called *measurable sets*.

---

[a] The prefix σ is always connected with the notion of the countability.

The set of all algebras (σ–algebras) of a given space may be partially ordered accounting that the algebra (respectively σ–algebra) $\mathcal{A}_1$ is smaller than the algebra $\mathcal{A}_2$ (respectively σ–algebra) if $\mathcal{A}_1$ is completely included in $\mathcal{A}_2$, $\mathcal{A}_1 \subset \mathcal{A}_2$.

The minimal algebra containing a given class of sets $\mathcal{C}$ is called the *algebra induced by the class of the sets* $\mathcal{C}$ and is often denoted by $\mathcal{A}(\mathcal{C})$. The minimal σ–algebra containing a given class of the sets $\mathcal{C}$ is called the *σ–algebra induced by the class of the sets* $\mathcal{C}$ and is often denoted by $\sigma(\mathcal{C})$.

It is evident that the intersection of any set of σ–algebras of the given space is also the σ–algebra as if the sets A_k ($k = 1, 2, \ldots$) simultaneously belong to all σ–algebras $\mathcal{A}_\alpha$. According to the definition of a σ–algebra their union $\bigcup A_k$ also belongs to all σ–algebras $\mathcal{A}_\alpha$, i.e. from $A_k \in \bigcap \mathcal{A}_\alpha$ follows $\bigcup A_k \in \bigcap \mathcal{A}_\alpha$. It is true also for algebras of sets: the intersection of any algebras of sets of the given space is also an algebra.

Theorem 2.2. *For any class of the sets $\mathcal{C}$ there exists σ–algebra $\sigma(\mathcal{C})$ induced by it.*

▷ There always exist σ–algebras containing a given class of sets $\mathcal{C}$, for instance, a σ–algebra of all sets of a given space. The intersection of all σ–algebras containing $\mathcal{C}$ will be the minimal σ–algebra $\sigma(\mathcal{C})$ which contains $\mathcal{C}$. ◁

R e m a r k. This theorem is also valid for the algebras: for any class of sets $\mathcal{C}$ there exists the algebra $\mathcal{A}(\mathcal{C})$ induced by it.

E x a m p l e 2.3. The class of all finite unions of intervals of the real axis R represents an algebra.

E x a m p l e 2.4. The class of all finite unions of rectangles of the space R^n (including the rectangles with infinite sides) represents an algebra.

E x a m p l e 2.5. The class of all intervals of the real axis R (the rectangles of the space R^n) induces the σ–algebra which is called a *σ–algebra of Borel sets*. In the case R^n, $n > 1$ this σ–algebra coincides with the σ–algebra induced by the set of all the balls.

2.1.4. Construction of a Semi–Algebra and an Algebra Induced by a Given Class of Sets

Let $\mathcal{C}_0$ be an arbitrary class of sets of the space X. We set up the problem to find algebra of sets induced by the class of the sets $\mathcal{C}_0$.

We complement $\mathcal{C}_0$ by the empty set $\varnothing$ and after that we complement the received class of sets by the complement of all the sets which are included in it. The class of sets obtained in such a way we denote by C_1. This class of sets contains the empty set $\emptyset$ and is closed under the complement operation. Besides this $C_0 \subset C_1$.

Let C_2 be a set of all finite intersections of sets of the class C_1. We shall prove that C_2 is a semi–algebra.

▷ It is evident that any set of the class C_1 also enters in C_2, $C_1 \subset C_2$. Consequently, C_2 contains the empty set $\emptyset$ and the whole space X. It is also evident that any finite intersection of finite intersections of sets of the class C_1 in its turn is a finite intersection of sets of the class C_2. Thus the class of sets C_2 is closed under the operation of finite intersection.

As any set A of the class C_2 may be presented in the form

$$A = \bigcap_{k=1}^{n} A_k, \quad A_1, \ldots, A_n \in C_1,$$

then

$$\bar{A} = \bigcup_{k=1}^{n} \bar{A}_k = \bar{A}_1 \bigcup \bar{A}_2 A_1 \bigcup \ldots \bigcup \bar{A}_n A_1 \ldots A_{n-1}.$$

But the class of sets C_1 is closed under the complement operation. Consequently, $\bar{A}_1, \bar{A}_2, \ldots, \bar{A}_n \in C_1$ and $\bar{A}_1, \bar{A}_2 A_1, \ldots, \bar{A}_n A_1 \ldots A_{n-1} \in C_2$. Thus the complement of any set of the class C_2 may be presented in the form of finite union of pairwise nonintersecting sets from C_2. ◁

Let determine now the class of sets C_3 as the set of all finite unions of sets of the class C_2. We shall prove that C_3 represents an algebra of sets. It is sufficient for this purpose to show that C_3 contains the empty set $\emptyset$ and is closed under the operations of finite intersection and complement.

▷ First it is clear that any set from the semi–algebra C_2 also enters into C_3, $C_2 \subset C_3$. Consequently, $\emptyset, X \in C_3$.

If $A, B \in C_3$ then

$$A = \bigcup_{k=1}^{n} A_k, \quad B = \bigcup_{l=1}^{m} B_l, \quad A_1, \ldots, A_n, B_1, \ldots, B_m \in C_2$$

and consequently,

$$AB = \bigcup_{k=1}^{n} \bigcup_{l=1}^{m} A_k B_l \in C_3$$

by virtue of the fact that $A_k B_l \in C_2$.

If $C \in C_2 \subset C_3$ then in view of C_2 is a semi–algebra,

$$\bar{C} = \bigcup_{p=1}^{N} C_p, \quad C_1, \ldots, C_N \in C_2.$$

So, the complement of any set from the semi–algebra C_2 belongs to C_3 as the finite union of the sets of the class C_2. If A is an arbitrary set of the class C_3 then

$$A = \bigcup_{k=1}^{n} A_k, \quad A_1, \ldots, A_n \in C_2,$$

and as a result we have

$$\bar{A} = \bigcap_{k=1}^{n} \bar{A}_k \in C_3,$$

as from $A_1, \ldots, A_n \in C_2$ follows that $\bar{A}_1, \ldots, \bar{A}_n \in C_3$ and the class C_3 is closed under the operation of finite intersection. ◁

Thus the constructed class of sets C_3 represents an algebra. According to the construction $C_0 \subset C_1 \subset C_2 \subset C_3$ this algebra contains initial class of sets.

It remains to prove that algebra C_3 is in any other algebra which contains the class of the sets C_0. Let $\mathcal{A}$ be an arbitrary algebra which contains the class of the sets C_0.

▷ In view of the properties of an algebra the algebra $\mathcal{A}$ contains $\varnothing$, X and all complements of the sets from C_0, i.e. $C_1 \subset \mathcal{A}$; $\mathcal{A}$ contains all finite intersections of the sets from C_1, i.e. $C_2 \subset \mathcal{A}$; $\mathcal{A}$ contains all finite unions of the sets from C_2, i.e. $C_3 \subset \mathcal{A}$. ◁

So, we proved that the class of sets C_3 constructed by the stated way represents an algebra induced by the class of the sets C_0. In passing we proved that C_3 represents an algebra of the sets induced by the semi–algebra C_3. The latter proposition may be formulated in the form of the following theorem.

Theorem 2.3. *An algebra of sets induced by a semi–algebra represents a set of all finite unions of sets of this semi–algebra.*

It follows also the theorem.

Theorem 2.4. *A countable class of sets induces a countable algebra of sets.*

▷ If C_0 is a countable class of the sets then C_1 will be also a countable class of the sets. The semi–algebra C_2 as the set of all finite intersections of the sets of the countable class C_1 will be also a countable one. Finally, the algebra C_3 as the set of all finite unions of the sets of the countable class C_2 will be also a countable one. ◁

2.1.5. Construction of a σ-Algebra Induced by a Given Sets Algebra

It was shown in Subsection 2.1.4 how on the basis of an arbitrary class of the sets we may construct a semi-algebra and an algebra of the sets induced by this class. In this case it was proved Theorem 2.3 which states that the algebra of the sets induced by a given semi-algebra represents a set of all finite unions of the sets of this semi-algebra. Naturally, this brings up the question how to construct a σ-algebra induced by a given class of the sets? Now we shall show how to do it.

▷ Let $\mathcal{B}$ be an algebra of the sets. We consider a class of the sets $\mathcal{B}_1$ which represents an aggregate of all finite and countable unions and intersections of the sets of the algebra $\mathcal{B}$. This class of the sets is not yet a σ-algebra as it does not contain, for instance, countable unions of the sets which are included in it representing countable intersections of the sets of algebra $\mathcal{B}$. Therefore we shall construct a class of sets $\mathcal{B}_2$ which represents an aggregate of all finite and countable unions and intersections of the sets of the class $\mathcal{B}_1$. Continuing this process we shall obtain the sequence of the classes of sets $\{\mathcal{B}_n\}$. In this case every class $\mathcal{B}_n$ represents a set of all finite and countable unions and intersections of the class $\mathcal{B}_{n-1}$ and consequently, contains $\mathcal{B}_{n-1}$, $\mathcal{B}_n \supset \mathcal{B}_{n-1}$ $(n = 1, 2, \ldots; \mathcal{B}_0 = \mathcal{B})$. As a result the obtained limit class of sets $\mathcal{A} = \bigcup \mathcal{B}_n$ possesses such a property that every set $A \in \mathcal{A}$ is derived from the sets of initial algebra $\mathcal{B}$ by nothing more than countable set of operations of union and intersection. In this case $\mathcal{A}$ contains all the sets which may be received from the sets of the algebra $\mathcal{B}$ by finite or countable set of the operations of union and intersection. Consequently, the class of the sets $\mathcal{A}$ represents a σ-algebra. As any σ-algebra containing the algebra $\mathcal{B}$ also contains all the sets which may be received from the sets of algebra $\mathcal{B}$ by finite or countable set of the operations of union or intersection then $\mathcal{A}$ represents minimal σ-algebra which contains the algebra $\mathcal{B}$.

It is clear that if the algebra $\mathcal{B}$ is induced by some class of the sets $\mathcal{C}$ then the constructed σ-algebra $\mathcal{A}$ will be the σ-algebra induced by the class of the sets $\mathcal{C}$.

2.1.6. Rings and Semi-Rings of Sets

Besides the algebras and semi-algebras there are other classes of sets in sets theory, in particular, the rings and semi-rings of sets.

The class of sets $\mathcal{R}$ of the space X is called *a ring* if together with any sets A, $B \in \mathcal{R}$ it contains an intersection AB and a symmetric difference $A \triangle B$.

As $A \bigcup B = AB \triangle (A \triangle B)$, $A \setminus B = A \triangle AB$ then from A, $B \in \mathcal{R}$ follows $A \bigcup B \in \mathcal{R}$, $A \setminus B \in \mathcal{R}$. As $A \setminus A = \emptyset$ then any ring contains the empty set. Thus the ring is closed relative to the operations of finite intersections, union and difference. In the special case if a ring contains the whole space X then it represents an algebra (Subsection 2.1.3).

The class of sets $\mathcal{C}$ is called *a semi–ring* if it contains the empty set $\emptyset$, the intersections of any two sets which are included in it and from A, $A_1 \in \mathcal{C}$, $A_1 \subset A$ follows that

$$A = \bigcup_{k=1}^{n} A_k \,,$$

where $A_1, \ldots, A_n$ are pairwise nonintersecting sets of the class $\mathcal{C}$, A_1, $\ldots, A_n \in \mathcal{C}$, $A_k A_l = \emptyset$ at $k \neq l$.

In the special case if a semi–ring contains the whole space X then it is a semi–algebra (Subsection 2.1.2).

Thus an algebra and a semi–algebra represent a ring and a semi–ring with the unit (the whole space X is assumed as *the unit*).

R e m a r k. We shall not study here the rings and semi–rings of sets as we shall not meet with them in chapters that follow.

2.1.7. Monotone Classes of Sets

The sequence of sets $\{A_n\}$ is called *an increasing (decreasing)* one if $A_p \subset A_q$ (correspondingly $A_p \supset A_q$) at any p and $q > p$. Increasing and decreasing sequences are called *monotone*.

The limit of the increasing sequence of sets $\{A_n\}$ is called the union of all these sets

$$\lim_{n \to \infty} A_n = \bigcup_{n=1}^{\infty} A_n \,.$$

In the special case $\lim_{n \to \infty} A_n$ may coincide with the whole space X.

The limit of decreasing sequence of the sets $\{A_n\}$ is called the intersection of all these sets

$$\lim_{n \to \infty} A_n = \bigcap_{n=1}^{\infty} A_n \,.$$

In the special case $\lim_{n \to \infty} A_n$ may be the empty set $\emptyset$.

Let $\{A_n\}$ be any sequence of sets. It is evident that the sequences $\{B_n\}$ and $\{C_n\}$,

$$B_n = \bigcup_{k=1}^{n} A_k, \quad C_n = \bigcap_{k=n}^{\infty} A_k,$$

are increasing and the sequences$\{D_n\}$ and $\{E_n\}$,

$$D_n = \bigcup_{k=n}^{\infty} A_k, \quad E_n = \bigcap_{k=1}^{n} A_k,$$

are decreasing ones. Thus from any sequence of sets we may construct monotone sequences by means of the operations of union and intersection.

A *monotone class* of sets is called such a class of sets which contains the limits of all monotone sequences of sets included in it. Evidently an intersection of monotone classes is also a monotone class. Therefore for any class of sets $\mathcal{C}$ there exists a minimal monotone class containing $\mathcal{C}$. This minimal class is called *a monotone class induced by a class of sets* $\mathcal{C}$ and is denoted by $\mathcal{M}(\mathcal{C})$.

Theorem 2.5. *Any σ–algebra represents a monotone class of sets.*

▷ As σ–algebra contains the limits of monotone sequences of sets which are included in it as countable unions and intersections then any σ–algebra is a monotone class. ◁

Theorem 2.6. *If monotone class $\mathcal{M}$ is an algebra then it represents a σ–algebra.*

▷ Really, let $\{A_n\}$ be any sequence of sets of $\mathcal{M}$, $A_n \in \mathcal{M}$. From the fact that $\mathcal{M}$ is an algebra follows that $B_n = \bigcup_{k=1}^{n} A_k \in \mathcal{M}$ therewith the sets B_n form a monotone increasing sequence. And as $\mathcal{M}$ is a monotone class then it also contains a limit of the sequence B_n,

$$\lim_{n \to \infty} B_n = \bigcup_{n=1}^{\infty} B_n = \bigcup_{k=1}^{\infty} A_k \in \mathcal{M}.$$

Consequently, all countable unions of sets from $\mathcal{M}$ belong to $\mathcal{M}$, i.e. $\mathcal{M}$ is a σ–algebra. ◁

Theorem 2.7. *A monotone class $\mathcal{M}$ induced by an algebra of sets $\mathcal{C}$ coincides with σ–algebra $\mathcal{A}$ generated by the same algebra $\mathcal{C}$.*

▷ As σ–algebra is a monotone class and $\mathcal{M}$ is a minimal monotone class which contains algebra $\mathcal{C}$ then $\mathcal{M} \subset \mathcal{A}$. If we prove that $\mathcal{M}$ represents an algebra then according to Theorem 2.6 $\mathcal{M}$ will be a σ–algebra. And from the fact that $\mathcal{A}$ is minimal algebra containing $\mathcal{C}$ it will follow $\mathcal{A} \subset \mathcal{M}$. Necessary result $\mathcal{A} = \mathcal{M}$ will follow from obtained two inclusions. Thus it remains to prove that $\mathcal{M}$ is an algebra.

As $\mathcal{M}$ contains algebra $\mathcal{C}$ then $\varnothing \in \mathcal{M}$, $X \in \mathcal{M}$ and it remains to prove that from $A \in \mathcal{M}$ follows $\bar{A} \in \mathcal{M}$ and from A, $B \in \mathcal{M}$ follows that $AB \in \mathcal{M}$.

Let us assume that $\mathcal{M}' = \{A : A \in \mathcal{M}, \bar{A} \in \mathcal{M}\}$. It is evident that $\mathcal{C} \subset \mathcal{M}'$. Let $\{A_n\}$ any monotone sequence of sets of the class $\mathcal{M}'$. As $\mathcal{M}' \subset \mathcal{M}$, then $A_n \in \mathcal{M}$ and $A = \lim A_n \in \mathcal{M}$. But $\bar{A}_n \in \mathcal{M}$ and the sequence of sets $\{\bar{A}_n\}$ is also monotone. Consequently, $\bar{A} = \lim \bar{A}_n \in \mathcal{M}$ and $A = \lim A_n \in \mathcal{M}'$. Thus $\mathcal{M}'$ is a monotone class containing the algebra $\mathcal{C}$. As $\mathcal{M}$ is a minimal monotone class containing $\mathcal{C}$ then $\mathcal{M}' = \mathcal{M}$, i.e. $\mathcal{M}$ is closed relative to the complement operation.

In order to prove the completeness of $\mathcal{M}$ relative to finite intersections we determine for any set $A \in \mathcal{M}$ the class of sets $\mathcal{M}_A = \{B : B \in \mathcal{M}, AB \in \mathcal{M}\}$. In the same way as in the case of $\mathcal{M}'$ it is proved that $\mathcal{M}$ is a monotone class. It is evident that if $B \in \mathcal{M}_A$ then $A \in \mathcal{M}_B$. But if $A \in \mathcal{C}$ then $\mathcal{C} \subset \mathcal{M}_A$ and consequently, $\mathcal{M}_A = \mathcal{M}$ as $\mathcal{M}$ is a minimal monotone class containing $\mathcal{C}$. Therefore any set $B \in \mathcal{M}$ also belongs to the class $\mathcal{M}_A$, $B \in \mathcal{M}_A$ if $A \in \mathcal{C}$. As $A \in \mathcal{M}_B$ then $\mathcal{C} \subset \mathcal{M}_B$ and consequently, $\mathcal{M}_B = \mathcal{M}$ for any set $B \in \mathcal{M}$. Thus $\mathcal{M}$ is closed relative to the operations of complement and finite intersections and contains $\varnothing$ and X, i.e. represents an algebra. ◁

2.1.8. Product of Two Spaces

Let us consider two spaces X and Y. The set Z of all ordered pairs $z = \{x, y\}$, $x \in X$, $y \in Y$ is called *a direct product* or *Cartesian product* of the spaces X and Y and is denoted $Z = X \times Y$ (Subsection 1.2.2). Hence and the point $x \in X$ is called a *projection of a point $z = \{x, y\} \in Z$ on the space X* and the point $y \in Y$ is called *a projection of the point $z = \{x, y\} \in Z$ on the space Y*. Let $A \subset X$ and $B \subset Y$ be any arbitrary sets. The aggregate of all the pairs $z = \{x, y\}$, $x \in A$, $y \in B$ is called *a rectangle* with the sides A and B in the space $Z = X \times Y$ and is denoted $A \times B$.

So, for instance, if X and Y are perpendicular real axes on a plane then a direct product $X \times Y$ represents a plane, and the rectangles in

$X \times Y$ are ordinary rectangles on a plane if A and B are the intervals on lines X and Y.

Let $(X, \mathcal{A})$ and $(Y, \mathcal{B})$ be two measurable spaces. The aggregate of all the rectangles $A \times B$, $A \in \mathcal{A}$, $B \in \mathcal{B}$ being *measurable rectangles* represents a semi–algebra. Really, $\emptyset = \emptyset \times \emptyset$, $X \times Y$ are the rectangles. The intersection of two rectangles $(A \times B)(C \times D) = AC \times BD$ is a rectangle. The complement of the rectangle $A \times B$ represents an union of three pairwise nonintersecting rectangles

$$\overline{A \times B} = (\bar{A} \times B) \bigcup (A \times \bar{B}) \bigcup (\bar{A} \times \bar{B}) .$$

A minimal σ–algebra $\mathcal{C}$ which contains all countable rectangles (σ–algebra generated by a semi–algebra of countable rectangles $A \times B$, $A \in \mathcal{A}$, $B \in \mathcal{B}$) is called a *product of σ–algebras* $\mathcal{A}$ and $\mathcal{B}$ and is denoted $\mathcal{C} = \mathcal{A} \times \mathcal{B}$.

The product $X \times Y$ with a σ–algebra in it $\mathcal{C} = \mathcal{A} \times \mathcal{B}$, $(Z, \mathcal{C}) = (X \times Y, \mathcal{A} \times \mathcal{B})$ is called *a product of two measurable spaces* $(X, \mathcal{A})$ and $(Y, \mathcal{B})$.

Let take an arbitrary set $C \subset Z$ and choose from it the points $z = \{x, y\}$ correspodent to some fixed x. Then we obtain *the section* C_x of the set C in the point x. This definition of a section in the special case when X and Y are the real axes, and $Z = X \times Y$ is a plane gives a section of a plane figure by a line parallel to the axis y.

Theorem 2.8. *The sections of measurable sets in the product of measurable spaces are measurable.*

▷ Let separate from the σ–algebra $\mathcal{C} = \mathcal{A} \times \mathcal{B}$ the class of sets $\mathcal{C}'$ whose all sections are measurable, $\mathcal{C}' \subset \mathcal{C}$ and prove that $\mathcal{C}' = \mathcal{C}$. For this purpose we notice that according to the definition of measurable rectangle all its sections are measurable. Consequently, $\mathcal{C}'$ contains all measurable rectangles. Further $\mathcal{C}'$ is a σ–algebra as any countable unions of the sets with countable sections and any their countable intersections have measurable sections from the fact that $\mathcal{A}$ and $\mathcal{B}$ are σ–algebras. Thus $\mathcal{C}'$ is a σ–algebra which contains all measurable rectangles. But $\mathcal{C}'$ is a minimal σ–algebra containing all measurable rectangles. So, $\mathcal{C} \subset \mathcal{C}'$, and consequently, $\mathcal{C}' = \mathcal{C}$. ◁

2.1.9. Product of a Set of Spaces

Let take an arbitrary set of values T of some parameter t. To each $t \in T$ corresponds the space X_t. We shall choose for each $t \in T$ an arbitrary element $x_t \in X_t$ and consider a set $\{x_t : t \in T\}^a$. The set of all sets $\{x_t : t \in T\}$ is called *a product (direct product) of the spaces* X_t and is denoted by

$$X^T = \prod_{t \in T} X_t \, .$$

Here the element $x_t \in X_t$ at a given t is called *a projection of the point* $x_t \in X_t$ $x = \{x_t : t \in T\} \in X^T$ on the space X_t.

Let, for instance, T be an interval $(0, T)$ of the real axis R and X_t at each t, $0 < t < T$ be the real axis. If we choose the number $x_t \in X_t$ for each t then the set of numbers x_t correspondent to all t, $0 < t < T$, i.e. the set $\{x_t : t \in T\}$ will represent the function $x(t)$ and a product of the spaces X_t will represent a space of all functions determined at the interval $(0, T)$. Similarly, if T is a set of the space R^n and X_t at each $t \in T$ represents the space R^m then X^T is a space of all m–dimensional vector functions of n–dimensional vector argument $t \in T$.

Let take an arbitrary set $S \subset T$ and the correspondent product of the spaces

$$X^S = \prod_{t \in S} X_t \, .$$

The received space X^S is called a *subspace* of the space X^T and the points of the space X^S are called the *projections* of X^S of the correspondent points $\{x_t : t \in T\}$ of the space X^T.

Let A_S be an arbitrary set of the space X^S, $A_S \subset X^S$. The set of the points of the space X^T whose projections of X^S belong to the set A_S is called a *cylinder* in X^T with the base A_S in X^S. If S is a finite subset of the set T, $S = \{t_1, \ldots, t_n\}$ the cylinder will have finite–dimensional base A_S in finite product of the spaces $X_{t_1} \times \cdots \times X_{t_n}$. If

$$A_S = \prod_{k=1}^{n} A_k \, , \quad A_k \subset X_{t_k},$$

then the cylinder with the base A_S is called *a rectangle* in X^T with the sides $A_1, \ldots, A_n$ in the subspaces $X_{t_1}, \ldots, X_{t_n}$.

So, for example, the aggregate of all the functions whose values in the points $t_1, \ldots, t_n$ belong correspondingly to the sets $A_1, \ldots, A_n$ will be a rectangle with the sides $A_1, \ldots, A_n$ in the space X^T of all numerical functions $x(t)$ given at the interval $(0, T)$.

---

[a] It is possible on the basis of axiom of choice (Subsection 1.1.9).

If in each space X_t, $t \in T$ σ–algebra of the sets A_t is chosen then σ–algebra generated by a class of all measurable rectangles (i.e. the rectangles with measurable sides) in the space X^T is called *a product of* σ–*algebras* A_t, $t \in T$ and is denoted by

$$A^T = \prod_{t \in T} A_t,$$

The corresponding measurable space $\left(X^T, A^T\right)$ is called *a product of measurable spaces* $\left(X_t, A_t\right)$, $t \in T$.

Let C be an arbitrary set in the product of the spaces X^T, S be a subset of the set T, $S \subset T$. The set C_S of all the points $x \in C$ of the space X correspondent to fixed values $x_t \in X_t$, $t \in S$ is called *a section of the set* C *in the point* $x' = \left\{ x_t : t \in S \right\}$.

Let S be an arbitrary subset of the set T, $S \subset T$. Then to σ–algebra A^S in the product of the spaces X^S corresponds in X^T σ–algebra of cylinders with the bases in X^S. We shall denote this σ–algebra A^S similarly as σ–algebra of the bases of these cylinders. It is clear that $A^S \subset A^T$.

As for any subset S of the set T, $S \subset T$, $X^T = X^S \times X^{T \setminus S}$, $A^T = A^S \times A^{T \setminus S}$ then from Theorem 2.8 follows that in the product of any set of the spaces all the sections of measurable sets are measurable.

For further study the main classes of sets refer to (Halmos 1950, Xia Dao–Xing 1972).

Problems

2.1.1. Whether the set of all intervals of the real axis with end point coordinates m/n at fixed n be a semi–algebra? Consider two special cases: a) when the set of the fractions with the denominator n is complemented by $-\infty$ and $+\infty$; b) when this set is not complemented by $-\infty$ and $+\infty$.

2.1.2. Whether the set of all open intervals of the real axis be a semi–algebra? The same question concerning the cases of the set of all closed, right semi–closed and left semi–closed intervals is posed.

2.1.3. Whether the set of all finite unions of open intervals be an algebra? The same question concerning the cases of all closed, right semi–closed and left semi–closed intervals is raised.

2.1.4. Construct a semi–algebra and an algebra generated by the set of all open intervals of the real axis. Whether they coincide with the semi–algebra of all intervals and the algebra of finite union of all intervals?

2.1.5. Whether the set of all the balls in R^n be a semi–algebra?

2.1.6. Show that an algebra of sets may be defined as a class of sets containing complements and pairwise intersections of sets which belong to them.

2.1.7. Prove that the σ–algebra of Borel sets of the real axis coincides with the σ–algebra, induced by: a) the set of all the intervals with rational end points; b) the set of all open intervals.

2.1.8. Prove that the σ–algebra of Borel sets in R^n coincides with the σ–algebra induced by: a) the set of all the rectangles with rational coordinates of tops; b) the set of all open rectangles; c) the set of all balls; d) the set of all balls with rational radii and rational coordinats of their centres.

2.1.9. Prove that the set of all (Jordan) measurable sets in R^n is the algebra.

I n s t r u c t i o n. It is evident that from $B^- \subset A \subset B^+$ we have $\bar{B}^+ \subset \bar{A} \subset \bar{B}^-$ and $\bar{B}^- \setminus \bar{B}^+ = B^+ \setminus B^-$, where B^- and B^+ are cube unions which are contained in A and have nonempty intersections with A.

2.1.10. What is the difference between ordinary rectangles in the plane (in R^n) and measurable rectangles defined in Subsection 2.1.8? Prove that the σ–algebra of Borel sets in the plane (in R^n) induced by the semi–algebra of ordinary rectangles coincides with the σ–algebra induced by the semi–algebra of more general rectangles defined in Subsection 2.1.8.

2.1.11. Prove that a class of all measurable rectangles in the product of spaces $\prod_{t \in T} X_t$ is a semi–algebra.

I n s t r u c t i o n. It is clear that any two rectangles with sides $A_1 \times \cdots \times A_n$ and $B_{k+1} \times \cdots \times B_m$ in the product of spaces $X_{t_1} \times \cdots \times X_{t_n}$ and $X_{t_{k+1}} \times \cdots \times X_{t_m}$, $k < n < m$ may be considered as rectangles with bases $A_1 \times \cdots \times A_n \times X_{t_{n+1}} \times \cdots \times X_{t_m}$ and $X_{t_1} \times \cdots \times X_{t_k} \times B_{k+1} \times \cdots \times B_m$ in the same product of spaces $X_{t_1} \times \cdots \times X_{t_m}$ and also that the complement of the rectangle with the base $A_1 \times \cdots \times A_n$ in the product of spaces $X_{t_1} \times \cdots \times X_{t_n}$ is the union of $2^n - 1$ pairwise nonintersecting rectangles with bases $A_1 \times \cdots \times A_n$ where the sets $A_1, \ldots, A_n$ are replaced by their complements.

2.1.12. Prove that the union of σ–algebras $\mathcal{A}^S$ in the product of spaces X^T corresponding to all finite subsets S of the set T represents the algebra which induces the σ–algebra $\mathcal{A}^T$.

2.1.13. Prove that the union of σ–algebra $\mathcal{A}^S$ in the product of the spaces X^T corresponding to all countable subsets S of the set T is the σ–algebra $\mathcal{A}^T$.

2.1.14. Establish the inherent property of functions sets from $\mathcal{A}^T$ (Problem 2.1.13) when each of spaces X_t represents R^n (the same n for all $t \in T$).

2.1.15. Let $X_1, \ldots, X_n$ be linear normed spaces. Show that in the product of the spaces $X = X_1 \times \cdots \times X_n$ the norm of the element $x = \{x_1, \ldots, x_n\}$ may be defined by formula

$$\| x \| = \| x_1 \| + \cdots + \| x_n \| .$$

2.1.16. Under the conditions of Problem 2.1.15 when $X_1, \ldots, X_n$ are B–spaces the space X with the norm defined above will be also B–space.

2.1.17. Let $X_1, \ldots, X_n$ be H–spaces. Show that in the product of the spaces $X = X_1 \times \cdots \times X_n$ a scalar product of the elements $x = \{x_1, \ldots, x_n\}$ and $y = \{y_1, \cdots, y_n\}$ may be defined by following formula:

$$(x, y) = (x_1, y_1) + \cdots + (x_n, y_n) .$$

Will the space X with such a scalar product be H–space or not?

2.2. Set Functions and Measures

2.2.1. Additive Set Functions

All set functions which occur in the geometry and the physics possess one character property: while combining the finite number of pairwise noninterseting sets their values are summarized. This property of set functions is called *the additivity*. Therefore it is natural while studying the set functions to restrict ourselves to the functions possessing this property. But it is necessary for this purpose to restrict the class of possible spaces of the values of the set functions by linear spaces.

The set function $\varphi(A)$ determined at some class of sets $\mathcal{C}$ and taking the values from some linear space is called *an additive function* if for any pairwise nonintersecting sets $A_1, \ldots, A_n \in \mathcal{C}$, $A_k A_h = \emptyset$ at $h \neq k$, $\bigcup A_k \in \mathcal{C}$ we have

$$\varphi\left(\bigcup_{k=1}^{n} A_k\right) = \sum_{k=1}^{n} \varphi(A_k). \tag{2.2}$$

In order to underline that this equality is valid for any finite number of items the additive function $\varphi(A)$ is also called *a finite–additive*.

Theorem 2.9. *The additive function $\varphi(A)$ is equal to zero on the empty set*

$$\varphi(\emptyset) = 0 \tag{2.3}$$

(*certainly if its domain $\mathcal{C}$ contains $\emptyset$*).

▷ As $A = A \bigcup \varnothing$ for any set A,

$$\varphi(A) = \varphi(A \bigcup \varnothing) = \varphi(A) + \varphi(\varnothing) \quad \forall A \in \mathcal{C},$$

then whence it follows (2.3). ◁

If equality (2.2) is valid not only for any finite number of sets but for any sequence of pairwise nonintersecting sets $\{A_n\}$, $A_k A_h = \varnothing$ at $h \neq k$, $\bigcup A_k \in \mathcal{C}$,

$$\varphi\left(\bigcup_{k=1}^{\infty} A_k\right) = \sum_{k=1}^{\infty} \varphi(A_k), \qquad (2.4)$$

then the function $\varphi(A)$ is called *countable-additive* or shortly *σ-additive*.

It is clear that the notion of σ–additive function is applicable only to such functions in whose space of values the convergence is determined. Otherwise the infinite sum in the right–hand side of formula (2.4) has no sense. Therefore further we shall consider only the set functions with the values in normed linear spaces.

The additive and σ–additive functions are called *the measures*. Later on while speaking about the measures we shall always imply σ–additive measures where an additive measure is not concerned.

The natural measure domain is some σ–algebra. One of the reasons of this fact is the property of σ–algebra to contain all countable unions of sets which enter into it. The other reasons will be clear later on. But to prescribe a concrete measure immediately on σ–algebra of the sets is impossible. Therefore at first we have to determine the measure on more simple class of sets and after that to perform its extension on a chosen σ–algebra.

The space X with a given σ–algebra of its sets $\mathcal{A}$ and determined on $\mathcal{A}$ by the measure $\mu(A)$, $A \in \mathcal{A}$ is called *a space with a measure* and is denoted by $(X, \mathcal{A}, \mu)$.

The numerical measures are of great importance, i.e. the measures with the values on the real axis or on the complex plane.

Numerical additive function is called *a finite* one if it does not have infinite values on none of the domain set.

If it is granted that numerical additive functions $\varphi(A)$ given on the algebra of sets may take infinite values of different signs then we shall arrive to the uncertainty. Really, if there exist such sets A and B that $\varphi(A) = +\infty$, $\varphi(B) = -\infty$ and $\varphi(\bar{A}) \neq -\infty$, $\varphi(\bar{B}) \neq +\infty$ then we shall have

$$\varphi(X) = \varphi(A) + \varphi(\bar{A}) = +\infty, \quad \varphi(X) = \varphi(B) + \varphi(\bar{B}) = -\infty.$$

In order to avoid this uncertainty we have to assume that $\varphi(A)$ may take infinite values of only one sign. For definiteness we shall consider that $\varphi(A)$ does not take the value $-\infty$.

The numerical additive functions $\varphi(A)$ is called σ-*finite* if there exists such a partition of the space X into pairwise nonintersecting sets $X = \bigcup\limits_{k=1}^{\infty} X_k,\ X_k X_l = \emptyset$ at $k \neq l$, that $\varphi(A)$ is finite on all the sets X_k $(k = 1, 2, \ldots)$.

Theorem 2.10. *If a real function* $\varphi(A)$ *is additive and*

$$\varphi\left(\bigcup_{k=1}^{\infty} A_k\right) < \infty,$$

$A_k A_h = \emptyset$ *at* $h \neq k$ *then the series*

$$\sum_{k=1}^{\infty} \varphi(A_k)$$

converges absolutely.

▷ Really, let

$$A_k^+ = \begin{cases} A_k & \text{at} \quad \varphi(A_k) \geq 0, \\ \emptyset & \text{at} \quad \varphi(A_k) < 0, \end{cases} \quad A_k^- = \begin{cases} A_k & \text{at} \quad \varphi(A_k) < 0, \\ \emptyset & \text{at} \quad \varphi(A_k) \geq 0. \end{cases}$$

Then

$$\varphi\left(\bigcup_{k=1}^{\infty} A_k\right) = \varphi\left(\bigcup_{k=1}^{\infty} A_k^+\right) + \varphi\left(\bigcup_{k=1}^{\infty} A_k^-\right) < \infty$$

and as $\varphi(A)$ may not take the value $-\infty$, so,

$$-\infty < \varphi\left(\bigcup_{k=1}^{\infty} A_k^-\right) = \sum_{k=1}^{\infty} \varphi(A_k^-) \leq 0$$

$$0 \leq \varphi\left(\bigcup_{k=1}^{\infty} A_k^+\right) = \sum_{k=1}^{\infty} \varphi(A_k^+) < \infty.$$

It follows from these inequalities that the series with nonnegative terms $\sum |\varphi(A_k^-)|$ and $\sum \varphi(A_k^+)$ converge what proves our statement. ◁

E x a m p l e 2.6. The probability represents a nonnegative σ-additive set function, i.e. a nonnegative measure. This measure evidently is a finite (it cannot be more than 1).

E x a m p l e 2.7. The Riemann integral

$$\varphi(A) = \int\limits_A f(x)dx \,,$$

extended on the interval A of the real axis or on the domain A of a finite–dimensional space (certainly, Jordan measurable) is an additive set function. If an integral extended on the whole space represents a convergent improper integral then the function $\varphi(A)$ is finite.

E x a m p l e 2.8. The *Lebesgue measure* is determined on the intervals of the real axis R (on rectangular sets spaces R^n) as the length of the interval (the volume of a rectangular set)

$$l((a,\,b)) = l([a,\,b)) = l((a,\,b]) = l([a,\,b]) = b - a \,,$$
$$l(A) = (b_1 - a_1) \cdots (b_n - a_n) \,,$$

where $A = \{x \,:\, x_k \in (a_k,\,b_k),\, k = 1,\,\ldots,\,n\}$ or any rectangular set received from A by the change of the aggregate of open intervals $(a_k,\,b_k)$ by any combination of open, closed and semi–closed intervals. Determined in such a way on semi–algebra of the intervals (rectangles) the Lebesgue measure is uniquely determined according to the theorem about measure extension (Example 2.13) and on all Borel sets of the real axis R (space R^n).

E x a m p l e 2.9. The *Lebesgue–Stieltjes measure* is determined on the intervals of the real axis R by the formulae

$$s([a,\,b)) = F(b) - F(a) \,, \quad s([a,\,b]) = F(b+0) - F(a) \,,$$
$$s((a,\,b)) = F(b) - F(a+0) \,, \quad s((a,\,b]) = F(b+0) - F(a+0) \,,$$

where $F(x)$ is a continuous function from the left. Determined in such a way on semi–algebra of the intervals the Lebesgue–Stieltjes measure is uniquely determined according to the theorem about measure extension (Example 2.14) and on all Borel sets. The Lebesgue–Stieltjes measure is finite if the functions $F(x)$ is bounded.

By the same procedure we may determine the Lebesgue–Stieltjes measure in any finite–dimensional space R^n.

2.2.2. Continuous Set Functions

Set function $\varphi(A)$ is called *upper continuous* if for any monotone decreasing sequence of the sets $\{A_n\}$, $A_{n+1} \subset A_n$ $(n = 1,\, 2,\, \ldots)$,

$$\varphi(\lim A_n) = \varphi(\bigcap A_n) = \lim \varphi(A_n)$$

(in the case of numerical function φ must be $\varphi(A_n) < \infty$ $\forall n$). The set function $\varphi(A)$ is called *lower continuous* if for any increasing sequence of sets $\{A_n\}$, $A_{n+1} \supset A_n$, $(n = 1, 2, \ldots)$,

$$\varphi(\lim A_n) = \varphi(\bigcup A_n) = \lim \varphi(A_n).$$

The set function which is upper and lower continuous is called *a continuous* one.

2.2.3. General Properties of Measures

Let us study the general properties of the additive functions and measures.

Theorem 2.11. *The measure $\mu(A)$ determined on a σ-algebra of sets $\mathcal{A}$ is continuous.*

▷ Let $\{A_n\}$ be a monotone increasing sequence of sets $A = \lim\limits_{n \to \infty} A_n$, $A_n \in \mathcal{A}$. As $\lim A_n = \bigcup A_n$ and $\mathcal{A}$ is a σ-algebra then $A \in \mathcal{A}$. We put for generality $A_0 = \emptyset$. Then we shall have

$$A_n = \bigcup_{k=1}^{n} (A_k \backslash A_{k-1}), \quad \mu(A_n) = \sum_{k=1}^{n} \mu(A_k \backslash A_{k-1}),$$

$$A = \lim_{n \to \infty} A_n = \bigcup_{k=1}^{\infty} (A_k \backslash A_{k-1}),$$

where $A_k \backslash A_{k-1}$ are pairwise nonintersecting sets. Consequently, by virtue of the σ-additivity of measure

$$\mu(A) = \sum_{k=1}^{\infty} \mu(A_k \backslash A_{k-1}) = \lim_{n \to \infty} \sum_{k=1}^{n} \mu(A_k \backslash A_{k-1}) = \lim \mu(A_n).$$

Thus $\mu(A)$ is lower continuous.

Let $\{A_n\}$, $A_n \in \mathcal{A}$ be now a monotone decreasing sequence of sets. Then the sequence of sets $\{A_1 \backslash A_n\}$ will be monotone increasing and its limit will be $A_1 \backslash A$. From just proved lower continuity of $\mu(A)$ follows

$$\mu(A_1 \backslash A) = \lim \mu(A_1 \backslash A_n)$$

or

$$\mu(A_1) - \mu(A) = \mu(A_1) - \lim \mu(A_n).$$

Hence we get (in the case of a numerical measure should be $\mu(A_n)$ $< \infty$ $\forall n$)

$$\mu(A) = \lim \mu(A_n) \,,$$

i.e. $\mu(A)$ is upper continuous. Thus the measure $\mu(A)$ is upper and lower continuous, i.e. is continuous. ◁

The following two theorems may be considered as the theorems inverse of Theorem 2.11.

Theorem 2.12. *If the additive function* $\varphi(A)$ *determined on some class of sets* $\mathcal{C}$ *is lower continuous then it is* σ*-additive on* $\mathcal{C}$*.*

▷ Let $\{A_n\}$ be such arbitrary sequence of pairwise nonintersecting sets from $\mathcal{C}$, $A_n \in \mathcal{C}$, $A_k A_h = \varnothing$ at $h \neq k$ that all the sets

$$B_n = \bigcup_{k=1}^{n} A_k \quad (n = 1, 2, \ldots) \quad \text{and} \quad A = \bigcup_{k=1}^{\infty} A_k$$

belong to the class $\mathcal{C}$. It is evident that the sets B_n form a monotone increasing sequence whose limit is $A = \bigcup_{k=1}^{\infty} A_k = \bigcup_{n=1}^{\infty} B_n$. Consequently, by virtue of the lower continuous property of the function φ

$$\varphi(A) = \varphi\left(\bigcup_{k=1}^{\infty} A_k\right) = \lim_{n \to \infty} \varphi\left(\bigcup_{k=1}^{n} A_k\right),$$

and in view of its additivity

$$\varphi\left(\bigcup_{k=1}^{n} A_k\right) = \sum_{k=1}^{n} \varphi(A_k),$$

$$\varphi(A) = \lim_{n \to \infty} \sum_{k=1}^{n} \varphi(A_k) = \sum_{k=1}^{\infty} \varphi(A_k). \quad ◁$$

Theorem 2.13. *If the additive function* φ *determined on some class of sets* $\mathcal{C}$ *which contains the empty set is continuous at zero, i.e. if* $\lim \varphi(C_n) = 0$ *for any monotone decreasing sequence of sets* $\{C_n\} \subset \mathcal{C}$ *with the empty intersection* $\bigcap C_n = \varnothing$ *then it is* σ*-additive at* $\mathcal{C}$*.*

▷ Let $\{A_n\}$ be such an arbitrary sequence of pairwise nonintersecting sets from $\mathcal{C}$, $A_n \in \mathcal{C}$, $A_k A_h = \varnothing$ at $h \neq k$ that all the sets

$$C_n = \bigcup_{k=n+1}^{\infty} A_k \quad (n = 1, 2, \ldots) \quad \text{and} \quad A = \bigcup_{k=1}^{\infty} A_k$$

belong to the class $\mathcal{C}$. It is obvious that the sets C_n form a monotone decreasing sequence with the empty intersection. Therefore owing to the continuity of the function φ at zero

$$\lim_{n\to\infty} \varphi(C_n) = \varphi(\emptyset) = 0\,,$$

and by virtue of its additivity

$$\varphi(A) = \varphi\left(\bigcup_{k=1}^{\infty} A_k\right) = \sum_{k=1}^{n} \varphi(A_k) + \varphi\left(\bigcup_{k=n+1}^{\infty} A_k\right) = \sum_{k=1}^{n} \varphi(A_k) + \varphi(C_n)\,.$$

From these two equalities we receive the formula

$$\varphi(A) = \sum_{k=1}^{\infty} \varphi(A_k)\,. \quad \triangleleft$$

2.2.4. Properties of Nonnegative Measures

Let study now the peculiar properties of nonnegative additive functions and measures which are of great importance.

Theorem 2.14. *If a nonnegative additive function μ is determined on a semi–algebra of sets $\mathcal{C}$ them for any sets A, $B \in \mathcal{C}$, $A \subset B$*

$$\mu(A) \leq \mu(B)\,. \tag{2.5}$$

$\triangleright$ As $B = A \bigcup \bar{A}B$ and $\bar{A}$ represents a finite union of pairwise non-intersecting sets $C_1, \ldots, C_N \in \mathcal{C}$,

$$\bar{A} = \bigcup_{k=1}^{N} C_k\,,$$

then

$$B = A \bigcup BC_1 \bigcup \cdots \bigcup BC_N\,.$$

As by definition of a semi–algebra $BC_1, \ldots, BC_N \in \mathcal{C}$ then by virtue of the additivity of μ at $\mathcal{C}$

$$\mu(B) = \mu(A) + \sum_{k=1}^{N} \mu(BC_k) \geq \mu(A)\,. \quad \triangleleft$$

Theorem 2.15. *If a nonnegative additive function μ is determined on a semi–algebra of sets $\mathcal{C}$ then for any set $A \in \mathcal{C}$ and any pairwise nonintersecting subsets $A_1, \ldots, A_n \in \mathcal{C}$, $A_k A_h = \emptyset$ at $k \neq h$ the following inequality is valid:*

$$\sum_{k=1}^{n} \mu(A_k) \leq \mu(A). \tag{2.6}$$

▷ If the union of sets $A_1, \ldots, A_n$ belongs to $\mathcal{C}$ then (2.6) follows immediately from the additivity of μ and Theorem 2.14. Otherwise,

$$A \backslash \bigcup_{k=1}^{n} A_k = A\left(\overline{\bigcup_{k=1}^{n} A_k} \right) = A\left(\bigcap_{k=1}^{n} \bar{A}_k \right),$$

$$\bar{A}_k = \bigcup_{l=1}^{N_k} C_{kl}, \quad C_{kl} \in \mathcal{C}, \quad C_{kl} C_{kh} = \emptyset \quad \text{at} \quad k \neq h. \tag{2.7}$$

Putting $N = \max N_k$, $C_{kl} = \emptyset$ at $l > N_k$ we get

$$A \backslash \bigcup_{k=1}^{n} A_k = \bigcup_{l_1, \ldots, l_n = 1}^{N} A C_{1l_1} \cdots C_{nl_n}$$

and

$$A = \left(\bigcup_{k=1}^{n} A_k \right) \cup \left(A \backslash \bigcup_{k=1}^{n} A_k \right) = \left(\bigcup_{k=1}^{n} A_k \right) \cup \left(\bigcup_{l_1, \ldots, l_n = 1}^{N} A C_{1l_1} \cdots C_{nl_n} \right).$$

This formula expresses the set $A \in \mathcal{C}$ in the form of finite union of pairwise nonintersecting sets of a semi–algebra $\mathcal{C}$. Consequently, owing to the additivity of μ at $\mathcal{C}$ we have

$$\mu(A) = \sum_{k=1}^{n} \mu(A_k) + \sum_{l_1, \ldots, l_n = 1}^{N} \mu(A C_{1l_1} \cdots C_{nl_n}) \geq \sum_{k=1}^{n} \mu(A_k). \quad ◁$$

Theorem 2.16. *If a nonnegative additive function μ is determined at a semi–algebra of sets $\mathcal{C}$ then for any sets $A_1, \ldots, A_n \in \mathcal{C}$ for which*

$$A = \bigcup_{k=1}^{n} A_k \in \mathcal{C},$$

is valid the inequality

$$\mu(A) \le \sum_{k=1}^{n} \mu(A_k).\qquad(2.8)$$

▷ As

$$A = \bigcup_{k=1}^{n} A_k = A_1\bigcup A_2\bar{A}_1\bigcup\cdots\bigcup A_n\bar{A}_1\cdots\bar{A}_{n-1}\qquad(2.9)$$

and is valid formula (2.7) then similarly as at the proof of Theorem 2.15 we obtain

$$A_k\bar{A}_1\cdots\bar{A}_{k-1} = \bigcup_{l_1,\ldots,l_{k-1}=1}^{N} A_k C_{1l_1}\cdots C_{k-1,l_{k-1}} \quad (k = 2, \ldots, n).$$
$$(2.10)$$

Formulae (2.9) and (2.10) give the representation of the set A in the form of finite union of pairwise nonintersecting sets of a semi–algebra $\mathcal{C}$. Consequently, due to the additivity of μ at $\mathcal{C}$

$$\mu(A) = \mu(A_1) + \sum_{k=2}^{n} \sum_{l_1,\ldots,l_{k-1}=1}^{N} \mu\bigl(A_k C_{1l_1}\cdots C_{k-1,l_{k-1}}\bigr).\qquad(2.11)$$

As $A_k\bar{A}_1\ldots\bar{A}_{k-1} \subset A_k$ $(k = 2, \ldots, n)$ then by virtue of (2.10) from Theorem 2.15 follows

$$\sum_{l_1,\ldots,l_{k-1}=1}^{N} \mu\bigl(A_k C_{1l_1}\cdots C_{k-1,l_{k-1}}\bigr) \le \mu(A_k).\qquad(2.12)$$

This formula together with formula (2.11) leads to inequality (2.8). ◁

Theorem 2.17. *If μ is a nonnegative measure determined at a semi–algebra of sets $\mathcal{C}$ them for any sequence of the sets $\{A_n\} \subset \mathcal{C}$ for which*

$$A = \bigcup_{k=1}^{\infty} A_k \in \mathcal{C},$$

is valid the inequality

$$\mu(A) \le \sum_{k=1}^{\infty} \mu(A_k).\qquad(2.13)$$

This property of a nonnegative measure is called *the semi–additivity* of a measure.

▷ In this case formula (2.9) at $n = \infty$ and (2.10) give the representation of the set A in the form of countable union of pairwise nonintersecting sets of semi–algebra C and due to the σ–additivity of a measure formula (2.11) is valid at $n = \infty$. This formula together with (2.12) gives inequality (2.13). ◁

The following theorem which is in some sense inverse of Theorem 2.17 is often used while proving the σ–additivity of nonnegative additive function.

Theorem 2.18. *If a nonnegative additive function μ determined on a semi–algebra of sets C is semi–additive then it is σ–additive.*

▷ Let $\{A_n\} \subset C$ be any sequence of pairwise nonintersecting sets such as

$$A = \bigcup_{k=1}^{\infty} A_k \in C.$$

As at any n

$$\bigcup_{k=1}^{n} A_k \subset A,$$

then according to Theorem 2.16 and owing to the semi–additivity of the function μ

$$\sum_{k=1}^{n} \mu(A_k) \le \mu(A) \le \sum_{k=1}^{\infty} \mu(A_k) \qquad (2.14)$$

at any n. If $\mu(A) < \infty$ then the sequence of the sums in the left–hand side of these inequalities as a nondecreasing sequence of positive numbers upper bounded by the number $\mu(A)$ has the limit

$$\lim_{n \to \infty} \sum_{k=1}^{n} \mu(A_k) = \sum_{k=1}^{\infty} \mu(A_k).$$

Hence and from the validity of (2.14) at all n follows

$$\mu(A) = \sum_{k=1}^{\infty} \mu(A_k). \quad ◁$$

Theorems 2.14–2.18 are also valid in the case when C is an algebra or a σ–algebra of sets as any algebra of sets represents a semi–algebra.

E x a m p l e 2.10. The Lebesgue measure l is σ–additive on a semi–algebra of the intervals (rectangular sets in the case of the space R^n). It is evident that the that the Lebesgue measure is additive on $\mathcal{C}$: at any finite partition of an interval (a rectangle) into pairwise nonintersecting intervals the Lebesgue measure of a given interval is equal to the sum of measures of the intervals into which it is subdivided. In order to prove the σ–additivity of the Lebesgue measure on the basis of Theorem 2.18 it is sufficient to prove its semi–additivity on $\mathcal{C}$. For this purpose we shall take any finite interval (rectangle) A and divide it into countable set of pairwise nonintersecting intervals (rectangles)

$$A = \bigcup_{n=1}^{\infty} A_n \,.$$

Let set an arbitrary $\varepsilon > 0$ and take such a closed interval $B_\varepsilon \subset A$ that

$$l(A) < l(B_\varepsilon) + \varepsilon/2 \,.$$

We shall cover each interval A_n by such open interval $A_n^\varepsilon \supset A_n$ that

$$l(A_n^\varepsilon) < l(A_n) + \varepsilon/2^{n+1} \quad (n = 1, 2, \ldots) \,.$$

Then $\bigcup A_n^\varepsilon$ will represent a covering of a closed interval B_ε by open intervals

$$B_\varepsilon \subset A = \bigcup_{k=1}^{\infty} A_n \subset \bigcup_{k=1}^{\infty} A_n^\varepsilon \,.$$

And as a finite closed interval is compact then the set of intervals $\{A_n^\varepsilon\}$ contains a finite subset of the intervals covering the whole interval B_ε. We denote these intervals $A_{n_1}^\varepsilon, \ldots, A_{n_N}^\varepsilon$:

$$B_\varepsilon \subset \bigcup_{p=1}^{N} A_{n_p}^\varepsilon \,.$$

A finite union of the intervals (rectangles) in the right–hand side of this inclusion similarly as the intervals A and B_ε belongs to the algebra induced by semi–algebra $\mathcal{C}$ (Subsection 2.1.4). Therefore by virtue of Theorems 2.14 and 2.16

$$l(B_\varepsilon) \leq l\left(\bigcup_{p=1}^{N} A_{n_p}^\varepsilon\right) \leq \sum_{p=1}^{N} l(A_{n_p}^\varepsilon) \,.$$

Hence and from the definition of the intervals B_ε and A_n^ε we obtain

$$l(A) \leq \sum_{p=1}^{N} l(A_{n_p}^\varepsilon) + \tfrac{\varepsilon}{2} \leq \sum_{n=1}^{\infty} l(A_n^\varepsilon) + \tfrac{\varepsilon}{2} \leq \sum_{n=1}^{\infty} l(A_n) + \varepsilon \,.$$

From this inequality and due to the arbitrariness of $\varepsilon > 0$ follows the semi–additivity of the Lebesgue measure on a semi–algebra of all the intervals.

E x a m p l e 2.11. Analogously it is proved the σ–additivity of the nonnegative Lebesgue–Stieltjes measure on a semi–algebra of the intervals (the rectangles of the space R^n).

2.2.5. Representation of a Real Numeric Measure as a Difference of Nonnegative Measures

Any real measure $\mu(A)$ determined on a σ–algebra of sets $\mathcal{A}$ of the space X may be presented in the form of difference of two nonnegative measures

$$\mu(A) = \mu^+(A) - \mu^-(A). \tag{2.15}$$

This presentation of a real numeric measure is called *Jordan expansion*.

Theorem 2.19. *If a real measure $\mu(A)$ is defined on a σ–algebra $\mathcal{A}$ then it achieves its accurate lower bound on some set $D^- \in \mathcal{A}$, i.e. there exists such a set $D^- \in \mathcal{A}$ that*

$$\mu(D^-) = \inf_{A \in \mathcal{A}} \mu(A)^{\,a}.$$

▷ By definition of a lower bound there exists such a sequence of sets $\{A_n\}$, $A_n \in \mathcal{A}$ that $\lim \mu(A_n) = \inf \mu(A)$. We put $B = \bigcup A_k$. At any n the set B may be expanded into 2^n pairwise nonintersecting sets each of which represents the intersection of n sets – some ones from the sets $A_1, \dots, A_n$ and the complements till B of the rest sets $A_1, \dots, A_n$. As an illustration the case when $n = 2$ is shown on Fig 2.1. In this case

$$A_{12} = A_1 A_2, \quad A_{22} = (B \backslash A_1) A_2,$$

$$A_{32} = A_1(B \backslash A_2), \quad A_{42} = (B \backslash A_1)(B \backslash A_2).$$

Thus at each n

$$B = \bigcup_{m=1}^{2^n} A_{mn},$$

$$A_{mn} = \begin{cases} A_{m,n-1} A_n & (m = 1, \dots, 2^{n-1}), \\ A_{m-2^{n-1},n-1}(B \backslash A_n) & (m = 2^{n-1} + 1, \dots, 2^n), \end{cases}$$

---

^{a} For nonnegative measures this theorem is trivial. In this case $\emptyset \in \mathcal{A}$ and $\inf \mu(A) = \mu(\emptyset) = 0$.

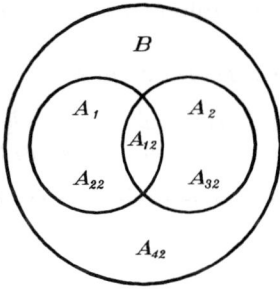

Fig. 2.1

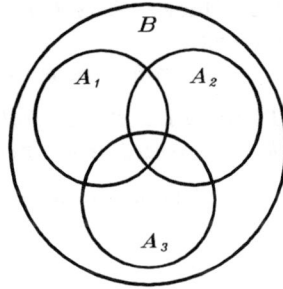

Fig. 2.2

(some sets A_{mn} may be empty). It is evident that each set A_{mn} represents an union of some sets A_{pq} correspondent to $q > n$, i.e. while increasing n the parts into which the set B is partitioned in its turn are divided into nonintersecting parts. This is illustrated on Fig.2.2 for the case when $n = 2$, $q = 3$. Let choose now at each n those sets A_{mn} for which $\mu(A_{mn}) < 0$. We shall mark these sets by the sign minus at the top and put $B_n = \bigcup_m A_{mn}^-$. Then we shall have

$$\mu\left(\bigcup_{p=n}^{\infty} B_p\right) \leq \mu(B_n) \leq \mu(A_n).$$

The sets $\bigcup_{p=n}^{\infty} B_p$ form at $n \to \infty$ a monotone decreasing sequence. We denote its limit by D^-:

$$D^- = \lim_{n\to\infty} \bigcup_{p=n}^{\infty} B_p = \bigcap_{n=1}^{\infty} \bigcup_{p=n}^{\infty} B_p \in \mathcal{A}.$$

Then accounting that the measure $\mu(A)$ is upper continuous we obtain from the previous inequality

$$\mu(D^-) = \lim_{n\to\infty} \mu\left(\bigcup_{p=n}^{\infty} B_p\right) \leq \lim_{n\to\infty} \mu(A_n) = \inf_{A\in\mathcal{A}} \mu(A).$$

And as at any $C \in \mathcal{A}$ $\quad \mu(C) \geq \inf \mu(A)$ then

$$\mu(D^-) = \inf_{A\in\mathcal{A}} \mu(A). \quad \triangleleft$$

Corollary. *A real measure $\mu(A)$ determined on a σ-algebra is lower bounded.*

▷ As the value $-\infty$ is excluded then $\mu(D^-) = \inf \mu(A) > -\infty$. ◁

Theorem 2.20. *A real measure μ determined on a σ-algebra $\mathcal{A}$ may be represented as a difference of two nonnegative measures* (2.15).

▷ Let $\mu(A)$ be a measure determined on a σ-algebra $\mathcal{A}$, $D^- \in \mathcal{A}$ be a set on which $\mu(A)$ achieves the lower bound $\mu(D^-) = \inf \mu(A)$. For any set $A \in \mathcal{A}$

$$\mu(D^-) = \mu(AD^-) + \mu(\bar{A}D^-), \quad \mu(D^- \bigcup AD^+) = \mu(D^-) + \mu(AD^+),$$

where $D^+ = \overline{D^-}$. By the definition of the set D^-

$$\mu(D^-) \leq \mu(\bar{A}D^-), \quad \mu(D^- \bigcup AD^+) \geq \mu(D^-).$$

Consequently, we get

$$\mu(AD^-) \leq 0, \quad \mu(AD^+) \geq 0$$

for any set $A \in \mathcal{A}$. Thus the space X is divided into two nonintersecting parts $X = D^- \bigcup D^+$ in such a way that on all subsets of the set D^- the measure $\mu(A)$ is negative or is equal to zero and on all subsets of the set D^+ it is nonnegative. This expansion of the space is called *Hahn expansion*.

Now we put for any set $A \in \mathcal{A}$

$$\mu^+(A) = \mu(AD^+), \quad \mu^-(A) = -\mu(AD^-). \tag{2.16}$$

Then we shall have

$$\mu(A) = \mu(AD^+) + \mu(AD^-) = \mu^+(A) - \mu^-(A).$$

It remains to prove that the functions $\mu^+(A)$ and $\mu^-(A)$ are the measures. For this purpose we notice that for any pairwise nonintersecting sets $A_1, A_2, \ldots \in \mathcal{A}$

$$\mu^+\left(\bigcup_{k=1}^{\infty} A_k\right) = \mu\left(\bigcup_{k=1}^{\infty} A_k D^+\right) = \sum_{k=1}^{\infty} \mu(A_k D^+) = \sum_{k=1}^{\infty} \mu^+(A_k),$$

$$\mu^-\left(\bigcup_{k=1}^{\infty} A_k\right) = -\mu\left(\bigcup_{k=1}^{\infty} A_k D^-\right) = -\sum_{k=1}^{\infty} \mu(A_k D^-) = \sum_{k=1}^{\infty} \mu^-(A_k).$$

Thus $\mu^+(A)$ and $\mu^-(A)$ are the nonnegative σ–additive functions, i.e. the nonnegative measures. ◁

Notice that Hahn expansion $X = D^- \bigcup D^+$ is not unique whereas Jordan expansion (2.15) is unique.

▷ Really, supposing that

$$X = D_1^- \bigcup D_1^+ = D_2^- \bigcup D_2^+ , \qquad (2.17)$$

we have $D_1^+ D_2^- \subset D_2^-$, as a result

$$\mu(D_1^+ D_2^-) \leq 0 .$$

On the other hand, $D_1^+ D_2^- \subset D_1^+$ as a result

$$\mu(D_1^+ D_2^-) \geq 0 .$$

These opposite inequalities give $\mu(D_1^+ D_2^-) = 0$ and in the same way $\mu(D_1^- D_2^+) = 0$. Consequently, two Hahn expansions (2.16) may exist only in such a case when

$$\mu(D_1^+ D_2^-) = \mu(D_1^- D_2^+) = 0 .$$

But in this case for any set $A \in \mathcal{A}$

$$\mu(AD_1^+) = \mu(AD_1^+ D_2^-) + \mu(AD_1^+ D_2^+) = \mu(AD_1^+ D_2^+)$$
$$= \mu(AD_1^+ D_2^+) + \mu(AD_1^- D_2^+) = \mu(AD_2^+) ,$$

$$\mu(AD_1^-) = \mu(AD_1^- D_2^+) + \mu(AD_1^- D_2^-) = \mu(AD_1^- D_2^-)$$
$$= \mu(AD_1^- D_2^-) + \mu(AD_1^+ D_2^-) = \mu(AD_2^-) . \quad ◁$$

2.2.6. Total Variation of a Measure

Let $\mu(A)$ be a measure (or an additive measure) with the values in a normed linear space Y determined on a class of sets $\mathcal{A}$. Real nonnegative set function

$$| \mu | (A) = \sup \sum \| \mu(A_k) \| , \qquad (2.18)$$

where the upper bound is taken on all finite aggregates of pairwise nonintersecting subsets $A_k \in \mathcal{A}$ of the set A, $A_k \subset A$ is called *a complete variation* of a measure (an additive measure) μ.

Theorem 2.21. *If a measure or an additive measure μ is determined on an algebra or a σ-algebra of sets $\mathcal{A}$ then its complete variation is additive.*

▷ Let $A_1, \ldots, A_N$ be arbitrary pairwise nonintersecting sets $A_k \in \mathcal{A}$, $A = \bigcup A_k$. Then for any pairwise nonintersecting sets $B_1, \ldots, B_n$, $B_p \subset A$ we have

$$\sum_{p=1}^{n} \| \mu(B_p) \| \leq \sum_{p=1}^{n} \sum_{k=1}^{N} \| \mu(B_p A_k) \| = \sum_{k=1}^{N} \sum_{p=1}^{n} \| \mu(B_p A_k) \| . \qquad (2.19)$$

As $B_p A_k \subset A_k$ then according to the definition of a complete variation it follows from (2.19)

$$| \mu | (A) \leq \sum_{k=1}^{N} | \mu | (A_k). \qquad (2.20)$$

If $| \mu | (A) < \infty$ then for any $\varepsilon > 0$ there exist such numbers n_k and pairwise nonintersecting subsets B_{kp} of the sets A_k, $B_{kp} \subset A_k$ that

$$| \mu | (A_k) \leq \sum_{p=1}^{n_k} \| \mu(B_{kp}) \| + \frac{\varepsilon}{N} ,$$

$$\sum_{k=1}^{N} | \mu | (A_k) \leq \sum_{k=1}^{N} \sum_{p=1}^{n_k} \| \mu(B_{kp}) \| + \varepsilon \leq | \mu | (A) + \varepsilon .$$

From this inequality and inequality (2.20) by virtue of the arbitrariness of ε follows

$$| \mu | \left(\bigcup_{k=1}^{N} A_k \right) = \sum_{k=1}^{N} | \mu | (A_k) ,$$

what proves the additivity of $| \mu |$. If $| \mu | (A) = \infty$ then in inequality (2.20) holds the sign of equality. So in this case the condition of the additivity $| \mu |$ is fulfilled. ◁

Let consider now the special case of a real numeric additive measure μ. Introducing the functions

$$\mu^+(A) = \frac{1}{2}[| \mu | (A) + \mu(A)], \quad \mu^-(A) = \frac{1}{2}[| \mu | (A) - \mu(A)] , \qquad (2.21)$$

we shall have

$$\mu(A) = \mu^+(A) - \mu^-(A), \quad |\mu|(A) = \mu^+(A) + \mu^-(A). \qquad (2.22)$$

It is evident that the functions $\mu^+(A)$ and $\mu^-(A)$ represent nonnegative additive measures. The obtained results set the validity of the following theorem.

Theorem 2.22. *Any real additive measure may be represented as a difference of two nonnegative additive measures. Hence its complete variation is equal to the sum of these nonnegative additive measures.*

Corollary. *Any complex additive measure μ may be expressed in terms of nonnegative additive measures by formula*

$$\mu(A) = \mu_R^+(A) - \mu_R^-(A) + i\left[\mu_I^+(A) - \mu_I^-(A)\right]. \qquad (2.23)$$

Theorem 2.23. *Nonnegative measures μ^+ and μ^- in Jordan expansion* (2.15) *of a real measure μ coincide with nonnegative additive measures determined by formulae* (2.21).

▷ It is sufficient to prove that a complete variation of a measure μ determined on a σ–algebra of sets $\mathcal{A}$ is expressed in terms of nonnegative measures μ^+ and μ^- in its Jordan expansion (2.15) by the second formula (2.22). Then from (2.22) will follow formulae (2.21) for μ^+ and μ^-. As $AD \subset A$, $A\bar{D} \subset A$ for any sets A, $D \subset \mathcal{A}$ then after taking the set $D = D^+$ from Hahn expansion $X = D^- \bigcup D^+$ we get

$$\mu^+(A) + \mu^-(A) = \mu(AD^+) - \mu(AD^-) = |\mu(AD^+)| + |\mu(AD^-)|,$$

whence

$$\mu^+(A) + \mu^-(A) \le |\mu|(A). \qquad (2.24)$$

On the other hand, for any pairwise nonintersecting subspaces A_1, ..., A_n of the set A according to Theorem 2.15 we have

$$\mu^+(A) \ge \sum_{k=1}^{n} \mu^+(A_k) = \sum_{k=1}^{n} \mu(A_k D^+) = \sum_{k=1}^{n} |\mu(A_k D^+)|,$$

$$\mu^-(A) \ge \sum_{k=1}^{n} \mu^-(A_k) = -\sum_{k=1}^{n} \mu(A_k D^-) = \sum_{k=1}^{n} |\mu(A_k D^-)|,$$

$$\mu^+(A) + \mu^-(A) \geq \sum_{k=1}^{n} \left\{ | \mu(A_k D^+) | + | \mu(A_k D^-) | \right\} \geq \sum_{k=1}^{n} | \mu(A_k) | ,$$

whence

$$\mu^+(A) + \mu^-(A) \geq | \mu | (A) .$$

Comparing this inequality with (2.24) we obtain

$$| \mu | (A) = \mu^+(A) + \mu^-(A) .$$

This is just the second formula (2.22). ◁

Corollary. *A complete variation of a real measure determined on a σ-algebra of sets is σ-additive.*

The measures μ^+ and μ^- are called *the positive and the negative* variation of the measure $\mu(A)$ correspondingly.

It is evident that a complete variation of the nonnegative measure μ coincides with this measure $| \mu | (A) = \mu(A)$.

E x a m p l e 2.12. A complete variation of the Lebesgue–Stieltjes measure on the real axis R

$$\mu([a, b)) = F(b) - F(a) ,$$

is determined by formula

$$| \mu | ([a, b)) = \sup \sum_{k=1}^{n} | F(x_k) - F(x_{k-1}) | ,$$

where supremum is taken over all finite partitions of the interval $[a, b)$ into pairwise nonintersecting parts

$$[a, b) = \bigcup_{k=1}^{n} [x_{k-1}, x_k) .$$

If $| \mu | (R) = | \mu | ((-\infty, \infty)) < \infty$ then the function $F(x) = \mu((-\infty, x))$ is called *a function of bounded variation*. After constructing Jordan expansion (2.15) for the Lebesgue–Stieltjes measure we ensure that any left–hand side continuous function may be represented as a difference of two nondecreasing functions.

More detailed set functions theory is in (Edwards 1965, Halmos 1950). Probability applications are given in (Cramer and Leadbetter 1967, Kolmogorov 1933, Parthasarathy 1980, Pugachev and Sinitsyn 1987).

Problems

2.2.1 The Lebesgue–Stieltjes measure on the real axis R is determined by formula

$$s(A) = \int\limits_A f(x)dx \,,$$

where $f(x)$ is bounded integrable (in Riemann) function. Define the corresponding function $F(x)$ in formulae of Example 2.9.

2.2.2. The Lebesgue–Stieltjes measure on the real axis R is determined by formula

$$s(A) = \int\limits_A f(x)dx + \sum_{\{k\,:\,x_k \in A\}} p_k \,,$$

where $\{x_n\}$ and $\{p_k\}$ are sequences of real numbers, $f(x)$ is bounded integrable (in Riemann) function. Define the function $F(x)$ in formulae of Example 2.9.

2.2.3. Prove the σ–additivity of measures in Problems 2.2.1 and 2.2.2.

2.2.4. Show explicitly how to choose the intervals B_ε and A_n^ε in Example 2.11 accounting the possible points of discontinuity of function $F(x)$ which in this case must be nondecreasing.

2.2.5. Find the sets D^- and D^+ of Theorems 2.19 and 2.2 for the Lebesgue–Stieltjes measure in Problem 2.2.4 in the case of piecewise continuous function $F(x)$ with finite number of discontinuity points and extremuma. Find $F(x)$ expansion corresponding the Jordan expansion of the Lebesgue–Stieltjes measure.

2.2.6.   Find the complete, positive and negative variations of the Lebesgue–Stieltjes measure in Example 2.12.

2.2.7. The set function is determined on intervals by formula

$$\mu((a,\,b)) = \mu([a,\,b]) = \mu([a,\,b)) = \mu((a,\,b]) = \int\limits_a^b f(x)dx \,,$$

where $f(x)$ is bounded continuous function for which its improper intergral

$$\int\limits_{-\infty}^{\infty} |\,f(x)\,|\,dx$$

converges. Prove its additivity and σ–additivity on a semi–algebra of intervals. Find its complete, positive and negative variations.

2.2.8. The set function is determined on a semi–algebra of rectangles of the space R^n by formula

$$\mu(A) = \int\limits_{a_1}^{b_1} dx_1 \cdots \int\limits_{a_n}^{b_n} f(x_1,\,\ldots,\,x_n)dx_n \,,$$

where $f(x_1, \ldots, x_n)$ is continuous bounded function for which its improper integral

$$\int\limits_{-\infty}^{\infty} dx_1 \cdots \int\limits_{-\infty}^{\infty} |f(x_1, \ldots, x_n)| \, dx_n \, .$$

converges. Prove its additivity and σ–additivity on a semi–algebra of rectangles. Find its complete, positive and negative variations.

2.2.9. Prove Theorems 2.14–2.18 for the case when $\mathcal{C}$ is an algebra of sets.

2.2.10. Prove that the set of discontinuity points of a function of bounded variation on the interval $[a, b]$ is not more than countable and it consists of the first order discontinuity points.

I n s t r u c t i o n. Use the monotone property of corresponding functions on the interval $[a, b]$.

2.2.11. Prove that a function with a bounded derivative defined on the interval $[a, b]$ is a function of bounded variation.

2.2.12. Define the variation by formula

$$\mathrm{Var}_a^b(f) = \int\limits_a^b |f'(x)| \, dx \, ,$$

f being integrable with the first derivative (in Riemann) on the interval $[a, b]$. Find variations: 1) $\mathrm{Var}_0^{100}(e^x)$, 2) $\mathrm{Var}_2^4(\ln x)$, 3) $\mathrm{Var}_0^{2\pi}(\cos x)$, 4) $\mathrm{Var}_0^{2\pi}(\sin x)$.

2.2.13. The random function $Z(A)$ of set of the points of an N–dimensional space R^N is called *stochastic measure* defined on some class of sets $\mathcal{A}$, it is equal to zero on the empty set, $Z(\varnothing) = 0$, the random variables $Z_A = Z(A)$ and $Z_B = Z(B)$ are uncorrelated for any nonintersecting sets A, $B \in \mathcal{A}$ and for any sequence of pairwise nonintersecting sets $\{A_n\}$, $A_n \in \mathcal{A}$, $A_n A_m = \varnothing$ at $m \neq n$ $\bigcup A_n \in \mathcal{A}$,

$$Z\left(\bigcup_{n=1}^{\infty} A_n\right) = \sum_{n=1}^{\infty} Z(A_n) \, .$$

This equality represents the property of the σ–additivity of a stochastic measure. Prove that the σ–additive random scalar function of set $Z(A)$ with a finite second order moment, $Z(\varnothing) = 0$, is a stochastic measure if and only if its covariance function $K_Z(A, B)$ represents a function of the intersection AB,

$$K_Z(A, B) = k(AB)$$

where $k(\mathcal{C})$ is a nonnegative, nondecreasing function, $0 \leq k(\mathcal{C}_1) \leq k(\mathcal{C}_2)$ at $\mathcal{C}_1 \subset \mathcal{C}_2$ $k(\varnothing) = 0$.

2.2.14. Prove that in the case of vector Z (Problem 2.2.13) $k(\mathcal{C})$ represents a matrix and the nonnegativity of $k(\mathcal{C})$ represents the nonnegative definiteness, and nondecreasing represents the nonnegative definiteness of $k(\mathcal{C}_2) - k(\mathcal{C}_1)$ at $\mathcal{C}_1 \subset \mathcal{C}_2$.

2.3. Extension of a Measure

2.3.1. Extension and Contraction of a Function

In many cases there arises a problem of extension of a function with a given domain on a more wider domain. A typical example of such a problem is a problem of a measure determination on all Borel sets of the real axis R (or finite–dimensional space R^n) initially given only on a semi–algebra of the intervals (on rectangular sets — rectangles respectively). Whatsoever any numeric measure is usually determined on some semi–algebra of sets and a problem of its extension on a wider class of sets arises.

If the function $f(x)$ is given in the domain D_f and the $f_1(x)$ is given in D_{f_1}, $D_f \subset D_{f_1}$ and $f_1(x) = f(x)$ at $x \in D_f$ then the function $f_1(x)$ is called *an extension* of the function $f(x)$ on the domain D_{f_1}, the function $f(x)$ is called *a contraction* of the function $f_1(x)$ on the domain D_f.

2.3.2. Problem of Extension of a Numeric Measure

Suppose that a nonnegative σ–additive σ–finite measure μ is given on a some semi–algebra of sets $\mathcal{C}$ and it is required to extend it over more wider class of sets (certaintly, which contains a semi–algebra $\mathcal{C}$).

In order to find the way of the solution of this problem we may use the method applied by measure theory founder Lebesgue for the extension of a measure on the real axis given on the intervals. We shall show the main idea of this method at the example of a measure extension of the plane given on a semi–algebra of all rectangles. So, in order to determine the extension μ^* of the measure μ on a class of all the sets of the plane points it is natural to consider all possible coverings of the set A by the rectangles and to take the lower bound of measures sums μ of the rectangles on all these coverings.

2.3.3. Exterior Measure

Following Lebesgue method we determine for any set $A \subset X$ a set function

$$\mu^*(A) = \inf \sum_k \mu(C_k), \tag{2.25}$$

where the lower bound is taken on all possible countable coverings of the set A by the sets of the class C : $A \subset \bigcup_k C_k, C_k \in C$. This set function is called *an exterior measure*. By virtue of Theorem 2.1 without the loss of generality we may consider the sets C_k in (2.25) as pairwise nonintersecting ones $C_k C_l = \emptyset$ at $l \neq k$.

Let study the main properties of an exterior measure.

Theorem 2.24. *If $A \in C$ then $\mu^*(A) = \mu(A)$.*

▷ For any countable covering of the set A by pairwise nonintersecting sets from C $A = \bigcup_k AC_k$ and as $A \in C, AC_k \in C$ then due to σ–additivity of the measure μ on C

$$\mu(A) = \sum_k \mu(AC_k) \leq \sum_k \mu(C_k).$$

Hence it follows the inequality

$$\mu(A) \leq \mu^*(A)$$

and as $\mu^*(A)$ is the lower bound of the sum $\sum_k \mu(C_k)$ on all coverings of the set A by sets of the class C, and $A \in C$ then

$$\mu^*(A) \leq \mu(A).$$

From two opposite inequalites follows

$$\mu^*(A) = \mu(A). \quad ◁$$

In particular, hence it follows that $\mu^*(X) = \mu(X)$ and $\mu^*(\emptyset) = \mu(\emptyset) = 0$.

Thus an exterior measure $\mu^*(A)$ formally is the extension of the measure μ over the class $\mathcal{P}$ on all the sets of the space X. However we are interested only in the extension of a measure with retaining the main property of a measure its σ–additivity. Therefore it is necessary to find such a class of sets on which exterior measure μ^* is σ–additive.

Theorem 2.25. *If a set A represents a countable union of pairwise nonintersecting sets of the class C, $A = \bigcup_p A_p$, $A_p \in C$, $A_p A_q = \emptyset$ at $q \neq p$, then*

$$\mu^*(A) = \sum_p \mu(A_p).$$

▷ For any covering of the set A by the sets of the class $\mathcal{C}$, $A = \bigcup_p A_p \subset \bigcup_k C_k, C_k \in \mathcal{C}, C_k C_l = \emptyset$ at $l \neq k$, we have $A_p = \bigcup_k A_p C_k$. Hence as a result of the σ–additivity of the measure μ on $\mathcal{C}$ $(A_p C_k \in \mathcal{C}$ as $A_p \in \mathcal{C}, \ C_k \in \mathcal{C})$ follows

$$\mu(A_p) = \sum_k \mu(A_p C_k), \ (p = 1, 2, \ldots), \ \sum_p \mu(A_p) = \sum_{p,k} \mu(A_p C_k).$$

But as $\bigcup_p A_p C_k \subset C_k$, then $\sum_p \mu(A_p C_k) \leq \mu(C_k)$. Consequently,

$$\sum_p \mu(A_p) \leq \sum_k \mu(C_k),$$

whence

$$\sum_p \mu(A_p) \leq \mu^*(A).$$

And as $\bigcup_p A_p = A$ also represents a covering of the set A by the sets of the class $\mathcal{C}$, then

$$\mu^*(A) \leq \sum_p \mu(A_p).$$

From obtained two opposite inclusions follows the required result

$$\mu^*(A) = \sum_p \mu(A_p). \quad \triangleleft \tag{2.26}$$

Theorem 2.26. *If $A \subset B$ then*

$$\mu^*(A) \leq \mu^*(B). \tag{2.27}$$

▷ It follows from the fact that any covering of the set B is also covering of the set A and not the reverse (i.e. the class of the coverings of the sets B is the part of the class of the coverings of the set A).◁

Theorem 2.27. *An exterior measure μ^* is semi–additive.*

▷ Let $\{A_p\}$ be an arbitrary countable system of sets. According to the definition of an exterior measure μ^* for any A_p and any $\varepsilon >$

0 there exists such a covering of the set A_p by the sets from $\mathcal{C}$, A_p $\subset \bigcup_k C_{pk}$, $C_{pk} \in \mathcal{C}$ that

$$\sum_k \mu(C_{pk}) \leq \mu^*(A_p) + \frac{\varepsilon}{2^p} \qquad (p = 1, 2, \ldots).$$

Hence by summing over p we get

$$\sum_{p,k} \mu(C_{pk}) \leq \sum_p \mu^*(A_p) + \varepsilon.$$

On the other hand, as $\bigcup_p A_p \subset \bigcup_{p,k} C_{pk}$ then

$$\mu^*\left(\bigcup_p A_p\right) \leq \sum_{p,k} \mu(C_{pk}), \quad \mu^*\left(\bigcup_p A_p\right) \leq \sum_p \mu^*(A_p) + \varepsilon.$$

And as $\varepsilon > 0$ is arbitrary then

$$\mu^*\left(\bigcup_p A_p\right) \leq \sum_p \mu^*(A_p), \tag{2.28}$$

what proves the semi–additivity of μ^*. ◁

Corollary. *For any sets $A, D \in \mathcal{P}$*

$$\mu^*(D) \leq \mu^*(AD) + \mu^*(\bar{A}D). \tag{2.29}$$

Theorem 2.28. *Any σ–additive extension φ of the measure μ on some σ–algebra $\mathcal{D}_\varphi \supset \mathcal{C}$ satisfies the inequalities*

$$\mu^*(D) - \mu^*(\bar{A}D) \leq \varphi(A) \leq \mu^*(A), \tag{2.30}$$

where $A \in \mathcal{D}_\varphi$, and D is an arbitrary countable covering of the set A by pairwise nonintersecting sets from the semi–algebra $\mathcal{C}$, $A \subset D$ $= \bigcup_k C_k$, $C_k \in \mathcal{C}$, $C_k C_h = \varnothing$ at $h \neq k$.

▷ As $C_k, D \in \mathcal{D}_\varphi$ and $\varphi(C) = \mu(C)$ at $C \in \mathcal{C}$ then

$$\varphi(A) \leq \varphi(D) = \sum_k \varphi(C_k) = \sum_k \mu(C_k).$$

Hence and from formula (2.25) follows

$$\varphi(A) \leq \mu^*(A) \quad \text{for any} \quad A \in \mathcal{D}_\varphi. \tag{2.31}$$

In particular, as $\mathcal{D}_\varphi$ is a σ–algebra and $\bar{A}, \bar{A}D \in \mathcal{D}_\varphi$ then

$$\varphi(\bar{A}D) \leq \mu^*(\bar{A}D). \tag{2.32}$$

But $\varphi(\bar{A}D) = \varphi(D) - \varphi(A)$ as far as $A = AD$ and owing to $D = \bigcup_k C_k$, $C_k \in \mathcal{C}$ according to Theorem 2.25 $\mu^*(D) = \sum_k \mu(C_k)$ $= \sum_k \varphi(C_k) = \varphi(D)$. Consequently, $\varphi(\bar{A}D) = \mu^*(D) - \varphi(A)$ and inequality (2.32) gives

$$\mu^*(D) - \varphi(A) \leq \mu^*(\bar{A}D). \tag{2.33}$$

From inequalities (2.31) and (2.33) follows inequality (2.30). ◁

Corollary 1. *For any set $A \in \mathcal{D}_\varphi$ and countable union D of the sets from the semi–algebra $\mathcal{C}$ which has countable exterior measure the following inequalities are valid:*

$$\mu^*(D) - \mu^*(\bar{A}D) \leq \varphi(AD) \leq \mu^*(AD). \tag{2.34}$$

▷ This follows directly from inequality (2.30) if we account that $AD \in \mathcal{D}_\varphi$, $AD \subset D$ and $\mu^*(AD) \leq \mu^*(D) < \infty$. ◁

Corollary 2. *The extension φ of the measure μ will be the unique σ–additive extension μ if*

$$\mathcal{D}_\varphi \subset \{A : \mu^*(D) = \mu^*(AD) + \mu^*(\bar{A}D)$$

$$\forall D = \bigcup_k C_k, C_k \in \mathcal{C}, \mu^*(D) < \infty\}.$$

From inequalities (2.34) and formula (2.31) follows that this extension of a measure μ must coincide with an exterior measure μ^*.

On the basis of Corollary 2 for finding a single–valued extension of a measure μ it is necessary to study the class $\mathcal{L}$ of the sets A which satisfy the condition

$$\mu^*(D) = \mu^*(AD) + \mu^*(\bar{A}D) \tag{2.35}$$

for any set D representing a countable union of the sets from the semi–algebra $\mathcal{C}$ for which $\mu^*(D) < \infty$ and to establish the σ–additivity of the exterior measurement μ^* on this class of the sets $\mathcal{L}$.

2.3.4. Class of Lebesgue Measurable Sets

The sets of the class $\mathcal{L}$ are called *Lebesgue measurable sets*[a].

Theorem 2.29. *The class of the sets $\mathcal{L}$ contains the semi-algebra $\mathcal{C}$.*

▷ For any set $A \in \mathcal{C}$ according to the definition of a semi-algebra

$$\bar{A} = \bigcup_{p=1}^{n} A_p, \quad A_p \in \mathcal{C}, \quad A_p A_q = \emptyset \quad \text{at} \quad q \neq p.$$

Let $D = \bigcup_k C_k$, $C_k \in \mathcal{C}$, $C_k C_l = \emptyset$ at $l \neq k$. Then accounting that

$$C_k = X C_k = A C_k \bigcup \left(\bigcup_{p=1}^{n} A_p C_k \right)$$

and that $A C_k, A_p C_k \in \mathcal{C}$ as a result of the σ-additivity of a measure μ in $\mathcal{C}$ we may write

$$\mu(C_k) = \mu(A C_k) + \sum_{p=1}^{n} \mu(A_p C_k) \quad (k = 1, 2, \ldots),$$

$$\mu^*(D) = \sum_k \mu(C_k) = \sum_k \left[\mu(A C_k) + \sum_{p=1}^{n} \mu(A_p C_k) \right].$$

On the other hand,

$$\mu^*(AD) = \sum_k \mu(A C_k), \quad \mu^*(\bar{A}D) = \sum_k \sum_{p=1}^{n} \mu(A_p C_k).$$

Comparing the received equalities we ensure that any set $A \in \mathcal{C}$ satisfies condition (2.35) , and consequently, belongs to the class of the sets $\mathcal{L}$. ◁

---

[a] The sets of the class $\mathcal{L}$ sometimes are called *Caratheodory measurable* and *Lebesgue measurable* sets are called the sets A for which equality (2.35) is satisfied only for $D = X_n$ where $\bigcup_n X_n$ is a countable partition of the space X into pairwise nonintersecting sets $X_n \in \mathcal{C}$ of a finite measure $\mu, \mu(X_n) < \infty$ $(n = 1, 2, \ldots)$. But the class of Caratheodory measurable sets $\mathcal{L}$ coincides with the class of Lebesgue measurable sets $\mathcal{L}_1$. Therefore we may not distinguish these two classes of the sets.

Theorem 2.30. *The set A belongs to the class $\mathcal{L}$ if and only if it satisfies condition* (2.35) *for any set $D \subset X$ of a finite exterior measure $D \in \mathcal{P}$, $\mu^*(D) < \infty$.*

▷ The sufficiency of this condition is evident. In order to prove its necessity we notice that by the definition of an exterior measure for any set $D \in \mathcal{P}$, $\mu^*(D) < \infty$ at any $\varepsilon > 0$ there exists such a covering by the sets from the semi–algebra $\mathcal{C}$

$$D^+ = \bigcup_k C_k \supset D, \quad C_k \in \mathcal{C},$$

that

$$\mu^*(D) > \sum_k \mu(C_k) - \varepsilon = \mu^*(D^+) - \varepsilon.$$

For any set $A \in \mathcal{L}$ condition (2.35) is satisfied if we replace in it the set D by the set D^+:

$$\mu^*(D^+) = \mu^*(AD^+) + \mu^*(\bar{A}D^+).$$

Substituting this expression into the previous inequality and taking into account that $AD^+ \supset AD$, $\bar{A}D^+ \supset \bar{A}D$ and as result $\mu^*(AD^+) \geq \mu^*(AD)$, $\mu^*(\bar{A}D^+) \geq \mu^*(\bar{A}D)$, we get

$$\mu^*(D) > \mu^*(AD) + \mu^*(\bar{A}D) - \varepsilon.$$

As this inequality is valid at any $\varepsilon > 0$ then

$$\mu^*(D) \geq \mu^*(AD) + \mu^*(\bar{A}D).$$

Hence owing to the semi–additivity of the exterior measure μ^* we obtain condition (2.35). ◁

On the basis of this theorem the Lebesgue measurable sets may be defined as the sets A satisfying condition (2.35) for any set $D \subset X$, $\mu^*(D) < \infty$.

Theorem 2.31. *The class of the sets $\mathcal{L}$ represents an algebra and its exterior measure μ^* is additive on $\mathcal{L}$.*

▷ Due to the symmetry of condition (2.35) the class $\mathcal{L}$ together with any set A which enters into it also contains its complement $\bar{A}$.

Further the class $\mathcal{L}$ contains the whole space X and an empty set $\emptyset$ as $\mu^*(XD) = \mu^*(D)$, $\mu^*(\emptyset D) = \mu^*(\emptyset) = 0$, and consequently,

$$\mu^*(D) = \mu^*(XD) + \mu^*(\emptyset D).$$

It remains to prove that the class $\mathcal{L}$ together with any sets A and B which enter into it contains their intersection AB. If $A, B \in \mathcal{L}$ then for any set $D, \mu^*(D) < \infty$,

$$\mu^*(D) = \mu^*(AD) + \mu^*(\bar{A}D), \ \ \mu^*(AD) = \mu^*(ABD) + \mu^*(A\bar{B}D)$$

and consequently,

$$\mu^*(D) = \mu^*(ABD) + \mu^*(A\bar{B}D) + \mu^*(\bar{A}D). \tag{2.36}$$

On the other hand, from $A \in \mathcal{L}$ follows

$$\mu^*(\overline{AB}D) = \mu^*(A\overline{AB}D) + \mu^*(\bar{A}\,\overline{AB}D)$$

or as $A\overline{AB} = A(\bar{A}\bigcup\bar{B}) = A\bar{B}, \bar{A}\,\overline{AB} = \bar{A}(\bar{A}\bigcup\bar{B}) = \bar{A}$ then

$$\mu^*(\overline{AB}D) = \mu^*(A\bar{B}D) + \mu^*(\bar{A}D).$$

Hence and from formula (2.36) we find

$$\mu^*(D) = \mu^*(ABD) + \mu^*(\overline{AB}D).$$

This equality shows that $AB \in \mathcal{L}$. From the assigned properties of the class $\mathcal{L}$ follows that it represents an algebra of sets.

Now let A be any set of the class $\mathcal{L}$, B be any set nonintersecting with A, C be any set. Putting in condition (2.35) $D = (A\bigcup B)C$ we get

$$\mu^*((A\bigcup B)C) = \mu^*(A(A\bigcup B)C) + \mu^*(\bar{A}(A\bigcup B)C).$$

But $A(A\bigcup B) = A, \ \bar{A}(A\bigcup B) = B$. Consequently,

$$\mu^*((A\bigcup B)C) = \mu^*(AC) + \mu^*(BC). \tag{2.37}$$

This equality is valid for any nonintersecting sets A, B and any set C at the unique condition that $A \in \mathcal{L}$ (or $B \in \mathcal{L}$). In particular, it is valid at any $A, B \in \mathcal{L}$. Hence by induction follows that for any pairwise nonintersecting sets $A_1, \ldots, A_n \in \mathcal{L}$ and for any set $C \in \mathcal{P}$

$$\mu^*\left(\bigcup_{p=1}^{n} A_p C\right) = \sum_{p=1}^{n} \mu^*(A_p C). \tag{2.38}$$

Putting in formula (2.38) $C = X$ we obtain

$$\mu^* \left(\bigcup_{p=1}^{n} A_p \right) = \sum_{p=1}^{n} \mu^*(A_p) \tag{2.39}$$

for any pairwise nonintersecting sets from $\mathcal{L}$. Thus an exterior measure μ^* is additive on the algebra of sets $\mathcal{L}$. ◁

Corollary. *The class of sets $\mathcal{L}$ represents an algebra which contains a semi–algebra $\mathcal{C}$. Consequently, $\mathcal{L}$ contains an algebra of the sets $\mathcal{B}$ induced by a semi–algebra $\mathcal{C}$, $\mathcal{B} = A(\mathcal{C}) \subset \mathcal{L}$.*

Theorem 2.32. *The class of the sets $\mathcal{L}$ represents a σ–algebra and its exterior measure μ^* is σ–additive on $\mathcal{L}$.*

▷ Let $\{A_n\}$ be an arbitrary sequence of pairwise nonintersecting sets of the class $\mathcal{L}$, $A_n \in \mathcal{L}$, $A_n A_m = \emptyset$ at $m \neq n$. We put

$$B_n = \bigcup_{p=1}^{n} A_p, \quad B = \lim_{n \to \infty} B_n = \bigcup_{p=1}^{\infty} A_p.$$

As according to proved $B_n \in \mathcal{L}$ then at any finite n for any set D the equality is valid

$$\mu^*(D) = \mu^*(B_n D) + \mu^*(\bar{B}_n D).$$

But by virtue of formula (2.38)

$$\mu^*(B_n D) = \sum_{p=1}^{n} \mu^*(A_p D).$$

Further as $B_n \subset B$, and consequently, $\bar{B}_n \supset \bar{B}$ then $\mu^*(\bar{B}_n D) \geq \mu^*(\bar{B}D)$. Therefore

$$\mu^*(D) \geq \sum_{p=1}^{n} \mu^*(A_p D) + \mu^*(\bar{B}D).$$

This inequality is valid at any n. Consequently, it is also valid in the limit at $n \to \infty$:

$$\mu^*(D) \geq \sum_{p=1}^{\infty} \mu^*(A_p D) + \mu^*(\bar{B}D).$$

As $\displaystyle\bigcup_{p=1}^{\infty} A_p D = BD$ then due to the semi–additivity of μ^*

$$\sum_{p=1}^{\infty} \mu^*(A_p D) \geq \mu^*(BD).$$

Therefore

$$\mu^*(D) \geq \sum_{p=1}^{\infty} \mu^*(A_p D) + \mu^*(\bar{B}D) \geq \mu^*(BD) + \mu^*(\bar{B}D). \qquad (2.40)$$

Hence accounting the semi–additivity of μ^* we get

$$\mu^*(D) = \mu^*(BD) + \mu^*(\bar{B}D). \qquad (2.41)$$

Consequently,

$$B = \bigcup_{p=1}^{\infty} A_p \in \mathcal{L}.$$

But any countable union of sets $A_1', A_2', \ldots$ of the class $\mathcal{L}$ may be represented as a countable union of pairwise nonintersecting sets $A_1 = A_1'$, $A_2 = A_2' \bar{A}_1', \ldots$, $A_n = A_n' \bar{A}_1' \ldots \bar{A}_{n-1}$, which belong to the class $\mathcal{L}$ by virtue of Theorem 2.31 $\mathcal{L}$ is an algebra. Thus the class of the sets $\mathcal{L}$ contains all countable unions of the sets which enter into it, i.e. represents a σ–algebra. To prove the σ–additivity of an exterior measure μ^* on $\mathcal{L}$ it is sufficient to notice that by virtue of inequality (2.40) and formula (2.41)

$$\mu^*(BD) = \mu^*\left(\bigcup_{p=1}^{\infty} A_p D\right) = \sum_{p=1}^{\infty} \mu^*(A_p D) \qquad (2.42)$$

for any set $D \in \mathcal{P}, \mu^*(D) < \infty$. From formula (2.42) at $D = B = \displaystyle\bigcup_{p=1}^{\infty} A_p$ we obtain

$$\mu^*\left(\bigcup_{p=1}^{\infty} A_p\right) = \sum_{p=1}^{\infty} \mu^*(A_p) \qquad (2.43)$$

for any pairwise nonintersecting sets $A_p \in \mathcal{L}$. We notice that the σ–additivity of μ^* on $\mathcal{L}$ also follows from Theorem 2.18 as μ^* is semi–additive and in accordance with Theorem 2.31 is additive on $\mathcal{L}$. ◁

R e m a r k. Summarizing all stated facts we come to the conclusion that an exterior measure μ^* represents the unique σ-additive extension of a measure μ determined on a semi-algebra of the sets C on the σ-algebra $\mathcal{L}$ which contains C.

2.3.5. A Property of the Class Lebesgue Measurable Sets

From the fact that the σ-algebra $\mathcal{L}$ of Lebesgue measurable sets contains a semi-algebra C, $C \subset \mathcal{L}$ (Theorem 2.29) follows that it contains a σ-algebra $\mathcal{A} = \sigma(C)$ induced by a semi-algebra $C : \mathcal{A} \subset \mathcal{L}$ ($\mathcal{A}$ by the definition is the minimum σ-algebra which contains C). Naturally, the question arises how much the σ-algebra $\mathcal{L}$ is larger than the σ-algebra $\mathcal{A}$ induced by a semi-algebra C. In order to answer this question let establish one more property of the σ-algebra $\mathcal{L}$.

Theorem 2.33. *The σ-algebra $\mathcal{L}$ contains all the subsets of the sets of zero measure μ^*.*

▷ An exterior measure μ^* is determined for all the sets. Therefore if $\mu^*(N) = 0$ and $N_1 \subset N$ then $\mu^*(N_1) = 0$ by virtue of Theorem 2.26. But if $\mu^*(N) = 0$ then due to the same Theorem 2.26 $\mu^*(ND) = 0$ and $\mu^*(\bar{N}D) \leq \mu^*(D)$ for any set $D \in \mathcal{P}$. Consequently, for any set N of zero measure μ^*

$$\mu^*(ND) + \mu^*(\bar{N}D) \leq \mu^*(D).$$

Together with property (2.29) of the measure μ^* it gives

$$\mu^*(ND) + \mu^*(\bar{N}D) = \mu^*(D),$$

what proves the belonging of any set of zero measure μ^* to the σ-algebra $\mathcal{L}$ of Lebesgue measurable sets. ◁

2.3.6. Complete σ-Algebra

Let $(X, \mathcal{A}, \mu)$ be the space with a nonnegative measure. A σ-algebra $\mathcal{A}$ is called *complete relatively to a measure μ* if it contains all the subsets of the sets of a zero measure μ which enter into it, i.e. if from $N \in \mathcal{A}$, $\mu(N) = 0$, $N_1 \subset N$ follows $N_1 \in \mathcal{A}$ (and certainly $\mu(N_1) = 0$). Theorem 2.33 states that a σ-algebra $\mathcal{L}$ of Lebesgue measurable sets is complete relatively to an exterior measure μ^*.

Theorem 2.34. *Any σ-algebra may be completed.*

▷ Let $\mathcal{N}$ be a class of all subsets of the sets from a σ-algebra $\mathcal{A}$ which have zero measure μ. Consider a class of the sets

$$\mathcal{A}^* = \{C : C = A \bigcup N, A \in \mathcal{A}, N \in \mathcal{N}\}.$$

It is clear that $\mathcal{A} \subset \mathcal{A}^*$ and $\mathcal{N} \subset \mathcal{A}^*$ as from $A \in \mathcal{A}, N \in \mathcal{N}$ follows $A = A \bigcup \emptyset \in \mathcal{A}^*$ and $N = \emptyset \bigcup N \in \mathcal{A}^*$. We shall prove that $\mathcal{A}^*$ is a σ-algebra. Let $N \subset N_0$, $N_0 \in \mathcal{A}$, $\mu(N_0) = 0$. Then for any $A \in \mathcal{A}$

$$\overline{A \bigcup N} = \bar{A} \bar{N} = \bar{A} \bar{N_0} \bigcup \bar{A}(N_0 \backslash N)$$

and $\bar{A} \bar{N_0} \in \mathcal{A}$, $\bar{A}(N_0 \backslash N) \in \mathcal{N}$ as $\bar{A}(N_0 \backslash N) \subset N_0$. Consequently, the class of the sets $\mathcal{A}^*$ is closed relatively to the operation of the complement. After taking an arbitrary sequence of the sets $\{C_n\}$ from $\mathcal{A}^*$, $C_n = A_n \bigcup N_n$, $A_n \in \mathcal{A}$, $N_n \in \mathcal{N}$ we receive

$$\bigcup_n (A_n \bigcup N_n) = \left(\bigcup_n A_n\right) \cup \left(\bigcup_n N_n\right) \in \mathcal{A}^*,$$

as $\bigcup_n A_n \in \mathcal{A}$, $\bigcup_n N_n \in \mathcal{N}$. Consequently, the class of sets $\mathcal{A}^*$ is closed relatively to the operation of countable union. The established properties of the class $\mathcal{A}^*$ ensure us in the fact that $\mathcal{A}^*$ is complete σ-algebra which contains all the subsets of all sets of σ-algebra $\mathcal{A}$ of zero measure μ.

Putting

$$\mu(C) = \mu(A \bigcup N) = \mu(A)$$

for any $A \in \mathcal{A}, N \in \mathcal{N}$ we obtain the unique extension of a measure μ on a complete σ-algebra $\mathcal{A}^*$. Really, if $A_1 \bigcup N_1 = A_2 \bigcup N_2$, $A_1, A_2 \in \mathcal{A}$, $N_1, N_2 \in \mathcal{N}$ then $A_1 \backslash A_2 = A_1 \bar{A}_2 \subset N_2$ and $A_2 \backslash A_1 = A_2 \bar{A}_1 \subset N_1$ as a result $\mu(A_1 \bar{A}_2) = \mu(A_2 \bar{A}_1) = 0$ and $\mu(A_2) = \mu(A_1 A_2) + \mu(A_2 \bar{A}_1)$ $= \mu(A_1 A_2) + \mu(A_1 \bar{A}_2) = \mu(A_1)$. ◁

Measure μ determined on a complete σ-algebra relative to it is also called *a complete measure*, and the correspondent space with a measure $(X, \mathcal{A}, \mu)$ is called *a complete space*.

2.3.7. Coincidence of the σ-Algebra of Lebesgue Measurable Sets with the Minimal Complete σ-Algebra Containing the Semi-Algebra $\mathcal{C}$

Theorem 2.33 asserts that a σ-algebra $\mathcal{L}$ of Lebesgue measurable sets is complete. At the same time a σ-algebra $\mathcal{A}$ induced by a semi-algebra $\mathcal{C}$ may be not complete. It turns out that it is the only distinction between σ–algebra $\mathcal{A}$ and $\mathcal{L}$. If a σ–algebra $\mathcal{A}$ induced by a semi–algebra $\mathcal{C}$ is completed by all subsets of all its sets of zero measure μ^* then the obtained minimal complete σ–algebra $\mathcal{A}^*$ containing $\mathcal{C}$ coincides with $\mathcal{L}$.

Theorem 2.35 *The σ–algebra $\mathcal{L}$ of Lebesgue measurable sets coincides with the minimal σ–algebra $\mathcal{A}^*$ containing the semi–algebra $\mathcal{C}$.*

▷ It is evident that the σ–algebra $\mathcal{A}^*$ contains all countable unions of the sets of a semi–algebra $\mathcal{C}$. Hence from definition (2.25) of the measure μ^* and from Theorem 2.25 follows that for any set $A \in \mathcal{P}$ of the finite measure μ^*, $\mu^*(A) < \infty$ and for any $\varepsilon > 0$ in the σ–algebra $\mathcal{A}^*$ there exists such a sequence of the sets $\{B_n\}$, $B_n \in \mathcal{A}^*$ that $A \subset B_n$ and

$$\mu^*(B_n) < \mu^*(A) + \varepsilon/2^n \quad (n = 1, 2, \ldots). \tag{2.44}$$

In this case without the loss of the generality we may assume the sequence of the sets $\{B_n\}$ as a monotone decreasing one as otherwise the sets B_n, $n \geq 2$ may be subtituted by the sets

$$B_n' = \bigcap_{k=1}^{n} B_k \subset B_n,$$

for which inequalities (2.44) will be also valid. From inequalities (2.44) and accounting that $\mu^*(A) \leq \mu^*(B_n)$ we obtain

$$\lim_{n \to \infty} \mu^*(B_n) = \mu^*(A). \tag{2.45}$$

On the other hand, putting

$$B = \lim_{n \to \infty} B_n = \bigcap_{n=1}^{\infty} B_n,$$

by virtue of the continuity of the measure μ^* on $\mathcal{L}$ (Theorem 2.11) we get

$$\lim_{n \to \infty} \mu^*(B_n) = \mu^* \left(\lim_{n \to \infty} B_n \right) = \mu^*(B). \tag{2.46}$$

From (2.44) and (2.45) in view of the fact that $A \subset B$, $B \in \mathcal{A}^*$ follows that any set $A \in \mathcal{P}$ of the finite measure μ^* is a subset of some set $B \in \mathcal{A}^*$ of the same measure μ^*. In particular, any set $A \in \mathcal{P}$ of a zero measure μ^* is a subset of some set $B \in \mathcal{A}^*$ of a zero measure μ^*. Hence due to the completeness of a σ–algebra $\mathcal{A}^*$ follows that σ–algebra $\mathcal{A}^*$ contains all the sets of a zero measure μ^*.

Let now A be any set from σ–algebra $\mathcal{L}$ of a finite measure μ^*, $\mu^*(A) < \infty$, B be a set from σ–algebra $\mathcal{A}^*$ of the same measure μ^* containing A, $A \subset B$, $\mu^*(B) = \mu^*(A)$. As $\mathcal{A}^* \subset \mathcal{L}$ then by virtue of the additivity of μ^* on $\mathcal{L}$ (Theorem 2.31) $\mu^*(B) = \mu^*(A \bigcup (B \backslash A))$ $= \mu^*(A) + \mu^*(B \backslash A)$. After comparing this equality with the previous one we obtain $\mu^*(B \backslash A) = 0$ and consequently, $B \backslash A \in \mathcal{A}^*$. From $B, B \backslash A \in \mathcal{A}^*$ follows that $A = B \backslash (B \backslash A) = B(\overline{B \backslash A}) \in \mathcal{A}^*$.

If $A \in \mathcal{L}$, $\mu^*(A) = \infty$ then the set A may be represented as a countable union of pairwise nonintersecting sets $A_n \in \mathcal{L}$ of a finite measure $\mu^*(A_n) < \infty$. We may realize it, for instance, putting $A_n = AX_n$ where $X = \bigcup_{n=1}^{\infty} X_n$ is the partition of the space X into pairwise nonintersecting sets of a finite measure $X_n \in \mathcal{C}$, $\mu(X_n) < \infty$. According to the proved $A_n \in \mathcal{A}^*$ $(n = 1, 2 \ldots)$, and consequently, $A = \bigcup_{n=1}^{\infty} A_n \in \mathcal{A}^*$. Thus any set $A \in \mathcal{L}$ belongs to a σ–algebra $\mathcal{A}^*$. Consequently, $\mathcal{L} \subset \mathcal{A}^*$. And as $\mathcal{A}^* \subset \mathcal{L}$ then $\mathcal{L} = \mathcal{A}^*$. ◁

In particular, a σ–algebra of Lebesgue measurable sets of the real axis R or the finite–dimensional space R^n represents a σ–algebra of Borel sets completed by all subsets of all the sets of zero Lebesgue measure.

2.3.8. Approximation Property of the Algebra Inducing a σ–Algebra

Let establish one more important property of a σ–algebra of Lebesgue measurable sets and measure μ^*.

Theorem 2.36. *In order that the set A of the finite exterior measure $\mu^*(A) < \infty$ be Lebesgue measurable it is necessary and sufficient that for any $\varepsilon > 0$ in the algebra $\mathcal{B} = \mathcal{A}(\mathcal{C})$ induced by the semi–algebra $\mathcal{C}$ will occur such a set B for which*

$$\mu^*(A \triangle B) < \varepsilon. \tag{2.47}$$

▷ According to the definition of an exterior measure μ^* at any $\varepsilon > 0$ it will appear such a covering of the set $A \in \mathcal{L}$, $\mu^*(A) < \infty$ by pairwise nonintersecting sets from a semi–algebra $\mathcal{C}$

$$A \subset \bigcup_k C_k, \quad C_k \in \mathcal{C}, \quad C_k C_l = \emptyset \quad \text{at} \quad l \neq k,$$

that

$$\sum_k \mu(C_k) < \mu^*(A) + \varepsilon/2. \tag{2.48}$$

Let choose such a natural n that

$$\sum_{k>n} \mu(C_k) < \varepsilon/2, \tag{2.49}$$

and put

$$B = \bigcup_{k=1}^{n} C_k, \ E = \bigcup_{k>n} C_k, \ D = \bigcup_k C_k = B \bigcup E.$$

By Theorem 2.3 we have $B \in \mathcal{B}$. From $A \subset D$ follows that $A \backslash B \subset D \backslash B = E$ as a result

$$\mu^*(A \backslash B) \le \mu^*(E) = \sum_{k>n} \mu(C_k) < \varepsilon/2. \tag{2.50}$$

On the other hand, from $B \subset D$ follows $B \backslash A = B\bar{A} \subset D\bar{A}$ and

$$\mu^*(B \backslash A) \le \mu^*(D\bar{A}). \tag{2.51}$$

If $A \in \mathcal{L}$ then by virtue of $A \subset D$

$$\mu^*(D) = \mu^*(AD) + \mu^*(\bar{A}D) = \mu^*(A) + \mu^*(\bar{A}D).$$

Hence from Theorem 2.25 and inequality (2.48) follows

$$\mu^*(\bar{A}D) = \mu^*(D) - \mu^*(A) = \sum_k \mu(C_k) - \mu^*(A) < \varepsilon/2.$$

This inequality together with inequality (2.51) gives

$$\mu^*(B \backslash A) < \varepsilon/2. \tag{2.52}$$

From inequalities (2.50) and (2.52) follows that

$$\mu^*(A \triangle B) = \mu^*(A \backslash B) + \mu^*(B \backslash A) < \varepsilon.$$

Thus condition (2.47) is necessary.

For proving the sufficiency of condition (2.47) we notice that from inclusions

$$AD \triangle BD \subset A \triangle B, \ \bar{A}D \triangle \bar{B}D \subset \bar{A} \triangle \bar{B} = A \triangle B$$

and from condition (2.47) follows that

$$\mu^*(AD\Delta BD) < \varepsilon, \quad \mu^*(\bar{A}D\Delta\bar{B}D) < \varepsilon \qquad (2.53)$$

for any set $D \in \mathcal{P}$. But for any sets E, $F \in \mathcal{P}$, $\mu^*(E) < \infty$, $\mu^*(F) < \infty$, because inclusions of $E \subset F \bigcup(E\backslash F)$, $F \subset E \bigcup(F\backslash E)$ and the semi–additivity μ^*

$$\mu^*(E) - \mu^*(F) \le \mu^*(E\backslash F) \le \mu^*(E\Delta F),$$

$$\mu^*(F) - \mu^*(E) \le \mu^*(F\backslash E) \le \mu^*(E\Delta F).$$

Therefore inequalities (2.53) give

$$|\,\mu^*(AD) - \mu^*(BD)\,| \le \mu^*(AD\Delta BD) < \varepsilon,$$

$$|\,\mu^*(\bar{A}D) - \mu^*(\bar{B}D)\,| \le \mu^*(\bar{A}D\Delta\bar{B}D) < \varepsilon \qquad (2.54)$$

for any set D of the finite measure μ^*, $\mu^*(D) < \infty$. As $B \in \mathcal{B} \subset \mathcal{L}$ then for any set D of the finite measure $\mu^*(D) < \infty$

$$\mu^*(BD) + \mu^*(\bar{B}D) = \mu^*(D). \qquad (2.55)$$

From inequalities (2.54) and (2.55) follows

$$|\,\mu^*(AD) + \mu^*(\bar{A}D) - \mu^*(D)\,|$$
$$\le |\,\mu^*(AD) - \mu^*(BD)\,| + |\,\mu^*(\bar{A}D) - \mu^*(\bar{B}D)\,| < 2\varepsilon.$$

As it is valid at any $\varepsilon > 0$ then

$$\mu^*(AD) + \mu^*(\bar{A}D) = \mu^*(D),$$

what proves the belonging of the set A to the σ–algebra $\mathcal{L}$ of Lebesgue measurable sets. ◁

Corollary. *If the σ–algebra $\mathcal{A}$ is induced by the algebra $\mathcal{B}$ and a nonnegative measure μ is given on $\mathcal{A}$ then for any $A \in \mathcal{A}$ of the finite measure $\mu(A) < \infty$ and any $\varepsilon > 0$ there will appear such a set $B \in \mathcal{B}$ that*

$$\mu(A\Delta B) < \varepsilon.$$

▷ In this case an exterior measure μ^* coincides with μ on $\mathcal{A}$ as μ^* is the unique σ–additive extension of μ from the algebra $\mathcal{B}$ on the σ–algebra $\mathcal{A}$ induced by the algebra $\mathcal{B}$. ◁

2.3.9. Coincidence of Classes of Sets $\mathcal{L}$ and $\mathcal{L}_1$

Let μ be a nonnegative σ–finite measure determined on the semi–algebra of the sets $\mathcal{C}$, μ^* be the correspondent exterior measure, $X = \bigcup_k X_n$ be the partition of the space X into pairwise noninter-secting sets of the finite measure $X_n \in \mathcal{C}$, $\mu(X_n) < \infty$.

Theorem 2.37. *The classes of the sets*

$$\mathcal{L} = \left\{ A : \mu^*(AD) + \mu^*(\bar{A}D) = \mu^*(D), \quad D \in \mathcal{P}, \, \mu^*(D) < \infty \right\},$$

$$\mathcal{L}_1 = \left\{ A : \mu^*(AX_n) + \mu^*(\bar{A}X_n) = \mu^*(X_n) \quad (n = 1, 2 \ldots) \right\}$$

coincide.

▷ It is evident that $\mathcal{L} \subset \mathcal{L}_1$. In order to prove the opposite inclusion it is sufficient to show that for any set $A \in \mathcal{L}_1$ at any $\varepsilon > 0$ it will occur such a set $B_n \in \mathcal{B} = \mathcal{A}(\mathcal{C})$ that $\mu^*(AX_n \triangle B_n) < \varepsilon$. Then from Theorem 2.36 it will follow that $AX_n \in \mathcal{L}$ and $A = \bigcup_n AX_n \in \mathcal{L}$, i.e. $\mathcal{L}_1 \subset \mathcal{L}$.

Let $C = \bigcup_k C_k$ and $D = \bigcup_l D_l$ be such coverings of the sets AX_n and $\bar{A}X_n$ by the sets from the semi–algebra $\mathcal{C}$

$$AX_n \subset \bigcup_k C_k, \quad \bar{A}X_n \subset \bigcup_l D_l,$$

$C_k, D_l \in \mathcal{C}, \; C_k, D_l \subset X_n$ that

$$\sum_k \mu(C_k) < \mu^*(AX_n) + \frac{\varepsilon}{3}, \quad \sum_l \mu(D_l) < \mu^*(\bar{A}X_n) + \frac{\varepsilon}{3}. \qquad (2.56)$$

We choose natural h in such a way that

$$\sum_{k>h} \mu(C_k) < \varepsilon/3,$$

and put

$$B_n = \bigcup_{k=1}^h C_k, \quad E_n = \bigcup_{k>h} C_k = C \backslash B_n.$$

Then from $AX_n \subset C$ will follow $AX_n \backslash B_n \subset C \backslash B_n = E_n$ and

$$\mu^*(AX_n \backslash B_n) \le \mu^*(E_n) = \sum_{k>h} \mu(C_k) < \varepsilon/3. \qquad (2.57)$$

Further from $AX_n \subset C$, $\bar{A}X_n \subset D$, $D = DC \bigcup D\bar{C}$ and $C \bigcup D$
$= C \bigcup D\bar{C} = X_n$ follows

$$\mu^*(D) = \mu^*(DC) + \mu^*(D\bar{C}), \quad \mu^*(C) + \mu^*(D\bar{C}) = \mu^*(X_n),$$

$$\mu^*(C) + \mu^*(D) = \mu^*(C) + \mu^*(D\bar{C}) + \mu^*(DC) = \mu^*(X_n) + \mu^*(DC). \quad (2.58)$$

On the other hand, from Theorem 2.25 and inequality (2.56) follows

$$\mu^*(C) + \mu^*(D) = \sum_k \mu(C_k) + \sum_l \mu(D_l) < \mu^*(AX_n)$$

$$+ \mu^*(\bar{A}X_n) + 2\varepsilon/3 = \mu^*(X_n) + 2\varepsilon/3 \,.$$

Hence and from inequality (2.58) we find

$$\mu^*(DC) < 2\varepsilon/3 \,.$$

As $B_n \backslash AX_n = B_n \bar{A}X_n \subset CD$ owing to $B_n \subset C$, $\bar{A}X_n \subset D$ then

$$\mu^*(B_n \backslash AX_n) \le \mu^*(DC) < 2\varepsilon/3 \,.$$

This inequality together with (2.57) gives

$$\mu^*(AX_n \vartriangle B_n) = \mu^*(AX_n \backslash B_n) + \mu^*(B_n \backslash AX_n) < \varepsilon \,. \triangleleft$$

2.3.10. General Extension of Numeric Measure Theorem

Summarizing all the obtained results we may formulate the following general theorem about measure extension.

Theorem 2.38. *Any nonnegative σ–finite σ–additive measure given on the semi–algebra of the sets C (or the algebra of the sets B) may be uniquely extended on the minimal complete σ–algebra A^* containing the semi–algebra C (the algebra B).*

E x a m p l e 2.13. It was shown in Example 2.10 that the Lebesgue measure l is σ–additive on a semi–algebra of all the intervals (of the rectangles in the case of the space R^n). According to Theorem 2.38 it may be uniquely extended on a σ–algebra of Borel sets completed by the subsets of all Borel sets of zero measure.

E x a m p l e 2.14. It was shown in Example 2.11 that the Lebesgue–Stieltjes measure on the real axis is σ–additive on a semi–algebra of all the intervals. By Theorem 2.38 it may be uniquely extended on a σ-algebra of Borel sets completed by the subsets of all Borel sets of zero measure.

E x a m p l e 2.15. The Wiener measure in the space of all scalar functions of the scalar argument $t \in [0, \infty)$ is determined on the semi-algebra of measurable rectangles of finite product of measurable spaces $(X^T, \mathcal{A}^T)$ (Section 2.1.9)

$$X^T = \prod_{t \in T} X_t, \quad \mathcal{A}^T = \prod_{t \in T} \mathcal{A}_t, \quad X_t = R, \quad \mathcal{A}_t = \mathcal{B}, \quad T = [0, \infty),$$

i.e. the space of all the functions $x(t)$ with the domain $T = [0, \infty)$ by the following formula:

$$\mu_W (A_1 \times \cdots \times A_n) = \frac{1}{\sqrt{(2\pi)^n t_1 (t_2 - t_1) \ldots (t_n - t_{n-1})}}$$

$$\times \int_{A_1} \cdots \int_{A_n} \exp\left\{ -\frac{x_1^2}{2t_1} - \frac{1}{2} \sum_{k=2}^{n} \frac{(x_k - x_{k-1})^2}{t_k - t_{k-1}} \right\} dx_1 \cdots dx_n,$$

where $A_1, \ldots, A_n$ are the intervals of the real axis $t_1, \ldots, t_n \in [0, \infty)$, $t_1 < t_2 < \cdots < t_n$, $(n = 1, 2, \ldots)$, $\mathcal{B}$ is a σ-algebra of Borel sets on R. It is evident that the measure μ_W is additive on the semi-algebra of the rectangles of the space X^T. Notice that the measure μ_W may be also considered as a measure on any subspace $(X^{[a,b]}, \mathcal{A}^{[a,b]})$ of the space $(X^T, \mathcal{A}^T)$. The Wiener measure may serve as an example of a measure in an infinite-dimensional space.

R e m a r k. Using the Jordan expansion we may extend Theorem 2.38 on any numeric measure. Thus any numeric measure (real or complex) given on a semi-algebra of the sets may be uniquely extended on a complete σ-algebra of the sets induced by its algebra.

2.3.11. Jordan Measure Extension

In the course of standart mathematical analysis while generalizing the notion of the volume another way of the measure extension (volume) is used. It is determined by a natural way on the semi-algebra of the rectangles $\mathcal{C}$, i.e. by the Jordan extension. Firstly, in the definition of an exterior measure by Jordan the lower bound in definition (2.25) is taken only on all *finite* coverings of the set A by pairwise nonintersecting sets of the class $\mathcal{C}$. And as according to Theorem 2.3 the set of all finite unions of the sets from a semi-algebra $\mathcal{C}$ represents an algebra

$\mathcal{B} = \mathcal{A}(\mathcal{C})$ induced by a semi–algebra $\mathcal{C}$ then it leads to the following definition of an exterior measure of the set A:

$$\bar{\mu}^*(A) = \inf \bar{\mu}(B), \qquad\qquad (2.59)$$

where the lower bound is taken on all $B \in \mathcal{B}$, $B \supset A$ and $\bar{\mu}(B)$ $= \sum_{k=1}^{N} \mu(C_k)$ at $B = \bigcup_{k=1}^{N} C_k$, $C_1, \ldots, C_N \in \mathcal{C}$. It is clear that $\mu^*(A) \leq \bar{\mu}^*(A)$ by virtue of the fact that in equality (2.25) the lower bound is taken on more wide class of the sets. Secondly, together with the exterior measure the interior measure is determined by the following equality:

$$\bar{\mu}_*(A) = \sup \bar{\mu}(B), \qquad\qquad (2.60)$$

where the upper bound is taken on all $B \in \mathcal{B}$, $B \subset A$. It is clear that $\bar{\mu}_*(A) \leq \bar{\mu}^*(A)$.

The set A is called *Jordan measurable* if $\bar{\mu}^* = \bar{\mu}_*(A)$ and this common value of the exterior and the interior measures is called the *Jordan measure* of the set A.

It is easy to see that the class of Jordan measurable sets is contained in the class of Lebesgue measurable sets $\mathcal{L}$ $J \subset \mathcal{L}$ and the Jordan measure coincides with μ^* on J. For the proof it is sufficient to notice that if $A \in J$ than at any $\varepsilon > 0$ there exist such sets B^-, $B^+ \in \mathcal{B}$, $B^- \subset A$ $\subset B^+$ that

$$\bar{\mu}(B^+ \backslash B^-) = \bar{\mu}(B^+) - \bar{\mu}(B^-) < \varepsilon.$$

And as $\bar{\mu}(B) = \mu^*(B)$ for any $B \in \mathcal{B}$ (Theorem 2.25) and $A \backslash B^-$, $B^+ \backslash A \subset B^+ \backslash B^-$ and $A \backslash B^+ = B^- \backslash A = \emptyset$ then

$$\mu^*(A \triangle B^-) < \varepsilon, \quad \mu^*(A \triangle B^+) < \varepsilon.$$

Hence on the basis of Theorem 2.36 follows that $A \in \mathcal{L}$. Finally, from inequalities $\bar{\mu}_*(A) \leq \mu^*(A) \leq \bar{\mu}^*(A)$ follows that the Jordan measure coincides with the exterior measure μ^* on J.

Thus the Lebesgue measure extension is the extension on more wide class of the sets than the Jordan extension. It may be proved that the class of Jordan measurable sets J represents an algebra of sets. From the coincidence of the Jordan measure $\bar{\mu}$ with the exterior measure μ^* on J and from the σ–additivity of μ^* (Theorem 2.32) follows the σ–additivity of the Jordan measure $\bar{\mu}$ on J.

For deep study of measure theory refer to (Edwards 1965, Halmos 1950, Xia Dao–Hing 1972). Some applications in probability theory and statistics are in (Cramer and Leadbetter 1967, Kolmogorov 1933, Parthasarathy 1980), in mathematical physics (Reed and Simon 1972) and in control theory (Balakrishnan 1976, Pugachev and Sinitsyn 1987).

Problems

2.3.1. Prove that the Borel sets on the real axis R are Lebesgue measurable.

2.3.2. Prove that any Lebesgue measurable set on the real axis R is the union of the Borel sets and the set with zero measure.

S o l u t i o n. Let $A \subset R$. The set $A \subset [o, 1]$ is Lebesgue measurable if and only if $\mu^*(A) = 1 - \mu^*([0, 1]/A)$. For any $\varepsilon > 0$ there exists such a closed set $B_\varepsilon \subset A$ that $\mu^*(A/B) < \varepsilon$. Then $\bigcap\limits_{n=1}^{\infty} B_{1/n}$ is the required Borel set.

2.3.3. Find a Lebesgue nonmeasurable set on the real axis R.

S o l u t i o n. Define on the interval $[0, 1]$ the relation of equivalence $x \sim y$ if $x - y \in Q$, Q being the set of all rational numbers. Let A be a subset of interval $(0, 1]$ containing at least one element from every quotient. For $r \in (0, 1]$ define the set $A_r \subset (0, 1]$ received from A by the shift r:

$$A_r \equiv ([r + A]\bigcup[(r - 1) + A])\bigcup(0, 1].$$

Interval $(0, 1]$ presents the union of all pairwise nonintersecting sets $\{A_r\}$ where $r \in Q\bigcup(0, 1]$. So, the set A is not measurable.

2.3.4. Let the set $A \subset [0, 1)$ be nonmeasurable. Prove that the set on the plane

$$\{A \times \{0\}\}\bigcup\{\{0\} \times A\} \subset [0, 1] \times [0, 1]$$

is measurable.

2.3.5. Prove that the countable sets on the real axis R have zero Lebesgue measure.

2.3.6. Prove that the Wiener measure (Example 2.15) is finite and its value in the whole space X^T is equal to 1, $\mu_W(X^T) = 1$.

I n s t r u c t i o n. It is evident that the whole space X^T is the rectangle with infinite sides of the base in any finite product of the spaces $X_{t_1} \times \cdots \times X_{t_n}$ $(n = 1, 2, \ldots)$. Then apply the known formula for Poisson integral (Appendix 2):

$$\int\limits_{-\infty}^{\infty} e^{-x^2} dx = \sqrt{\pi}.$$

2.3.7. Prove that the Wiener measure μ_W for the set of functions from X^T for which its increment on the interval $(t,\ t + \tau)$ is no less than $\varepsilon > 0$ tends to zero at $\tau \to 0$ at any fixed t.

I n s t r u c t i o n. For calculation of μ_W of the set of functions for which $|\ x(t + \tau) - x(t)\ | \geq \varepsilon$ use the functions (Appendix 2)

$$\Phi(u) = \frac{1}{\sqrt{\pi}} \int_0^u e^{-x^2/2}\, dx, \quad \Phi(\infty) = \tfrac{1}{2}, \quad \Phi(-u) = -\Phi(u)\,. \qquad (2.61)$$

As a result we get

$$\mu_W(\{x(t)\ :\ |\ x(t + \tau) - x(t)\ | \geq \varepsilon\})$$

$$= \frac{1}{2\pi\sqrt{t\tau}} \int_{-\infty}^{\infty} dx_1 \int_{|x_2 - x_1| \geq \varepsilon} \exp\Big\{-\frac{x_1^2}{2t} - \frac{(x_2 - x_1)^2}{2\tau}\Big\} dx_2$$

$$= \frac{2}{\pi\sqrt{t\tau}} \int_0^{\infty} \exp\Big\{-\frac{x_1^2}{2t}\Big\} dx_1 \int_\varepsilon^{\infty} \exp\Big\{-\frac{x^2}{2\tau}\Big\} dx$$

$$= 1 - 2\Phi(\varepsilon/\sqrt{\tau}) \to 0 \quad \text{at} \quad \tau \to 0 \quad \forall \varepsilon > 0\,. \qquad (2.62)$$

Thus for any sequence $\{\tau_n\}$, $\tau_n > 0$ that converges to zero the sequence of functionals form function $x(t) \in X^T$ and $f_n(x) = x(t + \tau_n) - x(t)$ $(n = 1, 2, \ldots)$ converges to zero by the measure μ_W.

2.3.8. On the basis of the result of Problem 2.3.7 prove that the Wiener measure μ_W may be considered as concentrated on the subspace C of continuous functions from the space X^T of all scalar functions of the variable $t \in [0, \infty)$.

S o l u t i o n. Let denote by S the set of binary–rational numbers $m2^{-p}$ $(m = 0, 1, 2, \ldots\,;\ p = 1, 2, \ldots)$. Then applying formula (2.62) and the inequality:

$$1 - 2\Phi(u) < k!2^{2k+1}u^{-2k-1}/\sqrt{2\pi}\,, (k = 1, 2, \ldots)\,, \qquad (2.63)$$

valid for any natural number k, we get

$$\mu_W(\{x(t)\ :\ |\ x((m + 1)2^{-p}) - x(m2^{-p})\ | \geq \varepsilon\})$$

$$< k!2^{2k+1}(\varepsilon 2^{p/2})^{-2k-1}/\sqrt{2\pi} \quad \forall p,\ m\,. \qquad (2.64)$$

Let put

$$z_p = \sup_m |\ x((m + 1)2^{-p}) - x(m2^{-p})\ |\,, \qquad (2.65)$$

where the upper bound is taken for all m satisfying the condition $m2^{-p}$ $< a$, $a > 0$ is an arbitrary number. Then we have

$$\{x(t) \ : \ z_p \geq \varepsilon\} \subset \bigcup_{m=0}^{[a2^p]} \{x(t) \ : \ | \ x((m+1)2^{-p}) - x(m2^{-p}) \ | \geq \varepsilon\} \ .$$

Putting $\varepsilon = A2^{-p\alpha}$, $0 < \alpha < 1/2$, $A > 0$, $c = (a+1)k!2^{2k+1}/\sqrt{2\pi}A^{2k+1}$ we receive from inequality (2.64)

$$\mu_W\left(\{x(t) \ : \ z_p \geq A2^{-p\alpha}\}\right) < c2^{-p(\beta-1)} \ , \tag{2.66}$$

where $\beta = (1/2 - \alpha)(2k+1)$. Define the sets

$$A_p = \{x(t) \ : \ z_p \geq A2^{-p\alpha}\}, \quad B_n = \bigcup_{p=n}^{\infty} A_p, \quad N = \bigcap_{n=1}^{\infty} B_n \ .$$

It is clear that $\mu_W(N) \leq \mu_W(B_n) \forall n$ and using inequality (2.66)

$$\mu_W(B_n) \leq \sum_{p=n}^{\infty} \mu_W(A_p) < c \sum_{p=n}^{\infty} 2^{-p(\beta-1)} \ .$$

After choosing k in such a way that $\beta > 1$ we get

$$\mu_W(B_n) < c2^{-n(\beta-1)}/(1 - 2^{-\beta+1}) \quad \forall n \ ,$$

$$\mu_W(N) \leq \mu_W(B_n) < c'2^{-n(\beta-1)} \ ,$$

where $c' = c/(1 - 2^{-\beta+1})$. As this inequality is valid for all n then $\mu_W(N) = 0$.

Any function $x(t) \in \bar{N}$ belongs to any set $\bar{B}_n$, and consequently, to all sets $\bar{A}_p$, $p \geq n$. So,

$$z_p = \sup_m | \ x((m+1)2^{-p}) - x(m2^{-p}) \ |< A2^{-p\alpha} \tag{2.67}$$

for all sufficiently large p. Let s, $s' \in S$ be arbitrary binary–rational numbers, q be the least natural number satisfying the condition $| \ s' - s \ |< 2^{-q+1}$. In this case there exists a point $s_0 = m2^{-q}$ between the points s and s' for some m and

$$s - s_0 = \pm \sum_{p=1}^{l} k_p 2^{-q-p} \ ,$$

where the numbers k_l are equal to zero or unit. If we realize the transfer from the point s_0 to the point s sequentially by the steps of the length $k_1 2^{-q-1}$, $\ldots, k_l 2^{-q-l}$ then due to (2.67) we get (2.68)

$$| x(s) - x(s_0) | < A \sum_{p=1}^{l} 2^{-(q-p)\alpha} < A2^{-q\alpha} \sum_{p=1}^{\infty} 2^{-p\alpha} = \frac{A2^{-(q+1)\alpha}}{1-2^{-\alpha}},$$

$$| x(s') - x(s) | < \frac{2A2^{-(q+1)\alpha}}{1-2^{-\alpha}}, \qquad \sup_{\substack{s,s' \in S \\ |s'-s| < 2^{-q}}} | x(s') - x(s) | < \frac{2A2^{-(q+1)\alpha}}{1-2^{-\alpha}}.$$

It is clear that $\sup | x(s') - x(s) | \to 0$ at $s' \to s$, s, $s' \in S[0, a]$. This proves that all the functions $x(t) \in X^T$ excluding the set N of zero measure μ_W are continuous on the set S of binary–rational numbers for any finite interval $[0, a]$. Now let take an arbitrary $t \in [0, a)$ and the sequence $\{s_n\}$ of binary–rational numbers converging to t and $| s_p - t | < 2^{-p}$. Define the sets

$$A_p = \left\{ x(t') : | x(t) - x(s_p) | \geq A2^{-p\alpha} \right\},$$

$$B_n = \bigcup_{p=n}^{\infty} A_p, \quad N = \bigcap_{n=1}^{\infty} B_n.$$

Analogously we prove that $\mu_W(N) = 0$ and that the sequence $\{x(s_p)\}$ converges to $x(t)$ beyond the set N. Thus at any fixed $t \in [0, a)$ almost all concerning the measure μ_W functions from the space X^T are continuous at the point t for any a. This result and the results of Problems 2.1.13 and 2.1.14 make possible to extend the measure μ_W on the σ–algebra $A^T C = \{B : B = AC, C \in A^T\}$ of the space C of continuous functions from the space of all functions X^T putting $\mu_W(B) = \mu_W(A)$, $B \in A^T C$.

As it is known from the probability theory any function from the space X^T without changing the values of measure on rectangle sets from the space X^T and also on all sets of an induced σ–algebra may be replaced by so called a *separable* function. Such a function is almostly defined for all t by its values on some countable set called a *separant*. In our case separant in the set of all binary–rational numbers and separable function is the set of all continuous functions.

Notice that the Lebesgue extention of the measure μ_W initially defined on the sets of the rectangles of all functions X^T may be determined on the space C of all continuous functions.

2.3.9. Prove that the σ–algebra $A^T C$ in Problem 2.3.8 coincides with the σ–algebra C of the space $C = C([0, \infty))$ of bounded continuous functions with supnorm (Subsection 1.3.12) induced by the set of all balls.

2.3.10. Prove that almost all concerning the Wiener measure μ_W functions of the space $C \subset X^T$ are equal to zero at $t = 0$.

CHAPTER 3
INTEGRALS

Chapter 3 contains the theory of integrals. In Section 3.1 the notion of a measurable function is introduced and general properties of measurable functions are studied. In Section 3.2 the main modes of convergence of sequences of functions are considered, i.e. almost everywhere convergence, convergence in measure, almost uniform convergence. The general definition of an integral of a function with values in a separable B-space (Bochner integral) is given and its general properties are studied in Section 3.3. The abstract Lebesgue integral, Lebesgue integral by Lebesgue measure and Lebesgue–Stieltjes integral are considered as special cases in Section 3.4. The relation between the Lebesgue integral by Lebesgue measure and the Riemann integral as well as between Lebesgue–Stieltjes integral and Riemann–Stieltjes integral is established. The Riemann integral of a function with values in a separable B-space is defined and its relation to Bochner integral is established. General theorems on passing to the limit under the integral sign are proved in Section 3.5. The notions of absolute continuity and singularity of measures are introduced. The definition of the Radon-Nikodym derivative is given in Section 3.6. Then in Section 3.7 the definition of Lebesgue spaces is given which constitute one of the most important classes of spaces. After that Sobolev spaces are defined. In the last Section 3.8 some methods of constructing and studying measures in product spaces are outlined and the theory of multiple and iterated integrals is given.

3.1. Measurable Functions

3.1.1. Definition of a Measurable Function

Let $(X, \mathcal{A})$ and $(Y, \mathcal{B})$ be two measurable spaces. The function $y = f(x)$ mapping the space X onto the space Y is called *measurable relatively to σ–algebra $\mathcal{A}$ and $\mathcal{B}$* or $(\mathcal{A}, \mathcal{B})$–*measurable* if the inverse images of all the sets $B \in \mathcal{B}$ are measurable, i.e. from $B \in \mathcal{B}$ follows $f^{-1}(B) \in \mathcal{A}$. In those cases when σ–algebras $\mathcal{A}$ and $\mathcal{B}$ are known $(\mathcal{A}, \mathcal{B})$–measurable function is shortly called a *measurable* one without specifying the σ–algebras in the spaces X and Y with respect to which it is measurable.

It follows exactly from the definition that the inverse image of σ–algebra $\mathcal{B}$ determined by $(\mathcal{A}, \mathcal{B})$–measurable function $y = f(x)$ is completely contained in the σ–algebra $\mathcal{A}$:

$$f^{-1}(\mathcal{B}) \subset \mathcal{A}.$$

By the properties of the inverse mappings the class of the sets $f^{-1}(\mathcal{B})$ represents the σ–algebra. This σ–algebra is called the *σ–algebra induced in the space X by the function f* and is denoted $\mathcal{A}_f$, $\mathcal{A}_f = f^{-1}(\mathcal{B})$. Thus $\mathcal{A}_f \subset \mathcal{A}$. But $\mathcal{A}_f$ may not coincide with $\mathcal{A}$. It is clear that different measurable functions induce in the space X different σ–algebras.

If the function f is not measurable then the σ–algebra $\mathcal{A}_f$ induced by it does not enter into the σ–algebra $\mathcal{A}$.

Theorem 3.1. *Let $(X, \mathcal{A})$, $(Y, \mathcal{B})$ and $(Z, \mathcal{C})$ be measurable spaces. If the function $y = f(x)$, $f : X \rightarrow Y$, is $(\mathcal{A}, \mathcal{B})$-measurable and the function $z = g(y)$, $g : Y \rightarrow Z$, is $(\mathcal{B}, \mathcal{C})$-measurable then the composite function (the composition of the mappings) $gf(x) = g(f(x))$ is $(\mathcal{A}, \mathcal{C})$-measurable.*

▷ It is clear that as the function g is measurable then the inverse image of any set $C \in \mathcal{C}$ belongs to the σ–algebra $\mathcal{B}$, $g^{-1}(C) \in \mathcal{B}$. Analogously as the function f is measurable than the inverse image of any set $B \in \mathcal{B}$, in particular, the set $g^{-1}(C) \in \mathcal{B}$ belongs to σ–algebra $\mathcal{A}$, $f^{-1}(g^{-1}(C)) = (gf)^{-1}(C) \in \mathcal{A}$. ◁

For the theory of integral the measurable functions with the values in a separable B–space Y are of prime importance. In this case as the σ–algebra $\mathcal{B}$ in Y we usually take the σ–algebra induced by the class of all open balls which is called similarly as in the case of the finite–dimensional space R^n a *σ–algebra of Borel sets*. Here the measurable space $(X, \mathcal{A})$ may be an arbitrary one.

E x a m p l e 3.1. Let us consider the function $f(x) = \varphi_1(s)x^2(t_1) + \varphi_2(s)x^2(t_2)$ where $\varphi_1(s), \varphi_2(s)$ are the continuous functions on the bounded closed interval S, i.e. the elements of the B–space $C(S)$, and $x(t)$ is a scalar function belonging to the space X^T with the σ–algebra $\mathcal{A}^T$, $T = [0, \infty)$ of Example 2.15. The function $f(x)$ maps the measurable space $(X^T, \mathcal{A}^T)$ into the separable B–space $C(S)$. This function is $(\mathcal{A}^T, \mathcal{B})$–measurable one as the inverse image of any ball of the space $C(S)$

$$f^{-1}(S_r(\psi)) = \{x(t) : \psi(s) - r < \varphi_1(s)\, x^2(t_1) + \varphi_2(s)\, x^2(t_2)$$

$$< \psi(s) + r, \ \forall s \in S\}$$

may be presented as a countable intersection of the cylinders in X^T with two–dimensional bases $\psi(s) - r < \varphi_1(s) x_1^2 + \varphi_2(s) x_2^2 < \psi(s) + r$ in the product of the spaces $X_{t_1} \times X_{t_2}$ correspondent to all rational $s \in S$.

Analogously the $(\mathcal{A}^T, \mathcal{B})$–measurability of the function

$$f(x) = \sum_{k=1}^{n} \varphi_k(s) x^2(t_k),$$

mapping the space X^T of Example 2.15 in the separable B–space $C(S)$ is proved.

3.1.2. Properties of Measurable Functions

Let study the main properties of the measurable functions with the values in a separable B–space.

Theorem 3.2. *In order that the function $f(x)$ should be measurable it is necessary and sufficient that the inverse images of all the balls of the space Y should be measurable sets, i.e. should belong to σ–algebra $\mathcal{A}$.*

▷ The necessity is clear. If the function $f(x)$ is measurable then due to the fact that all the balls belong to the σ–algebra $\mathcal{B}$ the inverse images of all the balls belong to the σ–algebra $\mathcal{A}$. For proving the sufficiency we notice that by virtue of the properties of inverse mappings and the definition of the σ–algebra $\mathcal{B}$ the inverse image $f^{-1}(\mathcal{B})$ represents the minimal σ–algebra containing the inverse images of all the balls which under the condition also belong to σ–algebra $\mathcal{A}$. Consequently, $f^{-1}(\mathcal{B}) \subset \mathcal{A}$, i.e. $f^{-1}(B) \in \mathcal{A}$ for any set $B \in \mathcal{B}$. This proves that the function $f(x)$ is a measurable one. ◁

Theorem 3.3. *The limit of the sequence of measurable functions represents a measurable function.*

▷ Let $\{f_n(x)\}$ be a sequence of measurable functions converging to the function $f(x)$, $\lim f_n(x) = f(x)$. For proving that the function $f(x)$ is measurable on the basis of Theorem 3.2 it is sufficient to show that the inverse image of any ball determined by the function $f(x)$ is measurable

$$S = f^{-1}(S_r(b)) = \{x : \| f(x) - b \| < r\} \in \mathcal{A}.$$

In order to do this we shall show that S may be obtained by the countable intersections and unions of the sets from $\mathcal{A}$. We take an arbitrary sequence of positive numbers $\{\varepsilon_m\}$ converging to zero, $0 < \varepsilon_m < r, \varepsilon_m \to 0$. From the property of measurable function $f_n(x)$ follows that all the sets

$$A_n^m = \{x : \| f_n(x) - b \| < r - \varepsilon_m\}$$

are measurable, $A_n^m \in \mathcal{A}$ as they represent the inverse images of the balls $S_{r-\varepsilon_m}(b)$, $\{x : \| f_n(x) - b \| < r - \varepsilon_m\} = f_n^{-1}(S_{r-\varepsilon_m}(b))$. We put

$$B_p^m = \bigcap_{n=p}^{\infty} A_n^m, \quad C = \bigcup_{p,m=1}^{\infty} B_p^m.$$

These sets are also measurable $B_p^m, C \in \mathcal{A}$ as they are obtained from A_n^m by the countable intersections and unions. If we prove that $S = C$ then we prove that the set S is measurable.

If $x \in C$ then $x \in B_p^m$ for some m and p. Consequently, the point x belongs to all the sets A_n^m whose intersection is the given set B_p^m and for all $n \geq p$ there exists the inequality $\| f_n(x) - b \| < r - \varepsilon_m$. In particular, this inequality is valid at such n at which $\| f_n(x) - f(x) \| < \varepsilon_m$. From the triangle inequality we get that in such a case

$$\| f(x) - b \| \leq \| f_n(x) - b \| + \| f_n(x) - f(x) \| < r.$$

Thus any point of the set C belongs to the set S, i.e. $C \subset S$.

If $x \in S$, i.e. $\| f(x) - b \| < r$ then there exists such $\delta > 0, \delta < r$, that $\| f(x) - b \| < r - \delta$. By virtue of the convergence of the sequence $\{f_n(x)\}$ to $f(x)$ at all sufficiently large n and at all m at which $\varepsilon_m < \delta$ the inequality $\| f_n(x) - f(x) \| < \delta - \varepsilon_m$ is true. So, by virtue of the triangle inequality also the inequality $\| f_n(x) - b \| < r - \varepsilon_m$ is valid. Thus any point x of the set S belongs to all the sets A_n^m correspondent to sufficiently large n and m and consequently, to all the sets B_p^m at sufficiently large p and m. But in this case $x \in C$. Consequently, any point of the set S belongs to the set C, i.e. $S \subset C$. From the obtained two inclusions follows that $S = C$. Thus the inverse image of any ball determined by the function $f(x)$ is measurable. Hence according to Theorem 3.2 we conclude that the function $f(x)$ is measurable. ◁

R e m a r k. So, we proved that the class of measurable functions with the values in a separable B–space is closed relatively to the operation of the passage to the limit. The limit of any sequence of measurable functions is always a measurable function.

3.1.3. Simple and Elementary Functions

If the range R_f of the function $f(x)$ consists of the finite number of the points then the function $f(x)$ is called *finite–valued*. If R_f is a countable set of the points then the function $f(x)$ is called *countable–valued*. As an example of a finite–valued or countable–valued function may serve any step–function.

Let $f(x)$ be a finite–valued or countable–valued function which takes the values $y_1, y_2, \ldots$ correspondingly on pairwise nonintersecting set $E_1, E_2, \ldots, E_k E_h = \emptyset$ at $k \neq h$:

$$f(x) = y_n \quad \text{at} \quad x \in E_n \ (n = 1, 2, \ldots).$$

The domain D_f of this function represents the union $\bigcup_k E_k$ and the range R_f of this function is a finite (correspondingly, countable) set of the points $\{y_n\}$. It is easy to see that such a function may be expressed by the formula

$$f(x) = \sum_n y_n \mathbf{1}_{E_n}(x), \tag{3.1}$$

where $\mathbf{1}_{E_n}(x)$ is an indicator of the set E_n (Subsection 1.1.5).

A measurable finite–valued function is called *a simple* one. A measurable countable–valued function is called an *elementary one*.

Theorem 3.4. *In order that the finite–valued or countable–valued function $f(x)$ should be measurable it is necessary and sufficient that all the sets $E_1, E_2, \ldots$ in formula (3.1) should be measurable, $E_n \in \mathcal{A}$.*

▷ As for any set B

$$f^{-1}(B) = \bigcup_{\{k : y_k \in B\}} E_k$$

and, in particular, $f^{-1}(B) = E_p$ if y_p is the unique point from the points $y_1, y_2, \ldots$ in B then $f^{-1}(B) \in \mathcal{A}$ at all $B \in \mathcal{B}$ if and only if all the sets $E_1, E_2, \ldots$ are measurable. ◁

3.1.4. Measurable Functions as Limits of Sequences of Elementary Functions

Let us continue studying the properties of the measurable functions with the values in a separable B–space.

Theorem 3.5. *The funcion $f(x)$ is measurable if and only if it represents the limit of the uniformly convergent sequence of elementary functions.*

▷ The sufficiency of this condition follows directly from Theorem 3.3. For proving the necessity we suppose that the function $f(x)$ is measurable. Let $\{y_n\}$ be a countable set dense in Y. We take the sequence of positive numbers convergent to zero $\{\varepsilon_m\}, \varepsilon_m > 0$, $\varepsilon_m \to 0$ at $m \to \infty$. We form the sets

$$A_n^m = \{x : \| f(x) - y_n \| < \varepsilon_m\} \quad (n = 1, 2, \ldots),$$

$$E_1^m = A_1^m, \quad E_n^m = A_n^m \bigcap_{p=1}^{n-1} \overline{A_p^m} \quad (n = 2, 3, \ldots)$$

and construct the step functions

$$f_m(x) = \sum_n y_n \, 1_{E_n^m}(x) \quad (m = 1, 2, \ldots).$$

All the sets A_n^m and E_n^m are measurable as the function $f(x)$ is measurable. Consequently, according to Theorem 3.4 $\{f_m(x)\}$ represents the sequence of the elementary functions. It is evident that at any x from the domain D_f of the function $f(x)$

$$\| f_m(x) - f(x) \| < \varepsilon_m .$$

Thus the sequence of the elementary functions $\{f_m(x)\}$ uniformly converges to the function $f(x)$. ◁

In particular, in order that the numeric function (functional) should be measurable it is necessary and sufficient that it should be presented as the limit of the uniformly convergent sequence of elementary functions. For proving it is sufficient to notice that the real axis R and the complex plane C are separable (the set of the rational numbers is countable and dense in R and C).

Corollary 1. *If the functions $f_1(x), \ldots, f_n(x)$ are measurable then their sum*

$$f(x) = \sum_{k=1}^{n} f_k(x)$$

represents a measurable function.

Corollary 2. *If the function $f(x)$ and the numeric function $g(x)$ are measurable then their product $f(x)\,g(x)$ represents a measurable function.*

Corollary 3. *If the function $f(x)$ and the numeric function $g(x)$ are measurable and $g(x)$ nowhere vanishes then the function $f(x)/g(x)$ is measurable.*

Corollary 4. *If the functions $f_1(x)$, ... , $f_n(x)$ are measurable and the sets E_1, ... , E_n are measurable then the function*

$$f(x) = \sum_{k=1}^{n} f_k(x) 1_{E_k}(x)$$

is measurable.

Corollary 5. *If the numeric functions $f_1(x)$, ... , $f_n(x)$ are measurable then the functions*

$$f(x) = \max_k f_k(x), \quad g(x) = \min_k f_k(x)$$

are measurable.

▷ To prove all these Corollaries it is sufficient to represent all measurable functions as the limits of the uniformly convergent sequences of elementary functions and use the general theorems about the limits, and for proving Corollary 5 take also into account that according to Corollary 1 the functions $f_k(x) - f_h(x)$ are measurable at all k, h. As a result all the sets $\{x : f_k(x) > f_h(x)\}$ are measurable at all k, h. ◁

R e m a r k. Corollaries 1, 4 and 5 are also extended to the countable sets of the measurable functions $\{f_n(x)\}$ but in Corollary 5 it is essential to take instead of max and min the signs sup and inf correspondingly. It is evident for Corollaries 1 and 4. In the case of Corollary 5 this follows from

$$\sup_k f_k(x) = \lim_{n \to \infty} \max\{f_1(x), \ldots, f_n(x)\},$$

$$\inf_k f_k(x) = \lim_{n \to \infty} \min\{f_1(x), \ldots, f_n(x)\},$$

and from Theorem 3.3.

3.1.5. Almost Everywhere Convergence

Let $(X, \mathcal{A}, \mu)$ be the space with nonnegative measure μ. If any statement is true at all the points of the space X (or set $A \subset X$) except the points belonging to some set $E \in \mathcal{A}$ of zero measure, $\mu(E) = 0$, then it is assumed that this statement is true *almost everywhere* (respectively *almost everywhere on the set A*). The sequence of functions $\{f_n(x)\}$ mapping the space X in B–space Y is called *convergent almost everywhere* (respectively *almost everywhere on the set A*) to the function $f(x)$ if it converges at all x (respectively at all $x \in A$) besides may be the points of the sets $E \in \mathcal{A}$ of zero measure, $\mu(E) = 0$. It is written as $f_n(x) \xrightarrow{\text{a.e.}} f(x)$.

The sequence of the functions $\{f_n(x)\}$ is called *fundamental almost everywhere* or a *Chauchy sequence almost everywhere* (or *almost everywhere on the set A*) if

$$f_n(x) - f_m(x) \xrightarrow{\text{a.e.}} 0 \quad \text{at any} \quad n, \, m \to \infty,$$

i.e. if for any point x except $x \in E$, $\mu(E) = 0$, and any $\varepsilon > 0$ there exists such a number $N = N(\varepsilon)$ that

$$\| f_n(x) - f_m(x) \| < \varepsilon \quad \text{at all} \quad n, \, m > N.$$

Theorem 3.6. *The sequence of the functions $\{f_n(x)\}$ with the values in a B–space converges almost everywhere if and only if it is a fundamental sequence almost everywhere.*

▷ By virtue of the completeness of a B–space the sequence $\{f_n(x)\}$ converges at a given x if and only if it is a fundamental sequence at this x. Consequently, the set of the points of the sequence $\{f_n(x)\}$ convergence coincides with the set of the points of its being a fundamental sequence. ◁

R e m a r k. Theorem 3.3 is also extended on the sequences of the functions $\{f_n(x)\}$ convergent to the function $f(x)$ almost everywhere if the σ–algebra $\mathcal{A}$ is complete (Subsection 2.3.6).

Theorem 3.7. *If the sequence of the measurable functions $\{f_n(x)\}$ converges to the function $f(x)$ almost everywhere relatively to the measure μ and the σ–algebra $\mathcal{A}$ is complete relatively to the measure μ then the limit function $f(x)$ is measurable.*

▷ The proof is quite the same as in Theorem 3.3 with changing the sets A_n^m, B_p^m, C and S by the sets $A_n^m \setminus E$, $B_p^m \setminus E$, $C \setminus E$ and $S \setminus E$ respectively where E is a set of the points of the divergence of the sequence $\{f_n(x)\}$, $E \in \mathcal{A}$, $\mu(E) = 0$. The capacity of S to be measurable follows from the same property of $S \setminus E$ and SE as $S = (S \setminus E) \bigcup SE$ and the set SE is measurable by virtue of the completeness of the σ–algebra $\mathcal{A}$. ◁

Corollary. *If the σ–algebra $\mathcal{A}$ is not complete and $\mathcal{A}^*$ is its complement relative to the measure μ then the limit $f(x)$ of almost everewhere convergent sequence $\{f_n(x)\}$ of $(\mathcal{A}, \mathcal{B})$–measurable functions represents the $(\mathcal{A}^*, \mathcal{B})$–measurable function.*

▷ For proving it is sufficient to notice that any $(\mathcal{A}, \mathcal{B})$–measurable function is $(\mathcal{A}^*, \mathcal{B})$–measurable. ◁

The notion of a measurable function is not associated with any measure property. But in the case where X is a space with the measure $(X, \mathcal{A}, \mu)$ except the notion of $(\mathcal{A}, \mathcal{B})$–measurable naturally arises the notion of $(\mathcal{A}^*, \mathcal{B})$–measurable, and $\mathcal{A}^*$ is the complement of the σ–algebra $\mathcal{A}$ relative to the measure μ. It links the notion to be measurable with a measure. Therefore $(\mathcal{A}^*, \mathcal{B})$–measurable functions are usually called *measurable relative to the measure* μ or shortly, μ–*measurable*. It is clear that any $(\mathcal{A}, \mathcal{B})$–measurable function is measurable relative to any measure μ determined on the σ–algebra $\mathcal{A}$. But not every μ–measurable function is $(\mathcal{A}, \mathcal{B})$–measurable if the σ–algebra $\mathcal{A}$ is not complete.

E x a m p l e 3.2. Let consider a function

$$f(x) = f_x(s) = \int_a^b \varphi(s, t)\, x^2(t)\, dt\,,$$

where $\varphi(s, t)$ is a continuous function $s, t, s \in S$, S is a bounded closed set. This function maps the space of all the functions $\mathcal{A}^T$ of the variable $t \in T = [0, \infty)$ into the separable $\mathcal{B}$–space $C(S)$. In accordance with the result of Problem 2.3.8 the function $f(x)$ is determined almost everywhere in X^T relative to the Wiener measure μ_W. Therefore almost for all the functions $x(t) \in X^T$ there exists the limit

$$f(x) = \lim_{n \to \infty} f_n(x) = \lim_{n \to \infty} \tfrac{b-a}{n-1} \sum_{k=0}^{n-1} \varphi(s, t_k)\, x^2(t_k)\,,$$

$t_k = a + k\,(b-a)/(n-1)$. But all the functions $f_n(x)$ are $(\mathcal{A}^T, \mathcal{B})$–measurable according to the result of Example 3.1. Consequently, $f(x)$ represents the limit almost everywhere relative to the measure μ_W of the convergent sequence of $(\mathcal{A}^T, \mathcal{B})$–measurable functions. In conformity with Corollary of Theorem 3.7 the function $f(x)$ is $(\mathcal{A}_W^T, \mathcal{B})$–measurable where $\mathcal{A}_W^T$ is the complement of the σ–algebra $\mathcal{A}^T$ relative to the measure μ_W, i.e. μ_W is measurable.

Theorem 3.8. *Let* $(X, \mathcal{A}, \mu)$ *be the space with nonnegative measure,* $(Y, \mathcal{B})$–*measurable space. If the function* $y = f(x)$ *is* $(\mathcal{A}, \mathcal{B})$–*measurable* $g(x) = f(x)$ *almost everywhere and the* σ–*algebra* $\mathcal{A}$ *is complete then the function* $y = g(x)$ *is* $(\mathcal{A}, \mathcal{B})$–*measurable.*

▷ As $g(x) = f(x)$ almost everywhere then the sets $f^{-1}(B) = \{\, x : f(x) \in B\,\}$ and $g^{-1}(B) = \{\, x : g(x) \in B\,\}$ may differ one from another only by the subset E_B of the set of zero measure μ. Due to the completeness of the σ–algebra $\mathcal{A}$ the set E_B is measurable. Therefore the sets $f^{-1}(B)$ and $g^{-1}(B)$ are measurable or nonmeasurable simultaneously. ◁

The functions equal to each other almost everywhere are called *equivalent*.

3.1.6. Measurable Functions in Product Spaces

Let $f(z) = f(x, y)$ be the function mapping the product of the spaces $Z = X \times Y$ into some space U. The function $f_x(y) = f(x, y)$ of the variable y at a fixed value x is called *a section* of the function $f(z)$ in the point x. In the special case when $X = Y = U = R$ the section $f_x(y)$ of the function $f(z)$ represents a curve, i.e. the section of the surface $u = f(x, y)$ by the plane parallel to the ordinate axis at a given x.

Theorem 3.9. *All sections of a measurable function are measurable.*

▷ Let $\mathcal{D}$ be the σ–algebra in the space U, $u = f(z)$ be $(\mathcal{C}, \mathcal{D})$–measurable function mapping the space $Z = X \times Y$ into U. As the function $f(z)$ is measurable it means that $f^{-1}(D) \in \mathcal{C}$ at any $D \in \mathcal{D}$. But an inverse image of the set D in Y at a fixed x evidently is the correspondent section $f_x^{-1}(D)$ of the set $f^{-1}(D)$. From the capacity to be measured $f_x^{-1}(D)$ according to Theorem 2.8 follows the same property of $f_x^{-1}(D)$, $f_x^{-1}(D) \in \mathcal{B}$ what proves that the function $u = f_x(y)$ at a given x is $(\mathcal{B}, \mathcal{D})$–measurable. ◁

Consider now the functions in the product of any set of the spaces. Let $u = f(x)$ be the function mapping the product of the space X^T (Subsection 2.1.9) into some space U, S be any subset of the set T, $S \subset T$, $x' = \{x_t : t \in S\}$, $x'' = \{x_t : t \in S \backslash T\}$ be the projections of the point $x = \{x_t : t \in T\} \in X^T$ on the subset X^S and $X^{S \backslash T}$ correspondingly. The function $f_S(x'') = f(x) = f(x', x'')$ of the variable x'' at a fixed value x' (i.e. at fixed values $x_t \in X_t$, $t \in S$) is called *a section* of the function $f(x)$ in the point $x' = \{x_t : t \in S\}$.

From evident relations $X^T = X^S \times X^{T \backslash S}$ and $A^T = A^S \times A^{T \backslash S}$ and from Theorem 3.9 follows that all the sections of a measurable function in the product of any set of the spaces are measurable.

3.1.7. Measures Induced by Measurable Functions

Let $(X, \mathcal{A}, \mu)$ be a space with a measure, $(Y, \mathcal{B})$ be a measurable space, $y = f(x)$ be $(\mathcal{A}, \mathcal{B})$–measurable function. It is natural to determine on the σ–algebra $\mathcal{B}$ in the space $(Y, \mathcal{B})$ the measure

$$\nu(B) = \mu\left(f^{-1}(B)\right). \tag{3.2}$$

In accordance with this formula the measurable function $f(x)$ transfers the measure μ into the space $(Y, \mathcal{B})$.

The measure μ determined by formula (3.2) is called *a measure induced* in the space $(Y, \mathcal{B})$ by the function $y = f(x)$.

R e m a r k. Notice that formula (3.2) determines in the space $(Y, \mathcal{B})$ the measure μ only in the case when the function $y = f(x)$ is $(\mathcal{A}, \mathcal{B})$–measurable.

For detailed study of the measurable functions and probability applications refer to (Halmos 1950, Parthasarathy 1980).

P r o b l e m s

3.1.1. Prove that the numeric function $f(x)$ is measurable if and only if the set $\{x : f(x) < c\}$ at any c is measurable.

3.1.2. Prove that if the numeric function $f(x)$ is measurable then the function $|f(x)|$ is measurable also but vice versa. If the set A in the space X is not measurable then the function

$$f(x) = \begin{cases} \varphi(x) & \text{at } x \in A, \\ -\varphi(x) & \text{at } x \in \bar{A} \end{cases}$$

is nonmeasurable for any measurable funtion φ whereas the function $|\,f(x)\,| = |\,\varphi(x)\,|$ is measurable.

3.1.3. Let $g(x)$ be a measurable function defined on the real axis R and f is a real continuous function. Show that in general case the function $h(x) = g(f(x))$ is nonmeasurable.

3.1.4. Let X be separable metric space with σ–algebra $\mathcal{A}$ induced by the set of all the balls. Prove that the continuous function $f(x)$ mapping the whole space X into the real axis R is $(\mathcal{A}, \mathcal{B})$–measurable.

I n s t r u c t i o n. Apply the result of Problem 1.2.12.

3.1.5. Let $y = f(x)$ is $(\mathcal{A}, \mathcal{C})$–measurable function mapping any whole space X into separable metric space Y with σ–algebra $\mathcal{C}$ induced by the set of all the balls and $z = g(y)$ is continuous function mapping Y into the real axis R (with Borel σ–algebra $\mathcal{B}$). Prove that the composite function $z = g(f(x))$ (composition of functions f and g) is $(\mathcal{A}, \mathcal{B})$–measurable.

3.1.6. Evidently the function of scalar argument $f(x) = [x]$ is measurable. Describe the σ–algebra $\mathcal{A}_f$ induced by this function. Whether the function $g([x])$ be measurable relative to σ–algebra $\mathcal{A}_f$ if: a) $g(y)$ is measurable; b) $g(y)$ is continuous?

3.1.7. For numeric function of a scalar agrument there exist Borel–measurable functions (relatively to σ–algebra of Borel set $\mathcal{B}$ on the real axis) and Lebesgue–measurable functions (relatively to σ–algebra $\mathcal{B}^*$ of Lebesgue measurable sets). Borel–measurable functions are called *Borel functions.* Is any Borel function Lebesgue–measurable? Is any Lebesgue–measurable function a Borel function? Show that any Lebesgue–measurable function may be replaced by corresponding Borel function by changing its values on the set with zero Lebesgue measure.

I n s t r u c t i o n. Apply Theorems 2.33 and 2.34. Take into account that in order the function $f(x)$ be a Borel function it is sufficient that the sets $\{\, x : f(x) < c \,\}$ be Borel sets for rational c.

3.1.8. For any m and n, $(\varphi_1, \ldots, \varphi_n \in C(S))$ prove that the function

$$f(x) = \sum_{k=1}^n \varphi_k(s)\, x^m\,(t_k)$$

is $(\mathcal{A}^T, \mathcal{B})$–measurable.

3.1.9. Generalizing Problem 3.1.8 prove that the function

$$f(x) = \varphi\left(s,\, t_1,\, \ldots,\, t_n,\; x(t_1),\, \ldots,\, x(t_n)\right)$$

is $(\mathcal{A}^T, \mathcal{B})$–measurable for any n where $\varphi\left(s, t_1, \ldots, t_n, x_1, \ldots, x_n\right)$ is a continuous function $s \in S,\, t_1, \ldots, t_n \in T = [0, \infty),\, x_1, \ldots, x_n \in R$.

3.1.10. Let

$$f_1(x) = \max_{t\in[a,b]} x(t), \quad f_2(x) = \min_{t\in[a,b]} x(t),$$

be functionals defined on the subspace of continuous functions $C([a, b])$ of the space X^T. Prove that these functionals are $(\mathcal{A}^T, \mathcal{B})$–measurable.

I n s t r u c t i o n. It is evident that $f_1(x) = \sup x(t_n),\quad f_2(x) = \inf x(t_n)$, where $\{t_n\}$ is the set of rational numbers of interval $[a, b]$. Apply Corollary 5 of Theorem 3.5 to the sequence of functions $\{\varphi_n(x)\}$, $\varphi_n(x) = x(t_n)$.

3.1.11. Prove that the following functions:

$$f(x) = \exp\left\{\int_a^b \varphi\,(s,t)\, x^m(t)\, dt\,\right\},$$

$$f(x) = \int_a^b \psi\,(s, t, x(t))\, dt\,,$$

are $(\mathcal{A}_W^T, \mathcal{B})$–measurable, where $\varphi(s,t)$ and $\psi\,(s, t, x)$ are continuous functions, $s \in S,\, t \in [0, \infty),\, x \in R$.

I n s t r u c t i o n. Apply Example 3.2.

3.1.12. Let $f(x)$ be $(\mathcal{A}^T, \mathcal{B})$–measurable function mapping the product of the spaces X^T with σ–algebra $\mathcal{A}^T$ into some space Y with σ–algebra $\mathcal{B}$. Prove that if $\mathcal{B}$ is induced by a countable class of sets $\{\mathcal{B}_n\}$ then there exists such a countable subset S of the set T, $S \subset T$ that $f(x)$ is $(\mathcal{A}^S, \mathcal{B})$–measurable.

S o l u t i o n. If σ–algebra $\mathcal{B}$ is induced by a countable class of sets $\mathcal{B}_n$ then induced by the function $f(x)$ σ–algebra $\mathcal{A}_f = f^{-1}(\mathcal{B})$ due to the properties of inverse mappings is induced by a countable class of the sets $\{f^{-1}(\mathcal{B}_n)\}$ $\subset \mathcal{A}^T$. On the basis of the results of Problem 2.1.13 σ–algebra presents the union of σ–algebras $\mathcal{A}^S$ corresponding to all countable sets $S \subset T$:

$$\mathcal{A}^T = \bigcup \mathcal{A}^S.$$

Consequently, for any set $A \in \mathcal{A}^T$ there exists countable set $S \subset T$, that $A \in \mathcal{A}^S$. Let $S_n \subset T$ be such countable subsets that $f^{-1}(\mathcal{B}_n) \in \mathcal{A}^{S_n}$, $S = \bigcup_{n=1}^{\infty} S_n$. It is evident that all sets $f^{-1}(\mathcal{B}_n)$ belong to σ–algebra $\mathcal{A}^S$. Consequently, σ–algebra $\mathcal{A}_f = f^{-1}(\mathcal{B})$ induced by the class of sets $\{f^{-1}(\mathcal{B}_n)\}$ as the minimal σ–algebra containing this class of sets is included into σ–algebra $\mathcal{A}^S$, $\mathcal{A}_f = f^{-1}(\mathcal{B}) \subset \mathcal{A}^S$. This proves that $f(x)$ is $(\mathcal{A}^S \mathcal{B})$–measurable.

This result shows that any $(\mathcal{A}^T, \mathcal{B})$–measurable function may be dependent upon the values of function $x(t) \in X^T$ not more than in countable set of points. So, the integrals of Example 3.2 and Problem 3.1.11 cannot be $(\mathcal{A}^T, \mathcal{B})$–measurable because they depend upon all values of function $x(t)$ in the interval $[a, b]$. Analogously the integrals where the functions $\varphi(s, t)$, $\psi(s, t, x)$ are replaced by $\varphi(t)$, $\psi(t, x)$ with values in any separable $\mathcal{B}$–space cannot be measurable. It follows from the fact that the σ–algebra $\mathcal{B}$ in such a space is induced by countable set of open balls with rational radii and with centres in all points of a dense countable set. See also Subsection 3.4.6 concerning Riemann integrals with values in a separable $\mathcal{B}$–space.

3.2. Convergence in Measure. Almost Uniform Convergence

3.2.1. Convergence in Measure

Let $(X, \mathcal{A}, \mu)$ be the space with the nonnegative measure μ. Without the loss of the generality we may consider the σ–algebra $\mathcal{A}$ complete relative to the measure μ (Theorem 2.34).

The sequence of the measurable functions $\{f_n(x)\}$ with the values in the separable $\mathcal{B}$–space Y is called *convergent in the measure* μ to the measurable function $f(x)$ on the set A if at any $\varepsilon > 0$

$$\lim_{n \to \infty} \mu(\{x : \| f_n(x) - f(x) \| \ge \varepsilon\}A) = 0. \tag{3.3}$$

It is written in the form $f_n(x) \xrightarrow{\mu} f(x)$.

The sequence of the measurable functions $\{f_n(x)\}$ with the values in the separable B–space Y is called *a fundamental sequence in the measure* μ on the set A if

$$f_n(x) - f_m(x) \xrightarrow{\mu} 0, \quad x \in A \quad \text{at } n, m \to \infty,$$

i.e. if at any $\varepsilon, \delta > 0$ there exists such a number $N = N(\varepsilon, \delta)$ that

$$\mu(\{x : \| f_n(x) - f_m(x) \| \geq \varepsilon\}A) < \delta \quad \text{at all } n, m > N. \qquad (3.4)$$

If the sequence $\{f_n(x)\}$ converges to $f(x)$ or is a fundamental sequence in the whole domain of the functions $f_n(x)$ and $f(x)$ then conditions (3.3) and (3.4) are written in the form

$$\lim_{n \to \infty} \mu(\{x : \| f_n(x) - f(x) \| \geq \varepsilon\}) = 0, \qquad (3.5)$$

$$\mu(\{x : \| f_n(x) - f_m(x) \| \geq \varepsilon\}) < \delta \quad \text{at all } n, m > N. \qquad (3.6)$$

These definitions are correct as according to the theorems of Section 3.1 the functions $f_n(x) - f(x)$ and $f_n(x) - f_m(x)$ are measurable at all n, m. With understanding that further we consider only measurable functions with the values in a separable B–space.

Theorem 3.10. *Any sequence convergent in the measure μ is a fundamental sequence in the measure μ.*

▷ Really,

$$\| f_n(x) - f_m(x) \| \leq \| f_n(x) - f(x) \| + \| f_m(x) - f(x) \|,$$

as a result

$$\{x : \| f_n(x) - f_m(x) \| \geq \varepsilon\}$$
$$\subset \{x : \| f_n(x) - f(x) \| \geq \varepsilon/2\} \bigcup \{x : \| f_m(x) - f(x) \| \geq \varepsilon/2\}$$

and

$$\mu(\{x : \| f_n(x) - f_m(x) \| \geq \varepsilon\})$$
$$\leq \mu(\{x : \| f_n(x) - f(x) \| \geq \varepsilon/2\})$$
$$+ \mu(\{x : \| f_m(x) - f(x) \| \geq \varepsilon/2\}). \quad ◁$$

For proving the inverse proposal it is necessary at first to compare the convergence in the measure with the almost everywhere convergence and reveal the relation between these two types of the convergence.

Theorem 3.11. *The limit of the sequence of the functions in the measure is the unique accurate to the equivalence.*

▷ Suppose that the sequence of the functions $\{f_n(x)\}$ has two different limits in the measure $f(x)$ and $f'(x)$. Then at any $\varepsilon, \delta > 0$

$$\mu(\{\, x : \|\, f_n(x) - f(x)\, \| \geq \varepsilon/2\,\}) < \delta/2,$$

$$\mu(\{\, x : \|\, f_n(x) - f'(x)\, \| \geq \varepsilon/2\,\}) < \delta/2 \qquad (3.7)$$

at all sufficiently large n. But

$$\|\, f(x) - f'(x)\, \| \leq \|\, f_n(x) - f(x)\, \| + \|\, f_n(x) - f'(x)\, \|$$

and

$$\{\, x : \|\, f(x) - f'(x)\, \| \geq \varepsilon\,\}$$
$$\subset \{\, x : \|\, f_n(x) - f(x)\, \| \geq \varepsilon/2\,\} \bigcup \{\, x : \|\, f_n(x) - f'(x)\, \| \geq \varepsilon/2\,\},$$

as a result formulae (3.7) give

$$\mu(\{\, x : \|\, f(x) - f'(x)\, \| \geq \varepsilon\,\})$$
$$\leq \mu(\{\, x : \|\, f_n(x) - f(x)\, \| \geq \varepsilon/2\,\})$$
$$+ \mu(\{\, x : \|\, f_n(x) - f'(x)\, \| \geq \varepsilon/2\,\}) < \delta.$$

And as it is valid at all $\delta > 0$ then

$$\mu(\{\, x : \|\, f(x) - f'(x)\, \| \geq \varepsilon\,\}) = 0 \quad \forall \varepsilon > 0.$$

Thus $f(x)$ and $f'(x)$ may not coincide only on the set of zero measure μ, i.e. are equivalent. ◁

3.2.2. Almost Everywhere Convergence and Convergence in Measure

Theorem 3.12. *Any sequence of the measurable functions $\{f_n(x)\}$ convergent almost everywhere to the function $f(x)$ on the set A of a finite measure also converges to $f(x)$ in measure.*

▷ Let us prescribe an arbitrary $\varepsilon > 0$ and determine the sets

$$A_n = \{\, x : \|\, f_n(x) - f(x)\, \| \geq \varepsilon\,\}A \quad (n = 1, 2, \ldots),$$

$$B_n = \bigcup_{p=n}^{\infty} A_p, \quad C = \lim_{n \to \infty} B_n = \bigcap_{n=1}^{\infty} B_n.$$

By virtue of the measurable property of the functions $f_n(x)$ and $f(x)$ all the sets A_n, B_n, C are measurable, $A_n, B_n, C \in \mathcal{A}$. Due to the continuity of the measure given on a σ–algebra, $\mu(C) = \lim \mu(B_n)$ (Theorem 2.11). It is easy to understand that at any point $x \in C$ the sequence $\{f_n(x)\}$ converges. Really, if $f_n(x) \to f(x)$ then at any $\varepsilon > 0$ there exists such n_ε that $\| f_p(x) - f(x) \| < \varepsilon$ at all $p \geq n_\varepsilon$, i.e. the point x does not belong to none of the sets A_p at $p \geq n_\varepsilon$. Consequently, the point x does not belong to none of the sets $n \geq n_\varepsilon$, i.e. does not belong to C. Thus none of the points of the convergence of the sequence $f_n(x)$ can belong to the set C. In other words, C is the subset of the set of the sequence convergence $\{f_n(x)\}$. But the sequence $\{f_n(x)\}$ by the condition converges almost everywhere. Consequently, owing to the completeness of a σ–algebra $\mathcal{A}$, $\mu(C) = \lim \mu(B_n) = 0$. As at any n, $A_n \subset B_n$, and consequently, $\mu(A_n) \leq \mu(B_n)$, then $\lim \mu(A_n) = 0$, i.e. the sequence $\{f_n(x)\}$ converges in measure μ to $f(x)$. ◁

R e m a r k. The opposite statement of the theorem generally speaking is not true. Not every sequence convergent in measure is convergent almost everywhere. We shall prove some later on a more general supposition. But at first let us consider an example.

E x a m p l e 3.3. The sequence of the functions

$$f_n(x) = \frac{nx}{n^2 + x^2}$$

converges to zero at any $x \in R$ (i.e. everywhere). But it does not converge in Lebesgue measure, as the Lebesgue measure of the set $\{x : |f_n(x)| \geq \varepsilon\}$ equal to $2n\sqrt{1 - 4\varepsilon^2}/\varepsilon \to \infty$ at $n \to \infty$. This example shows that the sequence of the functions convergent almost everywhere (and even everywhere) on the set of the infinite measure may not converge in measure.

Theorem 3.13. *Any sequence of measurable functions* $\{f_n(x)\}$ *being a fundamental sequence in the measure* μ *contains the subsequence* $\{f_{n_k}(x)\}$ *which converges also almost everywhere in the measure.*

▷ Take an arbitrary $\varepsilon > 0$ и $\delta > 0$. As the sequence $\{f_n(x)\}$ is a fundamental sequence in measure it follows that for any $\eta > 0$, $\zeta > 0$ there exists such a natural number $N = N(\eta, \zeta)$ that

$$\mu(\{ x : \| f_n(x) - f_m(x) \| \geq \eta \}) < \zeta \quad \text{at all } n, m > N.$$

Therefore we may choose such a natural number n_1 that

$$\mu(\{ x : \| f_{n_1}(x) - f_m(x) \| \geq \varepsilon/2 \}) < \delta/2 \quad \text{at all } m > n_1.$$

After this we may choose such $n_2 > n_1$ that

$$\mu(\{\, x : \|\, f_{n_2}(x) - f_m(x)\,\| \geq \varepsilon/2^2 \,\}) < \delta/2^2 \quad \text{at all } m > n_2.$$

And so on we may choose such $n_k > n_{k-1}$ that

$$\mu(\{\, x : \|\, f_{n_k}(x) - f_m(x)\,\| \geq \varepsilon/2^k \,\}) < \delta/2^k \text{ at all } m > n_k \ (k = 1, 2, \ldots).$$

Now we construct the sets

$$F_k = \{\, x : \|\, f_{n_k}(x) - f_{n_{k+1}}(x)\,\| \geq \varepsilon/2^k \,\} \quad (k = 1, 2, \ldots),$$

$$G_p = \bigcup_{k=p}^{\infty} F_k, \quad E = \lim G_p = \bigcap_{p=1}^{\infty} G_p.$$

All these sets are measurable because of the fact that the functions $f_n(x)$ are measurable. As $E \subset G_p$ at all p and the measure μ is a semi–additive then

$$\mu(E) \leq \mu(G_p) \leq \sum_{k=p}^{\infty} \mu(F_k) < \sum_{k=p}^{\infty} \frac{\delta}{2^k} = \frac{\delta}{2^{p-1}}. \tag{3.8}$$

This inequality is valid at any p. Consequently, the right–hand side of the obtained inequality is an arbitrarily small. Therefore, $\mu(E) = 0$. We shall prove that the constructed subsequence of the functions $\{f_{n_k}(x)\}$ converges everywhere outside of the set E. It will mean that it converges almost everywhere. As $\bar{E} = \bigcup_{p=1}^{\infty} \bar{G}_p$ it is sufficient to show that the subsequence $\{f_{n_k}(x)\}$ converges everywhere outside of any set G_p. Let x be any point of the set $\bar{G}_p$. As $\bar{G}_p = \bigcap_{k=p}^{\infty} \bar{F}_k$ then $x \in \bar{F}_k$ at any $k \geq p$. Consequently,

$$\|\, f_{n_k}(x) - f_{n_{k+1}}(x)\,\| < \varepsilon/2^k$$

and at all $k, l, l > k \geq p$

$$\|\, f_{n_k}(x) - f_{n_l}(x)\,\| \leq \sum_{q=k}^{l-1} \|\, f_{n_q}(x) - f_{n_{q+1}}(x)\,\| < \sum_{q=k}^{l-1} \frac{\varepsilon}{2^q} < \frac{\varepsilon}{2^{k-1}}.$$

After taking $\eta > 0$ and $\eta_1 < \eta$ we obtain at all $k, l, l > k \geq p, \varepsilon/2^{k-1} < \eta_1$

$$\|\, f_{n_k}(x) - f_{n_l}(x)\,\| < \eta_1. \tag{3.9}$$

Thus the subsequence $\{f_{n_k}(x)\}$ is a fundamental sequence on any set $\bar{G}_p$ and consequently, on $\bar{E}$. As a B–space is complete then from the property of a fundamental sequence $\{f_{n_k}(x)\}$ follows the existence of the limit $f(x)$ (certainly dependent on x). Thus the constructed subsequence $\{f_{n_k}(x)\}$ converges to $f(x)$ at all x which do not belong to the set E of the zero measure μ, i.e. converges to $f(x)$ almost everywhere. By Theorem 3.7 due to the completeness of the σ–algebra $\mathcal{A}$ the limit function $f(x)$ is measurable. In order to prove the convergence of the constructed sequence to $f(x)$ in measure let pass in inequality (3.9) to the limit at $l \to \infty$. Then we obtain at all $x \in \bar{G}_p$ and $k \geq p, \varepsilon/2^{k-1} < \eta_1$

$$\| f_{n_k}(x) - f(x) \| \leq \eta_1 < \eta.$$

Together with inequalities (3.8) it gives

$$\mu(\{x : \| f_{n_k}(x) - f(x) \| \geq \eta\}) \leq \mu(G_p) < \delta/2^{p-1}.$$

Hence as p is arbitrary then follows the convergence of the subsequence $\{f_{n_k}(x)\}$ to $f(x)$ in measure. ◁

Corollary. *Any sequence $\{f_n(x)\}$ convergent in measure to $f(x)$ contains the subsequence convergent to $f(x)$ almost everywhere.*

▷ By Theorem 3.10 the sequence $\{f_n(x)\}$ is a fundamental sequence in measure. According to Theorem 3.13 it contains the subsequence convergent almost everywhere in measure to some function which may differ from $f(x)$ only on the set of zero measure μ by virtue of Theorem 3.11 about the uniqueness of the limit in measure. ◁

3.2.3. Convergence in Measure of a Fundamental Sequence

Theorem 3.14. *Any sequence of the functions with the values in a separable B–space which is a fundamental sequence in measure μ converges in measure μ to some function.*

▷ From Theorem 3.13 follows that any fundamental sequence in measure $\{f_n(x)\}$ contains the subsequence $\{f_n(x)\}$ convergent in measure to some limit function $f(x)$. As

$$\| f_n(x) - f(x) \| \leq \| f_n(x) - f_{n_k}(x) \| + \| f_{n_k}(x) - f(x) \|$$

and the limit function $f(x)$ is measurable, then

$$\mu(\{x : \| f_n(x) - f(x) \| \geq \varepsilon\})$$

$$\leq \mu(\{\, x : \|\, f_n(x) - f_{n_k}(x)\,\| \geq \varepsilon/2\,\}) + \mu(\{\, x : \|\, f_{n_k}(x) - f(x)\,\| \geq \varepsilon/2\,\}).$$

The first item in the right–hand side tends to zero at $n, k \to \infty$ as the sequence $\{f_n(x)\}$ is a fundamental sequence in measure μ, and the second item – as a result of the convegence in measure μ of the subsequence $\{f_n(x)\}$ what proves the convergence in measure μ of the whole sequence $\{f_n(x)\}$. ◁

From Theorems 3.10 and 3.14 follows that for the convergence of the sequence in measure it is necessary and sufficient that the sequence be a fundamental sequence in measure.

3.2.4. Almost Uniform Convergence

The sequence of the functions $\{f_n(x)\}$ convergent to some function $f(x)$ on the set A uniformly except the points x which belong to the set of an arbitrarily small measure is called *convergent almost uniformly to* $f(x)$.

Theorem 3.15. *If the sequence of measurable functions $\{f_n(x)\}$ converges to $f(x)$ almost everywhere on the set $A \in \mathcal{A}$ of the finite measure $\mu(A) < \infty$ then it converges to $f(x)$ almost uniformly* (Egorov theorem).

▷ Let prescribe the sequence of positive numbers $\{\varepsilon_m\}$ convergent to zero, $\varepsilon_m > 0$, $\varepsilon_m \to 0$ at $m \to \infty$ and determine the sets

$$A_m^n = \bigcap_{p=n}^{\infty} \{\, x : \|\, f_p(x) - f(x)\,\| < \varepsilon_m\,\} A.$$

As the functions $f(x)$ and $f_p(x) - f(x)$ according to Theorem 3.7 and Corollary 1 of Theorem 3.5 are measurable then at any p and $\varepsilon > 0$ the set $\{\, x : \|\, f_p(x) - f(x)\,\| < \varepsilon\,\}$ is measurable, and consequently, all the sets A_m^n are measurable, i.e. $A_m^n \in \mathcal{A}$. At any fixed m the sets A_m^n form a monotone increasing sequence whose limit is determined by formula

$$A_m = \lim_{n \to \infty} A_m^n = \bigcup_{n=1}^{\infty} A_m^n.$$

From the property of all the sets A_m^n to be measurable follows the same property of the set A_m. Besides that as $A_m^n \subset A$ then $A_m \subset A$. Consequently,

$$\mu(A_m) \leq \mu(A) < \infty.$$

By virtue of the continuity of the measure given on the σ–algebra

$$\mu(A_m) = \lim_{n \to \infty} \mu(A_m^n)$$

and consequently, for any $\delta > 0$ we may find such $n = n_m = n_m(\delta)$ that

$$\mu(A_m \backslash A_m^{n_m}) < \delta/2^m \quad (m = 1, 2, \ldots). \tag{3.10}$$

We determine the sets

$$A_\delta = \bigcap_{m=1}^{\infty} A_m^{n_m}, \quad E_\delta = A \backslash A_\delta$$

and estimate $\mu(E_\delta)$:

$$\mu(E_\delta) = \mu(A \backslash \bigcap_{m=1}^{\infty} A_m^{n_m}) = \mu(\bigcup_{m=1}^{\infty} (A \backslash A_m^{n_m})).$$

But by virtue of the semi–additivity of the nonnegative measure

$$\mu(\bigcup_{m=1}^{\infty} (A \backslash A_m^{n_m})) \le \sum_{m=1}^{\infty} \mu(A \backslash A_m^{n_m}).$$

As $A \backslash A_m^{n_m} = (A \backslash A_m) \bigcup (A_m \backslash A_m^{n_m})$ then

$$\mu(A \backslash A_m^{n_m}) = \mu(A \backslash A_m) + \mu(A_m \backslash A_m^{n_m}). \tag{3.11}$$

We notice now that the sequence $\{f_n(x)\}$ diverges in all the points of the set $A \backslash A_m$. Really, it follows from the definition of the set A_m that if $x \in A \backslash A_m$ then the point x cannot belong to any one of the sets A_m^n correspondent to a given m. By the definition of the set A_m^n in this case at any n follows that it will be found such $p \ge n$ that

$$\| f_p(x) - f(x) \| \ge \varepsilon_m.$$

Hence it follows that the sequence $\{f_n(x)\}$ diverges in all the points of the set $A \backslash A_m$. As the sequence $\{f_n(x)\}$ converges almost everywhere on A then the measure of the set $A \backslash A_m$ cannot be different from zero, $\mu(A \backslash A_m) = 0$. Accounting inequality (3.10) we get from formula (3.11)

$$\mu(A \backslash A_m^{n_m}) = \mu(A_m \backslash A_m^{n_m}) < \delta/2^m \quad (m = 1, 2, \ldots).$$

Consequently,

$$\mu(E_\delta) \le \sum_{m=1}^{\infty} \mu(A\backslash A_m^{n_m}) < \delta \sum_{m=1}^{\infty} \frac{1}{2^m} = \delta.$$

Thus at any $\delta > 0$ the measure of the correspondent set E_δ is smaller than δ, i.e. may be arbitrarily small. It remains to prove that the sequence $\{f_n(x)\}$ converges uniformly on the remainder part A_δ of the set A. But it is clear from the definition of the set A_δ; if $x \in A_\delta$ then x belongs to all $A_m^{n_m}$ $(m = 1, 2, \ldots)$. Consequently, no matter what $\varepsilon > 0$ may be there will appear such m that $\varepsilon_m < \varepsilon, x \in A_m^{n_m}$, and consequently, at all $p \ge n_m$ the following inequality will take place:

$$\| f_p(x) - f(x) \| < \varepsilon_m < \varepsilon.$$

As n_m does not depend on x it means that the sequence $\{f_n(x)\}$ converges uniformly on the set A_δ which differs from A by the set E_δ on an arbitrarily small measure. ◁

An inverse more strong suggestion is also valid.

Theorem 3.16. *Any sequence of the functions $\{f_n(x)\}$ with the values in a B–space almost uniformly convergent to the function $f(x)$ on the set $A \in \mathcal{A}$ converges on A almost everywhere. Hence it is not obligatory for the functions f_n to be measurable, the space of their values may be not separable, and the set A may also have an infinite measure.*

▷ For proving we take an arbitrary sequence of positive numbers $\{\delta_m\}$ convergent to zero $\delta_m > 0$, $\delta_m \to 0$ at $m \to \infty$. As by the condition the sequence $\{f_n(x)\}$ converges almost uniformly on A than to any δ_m corresponds such a measurable set $E_{\delta_m} \subset A$, that $\mu(E_{\delta_m}) < \delta_m$ and the sequence $\{f_n(x)\}$ converges uniformly on the set $A_{\delta_m} = A\backslash E_{\delta_m}$. It is evident that in this case the sequence $\{f_n(x)\}$ converges in all the points of the set

$$A_0 = \bigcup_{m=1}^{\infty} A_{\delta_m},$$

as any point $x \in A_0$ obligatory belongs to some set A_{δ_m}. It remains to prove that the measure of the set $E_0 = A\backslash A_0$ is equal to zero. For this purpose we notice that

$$E_0 = A\backslash A_0 = A\backslash \bigcup_{m=1}^{\infty} A_{\delta_m} = \bigcap_{m=1}^{\infty} (A\backslash A_{\delta_m}) = \bigcap_{m=1}^{\infty} E_{\delta_m}.$$

Consequently, E_0 is measurable, $E_0 \subset E_{\delta_m}$ at any m and

$$\mu(E_0) \leq \mu(E_{\delta_m}) < \delta_m.$$

As it is valid at any m and $\delta_m \to 0$ at $m \to \infty$ then $\mu(E_0) = 0$. Thus the sequence $\{f_n(x)\}$ converges everywhere on A except the points of the set E_0 of zero measure, i.e. converges almost everywhere. ◁

3.2.5. Measurable Functions as Limits of Sequences of Simple Functions

Theorem 3.5 states that any measurable function represents the limit of the sequence of the elementary functions. For constructing the theory of an integral it is expedient to consider the measurable functions as the limits of the sequences of simple functions. The possibility of this fact is defined by the following theorem.

Theorem 3.17. *Any measurable function $f(x)$ with the values in a separable B–space Y may be presented on any set of a finite measure as the limit of almost everywhere convergent sequence of simple (i.e. measurable finite–valued) functions.*

▷ Let $A \subset D_f$ be a set of a finite measure, $\{y_k\}$ be a countable set which is dense in Y. We assign two sequences of positive numbers $\{\varepsilon_n\}$ and $\{\delta_n\}$ convergent to zero, $\varepsilon_n, \delta_n > 0$; $\varepsilon_n, \delta_n \to 0$ at $n \to \infty$ and form the sets

$$A_k^n = A\{\, x : \| f(x) - y_k \| < \varepsilon_n \,\} \quad (k, n = 1, 2, \ldots),$$

$$E_1^n = A_1^n, \quad E_k^n = A_k^n \bigcap_{p=1}^{k-1} \overline{A_p^n} \quad (k = 2, 3, \ldots; n = 1, 2, \ldots).$$

All these sets are measurable because the set A and the function $f(x)$ are measurable. As

$$\bigcup_{k=1}^{\infty} E_k^n = A \qquad \mu(A) < \infty,$$

then at any n and any $\delta > 0$ by virtue of the continuity of the measure it will occur such a natural number $N = N(\delta)$ that

$$\mu\Big(\bigcup_{k=N}^{\infty} E_k^n \Big) < \delta.$$

Therefore we may determine such an infinitely increasing sequence of natural numbers $\{N_n\}, N_n \to \infty$ that

$$\mu\left(\bigcup_{k=N_n}^{\infty} E_k^n\right) < \delta_n \quad (n = 1, 2, \ldots). \tag{3.12}$$

We construct the simple functions

$$f^n(x) = \sum_{k=1}^{N_n-1} y_k \mathbf{1}_{E_k^n}(x) \quad (n = 1, 2, \ldots).$$

By the definitions of the sets E_k^n

$$\| f^n(x) - f(x) \| < \varepsilon_n \quad \text{at all } x \in \bigcup_{k=1}^{N_n-1} E_k^n. \tag{3.13}$$

Now we set an arbitrary number $\delta > 0$ and choose such m that $\delta_m < \delta$. We introduce a set

$$E_\delta = \bigcup_{k=N_m}^{\infty} E_k^m.$$

Then by virtue of inequality (3.12) we shall have $\mu(E_\delta) < \delta$. After taking an arbitrary $\varepsilon > 0$ we shall get from inequality (3.13) for all $n > m$ for which $\varepsilon_n < \varepsilon$

$$\| f^n(x) - f(x) \| < \varepsilon \quad \text{at all } x \in A \backslash E_\delta.$$

This inequality proves an uniform convergence of the sequence of simple functions $\{f^n(x)\}$ to the function $f(x)$ on the set $A \backslash E_\delta$. Hence owing to the arbitrariness of $\delta > 0$ follows almost uniform convergence of the sequence $\{f^n(x)\}$ to $f(x)$ on A. According to Theorem 3.16 the sequence $\{f^n(x)\}$ converges to $f(x)$ almost everywhere on A. ◁

3.2.6. Property of Functions Measurable with Respect to σ-Algebra Induced by Another Function

Let $(X, \mathcal{A}, \mu)$ be a space with nonnegative σ-finite measure, $(Y, \mathcal{B})$ and $(Z, \mathcal{C})$ be measurable spaces, Z be a separable B-space, and $\mathcal{C}$ as usual in this case is the σ-algebra induced by the set of all open balls in Z, $y = f(x)$ be $(\mathcal{A}, \mathcal{B})$-measurable function, $z = g(x)$ be $(\mathcal{A}, \mathcal{C})$-measurable function, $\mathcal{A}_f = f^{-1}(\mathcal{B}) \subset \mathcal{A}$ be σ-algebra induced in X by the function $f(x)$. The following theorem is valid.

Theorem 3.18. *The function $g(x)$ is $(\mathcal{A}_f, \mathcal{C})$-measurable if and only if there exists such $(\mathcal{B}, \mathcal{C})$-measurable function $z = \varphi(y)$ that*

$$g(x) = \varphi(f(x)) \quad \text{almost everywhere.} \tag{3.14}$$

▷ If condition (3.14) is valid then the property of the function $g(x)$ to be $(\mathcal{A}_f, \mathcal{C})$-measurable follows directly from the same property of the function $f(x)$ to be $(\mathcal{A}_f, \mathcal{B})$-measurable, of the function $\varphi(y)$ to be $(\mathcal{B}, \mathcal{C})$-measurable and from Theorems 3.1 and 3.8 and even whether Z is a separable B–space or not.

Suppose now that the function $g(x)$ is $(\mathcal{A}_f, \mathcal{C})$-measurable. At first we admit that $g(x)$ is a simple function:

$$g(x) = \sum_{k=1}^{N} z_k \mathbf{1}_{E_k}(x),$$

where $E_1, \ldots, E_N \in \mathcal{A}_f$ are pairwise nonintersecting sets. From $E_k \in \mathcal{A}_f$ follows that in the range R_f of the function $f(x)$ there exists a set $B_k \in \mathcal{B}$ whose inverse image coincides with E_k, $f^{-1}(B_k) = E_k$ $(k = 1, \ldots, N)$. It is evident that the sets $B_1, \ldots, B_N$ do not pairwise intersect as by virtue of the properties of inverse maps $f^{-1}(B_k B_h) = E_k E_h = \emptyset$ at $k \neq h$ what is possible only if $B_k B_h = \emptyset$ as $B_k, B_h \subset R_f$. Let us determine in the space $(Y, \mathcal{B})$ a simple function

$$\varphi(y) = \sum_{k=1}^{N} z_k \mathbf{1}_{B_k}(y).$$

As the function $y = f(x)$ takes the values from B_k at $x \in E_k$ then $g(x) = \varphi(f(x))$ $\forall x$. Thus the statement of the theorem is valid for all simple functions.

Now let $g(x)$ be an arbitrary $(\mathcal{A}_f, \mathcal{C})$-measurable function. Without loss of the generality the σ–algebras $\mathcal{A}_f$ and $\mathcal{B}$ may be considered as complete on the basis of Theorem 2.34. In this case according to Theorem 3.17 any $(\mathcal{A}_f, \mathcal{C})$-measurable function may be presented on any set of the finite measure μ as the limit of the sequence of simple functions (certainly, $(\mathcal{A}_f, \mathcal{C})$-measurable). As the measure μ is σ–finite then there exists such a partition of the space X into pairwise nonintersecting sets

$$X = \bigcup_{k=1}^{\infty} X_k, \quad X_k \in \mathcal{A}, \tag{3.15}$$

that the measure μ is finite at every set X_k. Let $\{g^n(x)\}$ be the sequence of the simple functions convergent almost everywhere to $g(x)$ on the set X_k. As it was already proved to every simple function $g^n(x)$ corresponds such a simple function $z = \varphi^n(y)$ that $g^n(x) = \varphi^n(f(x))$. From the convergence of the sequence $\{g^n(x)\}$ at the point x follows the sequence $\{\varphi^n(y)\}$ which is a fundamental sequence at the correspondent point $y = f(x)$. Hence owing to the completeness of the B–space Z follows the existence of the limit function $\varphi(y) = \lim \varphi^n(y)$ at all the points y which correspond to the points of the convergence x of the sequence $\{g^n(x)\}$. It is clear that the inverse image of the set of the divergence points y of the sequence $\{\varphi^n(y)\}$ determined by the function $y = f(x)$ is the subset of the set of the divergence points x of the sequence $\{g^n(x)\}$. Consequently, by virtue of formula (3.2) the set of the divergence points y of the sequence has zero measure μ induced into the space $(Y, \mathcal{B})$ by the function $y = f(x)$. Thus the sequence of simple functions $\{\varphi^n(y)\}$ converges to $\varphi(y)$ almost everywhere with respect to the measure μ. According to Theorem 3.7 the limit function $\varphi(y)$ is $(Y, \mathcal{B})$–measurable. Finally, the limit passage in the equality $g^n(x) = \varphi^n(f(x))$ make us sure in the validity of equality (3.14) almost everywhere with respect to the measure μ. It remains to notice that the validity of Theorem 3.18 for each set X_k of the finite measure μ in formula (3.15) means its validity for the whole space X. ◁

For deep study of different modes of convergence and their applications in probability theory refer to (Edwards 1965, Halmos 1950, Parthasarathy 1980, Xia Dao–Hing 1972).

Problems

3.2.1. Prove that the sequence of functions $\{\sin(x/n)\}$: a) converges to zero at all x; b) converges to zero in the Lebesgue measure for any finite interval (a, b); c) diverges in the Lebesgue measure for the whole real axis.

3.2.2. Whether the sequence of functions $\{1/nx\}$ be convergent almost uniformly? If so then define for every $\delta > 0$ the corresponding set E_δ, where outside of E_δ the sequence of functions is convergent uniformly. Whether this sequence be convergent in the Lebesgue measure almost everywhere?

3.2.3. Prove that the sequence of functions in Example 3.3 is convergent in the Lebesgue measure for any finite interval.

3.2.4. The sequence of integral sums in Examples 3.2 is convergent almost everywhere relative to the Wiener measure μ_W. Is this sequence convergent: a) in μ_W; b) almost uniformly?

3.2.5. Let $f_n \xrightarrow{\mu} h$ and $f_n \xrightarrow{\mu} g$. Prove that f and g are equivalent in measure.

3.3. Bochner Integral

3.3.1. Integration of Simple Functions

Let X be a measurable space with a σ–algebra of the sets $\mathcal{A}$ and the nonnegative measure μ given on $\mathcal{A}$. A simple function with the values in an arbitrary linear space Y

$$f(x) = \sum_{k=1}^{N} y_k \mathbf{1}_{E_k}(x)$$

is called μ–integrable if it is different from zero only on the set of the finite measure, i.e. if $\mu(E_k) < \infty$ at all k.

An integral of the simple function $f(x)$ over the set $A \in \mathcal{A}$ and over the whole space X is determined by formulae

$$\int_A f(x)\mu(dx) = \int_A f d\mu = \sum_{k=1}^{N} y_k \mu(E_k A) \,, \tag{3.16}$$

$$\int f(x)\mu(dx) = \int f d\mu = \sum_{k=1}^{N} y_k \mu(E_k) \,. \tag{3.17}$$

Hence the region of the integration is not indicated in (3.17). Thus the integrals without the indication of the integration region are always considered as the integrals over the whole space X.

3.3.2. Properties of an Integral of a Simple Function

It follows directly from definition (3.16) that at any complex α

$$\int_A \alpha f \, d\mu = \alpha \int_A f \, d\mu \,. \tag{I}$$

The sum of two μ–integrable simple functions

$$f_1(x) = \sum_{i=1}^{N} y_i \mathbf{1}_{E_i}(x), \quad f_2(x) = \sum_{j=1}^{M} z_j \mathbf{1}_{F_j}(x)$$

represents a μ–integrable simple function

$$f_1(x) + f_2(x) = \sum_{i=0}^{N}\sum_{j=0}^{M}(y_i + z_j)\mathbf{1}_{E_i F_j}(x),$$

where $E_0 = \bigcap_{i=1}^{N} \bar{E}_i$, $F_0 = \bigcap_{j=1}^{M} \bar{F}_j$, $y_0 = z_0 = 0$. The integral over the sum of the functions $f_1(x)$ and $f_2(x)$ according to formula (3.16) is determined by formula

$$\int_A (f_1 + f_2)d\mu = \sum_{i=0}^{N}\sum_{j=0}^{M}(y_i + z_j)\mu(E_i F_j A)$$

$$= \sum_{i=1}^{N} y_i\mu(E_i A) + \sum_{j=1}^{M} z_j\mu(F_j A) = \int_A f_1 d\mu + \int_A f_2 d\mu.$$

Thus the integral over the sum of two simple functions is equal to the sum of the integrals over these functions:

$$\int_A (f_1 + f_2)d\mu = \int_A f_1 d\mu + \int_A f_2 d\mu. \tag{II}$$

From (I) and (II) follows that any finite linear combination of μ–integrable simple functions is μ–integrable function and

$$\int_A \sum_{k=1}^{n} \alpha_k f_k d\mu = \sum_{k=1}^{n} \alpha_k \int_A f_k d\mu. \tag{III}$$

Putting here $A = \bigcup_{k=1}^{n} A_k$, $A_k \in \mathcal{A}$, $A_k A_h = \emptyset$ at $h \neq k$, $\alpha_k = 1$, $f_k(x) = f(x)\mathbf{1}_{A_k}(x)$ we get

$$\int_A f d\mu = \sum_{k=1}^{n} \int_{A_k} f d\mu. \tag{IV}$$

Thus the integral of a simple function represents an additive set function.

Let T be a linear operator mapping the linear space Y into another linear space Z. The function

$$g(x) = Tf(x) = \sum_{k=1}^{N} Ty_k \mathbf{1}_{E_k}(x),$$

evidently represents a μ–integrable simple function with the values in the space Z by virtue of the linearity of linear operator T

$$\int_A Tf d\mu = \sum_{k=1}^{N} Ty_k \mu(E_k A) = T \sum_{k=1}^{N} y_k \mu(E_k A) = T \int_A f d\mu.$$

Thus for any μ–integrable function $f(x)$ and any linear operator T the simple function $Tf(x)$ is μ–integrable and

$$\int_A Tf d\mu = T \int_A f d\mu. \tag{V}$$

The following properties of an integral of a simple function are evident:

$$\int_A f d\mu = 0, \quad \text{if} \quad \mu(A) = 0, \tag{VI}$$

$$\int_A f d\mu = \int_A g d\mu, \quad \text{if} \quad f(x) \overset{\text{a.e.}}{=} g(x). \tag{VII}$$

If the space of the values Y of the simple function $f(x)$ is normed then the simple function $\| f(x) \|$ is μ–integrable if and only if $f(x)$ is μ–integrable. And from the inequality

$$\left\| \sum_{k=1}^{N} y_k \mu(E_k A) \right\| \leq \sum_{k=1}^{N} \| y_k \| \mu(E_k A)$$

follows

$$\left\| \int_A f d\mu \right\| \leq \int_A \| f \| d\mu. \tag{VIII}$$

Due to the fact that a μ–integrable simple function is always bounded $\| f(x) \| \leq M < \infty$, from (VIII) follows that

$$\left\| \int_A f d\mu \right\| \leq M \mu(A). \tag{IX}$$

We may assume $M = \max\limits_{k} \| y_k \|$. Hence it follows that at any $\varepsilon > 0$ for all $A \in \mathcal{A}$ for which $\mu(A) < \delta = \varepsilon/M$

$$\left\| \int\limits_A f d\mu \right\| < \varepsilon. \tag{X}$$

If $f(x)$ is a μ–integrable simple function with the values on the real axis R and $f(x) > 0$ almost everywhere on A, $\mu(A) > 0$ then

$$\int\limits_A f d\mu > 0. \tag{XI}$$

Hence it is clear that for any μ–integrable simple function $f(x)$ with the values in a normed linear space almost everywhere

$$f(x) \overset{\text{a.e.}}{=} 0, \quad x \in A, \quad \text{if} \quad \int\limits_A \| f \| d\mu = 0 \quad \text{and} \quad \mu(A) > 0. \tag{XII}$$

From (XI) also follows that if $f(x)$ and $g(x)$ are μ–integrable simple functions with the values on the real axis R and $f(x) \leq g(x)$ almost everywhere on A then

$$\int\limits_A f d\mu \leq \int\limits_A g d\mu, \tag{XIII}$$

and the sign of the equality takes place if and only if $f(x) = g(x)$ almost everywhere.

If $A_1, A_2 \in \mathcal{A}$, $A_1 \subset A_2$ and $f(x) \geq 0$ almost everywhere on A_2 then (XIII) gives

$$\int\limits_{A_1} f d\mu \leq \int\limits_{A_2} f d\mu, \tag{XIV}$$

as

$$\int\limits_{A_1} f d\mu = \int\limits_{A_2} f \mathbf{1}_{A_1} d\mu \leq \int\limits_{A_2} f d\mu,$$

because $f(x) \mathbf{1}_{A_1}(x) \leq f(x)$.

We underline that the integrals of the simple functions with the values in any linear space possess properties (I)–(VII), the integrals of simple functions with the values in a normed linear space possess properties (VIII)–(X) and (XII) and only the integrals of simple functions with the values on the real axis possess properties (XI), (XIII) and (XIV).

3.3.3. Integration of B–Space–Valued Functions

The measurable function $f(x)$ with the values in the separable B–space Y is called μ–*integrable* if there exists such a sequence of μ–integrable simple functions $\{f^n(x)\}$ convergent almost everywhere to $f(x)$ that

$$\int \| f^n - f^m \| \, d\mu \to 0 \quad \text{at} \quad n, m \to \infty . \tag{3.18}$$

Such a sequence is called the *determining* sequence for the function $f(x)$.

An *integral* of the function $f(x)$ over the set $A \in \mathcal{A}$ is determined by formula

$$\int\limits_A f(x)\mu(dx) = \int\limits_A f d\mu = \lim_{n \to \infty} \int\limits_A f^n d\mu . \tag{3.19}$$

This integral is called *Bochner integral*. In the special case of the real numeric measurable function $f(x)$ Bochner integral is called *Lebesgue integral* (sometimes *abstract Lebesgue integral* in order to distinguish it from the integral introduced for the first time by Lebesgue).

Above all we show that the limit in formula (3.19) exists. For this purpose we notice that from inequalities (VIII) and (XIV) follows

$$\left\| \int\limits_A f^n d\mu - \int\limits_A f^m d\mu \right\| \leq \int\limits_A \| f^n - f^m \| \, d\mu \leq \int \| f^n - f^m \| \, d\mu . \tag{3.20}$$

Hence and from condition (3.18) follows that a sequence of the integrals $\{\int\limits_A f d\mu\}$ is a fundamental sequence. As a B–space is complete then this sequence has the limit in it to which it converges uniformly relatively to A. For proving the uniform convergence it is sufficient to notice that from condition (3.18) follows the existence at any $\varepsilon > 0$ such natural N_ε that

$$\int \| f^n - f^m \| \, d\mu < \varepsilon \quad \forall n, m > N_\varepsilon .$$

Hence and from inequalities (3.20) follows

$$\left\| \int\limits_A f^n d\mu - \int\limits_A f^m d\mu \right\| < \varepsilon \quad \forall A \in \mathcal{A}, \, n, m > N_\varepsilon .$$

Passing to the limit at $m \to \infty$ we obtain

$$\left\| \int_A f^n d\mu - \int_A f d\mu \right\| < \varepsilon \quad \forall A \in \mathcal{A}, \, n > N_\varepsilon \, .$$

Due to the independence of N_ε from A follows an uniform relative to A convergence of the sequence of the integrals of simple functions f^n to the integral of the function f.

Generalizing the definition of a μ–integrable function to the case when the function is not a μ–integrable over the whole space X we shall call the function $f(x)$ a μ-integrable on the set $B \subset X$, $B \in \mathcal{A}$ if there exists such a sequence of μ–integrable on B simple functions $\{f^n(x)\}$ convergent almost everywhere to $f(x)$ on B that

$$\int_B \| f^n - f^m \| \, d\mu \to 0 \quad \text{at} \quad n, m \to \infty \, . \tag{3.21}$$

In this case the sequence of the integrals $\int\limits_A f^n d\mu$ converges to the integral $\int\limits_A f d\mu$ uniformly relative to $A \in \mathcal{A}$, $A \subset B$.

3.3.4. Correctness of Definition of an Integral

It remains to prove the correctness of definition of an integral. For this purpose we shall prove the following suggestion.

Theorem 3.19. *An integral does not depend on the choice of the sequence of simple functions determining the integrand.*

▷ Let $\{f_1^n(x)\}$ and $\{f_2^n(x)\}$ be two sequences which determine the function $f(x)$ with the values in a separable B–space: $f_1^n(x) \xrightarrow{\text{a.e.}} f(x)$, $f_2^n(x) \xrightarrow{\text{a.e.}} f(x)$, and

$$\int \| f_1^n - f_1^m \| \, d\mu \to 0, \quad \int \| f_2^n - f_2^m \| \, d\mu \to 0 \tag{3.22}$$

at $n, m \to \infty$. We put

$$g^n(x) = \| f_1^n(x) - f_2^n(x) \| \quad (n = 1, 2, \ldots) \, .$$

As

$$\left| g^n - g^m \right| = \left| \| f_1^n(x) - f_2^n(x) \| - \| f_1^m(x) - f_2^m(x) \| \right|$$

$$\leq \| f_1^n - f_2^n - f_1^m + f_2^m \| \leq \| f_1^n - f_1^m \| + \| f_2^n - f_2^m \|$$

and both sequences $\{f_1^n\}$ and $\{f_2^n\}$ satisfy condition (3.22) then also the sequence of simple functions $\{g^n(x)\}$ convergent almost everywhere to zero satisfies condition (3.22):

$$\int |g^n - g^m| \, d\mu \to 0 \quad \text{at} \quad n, m \to \infty.$$

Consequently, $\{g^n\}$ defines zero and the sequence of the integrals $\{\int_A g^n d\mu\}$ converges uniformly on $A \in \mathcal{A}$. Our task is to prove that it converges to zero. It is sufficient for this purpose to show at any $\varepsilon > 0$

$$\int g^p \, d\mu < \varepsilon \tag{3.23}$$

at all sufficiently large p (recall that $g^p \geq 0$ at all p).

From the convergence of the sequence $\{\int_A g^n d\mu\}$ follows its property to be a fundamental sequence. Consequently, at any $\eta > 0$ there exists such natural number N_η that

$$\left| \int_A g^p \, d\mu - \int_A g^n \, d\mu \right| < \eta \quad \text{at all} \quad p > n > N_\eta \tag{3.24}$$

and at all $A \in \mathcal{A}$. We fix $n > N_\eta$. The function g^n as a μ–integrable function is different from zero only on some set $B \in \mathcal{A}$ of a finite measure $\mu(B) < \infty$. From $g^n(x) = 0$ at $x \in \bar{B}$ follows

$$\int_{\bar{B}} g^n \, d\mu = 0.$$

But then from inequality (3.24) follows the inequality

$$\int_{\bar{B}} g^p \, d\mu < \eta \quad \text{at all} \quad p > n. \tag{3.25}$$

It remains to estimate $\int_B g^p d\mu$. At first we shall choose such $\delta > 0$ that

$$\int_C g^n \, d\mu < \eta \quad \forall C \in \mathcal{A}, \quad \mu(C) < \delta. \tag{3.26}$$

It is possible on the basis of property (X) of an integral of a simple function. As $\mu(B) < \infty$ then according to Theorem 3.12 the sequence of the functions $\{g^p\}$ converges to zero in the measure on B. Therefore at any $\delta > 0$, $\zeta > 0$ there exists such natural $N_{\delta,\zeta}$ that

$$\mu(\{x \ : \ g^p(x) \geq \zeta\}B) < \delta \quad \forall p > N_{\delta,\zeta}\,.$$

After choosing some $m > \max(N_\eta, N_{\delta,\zeta})$ and putting $C = \{x \ : \ g^m(x) \geq \zeta\}B$ we shall have $\mu(C) < \delta$, $g^m(x) < \zeta \quad \forall x \in B\backslash C$ and

$$\int\limits_{B\backslash C} g^m\,d\mu < \zeta\mu(B\backslash C) < \zeta\mu(B)\,. \tag{3.27}$$

From inequalities (3.24)–(3.27) follows that at all $p > \max(N_\eta, N_{\delta,\zeta})$

$$0 \leq \int g^p\,d\mu = \int\limits_{\bar{B}} g^p\,d\mu + \int\limits_{B\backslash C} g^m\,d\mu + \left(\int\limits_{B\backslash C} g^p\,d\mu\right.$$
$$\left. - \int\limits_{B\backslash C} g^m\,d\mu\right) + \int\limits_{C} g^n\,d\mu + \left(\int\limits_{C} g^p\,d\mu - \int\limits_{C} g^n\,d\mu\right)$$
$$< \eta + \zeta\mu(B) + 3\eta = 4\eta + \zeta\mu(B)\,.$$

After setting an arbitrary $\varepsilon > 0$ and putting $\eta = \varepsilon/5$, $\zeta = \varepsilon/5\mu(B)$ we get

$$0 \leq \int g^p\,d\mu < \varepsilon\,.$$

Hence it follows that for any $A \in \mathcal{A}$ we get

$$\int\limits_{A} \| f_1^n - f_2^n \|\,d\mu \to 0 \quad \text{at} \quad n \to \infty\,,$$

and

$$\left\| \int\limits_{A} f_1^n\,d\mu - \int\limits_{A} f_2^n\,d\mu \right\| \leq \int\limits_{A} \| f_1^n - f_2^n \|\,d\mu \to 0\,.$$

Thus

$$\lim \int\limits_{A} f_1^n\,d\mu = \lim \int\limits_{A} f_2^n\,d\mu\,, \tag{3.28}$$

what proves the independence of the limit in definition (3.19) of the integral of the choice of the sequence of simple functions determining the integrand $f(x)$. ◁

E x a m p l e   3.4.   To calculate the integral of the function $f(x)$ $= \varphi_1(s)x^2(t_1) + \varphi_2(s)x^2(t_2)$ over the whole space of the functions X^T on the Wiener measure μ_W (Example 2.15) if $\varphi_1(s)$ and $\varphi_2(s)$ are continuous functions on bounded closed interval S of the real axis.

In a given case as an argument of the function $f(x)$ serves the function $x(t)$ from the space X^T of scalar functions $t \in [0, \infty)$ on which the Wiener measure μ_W is determined. The space of the values of the function $f(x)$ serves the B–space of continuous functions $C(S)$. As the function $f(x)$ depends only on the values of the function $x(t)$ in two fixed points t_1, t_2 then for calculating the integral $\int f(x)\mu_W(dx)$ we may use the expression of Example 2.15 of the measure μ_W on rectangular sets at $n = 2$. As a result we receive

$$\int f(x)\mu_W(dx) = \frac{1}{2\pi\sqrt{t_1(t_2-t_1)}} \int\limits_{-\infty}^{\infty} \int\limits_{-\infty}^{\infty} [\varphi_1(s)x_1^2$$

$$+\varphi_2(s)x_2^2]\exp\left\{-\frac{x_1^2}{2t_1} - \frac{(x_2-x_1)^2}{2(t_2-t_1)}\right\}dx_1 dx_2 .$$

For calculating the integrals over x_1 and over x_2 we use formulae (Appendix 2)

$$\int\limits_{-\infty}^{\infty} e^{\eta x - cx^2/2}dx = \sqrt{\frac{2\pi}{c}}e^{\eta^2/2c} , \tag{$*$}$$

$$\int\limits_{-\infty}^{\infty} e^{-cx^2/2}dx = \sqrt{\frac{2\pi}{c}} ,$$

$$\int\limits_{-\infty}^{\infty} xe^{-cx^2/2}dx = 0 , \quad \int\limits_{-\infty}^{\infty} x^2 e^{-cx^2/2}dx = \frac{1}{c}\sqrt{\frac{2\pi}{c}} .$$

So, we find

$$\int f(x)\mu_W(dx) = \varphi_1(s)t_1 + \varphi_2(s)t_2 .$$

3.3.5. Properties of Integrals

It is easy to be sure that properties (I)–(IV), (VI), (VII), (IX) of an integral of a simple function are also typical for an integral of a measurable function with the values in a B-space. Property (III) of an integral

$$\int\limits_A \sum_{k=1}^n \alpha_k f_k d\mu = \sum_{k=1}^n \alpha_k \int\limits_A f_k d\mu$$

for any μ-integrable functions $f_1, \ldots, f_n$ and any numbers $\alpha_1, \ldots, \alpha_n$ shows that an integral represents a linear operator mapping the space of μ-integrable functions into the space of their values. Property (IV) of an integral shows that an integral represents an additive measure.

▷ For proving property (VIII) it is sufficient to notice that from the inequality $||| f^n || - || f^m ||| \leq || f^n - f^m ||$ follows that the sequence of μ-integrable simple functions $\{|| f^n(x) ||\}$ determines the function $|| f(x) ||$. Thus from the μ-integrability of $f(x)$ follows the μ-integrability $|| f(x) ||$. Further from inequality (VIII) for all simple functions

$$\left\| \int\limits_A f^n d\mu \right\| \leq \int\limits_A || f^n || d\mu,$$

follows that inequality (VIII) is also valid for the limit function $f(x)$. Thus for the μ-integrability of the function $f(x)$ it is necessary the μ-integrability of its norm $\|f(x)\|$. The sufficiency of the μ-integrability of the norm $\|f(x)\|$ for the μ-integrability of the function $f(x)$ we shall prove in Subsection 3.3.6. ◁

▷ In order to generalize property (X) on the integrals of the measurable functions we suppose that $f(x)$ is a μ-integrable function and $\{f^n(x)\}$ is its determining sequence of the μ-integrable simple functions. According to the proved in Subsection 3.3.3 for any $\varepsilon > 0$ there exists such a natural $N = N(\varepsilon)$, that $\| \int_A f^n d\mu - \int_A f d\mu \| < \frac{\varepsilon}{2}$ for all $A \in \mathcal{A}$ and $n \geq N$. We choose any $n \geq N$. As f^n is the μ-integrable simple function then on the basis of inequality (X) for the integrals of simple functions f^n at any $\varepsilon > 0$ there exists such $\delta > 0$ that

$$\left\| \int\limits_A f^n d\mu \right\| < \frac{\varepsilon}{2} \quad \text{for all} \quad A \in \mathcal{A}, \ \mu(A) < \delta.$$

From two obtained inequalities follows

$$\left\| \int\limits_A f d\mu \right\| < \varepsilon \quad \text{for all} \quad A \in \mathcal{A}, \ \mu(A) < \delta. \qquad \text{(X)} \ ◁$$

▷ For proving property (XII) we notice that for an arbitrary $\varepsilon > 0$, if $\| f(x) \| \geq \varepsilon$ on the set $B_\varepsilon \in \mathcal{A}$, $B_\varepsilon \subset A$ then from (XIII) and (XIV) follows

$$\int_A \| f \| \, d\mu \geq \int_{B_\varepsilon} \| f \| \, d\mu \geq \varepsilon \mu(B_\varepsilon) \,. \qquad (3.29)$$

Therefore if

$$\int_A \| f \| \, d\mu = 0 \,,$$

then $\mu(B_\varepsilon) = 0$. As $\varepsilon > 0$ is an arbitrary one then $f(x)$ may differ from zero only on the subset of zero measure of the set A, i.e. $f(x) = 0$ almost everywhere on A. ◁

From inequalities (3.29) follows that at any $\varepsilon > 0$ and $A \in \mathcal{A}$

$$\mu(\{x : \| f(x) \| > \varepsilon\} A) \leq \frac{1}{\varepsilon} \int_A \| f \| \, d\mu \,. \qquad (3.30)$$

Putting here, in particular $A = X$, we receive

$$\mu(\{x : \| f(x) \| \geq \varepsilon\}) \leq \frac{1}{\varepsilon} \int \| f \| \, d\mu \,. \qquad (3.31)$$

Inequalities (3.30) and (3.31) are called *Chebyshev inequalities*.

Let state two more important properties of an integral. We choose an arbitrary $\varepsilon > 0$ and some natural n for which

$$\left\| \int_A f d\mu - \int_A f^n d\mu \right\| < \varepsilon \quad \text{for all} \quad A \in \mathcal{A} \,.$$

The simple function f^n is different from zero only on some set A_ε of the finite measure $A_\varepsilon \in \mathcal{A}$, $\mu(A_\varepsilon) < \infty$. On the complement of this set $\bar{A}_\varepsilon$ the previous inequality gives

$$\left\| \int_{\bar{A}_\varepsilon} f d\mu \right\| < \varepsilon \,. \qquad (XV)$$

Thus at any $\varepsilon > 0$ there exists the division of the space X into additional sets A_ε and $\bar{A}_\varepsilon$ one of which has the finite measure and on the other one the norm of the integral is smaller than ε.

While extending the measure μ on the σ-algebra $\mathcal{A}_1 \supset \mathcal{A}$ the integral of $(\mathcal{A}, \mathcal{B})$-measurable function over the set $A \in \mathcal{A}$ does not change:

$$\int\limits_A f(x)\mu(dx) = \int\limits_A f(x)\mu_1(dx), \qquad \text{(XVI)}$$

where μ_1 is the extension of the measure μ. For proving it is sufficient to notice that any $(\mathcal{A}, \mathcal{B})$-measurable function is $(\mathcal{A}_1, \mathcal{B})$-measurable and $\mu_1(A) = \mu(A)$ for any set $A \in \mathcal{A}$.

3.3.6. Integrability of a Function whose Norm is Bounded by an Integrable Function

Let contunue studying the properties of the integrals and integrable functions. In particular, we prove the sufficiency of the integrability of the norm of the function for the integrability of the function itself. The following general theorem is of great importance.

Theorem 3.20. *If the function $f(x)$ with the values in a separable B-space is measurable and $\| f(x) \| \le g(x)$ where $g(x)$ is μ-integrable function then the function $f(x)$ is μ-integrable.*

▷ For proving μ-integrability of the function $f(x)$ it is sufficient to construct the sequence of simple functions determining $f(x)$. By Theorem 3.17 the function $f(x)$ may be presented on any set of the finite measure as the limit of the sequence of simple functions convergent almost everywhere. In order to prove the theorem in general case of σ-finite measure let construct at first the sequence of the functions $\{f_m(x)\}$ convergent to $f(x)$ in measure. Each is different from zero only on the set of the finite measure. Further on applying to each function $f_m(x)$ Theorem 3.17 we construct the sequence of simple functions $\{z^m(x)\}$ convergent to $f(x)$ in measure μ. After this we find the sequence of the simple functions $\{f^n(x)\}$ which determines $f(x)$.

By virtue of Chebyshev inequality (3.31) for any μ-integrable function $h(x)$ and for any $\varepsilon > 0$ the measure of the set $\{x : \| h(x) \| \ge \varepsilon\}$ is finite. Therefore after taking any sequence of positive numbers $\{\varepsilon_m\}$ convergent to zero we obtain the sequence of the sets of the finite measure $\{A_m\}$,

$$A_m = \{x : g(x) \ge \varepsilon_m/2\}, \quad \mu(A_m) < \infty.$$

Then we shall have $A_m \subset A_{m+1}$, $\lim A_m = \bigcup A_m = X$ and

$$\| f(x) \| \le g(x) < \varepsilon_m/2 \quad \text{at} \quad x \in \bar{A}_m.$$

After determining the functions

$$f_m(x) = \begin{cases} f(x) & \text{at} \quad x \in A_m, \\ 0 & \text{at} \quad x \in \bar{A}_m, \end{cases} \quad (m = 1, 2, \ldots), \qquad (3.32)$$

we get the sequence of the functions $\{f_m(x)\}$ convergent to $f(x)$ in measure μ. Really, from the definition of the function $f_m(x)$ it is clear that at all x

$$\| f_m(x) - f(x) \| < \varepsilon_m/2 \quad (m = 1, 2, \ldots).$$

As a result we get

$$\mu(\{x : \| f_m(x) - f(x) \| \geq \varepsilon_m/2\}) = 0 \quad (m = 1, 2, \ldots). \qquad (3.33)$$

Consequently, at any $\varepsilon > 0$

$$\mu(\{x : \| f_m(x) - f(x) \| \geq \varepsilon\}) = 0$$

at all m at which $\varepsilon_m/2 < \varepsilon$.

We notice now that each function $f_m(x)$ is different from zero only on the set A_m of the finite measure. As a result according to Theorem 3.17 it may be presented as the limit of the sequence of simple functions $\{f_m^k(x)\}_k$ (equal to 0 at $x \in \bar{A}_m$) convergent to it almost everywhere, and consequently, in measure. Let take an arbitrary sequence of the numbers $\{\delta_m\}$, $\delta_m > 0$ convergent to zero and choose for each m a natural k_m that

$$\mu(\{x : \| f_m^{k_m}(x) - f_m(x) \| \geq \varepsilon_m/2\}) < \delta_m . \qquad (3.34)$$

As a result we get the sequence of simple functions $\{z^m(x)\}$, $z^m(x) = f_m^{k_m}(x)$ convergent in measure μ to the function $f(x)$. Really, as

$$\| z^m(x) - f(x) \| \leq \| z^m(x) - f_m(x) \| + \| f_m(x) - f(x) \| ,$$

then

$$\mu(\{x : \| z^m(x) - f(x) \| \geq \varepsilon_m\})$$
$$\leq \mu(\{x : \| z^m(x) - f_m(x) \| \geq \varepsilon_m/2\})$$
$$+ \mu(\{x : \| f_m(x) - f(x) \| \geq \varepsilon_m/2\}) .$$

Hence due to (3.33) and (3.34) we obtain at all m

$$\mu(\{x : \| z^m(x) - f(x) \| \geq \varepsilon_m\}) < \delta_m . \qquad (3.35)$$

Therefore at any ε, $\delta > 0$

$$\mu(\{x : \parallel z^m(x) - f(x) \parallel \geq \varepsilon\}) < \delta$$

at all m for which $\varepsilon_m < \varepsilon$, $\delta_m < \delta$ what proves the convergence of the sequence of simple functions $\{z^m(x)\} \to f(x)$ in measure μ.

In order to get from $\{z^m(x)\}$ the sequence determining $f(x)$ we "improve" each function $z^m(x)$ replacing it by zero in the set

$$E_m = \{x : \parallel z^m(x) - f(x) \parallel \geq \varepsilon_m\} \qquad (3.36)$$

and on that part of the set $\bar{E}_m$ where $\parallel z^m(x) \parallel \leq 2\varepsilon_m$. As a result we obtain the sequence of simple functions $\{u^m(x)\}$

$$u^m(x) = \begin{cases} z^m(x) & \text{at } x \in \bar{E}_m \quad \text{and} \quad \parallel z^m(x) \parallel > 2\varepsilon_m, \\ 0 & \text{at } x \in E_m \quad \text{or} \quad \parallel z^m(x) \parallel \leq 2\varepsilon_m. \end{cases}$$

It is evident that at all $x \in \bar{E}_m$

$$\parallel u^m(x) - f(x) \parallel < 3\varepsilon_m ,$$

as

$$\parallel u^m(x) - f(x) \parallel < \varepsilon_m \quad \text{everywhere} \quad u^m(x) = z^m(x) ,$$

and

$$\parallel u^m(x) - f(x) \parallel = \parallel f(x) \parallel \leq \parallel z^m(x) \parallel + \parallel z^m(x) - f(x) \parallel < 3\varepsilon_m ,$$

at all $x \in \bar{E}_m$ at which $\parallel z^m(x) \parallel \leq 2\varepsilon_m$. Putting arbitrary ε, $\delta > 0$ we shall have by virtue of (3.36) and (3.35) at all m at which $3\varepsilon_m < \varepsilon$, $\delta_m < \delta$

$$\mu(\{x : \parallel u^m(x) - f(x) \parallel \geq \varepsilon\}) \leq \mu(E_m) < \delta .$$

Consequently, the sequence $\{u^m(x)\}$similarly as $\{z^m(x)\}$ converges to $f(x)$ in measure. But the functions $u^m(x)$ have an advantage over $z^m(x)$ that they are bounded in measure by an integrable function. Really, at all x at which $u^m(x) \neq 0$, $\parallel u^m(x) \parallel > 2\varepsilon_m$ and $\parallel u^m(x) - f(x) \parallel < \varepsilon_m$. As a result we have

$$\parallel u^m(x) \parallel \leq \parallel f(x) \parallel + \parallel u^m(x) - f(x) \parallel$$

$$< \parallel f(x) \parallel + \varepsilon_m < \parallel f(x) \parallel + \frac{1}{2} \parallel u^m(x) \parallel .$$

Hence

$$\| u^m(x) \| < 2 \| f(x) \| \le 2g(x), \tag{3.37}$$

at all x. Inequality (3.37) shows that all simple functions $u^m(x)$ are μ-integrable by virtue of the μ-integrability of $g(x)$. Now we separate out the sequence $\{u^m(x)\}$ convergent to $f(x)$ in measure the subsequence $\{f^n(x)\}$, $f^n(x) = u^{m_n}(x)$ convergent almost everywhere. Then by virtue of inequality (3.37) we get

$$\| f^n(x) - f^m(x) \| \le \| f^n(x) \| + \| f^m(x) \| < 4g(x)$$

and

$$\int \| f^n - f^m \| \, d\mu < 4 \int g d\mu. \tag{3.38}$$

On the basis of properties (X) and (XV) of an integral at any $\eta > 0$ there exists such number $\delta > 0$ and the set of the finite measure B, $\mu(B) < \infty$, that

$$\int_C g d\mu < \eta/4 \quad \forall C, \quad \mu(C) < \delta, \quad \int_{\bar{B}} g d\mu < \eta/4.$$

Hence and from inequality (3.38) it follows

$$\int_C \| f^n - f^m \| \, d\mu < \eta \quad \forall C, \quad \mu(C) < \delta, \quad \forall n, m, \tag{3.39}$$

$$\int_{\bar{B}} \| f^n - f^m \| \, d\mu < \eta \quad \forall n, m. \tag{3.40}$$

From the convergence of the sequence $\{f^n(x)\}$ in measure follows its property to be a fundamental sequence in measure (Theorem 3.10). Therefore at any $\delta > 0$, $\zeta > 0$ there exists such natural N that

$$\mu(\{x : \| f^n(x) - f^m(x) \| \ge \zeta\}) < \delta \quad \forall n, m > N.$$

After fixing $n, m > N$ and putting $C = \{x : \| f^n(x) - f^m(x) \| \ge \zeta\}$ we shall have $\mu(C) < \delta$, $\| f^n(x) - f^m(x) \| < \zeta$, $\forall x \in \bar{C}$ and

$$\int_{B \backslash C} \| f^n - f^m \| \, d\mu < \zeta \mu(B \backslash C) \le \zeta \mu(B). \tag{3.41}$$

From (3.39)–(3.41) follows

$$\int \| f^n - f^m \| \, d\mu = \int_{\bar{B}} \| f^n - f^m \| \, d\mu + \int_{BC} \| f^n - f^m \| \, d\mu$$

$$+ \int_{B\bar{C}} \| f^n - f^m \| \, d\mu < 2\eta + \zeta\mu(B).$$

After setting an arbitrary $\varepsilon > 0$ and putting $\eta = \varepsilon/3$, $\zeta = \varepsilon/3\mu(B)$ we have

$$\int \| f^n - f^m \| \, d\mu < \varepsilon \quad \forall n, m > N.$$

Thus the constructed sequence of the μ–integrable simple functions $\{f^n(x)\}$ determines $f(x)$ what proves the μ–integrability of $f(x)$. ◁

Corollary 1. *From the μ–integrability of $\| f(x) \|$ follows the μ–integrability of $f(x)$.*

▷ For proving it is sufficient to put $g(x) = \| f(x) \|$. ◁

Recalling that the μ–integrability of $f(x)$ implies the μ–integrability of $\| f(x) \|$ we come to the conclusion that for the μ–integrability of the measurable function $f(x)$ with the values in a separable B–space it is necessary and sufficient that its norm $\| f(x) \|$ be μ–integrable.

Corollary 2. *From any sequence of simple functions $\{f_1^n(x)\}$ which determines μ–integrable function $f(x)$ we may obtain another sequence of simple functions $\{f^n(x)\}$ which determines $f(x)$ and satisfies the condition $\| f^n(x) \| < 2 \| f(x) \|$.*

▷ For proving it is sufficient to apply to $\{f_1^n(x)\}$ the same method as was applied to the sequence $\{z^m(x)\}$ in Theorem 3.20. ◁

3.3.7. Changing Variables

In many problems it is important to know how to pass in an integral from one variable of the integration to another one. Let $(X, \mathcal{A}, \mu)$ be the space with nonnegative measure, $(U, \mathcal{D})$ be a measurable space, $u = \varphi(x)$ be $(\mathcal{A}, \mathcal{D})$–measurable function mapping X into U,

$$\nu(D) = \mu(\varphi^{-1}(D)), \quad D \in \mathcal{D} \tag{3.42}$$

be a measure induced in the space U by the function $u = \varphi(x)$ (Subsection 3.1.7). Let us consider a ν–integrable function $y = g(u)$ which maps U into a separable B–space Y.

Theorem 3.21. *The following formula of changing variables in an integral is valid:*

$$\int_D g(u)\nu(du) = \int_{\varphi^{-1}(D)} g(\varphi(x))\mu(dx).$$ (3.43)

▷ Let $\{g^n(u)\}$ be a sequence of ν–integrable simple functions determining the function $g(u)$ almost everywhere:

$$g^n(u) \xrightarrow{\text{a.e.}} g(u), \int \parallel g^n - g^m \parallel d\nu \to 0 \quad \text{at} \quad n, m \to \infty.$$ (3.44)

If

$$g^n(u) = \sum_{k=1}^{N} y_k \mathbf{1}_{F_k}(u),$$

then on the basis of formula (3.42)

$$\int_D g^n d\nu = \sum_{k=1}^{N} y_k \nu(F_k D) = \sum_{k=1}^{N} y_k \mu(\varphi^{-1}(F_k)\varphi^{-1}(D))$$

$$= \int_{\varphi^{-1}(D)} g^n(\varphi(x))\mu(dx)$$ (3.45)

and

$$\int_D \parallel g^n - g^m \parallel d\nu = \int_{\varphi^{-1}(D)} \parallel g^n(\varphi(x)) - g^m(\varphi(x)) \parallel \mu(dx).$$

Hence at $D = U$ and from condition (3.44) is clear that the simple functions

$$f^n(x) = g^n(\varphi(x))$$

are μ–integrable and the sequence $\{f^n(x)\}$ determines the function

$$f(x) = g(\varphi(x)).$$

Consequently, from equality (3.45) follows formula (3.43). ◁

At $g(u) = u$, $D = U$, $D_\varphi = X$ formula (3.43) takes the form

$$\int uv(du) = \int \varphi(x)\mu(dx).$$ (3.46)

This formula is often used in the probability theory.

R e m a r k. The stated integral theory is valid in the case when $\mathcal{A}$ is an algebra of sets and μ is an additive measure (Dunford and Schwarz 1958). We shall meet with the integrals over the additive measure in Section 6.4.

3.3.8. Integrals by a Complex Measure

Jordan expansion (2.15) gives the possibility to determine Bochner and Lebesgue integrals by any real and complex measures. From (2.15) follows that any complex measure μ may be presented in the form

$$\mu(A) = \mu_R^+(A) - \mu_R^-(A) + i[\mu_I^+(A) - \mu_I^-(A)],$$ (3.47)

where $\mu_R^+(A)$, $\mu_R^-(A)$, $\mu_I^+(A)$ and $\mu_I^-(A)$ are nonnegative measures. Correspondingly the integral by the function $f(x)$ with the values in a B–space (in particular, with the values in the set of complex numbers C) by the set $A \in \mathcal{A}$ naturally will be determined by the formula

$$\int\limits_A f(x)\mu(dx) = \int\limits_A fd\mu = \int\limits_A fd\mu_R^+ - \int\limits_A fd\mu_R^-$$

$$+i\left\{ \int\limits_A fd\mu_I^+ - \int\limits_A fd\mu_I^- \right\}.$$ (3.48)

3.3.9. Integrals of Numeric Functions by a B–Space–Valued Measure

Similarly as in Subsection 3.3.3 we may determine an integral of the numeric function $f(x)$ in the measure μ with the values in a B–space

$$\int\limits_A f(x)\mu(dx) = \int\limits_A fd\mu.$$

Hence the general theory stated in Subsections 3.3.3–3.3.6 may be also applied to such integrals. Only nonnegative integrals of the form

$$\int\limits_A \| f \| d\mu = \int\limits_A \| f(x) \| \mu(dx)$$

are everywhere replaced by the integrals of the form

$$\int\limits_A |f|\,d\,|\,\mu\,| = \int\limits_A |f(x)|\,|\,\mu\,|\,(dx),$$

and the almost everywhere convergence relative to the measure μ is replaced by the almost everywhere convergence relative to complete variation $|\,\mu\,|$ of the measure μ.

For deep study of the Bochner integral refer to (Dunford and Schwarz 1958).

Problems

3.3.1. Using Example 3.4 calculate the integral of the function

$$f(x) = \sum_{k=1}^{n} \varphi_k(s)x^2(t_k), \quad \varphi_k = C(S), \quad (k = 1, 2, \ldots, n)$$

over the whole space of functions X^T in the Wiener measure μ_W.

3.3.2. Show that the integral of the function

$$f(x) = \sum_{k=1}^{n} \varphi_k(s)x^m(t_k)$$

over the whole space of functions X^T in the Wiener measure is equal to zero at odd $m = 2r + 1$ and at even $m = 2r$

$$\int f(x)\mu_W(dx) = (2r - 1)!! \sum_{k=1}^{n} \varphi_k(s)t_k^r\,.$$

Instruction. Apply formula $(*)$ of Example 3.4 and corresponding formulae received by differention $(*)$ with respect to η.

3.3.3. Calculate the integral in the Wiener measure of the function

$$f(x) = \varphi_1(s)x(t_1) + \varphi_2(s)x(t_2), \quad \varphi_1, \varphi_2 \in C(S),$$

over the rectangle set of space X^T with the base $A = (a_1, b_1) \times (a_2, b_2)$ in the product of the spaces $X_{t_1} \times X_{t_2}$.

Instruction. Apply the function (Appendix 2)

$$\Phi(u) = \frac{1}{\sqrt{2\pi}} \int\limits_0^u e^{-x^2/2}dx$$

and integrate by parts.

3.3.4. Show that the following formula is valid

$$\int f(x)\mu_W(dx) = \exp\Big\{ \sum_{k,l=1}^{n} \varphi_k(s)\varphi_l(s)\min(t_k,\,t_l) \Big\}$$

for the integral over the whole space X^T in the Wiener measure of the function

$$f(x) = \exp\Big\{ \sum_{k=1}^{n} \varphi_k(s)x(t_k) \Big\}.$$

I n s t r u c t i o n. Apply the formula (Appendix 2)

$$\int_{-\infty}^{\infty} \exp\{\eta^T x - x^T C x/2\}\, dx = \sqrt{\tfrac{(2\pi)^n}{|C|}}\, \exp\{\eta^T C^{-1}\eta/2\},$$

where n is the dimension of vectors x and η, $|\,C\,|$ is the determinant of some positively defined matrix C.

3.4. Lebesgue, Lebesgue–Stieltjes and Riemann Integrals

3.4.1. Lebesgue Integral

In the special case of the real numeric function $f(x)$, $f : X \to R$ Bochner integral (3.19) is called the *Lebesgue integral* which was named for the founder of a modern measure and integration theory French scientist Lebesgue. So, we have

$$f(x) = f^+(x) - f^-(x), \qquad (3.49)$$

$$f^+(x) = \max\{\,0, f(x)\,\},$$
$$f^-(x) = -\min\{\,0, f(x)\,\}.$$

Thus any real function may be presented as the difference of two nonnegative functions. It gives grounds to determine the Lebesgue integral only for nonnegative functions. Then from property (III) of the Bochner integral will follow the definition of the Lebesgue integral for any real function.

Let $f(x)$ be a measurable nonnegative function. It is evident that the nondecreasing sequence of the simple functions $\{f^n(x)\}$

$$f^n(x) = \begin{cases} k2^{-n} & \text{at } k2^{-n} \le f(x) < (k+1)2^{-n} \ (k = 0,1,\ldots,2^{2n}-1) \\ 2^n & \text{at } f(x) \ge 2^n \ (n = 1,2,\ldots) \end{cases} \qquad (3.50)$$

converges to $f(x)$ at all x. The property to be measurable of all the functions f^n follows from the fact that the sets $\{x : k2^{-n} \le f(x) < (k+1)2^{-n}\}$ represent the inverse images of the intervals $[k2^{-n}, (k+1)2^{-n})$.

Theorem 3.22. *If the sequence $\{\int_A f^n d\mu\}$ is bounded*

$$\int_A f^n d\mu < c < \infty, \tag{3.51}$$

then the sequence $\{f^n(x)\}$ determines the function $f(x)$.

▷ From (3.51) follows that the nondecreasing sequence of the integrals $\{\int_A f^n d\mu\}$ converges, and consequently, is a fundamental sequence

$$\int_A f^n d\mu - \int_A f^m d\mu \to 0 \qquad \text{at} \quad n, m \to \infty.$$

But by virtue of the nonnegativity of functions f^n and nondecrease of the sequence $\{f^n\}$

$$\int_A |f^n - f^m| \, d\mu = \left| \int_A f^n d\mu - \int_A f^m d\mu \right| \to 0 \qquad \text{at} \quad n, m \to \infty.$$

In particular, this is valid at $A = X$. It remains to prove that the sequence $\{f^n(x)\}$ converges to $f(x)$ almost everywhere. For this purpose we notice that according to definition (3.50) the sequence $\{f^n\}$ converges to $f(x)$ uniformly on any set $B_n = \{x : f(x) < 2^n\}$. For estimating the measure of the set $A \backslash B_n = A\{x : f(x) \ge 2^n\} = A\{x : f^n(x) = 2^n\}$ we notice that according to (3.50) and (3.51)

$$2^n \mu(A \backslash B_n) = \int_{A \backslash B_n} f^n d\mu \le \int_A f^n d\mu < c,$$

whence we find $\mu(A \backslash B_n) < c2^{--n}$. After taking an arbitrary $\delta > 0$ and n satisfying the inequality $2^n > c/\delta$ we get $\mu(A \backslash B_n) < \delta$. Consequently, the sequence $\{f^n(x)\}$ converges to $f(x)$ almost uniformly. By Theorem 3.16 it converges to $f(x)$ almost everywhere. ◁

If the sequence of the integrals $\{\int_A f^n \, d\mu\}$ is unbounded then the sequence of the simple functions $\{f^n\}$ does not determine the function f and the function f is not μ–integrable.

Thus condition (3.51) is necessary and sufficient for the μ–integrability of the nonnegative function $f(x)$. The μ–integrability of $f(x)$ in this case also follows directly from the convergence of the sequence of the integrals $\{\int_A f^n \, d\mu\}$.

From (3.49) and property (III) follows that for an arbitrary real function $f(x)$ the Lebesgue integral is determined as the difference of the integrals of the nonnegative functions $f^+(x)$ and $f^-(x)$:

$$\int_A f \, d\mu = \int_A f^+ \, d\mu - \int_A f^- \, d\mu. \tag{3.52}$$

Hence it follows that for the μ–integrability of the function $f(x)$ it is necessary and sufficient the μ–integrability of its modulus, i.e. the existence of the integral

$$\int_A |f| \, d\mu = \int_A f^+ \, d\mu + \int_A f^- \, d\mu.$$

For Lebesgue integrals the infinite values are also admissible, i.e. the values on the *expanded* real axis $[-\infty, \infty]$ (except the case when $\int f^+ \, d\mu = \infty$ and $\int f^- \, d\mu = \infty$). Thus we may consider Lebesgue integrals for any measurable (not obligatory integrable) functions.

3.4.2. Lebesgue Integral by Lebesgue Measure

Initially the definition of the Lebesgue integral was given for the case when the space X represents the real axis R, and the measure μ represents the Lebesgue measure on R. This definition can be trivially extended to the case of any finite–dimensional space $X = R^n$ and the Lebesgue measure μ in this space. In these cases the Lebesgue integral is denoted quite in the same way as the Riemann integral

$$\int_A f(x) \, dx.$$

Hence in the case of n–dimensional vector argument x the integral is assumed as a multiple integral

$$\int\limits_A f(x)\,dx = \int\limits_A \ldots \int f(x_1,\ldots,x_n)\,dx_1\ldots dx_n.$$

R e m a r k. Later on the notion of the Lebesgue integral was generalized to the case of any space with the measure $(X,\mathcal{A},\mu)$, and Bochner has also extended it to the integrands with the values in a B–space. For the case of the finite–dimensional space $Y = R^m$ the Lebesgue integral is determined quite natural as the vector whose coordinates serve Lebesgue integrals of the correspondent coordinates of the vector integrand.

3.4.3. Lebesgue and Riemann Integrals

As it is known the Riemann integral of the bounded real function $f(x)$ along bounded region A of the n–dimensional space R^n is determined by formula

$$\int\limits_A f(x)\,dx = \lim_{p\to\infty} \sum_{k=1}^{N_p} f(\xi_k^{(p)})v(A_k^{(p)}), \qquad (3.53)$$

where $\{P_p\}$

$$P_p : A = \bigcup_{k=1}^{N_p} A_k^{(p)} \quad (p = 1,2,\ldots)$$

is such a sequence of the partitions of the region A that

$$\Delta_p = \max_k \sup_{x,x'\in A_k^{(p)}} |x - x'| \to 0 \qquad \text{at} \quad p \to \infty,$$

$\xi_k^{(p)}$ is an arbitrary point of the region $A_k^{(p)}$, $\xi_k^{(p)} \in A_k^{(p)}$ and $v(A_k^{(p)})$ is the volume (in the special case the length, the square) of the region $A_k^{(p)}$ (the Lebesgue measure of the region $A_k^{(p)}$).

It is easy to notice the difference between the definiton of the Riemann integral and the defintion of the Lebesgue integral by the Lebesgue measure in the finite–dimensions space. While determining the Riemann integral *the integration region* is partitioned into parts and the value of the function in an arbitrary point is multiplied by the volume of this part. While determining the Lebesgue integral *the domain of the values of the integrand* is partitioned into parts – the intervals $[k2^{-n}, (k+1)2^{-n}]$ and the value of the integrand in the left end of each interval is multiplied by the volume of the inverse image of this interval. It gives the opportunity to determine the integral for much more wide class of the functions. So, for instance, the Riemann integral of Dirichlet function $f(x)$ equal to 1 in the rational points and equal to 0 in all other points of the real axis over any finite interval (a, b) does not exist, but Lebesgue integral exists and is equal to zero as Dirichlet function $f(x)$ represents a simple function with two values and the Lebesgue measure of the set of the rational points on which $f(x) = 1$ is equal to 0 and the Lebesgue measure of the set on which $f(x) = 0$ is equal to $b - a$.

Theorem 3.23. *If there exists the Riemann integral of the function $f(x)$ over the bounded region A then there also exists the Lebesgue integral of this function which is equal to the Riemaun integral.*

▷ Let consider the Riemann integral sum

$$S_p = \sum_{k=1}^{N_p} f(\xi_k^{(p)}) v(A_k^{(p)}).$$

It represents an integral of the integrable (in the Lebesgue measure) simple function

$$f^p(x) = \sum_{k=1}^{N_p} f(\xi_k^{(p)}) 1_{A_k^{(p)}}(x),$$

$$S_p = \int_a^b f^p(x) \, dx.$$

We shall prove that the sequence of the simple functions $\{f^p\}$ determines the function f. For this purpose we denote by $m_k^{(p)}$ and $M_k^{(p)}$ the lower and the upper bounds of the function $f(x)$ in the region $A_k^{(p)}$ correspondingly:

$$m_k^{(p)} \le f(x) \le M_k^{(p)} \qquad \text{at} \quad x \in A_k^{(p)}.$$

Due to the integrability of the function $f(x)$ by Riemann the limit of the sequence of the sums S_p does not depend on the choice of the sequence of the partitions of the region A. Therefore we may always take such a sequence of the partitions that at any $q > p$ each region $A_l^{(q)}$ will completely occur in some region $A_k^{(p)}$. Then we shall have at all p and $q > p$

$$|f^p(x) - f^q(x)| \le M_k^{(p)} - m_k^{(p)} \quad \text{at} \quad x \in A_k^{(p)},$$

and by property (IX) of the integral of a simple function

$$\int_A |f^p(x) - f^q(x)| \, dx \le \sum_{k=1}^{N_p} (M_k^{(p)} - m_k^{(p)}) v(A_k^{(p)}).$$

But the Riemann integral of $f(x)$ exists if and only if the sequences of the upper and lower Darboux sums

$$\sum_{k=1}^{N_p} M_k^{(p)} v(A_k^{(p)}) \quad \text{and} \quad \sum_{k=1}^{N_p} m_k^{(p)} v(A_k^{(p)})$$

converge to one and the same limit. Therefore

$$\int_A |f^p(x) - f^q(x)| \, dx \to 0 \quad \text{at} \quad p, q \to \infty.$$

Consequently, the function f is integrable by the Lebesgue measure of the region A, i.e. there exists the Lebesgue integral

$$\int_A f(x) \, dx = \lim_{p \to \infty} S_p = \lim_{p \to \infty} \sum_{k=1}^{N_p} f(\xi_k^{(p)}) v(A_k^{(p)}).$$

But by the definition of Riemann integral (3.53) the same limit represents the Riemann integral of $f(x)$. ◁

3.4.4. Riemann Improper Integral

We shall show now how Theorem 3.23 is extended to improper integrals. Suppose that the function $f(x)$ has the discontinuities of the second order at the points of some set $E \subset A$ when $f(x)$ unboundedly increases in modulus at $x \to \xi \in E$ and the region A is unbounded. Let A_δ be the region obtained from A by throwing out balls vicinities of the radius δ of all the points of the set E and that part of A which is arranged outside of the ball of radius $1/\delta$ with the center in the origin of coordinates. The Riemann improper integral is determined by formula

$$\int\limits_A f(x)\,dx = \lim_{\delta \to 0} \int\limits_{A_\delta} f(x)\,dx. \tag{3.54}$$

Theorem 3.24. *If there exists absolutely convergent improper Riemann integral (3.54), i.e. there exists integral*

$$\int\limits_A |f(x)|\,dx,$$

then there also exists Lebesgue integral of the function $f(x)$ over the region A which coincides with Riemann integral (3.54).

▷ Evidently it is sufficient to consider the case of nonnegative function $f(x)$. In this case we may take increasing sequence (3.50) of the simple functions $\{f^n\}$ convergent almost everywhere to f. But in accordance with the definition of the integrability of a simple function in Subsection 3.3.1 we have to multiply each function f^n from (3.50) by the indicator of some bounded region B_n, $B_n \subset B_{n+1}$, $\lim B_n = \bigcup B_n = A$. In order to distinguish Riemann and Lebesgue integrals before the moment when their coincidence be established we shall denote them for a while by

$$(R) \int\limits_A f(x)\,dx \qquad \text{and} \qquad (L) \int\limits_A f(x)\,dx.$$

respectively. As Riemann and Lebesgue integrals of the simple functions f^n coincide then from property (XIII) of the integral which is also inherent in Riemann integrals follows

$$\int\limits_A f^n(x)\,dx < (R) \int\limits_A f(x)\,dx. \tag{3.55}$$

Thus all the integrals of the simple functions f^n are bounded, and consequently, according to Theorem 3.22 there exists the Lebesgue integral

$$(L) \int\limits_A f(x)\,dx = \lim_{n \to \infty} \int\limits_A f^n(x)\,dx. \tag{3.56}$$

From (3.55) and (3.56) follows the inequality

$$(L) \int_A f(x)\,dx \le (R) \int_A f(x)\,dx. \qquad (3.57)$$

On the other hand, as Riemann and Lebesgue integrals over the region A_δ by Theorem 3.23 coincide then according to property (XIV) of the Lebesgue integral at all $\delta > 0$

$$(L) \int_A f(x)\,dx \ge \int_{A_\delta} f(x)\,dx.$$

But in accordance with formula (3.54)

$$(R) \int_A f(x)\,dx = \lim_{\delta \to 0} \int_{A_\delta} f(x)\,dx.$$

Consequently,

$$(L) \int_A f(x)\,dx \ge (R) \int_A f(x)\,dx. \qquad (3.58)$$

From inequalities (3.57) and (3.58) we get

$$(L) \int_A f(x)\,dx = (R) \int_A f(x)\,dx = \int_A f(x)\,dx. \quad \triangleleft$$

R e m a r k. Thus in all the cases when the modulus of the function f is Riemann–integrable, the function f is also Lebesgue–integrable even Riemann and Lebesgue integrals of this function coincide. But as we have already seen Lebesgue integral may also exist when the Riemann integral does not exist. On the other hand, if the Riemann integral

$$\int_A |f(x)|\,dx$$

does not exist then the function $f(x)$ is not Lebesgue–integrable while the improper Riemann integral

$$\int_A f(x)\,dx$$

may exist as the known example shows

$$\int\limits_0^\infty \frac{\sin x}{x}\, dx = \pi/2.$$

3.4.5. Lebesgue–Stieltjes and Riemann–Stieltjes Integrals

In the case of the finite dimensional space $X = R^n$ the common Lebesgue integral (3.52) is called the *Lebesgue–Stieltjes integral*. In particular, at $X = R$ we may establish the correspondence between any finite nonnegative measure μ and bounded nonincreasing function

$$F(x) = \mu((-\infty, x)). \qquad (3.59)$$

This function determines the Lebesgue–Stieltjes measure on R (Example 2.9) and for any interval of the form $[\alpha, \beta)$

$$\mu([\alpha, \beta)) = F(\beta) - F(\alpha). \qquad (3.60)$$

On the basis of Theorem 2.11 about the continuity of a measure given on the σ–algebra the function $F(x)$ is continuous on the left (namely, for this reason while determining the Lebesgue–Stieltjes measure on R in Example 2.9 we took the function $F(x)$ which is continuous on the left).

If the measure μ is σ–finite then after determining the function $F(x)$ by formula

$$F(x) = \begin{cases} \mu([0, x)) & \text{at} \quad x > 0, \\ 0 & \text{at} \quad x = 0, \\ -\mu([x, 0)) & \text{at} \quad x < 0, \end{cases}$$

we get again formula (3.60) for $\mu([\alpha, \beta))$ and $F(x)$ will be continuous on the left nonincreasing function.

Thus any measure on R represents the Lebesgue–Stieltjes measure. The formulae of Example 2.9 for the values of this measure on the intervals $[\alpha, \beta]$, $(\alpha, \beta]$ and (α, β) are obtained on the basis of Theorem 2.11 about the measure continuity. From the results of Examples 2.11 and 2.14 follows that any function $F(x)$ continuous on left determines the measure on R. Analogously we may show that any measure on R^n may be expressed in terms of some function $F(x)$. But at $n > 1$ it is unlikely expedient.

On the basis of the stated one–to–one correspondence between the measures on R and the functions continuous on the left the Lebesgue–Stieltjes integral over the interval $[a, b)$ is written in the form

$$\int_a^b f(x)\, dF(x). \qquad (3.61)$$

The same formula determines the Lebesgue–Stieltjes integral over the intervals $[a, b]$, (a, b) and $(a, b]$ if the function $F(x)$ is continuous at the points a and b. If the function $F(x)$ has the discontinuities of the first order with jumps $F(a + 0) - F(a)$ and $F(b + 0) - F(b)$ then Lebesgue–Stieltjes integrals over the intervals $[a, b]$, (a, b) and $(a, b]$ should be written in the form respectively

$$\int_a^{b+0} f(x)\, dF(x), \qquad \int_{a+0}^{b} f(x)\, dF(x) \quad \text{and} \quad \int_{a+0}^{b+0} f(x)\, dF(x).$$

Usually Lebesgue–Stieltjes integrals (3.61) are considered in the case of the function $F(x)$ of bounded variation (Example 2.12).

Riemann–Stieltjes integral of the function f over the function F is determined by the formula analogous to (3.53):

$$\int_a^b f(x)\, dF(x) = \lim_{p \to \infty} \sum_{k=1}^{N_p} f(\xi_k^{(p)})[F(x_k^{(p)}) - F(x_{k-1}^{(p)})], \qquad (3.62)$$

where $\{ P_p \} : a = x_0^{(p)} < x_1^{(p)} < \ldots < x_{N_p-1}^{(p)} < x_{N_p}^{(p)} = b \ (p = 1, 2, \ldots)$ is such a sequence of the partitions of the interval $[a, b)$ that

$$\lim_{p \to \infty} \Delta_p = \lim_{p \to \infty} \max_k (x_k^{(p)} - x_{k-1}^{(p)}) \to 0.$$

In exactly the same way as in Theorems 3.23 and 3.24 the following theorem is derived.

Theorem 3.25. *If the Riemann–Stieltjes integral*

$$\int_a^b |f(x)|\, dF(x) \qquad (3.63)$$

exists then there exists the Stieltjes–Lebesgue integral of the function $f(x)$
over $F(x)$ on the interval $[a, b)$ (or on the intervals $[a, b]$, $(a, b]$, (a, b)),
coinciding with the Riemann–Stieltjes integral (may be $a = -\infty$, $b = \infty$).

And similarly as in the case of the Lebesgue integral over the Lebesgue measure the Lebesgue–Stieltjes integral of the function f may exist while the Riemann–Stieltjes integral does not exist. And vice versa, if integral (3.63) does not exist then the Lebesgue–Stieltjes integral of $f(x)$ also does not exist while the Riemann–Stieltjes integral may exist (in particular, as the example of the integral of $\sin x/x$ from 0 till ∞ at $F(x) = x - a$).

Analogously we may determine the Lebesgue–Stieltjes and the Riemann–Stieltjes integral of the function $f(x)$ of a scalar variable x over any continuous on the left complex function $F(x)$ after presenting it in the form (Example 2.12)

$$F(x) = F_1(x) - F_2(x) + i[F_3(x) - F_4(x)],$$

where F_1, F_2, F_3, F_4 are any nondecreasing functions continuous on the left.

E x a m p l e 3.5 Using the formula of the change of variables (3.46) we may reduce the Lebesgue integral to the Lebesgue–Stieltjes integral

$$\int\limits_a^b f(x)\,dx = \int\limits_{-\infty}^{\infty} u\,dF(u),$$

putting $u = f(x)$, $F(y) = l(\{\, x : f(x) < y \,\}(a, b))$ according to the definition (3.59) of nondecreasing function continuous on the left which induces the Lebesgue–Stieltjes measure (l is the Lebesgue measure on the real axis R).

3.4.6. Integrals of Functions with the Values in a B–Space in Lebesgue Measure

The general definition of the Bochner integral (Subsection 3.3.3) gives, in particular, the definition of the Bochner integral in the Lebesgue measure in the space R^n. We may also determine the Riemann integral of the function with the values in a separable B–space. Let consider the sequence of the partitions of the bounded closed region (compact) A

$$\{\, P_n \,\} : A = \bigcup\limits_{k=1}^{N_n} A_k^{(n)}, \qquad \Delta_n = \max_k \sup_{x, x' \in A_k^{(n)}} |x' - x| \to 0$$

and determine the correspondent sequence of the integral sums

$$S_n = \sum_{k=1}^{N_n} f(\xi_k^{(n)}) v(A_k^{(n)}) \qquad (n = 1, 2, \ldots), \qquad (3.64)$$

where $\xi_k^{(n)}$ is an arbitrary point of the region $A_k^{(n)}$, $\xi_k^{(n)} \in A_k^{(n)}$ and $v(A_k^{(n)})$ is the Lebesgue measure of the region $A_k^{(n)}$. The limit of the sequence of the integral sums S_n (if it exists) is called the Riemann integral of the function $f(x)$ with the values in a separable B–space

$$\int_A f(x)\, dx = \lim_{n \to \infty} \sum_{k=1}^{N_n} f(\xi_k^{(n)}) v(A_k^{(n)}). \qquad (3.65)$$

Theorem 3.26. *If the function $f(x)$ is continuous in the bounded closed region A then there exist equal to each other Riemann–Bochner integrals of the function $f(x)$ over the region A.*

▷ The proof is very similarly to the proof of Theorem 3.23 with the only difference that for the integrals of simple function f^n of Theorem 3.23 the following relation is received:

$$\int_A \| f^p(x) - f^q(x) \|\, dx \leq \sum_{l=1}^{N_q} \sup_{A_l^{(q)}} \| f(\xi_k^{(p)}) - f(\xi_l^{(q)}) \|\, v(A_l^{(q)}),$$

where one and the same point $\xi_k^{(p)}$ corresponds to all l for which

$$\bigcup A_l^{(q)} = A_k^{(p)}.$$

From the continuity of the function $f(x)$ in the bounded closed region A follows its uniform continuity. Therefore at any $\varepsilon > 0$ there exists such $\delta > 0$ that

$$\sup_{1 \leq l \leq N_q} \sup_{A_l^{(q)}} \| f(\xi_k^{(p)}) - f(\xi_l^{(q)}) \| < \varepsilon$$

at $\Delta_p < \delta$. Therefore

$$\int_A \| f^p(x) - f^q(x) \|\, dx < \varepsilon v(A)$$

at all sufficiently large p and $q > p$. This proves that the sequence of the simple functions $\{f^n(x)\}$ determines the function $f(x)$. Consequently, Bochner integral exists and

$$\int\limits_A f(x)\,dx = \lim_{n\to\infty} \int\limits_A f^n(x)\,dx = \lim_{n\to\infty} S_n\,.$$

But this limit is equal to the Riemann integral. The coincidence of the Riemann integral with the Bochner integral by virtue of Theorem 3.19 proves the independence of the limit in (3.65) of the sequence choice of the partitions of the region A. ◁

It is clear that Riemann integral (3.65) of continuous function with the values in a B–space possesses all the properties of the Bochner integral which were established in Subsections 3.3.2 and 3.3.5.

Riemann integrals of the functions with the values in a B–space are of great importance for the theory of differential equations in the B–spaces (Ladas and Lakshmikantam 1972). Detailed treatment of Lebesgue, Lebesgue–Stieltjes and Riemann integrals is given in (Dunford and Schwarz 1958).

Problems

3.4.1. Prove the existence of the Riemann–Stieltjes integral in the case of the continuous function $f(x)$ and the function of bounded variation $F(x)$.

3.4.2. Calculate the Lebesgue integral over the interval $[-1, 1]$ of the functions

$$f_1(x) = \begin{cases} x^n & \text{at irrational } x, \\ \ln^2 x & \text{at rational } x \neq 0; \end{cases}$$

$$f_2(x) = \begin{cases} 1/(x - a) & \text{at rational } x,\ a \text{ is irrational}, \\ \sin x & \text{at irrational } x; \end{cases}$$

$$f_3(x) = \begin{cases} 1/\sqrt{(1 - x^2)(1 - k^2 x^2)} & \text{at rational } x, \\ 1/\sqrt{1 - x^2} & \text{at irrational } x; \end{cases}$$

$$f_4(x) = \begin{cases} \ln|x| & \text{at } x = \pm n^{-1}, (n = 1, 2, \ldots), \\ \sin x + bx^m & \text{at } x \neq \pm n^{-1}. \end{cases}$$

Are these functions measurable?

3.4.3. Calculate the Riemann–Stieltjes integral of the functions

$$f_1(x) = x^{-2}, \quad f_2(x) = a^x,\ a \in (0, 1), \quad f_3(x) = \Gamma(x + 1)$$

in the case when the Lebesgue–Stieltjes measure is determined by the function $F(x) = -[-x] - 1$ over the interval $(0, \infty)$ (the function $-[-x] - 1$ coincides with $[x]$ everywhere except the points $x = 0, \pm 1, \pm 2, \ldots$ but contrary to $[x]$ it is continuous not on the right but on the left: $F(x) = n$ at $x \in (n, n+1]$).

    3.4.4. Calculate the Riemann–Stieltjes integral over the interval $[-2, 2]$ in measure determined by the function

$$F(x) = \begin{cases} -x^{-2} & \text{at} \quad x \in (-\infty, -1], \\ 2x^3 & \text{at} \quad x \in (-1, 0], \\ x + 5 & \text{at} \quad x \in (0, 1], \\ e^x & \text{at} \quad x \in (1, \infty), \end{cases}$$

of the functions

$$f_1(x) = \tfrac{x}{1+x^2}, \qquad f_2(x) = \sin x.$$

    3.4.5. Calculate the Lebesgue–Stieltjes integral of the functions:

$$1) f_1(x) = x^n, \quad n \text{ (natural)}, \quad 2) f_2(x) = e^{ax}$$

over the function $F(x)$ of Example 3.9.

    I n s t r u c t i o n. Use the fact that by virtue of the construction of the function $F(x)$ (Example 3.9)

$$\int\limits_{-\infty}^{\infty} f(x)\, dF(x) = \int\limits_{0}^{1/3} f(x)\, dF(x) + \int\limits_{2/3}^{1} f(x)\, dF(x)$$

and $F(x) = F(3x)/2$ at $x \in (0, 1/3)$, $F(x) = [1 + F(3x - 2)]/2$ at $x \in (2/3, 1)$. As a result the change of variables $x = y/3$ in the first integral and $x = (2 + y)/3$ in the second integral gives

$$\int\limits_{-\infty}^{\infty} f(x)\, dF(x) = \int\limits_{0}^{1} f(x)\, dF(x)$$

$$= \tfrac{1}{2} \int\limits_{0}^{1} f\left(\tfrac{y}{3}\right) dF(y) + \tfrac{1}{2} \int\limits_{0}^{1} f\left(\tfrac{2+y}{3}\right) dF(y).$$

This relation gives recurrent formulae for integrals. As at $m = 0$, $a = 0$ these integrals are equal $\int dF(x) = 1$ then these recurrent formulae solve completely the problem.

    3.4.6. Taking the function $f(x)$ continuous at the points a and b derive the formula of the integration by parts for the Riemann–Stieltjes integral:

$$\int\limits_{a}^{b} f(x)\, dF(x) = f(b)F(b) - f(a)F(a) - \int\limits_{a}^{b} F(x)\, d f(x).$$

3.4.7. Prove that in the case of continuous piecewise differentiable function $F(x)$ the Riemann–Stieltjes integral is reduced to the Riemann integral

$$\int\limits_a^b f(x)\,dF(x) = \int\limits_a^b f(x)F'(x)\,dx.$$

3.4.8. Generalize definition (3.65) of the Riemann integral of the function with the values in a separable B–space to the case of the integral over the arbitrary finite measure μ and prove Theorem 3.26 for this case.

3.5. Passage to the Limit under the Integral Sign

3.5.1. Monotone Convergence Theorem

One of the most important questions of the integral theory is the question about the possibility of the passage to the limit under the integral sign. Therefore let study this question beginning with the integrals of the real functions.

Theorem 3.27. *If $\{f_n(x)\}$ is nondecresing sequence of non-negative measurable functions convergent almost everywhere to the function $f(x)$ then*

$$\lim_{n\to\infty} \int\limits_A f_n\,d\mu = \int\limits_A f\,d\mu. \tag{3.66}$$

▷ We represent each function $f_n(x)$ as the limit of nondecreasing sequence of the simple functions $\{f_n^k(x)\}$, $f_n(x) = \lim\limits_{n\to\infty} f_n^k(x)$. As we have already seen in Subsection 3.4.1 we may do it, for instance, assuming

$$f_n^k(x) = h2^{-k} \text{ at } h2^{-k} \le f_n(x) < (h+1)2^{-k}, \quad h = 0,1,\ldots,2^{2k}-1,$$
$$f_n^k(x) = 2^k \text{ at } f_n(x) \ge 2^k.$$

We determine the simple functions

$$g^n(x) = \max\{f_1^n(x),\ldots,f_n^n(x)\} \qquad (n = 1,2,\ldots).$$

Accounting that $f_m^n(x) \le f_m(x)$ at all m,n and $f_m(x) \le f_n(x)$ at all $m < n$ we obtain

$$f_m^n(x) \le g^n(x) \le f_n(x) \qquad \text{at } m \le n. \tag{3.67}$$

Hence by virtue of property (XIII) of the integral follows

$$\int_A f_m^n \, d\mu \leq \int_A g^n \, d\mu \leq \int_A f_n \, d\mu. \tag{3.68}$$

Let us pass now to the limit at $n \to \infty$. Then (3.67) will take the form

$$f_m(x) \leq \lim_{n \to \infty} g^n(x) \leq f(x).$$

Here the limit exists as $\{g^n(x)\}$ is a nondecreasing sequence upper bounded by the function $f(x)$. Hence taking into account that $f_m(x) \xrightarrow{\text{a.e.}} f(x)$ at $m \to \infty$ we get $\lim g^n(x) \stackrel{\text{a.e.}}{=} f(x)$, i.e. the sequence of the simple functions $\{g^n(x)\}$ converges almost everywhere to $f(x)$. Therefore accounting that $f_m^n(x)$ are also the simple functions on the basis of the integral definition (3.19) from (3.68) at $n \to \infty$ we obtain

$$\int_A f_m \, d\mu \leq \int_A f \, d\mu \leq \lim_{n \to \infty} \int_A f_n \, d\mu.$$

Hence as a result that

$$\int_A f_m \, d\mu \to \lim_{n \to \infty} \int_A f_n \, d\mu \qquad \text{at } m \to \infty,$$

we get (3.66). ◁

This preposition is known as the *monotone convergence theorem*.

Corollary. *If for nondecreasing sequence of μ–integrable non-negative functions $\{f_n(x)\}$ the sequence of the integrals $\{\int_A f_n d\mu\}$ is bounded then there exists the function $f(x) \stackrel{\text{a.e.}}{=} \lim_{n \to \infty} f_n(x)$ and formula (3.66) of the limit passage under the integral sign is valid (Levi theorem).*

▷ It is sufficient to prove the existence of the limit function $f(x)$ to which the sequence $\{f_n(x)\}$ converges almost everywhere:

$$f(x) \stackrel{\text{a.e.}}{=} \lim_{n \to \infty} f_n(x).$$

Then the validity of formula (3.66) will follow from Theorem 3.27. For proving the existence of the limit of $f(x)$ we determine the sets

$$A_n^r = \{x : f_n(x) \geq r\} A \quad (n, r = 1, 2, \ldots).$$

By virtue of the nondecrease of the sequence $\{f_n\}$ at any r

$$\{x : f_n(x) \geq r\} \subset \{x : f_{n+1}(x) \geq r\} \quad (n = 1, 2, \ldots).$$

Consequently, at any fixed r the sets A_n^r form a monotone increasing sequence and

$$B_r = \lim_{n \to \infty} A_n^r = \bigcup_{n=1}^{\infty} A_n^r \quad (r = 1, 2, \ldots).$$

At any fixed n the sets A_n^r form a monotone decreasing sequence as

$$\{x : f_n(x) \geq r + 1\} \subset \{x : f_n(x) \geq r\}.$$

Consequently,

$$C = \lim_{r \to \infty} B_r = \bigcap_{r=1}^{\infty} B_r.$$

From Chebyshev inequality (3.30) follows that at all n and r

$$\mu(A_n^r) \leq \int_A f_n \, d\mu / r < c/r,$$

if

$$\int_A f_n \, d\mu < c \quad \text{at all } n,$$

and by virtue of the continuity of the measure given on the σ–algebra

$$\mu(B_r) = \lim_{n \to \infty} \mu(A_n^r) \leq c/r,$$

and

$$\mu(C) \leq \mu(B_r) \leq c/r \quad \text{at any } r.$$

Hence it follows that $\mu(C) = 0$. But at any $x \in \overline{C}$, $x \in \overline{B}_r$ at some r, and consequently, r, $x \in \overline{A}_n^r$ at all n. So, $f_n(x) < r$ at all n and there exists the limit $f(x) = \lim_{n \to \infty} f_n(x)$ at all $x \in \overline{C}$. Thus the sequence $\{f_n(x)\}$ converges almost everywhere to some function $f(x)$. ◁

R e m a r k. If $\{f_n(x)\}$ is a nondecreasing sequence of real functions convergent almost everywhere to $f(x)$ and $f_n(x) \geq v(x)$ where $v(x)$ is a μ–integrable real function then after applying Theorem 3.27 to the functions $f_n(x) - v(x)$ we ensure that in this case also the limit passage under the integral sign is possible. This fact extends Theorem 3.27 to any real functions lower bounded by a μ–integrable function.

3.5.2. Termwise Integration of Series

From Theorem 3.27 it is easy to derive the theorem about the possibility of the termwise integration almost everywhere absolutely convergent series of measurable real functions.

Let $\{g_p(x)\}$ be an arbitrary sequence of real measurable functions such as the series $\sum\limits_{p=1}^{\infty} g_p(x)$ converges almost everywhere to some function $f(x)$. At first we suppose that all the functions $g_p(x)$ are nonnegative and determine the functions

$$f_n(x) = \sum_{p=1}^{n} g_p(x) \qquad (n = 1, 2, \ldots).$$

These functions form a nondecreasing sequence convergent almost everywhere to the function $f(x)$. Therefore Theorem 3.27 is applicable to them. This theorem gives

$$\int_A \sum_{p=1}^{\infty} g_p \, d\mu = \sum_{p=1}^{\infty} \int_A g_p \, d\mu. \qquad (3.69)$$

For arbitrary real functions $g_p(x)$ we determine corresponding nonnegative functions $g_p^+(x) = \max(g_p(x), 0)$ and $g_p^-(x) = -\min(g_p(x), 0)$. Then we shall have $g_p(x) = g_p^+(x) - g_p^-(x)$. Using (3.69) separately to the functions $g_p^+(x)$ and $g_p^-(x)$ we ensure that formula (3.69) is also valid in this case under the condition that the series $\sum\limits_{p=1}^{\infty} g_p(x)$ converges almost everywhere absolutely. Thus the following preposition is valid.

Theorem 3.28. *The series absolutely convergent almost everywhere may be termwise integrated.*

Let $f(x)$ be an arbitrary μ–integrable real function and $\{A_n\}$ be any sequence of pairwise nonintersecting sets $A_n \in \mathcal{A}$. Putting in (3.69) $A = \bigcup\limits_{p=1}^{\infty} A_p$, $g_p(x) = f(x)\mathbf{1}_{A_p}(x)$ we obtain

$$\int_A f d\mu = \sum_{p=1}^{\infty} \int_{A_p} f \, d\mu. \qquad (3.70)$$

Thus the integral of real numeric function represents the σ–additive function of the set.

3.5.3. Fatou Lemma

Let us consider now an arbitrary sequence of nonnegative measurable functions $\{f_n(x)\}$. Determine the functions

$$g_n(x) = \inf_{k \geq n} f_k(x) \qquad (n = 1, 2, \ldots).$$

It is evident that the sequence $\{g_n(x)\}$ is a nondecreasing sequence of nonnegative measurable functions convergent to $\underline{\lim} f_n(x)$. Therefore on the basis of Theorem 3.27

$$\lim \int_A g_n \, d\mu = \int_A \underline{\lim} f_n \, d\mu.$$

On the other hand, $g_n(x) \leq f_n(x)$ at all x, and consequently,

$$\int_A g_n \, d\mu \leq \int_A f_n \, d\mu.$$

Hence we derive

$$\lim \int_A g_n \, d\mu \leq \underline{\lim} \int_A f_n \, d\mu.$$

From this inequality and the equality obtained above follows

$$\int_A \underline{\lim} f_n \, d\mu \leq \underline{\lim} \int_A f_n \, d\mu. \qquad (3.71)$$

This inequality is the essence of *Fatou lemma*.

R e m a r k. Inequality (3.71) is also valid for any sequence of real functions $\{f_n(x)\}$ lower bounded by the μ–integrable function $v(x)$, $f_n(x) \geq v(x)$. To be sure that it is sufficient to apply (3.71) to the sequence of nonnegative functions $\{f_n(x) - v(x)\}$ and take into account that $\underline{\lim}(a_n - b) = \underline{\lim} a_n - b$ for any numeric sequence $\{a_n\}$ and any finite number b.

If the sequence $\{f_n(x)\}$ is upper bounded by the μ–integrable function $v(x)$ then from (3.71) follows

$$\overline{\lim} \int\limits_A f_n \, d\mu \leq \int\limits_A \overline{\lim} \, f_n \, d\mu. \qquad (3.72)$$

▷ For deriving this inequality it is sufficient to apply (3.71) to the sequence of nonnegative functions $\{v(x) - f_n(x)\}$ and account that $\underline{\lim}(b - a_n) = b - \overline{\lim} \, a_n$ for any numeric sequence $\{a_n\}$ and any finite number b. ◁

3.5.4. Magorizable Sequence Theorem

From inequalities (3.71) and (3.72) follows the main integration theorem of almost everywhere convergent sequences of real functions.

Theorem 3.29. *If the sequence of real measurable functions $\{f_n(x)\}$ converges almost everywhere to the function $f(x)$ and is lower and upper bounded by the μ–integrable functions $v(x)$ and $w(x)$ respectively*

$$v(x) \leq f_n(x) \leq w(x),$$

then all the functions $f_n(x)$ and $f(x)$ are μ–integrable and equality (3.66) is valid:

$$\lim_{n \to \infty} \int\limits_A f_n \, d\mu = \int\limits_A f \, d\mu.$$

▷ Similarly as while proving Theorem 3.27 we represent each of the functions $f_n(x)$ as the limit of nondecreasing sequence of the simple functions $\{f_n^k(x)\}$:

$$f_n(x) = \lim_{k \to \infty} f_n^k(x), \quad f_n^p(x) \leq f_n^q(x) \leq f_n(x) \qquad \text{at } p < q.$$

Then we get a nondecreasing sequence of real numbers $\{\int f_n^k \, d\mu\}$ upper bounded by the number $\int w \, d\mu$. According to Theorem 3.22 the sequence of the simple functions $\{f_n^k(x)\}$ at fixed n determines the function $f_n(x)$. Consequently, $f_n(x)$ is μ–integrable.

Now after applying to the functions $f_n(x)$ inequalities (3.71) and (3.72) and accounting that in this case $\underline{\lim} f_n(x) = \overline{\lim} f_n(x) = \lim f_n(x) = f(x)$ we obtain

$$\overline{\lim} \int\limits_A f_n(x) \, d\mu \leq \int\limits_A f \, d\mu \leq \underline{\lim} \int\limits_A f_n \, d\mu.$$

Hence it is clear that there exists the limit

$$\lim \int\limits_A f_n \, d\mu = \int\limits_A f \, d\mu. \quad \lhd$$

Thus in this case it is also possible the limit passage under the integral sign. The proved statement is usually called *the theorem of majorizable sequence.*

R e m a r k. For the validity of all proved theorems it is sufficient that their conditions be fulfilled almost everywhere.

3.5.5. General Lebesgue Theorem on Passage to the Limit under the Integral Sign

From Theorem 3.20 follows general theorem on passage to the limit under the integral sign.

Theorem 3.30. *If the sequence of measurable functions $\{f_n(x)\}$ with the values in the separable B–space converges almost everywhere to the function $f(x)$ and all the functions $f_n(x)$ are bounded in the norm by μ–integrable function $g(x), \| f_n(x) \| \leq g(x)$ then the function $f(x)$ is μ–integrable and*

$$\int\limits_A f \, d\mu = \lim_{n \to \infty} \int\limits_A f_n \, d\mu \qquad (3.73)$$

(Lebesgue theorem).

▷ By Theorem 3.20 all the functions $f_n(x)$ are μ–integrable. From the fact that $\| f_n(x) \| \leq g(x)$ and $f_n(x) \xrightarrow{\text{a.e.}} f(x)$ follows that $\| f(x) \| \leq g(x)$ almost everywhere. Therefore the function $f(x)$ is also μ–integable. Besides as $0 \leq \| f_n(x) - f(x) \| \leq 2g(x)$ almost everywhere then all nonnegative functions $\|f_n(x) - f(x)\|$ are also μ–integrable and according to Theorem 3.29

$$\lim_{n \to \infty} \int \| f_n - f \| \, d\mu = 0.$$

Hence by virtue of the inequality

$$\left\| \int\limits_A f_n \, d\mu - \int\limits_A f \, d\mu \right\| \leq \int \| f_n - f \| \, d\mu,$$

follows (3.73), and the sequence of the integrals $\{\int_A f_n \, d\mu\}$ converges to $\int_A f \, d\mu$ uniformly relative to $A \in \mathcal{A}$. ◁

  Corollary. *If the sequence of measurable functions* $\{f_n(x)\}$ *with the values in a separable B–space converges in measure* μ *to the function* $f(x)$ *and all the functions* $f_n(x)$ *are bounded in the norm by the* μ*–integrable function* $g(x)$, $\| f_n(x) \| \le g(x)$ *then the function* $f(x)$ *is* μ*–integrable and formula (3.73) is valid.*

  ▷ As the sequence $\{f_n(x)\}$ convergent in measure μ to the function $f(x)$ contains the subsequence convergent to $f(x)$ almost everywhere then by Theorem 3.30 the function $f(x)$ is μ–integrable. We put

$$\alpha = \overline{\lim} \int \| f_n - f \| \, d\mu.$$

From the definition of the upper limit follows that from the sequence $\{f_n(x)\}$ we may select such subsequence $\{f_{n_k}(x)\}$ that

$$\lim_{k \to \infty} \int \| f_{n_k} - f \| \, d\mu = \alpha.$$

This subsequence converges in measure to the same function $f(x)$. We put for brevity $g_k(x) = f_{n_k}(x)$ and select from $\{g_k(x)\}$ the subsequence $\{g_{k_p}(x)\}$ convergent to $f(x)$ almost everywhere. Then by Theorem 3.30 we shall have

$$\lim_{p \to \infty} \int_A \| g_{k_p} - f \| \, d\mu = 0.$$

On the other hand, as the subsequence always converges to the same limit as the sequence from which it is selected then

$$\lim_{p \to \infty} \int_A \| g_{k_p} - f \| \, d\mu = \alpha.$$

Consequently, $\alpha = 0$ there exists the limit

$$\lim \int_A \| f_n - f \| \, d\mu = 0.$$

Hence similarly as while proving Theorem 3.30 follows (3.73). ◁

  R e m a r k. Thus we have proved the possibility of the limit passage under the integral sign for any sequence of the functions bounded in the norm by the μ–integrable function and convergent to some function almost everywhere or in the measure.

3.5.6. General Theorem on Termwise Integration of Series

From the proved theorems in Subsections 3.5.2–3.5.5 it is easy to derive a theorem about the possibility of termwise integration of the convergent almost everywhere or in the measure series.

Theorem 3.31. *If the series*

$$\sum_{p=1}^{\infty} g_p(x)$$

converges almost everywhere or in the measure to some function $f(x)$ and all its finite segments

$$f_n(x) = \sum_{p=1}^{n} g_p(x) \qquad (n = 1, 2, \ldots)$$

are bounded in the measure by μ–integrable function $g(x)$ then the function $f(x)$ is μ–integrable and

$$\int_A f \, d\mu = \int_A \sum_{p=1}^{\infty} g_p \, d\mu = \sum_{p=1}^{\infty} \int_A g_p \, d\mu. \qquad (3.74)$$

▷ The validity of this theorem follows from Theorem 3.30 and the Corollary from it if we notice that the sequence $\{f_n(x)\}$ converges almost everywhere or in the measure to $f(x)$ and all the functions $f_n(x)$ are bounded in the measure by the μ–integrable function $g(x)$. ◁

Thus the series convergent almost everywhere or in measure may be integrated termwise under the condition that all its finite segments are bounded in measure by the integrable function.

Corollary. *The integral of the function with the values in a separable B–space represents the σ–additive set function, i.e. the measure with the values in the same B–space.*

▷ Let $f(x)$ be a μ–integrable function with the values in a separable B–space, $A = \bigcup_{p=1}^{\infty} A_p, A_p \in \mathcal{A}, A_p A_q = \emptyset$ at $q \neq p$. Putting in (3.74) $g_p(x) = f(x)\mathbf{1}_{A_p}(x)$ we get

$$\int_A f \, d\mu = \sum_{p=1}^{\infty} \int_{A_p} f \, d\mu. \quad ◁ \qquad (3.75)$$

E x a m p l e 3.6. We calculate the integral

$$\int f(x)\mu_W(dx), \quad f(x) = \int_a^b \varphi(s,t)x^2(t)\, dt.$$

In Example 3.2 it was evaluated that the function $f(x)$ is $(\mathcal{A}_W^T, \mathcal{B})$–measurable, and in Example 3.4 was calculated the integral for the function obtained from $f(x)$ by the substitution of the integral by the sum. On the basis of these results putting as in Example 3.2

$$f_n(x) = \tfrac{b-a}{n-1} \sum_{k=0}^{n-1} \varphi(s,t_k)x^2(t_k),$$

we get

$$\int f_n(x)\mu_W(dx) = \tfrac{b-a}{n-1} \sum_{k=0}^{n-1} \varphi(s,t_k)t_k$$

and

$$\lim_{n\to\infty} \int f_n(x)\mu_W(dx) = \int_a^b \varphi(s,t)t\, dt. \qquad (*)$$

It remains to evaluate the possibility of the limit passage under the integral sign. For this purpose we notice that

$$\| f_n(x) \| \le g_n(x) = \tfrac{b-a}{n-1} \sum_{k=1}^{n-1} \sup_{s\in S} |\varphi(s,t_k)| x^2(t_k).$$

Putting

$$g(x) = \lim_{n\to\infty} g_n(x) = \int_a^b \sup_{s\in S} |\varphi(s,t)| x^2(t)\, dt,$$

we obtain similarly as earlier

$$\lim_{n\to\infty} \int g_n(x)\mu_W(dx) = \int_a^b \sup_{s\in S} |\varphi(s,t)| t\, dt.$$

But in accordance with Fatou lemma (Subsection 3.5.3)

$$\int g(x)\mu_W(dx) \le \lim_{n\to\infty} \int g_n(x)\mu_W(dx).$$

Thus the function $g(x)$ is μ_W–integrable and the functions f_n are bounded in the measure by μ_W–integrable function $g(x) + c$ at some $c > 0$. As $f_n(x) \to f(x)$ almost everywhere relative to the measure μ_W then by Theorem 3.30 we get

$$\int f(x)\mu_W(dx) = \lim_{n\to\infty} \int f_n(x)\mu_W(dx) = \int_a^b \varphi(s,t)t\, dt.$$

For more detailed treatment of passage to the limit under the integral sign refer to (Dunford and Schwarz 1958).

P r o b l e m s

3.5.1.   Using the procedure of Example 3.6 prove that the integral of the function

$$f(x) = \int_a^b \varphi(s,t) x^{2r}(t)\, dt$$

over the Wiener measure μ_W extended on the whole space X^T is determined by

$$\int f(x)\mu_W(dx) = (2r-1)!! \int_a^b \varphi(s,t) t^r\, dt.$$

3.5.2. Prove that the integral of the function

$$f(x) = \exp\left\{ \int_a^b \varphi(s,t) x(t)\, dt \right\}$$

over the Wiener measure μ_W extended on the whole space X^T is determined by

$$\int f(x)\mu_W(dx) = \exp\left\{ \int_a^b\!\!\int_a^b \varphi(s,t_1)\varphi(s,t_2) \min(t_1,t_2)\, dt_1\, dt_2 \right\}. \qquad (*)$$

S o l u t i o n. Introducing the functions

$$f_n(x) = \exp\left\{ \frac{b-a}{n-1} \sum_{k=0}^{n-1} \varphi(s,t_k) x(t_k) \right\}$$

and using the result of Example 3.4 we find similarly as in Example 3.6

$$\lim_{n\to\infty} \int f_n(x)\mu_W(dx) = \exp\left\{ \int_a^b\!\!\int_a^b \varphi(s,t_1)\varphi(s,t_2) \min(t_1,t_2)\, dt_1\, dt_2 \right\}.$$
$$(**)$$

It remains to prove the possiblity of the limit passage under the integral sign. We notice for this purpose that according to Fatou lemma (Subsection 3.5.3) which may be used as a result of nonnegativity of all the functions $f_n(x)$ at each $s \in S$,

$$\int f(x)\mu_W(dx) = \int \lim f_n(x)\mu_W(dx) \le \lim \int f_n(x)\mu_W(dx).$$

Hence and from $(**)$ it is clear that the function $f(x)$ is μ_W–integrable. Therefore its norm is equal to

$$\| f(x) \| = \sup_{s \in S} \exp \left\{ \int_a^b \varphi(s,t)x(t)\, dt \right\}.$$

And as the sequence $\{f_n(x)\}$ converges to $f(x)$ almost everywhere relative to the measure μ_W then

$$\| f_n(x) \| = \sup_{s \in S} \exp \left\{ \frac{b-a}{n-1} \sum_{k=0}^{n-1} \varphi(s,t_k)x(t_k) \right\} \leq C \, \| f(x) \|$$

then at some $C > 1$. Thus the sequence of the functions $\{f_n(x)\}$ convergent almost everywhere relative to μ_W is bounded in the norm by μ_W–integrable function. According to Lebesgue theorem of Subsection 3.5.4 hence it follows

$$\int f(x)\mu_W(dx) = \lim_{n \to \infty} \int f_n(x)\mu_W(dx),$$

and we obtain the required formula $(*)$.

The received formulae and the conditions of the existence of Bochner integral obtained as a result of solution of Problems 3.5.1 and 3.5.2 show that the functions

$$f_1(x) = \int_0^\infty \varphi(s,t)x^{2r}(t)\, dt,$$

$$f_2(x) = \exp \left\{ \int_0^\infty \varphi(s,t)x(t)\, dt \right\}$$

are μ_W–integrable if and only if improper Rieman integrals

$$\int_0^\infty \sup_{s \in S} |\varphi(s,t)|t\, dt, \quad \int_0^\infty \int_0^\infty \sup_{s \in S} |\varphi(s,t_1)\varphi(s,t_2)| \min(t_1,t_2)\, dt_1\, dt_2$$

converge. Notice that in Example 3.4 and in Problems 3.3.1–3.3.4, 3.5.1 and 3.5.2 the conditions of the existence of Bochner integral – the integrability of the norm of the function $f(x)$ considered as the functions s, $s \in S$ from the space $C(S)$ are fulfilled.

3.5.3. Calculate the integrals in the Wiener measure μ_W extended over the whole space X^T of the following functions

$$f_1(x) = \sin \int_a^b \varphi(s,t)\, x(t)\, dt,$$

$$f_2(x) = \cos \int\limits_a^b \varphi(s,t)\, x(t)\, dt,$$

$$f_3(x) = \sinh \int\limits_a^b \varphi(s,t)\, x(t)\, dt,$$

$$f_4(x) = \cosh \int\limits_a^b \varphi(s,t)\, x(t)\, dt,$$

$$f_5(x) = \exp\left\{ \int\limits_a^b \varphi_1(s,t)\, x(t)\, dt \right\} \sin \int\limits_a^b \varphi_2(s,t)\, x(t)\, dt,$$

$$f_6(x) = \exp\left\{ \int\limits_a^b \varphi_1(s,t)\, x(t)\, dt \right\} \cos \int\limits_a^b \varphi_2(s,t)\, x(t)\, dt,$$

$$f_7(x) = \exp\left\{ \int\limits_a^b \varphi_1(s,t)\, x(t)\, dt \right\} \sinh \int\limits_a^b \varphi_2(s,t)\, x(t)\, dt,$$

$$f_8(x) = \exp\left\{ \int\limits_a^b \varphi_1(s,t)\, x(t)\, dt \right\} \cosh \int\limits_a^b \varphi_2(s,t)\, x(t)\, dt,$$

$$f_9(x) = \psi(x) \exp\left\{ \int\limits_a^b \varphi(s,t)\, x(t)\, dt \right\}, \quad \psi(x)$$

$$= \int\limits_a^b \ldots \int\limits_a^b \varphi(s,t_1) \ldots \varphi(s,t_m) x(t_1) \ldots x(t_m)\, dt_1 \ldots dt_m,$$

$$f_{10}(x) = \psi(x) \sin \int\limits_a^b \varphi(s,t)\, x(t)\, dt,$$

$$f_{11}(x) = \psi(x) \cos \int\limits_a^b \varphi(s,t)\, x(t)\, dt,$$

$$f_{12}(x) = \psi(x)\, \mathrm{sh} \int\limits_a^b \varphi(s,t)\, x(t)\, dt,$$

$$f_{13}(x) = \psi(x)\, \mathrm{ch} \int\limits_a^b \varphi(s,t)\, x(t)\, dt.$$

I n s t r u c t i o n. For calculating the integrals of f_9–f_{13} replace in the formulae of Problem 3.5.2 and in the formulae for the integrals of the functions $f_1, \ldots, f_4$ the function $\varphi(s,t)$ by the fucntion $\alpha\varphi(s,t)$, differentiate the obtained formulae with respect to α m times and put after that $\alpha = 1$.

3.5.4. Calculate the integral in the Wiener measure μ_W extended over the whole space X^T of the function

$$f(x) = \int\limits_a^b \ldots \int\limits_a^b \varphi(s,t_1,\ldots,t_m) x(t_1) \ldots x(t_m)\, dt_1 \ldots dt_m.$$

If the functions φ in all previous formulae do not depend on s then all the integrals in Example 3.4 and in Problems 3.3.1–3.3.4, 3.5.1–3.5.3 represent abstract Lebesgue integrals.

3.6. Absolute Continuity and Singularity of Measures

3.6.1. Definitions

Let $(X, \mathcal{A})$ be an arbitrary measurable space, $\mu(A)$ be a nonnegative measure determined on the σ–algebra $\mathcal{A}$, $\varphi(A)$ be the measure with the values in a separable B–space determined on the σ–algebra $\mathcal{A}$.

The measure $\varphi(A)$ is called an *absolutely continuous relative to the measure* μ or shortly, μ–*continuous* if $\varphi(A) = 0$ on any set $A \in \mathcal{A}$ of zero measure μ, $\mu(A) = 0$. In other words, the measure $\varphi(A)$ is μ–continuous if from $\mu(A) = 0$ follows $\varphi(A) = 0$.

The measure $\varphi(A)$ is called a *singular relative to the measure* μ or shortly, μ–*singular* if it is equal to zero everywhere outside some set $A \in \mathcal{A}$ of zero measure, i.e. if $\varphi(B) = 0$ at any $B \in \mathcal{A}$, $B \subset \bar{A}$, $\mu(A) = 0$.

It is evident that the measure simultaneously μ–continuous and μ–singular is identically equal to zero.

Two nonnegative measures $\mu(A)$ and $\nu(A)$ determined on one and the same σ–algebra $\mathcal{A}$ are called *equivalent* if the measure $\mu(A)$ is ν–continuous and the measure $\nu(A)$ is μ–continuous.

Two nonnegative measures $\mu(A)$ and $\nu(A)$ are called *orthogonal* if $\mu(A)$ is ν–singular and $\nu(A)$ is μ–singular.

E x a m p l e 3.7. A Lebesgue integral

$$P(A) = \int\limits_{A} f(x)\, dx$$

represents a measure absolutely continuous relative to the Lebesgue measure in the correspondent space R^n, $l(A) = v_A$ (the volume A).

E x a m p l e 3.8. The measure concentrated on the finite or countable set of the points x_k $(k = 1, 2, ...)$ of the space R^n is determined on the set A by the formula

$$P(A) = \sum\limits_{x_k \in A} p_k \, ,$$

where p_k is the value of the measure P on the single–point set $\{x_k\}$. It is singular relative to Lebesgue measure in R^n as $P(A) = 0$ on any set A which does not intersect with the set of the points $\{x_k\}$ possessing zero Lebesgue measure.

E x a m p l e 3.9. More complicated example of the singular measure relative to the Lebesgue measure represents the Lebesgue–Stieltjes measure which is determined by the formula

$$\varphi([a,\, b)) = F(b) -- F(a) \, ,$$

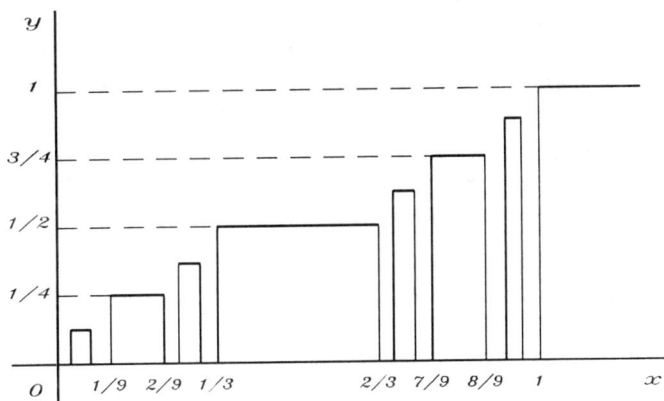

Fig. 3.1

where $F(x)$ is constructed by the following rule: it is assumed that $F(0) \leq 0$, $F(1) \geq 1$; after that the segment $[0, 1)$ is divided into three equal parts and it is divided into three equal parts and it is assummed everywhere on the middle part that $F(x) = \frac{1}{2}$; the remained two segments in their turn are divided into three equal parts each of them and it is assumed in the middle parts that $F(x) = \frac{1}{4}$ and $F(x) = \frac{3}{4}$. This process (Cantor ladder) lasts indefinitely and each segment on which the function $F(x)$ is not yet determined is divided into three equal parts and everywhere in the middle part $F(x)$ is assumed as constant, equal to the arithmetic mean of its values on the ends of divided segment (Fig. 3.1) It is easy to see that the function $F(x)$ is continuous as the difference of its values at any two points of the segment $[0, 1]$ is as small as one likes at sufficiently small distance between these points. On the other hand, the set E of the growing points of the function $F(x)$ has zero Lebesgue measure as the union measure of all the segments on which $F(x)$ is constant evidently is equal

$$\frac{1}{3} + \frac{2}{9} + \frac{4}{27} + \frac{8}{81} + \cdots = \sum_{n=1}^{\infty} \frac{2^{n-1}}{3^n} = \frac{1}{3}\frac{1}{1-\frac{2}{3}} = 1.$$

It is evident that $\varphi(A) = 0$ on any set A which contains none of the growing points of the function $F(x)$ i.e. everywhere outside the set E of zero Lebesgue measure.

From property (VI) of the integral and its σ–additivity follows that the integral represents μ–continuous measure. Naturally, the question arises whether the set of all μ–measurable measures is exhausted by the integrals of some functions. In other words, what kind of μ–continuous measure represents an integral of some function. The positive answer to this question for real measures is given by Theorems 3.32 and 3.33.

3.6.2. Representation of a Measure as the Sum of Absolutely Continuous and Singular Measures

Let $\varphi(A)$ be an arbitrary numeric measure determined on the same σ–algebra $\mathcal{A}$ that the nonnegative measure $\mu(A)$. Then the following theorem is valid.

Theorem 3.32. *The measure $\varphi(A)$ may be presented by the formula*

$$\varphi(A) = \alpha(A) + \beta(A), \quad A \in \mathcal{A}, \tag{3.76}$$

where $\alpha(A)$ is a μ–continuous measure, $\beta(A)$ is a μ–singular measure.

▷ At first we suppose that $\varphi(A)$ is nonnegative and both measures φ and μ are finite. Let consider the set Φ of all μ–integrable nonnegative functions $f(x)$ safisfying the condition

$$\int\limits_A f d\mu \le \varphi(A) \quad \text{at all} \quad A \in \mathcal{A}. \tag{3.77}$$

We put

$$\alpha = \sup_{f \in \Phi} \int f d\mu.$$

By the condition $\alpha \le \varphi(X) < \infty$. We shall prove that in Φ there exists the function $z(x)$ for which $\int z d\mu = \alpha$. According to the definition of the upper bound in Φ there exists such a sequence of the functions $\{f_n(x)\}$ that

$$\lim_{n \to \infty} \int f_n d\mu = \alpha.$$

We determine the functions

$$z_n(x) = \max\{f_1(x), \dots, f_n(x)\} \quad (n = 1, 2, \dots).$$

These functions evidently are μ–integrable and belong to the set Φ. Really, after determining the sets

$$E_k = A\{x : z_n(x) = f_k(x)\} \quad (k = 1, \dots, n),$$

$$A_1 = E_1, \quad A_k = E_k \bigcap_{p=1}^{k-1} \bar{E}_p \quad (k = 1, \ldots, n)$$

and taking into account that the sets $A_1, \ldots, A_n$ are pairwise non-intersecting and $\bigcup_{k=1}^{n} A_k = A$ we get

$$\int_A z_n d\mu = \sum_{k=1}^{n} \int_{A_k} f_k d\mu \leq \sum_{k=1}^{n} \varphi(A_k) = \varphi(A). \tag{3.78}$$

Thus the functions $z_n(x)$ satisfy condition (3.77), i.e. belong to Φ. Now we notice that $z_n(x) \geq f_n(x)$ and we come to the conclusion that

$$\int z_n d\mu \geq \int f_n d\mu.$$

Consequently,

$$\lim \int z_n d\mu = \alpha.$$

Evidently, the functions $z_n(x)$ form a nondecreasing sequence convergent to the function $z(x) = \sup f_n(x)$. On the basis of the monotone convergence theorem (Subsection 3.5.1)

$$\int z d\mu = \lim \int z_n d\mu = \alpha.$$

As inequality (3.78) is valid for all the functions $z_n(x)$ then it is also valid for the limit function $z(x)$. Consequently, the function $z(x)$ is μ–integrable and belongs to the set Φ.

Now we determine a μ–continuous measure on the σ–algebra $\mathcal{A}$

$$\alpha(A) = \int_A z d\mu \tag{3.79}$$

and study the difference

$$\beta(A) = \varphi(A) - \alpha(A). \tag{3.80}$$

By virtue of condition (3.77) to which satisfies $z(x)$ the function $\beta(A)$ represents a nonnegative measure. We shall prove that $\beta(A)$ is μ–singular. For this purpose we consider the measures

$$\gamma_n(A) = \beta(A) - \frac{1}{n}\mu(A) \quad (n = 1, 2, \ldots).$$

Let $X = D_n^- \bigcup D_n^+$ be Hahn expansion correspondent to the measure γ_n. Then we shall have at any $A \in \mathcal{A}$

$$\gamma_n(AD_n^-) \leq 0, \quad \gamma_n(AD_n^+) \geq 0.$$

We determine the set $D = \bigcap_{n=1}^{\infty} D_n^-$. As $D \subset D_n^-$ then for all n and $A \in \mathcal{A}$ we have $\gamma_n(AD) \leq 0$, whence it follows

$$\beta(AD) \leq \frac{1}{n}\mu(AD) \quad \text{for all} \quad n \quad \text{and} \quad A \in \mathcal{A}.$$

But it takes place only if $\beta(AD) = 0$ for all $A \in \mathcal{A}$. It remains to prove that the set $\bar{D}$ has zero measure μ. For this purpose we consider the functions

$$u_n(x) = z(x) + \frac{1}{n}1_{D_n^+}(x) \quad (n = 1, 2, \ldots).$$

All these functions belong to the set Φ as

$$\int_A u_n d\mu = \alpha(A) + \frac{1}{n}\mu(AD_n^+) = \alpha(A) + \beta(AD_n)$$

$$- \gamma_n(AD_n^+) \leq \alpha(A) + \beta(A) - \gamma_n(AD_n^+)$$
$$= \varphi(A) - \gamma_n(AD_n^+) \leq \varphi(A),$$

and $\gamma_n(AD_n^+) \geq 0$. On the other hand,

$$\int u_n d\mu = \alpha + \frac{1}{n}\mu(D_n^+),$$

what takes place only if $\mu(D_n^+) = 0$. Hence it follows

$$\mu(\bar{D}) = \mu\left(\bigcup_{n=1}^{\infty} D_n^+\right) \leq \sum_{n=1}^{\infty} \mu(D_n^+) = 0.$$

Thus the function $\beta(A)$ is equal to zero outside the set $\bar{D}$ of zero measure and consequently, is μ–singular.

As a result from (3.80) follows expansion (3.76):

$$\varphi(A) = \alpha(A) + \beta(A)$$

of an arbitrary $\varphi(A)$ in terms of μ–continuous and μ–singular parts, therewith the μ–continuous part may be presented in the form of integral (3.79).

We shall prove the uniqueness of expansion (3.76). We suppose that besides (3.76) there exists another expansion

$$\varphi(A) = \alpha_1(A) + \beta_1(A).$$

Then we get

$$\alpha(A) - \alpha_1(A) = \beta_1(A) - \beta(A).$$

The left part is μ–continuous here, the right part is μ–singular. Consequently, both of them are identically equal to zero, i.e. $\alpha_1(A) = \alpha(A)$, $\beta_1(A) = \beta(A)$.

For extending formula (3.76) to the case when $\varphi(A)$ is an arbitrary real or complex finite measure it is sufficient to recall that any real measure may be presented as the difference of two nonnegative measures. Finally, to extend formula (3.76) to the σ–finite measures $\varphi(A)$ and $\mu(A)$ it is sufficient to expand the space X into countable set of pairwise non-intersecting parts (φ and μ are finite on each of them) and represent $\varphi(A)$ by formula (3.76) on each of these parts. ◁

3.6.3. Radon–Nikodym Theorem

From formula (3.76) follows.

Theorem 3.33. *Any real or complex μ–continuous measure $\varphi(A)$ represents an integral of some (μ–integrable) function $z(x)$:*

$$\varphi(A) = \int_A z \, d\mu \tag{3.81}$$

(Radon–Nikodym theorem).

A function $z(x)$ in (3.81) is called a *Radon–Nikodym derivative* of the measure $\varphi(A)$ with respect to the measure $\mu(A)$ and is denoted by

$$z(x) = \frac{d\varphi}{d\mu}(x) = \frac{d\varphi}{d\mu} = \varphi'_\mu(x).$$

As the integral over the measure μ does not change at any integrand on any set of zero measure μ then a Radon–Nikodym derivative represents a class of equivalent functions which differ one from another on the set of zero measure μ.

Theorem 3.34. *If a real measure λ determined on the same σ-algebra of the sets A that the measure μ is μ-continuous and a real function $f(x)$ is λ-integrable then*

$$\varphi(A) = \int\limits_A f d\lambda = \int\limits_A f \lambda'_\mu d\mu . \tag{3.82}$$

▷ Really, this equality is valid for any simple function

$$f(x) = \sum_{k=1}^n y_k \mathbf{1}_{E_k}(x) ,$$

as in this case

$$\varphi(A) = \sum_{k=1}^n y_k \lambda(E_k A) = \sum_{k=1}^n y_k \int\limits_{E_k A} \lambda'_\mu d\mu = \int\limits_A f \lambda'_\mu d\mu .$$

After presenting any nonnegative λ-integrable function $f(x)$ as the limit of nondecreasing sequence of the simple functions which determines $f(x)$ and applying Theorem 3.27 and recalling the definition of Lebesgue integral for any real function $f(x)$ we ensure that formula (3.82) is valid in a general case. ◁

It follows from (3.82) that

$$\varphi'_\mu = \varphi'_\lambda \lambda'_\mu . \tag{3.83}$$

This formula represents the generalization of the differentiation rule of a composite function on the real measures.

E x a m p l e 3.10. The integrand $f(x)$ in Example 3.7 represents the Radon–Nikodym derivative of the measure $P(A)$ with respect to Lebesgue measure $l(A)$

$$f(x) = \frac{dP}{dl}(x) .$$

E x a m p l e 3.11. In Example 3.8 the function $f(x)$ equal to p_k at every point x_k represents the Radon–Nikodym derivative of the measure $P(A)$ with respect to the measure whose value on every set A is equal to the number of the points of the set $\{x_k\}$ belonging to A.

E x a m p l e 3.12. Let $x = \varphi(u)$ be the one–to–one differentiable mapping of the n–dimensional $U = R^n$ into another n–dimensional space $X = R^n$, $l_u(C)$ be the Lebesgue measure in U, and $l_x(A)$ be the Lebesgue measure in X (certainly, l_u and l_x are determined on the correspondent σ–algebras of Borel sets). From the formula

$$l_x(A) = \int \cdots \int_A dx_1 \ldots dx_n = \int \cdots \int_{\varphi^{-1}(A)} \left| \frac{\partial(\varphi_1, \ldots, \varphi_n)}{\partial(u_1, \ldots, u_n)} \right| du_1 \cdots du_n \,,$$

which may be rewritten in the form

$$\nu(C) = l_x(\varphi(C))$$

$$= \int \cdots \int_{\varphi(C)} dx_1 \ldots dx_n = \int \cdots \int_C \left| \frac{\partial(\varphi_1, \ldots, \varphi_n)}{\partial(u_1, \ldots, u_n)} \right| du_1 \ldots du_n \,, \qquad (3.84)$$

follows that the modulus of the functions Jacobian $\varphi_1(u_1, \ldots, u_n)$, $\ldots$, $\varphi_n(u_1, \ldots, u_n)$ represents the Radon–Nikodym derivative of the Lebesgue measure in the space X which is considered as a function of the correspondent sets in the space U with respect to the Lebesgue measure in the space U:

$$\left| \frac{\partial(\varphi_1, \ldots, \varphi_n)}{\partial(u_1, \ldots, u_n)} \right| = \frac{d\nu}{dl_u}(u) = \frac{d(l_x \varphi)}{dl_u}(u) \,.$$

Here $l_x \varphi$ is the composition of the mappings φ and l_x. The mapping φ establishes the correspondence between any set C of the space U and its image $\varphi(C)$ in the space X, and the other mapping l_x establishes the correspondence between the set $\varphi(C) \subset X$ and a nonnegative number – the Lebesgue measure $\varphi(C)$. In a given case Radon–Nikodym derivative may be determined by the formula

$$\frac{d\nu}{dl_u}(u) = \left| \frac{\partial(\varphi_1, \ldots, \varphi_n)}{\partial(u_1, \ldots, u_n)} \right| = \lim_{C \to \{u\}} \frac{\nu(C)}{l_u(C)} \,, \qquad (3.85)$$

which is easily derived by applying to the second integral in (3.84) the theorem about the mean value with succeeding passage to the limit while shrinking the set C to the point u.

E x a m p l e 3.13. Let φ be a measure on the real axis R, and l be the Lebesgue measure on R and $\varphi(A)$ is l–continuous. Then according to Radon–Nikodym theorem

$$F(x) = \varphi([a, x)) = \int_a^x f(u)du \,, \qquad (3.86)$$

where $f(u)$ is the Radon–Nikodym derivative with respect to Lebesgue measure l, $f(u) = \frac{d\varphi}{dl}(u)$. In this case the function $F(x)$ is called *absolutely conti-nuous* and the function $f(x)$ is its derivative.

Thus it follows from Radon–Nikodym theorem that any absolutely continuous function $F(x)$ has the derivative $f(x) = F'(x)$ in which terms it is expressed by ordinary formula (3.86). Hence the derivative $F'(x)$ of the function $F(x)$ may not exist in the ordinary sense.

It is clear that any piecewise differentiable in the ordinary sense continuous function is absolutely continuous. Thus the notion of absolutely continuous function in some sense generalizes the notion of the differentiable function.

E x a m p l e 3.14. For the Wiener measure μ_W (Example 2.15) and the modified Wiener measure μ'_W (Example 3.19) are equivalent. Really, they both are equal to zero on the set C if the integral over the measure $\nu_{t_1}(C; x_1)$ is equal to zero almost at all x_1 or if and only if the section of the set C at the point x_S, $S = T \backslash \{t\}$ has zero Lebesgue measure on R_t. But at the same time the calculation of Radon–Nikodym derivatives $d\mu_W / d\mu'_W$ and $d\mu'_W / d\mu_W$ is very difficult.

Detailed treatment of absolute continuity and singularity of measures including Radon–Nikodym derivatives is in (Dunford and Schwarz 1958). Some probability applications are given in (Parthasarathy 1980).

P r o b l e m s

3.6.1.   Whether the Lebesgue–Stieltjes measure induced by the function $\varphi(F(\psi(x)))$ be absolutely continuous (or singular) relative to the Lebesgue measure if $F(y)$ is the function of Example 3.9 (*Cantor ladder*) and φ and ψ are continuous functions?

3.6.2.   Let $\sum 2^n a_n$, $a_n > 0$ be convergent series. Show that the function

$$\Phi(x) = \sum_{n=1}^{\infty} a_n F_n(x) \,,$$

where $F_n(x)$ is the sum of the functions obtained from the function $F(x)$ of Example 3.9 by the mapping of the interval $[0, 1)$ into each from 2^{n-1} intervals of constancy $F(x)$ with the coordinates of the ends $k3^{-n}$, $(k + 1)3^{-n}$ is continuous, is not constant at none of the intervals $[a, b) \subset C(0, 1)$, is singular relative to the Lebesgue measure and its derivative is equal to zero almost everywhere.

3.6.3.   Is the measure

$$\nu(A) = \tfrac{1}{\sqrt{2\pi}} \int\limits_A e^{-t^2/2} dt$$

absolutely continuous or singular relative to Lebesgue measure on R? Are the measure $\nu(A)$ and Lebesgue measure $l(A)$ equivalent? At positive answer on the last question find Radon–Nikodym derivatives $d\mu/dl$ and $dl/d\mu$.

3.6.4. Are the Lebesgue and Lebesgue–Stieltjes measures induced by the function $F(x) = -[x^2] - 1$ absolutely continuous or singular relative one to another?

3.6.5. The same question (Problem 3.6.4) in the case of the Lebesgue–Stieltjes measures induced by the functions $F_1(x) = -[x^2] - 1$ and $F_2(x) = x F_1(x)$. In the case of absolute continuity find the correspondent Radon–Nikodym derivatives.

3.6.6. The same question (Problem 3.6.4) in the case of the functions $F_1(x) = -[x] - 1$ and $F_2(x) = -[x^2] - 1$.

3.6.7. Under what conditions one of the measures

$$\mu_1(A) = \int_A \varphi_1(x)\, dx, \quad \mu_2(A) = \int_A \varphi_2(x)\, dx$$

is absolutely continuous (singular) relative to the other measure? Under what conditions they are equivalent (orthogonal)?

3.6.8. The same question (Problem 3.6.7) in the case of the functions $F_1(x) = -[x] - 1$ and $f_2(x) = -[x + a] - 1$ at $0 < a < 1$.

3.6.9. Are the Lebesgue–Stieltjes measures induced by the functions $F_1(x) = -[x] - 1$ and $F_2(x) = -[x] - 1 + x$ absolutely continuous or singular relative to each other? In the case of absolute continuity find correspondent Radon–Nikodym derivatives. Find the Lebesgue expansion of the correspondent measure.

3.7. Lebesgue Spaces

3.7.1. Definition of a Lebesgue Space

The set of all measurable functions mapping the space X with the nonnegative measure μ into a separable B-space Y and satisfying condition (3.87) is called *Lebesgue space* L_p, $1 \le p < \infty$

$$\int \| f \|^p \, d\mu = \int \| f(x) \|^p \, \mu(dx) < \infty. \qquad (3.87)$$

As may be necessary this space is denoted by $L_p(X, \mathcal{A}, \mu)$ or $L_p(X, \mathcal{A}, \mu, Y)$ indicating the space X on which the functions are determined, the σ–algebra $\mathcal{A}$ in X and the measure μ determined in it and may be the space of the values of the function Y.

Let us introduce a notation

$$\| f \|_p = \left\{ \int \| f \|^p \, d\mu \right\}^{1/p}. \qquad (3.88)$$

Later on we show that the variable $\|f\|_p$ possesses all the properties of a norm. Therefore it may be assumed as the norm of the function $f(x)$

considered as an element of the space L_p (not to be confused with the norm $\|f(x)\|$ of the *value* of the function $f(x)$ at fixed x in the space Y). In the special case of scalar functions in the space R^n with the Lebesgue measure this space is denoted by $L_p(X)$ where X is the domain of the functions.

3.7.2. An Auxiliary Inequality

Lemma. *Let p, q, $r \geq 1$ and $p^{-1} + q^{-1} = r^{-1}$. If $f(x)$ belongs to L_p and numeric complex–valued function $\alpha(x)$ belongs to L_p then $\alpha(x)f(x) \in L_r$ and*

$$\| \alpha f \|_r \leq \| \alpha \|_q \| f \|_p \ . \tag{3.89}$$

▷ To prove Lemma we use the inequality

$$\frac{a^p}{p} + \frac{b^q}{q} \geq \frac{a^r b^r}{r} , \tag{3.90}$$

which is valid at any a, $b > 0$ and any p, q, $r \geq 1$ satisfying the condition $p^{-1} + q^{-1} = r^{-1}$ [a] .

Putting in (3.90) $a = \| f(x) \| / \| f \|_p$, $b = | \alpha(x) | / \| \alpha \|_q$ we get

$$\frac{1}{r} \frac{\| f(x) \|^r | \alpha(x) |^r}{\| f \|_p^r \| \alpha \|_q^r} \leq \frac{1}{p} \frac{\| f(x) \|^p}{\| f \|_p^p} + \frac{1}{q} \frac{| \alpha(x) |^q}{\| \alpha \|_q^q}$$

or

$$\| \alpha(x)f(x) \|^r \leq r \| \alpha \|_q^r \| f \|_p^r \left(\frac{1}{p} \frac{\| f(x) \|^p}{\| f \|_p^p} + \frac{1}{q} \frac{| \alpha(x) |^q}{\| \alpha \|_q^q} \right) . \tag{3.91}$$

---

[a] To prove this inequality we consider the function

$$\varphi(t) = \frac{t^p}{p} + \frac{t^{-q}}{q}$$

at $t > 0$. As $\varphi'(t) = t^{p-1} - t^{-q-1}$ then $\varphi(t)$ decreases at $0 < t < 1$ and increases at $t > 1$. Consequently, $\varphi(t)$ has the unique minimum $\varphi(1) = p^{-1} + q^{-1} = r^{-1}$ at the point 1 at $t > 0$ and

$$\frac{t^p}{p} + \frac{t^{-q}}{q} \geq \frac{1}{r} \quad \text{at all} \quad t > 0 .$$

Assuming that $t = a^{r/q} b^{-r/p}$, a, $b > 0$ we obtain inequality (3.90).

Hence according to Theorem 3.20 follows that the function $\| \, \alpha(x)f(x) \, \|^r$ is μ–integrable as $\| \, f(x) \, \|^p$ and $| \, \alpha(x) \, |^q$ are μ–integrable by virtue of $f \in L_p, \ \alpha \in L_q$. Thus $\alpha f \in L_r$. Integrating inequality (3.91) and taking into account definition(3.88) we receive

$$\| \, \alpha f \, \|_r^r = \int \| \, \alpha f \, \|^r \, d\mu \leq$$

$$\leq r \, \| \, \alpha \, \|_q^r \| \, f \, \|_p^r \left(\frac{1}{p} + \frac{1}{q} \right) = \| \, \alpha \, \|_q^r \| \, f \, \|_p^r \, ,$$

whence follows (3.89). ◁

In particular, at $r = 1$ from the proved Lemma follows that if $f \in L_p, \ \alpha \in L_q, \ p^{-1} + q^{-1} = 1$ then the product $\alpha(x)f(x)$ is μ–integrable and

$$\| \, \alpha f \, \|_1 = \int | \, \alpha \, | \| \, f \, \| \, d\mu \leq \| \, \alpha \, \|_q \| \, f \, \|_p \, . \qquad (3.92)$$

This inequality is called *Holder inequality*.

Corollary 1. *If a measure* μ *is finite,* $\mu(X) < \infty$, *then at any* $p > r \geq 1$

$$\frac{\| \, f \, \|_r}{\mu^{1/r}(X)} \leq \frac{\| \, f \, \|_p}{\mu^{1/p}(X)} . \qquad (3.93)$$

▷ For proving it is sufficient to put in (3.89) $\alpha(x) \equiv 1$ and account that $\| \, 1 \, \|_q = \mu^{1/q}(X) = \mu^{1/r - 1/p}(X)$. ◁

Corollary 2. *In the case of a finite measure* μ *from* $f \in L_p$ *follows* $f \in L_r$ *at all* $r < p$.

▷ It follows directly from (3.93). ◁

3.7.3. Norm in a Lebesgue Space

We shall prove now that the variable $\| \, f \, \|_p$ determined by formula (3.88) may be assumed as a norm in the space L_p.

Theorem 3.35. *The variable* $\| \, f \, \|_p$ *determined by formula (3.88) satisfies the axioms of a norm.*

▷ First of all it is evident that $\| \, f \, \|_p \geq 0$ and $\| \, f \, \|_p = 0$ if and only if $f(x) \stackrel{a.e.}{=} 0$. Taking the function $f(x) \stackrel{a.e.}{=} 0$ with arbitrary values on the set of zero measure μ as zero of the space L_p we make sure that $\| f \|_p$ safisfies the first axiom of a norm (Subsection 1.3.5) [a] .

Further directly from the integral properties follows that $\| \alpha f \|_p = | \alpha | \| f \|_p$ for any complex number α and any function $f \in L_p$. Thus $\| f \|_p$ satisfies the second axiom of the norm (Subsection 1.3.5).

From the inequality

$$\| f_1(x) + f_2(x) \|^p \leq (\| f_1(x) \| + \| f_2(x) \|)^p$$

$$\leq 2^{p-1}(\| f_1(x) \|^p + \| f_2(x) \|^p) \quad [b]$$

by Theorem 3.20 follows that if $f_1, f_2 \in L_p$ then $f_1 + f_2 \in L_p$. Therefore

$$\| f_1 + f_2 \|_p^p = \int \| f_1 + f_2 \|^p \, d\mu \leq \int (\| f_1 \| + \| f_2 \|) \| f_1 + f_2 \|^{p-1} \, d\mu .$$
$$(3.94)$$

It is evident that the numeric function $\alpha(x) = \| f_1(x) + f_2(x) \|^{p-1}$ belongs to the space L_q, $q = p/(p-1)$ and $p^{-1} + q^{-1} = 1$. Consequently, according to inequality (3.92)

$$\int \| f_k \| \| f_1 + f_2 \|^{p-1} \, d\mu = \int \alpha \| f_k \| \, d\mu$$

$$\leq \| \alpha \|_q \| f_k \|_p \quad (k = 1, 2) .$$

But $\| \alpha \|_q^q = \| f_1 + f_2 \|_p^p$, whence $\| \alpha \|_q = \| f_1 + f_2 \|_p^{p/q} = \| f_1 + f_2 \|_p^{p-1}$. Therefore

$$\int \| f_k \| \| f_1 + f_2 \|^{p-1} \, d\mu \leq \| f_k \|_p \| f_1 + f_2 \|_p^{p-1} \quad (k = 1, 2) ,$$

and from (3.94) we receive

$$\| f_1 + f_2 \|_p^p \leq (\| f_1 \|_p + \| f_2 \|_p) \| f_1 + f_2 \|_p^{p-1} .$$

---

[a] Taking as zero of the space L_p the function equal to zero almost everywhere hence we determine L_p as the space of the classes of μ-*equivalent* functions which differ one from another only on any set of zero measure μ. In other words, if as L_p^0 we assume the space of the functions satisfying condition (3.87) then denoting in terms of N the set of all the functions from L_p^0 which are equivalent to zero we determine Lebesgue space L_p as the $L_p = L_p^0/N$ (Subsection 1.3.4).

[b] In order to make sure in the validity of this inequality it is sufficient to show that the function $(1 + t)^p/(1 + t^p)$ at $t \geq 0$ has the unique maximum 2^{p-1} at $t = 1$.

Hence it follows that $\| f \|_p$ satisfies the third axiom of a norm $\| f_1 + f_2 \|_p \leq \| f_1 \|_p + \| f_2 \|_p$ (Subsection 1.3.5). ◁

Finally, we have proved that the variable $\| f \|_p$ possesses all the properties of a norm. If we assume $\| f \|_p$ as the norm of the element $f = f(x)$ of the space L_p then L_p will be a normed linear space.

The inequality of triangle for the norm L_p in which the norms are substituted by the correspondent integrals is called *Minkowski inequality*.

3.7.4. Convergence in Mean

The sequence of the functions $\{f_n(x)\} \subset L_p$ is called *convergent in mean of p degree*, or shortly in *p–mean* to the function $f(x) \in L_p$, $f_n(x) \xrightarrow{\text{P}} f(x)$ if $\| f_n - f \|_p \to 0$ at $n \to \infty$, i.e.

$$\int \| f_n - f \|^p \, d\mu \to 0 \quad \text{at} \quad n \to \infty.$$

In other words, the sequence $\{f_n(x)\}$ converges in p–mean to $f(x)$ if it converges to $f = f(x)$ in the metric of the space L_p generated by the norm $\| f \|_p$.

The sequence $\{f_n(x)\} \subset L_p$ is called *a fundamental in mean of p degree* or a *Chauchy sequence in mean of p degree*, or shortly, in *p–mean* if $\| f_n - f_m \|_p \to 0$ at $n, m \to \infty$, i.e. if at any $\varepsilon > 0$ $\| f_n - f_m \|_p < \varepsilon$ at all n, m exceeding some natural number $N = N(\varepsilon)$. So the sequence $\{f_n(x)\}$ is a fundamental sequence in p–mean if it is a fundamental sequence in the metric of the space L_p generated by the norm $\| f \|_p$.

Condition (3.18) in the definition of an integral represents nothing more but the condition of the sequence $\{f^n(x)\}$ to have a fundamental sequence property in the space $L_1(X, \mathcal{A}, \mu)$ which is evidently the space of μ–integrable functions.

From the general theorem about the property to be a fundamental sequence (Subsection 1.2.4) a convergent sequence in any metric (in particular, normed) space follows that any convergent in p–mean sequence of the functions is a fundamental sequence in p–mean.

Further from Chebyshev inequality (3.31) for the function $(f_n - f)^p$,

$$\mu(\{x : \| f_n(x) - f(x) \| \geq \varepsilon\}) \leq \int \| f_n - f \|^p \, d\mu / \varepsilon^p ,$$

follows directly.

Theorem 3.36. *Any convergent in p-mean sequence of functions (a fundamental sequence) converges (is a fundamental sequence) in the measure μ.*

According to Theorem 3.13 from any sequence of the functions convergent (a fundamental sequence) in the measure μ we may select a subsequence convergent almost everywhere relative to the measure μ. From this theorem and just proved preposition follows.

Theorem 3.37. *Any sequence of the functions convergent (a fundamental sequence) in p-mean contains a subsequence convergent almost everywhere.*

The following theorem establishes the sufficiency of a fundamental sequence in p-mean for its convergence in p-mean.

Theorem 3.38. *If the sequence $\{f_n(x)\} \subset L_p$ is a fundamental sequence in p-mean then it converges in p-mean to some function $f(x) \in L_p$.*

▷ Suppose that $\{f_n(x)\}$ is a fundamental sequence in p-mean and select from it the subsequence $\{f_{n_k}(x)\}$ convergent almost everywhere. We denote the limit of this subsequence in terms of $f(x)$. We shall prove that $f(x) \in L_p$. From $\left| \parallel f_n \parallel_p - \parallel f_m \parallel_p \right| \leq \parallel f_n - f_m \parallel_p$ and fundamental property in p-mean for sequence $\{f_n(x)\}$ follows the property of the numeric sequence of the norms $\{\parallel f_n \parallel_p\}$ to be a fundamental sequence and convergent. Consequently, there is such a number $c > 0$ that $\parallel f_n \parallel_p < c$ at all n. Therefore after applying to the sequence of the functions $\{\parallel f_{n_k}(x) \parallel^p\}$ Fatou lemma (Subsection 3.5.3) we get

$$\int \underline{\lim} \parallel f_{n_k} \parallel^p d\mu \leq \underline{\lim} \int \parallel f_{n_k} \parallel^p d\mu \leq c^p .$$

But $\underline{\lim}\parallel f_{n_k}(x)\parallel^p = \lim\parallel f_{n_k}(x)\parallel^p = \parallel f(x)\parallel^p$. So, $\parallel f(x)\parallel^p$ is μ-integrable, i.e. $f(x) \in L_p$. It remains to prove that the sequence $\{f_n(x)\}$ converges in p-mean to $f(x)$. As $f_n(x) - f(x) \overset{\text{a.e.}}{=} \lim_{k \to \infty} [f_n(x) - f_{n_k}(x)] = \lim_{k \to \infty} [f_n(x) - f_{n_k}(x)]$ then after applying to the sequence of the functions $\{f_n(x) - f_{n_k}(x)\}_{k=1}^{\infty}$ Fatou lemma we obtain

$$\parallel f_n - f \parallel_p^p \leq \underline{\lim}_{k \to \infty} \parallel f_n - f_{n_k} \parallel_p^p .$$

By virtue of a fundamental property for sequence $\{f_n(x)\}$ in p-mean at any $\varepsilon > 0$

$$\parallel f_n - f_{n_k} \parallel_p < \varepsilon$$

at all sufficiently large n and n_k. Therefore $\| f_n - f \|_p < \varepsilon$ at all n larger than some natural $N(\varepsilon)$ what proves the convergence of $\{f_n(x)\}$ to $f(x)$ in p–mean. ◁

Thus for the convergence of the sequence of the functions in p–mean it is necessary and sufficient its property to be a fundamental sequence in p–mean. It means that the space L_p is complete.

So, we proved that the space L_p is a complete normed linear space, i.e. a B–space.

We have seen that any sequence of the functions from L_p convergent in p–mean also converges in the measure μ and we may select from it the subsequence convergent almost everywhere. But not any sequence of the functions from L_p convergent in the measure or almost everywhere converges in p–mean. The following theorem gives a simple sufficient condition for the convergence in p–mean of the sequence convergent almost everywhere or in the measure.

Theorem 3.39. *If the sequence of the functions $\{f_n(x)\} \subset L_p$ converges almost everywhere or in the measure μ to $f(x)$ and is majorized in a norm by some function from L_p, $\| f_n(x) \| \leq \| g(x) \|$, $g(x) \in L_p$ then $f(x) \in L_p$ and the sequence $\{f_n(x)\}$ converges in p–mean to $f(x)$.*

▷ As $\| f_n(x) - f(x) \|^p \to 0$ and $\| f_n(x) \|_p^p - \| f(x) \|_p^p \to 0$ almost everywhere or in the measure μ then

$$\| f(x) \|^p \leq \| g(x) \|^p ,$$

and

$$0 \leq \| f_n(x) - f(x) \|^p \leq 2^p \| g(x) \|^p .$$

Hence by Theorem 3.30 follows that the functions $\| f(x) \|^p$ and $\| f_n(x) - f(x) \|^p$ are μ–integrable and

$$\| f_n - f \|_p^p = \int \| f_n - f \|^p \, d\mu \to 0 \quad \text{at} \quad n \to \infty ,$$

what proves our statement. ◁

3.7.5. Set of Simple Functions is Dense in a Lebesgue Space

For further study of the properties of L_p spaces we notice that all μ–integrable simple functions belong to L_p at any $p \geq 1$ as from μ–integrability of the simple function $f(x)$ immediately follows also μ–integrability of the simple function $\| f(x) \|^p$.

Theorem 3.40. *A set of the μ–integrable simple functions is dense in L_p.*

▷ For any function $f(x) \in L_p$ on the basis of Chebyshev inequality (3.31) at any $\varepsilon > 0$ the measure of the set $\{x : \| f(x) \| \geq \varepsilon\}$ is finite as the function $\| f(x) \|^p$ is μ–integrable. Therefore in the same way as at the proof of Theorem 3.20 we may construct the sequence of the μ–integrable simple functions $\{f^n(x)\}$ convergent almost everywhere and bounded in a norm by the function $2f(x) \in L_p$.

According to Theorem 3.39 the sequence $\{f^n(x)\}$ converges to $f(x)$ and in p–mean. Consequently, at any $\varepsilon > 0$ there exists such natural $N = N(\varepsilon)$ that

$$\| f^n - f \|_p < \varepsilon \quad \text{at all} \quad n > N .$$

Thus in any vicinity of any element f of the space L_p the elements f^n representing the μ–integrable simple functions are contained. This proves the density of a set of the μ–integrable simple functions in L_p. ◁

Corollary. *If the measure μ on the σ–algebra $\mathcal{A}$ may be approximated by its values on a countable class of sets $\mathcal{B} = \{B_n\}$, i.e. if for any $\varepsilon > 0$ and any $A \in \mathcal{A}$ will occur such set $B_k \in \mathcal{B}$ that $\mu(A \triangle B_k) < \varepsilon$ then the space L_p is a separable.*

▷ If $\mu(A \triangle B_k) < \varepsilon$ then $\| 1_A - 1_{B_k} \|_p = \mu^{1/p}(A \triangle B_k) < \varepsilon^{1/p}$. It means that the indicator of some set $B_k \in \mathcal{B}$ is contained in any vicinity of the indicator of any set $A \in \mathcal{A}$ in the space L_p.

Let $\{y_n\}$ be a dense countable set in a B–space Y. It follows from the proved that any μ–integrable simple function may be approximated in L_p with any degree of the accuracy by a simple function of the form

$$f(x) = \sum_{k=1}^{n} y_k \, 1_{B_k}(x) .$$

Consequently, a countable set of such functions is dense on the set of μ–integrable simple functions which in its turn is dense in the space L_p in accordance with Theorem 3.40. ◁

R e m a r k. If the σ–algebra $\mathcal{A}$ is induced by a countable algebra $\mathcal{B}$ then according to the Corollary of Theorem 2.36 the value of the measure μ on any set $A \in \mathcal{A}$ of a finite measure is approximated with any degree of the accuracy by its value on some set from countable class of the sets $\mathcal{B}$. By the Corollary of Theorem 3.40 in this case Lebesgue space $L_p(X, \mathcal{A}, \mu)$ is separable. In particular, the σ–algebra of Borel sets in any finite–dimensional space R^n is induced by a countable set of the triangles with rational coordinates of the tops (the intervals with rational ends in the case when $n = 1$). Therefore all Lebesgue spaces $L_p(X, \mathcal{A}, \mu)$ are separable at $X \subset R^n$ $(n = 1, 2, \ldots)$.

3.7.6. Hilbert Space of Scalar Functions with Integrable Square of Modulus

In the special case of the spaces $L_p(X, \mathcal{A}, \mu)$ of numeric functions $(Y = K)$ the norm $\| f \|_2$ in the space $L_2(X, \mathcal{A}, \mu)$ is generated by a scalar product

$$(f_1, f_2) = \int f_1 \, \bar{f_2} \, d\mu = \int f_1(x) \, \overline{f_2(x)} \mu(dx), \qquad (3.95)$$

as

$$(f, f) = \int |f|^2 \, d\mu.$$

Thus the space $L_2(X, \mathcal{A}, \mu)$ of scalar functions (the classes of equivalent functions) being *square of modulus integrable* represents a B–space in which the norm is generated by scalar product, i.e. a H–space. This space plays an important role in modern mathematics and its applications. In particular, the spaces $L_2(X)$, $X \subset R^n$ (Subsection 3.7.1) are of great importance. According to the Remark at the end of Subsection 3.7.5 they are all separable.

E x a m p l e 3.15. Consider the spaces $L_p(X^T, \mathcal{A}_W^T, \mu_W)$. According to the result of Problem 2.3.3 the measure μ_W may be extended to the space of the continuous functions $C \subset X^T$ with the σ–algebra $C\mathcal{A}^T$. Setting in the space C an ordinary norm $\| x \| = \sup | x(t) |$ we get the space $C(T)$. Show that the σ–algebra in the space $C(T)$ induced by a set of all open balls enters into the σ–algebra $C(T)\mathcal{A}^T$. For this purpose we notice that any open ball in $C(T)$ represents the limit of decreasing sequence of the rectangles:

$$\{x(t) : | x(t) - \varphi(t) | < r \quad \forall t \subset T\}$$

$$= \lim_{n \to \infty} \{x(t) : | x(s_k) - \varphi(s_k) | < r \quad (k = 1, \ldots, n)\},$$

where $\{s_n\}$ is a set of rational numbers of the interval $[0, \infty)$. Consequently, any ball in $C(T)$ belongs to the σ–algebra $C(T)\mathcal{A}^T$. It means that the whole σ–algebra $\mathcal{C}$ induced by a set of the balls in $C(T)$ inters into $C(T)\mathcal{A}^T$. But the σ–algebra $\mathcal{C}$ in $C(T)$ is induced by a countable set of the balls of rational radii with the centers at the points of a dense countable set (the space $C(T)$ is separable as a countable set of continuous functions with rational values in rational points is dense in $C(T)$).

Consequently, in accordance with the last result of Subsection 3.7.5 all the spaces $L_p(C[a, b], C, \mu_W)$ are separable. In particular, the H–space of numeric functions $L_2(C[a, b], C, \mu_W)$ is separable.

3.7.7. Set of Continuous Functions is Dense in $L_p(X)$

Let us consider Lebesgue space $L_p(X)$ of the numeric functions in the space R^n with integrable in the Lebesgue measure the p–th degree of the modulus, $p \geq 1$:

$$\| f \|_p = \int_X | f(x) |^p \ dx < \infty. \tag{3.96}$$

For the applications of functional analysis the following theorem has the great importance.

Theorem 3.41. *A set of continuous functions is dense in $L_p(X)$ at any $X \subset R^n$ (in particular, at $X = R^n$).*

▷ By virtue of Theorem 3.40 it is sufficient to prove that the indicator of any Borel set B in R^n may be approximated in $L_p(X)$ with any degree of accuracy by a continuous function. Let take an arbitrary $\varepsilon > 0$ and choose such an open set $G \supset B$ that the Lebesgue measure of the difference $G\backslash B$ will be smaller than ε, $l(G\backslash B) < \varepsilon$ [a] . Determine the function

$$f(x) = \frac{d(x, \bar{G})}{d(x, \bar{G}) + d(x, B)} ,$$

where

$$d(x, C) = \inf_{y \in C} | x - y |$$

is the distance from the point x till the set C. This function is continuous and is equal to 1 on the set B ($d(x, B) = 0$ at $x \in B$) and is equal to 0 everywhere outside the set G ($d(x, \bar{G}) = 0$ at $x \in \bar{G}$). Therefore $f(x) - \mathbf{1}_B(x) = 0$ at $x \in B$ and $x \in \bar{G}$ and $| f(x) - \mathbf{1}_B(x) | < 1$ at $x \in G\backslash B$. Consequently,

$$\int_X | f(x) - \mathbf{1}_B(x) |^p \ dx < l(G\backslash B) < \varepsilon.$$

---

[a] The existence of such an open set follows from the definition of the Lebesgue measure.

Hence due to the arbitrariness of $\varepsilon > 0$ follows the density of a set of continuous functions in the set of all simple functions which is dense in $L_p(X)$ by virtue of Theorem 3.40. ◁

It follows from this theorem that any function from $L_p(X)$ represents the limit of the sequence of continuous functions from $L_p(X)$. These functions also belong to the normed space of continuous functions $C_p(X)$ with the norm determined by formula (3.96) in which the integral is assumed as the Riemann integral (see Examples 1.13 and 1.14). Thus the space $L_p(X)$ represents the space $C_p(X)$ supplemented by the limits of all fundamental sequences of the functions from $C_p(X)$. In a given case the process of the supplement of the space is formally performed by the passage from the Riemann integral to the Lebesgue integral in (3.96).

3.7.8. Sobolev Spaces

The natural generalization of Lebesgue spaces $L_p(X)$ are Sobolev spaces which have great importance in applications of functional analysis. In order to approach to the definition of Sobolev space let consider a linear space of scalar functions continuous together with their derivatives till the order N inclusively on the bounded closed interval $[a, b]$. Introducing in this space the norm

$$\| f \| = \left\{ \int_a^b \sum_{k=0}^N | f^{(k)}(x) |^p \, dx \right\}^{1/p} , \quad p \geq 1 , \qquad (3.97)$$

where the integral naturally is assumed as Riemann integral we obtain a normed linear space which we denote by $C_p^N([a, b])$. This space is not complete what is proved analogously as in Example 1.14, but inclined rectilinear segments on Fig 1.17 should by replaced by the segments of smooth curves in order to obtain the functions $f_n(t)$ which are continuous together with their derivatives till the order N inclusively. For supplying $C_p^N([a, b])$ it is necessary to generalize the notion of the derivative. Let consider a fundamental sequence of the functions $\{f_n(x)\} \subset C_p^N([a, b])$. For this sequence

$$\| f_n - f_m \|^p = \int_a^b \sum_{k=0}^N |f_n^{(k)}(x) - f_m^{(k)}(x)|^p \, dx \to 0 \quad \text{at} \quad n, m \to 0 .$$

Hence follows that each of the sequences $\{f_n^{(k)}(x)\}$ $(k = 0, 1, \ldots, N)$ is a fundamental sequence in $L_p([a, b])$ and by virtue of the completeness of L_p has the limit $f^{(k)}(x)$ $(k = 0, 1, \ldots, N)$. The limit functions $f'(x), \ldots, f^{(N)}(x) \in L_p([a, b])$ are called the *generalized derivatives of the limit function* $f(x) \in L_p([a, b])$ *in Sobolev sense*.

The set of all the functions from $L_p([a, b])$ which have generalized derivatives in Sobolev sense till the N-th order inclusively with the norm determined by formula (3.97) in which the integral is assumed as the Lebesgue integral is called *Sobolev space* $W_p^N([a, b])$.

Similarly Sobolev space $W_p^N(X)$ for any closed bounded region X of the space R^n with sufficiently smooth bound is determined. In this case the norm of the element $f(x) \in W_p^N(X)$ is determined by formula

$$
\| f \| = \left\{ \int_X \sum_{k_1,\ldots,k_n=0}^{N} \left| \frac{\partial^{k_1+\cdots+k_n} f(x)}{\partial x_1^{k_1} \ldots \partial x_n^{k_n}} \right|^p dx \right\}^{1/p} . \tag{3.98}
$$

In the special case at $p = 2$ the space $W_2^N(X)$ represents the H-space with the scalar product

$$
(f, g) = \int_X \sum_{k_1,\ldots,k_n=0}^{N} \frac{\partial^{k_1+\cdots+k_n} f(x)}{\partial x_1^{k_1} \ldots \partial x_n^{k_n}} \cdot \frac{\partial^{k_1+\cdots+k_n} \overline{g(x)}}{\partial x_1^{k_1} \ldots \partial x_n^{k_n}} dx \tag{3.99}
$$

This Sobolev space is denoted by $H^N(X)$, $W_2^N(X) = H^N(X)$.

From the definition of Sobolev space follows directly that each function of the space $W_p^N(X)$ represents an element of the space $C_p^N(X)$[a] or the limit of a fundamental sequence of the element from $C_p^N(X)$. Thus the set $C_p^N(X)$ is dense in $W_p^N(X)$. Besides, every fundamental sequence from $C_p^N(X)$ according to the definition has the limit in $W_p^N(X)$. Consequently, by Theorem 1.8 the space $W_p^N(X)$ represents a supplement of the space $C_p^N(X)$. Thus we have the second example when the supplement of the space may be performed by immediate addition to it the limits of all its fundamental sequences. Formally it is reduced to the change of the derivatives by generalized derivatives and the passage from the Riemann integral to the Lebesgue integral.

For deep study of Lebesgue spaces refer to (Dunford and Schwarz 1958, Halmos 1950). Detailed theory and some applications of Sobolev spaces are given in (Adams 1975).

---

[a] Precisely, a class of the functions equivalent to some function from $C_p^N(X)$.

Problems

3.7.1. Whether the sequences of Example 3.3 and Problems 3.2.1 and 3.2.2 converge in p–mean at some $p \geq 1$?

3.7.2. Whether the function $f(x)$ of Problem 3.5.2 belongs to the space $L_p(C(R), C, \mu_W)$ at any $p \geq 1$ and whether the sequence of the functions $\{f_n(x)\}$ converges to $f(x)$ in p–mean?

3.7.3. Whether the sequence $\{\sin x / n x^\alpha\}$ converges in p–mean at any p in $L_p(R)$ $0 < \alpha < 1$? Whether it converges almost everywhere and in the measure (on the whole real axis R; on a finite interval)? Whether it converges in Sobolev space H^2?

3.7.4. Whether the space $L_2(X, \mathcal{A}, \mu, Y)$ of the functions with the values in a B–space Y be a H–space? To what condition must satisfy the space Y in order that $L_2(X, \mathcal{A}, \mu, Y)$ be a H–space? Determine a scalar product in this case.

3.8. Measures in Product Spaces. Multiple Integrals

3.8.1. Measures in Product of Two Spaces

Let $(X, \mathcal{A})$ and $(Y, \mathcal{B})$ be two measurable spaces, μ be nonnegative measure determined on the σ–algebra $\mathcal{A}$ in X, λ_x be a family of nonnegative measures determined on σ–algebra $\mathcal{B}$ in Y dependent on the parameter $x \in X$. In this case of such a family at any set $B \in \mathcal{B}$ the function $\lambda_x(B)$ of the variable x is measurable and μ–integrable.

Theorem 3.42. *Formula*

$$\nu(A \times B) = \int_A \lambda_x(B)\mu(dx) \qquad (3.100)$$

determines nonnegative measure ν on all measurable rectangles $A \times B$, $A \in \mathcal{A}$, $B \in \mathcal{B}$ of the measurable spaces $(X \times Y, \mathcal{A} \times \mathcal{B})$.

▷ The nonnegativity of the function of the set ν determined by formula (3.100) is evident. In order to prove its σ–additivity we shall show that it is additive and continuous in zero, i.e. that for any decreasing sequence of the rectangles $\{A_n \times B_n\}$, $A_n \in \mathcal{A}$, $B_n \in \mathcal{B}$, $A_{n+1} \subset A_n$, $B_{n+1} \subset B_n$ $(n = 1, 2, \ldots)$ with empty intersection $\bigcap(A_n \times B_n) = \varnothing$,

$$\lim_{n \to \infty} \nu(A_n \times B_n) = 0.$$

Hence by Theorem 2.13 will follow the σ–additivity ν on the class of measurable rectangles.

For any partition of the rectangle $A \times B$ on pairwise nonintersecting rectangles $A \times B = \bigcup\limits_{k=1}^{n} (A_k \times B_k)$,

$$A \times B = \bigcup_{p,q=1}^{2^n} (C_p \times D_q),$$

where $C_p = AA_{p1} \ldots A_{pn}$, $D_q = BB_{q1} \ldots B_{qn}$, $A_{pk} = A_k$ or $\bar{A}_k$, $B_{ql} = B_l$ or $\bar{B}_l$ are pairwise nonintersecting sets in the spaces X and Y respectively,

$$\bigcup_{p=1}^{2^n} C_p = A, \qquad \bigcup_{q=1}^{2^n} D_q = B.$$

By virtue of the additivity of the measure λ_x

$$\lambda_x(B) = \sum_{q=1}^{2^n} \lambda_x(D_q) \qquad (q = 1, \ldots, 2^n).$$

Substituting this expession into formula (3.100) we get

$$\nu(A \times B) = \sum_{q=1}^{2^n} \int_A \lambda_x(D_q)\mu(dx) = \sum_{q=1}^{2^n} \nu(A \times D_q). \qquad (3.101)$$

Due to the additivity of the integral

$$\nu(A \times D_q) = \sum_{p=1}^{2^n} \nu(C_p \times D_q). \qquad (3.102)$$

From expressions (3.101) and (3.102) follows

$$\nu(A \times B) = \sum_{p,q=1}^{2^n} \nu(C_p \times D_q). \qquad (3.103)$$

On the other hand, after applying this formula to the rectangles $A_k \times B_k$ we obtain

$$\nu(A_k \times B_k) = \sum_{\{r,s : C_{p_r} \subset A_k, D_{q_s} \subset B_k\}} \nu(C_{p_r} \times D_{q_s}).$$

From this formula and from (3.103) follows the additivity of the function ν on the rectangles;

$$\nu(A \times B) = \sum_{k=1}^{n} \nu(A_k \times B_k).$$

Let $\{A_n \times B_n\}$ be a monotone decreasing sequence of the rectangles with empty intersection. Then or $\bigcap A_n = \emptyset$ or $\bigcap B_n = \emptyset$ or both. As at any n $A_n \subset A_1$ then formula (3.100) for $A = A_n$, $B = B_n$ may be presented in the form

$$\nu(A_n \times B_n) = \int_{A_1} \lambda_x(B_n) \mathbf{1}_{A_n}(x) \mu(dx). \qquad (3.104)$$

If $\bigcap B_n = \lim B_n = \emptyset$ then by Theorem 2.9 by virtue of the σ–additivity of the measure λ_x

$$\lim \lambda_x(B_n) = \lambda_x(\lim B_n) = \lambda_x(\emptyset) = 0.$$

If $\bigcap A_n = \lim A_n = \emptyset$ then

$$\lim \mathbf{1}_{A_n}(x) = 0.$$

In both cases the sequence of the functions $\{\lambda_x(B_n)\mathbf{1}_{A_n}(x)\}$ converges at all x to zero. As at any n $A_n \subset A_1$, $B_n \subset B_1$, consequently, $\lambda_x(B_n) \leq \lambda_x(B_1)$, $\mathbf{1}_{A_n}(x) \leq \mathbf{1}_{A_1}(x)$ then this sequence is majorized by the μ–integrable function $\lambda_x(B_1)\mathbf{1}_{A_1}(x)$. Consequently, according to Theorem 3.30 in (3.104) we may pass to the limit under the integral sign. As a result we receive

$$\lim \nu(A_n \times B_n) = \int_{A_1} \lim \lambda_x(B_n)\mathbf{1}_{A_n}(x)\mu(dx) = 0. \quad \triangleleft$$

Thus ν is σ–additive on the class of measurable rectangles representing a semi–algebra. By Theorem 2.38 the measure ν has the unique extension on the σ–algebra $\mathcal{C} = \mathcal{A} \times \mathcal{B}$ induced by the class of measurable rectangles.

Theorem 3.43. *The extension of the measure ν on σ–algebra $\mathcal{C}$* $= \mathcal{A} \times \mathcal{B}$ *is determined by the formula*

$$\nu(C) = \int \lambda_x(C_x)\mu(dx) = \int \mu(dx) \int \mathbf{1}_C(x,y)\lambda_x(dy), \quad C \in \mathcal{C},$$
$$(3.105)$$

where C_x is the section C at point x.

▷ We denote by $\mathcal{M} \subset \mathcal{C}$ the class of all the sets C for which formula (3.105) is valid. On the basis of (3.100) $\mathcal{M}$ contains all the measurable rectangles, and consequently, all the finite unions of measurable rectangles, i.e. the algebra of sets $\bar{B}$ induced by the semi–algebra of all measurable rectangles $\mathcal{M} \supset \bar{B}$. On the basis of Theorem 3.27 about monotone convergence and Theorem 3.29 about majorized sequence $\mathcal{M}$ contains the limits of all monotone sequences of its entering sets. Really, if $\{C_n\}$ is a monotone sequence of the sets and $C = \lim C_n$ then $\{1_{C_n}(x,y)\}$ is a monotone sequence of the functions convergent to $1_C(x,y)$. Hence if $\{C_n\}$ is a decreasing sequence $C_{n+1} \subset C_n$ and at some natural k the integral

$$\int \mu(dx) \int 1_{C_k}(x,y)\lambda_x(dy)$$

is finite (such k exists, if $\nu(C) < \infty$) then the function $1_{C_k}(x,y)$ of the variable y is λ_x–integrable almost at all x, and the function $\int 1_{C_k}(x,y)\lambda_x(dx)$ of the variable x is μ–integrable. In consequence of this all the functions $1_{C_n}(x,y)$, $n > k$ almost at all x are lower and upper bounded by λ_x–integrable functions 0 and $1_{C_k}(x,y)$, $0 \le 1_{C_n}(x,y) \le 1_{C_k}(x,y)$ and all the functions $\int 1_{C_n}(x,y)\lambda_x(dy)$, $n > k$ are lower and upper bounded by the μ–integrable functions 0 and $\int 1_{C_k}(x,y)\lambda_x(dy)$:

$$0 \le \int 1_{C_n}(x,y)\lambda_x(dy) \le \int 1_{C_k}(x,y)\lambda_x(dy).$$

After twice applying Theorem 3.27 about monotone convergence in the case of increasing sequence of the sets $\{C_n\}$ and Theorem 3.29 about majorized sequence in the case of decreasing sequence of the sets $\{C_n\}$ we shall have

$$\lim \int \mu(dx) \int 1_{C_n}(x,y)\lambda_x(dy) = \int \mu(dx)\left\{\lim \int 1_{C_n}(x,y)\lambda_x(dy)\right\}$$

$$= \int \mu(dx) \int \lim 1_{C_n}(x,y)\lambda_x(dy) = \int \mu(dx) \int 1_C(x,y)\lambda_x(dy).$$

On the other hand, by virtue of the continuity of the measure ν

$$\lim \nu(C_n) = \nu(\lim C_n) = \nu(C).$$

Consequently, the class of the sets $\mathcal{M}$ for which formula (3.105) is valid contains the limits of all monotone sequences of its entering sets. Thus $\mathcal{M}$ is a monotone class containing the algebra of the sets $\bar{\mathcal{B}}$ which is formed by all finite unions of measurable rectangles. But according to Theorem 2.7 the σ–algebra $\mathcal{C}$ represents a minimal monotone class which contains the algebra $\bar{\mathcal{B}}$. Consequently, $\mathcal{C} \subset \mathcal{M}$. From two opposite inclusions we get $\mathcal{M} = \mathcal{C}$. Thus formula (3.105) is valid for all measurable sets $C \in \mathcal{C}$ of the space $Z = X \times Y$. ◁

R e m a r k. We notice that the function $\lambda_x(B)$ may be considered as a measure with the values in the space of μ–integrable functions $L_1(X, \mathcal{A}, \mu)$. Thus we encountered the example of the measure with the values in a B–space.

3.8.2. Fubini Theorem

Let $(X, \mathcal{A})$ and $(Y, \mathcal{B})$ be measurable spaces, μ a nonnegative measure in $(X, \mathcal{A})$, λ_x be a family of nonnegative measures in $(Y, \mathcal{B})$, $\lambda_x(B)$ be a μ–integrable function x at any $B \in \mathcal{B}$, ν be the measure in $(Z, \mathcal{C}) = (X \times Y, \mathcal{A} \times \mathcal{B})$ determined by formula (3.105).

R e m a r k. Notice that the set $C \in \mathcal{C}$ has zero measure ν if and only if at all almost x relative to the measure μ the section C_x of the set C at the point x is the set of zero measure λ_x. Consequently, if any statement is valid almost everywhere relative to the measure ν in $X \times Y$ then it is valid almost at all x relative to μ almost at all y relative to correspondent measure λ_x.

Theorem 3.44. *Let $f(x, y)$ be ν–integrable function in the product of the spaces $Z = X \times Y$ with the values in some B–space U. Then the section $f_x(y) = f(x, y)$ of the function f is λ_x–integrable almost at all x (relative to the measure μ), the function*

$$w(x) = \int f(x, y)\lambda_x(dy)$$

μ–integrable and

$$\int f \, d\nu = \int w \, d\mu = \int \mu(dx) \int f(x, y)\lambda_x(dy) \qquad (3.106)$$

(Fubini theorem).

▷ Formula (3.105) shows that this statement is valid in the case when $f(x, y)$ represents an indicator of any measurable set C, $f(x, y)$

$= \mathbf{1}_C(x,y)$. Consequently, it is also valid for all ν–integrable simple functions. As any ν–integrable nonnegative function $f(x,y)$ may be presented as the limit of almost everywhere convergent nondecreasing sequence of ν–integrable simple functions then by Theorem 3.27 the statement is valid for all nonnegative ν–integrable functions. Therefore no matter what the ν–integrable function $f(x,y)$ may be the theorem for the correspondent nonnegative function $\|f(x,y)\|$ is valid. So, the function $f(x,y)$ is λ_x–integrable almost at all x relative to μ. As the function

$$w(x) = \int f\,d\lambda_x$$

is majorized in the measure by the μ–integrable function

$$v(x) = \int \|f\|\,d\lambda_x,$$

$\|w(x)\| \le v(x)$ then by Theorem 3.20 the function $w(x)$ is μ–integrable. Consequently, the integral in the right–hand side of (3.106) exists for any ν–integrable function $f(x,y)$. It remains to prove that this integral is equal to the integral in the left–hand side. Let $\{f^n(x,y)\}$ be the sequence of ν–integrable simple functions which determines the function $f(x,y)$. This sequence determines the section $f_x(y) = f(x,y)$ of the function f almost at all $x \in X$. Really, as it was shown while proving Theorem 3.20 the functions $f^n(x,y)$ may be chosen in such a way that they satisfy the condition $\|f^n(x,y)\| < 2\|f(x,y)\|$. Then the sequence of the functions $\|f^n(x,y) - f^m(x,y)\|$ $(n,m = 1,2,\ldots)$ convergent to zero almost everywhere will be upper and lower bounded by λ_x–integrable functions 0 and $4\|f(x,y)\|$. According to Theorem 3.29 we get

$$\lim_{n,m\to\infty} \int \|f^n - f^m\|\,d\lambda_x = \int \lim_{n,m\to\infty} \|f^n - f^m\|\,d\lambda_x = 0.$$

This proves that the sequence of λ_x–simple functions $f_x^n(y) = f^n(x,y)$ $(n = 1,2,\ldots)$ determines the section $f_x(y) = f(x,y)$ of the function f almost at all $x \in X$. Consequently,

$$w_n(x) = \int f^n\,d\lambda_x \to \int f\,d\lambda_x = w(x)$$

almost at all x. As all the functions $w_n(x)$ are majorized in the norm by μ–integrable function $2\int \|f\|\,d\lambda_x$ then by Theorem 3.30

$$\lim \int w_n\,d\mu = \int w\,d\mu = \int d\mu \int f\,d\lambda_x,$$

i.e.

$$\lim \int d\mu \int f^n \, d\lambda_x = \int d\mu \int f \, d\lambda_x.$$

But formula (3.106) is valid for the ν–integrable simple functions $f^n(x, y)$ and

$$\lim \int f^n \, d\nu = \int f \, d\nu.$$

Consequently, formula (3.106) is valid for any ν–integrable function $f(x, y)$ what terminates the proof of the theorem. ◁

R e m a r k. After substituting in formula (3.106) the function $f(x, y)$ by the product $f(x, y)\mathbf{1}_C(x, y)$, $C \in \mathcal{C} = \mathcal{A} \times \mathcal{B}$ we make sure that Fubini theorem is also valid for the integrals over any set $C \in \mathcal{C}$:

$$\int_C f \, d\nu = \int \mu(dx) \int_{C_x} f(x, y)\lambda_x(dy), \qquad (3.107)$$

where C_x has the same value as in formula (3.105).

3.8.3. Multiple and Iterated Integrals

The integrals over the spaces product are called *multiple integrals*. In particular, the integral over the product of two spaces in the left–hand side of formula (3.106) represents *a double integral*.

The integral in the right–hand side of formula (3.106) represents the result of sequential performance of two integrations: at first the integration over the space Y at the fixed value of x and next the integration of the received function of the variable x over the space Y. Such integrals are called *iterated integrals*.

Fubini theorem establishes the equality of a multiple integral to an iterated one.

In the special case when the measure λ_x is independent of x, $\lambda_x(B) = \lambda(B)$ at all $B \in \mathcal{B}$ and at all x the measures μ and λ in (3.105) may be interchanged and formula (3.105) takes a symmetric form:

$$\nu(C) = \int \mu(dx) \int \mathbf{1}_C(x, y)\lambda(dy) = \int \lambda(dy) \int \mathbf{1}_C(x, y)\mu(dx). \quad (3.108)$$

As a result of this Fubini theorem is also symmetric relative to the measures λ and μ and formula (3.106) takes the form

$$\int f \, d\nu = \int \mu(dx) \int f(x, y)\lambda(dy) = \int \lambda(dy) \int f(x, y)\mu(dx). \quad (3.109)$$

Then it follows that in the considered case a multiple integral does not depend on the integration order. The measure ν in this case is called *a product* of the measures λ and μ, $\nu = \lambda \times \mu$.

R e m a r k. Fubini theorem may be implemented to Bochner integrals and, in particular, to Lebesgue integrals. But on the basis of Theorems 3.23 and 3.24 it is also applicable to the proper and absolutely convergent improper Riemann integrals. In this case it follows from Fubini theorem as a special case the known theorem about the equality of a multiple integral to anyone of iterated integrals.

E x a m p l e 3.16. Let consider a double integral

$$I = \iint\limits_{x^2+y^2<c} (c - x^2 - y^2)\, dx\, dy.$$

It is evident that this integral exists as the integrand is bounded and the Lebesgue measure of the integration region is finite. Using Fubini theorem we find $I = \pi c^2/2$.

E x a m p l e 3.17. Similarly using Fubini theorem we find

$$\iint\limits_{[0,1]^2} \frac{xy}{x^2+y^2}\, dx\, dy = 2\ln 2.$$

Here the integral exists as $xy/(x^2 + y^2) < 1/2$ and the Lebesgue measure of the integration region is finite, and consequently, a multiple integral is equal to anyone from iterated integrals.

3.8.4. Measures in Finite Products of Spaces

In exactly the same way the measures in any finite products of the spaces are determined. Suppose that in the space $(X_1, \mathcal{A}_1)$ the measure $\lambda_1(A)$ is given, and in the spaces $(X_k, \mathcal{A}_k)$ $(k = 2, \ldots, n)$ the families of the measures $\lambda_k(A; x_1, \ldots, x_{k-1})$, $x_l \in X_l$ $(l = 1, \ldots, n-1)$ are given, though at each given $A \in \mathcal{A}_k$ $\lambda_k(A; x_1, \ldots, x_{k-1})$ represents the μ_{k-1}-integrable function $x_1, \ldots, x_{k-1}$,

$$\mu_1(A) = \lambda_1(A), \qquad A \in \mathcal{A}_1,$$

$$\mu_k(A_1 \times A_2) = \int\limits_{A_1} \lambda_k(A_2; x_1, \ldots, x_{k-1})\mu_{k-1}(dx_1 \times \cdots \times dx_{k-1}),$$

$$A_1 \in \mathcal{A}_1 \times \cdots \times \mathcal{A}_{k-1}, \quad A_2 \in \mathcal{A}_k \quad (k = 2, \ldots, n). \tag{3.110}$$

This formula determines each measure μ_k on the rectangular sets of the product of two spaces $(X_1 \times \cdots \times X_{k-1}, A_1 \times \cdots \times A_{k-1})$ and (X_k, A_k). Consequently, by Theorem 3.42 it is σ–additive on the semi–algebra of the measurable triangles and by Theorem 2.38 is uniquely extended on the σ–algebra $A_1 \times \cdots \times A_k$. In particular, at $k = n$ the measure $\mu = \mu_n$ is σ–additive and is uniquely determined on the σ–algebra $A_1 \times \cdots \times A_n$ of the product of the spaces $X_1 \times \ldots \times X_n$.

Now we notice that after expressing the measure μ_{k-1} in (3.110) by the same formula (3.110) in terms of the measures μ_{k-2} and λ_{k-1} we may apply to integral in (3.110) Fubini theorem in accordance of which the function $\lambda_k(A_2; x_1, \ldots, x_{k-1})$ represents the λ_{k-1}–integrable function x_{k-1} $(k = 2, \ldots, n)$ at almost all values $x_1, \ldots, x_{k-2}$ relative to the measure μ_{k-2}.

By Theorem 3.43 the measure $\mu = \mu_n$ is determined on any set $C \in A_1 \times \cdots \times A_n$ by formula

$$\mu(C) = \int \mu_{n-1}(dx^{(n-1)}) \int 1_C(x_1, \ldots, x_n) \lambda_n(dx_n; x^{(n-1)})$$

$$= \int \lambda_1(dx_1) \int \lambda_2(dx_2; x_1) \ldots \int 1_C(x_1, \ldots, x_n) \lambda_n(dx_n; x^{(n-1)}),$$

$$\tag{3.111}$$

where for brevity the element $\{x_1, \ldots, x_{n-1}\}$ of the product of the spaces $X_1 \times \cdots \times X_{n-1}$ is denoted in terms of $x^{(n-1)}$.

The function $\lambda_k(A; x_1, \ldots, x_{k-1})$ $(k = 2, \ldots, n)$ represents a measure with the values in a B–space of μ_{k-1}–integrable functions $L_1(X_1 \times \cdots \times X_{k-1}, A_1 \times \cdots \times A_{k-1}, \mu_{k-1})$.

Fubini theorem is extended by the induction on the integrals over any finite product of the spaces.

In the special case when the measures λ_k do not depend on x_1, $\ldots, x_{k-1}$ $(k = 2, \ldots, n)$ the measure μ determined by formula (3.111) is called *a product of the measures* $\lambda_1, \ldots, \lambda_n$. In this case from formula (3.109) by the induction we conclude that a multiple integral over any finite product of the measures is equal to each of the iterated integrals and is independent of the order of the integration.

3.8.5. Measures in Infinite Products of Spaces

Let us pass now to the measures in infinite products of the spaces. Let (X^T, A^T) be an infinite product of the measurable spaces (X_t, A_t)

(Subsection 2.1.9):

$$X^T = \prod_{t \in T} X_t, \qquad A^T = \prod_{t \in T} A_t.$$

Suppose that at all n, $t_1, \ldots, t_n \in T$ in the space (X_{t_1}, A_{t_1}) the finite nonnegative measure $\lambda_{t_1}(A)$ is determined and in the space (X_{t_k}, A_{t_k}) $(k = 2, \ldots, n)$ the families of the finite nonnegative measures $\lambda_{t_k; t_1, \ldots, t_{k-1}}(A; x_1, \ldots, x_{k-1})$, $x_l \in X_{t_l}$, $(l = 1, \ldots, n-1)$ are determined, though similarly as in Subsection 3.8.4 each function $\lambda_{t_k; t_1, \ldots, t_{k-1}}$ is $\mu_{t_1, \ldots, t_{k-1}}$–integrable at any $A \in A_{t_k}$,

$$\mu_{t_1}(A) = \lambda_{t_1}(A), \qquad A \in A_{t_1},$$

$$\mu_{t_1, \ldots, t_k}(A_1 \times A_2) = \int_{A_1} \lambda_{t_k; t_1, \ldots, t_{k-1}}(A_2; x_1, \ldots, x_{k-1})$$

$$\times \mu_{t_1, \ldots, t_{k-1}}(dx_1 \times \cdots \times dx_{k-1}),$$

$$A_1 \in A_{t_1} \times \cdots \times A_{t_{k-1}}, \quad A_2 \in A_{t_k} \qquad (k = 2, \ldots, n). \qquad (3.112)$$

As it was proved in Subsection 3.8.4 all the measures $\mu_{t_1, \ldots, t_k}$ are σ–additive on the correspondent σ–algebras $A_{t_1} \times \cdots \times A_{t_k}$. The aggregate of all the measures $\mu_{t_1, \ldots, t_n}$ at all n, $t_1, \ldots, t_n \in T$ determines the measure μ on the semi–algebra of the rectangles of the space X^T:

$$\mu(A_1 \times \cdots \times A_n) = \int_{A_1} \cdots \int_{A_n} \lambda_{t_1}(dx_1) \lambda_{t_2; t_1}(dx_2; x_1) \times$$

$$\cdots \times \lambda_{t_n; t_1, \ldots, t_{n-1}}(dx_n; x_1, \ldots, x_{n-1}), \quad A_k \in A_{t_k}, \qquad (3.113)$$

and this measure is σ–additive on each of σ–algebras $A_{t_1} \times \cdots \times A_{t_n}$, $t_1, \ldots, t_n \in T$. In order to extend the measure μ to the σ–algebra A^T of the space X^T it is necessary to prove its σ–additivity on the semi–algebra C of measurable rectangles of the space X^T. For this purpose we use Theorem 2.13 in accordance of which any additive function of a set continuous in zero is σ–additive.

Theorem 3.45. *The measure μ determined by formula (3.113) on the semi–algebra C of the measurable rectangles of the space (X^T, A^T), is σ–additive on C.*

▷ First of all we notice that μ is additive on C. Really, the union of the rectangles represents a rectangle if and only if their bases lay on one and the same finite product of the spaces. But then the additivity of μ follows from the additivity of the integrals. We set on the semi–algebra of the rectangles $\mathcal{C}$ the measures

$$\nu_{kn}(A_k \times \cdots \times A_n; x_1, \ldots, x_{k-1})$$

$$= \int_{A_k} \cdots \int_{A_n} \lambda_k(dx_k; x_1, \ldots, x_{k-1}) \ldots \lambda_n(dx_n; x_1, \ldots, x_{n-1}),$$

$$A_k \in \mathcal{A}_{t_k} \qquad (k = 2, \ldots, n; \ \ n = 2, 3, \ldots), \qquad (3.114)$$

where it is assumed $\lambda_k = \lambda_{t_k; t_1, \ldots, t_{k-1}}$. It is clear that $\nu_{1n} = \mu$, $\nu_{nn} = \lambda_n$ and

$$\nu_{kn}(A_k \times \cdots \times A_n; x_1, \ldots, x_{k-1}) = \int_{A_k} \nu_{k+1,n}(A_{k+1} \times \cdots \times A_n; x_1, \ldots, x_k)$$

$$\times \lambda_k(dx_k; x_1, \ldots, x_{k-1}). \qquad (3.115)$$

According to the proved in Subsection 3.8.4 all the measures ν_{kn} are σ–additive on the correspondent σ–algebras $\mathcal{A}_{t_k} \times \cdots \times \mathcal{A}_{t_n}$ ($k = 2, \ldots, n$; $n = 2, 3, \ldots$).

Let us take now an arbitrary monotone decreasing sequence of the rectangles $\{R_n\}$ with empty intersection and prove that $\lim \mu(R_n) = 0$. If the bases of all the rectangles R_n lay in one and the same finite product of the space $X_{t_1} \times \cdots \times X_{t_N}$ then $\lim \mu(R_n) = 0$ as the continuity of μ follows by Theorem 2.11 from its σ–additivity on $\mathcal{A}_{t_1} \times \cdots \times \mathcal{A}_{t_N}$. Otherwise, the bases of all the rectangles R_n lay on some countable product of the spaces X^S, $S = \{t_k\}$. Suppose that the basis of the rectangle R_n lays in the product of the spaces $X_{t_1} \times \cdots \times X_{t_{r_n}}$. It is clear that the numbers r_n form a nondecreasing sequence and $r_n \to \infty$ at $n \to \infty$. We denote the sides of the rectangle R_n in the correspondent spaces X_{t_k} in terms of $A_{n1}, \ldots, A_{nr_n}$ and put $B_n^{(k)} = A_{n,k+1} \times \cdots \times A_{nr_n}$. Then by formula (3.115) we find

$$\mu(R_n) = \nu_{1r_n}(R_n) = \int 1_{A_{n1}}(x_1) \nu_{2r_n}(B_n^{(1)}; x_1) \lambda_1(dx_1), \qquad (3.116)$$

$$\nu_{kr_n}(A_{nk} \times B_n^{(k)}; x_1, \ldots, x_{k-1})$$

$$= \int 1_{A_{nk}}(x_k) \nu_{k+1,r_n}(B_n^{(k)}; x_1, \ldots, x_{k+1}) \lambda_k(dx_k; x_1, \ldots, x_{k-1})$$

$$(k = 2, \ldots, r_n - 1; \quad n = 1, 2, \ldots). \tag{3.117}$$

From monotone decrease of $\{R_n\}$ follows $A_{n+1,k} \subset A_{nk}$ ($k = 1, \ldots, r_n$; $n = 1, 2, \ldots$). The sequences of the integrands in (3.116) and (3.117) monotonously decrease while n increases as $\mathbf{1}_{A_{nk}}(x) \le \mathbf{1}_{A_{n+1,k}}(x_k)$ and

$$\nu_{k+1,r_{n+1}}(B_{n+1}^{(k)}; x_1, \ldots, x_k) \le \nu_{k+1,r_{n+1}}(B_n^{(k)} \times X_{t_{r_n+1}} \times$$

$$\cdots \times X_{t_{r_n+1}}; x_1, \ldots, x_k) = \nu_{k+1,r_n}(B_n^{(k)}; x_1, \ldots, x_k)$$

are lower bounded by zero. Therefore at every point $\{x_1, \ldots, x_k\}$ there exists the limit

$$f_k(x_1, \ldots, x_k) = \lim_{n \to \infty} \mathbf{1}_{A_{nk}}(x_k) \nu_{k+1,r_n}(B_n^{(k)}; x_1, \ldots, x_k).$$

Thus the sequences of the integrands in (3.116) and (3.117) converge at all $x_k, \ldots, x_{k-1}$ and are lower and upper bounded by λ_k–integrable functions 0 and $\mathbf{1}_{A_{1k}} \nu_{k+1,r_1}(B_1^{(k)}; x_1, \ldots, x_k)$. Consequently, according to Theorem 3.29 in (3.116) and (3.117) we may pass to the limit at $n \to \infty$ under the integral sign. As a result we obtain

$$\lim_{n \to \infty} \mu(R_n) = \int f_1(x_1) \lambda_1(dx_1), \tag{3.118}$$

$$\lim_{n \to \infty} \nu_{kr_n}(A_{nk} \times B_n^{(k)}; x_1, \ldots, x_{k-1})$$

$$= \int f_k(x_1, \ldots, x_k) \lambda_k(dx_k; x_1, \ldots, x_{k-1}). \tag{3.119}$$

If $\mu(R_n)$ does not tend to 0 at $n \to \infty$ then there is such a number $\varepsilon > 0$ that $\mu(R_n) > \varepsilon \ \forall n$. Consequently, there exists such a point $\bar{x}_1 \in X_{t_1}$ that $f_1(\bar{x}_1) > \varepsilon_1 = \varepsilon/\lambda_1(A_{11})$. It is evident that $\bar{x}_1 \in A_{n1}$ at all n as $f_1(x_1) = 0$ at $x_1 \notin A_{n1}$. From the obtained inequality by virtue of nonincreasing sequence $\{\nu_{2r_n}(B_n^{(1)}; x_1)\}$ follows $\nu_{2r_n}(B_n^{(1)}; \bar{x}_1) > \varepsilon_1$ at all n. From this fact and (3.117) at $k = 2$, $x_1 = \bar{x}_1$ accounting that $B_n^{(1)} = A_{n2} \times B_n^{(2)}$ in the same way we come to the conclusion that there exists such a point $\bar{x}_2 \in X_{t_2}$ that $\bar{x}_2 \in A_{n2}$ and $\nu_{3r_n}(B_n^{(2)}; \bar{x}_1, \bar{x}_2) > \varepsilon_2$ $= \varepsilon_1/\lambda_2(A_{12}; \bar{x}_1)$ at all n. Proceeding in such a way we find at any p such a point $\bar{x}_p \in X_{t_p}$ that $\bar{x}_p \in A_{np}$ and $\nu_{p+1,r_n}(B_n^{(p)}; \bar{x}_1, \ldots, \bar{x}_p) > \varepsilon_p$ $= \varepsilon_{p-1}/\lambda_p(A_{1p}; \bar{x}_1, \ldots, \bar{x}_{p-1})$ at all n. Here the point $x^{(p)} = \{\bar{x}_1, \ldots, \bar{x}_p\} \in X_{t_1} \times \cdots \times X_{t_p}$ will belong to all the rectangles R_n at $r_n < p$.

In the limit we shall obtain the point $\{\bar{x}_k\}$ in the countable product of the spaces X^S, $S = \{t_k\}$ which belongs to all the rectangles R_n what contradicts the supposition that the intersection of the rectangles R_n is empty. This contradiction proves the continuity of the measure μ in zero and along with this its σ–additivity on the semi–algebra $\mathcal{C}$ of the measurable rectangles of the space X^T. ◁

By Theorem 2.38 the measure μ has the unique extension on the minimal complete σ–algebra $\mathcal{A}_\mu^T$ which contains all the measurable rectangles of the space $(X^T, \mathcal{A}^T)$.

R e m a r k. Notice that in passing we proved the σ–additivity of all the measures ν_{kr_n} determined by formula (3.117) and the possibility of their unique extension on the σ–algebra $\mathcal{A}^T$ at all $x_1, \ldots, x_{k-1}$ ($k = 2, \ldots, r_n - 1; n = 1, 2, \ldots$).

E x a m p l e 3.18. The Wiener measure μ_W introduced in Example 2.15 is determined by formula (3.113) at

$$\lambda_{t_1}(A) = \frac{1}{\sqrt{2\pi t_1}} \int_A \exp\{-\tfrac{x_1^2}{2t_1}\}\, dx_1,$$

$$\lambda_{t_k; t_1, \ldots, t_{k-1}}(A; x_1, \ldots, x_{k-1}) = \frac{1}{\sqrt{2\pi(t_k - t_{k-1})}} \int_A \exp\{-\tfrac{(x_k - x_{k-1})^2}{2(t_k - t_{k-1})}\}\, dx_k$$

$$(k = 2, \ldots, n). \tag{3.120}$$

According to Theorem 3.45 μ_W is σ–additive on the semi–algebra $\mathcal{C}$ of the measurable rectangles of the space $(X^T, \mathcal{A}^T)$ at $X_t = R$, $\mathcal{A}_t = \mathcal{B}$, $T = [0, \infty)$. Consequently, it may be uniquely extended on the σ–algebra $\mathcal{A}^T$ and on its supplement $\mathcal{A}_W^T$ relative to μ_W. Formula (3.114) which determines the measure ν_{2n} in a given case gives

$$\nu_{2n}(A_2 \times \cdots \times A_n; x_1) = \frac{1}{\sqrt{(2\pi)^{n-1}(t_2 - t_1)\ldots(t_n - t_{n-1})}}$$

$$\times \int_{A_2} \ldots \int_{A_n} \exp\{-\tfrac{1}{2} \sum_{k=2}^n \tfrac{(x_k - x_{k-1})^2}{t_k - t_{k-1}}\}\, dx_2 \ldots dx_n. \tag{3.121}$$

This measure according to the last remark before this example is also σ–additive and is uniquely extended on the σ–algebra $\mathcal{A}^{T_1}$, $T_1 = (t_1, \infty)$ at any $x_1 \in X_{t_1}$. We denote its extension in terms of $\nu_{t_1}(C; x_1)$. Then by (3.105) as a result of $X^{[t_1, \infty)} = X_{t_1} \times X^{(t_1, \infty)}$ the value of the measure μ_W on any set $C \in \mathcal{A}^{[t_1, \infty)}$ is expressed by the formula

$$\mu_W(C) = \frac{1}{\sqrt{2\pi t_1}} \int \exp\{-\tfrac{x^2}{2t_1}\}\, dx \int 1_C(x, y)\nu_{t_1}(dy; x) \tag{3.122}$$

at any fixed t_1. At all possible values $t_1 \in [0, \infty)$ formula (3.122) determines the measure μ_W on all the sets $C \in \mathcal{A}^T$.

E x a m p l e 3.19. Along with the Wiener measure μ_W the measure μ'_W determined of the rectangular sets of the space $(X^T, \mathcal{A}^T)$ by formula (3.123) is often used

$$\mu'_W(A_1 \times \cdots \times A_n) = \frac{1}{\sqrt{(2\pi)^{n-1}(t_2 - t_1) \ldots (t_n - t_{n-1})}}$$

$$\times \int_{A_1} \cdots \int_{A_n} \exp\left\{-\frac{1}{2} \sum_{k=2}^{n} \frac{(x_k - x_{k-1})^2}{t_k - t_1}\right\} dx_1 \ldots dx_n. \qquad (3.123)$$

We shall call this measure *a modified Wiener measure* in order to avoid confusion which may occur as the measure μ'_W is often called the Wiener measure. It is evident that the measure μ'_W differs from μ_W by the fact that instead of the measure λ_{t_1} determined by the first formula of (3.120) the Lebesgue measure on the real axis $X_{t_1} = R$ is taken. Therefore formula (3.122) is replaced by formula

$$\mu'_W(C) = \int dx \int \mathbf{1}_C(x, y) \nu_{t_1}(dy; x). \qquad (3.124)$$

The measures in finite products of the spaces are of great importance in modern probability theory. For further study of measure theory in product spaces and its some probability applications see (Balakrishnan 1976, Dunford and Schwarz 1958, Mc Shane 1974, Parthasarathy 1980, Pugachev and Sinitsyn 1987, Xia Dao–Xing 1972).

P r o b l e m s

3.8.1. Perform the calculations of the integrals

$$\iint_{[0,1]^2} \frac{xy \, dx \, dy}{(x^2 + y^2)^{3/2}} , \qquad \iint_{[0,1]^2} \frac{x - y}{(x+y)^{3/2}} \, dx \, dy.$$

Explain why in the second case the iterated integrals do not coincide.

3.8.2. Prove the σ–finiteness of the modified Wiener measure μ'_W of Example 3.19.

3.8.3. Prove that the modified Wiener measure μ'_W may be considered completely concentrated on the subspace of the continuous functions C of the space X^T.

I n s t r u c t i o n. Substitute the measure $\lambda_t(A)$ on R of Example 3.18 by the Lebesgue measure on the interval $[n, n+1)$, prove similarly as in Problem 2.3.3 that the measure μ'_W of the set of the functions from X^T which are not continuous on the set of binary rational numbers of any interval $[0, a]$ is equal to zero, and account that the space X^T may be represented as the union of countable set of pairwise nonintersecting spaces obtained from X^T by the replacement of the real axis $X_t = R_t$ by the intervals $[n, n+1)$ at any fixed t. After this terminate the problem similarly as Problem 2.3.3.

   3.8.4. Prove that the measure μ'_W of the set of the functions differentiable at a given point t is equal to zero. Accounting the result of Problem 3.8.3 make a conclusion that almost all the functions of the space C relative to the measure μ'_W are not differentiable at any point.

   I n s t r u c t i o n. The same as in Problem 3.8.3.

   3.8.5. Calculate integrals over the modified Wiener measure μ'_W with respect to the set of the continuous functions satisfying the condition $x(a) \in [\alpha, \beta]$ of the functions of Problems 3.3.2, 3.3.4 at $t_1 = a$ and Problems 3.5.1–3.5.4.

   I n s t r u c t i o n. Apply Fubini theorem after determining the measure μ'_W by formula (3.124) and use for calculating the integrals over the measure ν_{t_1} the formulae obtained in Problems 3.3.2, 3.3.4, 3.5.1–3.5.4.

   3.8.6. Let $(X^T, \mathcal{A}^T)$ be an infinite product of the measurable spaces

$$X^T = \prod_{t \in T} X_t, \quad \mathcal{A}^T = \prod_{t \in T} \mathcal{A}_t, \quad X_t = R^m, \quad \mathcal{A}_t = \mathcal{B}^m, \quad T = [0, \infty),$$

where $\mathcal{B}^m$ is the σ–algebra of Borel sets of the space R^m. It is evident that X^T represents a space of all m–dimensional vector functions with the domain T. We shall determine on the semi–algebra of the measurable rectangles of the space $(X^T, \mathcal{A}^T)$ the measure by formula

$$\mu(A_1 \times \cdots \times A_n) = \frac{1}{(2\pi)^m \sqrt{|K|}} \int_{A_1} \cdots \int_{A_n} \exp\{-x^T K^{-1} x/2\} \, dx_1 \, dx_2$$

$$\times \lambda_{t_3; t_1, t_2}(dx_3; x_1, x_2) \ldots \lambda_{t_n; t_1, \ldots, t_{n-1}}(dx_n; x_1, \ldots, x_n), \qquad (3.125)$$

where

$$x = \begin{bmatrix} x_1 \\ x_2 \end{bmatrix}, \quad K = \begin{bmatrix} K(t_1, t_1) & K(t_1, t_2) \\ K(t_2, t_1) & K(t_2, t_2) \end{bmatrix},$$

$K(s_1, s_2)$ is a continuous matrix function of the dimension $m \times m$ satisfying the following conditions:

   (i) $K(s_2, s_1) = K(s_1, s_2)^T$;

   (ii) $\sum\limits_{i,j=1}^{N} u_i^T K(s_i, s_j) u_j > 0$ at any N, $s_1, \ldots, s_N$ and m–dimensional vectors $u_1, \ldots, u_N$;

   (iii) $\operatorname{tr} \sigma(t_1, t_2) < B|t_1 - t_2|^\gamma$ at some $B, \gamma > 0$, where   (3.126)

$$\sigma(t_1, t_2) = K(t_1, t_1) - K(t_1, t_2) - K(t_2, t_1) + K(t_2, t_2),$$

$\lambda_{t_k;t_1,\ldots,t_{k-1}}(B;x_1,\ldots,x_{k-1})$ $(k=3,\ldots,n)$ are finite nonnegative measures in the correspondent spaces $(X_{t_k},\mathcal{A}_{t_k})$ representing at each $B \in \mathcal{A}_{t_k}$ $(\mathcal{A}_{t_1} \times \cdots \times \mathcal{A}_{t_{k-1}},B)$–measurable functions $x_1,\ldots,x_{k-1}$. In accordance with Theorem 3.45 the measure μ is σ–additive on the semi–algebra $\mathcal{C}$ of rectangles of the space X^T. By Theorem 2.38 it may be uniquely extended on the σ–algebra $\mathcal{A}_\mu^T$ received by means of the supplement $\mathcal{A}^T$ relative to μ.

Prove that the measure μ (3.125) is finite and the statements of Problems 2.3.3 and 2.3.5 are valid for it.

I n s t r u c t i o n. Apply the formula[a]

$$\mu(\{x(t):|x_k(t_2)-x_k(t_1)|\geq\eta\}$$

$$=\frac{1}{2\pi\sqrt{|K|}}\underset{|x_{2k}-x_{1k}|>\eta}{\int\int}\exp\{-x^T K^{-1}x/2\}\,dx_1\,dx_2$$

$$=1-2\Phi(\eta/\sqrt{\sigma_{kk}(t_1,t_2)})\quad(k=1,\ldots,m),$$

where $\sigma_{kk}(t_1,t_2)$ are diagonal elements of the matrix $\sigma(t_1,t_2)$ and inequality (2.63) and take into account that

$$|x(t_2)-x(t_1)|=\sqrt{\sum_{k=1}^m[x_k(t_2)-x_k(t_1)]^2}\leq\sum_{k=1}^m|x_k(t_2)-x_k(t_1)|,$$

as a result

$$\{x(t):|x(t_2)-x(t_1)|\geq\varepsilon\}\subset\bigcup_{k=1}^m\{x(t):|x_k(t_2)-x_k(t_1)|\geq\varepsilon/m\},$$

and take α in the interval $(0,\gamma/2)$. Then all further calculations of Problem 2.3.3 will remain without changes.

3.8.7. Prove that the statements of Problems 2.3.3 and 2.3.5 are also valid for the σ–finite measure determined on the measurable rectangles of the space $(X^T,\mathcal{A}^T)$ by formula

$$\mu(A_1\times\cdots\times A_n)=\frac{1}{\sqrt{(2\pi)^m|\sigma|}}\int_{A_1}\cdots\int_{A_n}\exp\{-(x_2^T-x_1^T)$$

$$\times\sigma^{-1}(x_2-x_1)\,dx_1\,dx_2\lambda_{t_3;t_1,t_2}(dx_3;x_1,x_2)$$

---

[a] $2m$–multiple integral in this formula analogously as $4m$–multiple integral in formula of Problem 3.8.9 is reduced by linear change of the variables to integral (2.61).

$$\ldots \lambda_{t_n;t_1,\ldots,t_{n-1}}(dx_n; x_1,\ldots, x_{n-1}), \qquad (3.127)$$

where $\sigma = \sigma(t_1, t_2)$ is positively determined continuous matrix function of the dimension $m \times m$ satisfying condition (3.126), and the measures $\lambda_{t_k;t_1,\ldots,t_{k-1}}$ are the same as in Problem 3.8.6 (the σ–finiteness of the measure μ will be proved).

3.8.8. Prove that the integral over the measure μ of Problem 3.8.6 of the function

$$f(x) = \sum_{h,l=1}^{n} \varphi_{hl}(t_1,\ldots,t_n)x(t_h)x(t_l)^T,$$

where φ_{hl} $(h, l = 1,\ldots, n)$ are continuous matrix functions of the dimension $m \times m$ with the values in a separable B–space extended to the whole space C is determined by formula

$$\int f(x)\mu(dx) = \sum_{h,l=1}^{n} \varphi_{hl}(t_1,\ldots,t_n)K(t_h, t_l).$$

I n s t r u c t i o n. Differentiate the formula given in Problem 3.3.4 with respect to the components of the vector η and after this put $\eta = 0$.

3.8.9. Under the conditions of Problem 3.8.6 determine the measure

$$\mu(A_1 \times \cdots \times A_n) = \frac{1}{(2\pi)^{2m}\sqrt{|K|}} \int_{A_1} \cdots \int_{A_n} \exp\{-x^T K^{-1}x/2\}\, dx_1$$

$$\ldots dx_4 \lambda_{t_5;t_1,\ldots,t_4}(dx_5; x_1,\ldots, x_4)\ldots \lambda_{t_n;t_1,\ldots,t_{n-1}}(dx_n; x_1,\ldots, x_{n-1}),$$
$$(3.128)$$

where

$$x = \begin{bmatrix} x_1 \\ x_2 \\ x_3 \\ x_4 \end{bmatrix}, \quad K = \begin{bmatrix} K(t_1,t_1) & K(t_1,t_2) & K(t_1,t_3) & K(t_1,t_4) \\ K(t_2,t_1) & K(t_2,t_2) & K(t_2,t_3) & K(t_2,t_4) \\ K(t_3,t_1) & K(t_3,t_2) & K(t_3,t_3) & K(t_3,t_4) \\ K(t_4,t_1) & K(t_4,t_2) & K(t_4,t_3) & K(t_4,t_4) \end{bmatrix},$$

$K(t_1, t_2)$ is a continuous together with its derivatives till the second order inclusively the matrix function of the dimension $m \times m$ possessing the first two properties from Problem 3.8.6 and the additional property

$$\operatorname{tr} \sigma(t_1, t_2, t_3, t_4) < B\tau^{\gamma}$$

at some $B > 0, \gamma > 2, \tau = \max_{i,j} |t_i - t_j|$,

$$\sigma(t_1, t_2, t_3, t_4) = K(t_1, t_1) + K(t_2, t_2) + K(t_3, t_3) + K(t_4, t_4)$$

$$-K(t_1, t_2) - K(t_2, t_1) - K(t_1, t_3) - K(t_3, t_1) + K(t_1, t_4) + K(t_4, t_1)$$
$$+K(t_2, t_3) + K(t_3, t_2) - K(t_2, t_4) - K(t_4, t_2) - K(t_3, t_4) - K(t_4, t_3),$$

and $\lambda_{t_k; t_1, \ldots, t_{k-1}}$ $(k = 5, \ldots, n)$ are the same measures as in Problem 3.8.6. Prove that almost all the functions of the space X^T relative to the measure μ are continuous and have continuous first derivatives.

    I n s t r u c t i o n. Use formula

$$\frac{1}{(2\pi)^{2m}\sqrt{|K|}} \underset{|x_{4k}-x_{3k}-x_{2k}+x_{1k}|\geq\eta}{\int\int\int\int} \exp\{-x^T K^{-1} x/2\}\, dx_1\, dx_2\, dx_3\, dx_4$$

$$= 1 - 2\Phi(\eta/\sqrt{\sigma_{kk}(t_1, t_2, t_3, t_4)}),$$

where $\sigma_{kk}(t_1, t_2, t_3, t_4)$ are the diagonal elements of the matrix $\sigma(t_1, t_2, t_3, t_4)$ use inequality (2.63), take α in the interval $(1, \gamma/2)$ and repeat in the same sense the calculations of Problem 2.3.3 for the variables

$$y(k2^{-p}) = [x((k+1)2^{-p}) - x(k2^{-p})]2^p.$$

    The obtained result gives the opportunity to transfer the measure μ on the σ–algebra $\mathcal{A}^T C^1 = \{B : B = AC^1, A \in \mathcal{A}^T\}$ in the space of the functions continuous together with their first derivatives putting $\mu(B) = \mu(A)$, $B \in \mathcal{A}^T C^1$.

    3.8.10. Prove that the integral over the measure μ of Problem 3.8.9 of the function

$$f(x) = \sum_{h,l=1}^{n} \varphi_{hl}(t_1, \ldots, t_n) x(t_h) x(t_l)^T$$

$$+ \sum_{h,l=1}^{n} \psi_{hl}(t_1, \ldots, t_n) x'(t_h) x(t_l)^T$$

$$+ \sum_{h,l=1}^{n} \omega_{hl}(t_1, \ldots, t_n) x'(t_h) x'(t_l)^T,$$

extended over the whole space C^1 where $\varphi_{hl}, \psi_{hl}, \omega_{hl}$ are continuous matrix functions with the values in a separable B–space is determined by the formula

$$\int f(x)\mu(dx) = \sum_{h,l=1}^{n} \varphi_{hl}(t_1, \ldots, t_n) K(t_h, t_l)$$

$$+ \sum_{h,l=1}^{n} \psi_{hl}(t_1, \ldots, t_n) \frac{\partial K(t_h, t_l)}{\partial t_h} + \sum_{h,l=1}^{n} \omega_{hl}(t_1, \ldots, t_n) L(t_h, t_l),$$

where

$$\frac{\partial K(t_h, t_h)}{\partial t_h} = \left[\frac{\partial K(t_h, t_l)}{\partial t_h}\right]_{t_l=t_h},$$

$$L(t, t') = \frac{\partial^2 K(t, t')}{\partial t\, \partial t'}, \quad L(t, t) = \left[\frac{\partial^2 K(t, t')}{\partial t\, \partial t'}\right]_{t'=t}.$$

    3.8.11. Solve Problems 3.8.6–3.8.10 for the case of r–dimensional vector argument t (after modifying Problem 3.8.10 as required).

CHAPTER 4
TOPOLOGICAL SPACES

This chapter contains the elements of topology and theory of topological spaces. In Section 4.1 general definitions are given of a topology and topological space, of basis and subbasis of a topology. Separability and countability axioms are introduced in Section 4.2. The general method for assigning a topology in a space is considered in Section 4.3. The notions of compact sets and spaces are defined and their general properties are studied in Sections 4.4 and 4.5. Theorems on compact sets in metric spaces are proved. Then in Section 4.6 the notion of a topological linear space is introduced and the general method for assigning a topology in a linear space is given. The notion of a weak topology in a topological linear space is introduced and its relation to the strong topology is considered in Section 4.7.

4.1. Basic Notions of Topology

4.1.1. Convergence and Continuity in Terms of Vicinities

The definitions of the convergence of a sequence and the continuity of a function in a metric space given in Subsection 1.2 may be formulated on the basis of the notion of the point vicinity without using explicitly the metric.

It is clear that the set $\{x' : d(x, x') < r\}$ represents a spherical vicinity of the point x. But any vicinity of the point x contains this point together with its some spherical vicinity. The condition of the convergence $d(x_n, x) < \varepsilon$ at all $n > N = N_\varepsilon$ means that any vicinity of the point x contains all the points x_n of the sequence $\{x_n\}$ beginning with someone. Thus the sequence of the points $\{x_n\}$ converges to the point x if any vicinity of the point x contains all the points x_n beginning with someone.

Similarly the condition of the continuity of the function $f(x)$ which maps the metric space (X, d) into the metric space (Y, r) at the point $x : r(f(x'), f(x)) < \varepsilon$ at all $x', d(x', x) < \delta = \delta_\varepsilon$ means that the inverse image of any vicinity of the point $y = f(x)$ (which also contains together with the point y its some spherical vicinity $\{y' : r(y', y) < \varepsilon\}$)

contains some vicinity of the point x (in a given case $\{x' \ : \ d(x', \, x) < \delta\}$) (Fig.4.1).

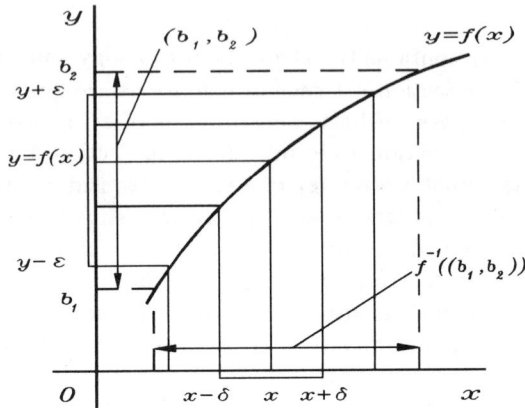

Fig.4.1

We see that the main role in the formation of the notions of the convergence (of a sequence) and the continuity (of a function) in a metric space plays the notion of the point vicinity. The metric plays only an auxiliary role as the vicinities are determined by means of the metric. Naturally, a question arises: is it possible to determine directly the vicinities of the points of the space without introducing into this space the metric. Especially it is expedient because there is no need to introduce the metric in applications. We shall see how to do it.

4.1.2. Topology

It is impossible to determine open sets in any space by natural way as it was done in the case of a metric space. Therefore in the general case we may arbitrarily select the definite class of the sets with understanding that only the sets of this class are open. It goes without saying that it is necessary to consider the fact that the class of open sets will possess the same properties that the class of open sets in a metric space which are determined by the theorems of Subsection 1.2.3.

The class of *open* sets in the space X may be assumed as an arbitrary chosen class of the sets τ which possesses the following properties:

(i) an empty set $\emptyset$ and the whole space X are open sets;

(ii) any union of open sets represents an open set;

(iii) an intersection of finite number of open sets represents an empty set.

A class τ of open sets of the space X is called *a topology* of this space. A space X with a given topology τ is called *a topological space* and is denoted by (X, τ). Properties (i)–(iii) of the topology are called (i)–(iii) *topology axioms*.

By virtue of the arbitrariness of the choice of the class of open sets in a given space X we may consider any number of different topological spaces. As a result we shall receive the different topological spaces.

The set of all possible topologies in a given space X similarly as the set of any class of the sets is partially ordered by the sign of the inclusion. If in the space X two topologies τ_1 and τ_2 are given and τ_1 is contained completely in τ_2 (i.e. any open set in τ_1 is an open set in τ_2) $\tau_1 \subset \tau_2$ then the topology τ_1 is *weaker* than the topology τ_2, and the topology τ_2 is stronger than the topology τ_1. The weakest topology in a given space X is the topology containing two sets – the whole space X and an empty set $\emptyset$. The strongest topology in the space X is the topology which contains all the sets of the space X. This topology is called *a discrete topology*. There are only two open sets in the weakest topology which are also closed at the same time (as the supplements of each other). In the discrete topology all the sets are simultaneously open and closed. It goes without saying that both these topologies are useless and the introduction of one of them into the space is equal to the absence of such an introduction.

It is easy to prove the following theorem.

Theorem 4.1. *An intersection of any set of topologies is a topology.*

The intersection of all the topologies which contains a given class of the sets $\mathcal{C}$ is the weakest topology containing $\mathcal{C}$. This topology is called *a topology induced by the class of the sets $\mathcal{C}$* and is often denoted by $\tau(\mathcal{C})$.

The supplement of any open set is called *a closed* set. On the basis of the duality principle the class of closed sets contains the whole space X and an empty set $\emptyset$, all the finite unions and any intersections of sets which are included in it. Thus an empty set and the whole space are simultaneously open and closed sets.

Theorem 4.2. *In any topological space (X, τ) the topology τ may be replaced by more weak topology $\tau_0 \subset \tau$ in which only an empty set $\emptyset$ and the whole space X will be simultaneously open and closed sets.*

▷ Suppose that the sets $A_\alpha \neq \emptyset, X, (\alpha \in S)$, and consequently, their supplements $\bar{A}_\alpha$ are simultaneously open and closed in the topology τ.

Let introduce a new, more weak topology τ_0 which coincides with τ
except the fact that each set A_α is open in τ_0 and its supplement $\bar{A}_\alpha$
is closed. It is evident that in the topology τ_0 only $\varnothing$ and X are
simultaneously open and closed. ◁

After choosing in a given space a topology we partition all the sets
of this space into three classes: open, closed and the other sets.

Any open set which contains a point x is called *a vicinity of the
point x*. Any open set containing a given set A is called *a vicinity of the
set A*.

4.1.3. Induced Topology

Let (X, τ) be a topological space, Y be an arbitrary set in X, Y
$\subset X$. It is evident that the class of the sets

$$\tau_Y = \{GY : G \in \tau\}$$

satisfies all the axioms of topology if we assume in addition that the
set Y is simultaneously open and closed in τ_Y. The topology τ_Y in
Y determined in such a way is called the *topology induced in Y by
the topology τ*. The set Y with induced topology τ_Y represents the
topological space (Y, τ_Y) which is called *a subspace* of the topological
space (X, τ).

R e m a r k. Notice that the sets $GY \in \tau_Y$ are open in the topology
τ (in space (X, τ)), $GY \in \tau$, if and only if the set Y is open in the
topology τ, $Y \in \tau$.

4.1.4. Internal Points, Adherent Points and Limit Points

A point x is called *an internal point* of the set A if there exists some
vicinity V_x of the point x which is completely contained in A, $V_x \subset A$.

A point x is called *an adherent point* of the set A if in any vicinity
of the point x at least only one point of the set A is contained. The
adherent point of the set A may belong or not to the set A.

A point x is called *a limit point* of the set A if in any vicinity of the
point x at least only one point of the set A different from x is contained.
The limit point of the set A may belong to the set A or not.

It is clear that any adherent point of the set is its limit point, and
any its limit point is its adherent point. The limit point of the set may
be not its adherent point, the adherent point may be neither limit nor

internal point. But any adherent point which does not belong to the set is its limit point.

Theorem 4.3. *The set A is open if and only if each of its points is an internal point.*

▷ If A is an open set then it is the vicinity of its any point, i.e. any point $x \in A$ is an internal point of A. On the contrary if the point $x \in A$ is an internal point of the set A then there exists the vicinity V_x of the point x which is completely contained in A. If all the points of the set A are internal then A represents an union of the vicinities of all the points of the set A which are contained in A,

$$A = \bigcup_{\substack{x \in A \\ V_x \subset A}} V_x \, .$$

Really, any point x of the set A belongs to its vicinity V_x which is contained in A, i.e. $x \in \bigcup V_x$ and $A \subset \bigcup V_x$. On the other hand, as all V_x in $\bigcup V_x$ belong to A then $\bigcup V_x \subset A$. Consequently, $A = \bigcup V_x$. But the union $\bigcup V_x$ of open sets is an open set. ◁

Theorem 4.4. *The set A is closed if and only if it contains all its adherent points.*

▷ If A is a closed set and x is its adherent point then x cannot belong to the open set $\bar{A}$ by virtue of Theorem 4.3. On the contrary if A contains all its adherent points then neither of them belongs to the supplement $\bar{A}$. Consequently, any point of the set $\bar{A}$ has the vicinity which does not contain the points of the set A, i.e. which is completely contained in $\bar{A}$. By Theorem 4.3 the set $\bar{A}$ is open, and A as the supplement of $\bar{A}$ is closed. ◁

Corollary. *The set A which has no limit points or which contains all its limit points is closed.*

Consider an arbitrary set A. The largest open set which is contained in A is called *an open kernel* of the set A and is denoted by A^0. The smallest closed set which is contained in A is called *a closure* of the set A and is denoted by $[A]$.

A^0 represents an union of all open sets which are contained in A and $[A]$ represents an intersection of all closed sets which contain A.

It is clear that an open kernel A^0 of the set A represents a set of all internal points of the set A, and the closure $[A]$ of the set A is the set of all adherent points of the set A.

The set $[A]\backslash A^0$ is called *a bound* of the set A. If A is an open set then by virtue of Theorem 4.3 $A^0 = A$. If A is a closed set then by

Theorem 4.4 $[A] = A$. Thus an open set does not contain any point of its bound, and a closed set completely contains its bound. All other sets contain the part of their bound.

The sets of topological space are called *separated* if none of them intersects with the closure of another. Pay attention that for the separability of two sets it is not sufficient their nonintersecting. For instance, the intervals $(0,1)$ and $[1,2]$ do not intersect but they are not separable while the intervals $(0,1)$ and $(1,2)$ are separable intervals.

A topological space X (or the set $A \subset X$) is called *coherent* if it may not be represented as the union of two separable sets.

If in the space X two topologies τ_1 and $\tau_2 \subset \tau_1$ are given then by virtue of the fact that the store of open, and consequently, the store of closed sets in τ_2 is less than in τ_1, $A^0_{\tau_2} \subset A^0_{\tau_1}$, $[A]_{\tau_1} \subset [A]_{\tau_2}$ for any set A (open kernels and the closures of the set A in the topologies τ_1 and τ_2 are indicated by the indexes τ_1 and τ_2 respectively). Thus the weaker is the topology in the space X the more "fuzzy" are the bounds of the set in X. As a result of this separable sets in the topology τ_1 may be not separabe in weaker topology τ_2 and correspondingly the coherent set (space) in topology τ_2 may be not coherent in stronger topology τ_1. Thus the notions of an open kernel, of a closure, of a bound and a coherency of a set are closely connected with the topology of the space and in the space without the topology have no sense.

4.1.5. Bases and Subbases

In order to set a topology in a given space there is no need to determine immediately all open sets. It is necessary to set only such aggregate of open sets from which all open sets may be received by the operations of any unions and finite intersections. In this connection appears a notion of the base of a topological space.

A *base of a topological space* or *a base of topology* of this space is called an aggregate of open sets by whose union of the sets may be received any open set. Thus the subclass $\mathcal{B}$ of open sets is the base of the topological space (X, τ) if any open set $G \in \tau$ represents an union of some sets $B_\alpha \in \mathcal{B}$, $G = \bigcup_\alpha B_\alpha$.

Theorem 4.5. *In order the subclass $\mathcal{B}$ of open sets be the base of a topological space it is necessary and sufficient that for any open set G and any its point $x \in G$ there should exist such a set $B_x \in \mathcal{B}$ that $x \in B_x \subset G$.*

▷ If $\mathcal{B}$ is a base of a topological space then any open set G is an union of some sets $B_\alpha \in \mathcal{B}$, $G = \bigcup B_\alpha$. Consequently, any point $x \in G$ belongs to some $B_\alpha \in \mathcal{B}$, and $x \in B_\alpha \subset G$ what proves the necessity of the condition. If for any G and $x \in G$ there exists such $B_x \in \mathcal{B}$ that $x \in B_x \subset G$, $G = \bigcup\limits_{x \in G} B_x$, i.e. any open set G is an union of some sets from $\mathcal{B}$. So, $\mathcal{B}$ is the base of a topological space what proves the sufficiency of the condition. ◁

It is easy to see that as the base of a topological space we may assume a set of all the vicinities of all the points of this space. Really, any open set G contains together with any point $x \in G$ some vicinity V_x of this point $x \in V_x \subset G$. Consequently, according to the proved theorem the aggregate of all the vicinities of all the points of the space is the base of this space.

In many cases for constructing the base of a topological space at first some auxiliary class of the sets from which we may receive a base is determined.

The set $\mathcal{C}$ of open sets whose all finite intersections form a base of a topological space is called *a subbase* of the topology of this space.

Theorem 4.6. *Any class of the sets $\mathcal{C}$ of a space X whose union of all the sets coincides with X may serve as a subbase of some topology in this space.*

▷ The class of the sets $\mathcal{C}$ will be a subbase if and only if the set of all the finite intersections of the sets from $\mathcal{C}$

$$\mathcal{B} = \left\{ B \ : \ B = \bigcap_{k=1}^{n} C_k, \quad C_1, \ldots, C_n \in \mathcal{C}, \quad n = 1, 2, \ldots \right\}$$

will be a base, i.e. when the set of all the unions of the sets from $\mathcal{B}$ will satisfy the topology axioms. Let consider the class $\mathcal{A}$ of all the unions of the sets from $\mathcal{B}$. Any set of the class $\mathcal{A}$ has the form

$$A = \bigcup_{\beta} B_\beta = \bigcup_{\beta} \bigcap_{k=1}^{n_\beta} C_{\beta k}, \quad C_{\beta k} \in \mathcal{C}.$$

Let $\{A_\alpha\}$ be a family of the sets of the class $\mathcal{A}$,

$$A_\alpha = \bigcup_{\beta} B_{\alpha\beta} = \bigcup_{\beta} \bigcap_{k=1}^{n_{\alpha\beta}} C_{\alpha\beta k}.$$

As the union

$$\bigcup_{\alpha} A_\alpha = \bigcup_{\alpha} \bigcup_{\beta} B_{\alpha\beta} = \bigcup_{\alpha,\beta} \bigcap_{k=1}^{n_{\alpha\beta}} C_{\alpha\beta k}$$

represents a set from $\mathcal{A}$ then $\mathcal{A}$ satisfies the second axiom of topology. As

$$\bigcap_{\alpha=1}^{N} \bigcup_{\beta} B_{\alpha\beta} = \bigcap_{\alpha=1}^{N} \bigcup_{\beta} \bigcap_{k=1}^{n_{\alpha\beta}} C_{\alpha\beta k} = \bigcup_{\beta} \bigcap_{\alpha=1}^{N} \bigcap_{k=1}^{n_{\alpha\beta}} C_{\alpha\beta k}$$

also represents a set of the class $\mathcal{A}$ then the class of the sets $\mathcal{A}$ safisfies the third axiom of the topology. Finally, as $X = \bigcup B_\beta \in \mathcal{A}$ and $\varnothing \in \mathcal{A}$ as the union of an empty family of the sets from $\mathcal{B}$ then the class of the sets also satisfies the first topology axiom. Thus the set $\mathcal{A}$ of all the unions of the sets from $\mathcal{B}$ represents the topology τ whose base serves $\mathcal{B}$. Consequently, the class of the sets $\mathcal{C}$ serves as the subbase of the topology τ. It is evident that the topology τ whose base serves $\mathcal{C}$ is the topology $\tau(\mathcal{C})$ induced by the subbase $\mathcal{C}$. Really, any topology containing $\mathcal{C}$ also contains all the sets

$$\bigcup_{\alpha} \bigcap_{k=1}^{n} C_{\alpha k}, \quad C_{\alpha k} \in \mathcal{C},$$

i.e. contains the topology $\mathcal{A} = \tau$. ◁

By Theorem 4.5 as the base of natural topology of a metric space generated by the metric serves the set of all spherical vicinities of all the points of this space as any open set contains its any point together with its some spherical vicinity. And most of all as the base of topology of a metric space serves also the set of the spherical vicinities of all the points of the space with the rational radii as any spherical vicinity of any point of a metric space contains a spherical vicinity of some smaller rational radius.

4.1.6. Tychonoff Product of Topological Spaces

Tychonoff product of two topological spaces (X, τ_x) and (Y, τ_y) is called a product of the spaces $Z = X \times Y$ with the topology τ_z whose base serves an aggregate of the sets $\{B : B = G_x \times G_y, G_x \in \tau_x, G_y \in \tau_y\}$. By means of the induction Tychonoff product of any finite set of the spaces is determined.

Let us consider now a product of any set of the topological spaces (X_t, τ_t), $t \in T$. Tychonoff product of the topological spaces (X_t, τ_t), $t \in T$ is called a product of the spaces

$$X^T = \prod_{t \in T} X_t$$

with the topology τ^T whose base serves the aggregate of the sets

$$B = \prod_{t \in T} G_t,$$

correspondent to all open sets $G_t \in \tau_t$ at the condition $G_t = X_t$ at all t except a finite set of the values t.

For more detailed study of basic notions of topology refer to (Kelly 1957).

P r o b l e m s

4.1.1. Prove that as the base of an ordinary topology on the plane generated by Euclidean metric may serve: a) a set of all open rectangles with the sides parallel to the coordinates axes; b) a set of all open rectangles with rational coordinates of the tops. Prove that in both cases we may restrict ourselves to the squares.

4.1.2. Generalize the result of Problem 4.1.1 on the space R^n with Euclidean metric.

4.1.3. May it serve as the base of an ordinary topology of the plane with Euclidean metric the set of the bands parallel to the coordinates axes whose bases are the open intervals on the correspondent coordinates axes? If not so, in what way we may obtain the base from these bands?

4.1.4. Generalize the result of Problem 4.1.3 on the spaces R^3 and R^n $n > 3$ after substituting the bands by the layers whose bases serve the open intervals on the correspondent coordinates axes.

4.1.5. Let (X, τ) be a topological space, $Y \subset X$ be a set in the space X, τ_Y be the topology on Y induced by the topoloby τ. Prove that: a) the set $A \subset Y$ is closed in the topology τ_Y if and only if $A = A_1 Y$ where A_1 is closed in the topology τ; b) the point $y \in Y$ is the limit point of the set $A \subset Y$ in the topology τ_Y if and only if it is the limit point of the set A in the topology τ; c) the closure of the set A in the topology τ_Y is the intersection Y with the closure A in the topology τ.

4.1.6. Prove that any subsets $B_1 \subset B$, $C_1 \subset C$ of the separated sets B and C of a topological space X are also the separated sets. In particular, for any set $A \subset X$ the sets AB and AC are separated.

4.1.7. Prove that the separated sets A and B of the topological space do not contain their common limit points.

4.1.8. Prove that any two nonintersecting sets are separated.

4.1.9. Let A and B be nonintersecting sets of the topological space (X, τ). Under what conditions the sets A and B will be open (closed) in the induced topology in $A \bigcup B$?

4.1.10. Prove that A and B represent the separated sets of the space X if and only if they are closed in the induced topology in $A \bigcup B$ and do not intersect.

4.1.11. Prove that the closure of the coherent sets is coherent.

4.1.12. Prove that any union of the coherent sets from which none of two are separated is coherent.

4.2. Separability and Countability Axioms

4.2.1. Separability Axioms

As a topology in any space may be given quite arbitrarily (merely the topology axioms will be satisfied) for constructing the theory it is worthwhile to choose a topology in such a way that to retain to a great extent such a property of a natural topology in a metric space that any point has a set of the vicinities which contracts to it. In connection with this one of four *separability axioms* is assumed.

The *first separability axiom* T_1. Either of two different points x, y of a topological space X has the vicinity which does not contain other point. In other words, if $x \neq y$ then the point x has the vicinity V_x which does not contain y, and the point y has the vicinity V_y which does not contain x. A topological space with the first axiom of the separability T_1 is called a T_1-*space*.

E x a m p l e 4.1. As an example of a topological space not satisfying the separability axiom T_1 may serve a plane on which as the open sets are assumed the bands parallel to ordinate axis whose intersections with abscissa axis represent the open sets of the real axis R (Fig.4.2). Such an aggregate of the open sets satisfies all the topology axioms as the aggregate of the open sets of the real axis R satisfies them. But none of the points a and b which lay on the same line parallel to the ordinate axis has no vicinity containing the other point.

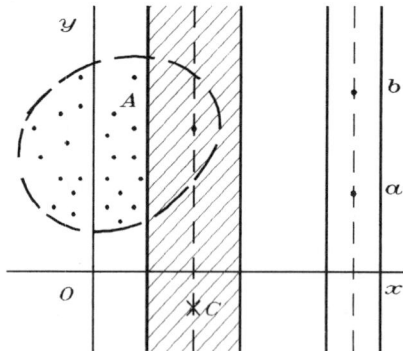

Fig. 4.2

Theorem 4.7. *The intersections of all the vicinities of any point* x *of a* T_1-*space represents the set* $\{x\}$ *consisting of one point* x *(a singleton).*

▷ As for any point $y \neq x$ there exists the vicinity V_x of the point x not containing the point y then the intersection of all the vicinities of the point x cannot contain any other point y of the space X. ◁

On the basis of this theorem any point of a T_1–space has a set of the vicinities contracting to this point.

Theorem 4.8. *Any finite set of the points of a T_1–space is closed.*

▷ Let x be any point of a T_1–space X. By virtue of the axiom T_1 any point y of the supplement $\overline{\{x\}} = X \backslash \{x\}$ of the singleton $\{x\}$ has the vicinity V_x which does not contain x, i.e. enters into $\overline{\{x\}}$ together its some vicinity V_y, $y \in V_y \subset \overline{\{x\}}$. Thus any point of the supplement $\overline{\{x\}}$ of the singleton $\{x\}$ is its internal point. By Theorem 4.3 follows that the set $\overline{\{x\}} = X \backslash \{x\}$ is open and it means that the singleton $\{x\}$ is closed. As any finite union of closed sets is closed then any finite set of the points of T_1–space is a closed set. ◁

It is easy to see that on the contrary any space in which all the finite sets of the points are closed is a T_1–space. Really, in this case at $x \neq y$ the supplement of the singleton $\{y\}$ represents the vicinity of the point x which does not contain the point y, and the supplemnt of the sigleton $\{x\}$ represents the vicinity of the point y which does not contain x.

Theorem 4.9. *In any vicinity of the limit point of the set in a T_1–space contains an infinite set of the points of this set.*

▷ Suppose that the vicinity V_x of the limit point x contains a finite set of the points $\{x_1, \ldots, x_n\}$ of the set A different from x. Let V_k be the vicinity of the point x which does not contain the point x_k ($k = 1, \ldots, n$). Then

$$V = \bigcap_{k=1}^{n} V_k$$

will be the vicinity of the point x which does not contain any of the points $x_1, \ldots, x_n$, and consequently, does not contain any of the points of the set A may be besides the point x itself. Thus the point x cannot be the limit point of the set A. ◁

In a topological space which does not satisfy the axiom T_1 the limit point of the set may have the vicinity which contains only the finite number of the points of this set. On the left part of Fig 4.2 it is shown a finite set of the points A. Any point of the plane laying on the same line parallel to the ordinate axis with any point of the set A is the limit point of the set A (for instance, the point C marked by a cross) as any vicinity of this point (its band parallel to the ordinate axis) contains at least one point of the set A. As the set A is finite then none of the vicinities of its limit point may contain more finite number of the points of the set A.

The *second separability axiom* T_2. Any two different points x, y of the topological space have nonintersecting vicinities. In other words, if $x \neq y$ then there exist such vicinities V_x, V_y of the points x, y that $V_x V_y = \varnothing$. The topological space with the second separability axiom T_2 is called a T_2–space or a *Hausdorff space* after German mathematician Hausdorff who served an outstanding role in the modern topology.

It is clear that any T_2–space is also a T_1–space but not the reverse. A T_1–space may be not a T_2–space.

E x a m p l e 4.2. Let us consider an arbitrary space X with uncountable set of the points in which only an empty set, the whole space and all the finite sets of the points are the closed sets. Each open set in such a space except an empty set and the whole space X represents the whole space with the pricked finite set of the points. This is a T_1–space as for any two points x and y, $x \neq y$ the whole space with the pricked point y represents the vicinity of the point x not containing y, and the whole space with the pricked point x is the vicinity of the point y not containing x. But it is not a T_2–space as any vicinities of two points x, y, $x \neq y$ have nonempty intersection.

The *third separability axiom* T_3. Each point x of the topological space and each closed set F which does not contain this point have non–intersecting vicinities. In other words, if $x \notin F$ where F is a closed set then there exist such vicinities V_x and V_F of the point x and the set F that $V_x V_F = \varnothing$. A T_1–space satisfying also the third separability axiom T_3 is called a T_3–space or *a regular space*.

Any T_3–space is a T_2–space but not the reverse, a T_2–space may be not regular.

E x a m p l e 4.3. It is easy to see that the plane of Example 4.1 (Fig.4.2) satisfies the axiom T_3 as any point on it and any closed band which does not contain this point parallel to the ordinate axis have nonintersecting vicinities. But as it was shown in Example 4.1 the plane is not a T_1–space, and consequently, is not a T_3–space.

E x a m p l e 4.4. Let X be an arbitrary T_2–space (whereas a T_1–space also), $\{x\}$ be an arbitrary sequence in X which has the unique limit point x_0. We change in this space the topology remaining invariant the vicinities of all the points except the point x_0 and as the vicinities of the point x_0 we assume all the open in the initial topology the vicinities of the point x_0 minus the set $\{x_n\}$. In other words, if V_0 is the vicinity of the point x_0 in the original topology then the correspondent vicinity of the point x_0 in new topology will be $V_0 \backslash \{x_n\}$. In new topology constructed in such a way the point x_0 will not be the limit point of the set $\{x_n\}$ as none of the vicinities of the point x_0 contains any point x_n. It is evident that in new topology similarly as well as in the original one any two different points have nonintersecting vicinities. Consequently, the space X with new topology is a T_2–space. But it does not satisfy the separability axiom T_3 as any vicinity of the closed (by the Corollary of Theorem 4.4) set $\{x_n\}$ intersects with any vicinity of the point x_0.

The fourth separability axiom T_4. Any two nonintersecting closed sets of the topological space have nonintersecting vicinities. In other words, if F_1, F_2 are two closed sets and $F_1 F_2 = \varnothing$ then there are exist such vicinities V_1, V_2 of the sets F_1, F_2 that $V_1 V_2 = \varnothing$. A T_1–space satisfying also the fourth separability axiom T_4 is called *a T_4–space* or *a normal space*.

Any T_4–space is a T_3–space but not inverse. A T_3–space may be not a T_4–space.

E x a m p l e 4.5. It is easy to see that the plane of Example 4.1 satisfies the axiom T_4 as any two nonintersecting closed bands (parallel to the ordinate axis) have nonintersecting vicinities. But this plane is not a T_4–space as it is not a T_1–space.

Theorem 4.10. *Any metric space is normal.*

▷ Really, let F_1 and F_2 be two nonintersecting closed sets in the metric space X. Each point $x \in F_1$ belongs to the supplement $\bar{F}_2$ of the set F_2, and consequently, it is contained in the open set $\bar{F}_2$ together with its some vicinity. Therefore any point $x \in F_1$ is at a distance different from zero away from the set F_2:

$$\rho_1(x) = \inf_{y \in F_2} d(x, y) > 0.$$

Analogously any point $y \in F_2$ is at a distance different from zero away from the set F_1:

$$\rho_2(y) = \inf_{x \in F_1} d(x, y) > 0.$$

The open sets

$$V_1 = \bigcup_{x \in F_1} S_{\rho_1(x)/2}(x), \quad V_2 = \bigcup_{y \in F_2} S_{\rho_2(y)/2}(y)$$

represent nonintersecting vicinities of the closed sets F_1 and F_2. Really, if $V_1 V_2 \neq \varnothing$ then for any point $z \in V_1 V_2$ will be found such a point

$x_z \in F_1$ that $z \in S_{\rho_1(x_z)/2}(x_z)$ and such point $y_z \in F_2$ that $z \in S_{\rho_2(y_z)/2}(y_z)$. Consequently,

$$d(x_z, y_z) \leq d(x_z, z) + d(y_z, z) < [\rho_1(x_z)$$
$$+ \rho_2(y_z)]/2 \leq \max[\rho_1(x_z), \rho_2(y_z)].$$

But by the definition $\rho_1(x)$ and $\rho_2(y)$ $d(x, y) \geq \max[\rho_1(x), \rho_2(y)]$ for any $x \in F_1$, $y \in F_2$. The received contradiction proves that the intersection $V_1 V_2 = \emptyset$. ◁

4.2.2. Base of Vicinities of a Point

The set $\mathcal{B}_x$ of the vicinities of the point x is called *a base of the vicinities* of the point x if any vicinity V_x of the point x contains some vicinity $B_x \in \mathcal{B}_x$, $B_x \subset V_x$.

From this definition and Theorem 4.7 follows that in a T_1–space the intersection of all the vicinities of the base of the vicinities of any point x consists of this one point x:

$$\bigcap_{B_x \in \mathcal{B}_x} B_x = \{x\}.$$

The following theorem is almost evident.

Theorem 4.11. *The union of the bases of the vicinities of all the points of a topological space represents a base of this space.*

E x a m p l e 4.6. In a metric space as the base of the vicinities of each point serves a countable set of all spherical vicinities of the rational radii.

4.2.3. Countability Axioms

Example 4.6 shows that in any metric space each point has countable base of the vicinities. In the general case of an arbitrary topological space such is not the case. Therefore it makes sense to select from the set of all topological spaces such spaces in which the points have countable bases of the vicinities. For this reason we introduce the countability axioms.

The first countability axiom. Each point of a topological space has a countable base of the vicinities.

It is evident that any metric space (Example 4.6) is a space with the first countability axiom.

The second countability axiom. A topological space has a countable base.

Theorem 4.12. *Any topological space with a countable base is a space with the first countability axiom.*

▷ Suppose that the topological space (X, τ) has the countable base $\mathcal{B} = \{B_n\}$. Then by Theorem 4.5 for any point x and its any vicinity V_x there exists such a set $B_k \in \mathcal{B}$ that $x \in B_k \subset V_x$. The aggregate of all the sets B_k which contain the point x represents a countable base of the vicinities of the point x:

$$\mathcal{B}_x = \{ B_k \ : \ B_k \in \mathcal{B}, \quad x \in B_k \}. \quad ◁$$

R e m a r k. The inverse in the general case is not true. A topological space with the first countability axiom may have not a countable base. In particular, a metric space (which is always a space with the first countability axiom) may have not a countable base.

4.2.4. Dense Sets. Separable Spaces

In Subsection 1.2.8 the notions of a dense set and the separability for the metric spaces were introduced. These notions are spread on any topological spaces.

The set A is called *a dense* in the set B (in the space X) if in any vicinity of any point of the set B (of the space X) are contained the points of the set A. In other words, the set A is dense in B if any point of the set B represents adherent point of the set A. It is evident that A is dense in B (in X) if and only if $[A] \supset B$ ($[A] = X$ respectively).

A topological space is called *separable* if it contains a countable dense set.

Theorem 4.13. *A topological space with a countable base is separable.*

▷ Let $\mathcal{B} = \{B_n\}$ be a countable base of the topological space X. We take in each set $B_n \in \mathcal{B}$ an arbitrary point x_n. By Theorem 4.5 for each point $x \in X$ and each its vicinity V_x there exists such a set $B_k \in \mathcal{B}$ that $x \in B_k \subset V_x$. Consequently, the vicinity V_x of the point x contains the point x_k from the countable set $\{x_n\}$. Thus in any vicinity of any point of the topological space X at least one point of the countable set $\{x_n\}$ is contained. It means that a countable set of the points $\{x_n\}$ is dense in X and the space X is separable. ◁

R e m a r k. The opposite statement in the general case is not valid. A separable topological space may have no countable base.

E x a m p l e 4.7. Let us consider the topological space X of Example 4.2 in which only an empty set, the whole space and all the finite sets of the points are closed. Any vicinity of any point of this space represents the whole space with the pricked finite set of the points. It is evident that any countable set of the points $\{x_n\}$ is dense in this space as any vicinity of any point may not contain only a finite set of the points x_n. Consequently, the space X is separable. But it has no countable set. Really, suppose that there exists the countable base $\mathcal{B} = \{B_n\}$ in it. Then according to Theorem 4.12 each point x of the space X has a countable base of the vicinities $\mathcal{B}_x = \{B_k \ : \ B_k \in \mathcal{B}, \ x \in B_k\}$. The intersection of all the vicinities of this base represents the whole space X with the pricked no more than countable set of the points. Each of the vicinities of B_k represents the whole space with pricked finite set of the points. But this contradicts the fact stated in Example 4.2 that X is the T_1–space. As a result the intersection of all the vicinities of the base of the vicinities of any point consists of only this point. Thus in a given case the separable space X cannot have a countable base.

The cited Example 4.7 shows that in the general case a separable topological space may have no countable base. But the following theorem is valid.

Theorem 4.14. *A separable metric space has a countable base.*

▷ Let $\{x_n\}$ be a dense countable set in a metric space X. The set of all open balls of the rational radii with the centers in the points x_n is countable. We shall prove that it represents a base of the space X. Let x be an arbitrary point of the space X, G be an arbitrary open set, $x \in G$. According to the definition of an open set in a metric space the set G contains the point x together with its some spherical vicinity $S_r(x)$, $S_r(x) \subset G$. By virtue of the density of the set $\{x_n\}$ in X the ball $S_{r/2}(x)$ contains any of the points x_n, $x_n \in S_{r/2}(x)$ and $d(x_n, x) < r/2$. We shall take such rational number ρ that $d(x_n, x) < \rho < r/2$. Then the ball $S_\rho(x_n)$ of the radius ρ with the center at the point x_n will contain the point x, $x \in S_\rho(x_n)$. On the other hand, it is evident that $S_\rho(x_n) \subset S_r(x) \subset G$. Thus $x \in S_\rho(x_n) \subset G$, i.e. for any point x and any open set G which is contained in it there exists a ball of the rational radius with the center in one of the points of the set $\{x_n\}$ which contains the point x and which is contained in the set G. By Theorem 4.5 hence it follows that a countable set of open balls of the rational radii with the centers in the points of a dense countable set represents a countable base of a metric space X. ◁

Thus in order a metric space be separable it is necessary and sufficient that it should have a countable base. In the general case of a topological space this condition by virtue of Theorem 4.13 is sufficient but not necessary. Example 4.7 shows that a topological space may be also separable even in the case when it has no countable base.

More detailed treatment of the separability and countability is given in (Kelly 1957).

P r o b l e m s

4.2.1. Prove that in a topological space for which the separability axiom T_1 is not valid the singleton $\{x\}$ is not closed. Give an example of an unclosed singleton.

4.2.2. Prove that a set of the vicinities of the points of the space $C([a, b])$

$$\{x(t) : | x(t) - x_0(t) | < \gamma_0(t), \quad t \in [a, b]\},$$

correspondent to all the functions $x_0(t) \in C([a, b])$, $\gamma_0(t) > 0$, $\gamma_0(t) \in C([a, b])$ represents a base of the topology of the space $C([a, b])$. Prove that $C([a, b])$ is the space which yields the first and the second countability axioms. Generalize the result on the space $C(T)$ where T is a closed bounded set in R^n.

4.2.3. Prove that a set of the vicinities of the points of the space of the functions which are continuous together with their derivatives till the n-th order inclusively $C^n([a, b])$

$$\{x(t) : | x(t) - x_0(t) | < \gamma_0(t), \quad t \in [a, b]\},$$

$$\{x(t) : | x^{(k)}(t) - x_0^{(k)}(t) | < \gamma_k(t), \quad t \in [a, b]\},$$

correspondent to all the functions $x_0(t) \in C^n([a, b])$, $\gamma_k(t) > 0$, $\gamma_k(t) \in C([a, b])$ and all $k = 1, \ldots, n$ represents a subbase of the topology of the space $C^n([a, b])$ generated by a norm. Prove that for the space $C^n([a, b])$ the first and the second countability axioms are valid. Is the space $C^n([a, b])$ separable or not?

4.2.4. What represent an open kernel, the closure and the set bound in the topological space of Example 4.2?

4.2.5. Prove that any space with the topology in which all the singletons are closed is a T_1-space.

4.2.6. Prove that in any space X the intersection of the topologies with the separability axiom T_1 represents a topology with the separability axiom T_1. Derive the existence and the uniqueness of the weakest topology with which X will be a T_1-space.

4.2.7. Prove that in any space X the weakest topology with which X is a T_1-space is the topology of Example 4.2 in which X is coherent and does not contain the separated sets.

4.2.8. Prove that the topology satisfies the separability axiom T_3 if and only if any vicinity of any point of this space contains the closure of some (the smallest) vicinity of this point.

4.2.9. Prove that the topology satisfies the separability axiom T_4 if and only if any vicinity of any closed set contains the closure of some vicinity of this set.

4.3. Convergence. Continuity of a Function

4.3.1. Convergence of a Sequence

The sequence of the points $\{x_n\}$ of a topological space is called *convergent in the point x*,

$$x_n \to x \quad \text{or} \quad x = \lim_{n \to \infty} x_n,$$

if any vicinity of the point x contains all the points x_n beginning with someone. In other words, the sequence $\{x_n\}$ converges to the point x if for any vicinity V_x of the point x there exists such natural $N = N(V_x)$ that $x_n \in V_x$ at all $n > N$.

Theorem 4.15. *If the sequence $\{x_n\}$ converges to x then any closed set F which does not contain the point x may contain no more than a finite set of the points of the sequence $\{x_n\}$.*

▷ Let F be a closed set which does not contain the point $x = \lim x_n$. Then $x \in \bar{F}$ and by virtue of the fact that $\bar{F}$ is an open set there exists the vicinity V_x of the point x which is completely contained in $\bar{F}$, and consequently, does not intersect with the set F, $V_x F = \varnothing$. But V_x contains all the points of the sequence, may be except a finite set of the first point. Consequently, the set F may contain no more than finite number of the first points of the sequence $\{x_n\}$. ◁

Theorem 4.16. *In a T_2-space the sequence may converge to no more than one point.*

▷ If the sequence of the points $\{x_n\}$ converges simultaneously to two different point x, y, $x \neq y$ then each of the nonintersecting vicinities V_x, V_y of these points $V_x V_y = \varnothing$ will contain all the points of the sequence $\{x_n\}$ beginning with someone. But it is impossible. ◁

Theorem 4.17. *In a T_1-space with the first countability axiom the point x may be the limit point of the set A if and only if in the set A there exists the sequence convergent to x.*

▷ If in the set A there is the sequence $\{x_n\}$ convergent to x then each vicinity of the point x contains the points $x_n \in A$ different from x. Consequently, x is the limit point of the set A. If x is the limit point then in each its vicinity will occur an infinite set of the points of the set A. Let $\mathcal{B}_x = \{B_n\}$ be the base of the vicinities of the point x. We determine the sets

$$C_n = \bigcap_{k=1}^{n} B_k \quad (n = 1, 2, \ldots)$$

and in each set C_n we take the point $x_n \in A$. It is possible as the sets C_n also represent the vicinities of the point x, and consequently, contain the points of the set A. As $C_{n+1} \subset C_n \subset B_n$ then each vicinity $B_n \in \mathcal{B}_x$ contains all the points $x_n, x_{n+1}, \ldots$. And as according to the definition of the base any vicinity V_x of the point x contains some vicinity B_n, $B_n \subset V_x$ then any vicinity of the point x contains all the points of the sequence $\{x_n\}$ beginning with someone. Consequently, the sequence $\{x_n\} \subset A$ converges to the point x. ◁

4.3.2. Continuity of a Function

The function $y = f(x)$ mapping a topological space X into a topological space Y is called *continuous at the point* x_0 if the inverse image of any vicinity W of the point $y_0 = f(x_0)$ contains some vicinity V of the point x_0, $V \subset f^{-1}(W)$. It is evident that this definition is equivalent to the following: the function $y = f(x)$ is called continuous at the point x_0 if any vicinity W of the point $y_0 = f(x_0)$ contains the image of some vicinity V of the point x_0, $f(V) \subset W$. The equivalence of these two definitions follows directly from the properties of inverse mappings.

The function $y = f(x)$ is called *continuous* if it is continuous at all the points of its domain D_f.

Theorem 4.18. *If the function* $y = f(x)$ *mapping a topological space* X *into a topological space* Y *is continuous at the point* x_0 *and the functions* $z = g(y)$ *mapping* Y *into a topological space* Z *is continuous at the point* $y_0 = f(x_0)$ *then the composite function* $gf(x) = g(f(x))$ *is continuous at the point* x_0.

▷ If the function g is continuous at the point $y_0 = f(x_0)$ then the inverse image of any vicinity W of the point $z_0 = g(y_0)$ contains some vicinity V of the point y_0, $V \subset g^{-1}(W)$. If f is continuous at the point x_0 then the inverse image of any vicinity of the point y_0, in particular, $f^{-1}(V)$ contains some vicinity U of the point x_0, $U \subset f^{-1}(V)$. From two obtained inclusions follows that

$$U \subset f^{-1}(V) \subset f^{-1}(g^{-1}(W)) = (gf)^{-1}(W).$$

Thus the inverse image of any vicinity of the point $z_0 = g(f(x_0))$ $= gf(x_0)$ contains some vicinity of the point x_0. Consequently, the composite function $gf(x) = g(f(x))$ is continuous. ◁

Theorem 4.19. *The function $y = f(x)$ mapping a topological space X into a topological space Y and determined on the whole space X is continuous if and only if the inverse image (which is determined by this functions) of any open set represents an open set.*

▷ Let B be an open set in Y, x be any point of the inverse image B, $x \in f^{-1}(B)$. If the function f is continuous then by virtue of the fact that B is the vicinity of the point $y = f(x)$ the inverse image of the set B $f^{-1}(B)$ contains the point x together with its some vicinity V_x, $x \in V_x \subset f^{-1}(B)$. As it is valid for any point $x \in f^{-1}(B)$ then by Theorem 4.3 the set $f^{-1}(B)$ is open.

If the inverse image of any open set determined by the function f represents an open set then for any $x \in X$ the inverse image $f^{-1}(V_y)$ of any vicinity V_y of the point $y = f(x)$ represents a vicinity of the point x, and consequently, the function f is continuous at the point x. As it is valid for any point $x \in X$ then the function f is continuous. ◁

From the proved theorem and the properties of inverse maps follows.

Corollary. *The function $y = f(x)$ mapping a topological space X into a topological space Y and determined on the whole space X is continuous if and only if the inverse image (which is determined by this function) of any closed set represents a closed set.*

R e m a r k. Theorem 4.19 is not valid if $D_f \neq X$ as the set $f^{-1}(Y) = D_f$ in the general case may be not simultaneously open and closed. But it is also valid in this case if we replace the topology τ_x in X by the induced topology $\tau_x D_f$ in D_f (Subsection 4.1.3).

Theorem 4.20. *If the function $f(x)$ is continuous then for any sequence $\{x_n\} \subset D_f$ convergent to $x \in D_f$ the sequence $\{y_n\}$, $y_n = f(x_n)$ converges to $y = f(x)$.*

▷ If $f(x)$ is continuous at the point x then any vicinity V_y of the point $y = f(x)$ contains an image of some vicinity V_x of the point x, $f(V_x) \subset V_y$. From the convergence of the sequence $\{x_n\}$ to x follows that all the points x_n beginning with someone are contained in V_x. Hence the correspondent points $y_n = f(x_n)$ are contained in $f(V_x) \subset V_y$. As it is valid for any vicinity V_y of the point $y = f(x)$ then the sequence $\{y_n\}$ converges to y. ◁

R e m a r k. The reverse in the general case is not true. But the following statement is valid.

Theorem 4.21. *If X is the space with the first countability axiom then from the convergence of the sequence $\{y_n\}$, $y_n = f(x_n)$, to $y = f(x)$ for any sequence $\{x_n\} \subset D_f$ convergent to $x \in D_f$ follows the continuity of the function $f(x)$.*

▷ If the function $f(x)$ is not continuous at the point x then the vicinity V_y of the point $y = f(x)$ whose inverse image $f^{-1}(V_y)$ will not contain anyone vicinity of the point x will be found. Let $\{V_n\}$ be a countable base of the vicinities of the point x. From the fact that anyone of the vicinities V_n is not contained in $f^{-1}(V_y)$ follows that in each V_n there exists the point $x_n \notin f^{-1}(V_y)$. The sequence $\{x_n\}$ converges to x. But none of the points $y_n = f(x_n)$ belongs to V_y, $y_n \notin V_y$ as $x_n \notin f^{-1}(V_y)$. Consequently, the sequence $\{y_n\}$, $y_n = f(x_n)$, does not converge to $y = f(x)$ what contradicts the condition of the theorem. ◁

Corollary. *In the space X with the first countability axiom for the continuity of the function $f(x) : X \to Y$ it is necessary and sufficient that for any convergent sequence $\{x_n\}$ the sequence $\{y_n\}$, $y_n = f(x_n)$, should converge to the correspondent value of the function $f(x)$.*

R e m a r k. It follows from Theorem 4.19 that the function $f(x)$ which is continuous in some topologies of the spaces X and Y may be not continuous in other topologies. In particular, at one and the same topology in the space Y the function $f(x)$ which is continuous in a given topology of the space X may be not continuous in more weak topology of the space X. And at one and the same topology in the space X the function $f(x)$ which is continuous in a given topology of the space Y is also continuous in more weak topology.

4.3.3. A Way to Assign a Topology

The simplest way of assigning a topology in a given space is the transfer of a topology from the space with already determined topology into a given space by means of direct and inverse mappings. Here by virtue of the fact that the inverse mappings hold all the relations between the sets the inverse mappings are more often used.

Let X be the space in which the topology is to be assigned, (Y, σ) be a topological space with the known topology σ, $y = f(x)$ be the mapping $X \to Y$. The inverse image $\tau = f^{-1}(\sigma)$ of the topology σ may be assumed as the topology in the space X as by virtue of the properties of inverse images the sets τ possess all the properties of open sets as the sets from σ also possess these properties.

But a topology in the composite space X transferred by means of the inverse mapping from the sufficiently simple space Y (only in the simple spaces there exist natural topologies) is often appeared too poor. Therefore we are not satisfied with one space with the known topology and one mapping. After taking the sufficient set of the spaces with the known topologies and the sufficient set of the mapping we may determine the sufficient strong topology in any space.

Let $\{(Y_\alpha, \sigma_\alpha) : \alpha \in A\}$ be the set of the topological spaces with the known topologies σ_α, $\{f_{\alpha\beta}(x) : \beta \in B\}$ be the set of the mappings of the space X into the space Y_α $(\alpha \in A)$. Then after using the stated way we shall obtain in the space X a set of the topologies

$$\tau_{\alpha\beta} = f_{\alpha\beta}^{-1}(\sigma_\alpha) \quad (\alpha \in A,\ \beta \in B).$$

By Theorem 4.6 the union of all these topologies may be assumed as the subbase of the topology τ in the space X:

$$\mathcal{C} = \bigcup_{\alpha,\beta} \tau_{\alpha\beta} = \bigcup_{\alpha,\beta} f_{\alpha\beta}^{-1}(\sigma_\alpha).$$

Usually as the spaces Y_α are assumed the real axis, the complex plane or more common metric spaces.

4.3.4. Measurable Topological Spaces

In the topological spaces the σ–algebra induced by a topology or the σ–algebra induced by a base of a topology are usually assigned. A topological space with the σ–algebra induced by a topology is called a *measurable topological space*.

The following theorems and corollaries are evident.

Theorem 4.22. *In a topological space the σ–algebra induced by a topology contains compeletely the σ–algebra induced by a base of a topology.*

Corollary. *Let $(X, \mathcal{A})$ be an arbitrary countable space, Y be a topological space, $\mathcal{B}$ be the σ–algebra in Y induced by a topology, $\mathcal{B}_0$ be the σ–algebra in Y induced by a base of a topology then any $(\mathcal{A}, \mathcal{B})$–measurable function $y = f(x) : X \to Y$ is $(\mathcal{A}, \mathcal{B}_0)$–measurable.*

Theorem 4.23. *In a topological space with a countable base the σ–algebra induced by a topology coincides with the σ–algebra induced by a base of a topology.*

Corollary. *In a separable metric space the σ–algebra induced by a topology coincides with the σ–algebra induced by a set of all open balls.*

In particular, in any finite–dimensional space R^n the σ–algebra induced by a topology coincides with the σ–algebra induced by a class of open balls (open intervals at $n = 1$).

Theorem 4.24. *All continuous functions mapping a measurable topological space into another measurable topological space are measurable.*

R e m a r k. Theorem 4.24 is of great importance as it establishes the measurability of all continuous functions. Namely for this purpose that the continuous functions will be measurable the σ–algebra in a topological space is always connected with a corresponding topology.

More detailed theory of convergence and function continuity is given in (Schaefer 1966).

P r o b l e m s

4.3.1. Let $f(x,\, y)$ be a continuous function of two variables. Under what conditions it is continuous in the topology of the plane xy of Example 4.1?

4.3.2. Let $f(x)$ be a continuous two–dimensional vector function mapping the space X into R^2. Under what conditions it is continuous in the topology of the plane xy of Example 4.1?

4.3.3. Determine a topology on the plane R^2 by means of the inverse mappings correspondent to the scalar functions $y = f^T(x - x_0)$ at all the vectors $x_0, f \in R^2$. Construct a subbase of such topology and the geometrical images of the sets of the subbases and bases. Prove that the topology determined in such a way coincides with an ordinary topology on the plane induced by Euclidean metric.

4.3.4. Generalize the result of Problem 4.3.3 on the space R^n.

4.3.5. Determine a topology on the real axis by means of the inverse mappings given by the function $y = [x]$. Will the function $y = x$ be continuous in such topology on the x–axis and in an ordinary topology on the y–axis?

4.3.6. Prove that at continuous mapping f of the whole topological space X into the topological space Y the inverse image of the closure of any set $B \subset Y$ contains the closure of the inverse image of the set B, $f^{-1}([B]) \supset [f^{-1}(B)]$. This theorem is also valid in the case when $D_f \neq X$ if we replace the topology in X by the induced topology in D_f.

I n s t r u c t i o n. Account that by virtue of Theorem 4.19 and the properties of the inverse mappings the inverse image of the closure of the set B is closed and represents the intersection of the inverse images of all closed sets which contain B, but the set of these inverse images may not contain all the closed sets which contain $f^{-1}(B)$.

4.3.7. Prove that the inverse images of the separated sets are separable.

4.3.8. Prove that the image of the coherent set at any continuous mapping is coherent.

4.3.9. Prove that the mapping P_x of Tychonoff product of the topological spaces (Subsection 4.1.6) $Z = X \times Y$ on the space X which establishes the correspondence between each point $z = \{x, y\}$ and its projection x on the space X is linear and continuous.

4.3.10. Let P_t be the mapping of Tychonoff product of the topological space (Subsection 4.1.6) $X^T = \prod_{t \in T} X_t$ on the space X_t which establishes the correspondence between each element $x = \{x_t : t \in T\}$ of the space X^T and the element x_t of the space X_t (the projection x on X_t). Prove that this mapping is linear and continuous.

I n s t r u c t i o n. Account that as the inverse image of any set $B \subset X_t$ in X^T serves the product of the sets

$$P_t^{-1}(B) = B \times \prod_{\substack{\tau \neq t \\ \tau \in T}} X_\tau .$$

4.3.11. Prove that the topology in Tychonoff product of the spaces is the weakest topology in the product of the spaces in which all the operators P_t $(t \in T)$ are continuous.

4.4. Compactness

4.4.1. Compact Sets and Spaces

The union of the sets $\bigcup G_\alpha$ is called *a covering* of the set A (the space X) if $A \subset \bigcup G_\alpha$ (correspondingly, $X = \bigcup G_\alpha$). If all the sets G_α are open then $\bigcup G_\alpha$ is called *an open covering* of the set A (the space X).

A set of a topological space (in particular, the space itself) is called *a compact* set if any open covering contains a finite covering (subcovering). A compact set of a T_2–space is called *a compact or compactum*.

As any metric space is normal (Theorem 4.10), and consequently, is a T_2–space then any compact set of the metric space is a compact.

A set of a topological space (in particular, the space itself) is called *a countable compact set* if its any countable open covering contains a finite covering.

A set of a topological space (in particular, the space itself) is called *a sequentially compact set* if its any sequence of its points contains convergent subsequence.

4.4.2. Centred System of Closed Sets

A system of closed sets is called *centred* if it does not contain the finite systems (subsystems) with an empty intersection. Immediatelly from the definition of a compact set follows the following statement.

Theorem 4.25. *The set A is a compact set if any centred system of the closed subsets of the set A has a nonempty intersection.*

We offer an equivalent formulation of this theorem. *The set A is a compact set if any system of its closed subsets with an empty intersection contains a finite system with an empty intersection.*

▷ Let $\{F_\alpha\}$ be a system of the closed subsets of the set A with an empty intersection $\bigcap F_\alpha = \emptyset$. Then the system of open sets $\{G_\alpha\}$, $G_\alpha = \overline{F_\alpha}$, forms an open covering of the set A, $\bigcup G_\alpha = \overline{\bigcap F_\alpha} = X \supset A$. If the set A is a compact set then the system $\{G_\alpha\}$ contains the finite system $\{G_{\alpha_1}, \ldots, G_{\alpha_n}\}$ which forms an open covering of the set A

$$\bigcup_{k=1}^{n} G_{\alpha_k} \supset A. \tag{4.1}$$

But then by virtue of the fact that $F_\alpha \subset A$, and consequently, $G_\alpha \supset \bar{A}$,

$$\bigcup_{k=1}^{n} G_{\alpha_k} \supset \bar{A}. \tag{4.2}$$

From (4.1) and (4.2) follows

$$\bigcup_{k=1}^{n} G_{\alpha_k} = X \quad \text{and} \quad \bigcap_{k=1}^{n} F_{\alpha_k} = \emptyset.$$

Thus the system $\{F_\alpha\}$ cannot be centred.

If the system $\{F_\alpha\}$ with an empty intersection is centred then an open covering of the A, $\bigcup G_\alpha = \overline{\bigcap F_\alpha} = X \supset A$, $G_\alpha = \overline{F_\alpha}$ cannot contain a finite covering, and consequently, the set A may be not a compact set. ◁

Analogous theorem is valid in the case of the countable compactness.

Theorem 4.26. *The set A is a countable–compact set if any countable centred system of the closed subsets of the set A has a nonempty intersection.*

The equivalent formulation of this theorem: *The set A is a countable–compact set if any countable system of closed subsets with an empty intersection contains a finite system with an empty intersection.*

4.4.3. Properties of Compact Sets and Spaces

It is clear from the definitions that any compact set (space) is a countable–compact set (space).

Theorem 4.27. *A closed subset of a compact set is a compact set.*

▷ Let A be a compact set, B be its closed subset, $B \subset A$, $\{F_\alpha\}$ be a centred system of the closed subsets of the set B, $F_\alpha \subset B$. As $F_\alpha = F_\alpha B$ then the sets F_α are closed subsets of the set A. By Theorem 4.25 from the compactness of A follows that the intersection of all the sets of the system $\{F_\alpha\}$ is not empty. Hence it follows in its turn that the set B is compact. ◁

Corollary. *A closed subset of a compact set is a compact set.*

The theorem analogous to Theorem 4.27 is also valid for a countable–compact set.

Theorem 4.28. *A closed subset of a countable–compact set is a countable–compact set.*

The statement which is inverse in the known sense to the Corollary of Theorem 4.27 is also valid.

Theorem 4.29. *A compact is closed in any T_2-space which is contained in it.*

▷ Let A be a compact in a T_2-space X, y be an arbitrary point of the supplement $\bar{A}$, $y \in \bar{A}$. Any point $x \in A$ and the point y have nonintersecting vicinities $V_x \ni x$, $U_x \ni y$, $V_x U_x = \varnothing$. The union of all such vicinities V_x correspondent to all the points $x \in A$ represents an open covering of the set A,

$$\bigcup_{x \in A} V_x \supset A \,.$$

By virtue of the compactness of A this covering contains a finite covering

$$\bigcup_{k=1}^{n} V_{x_k} \supset A \,.$$

But $\bigcap\limits_{k=1}^{n} U_{x_k}$ represents the vicinity of the point y which does not intersect with $\bigcup\limits_{k=1}^{n} V_{x_k}$, and consequently, also with A. It means that the point y

is an interval point of the set $\bar{A}$. As y is any point of the set $\bar{A}$ then $\bar{A}$ is an open set, and consequently, A is a closed set. ◁

Corollary 1. *A compact set of a metric space is closed as any metric space is normal, and consequently, represents a T_2-space.*

Corollary 2. *Any compact in a T_2-space and any point which does not belong to it have nonintersecting vicinities.*

▷ In the proof of Theorem 4.29 the vicinity $\bigcup\limits_{k=1}^{n} V_{x_k}$ of the compact A and the vicinity $\bigcap\limits_{k=1}^{n} U_{x_k}$ of the point y which does not belong to A do not intersect. ◁

Theorem 4.30. *Any compact is a normal space (T_4-space).*

▷ Let A be a compact, F_1, F_2 be its nonintersecting closed subsets, $F_1, F_2 \subset A$, $F_1 F_2 = \emptyset$. According to Theorem 4.27 F_1 and F_2 are the compacts. As they do not intersect then by Corollary 2 of Theorem 4.29 any point $x \in F_1$ and the compact F_2 have nonintersecting vicinities V_x and G_x, $x \in V_x$, $F_2 \subset G_x$, $V_x G_x = \emptyset$. As F_1 is a compact then its covering

$$\bigcup_{x \in F_1} V_x \supset F_1$$

contains a finite covering

$$\bigcup_{k=1}^{n} V_{x_k} \supset F_1 \,.$$

Here the open set $\bigcap\limits_{k=1}^{n} G_{x_k} \supset F_2$ will be the vicinity of the set F_2 not intersecting with the vicinity $\bigcup\limits_{k=1}^{n} V_{x_k}$ of the set F_1. ◁

Theorem 4.31. *In order a set should be a countable–compact set it is necessary that any its infinite subset should have the limit points.*

▷ Let A be a set, B be its infinite subset, $B \subset A$. If B has no limit points then none of its countable subset $\{x_n\}$ may have the limit points. But then by the Corollary of Theorem 4.4 the sets $F_n = \{x_n, x_{n+1}, \ldots\}$ ($n = 1, 2, \ldots$) are closed. As they form a countable centred system of the subsets of the set A with an empty intersection then by Theorem 4.26 the set A may be not a countable–compact. Suppose now that any infinite subset of the set A has the limit points. Let $\{F_n\}$ be an arbitrary countable centred system of the closed subsets of the set A. Determine the closed set

$$H_n = \bigcap_{k=1}^{n} F_k \quad (n = 1, 2, \dots).$$

It is evident that $H_{n+1} \subset H_n \subset F_n$ at any n and

$$\bigcap_{n=1}^{\infty} H_n = \bigcap_{k=1}^{\infty} F_k.$$

If at some N we have $H_p = H_N$ and all $p > N$ then

$$\bigcap_{k=1}^{\infty} F_k = \bigcap_{n=1}^{\infty} H_n = H_N \neq \emptyset.$$

If there is no such N then there exists an infinite set of nonempty sets $H_{n_k} \backslash H_{n_{k+1}}$. After taking in each set $H_{n_k} \backslash H_{n_{k+1}}$ the point x_k we obtain an infinite set of the points $\{x_k\}$ and each set H_n at $n \leq n_k$ will contain all the points $x_k, x_{k+1}, \dots$ as $H_n \supset H_{n_k}$ at $n \leq n_k$. Let x_0 be a limit point of the set $\{x_k\}$. It will be also the limit point of all the sets H_n. As the sets H_n are closed then $x_0 \in H_n$ at all n. Consequently,

$$\bigcap_{k=1}^{\infty} F_k = \bigcap_{n=1}^{\infty} H_n \neq \emptyset.$$

Thus if any infinite subset of the set A has the limit points then any countable centred system of its closed subsets has an nonempty intersection. By Theorem 4.26 the set A is countable–compact. ◁

Corollary 1. *Any infinite subset of a compact set has the limit points.*

▷ It follows from the proved theorem and the countable compactness of a compact set. ◁

Corollary 2. *Any sequentially compact set (space) is countable–compact set (space).*

Really, any infinite subset of a compact set contains the sequence which has the limit point – the limit of its convergent sequence.

R e m a r k. The inverse statement for the T_1–spaces with the first countability axiom is true as any countable–compact set is a sequentially compact set. It follows from Theorem 4.17 stating that any set in such a space which has the limit point contains the sequence convergent to this limit point. In particular, any sequence which has the limit point contains the subsequence convergent to this point.

Thus in the T_1–spaces with the first countability axiom the notion of the countable compactness and the sequential compactness coincide.

Theorem 4.32. *Any open covering of a topological space with a countable base contains a countable covering.*

▷ Let $\bigcup G_\alpha = X$ be an open covering of the space X, $\mathcal{B} = \{B_n\}$ be its countable base. According to Theorem 4.5 for each set G_α and its each point x there exists such a set $B_k \in \mathcal{B}$ that $x \in B_k \subset G_\alpha$. After choosing for each B_k someone set from the sets G_α, for instance, G_{α_k} we obtain a countable covering of the space X:

$$\bigcup_{k=1}^{\infty} G_{\alpha_k} \supset \bigcup_{k=1}^{\infty} B_k = X . \quad \triangleleft$$

Corollary. *If a set of a topological space with a countable base is a countable–compact set then it is a compact set .*

R e m a r k. Thus for the topological spaces with a countable base the notion of the compactness and countable compactness coincide.

Theorem 4.33. *Any T_3–space X with a countable base represents a T_4–space .*

▷ Let F_1 and F_2 be nonintersecting closed sets. Each point $x \in F_1$ has the vicinity U_x which does not intersect with some vicinity of the set F_2 and each point $y \in F_2$ has the vicinity which does not intersect with some vicinity of the set F_1. The union of all the vicinities U_x, $x \in F_1$, V_y, $y \in F_2$ and the supplements of the set $F_1 \bigcup F_2$ represents an open covering of the space X. By Theorem 4.32 it contains a countable covering of the space X. Let choose from this covering the countable covering $\bigcup U_n$ of the set F_1 and the countable covering $\bigcup V_m$ of the set F_2 and put

$$U'_n = U_n \backslash \bigcup_{k=1}^{n} [V_k], \quad V'_m = V_m \backslash \bigcup_{k=1}^{n} [U_k] .$$

It is evident that the set U'_n does not intersect with V_m and consequently, with V'_m at $m \leq n$, and the set V'_m does not intersect with U_n and consequently, with U'_n at $m \geq n$. Thus none of the sets U'_n intersects with none of the sets V'_m and the unions $\bigcup U'_n$ and $\bigcup V'_m$ represent non–intersecting vicinities of the sets F_1 and F_2. ◁

4.4.4. Precompact Sets

A set of a topological space is called *a precompact set* (*countable-precompact set*) or *a relative compact set* (*a relative–countable–compact set*) if its closure is a compact set (a countable compact set).

A topological space is called *a local compact set* if each its point has precompact vicinity.

4.4.5. Continuous Mappings of Compact Sets

The property of the compactness is retained at continuous mappings. Namely, the following statements are valid.

Theorem 4.34. *An image of a compact set at continuous mapping is a compact set.*

▷ Let $y = f(x)$ be a continuous mapping of the topological space X into the topological space Y, A be a compact set in X, $\bigcup H_\alpha$ be an open covering of the image B of the set A, $\bigcup H_\alpha \supset B = f(A)$. By virtue of the continuity of the function f and Theorem 4.19 the inverse images of the sets H_α represent the open sets $G_\alpha = f^{-1}(H_\alpha)$, and by virtue of the properties of the inverse mappings $\bigcup G_\alpha = \bigcup_\alpha f^{-1}(H_\alpha)$ represent an open covering of the set A. As A is a compact set then $\bigcup G_\alpha$ contains a finite covering $\bigcup_{k=1}^{n} G_{\alpha_k} \supset A$. But then $\bigcup_{k=1}^{n} H_{\alpha_k}$ will be a finite covering of the set $B = f(A)$. Thus any open covering of the set B contains a finite covering what proves the compactness of the set $B = f(A)$. ◁

Corollary. *An image of a compact at continuous mapping is a compact.*

▷ It is evident that Theorem 4.34 is also valid for the countable–compact set. ◁

Theorem 4.35. *An image of a compact set at continuous mapping is a countable–compact set.*

For formulating and proving the following theorem it is necessary to introduce one more notion.

An one–to–one and continuous mapping is called *a homeomorphism*. In other words, if the function $y = f(x)$ gives an one–to–one mapping of the topological space X in the topological space Y and both the function $y = f(x)$ and the inverse function $x = f^{-1}(y)$ are continuous then the function f is called the homeomorphism.

Theorem 4.36. *An one–to–one continuous mapping of a compact in a T_2–space is the homeomorphism.*

▷ Let $y = f(x)$ be an one–to–one continuous mapping of the compact A in a T_2–space Y. The theorem will be proved if we show that the inverse function $x = f^{-1}(y)$ is continuous. For this purpose we notice that by the Corollary of Theorem 4.27 any closed subset A_1 of the compact A is a compact. According to Corollary of Theorem 4.34 the image $B_1 = f(A_1)$ of the compact A_1 is a compact, and by Theorem 4.29 the compact B_1 is closed in a T_2–space Y. But B_1 is the inverse image of the set A_1 determined by the inverse function f^{-1}, $B_1 = (f^{-1})^{-1}(A_1)$. Thus the inverse image of any closed set A_1 determined by the function f^{-1} represents a closed set. Hence and from Corollary of Theorem 4.19 follows that f^{-1} is continuous. ◁

For detailed study of the compactness in metric spaces refer to (Kelly 1957).

Problems

4.4.1. Construct an open covering of the interval $(0, 1)$ which does not contain the finite coverings. Explain why each open covering of the interval $[0, 1]$ contains a finite covering while an open covering of smaller interval $(0, 1)$, $[0, 1)$ or $(0, 1]$ may not contain the finite covering.

4.4.2. Prove that a closed unit ball in an infinite–dimensional H–space is not a compact set.

I n s t r u c t i o n. Take an arbitrary orthogonal sequence of the vectors $\{x_n\}$ whose norms are equal to 1 and prove that this sequence has no limit points.

4.4.3. Using the result of Problem 4.3.8 prove that if a real function $y = f(x)$ takes at the coherent compact the values a and $b > a$ then it also takes any intermediate value $c \in (a, b)$.

4.5. Compactness in Metric Spaces

4.5.1. Totally Bounded Sets in a Metric Space

The set S is called an ε–net for the set A if for any point $x \in A$ the point $s \in S$ removed from x less than for ε, $d(x, s) < \varepsilon$ will be founded. Here some (or all) points of the set S may not belong to the set A. But any ε–net S for the set A may be replaced by the subset T of the set A which represents an 2ε–net for A. For this purpose it is sufficient to replace each point $s \in S$ by some point $t \in A$ for which $d(t, s) < \varepsilon$. If $s \in A$ then we may take $t = s$. If for a given point $s \in S$ in A there is no such a point t for which $d(t, s) < \varepsilon$ then this point s should be excluded. Then for any point $x \in A$ and any point $s \in S$ for which $d(x, s) < \varepsilon$ in the obtained set $T \subset A$ such a point t for which $d(x, t) \le d(x, s) + d(s, t) < 2\varepsilon$ will be founded.

It is evident that the set B which is dense in A serves as an ε–net for A at any $\varepsilon > 0$.

The set A of a metric space X is called *bounded* if there exists such number $c > 0$ that $d(x_1, x_2) < c$ for any points $x_1, x_2 \in A$.

The set A of a metric space X is called *totally bounded* if for each $\varepsilon > 0$ there exists the finite ε–net for A.

Theorem 4.37. *Any totally bounded set in a metric space is bounded.*

▷ Let A be totally bounded set, $S = \{s_1, \ldots, s_N\}$ be the finite ε–net for it at some $\varepsilon > 0$,

$$d = \max_{k,l} d(s_k, s_l).$$

Let us take any two points of the set A, $x_1, x_2 \in A$. Then such points $s_{k_1}, s_{k_2} \in S$ that $d(x_1, s_{k_1}) < \varepsilon$, $d(x_2, s_{k_2}) < \varepsilon$ will be found. As a result we get

$$d(x_1, x_2) \le d(x_1, s_{k_1}) + d(s_{k_1}, s_{k_2}) + d(x_2, s_{k_2}) < d + 2\varepsilon,$$

what proves the boundedness of the set A. ◁

In the finite dimensional space R^n the inverse statement is also true: any bounded set is totally bounded. To make sure it is sufficient to notice that at any $\varepsilon > 0$ the bounded set $A \in R^n$ may be covered by $N = (d\sqrt{n}/2\varepsilon)^n$ cubes where

$$d = \sup_{x_1, x_2 \in A} d(x_1, x_2).$$

The tops of these cubes will be a finite ε–net for A. In an infinite–dimensional space the bounded set may be not totally bounded. So, for instance, any ball in the infinite–dimensional B–space is bounded but not totally bounded.

R e m a r k. Thus the notions of the boundedness and the complete boundedness of the sets coincide for the finite–dimensional metric spaces and are different for the infinite–dimensional spaces.

4.5.2 Properties of Compact Sets in Metric Spaces

Let study the specific properties of compact and precompact sets in the metric spaces.

Theorem 4.38. *Any countable compact metric space is totally bounded.*

▷ If a metric space X is not totally bounded then at some $\varepsilon > 0$ in X such two points x_1, x_2 will be found that

$$d(x_1, x_2) \geq \varepsilon.$$

Otherwise any of the points x_1, x_2 will be the finite ε–net for X. After that a such point x_3 will be found in X that

$$d(x_1, x_3) \geq \varepsilon, \qquad d(x_2, x_3) \geq \varepsilon.$$

Continuing this process we add to the obtained points $x_1, \ldots, x_n$ such a point $x_{n+1} \in X$ that

$$d(x_1, x_{n+1}) \geq \varepsilon, \ldots, d(x_n, x_{n+1}) \geq \varepsilon.$$

Such point exists as in the opposite case the set $\{x_1, \ldots, x_n\}$ will be the finite ε–net for X. Thus if X is not totally bounded then there exists the sequence of the points $\{x_n\}$ in X which has no limit point as for any points x_r, x_s $d(x_r, x_s) \geq \varepsilon$. But then from Theorem 4.31 follows that X cannot be a countable–compact. ◁

Theorem 4.39. *Any totally bounded metric space has a countable base.*

▷ Let X be a metric space. By virtue of its complete boundedness at any $\varepsilon > 0$ there are the finite $\varepsilon/2^{n-1}$–nets S_n $(n = 1, 2, \ldots)$ in it. The union of these nets $\bigcup_n S_n$ represents a countable dense set in X. Consequently, X is separable. By Theorem 4.14 any separable metric space has a countable base. ◁

Theorem 4.40. *Any countable compact metric space is a compact space.*

▷ Let X be a countable compact metric space. On the basis of Theorems 4.38 and 4.39 it has a countable base. Consequently, according to Corollary of Theorem 4.32 the space X is a compact. ◁

R e m a r k. As the metric spaces are the T_1–spaces with the first countability axiom then on the basis of Corollary of Theorem 4.31 it is evident that any countable compact metric space is sequential compact. Thus for the metric spaces the notions of the compactness, countable compactness and the sequential compactness coincide.

On the basis of Theorems 4.17 and 4.31 a compact metric space (a set of a metric space) may be determined as a set whose each infinite subset contains a convergent sequence. So, we have come to the following statement.

Theorem 4.41. *A metric space is a compact if and only if it is complete and totally bounded.*

▷ Let X be a metric space. If it is not complete then there is the fundamental sequence $\{x_n\}$ which has no limit in X. Such sequence has no limit point in X. By Corollary 2 of Theorem 4.31 X may be not compact. Consequently, the completeness is necessary. The necessity of the complete boundedness follows from Theorem 4.38.

If X is totally bounded than at any $\varepsilon > 0$ the finite $\varepsilon/2^{n-1}$–nets S_n $(n = 1, 2, \ldots)$ will be found in X. Each such $\varepsilon/2^{n-1}$–net will be also finite $\varepsilon/2^{n-1}$–net for any set in the space X. Let $\{x_n\}$ be any sequence of the points in X. For each point s_{nk} of each finite $\varepsilon/2^{n-1}$–net S_n we take a ball vicinity of the radius $\varepsilon/2^{n-1}$. At least in one of the balls of the radius ε with the centers at the points of ε–net S_1, for instance, in B_1 an infinite set of the points x_p is contained. As the balls of the radius $\varepsilon/2$ with the centers at the points of the $\varepsilon/2$–net S_2 completely cover the ball B_1 then the intersection of the ball B_1 at least with one of these balls, for instance, with B_2 contains an infinite set of the points x_p. While continuing this process we get an infinite sequence of the balls $\{B_n\}$ and the correspondent sequence of open sets $C_n = B_1 \ldots B_n$, $C_{n+1} \subset C_n$, each of them contains an infinite set of the points x_p. After choosing in each set C_n the point x_{p_n} $(n = 1, 2, \ldots)$ we obtain the subsequence $\{x_{p_n}\}$. This subsequence is fundamental as all the points x_{p_n} beginning with x_{p_s} are contained in the set C_s. By virtue of the completeness of the space X this subsequence has the limit in X which is the limit point of the sequence $\{x_n\}$. Thus any sequence of the points of the space X has the limit points. By Corollary 1 of Theorem 4.31 the space X is countable compact and by virtue of Theorem 4.40 it is compact. ◁

Corollary 1. *A set in a complete metric space is precompact if and only if it is totally bounded.*

▷ For proving it is sufficient to notice that the closed set $[A]$ in a complete metric space is itself a complete metric space. ◁

Corollary 2. *Any precompact (compact)set in a complete metric space is bounded.*

▷ It follows from the previous Corollary 1 and Theorem 4.37. ◁

Corollary 3. *In the finite-dimensional space R^n any bounded set is precompact.*

Corollary 4. *An union of a finite set of the compacts in a complete metric space is a compact.*

▷ At any $\varepsilon > 0$ the union of finite ε–net $S_1, \ldots, S_n$ for the compacts $A_1, \ldots, A_n$,

$$S = \bigcup_{k=1}^{n} S_k,$$

is the finite ε–net for the union of these compacts

$$A = \bigcup_{k=1}^{n} A_k. \qquad ◁$$

4.5.3. Continuous Functions on a Compact

Let us pass now to the functions determined on the compacts in a metric space.

Theorem 4.42. *A continuous function mapping a compact of a metric space into a metric space is uniformly continuous.*

▷ Let $f(x)$be a continuous function mapping the compact A of the metric space (X, d) into the metric space (Y, r). If $f(x)$ is not uniformly continuous then for any $\delta > 0$ there exists such number $\varepsilon > 0$ and such pairs of the points $x_n, x_n' \in A$ ($n = 1, 2, \ldots$) that $d(x_n, x_n') < \delta/n$ and $r(f(x_n), f(x_n')) > \varepsilon$. As A is a compact then on the basis of Theorem 4.31 the set of the points $\{x_n\}$ has the limit point $x_0 \in A$ which is also evidently the limit point of the set $\{x_n'\}$.Consequently, the sequences $\{x_n\}$ and $\{x_n'\}$contain the subsequences $\{x_{n_p}\}$ and $\{x_{n_p}'\}$ convergent to x_0, $\lim\limits_{p \to \infty} x_{n_p} = \lim\limits_{p \to \infty} x_{n_p}' = x_0$. But at any p

$$\varepsilon < r(f(x_{n_p}), f(x_{n_p}')) \leq r(f(x_{n_p}), f(x_0)) + r(f(x_{n_p}'), f(x_0)),$$

and as a result at least one of the numbers $r(f(x_{n_p}), f(x_0))$, $r(f(x_{n_p}'), f(x_0))$ is more than $\varepsilon/2$. Thus at any $\delta > 0$ in the vicinity $S_\delta(x_0)$ of the point x_0 such a point x_{n_p} or x_{n_p}' will be found that

$$d(x_{n_p}, x_0) < \delta, \qquad r(f(x_{n_p}), f(x_0)) > \varepsilon/2$$

or
$$d(x'_{n_p}, x_0) < \delta, \qquad r(f(x'_{n_p}), f(x_0)) > \varepsilon/2.$$
This contradicts the continuity of the function $f(x)$. ◁

Theorem 4.43. *A continuous real function on a compact is bounded and achieves its smallest and largest values.*

▷ Let $f(x)$ be a continuous real function determined on the compact A. By Corollary of Theorem 4.34 the image $f(A)$ of the compact A in R which is determined by the function $f(x)$ is a compact. As the real axis R represents a complete metric space then on the basis of Corollary 2 of Theorem 4.41 the compact $f(A)$ represents a bounded set. Consequently, the function $f(x)$ is bounded. As by Theorem 4.29 the compact $f(A)$ is closed then it contains its bound. It proves that the function $f(x)$ achieves its smallest and largest values. ◁

4.5.4. A Criterion of Precompactness of a Set of Continuous Functions

Let X be a space of continuous functions $x(t)$ which map the metric compact (T, r) into the complete metric space (S, ρ). Introducing into the space X the metric
$$d(x_1, x_2) = \sup_{t \in T} \rho(x_1(t), x_2(t)), \qquad (4.3)$$
we obtain the metric space (X, d).

Theorem 4.44. *The space (X, d) is complete.*

▷ Let $\{x_n(t)\}$ be a fundamental sequence of the functions from (X, d). We take an arbitrary $\varepsilon > 0$. From $d(x_n, x_m) < \varepsilon$ at all $n, m > N = N_\varepsilon$ and formula (4.3) follows that $\rho(x_n(t), x_m(t)) < \varepsilon$ at all $n, m > N = N_\varepsilon$, $t \in T$. As the metric space (S, ρ) is complete then at each t there exists the limit $x(t) = \lim_{n \to \infty} x_n(t)$, and by virtue of the independence $N = N_\varepsilon$ of t the convergence of the sequence of the functions $\{x_n(t)\}$ to $x(t)$ is uniform on $t \in T$. The theorem will be proved if we show that the limit function $x(t)$ is continuous. At any $\varepsilon > 0$ we choose a natural number n in such a way that
$$\rho(x_n(t), x(t)) < \varepsilon/3 \qquad \forall t \in T. \qquad (4.4)$$

It is possible due to the uniform convergence of the sequence $\{x_n(t)\}$ to $x(t)$ on T. After this we shall choose the number $\delta = \delta_\varepsilon > 0$ in such a way that
$$\rho(x_n(t), x_n(t')) < \varepsilon/3 \quad \forall t, t', r(t, t') < \delta. \qquad (4.5)$$

It is possible as by Theorem 4.49 any continuous on the compact function is uniformly continuous. From (4.4) and (4.5) follows

$$\rho(x(t), x(t')) \leq \rho(x(t), x_n(t)) + \rho(x_n(t), x_n(t'))$$

$$+\rho(x_n(t'), x(t')) < \varepsilon \quad \forall t, t', r(t, t') < \delta,$$

what proves the continuity of the function $x(t)$. ◁

Theorem 4.44 generalizes Theorem 1.13 on the spaces of continuous functions with the values in any metric spaces. The restriction of the domain of the function by the condition of the compactness is easily removed by the modification of the proof of the continuity of the limit function $x(t)$ as it is performed while proving Theorem 1.13.

Theorem 4.45. *The set $A = \{x_\alpha(t)\}$ of the functions of a space X is precompact if and only if the union of the ranges of all the functions of the set A is precompact and all the functions of the set A are equicontinuous[a].*

▷ If A is precompact then at any $\varepsilon > 0$ there exists in it the finite ε-net $\Phi = \{\varphi_1(t), \ldots, \varphi_N(t)\}$ where all the functions $\varphi_1, \ldots, \varphi_N$ are continuous. Here for any function $x_\alpha(t) \in A$ such function $\varphi_k(t) \in \Phi$ will be found that

$$d(x_\alpha, \varphi_k) = \sup_{t \in T} \rho(x_\alpha(t), \varphi_k(t)) < \varepsilon. \tag{4.6}$$

Owing to the continuity of the functions $\varphi_1(t), \ldots, \varphi_N(t)$ and Theorem 4.34 all the sets $\varphi_1(T), \ldots, \varphi_N(T) \subset S$ are compact, and due to Corollary 4 of Theorem 4.41 their union $\bigcup_{k=1}^{N} \varphi_k(T)$ is compact. Consequently, in $\bigcup \varphi_k(T)$ there exists the finite ε-net. This net by virtue of inequality (4.6) is the finite 2ε-net for the union B of the ranges of all the functions $x_\alpha(t) \in A$. Thus the union B of the ranges of all the functions $x_\alpha(t) \in A$ is totally bounded. Hence on the basis of Corollary 1 of Theorem 4.41 follows that B is precompact. This proves the necessity of the first condition. For proving the necessity of the second condition notice

---

[a] The functions belonging to set $A = \{x_\alpha(t)\} \subset X$ are called *equicontinuous* if at any $\varepsilon > 0$ for all the functions $x_\alpha(t) \in A$ there exists one and the same such a number $\delta = \delta_\varepsilon > 0$ that $\rho(x_\alpha(t), x_\alpha(t')) < \varepsilon \; \forall t, t', r(t, t') < \delta$ (Subsection 1.2.8).

that by virtue of the continuity of the functions $\varphi_1(t), \ldots, \varphi_N(t)$ for any function $\varphi_l(t)$ at chosen $\varepsilon > 0$ there exists such a number $\delta_l > 0$ that

$$\rho(\varphi_l(t), \varphi_l(t')) < \varepsilon \qquad \forall t, t', r(t, t') < \delta_l. \tag{4.7}$$

Let $\delta = \min(\delta_1, \ldots, \delta_N)$. Then for any function $x_\alpha(t) \in A$ by virtue of inequalities (4.6) and (4.7)

$$\rho(x_\alpha(t), x_\alpha(t')) \leq \rho(x_\alpha(t), \varphi_k(t))$$

$$+\rho(\varphi_k(t), \varphi_k(t')) + \rho(\varphi_k(t'), x_\alpha(t')) < 3\varepsilon \tag{4.8}$$

at all $t, t' \in T$ satisfying the condition $r(t, t') < \delta$. As the number δ is the same for all the functions $x_\alpha(t) \in A$ then from inequality (4.8) follows that all the functions of the set A are equicontinuous. This proves the necessity of the second condition.

For proving the sufficiency of the conditions suppose that the union B of the ranges of all the functions of the set A is precompact and all the functions of the set A are equicontinuous. Then at any $\varepsilon > 0$ there exists such number $\delta = \delta_\varepsilon > 0$ that for all functions $x_\alpha(t) \in A$

$$\rho(x_\alpha(t), x_\alpha(t')) < \varepsilon \qquad \forall t, t', r(t, t') < \delta. \tag{4.9}$$

By virtue of the compactness of T there exists in it the finite δ–net $\{t_1, \ldots, t_m\}$ and due to the precompactness of B there exists in it the finite ε–net $\{s_1, \ldots, s_n\}$. Determine the sets

$$C_k = \{t \in T : r(t, t_k) < \delta\} \qquad (k = 1, \ldots, m),$$

$$D_1 = C_1, \; D_k = C_k \bar{C}_1 \ldots \bar{C}_{k-1} = D_{k-1} \bar{C}_{k-1} \quad (k = 2, \ldots, m)$$

and n^m finite–valued functions

$$\psi_{q_1, \ldots, q_m}(t) = \sum_{k=1}^{m} s_{q_k} \mathbf{1}_{D_k}(t) \quad (q_1, \ldots, q_m = 1, \ldots, n). \tag{4.10}$$

Then for any functions $x_\alpha(t) \in A$ such $s_{r_1}, \ldots, s_{r_m}$ in the ε–net $\{s_1, \ldots, s_n\}$ will be found that

$$\rho(x_\alpha(t_k), s_{r_k}) < \varepsilon \quad (k = 1, \ldots, m), \tag{4.11}$$

and at any $t \in D_k$ the correspondent function $\psi_{r_1, \ldots, r_m}(t)$ by virtue of (4.9) — (4.11) will satisfy the condition

$$\rho(x_\alpha(t), \psi_{r_1, \ldots, r_m}(t)) = \rho(x_\alpha(t), s_{r_k})$$

$$\leq \rho(x_\alpha(t), x_\alpha(t_k)) + \rho(x_\alpha(t_k), s_{r_k}) < 2\varepsilon.$$

As $\bigcup_{k=1}^{m} D_k = T$ then the inequality

$$\rho(x_\alpha(t), \psi_{r_1,\ldots,r_m}(t)) < 2\varepsilon$$

is valid at all $t \in T$, and consequently,

$$d(x_\alpha, \psi_{r_1,\ldots,r_m}) = \sup_{t\in T} \rho(x_\alpha(t), \psi_{r_1,\ldots,r_m}(t)) < 2\varepsilon. \qquad (4.12)$$

Hence owing to the arbitrariness of the choice of the function $x_\alpha(t) \in A$ follows that the functions $\psi_{q_1,\ldots,q_m}(t)$ form the finite 2ε–net for the set A. The proof may be ended at this point. But the functions $\psi_{q_1,\ldots,q_m}(t)$ do not belong to the set A but also to the space X while Theorem 4.41 and its Corollary 1 were proved only for the case when the correspondent ε–nets are the sets of the space X. Therefore we apply the approach mentioned at the beginning of Subsection 4.5.1 in order to substitute the functions $\psi_{q_1,\ldots,q_m}(t)$ by some functions of the set A.

Any function $\psi_{q_1,\ldots,q_m}(t)$ we replace by some function $x_{\alpha_{q_1,\ldots,q_m}}(t)$ $\in A$ for which

$$d(x_{\alpha_{q_1,\ldots,q_m}}, \psi_{q_1,\ldots,q_m}) < 2\varepsilon. \qquad (4.13)$$

If there is no such function $x_{\alpha_{q_1,\ldots,q_m}}(t)$ in the set A then we drop the correspondent function $\psi_{q_1,\ldots,q_m}(t)$ [a].

After establishing the unified numbering for the finite set of the function $x_{\alpha_{q_1,\ldots,q_m}}(t)$ we denote them in terms of $\varphi_1(t), \ldots, \varphi_N(t)$. Then for any function $x_\alpha(t) \in A$ after taking the correspondent function $\psi_{r_1,\ldots,r_m}(t)$ which satisfies condition (4.12) and the function $\varphi_k(t)$ $= x_{\alpha_{r_1,\ldots,r_m}}(t)$ which satisfies condition (4.13) we get

$$d(x_\alpha, \varphi_k) < 4\varepsilon.$$

It proves that the set of the functions $\{\varphi_1(t), \ldots, \varphi_N(t)\} \subset A$ represents the finite 4ε–net for the set A. Hence by virtue of $\varepsilon > 0$ follows that the

---

[a] It is evident that not for any function $\psi_{q_1,\ldots,q_m}$ there exists the function x_α $\in A$ satisfying condition (4.13). It is evident that the functions $x_\alpha(t) \in A$ which satisfy condition (4.13) exist only for such functions $\psi_{q_1,\ldots,q_m}(t)$ whose values on the neighbouring sets D_k differ no more than on 4ε, i.e. for which the condition $\rho(s_k, s_l) \leq 4\varepsilon$ for any k and l, $k < l$ for which $D_k \bar{C}_{k+1} \ldots \bar{C}_{l-1} C_l \neq \emptyset$.

set A is totally bounded. On the basis of Corollary 1 of Theorem 4.41 the set of the functions $A = \{x_\alpha(t)\}$ is precompact. ◁

In the special case of the finite–dimensional space $S = R^n$ or $S = C^n$ any precompact set in S may be enclosed into a ball of the finite radius $S_M(0) = \{s : |s| < M\}$ and Theorem 4.45 may be formulated in the following way.

Theorem 4.46. *The set* $A = \{x_\alpha(t)\}$ *of the functions of the space* $C(T, K^n)$ *of the continuous functions mapping the compact* T *into the space* K^n *is precompact if and only if all the functions of the set* A *are uniformly bounded and are equicontinuous* (Ascoli–Arzela theorem).

Detailed theory of compactness in the metric spaces is given in (Kelly 1957).

4.6. Topologies in Linear Spaces

4.6.1. Definition of a Topological Linear Space

Usually in a linear space a topology is determined in such a way that the operations of the vectors addition and the multiplication of a vector by a number are continuous. From the definition of the continuity of the function (Subsection 4.3.2) follows that at such determination of a topology for any two vectors x and y and for any vicinity V_z of the point $z = x + y$ there exist such vicinities V_x, V_y of the points x, y that $V_x + V_y \subset V_z$. In particular, for any vicinity of zero V there exists such vicinity of zero U that $U + U \subset V$. Besides, at such definition of a topology for any vector x, any number λ and any vicinity V_u of the point $u = \lambda x$ there exist such vicinities V_x, $V_\lambda = \{\lambda' : |\lambda' - \lambda| < \varepsilon\}$ of the point x and λ that $u' = \lambda' x' \in V_u$ at any $x' \in V_x$, $\lambda' \in V_\lambda$.

A linear space with a topology in which the operations of the vectors addition and the multiplication of a vector by a number are continuous is called *a topological linear space*.

If a topological linear space is a T_1–space then it is called *a separable* space.

Theorem 4.47. *If A is an open set of topological linear space X then for any vector y, any set B of a space X and any number $\lambda \neq 0$ the sets $A + y$, $A + B$ and λA are open.*

▷ At any fixed y the mapping $z = x - y$ of a space X on itself is continuous according to the definition, and from $z \in A$ follows $x \in A + y$, i.e. $A + y$ serves as the inverse image of the set A. Consequently, by Theorem 4.19 the set $A + y$ is open. As at any B

$$A + B = \bigcup_{y \in B} (A + y),$$

then the set $A + B$ is also open. Similarly at any fixed λ the set λA is open as the inverse image of an open set A at continuous mapping $z = x/\lambda$ of the space X on itself. ◁

Corollary. *In a topological linear space X any vicinity V_x of any point x is obtained by the shift of some vicinity of zero V_0 on the vector x.*

▷ By the proved theorem the set $V_0 = V_x - x$ is open. As it contains zero $x \in V_x$ then it is the vicinity of zero. Consequently, $V_x = V_0 + x$ where V_0 is the vicinity of zero. ◁

The set A of a topological linear space X is called *bounded* if for any vicinity of zero V there exists such a number $c > 0$ that $A \subset \lambda V$ at all λ, $|\lambda| > c$. Any finite set $\{x_1, \dots, x_N\}$ is bounded as by virtue of the continuity of the operation of vector multiplication by a number for any vicinity of zero V there exists such $\delta > 0$ that $\{\alpha x_1, \dots, \alpha x_N\} \subset V$ at all α, $|\alpha| < \delta$. Whence follows $\{x_1, \dots, x_N\} \subset \lambda V$ at all λ, $|\lambda| > \delta^{-1}$.

4.6.2. Fundamental Sequences

Unlike a general topological space we may introduce the notion of a fundamental or a Cauchy sequence in a topological linear space. The sequence of the points $\{x_n\}$ of a topological linear space X is called *fundamental* if any vicinity of zero contains all the points $x_n - x_m$ at $n, m > N$, where N is some natural number dependent on the chosen vicinity of zero. In other words, the sequence $\{x_n\}$ is fundamental if the sequence $\{x_n - x_m\}$ converges to zero.

Theorem 4.48. *Any convergent sequence in a topological linear space is a fundamental sequence.*

▷ If the sequence $\{x_n\}$ converges to x_0 then any vicinity of zero contains all the points $x_n - x_0$ beginning with someone. By virtue of the continuity of the addition operation (it means also the operation of the subtraction) in a topological linear space for any vicinity of zero V there exits such a vicinity of zero U that $U - U \subset V$. As U contains all the points $x_n - x_0$ at $n > N$ where N is some number dependent on U, and consequently, also on V then V contains all the points $x_n - x_m$ at $n, m > N$ what proves that the sequence $\{x_n\}$ is fundamental. ◁

R e m a r k. The inverse in the general case is not true. The same
is in the case of a metric space. A fundamental sequence in a topological
linear space may not converge to anyone limit (at none of the points of
this space).

A topological linear space is called *complete* if any fundamental
sequence has the limit in it.

4.6.3. Local Convexity of a Space

A topological linear space is called *locally convex* if any vicinity of
zero contains a convex vicinity of zero. From this definition and Corollary
of Theorem 4.47 follows that any vicinity of any point of a locally convex
topolgical linear space contains a convex vicinity of this point. Hence it
follows that any nonempty open set of a locally convex topological linear
space contains a nonempty convex open set. To ensure it is sufficient to
notice that any nonempty open set G is the vicinity of its any point x.
The convex vicinity V_x of the point x which is contained in G represents
a nonempty convex open set which is contained in G.

4.6.4. Ways to Assign Topology in a Linear Space

On the basis of the Corollary of Theorem 4.47 for the definition of
a topology in a linear space it is sufficient to assign a base of a vicinity
of zero. Then the base of the vicinities of any point x will be obtained
by a shift of a base of zero vicinities on the vector x.

In a normed linear space X the base of zero vicinities represents a
set of all open balls with the center in zero

$$S_\varepsilon(0) = \{x : \| x \| < \varepsilon\} \quad \forall \varepsilon > 0. \tag{4.14}$$

A normed linear space with such a topology similarly as any metric space
is a space with the first countability axiom.

It is easy to check that the operations of the vectors addition and
vector multiplication by a number are continuous in the topology determined by the vicinities of zero (4.14). Thus any normed linear space
represents a topological linear space.

A topology induced by a norm is determined by the single inverse
mapping correspondent to the mapping $z = \|x\|$ of a space X into the
real axis R. According to the general method of introducing the topology
in a given space we may use a set of inverse mappings from different
spaces. In particular, the topology in X by means of the norms set

may be determined. The important class of the topological linear spaces are the *countable normed spaces* in which the topology is determined by countable set of the norms satisfying some conditions.

In an arbitrary linear space the base of the vicinities of zero may be determined by means of the set of linear functionals by the transfer of the base of zero vicinities of the scalars field $|z| < \varepsilon$ into the space X by the inverse mappings (Subsection 4.3.3). Let F be any set of linear functionals on X. Then the base of zero vicinities in X may be determined as an aggregate of zero vicinities

$$\{x : |f_1 x| < \varepsilon, \ldots, |f_n x| < \varepsilon\}, \qquad (4.15)$$

correspondent to all n, $\varepsilon > 0$, $f_1, \ldots, f_n \in F$. Each such zero vicinity represents the intersection of the inverse images of zero vicinities $|z| < \varepsilon$ at the mappings $z = f_k x$ $(k = 1, \ldots, n)$ X in K.

On the basis of Theorem 4.19 all the functionals $f \in F$ are continuous in the topology τ_F induced by the base of zero vicinities (4.15).

Theorem 4.49. *The topology τ_F is the weakest topology in which all its functionals of the set F are continuous. If F is a linear space then any linear functional continuous in the topology τ_F belongs to F.*

▷ Let τ be any topology in which all the functionals of the set F are continuous. By Theorem 4.19 all the sets $\{x : |fx| < < \varepsilon\}$, $f \in F$ are open in τ as the inverse images of open sets at continuous mapping $z = fx$, i.e. belong to τ. Consequently, all the sets (4.15) are also open in τ as the finite intersections of open sets, i.e. τ contains the base of zero vicinities (4.15) of the topology τ_F, and here too, any open set from τ_F. Thus $\tau_F \subset \tau$.

Let φ be any linear functional continuous in the topology τ_F. According to Theorem 4.19 the set $\{x : |\varphi x| < 1\}$ which contains zero is open in the topology τ_F and consequently, represents zero vicinity in X as the inverse image of zero vicinity $\{z : |z| < 1\}$ of continuous mapping $z = \varphi x$. But any zero vicinity contains some zero vicinity which belongs to the base of zero vicinities. Therefore there are such $\varepsilon > 0$ and linear independent functionals $f_1, \ldots, f_n \in F$ that

$$\{x : |f_1 x| < \varepsilon, \ldots, |f_n x| < \varepsilon\} \subset \{x : |\varphi x| < 1\}.$$

In particular, the intersection A of the kernels of all the functionals $f_1, \ldots, f_n$ is contained in $\{x : |\varphi x| < 1\}$. Thus from $x \in A$ follows $|\varphi x| < 1$. As A represents the subspace of the space X then at any

$\delta > 0$ from $x \in A$ follows $x/\delta \in A$ and $|\varphi x/\delta| < 1$ whence we get $|\varphi x| < \delta$. Hence by virtue of the arbitrariness of $\delta > 0$ follows $\varphi x = 0$ for any $x \in A$, i.e. $A \subset \ker \varphi$. Now we notice that on the basis of Theorem 1.18 any vector x of the space X may be presented in the form

$$x = x_0 + \alpha_1 x_1 + \ldots + \alpha_n x_n,$$

where $x_0 \in A$ and $x_1, \ldots, x_n$ are linear independent vectors which do not belong to A. Consequently, for any $x \in X$

$$\varphi x = \alpha_1 \varphi x_1 + \ldots + \alpha_n \varphi x_n,$$
$$f_k x = \alpha_1 f_k x_1 + \ldots + \alpha_n f_k x_n \quad (k = 1, \ldots, n).$$

But the numbers $-1, \alpha_1, \ldots, \alpha_n$ which are not equal to zero simultaneously may satisfy the system of $n + 1$ homogeneous linear equations:

$$-\varphi x + \alpha_1 \varphi x_1 + \ldots + \alpha_n \varphi x_n = 0,$$
$$-f_k x + \alpha_1 f_k x_1 + \ldots + \alpha_n f_k x_n = 0 \quad (k = 1, \ldots, n)$$

only in the case when the determinant of this system is equal to zero:

$$\begin{vmatrix} \varphi x & \varphi x_1 & \ldots & \varphi x_n \\ f_1 x & f_1 x_1 & \ldots & f_1 x_n \\ \vdots & \vdots & \ddots & \vdots \\ f_n x & f_n x_1 & \ldots & f_n x_n \end{vmatrix} = 0.$$

After expanding the determinant in powers of the elements of the first column we obtain

$$c_0 \varphi x + c_1 f_1 x + \ldots + c_n f_n x = 0,$$

where the numbers $c_0, c_1, \ldots, c_n$ are independent of x. From the fact that this equality is valid at any x follows

$$c_0 \varphi + c_1 f_1 + \ldots + c_n f_n = 0,$$

and from linear independence of the functionals $f_1, \ldots, f_n$ follows that c_0 cannot be equal to zero. So the functional φ represents a linear combination of the functionals $f_1, \ldots, f_n \in F$. As F is a linear space then $\varphi \in F$. ◁

Thus the set of all linear functionals on X continuous in the topology τ_F coincides with F.

Theorem 4.50. *The operations of the vectors addition and vector multiplication by a number are continuous in the topology τ_F.*

▷ Let V_z be any vicinity of the point $z = x + y$. It contains some base vicinity of the point z:

$$V_z \supset \{z' : | f_1 z' - f_1 z | < \varepsilon, \ldots, | f_n z' - f_n z | < \varepsilon\}.$$

We determine the vicinities of the points x and y:

$$V_x = \{x' : | f_1 x' - f_1 x | < \varepsilon/2, \ldots, | f_n x' - f_n x | < \varepsilon/2\},$$

$$V_y = \{y' : | f_1 y' - f_1 y | < \varepsilon/2, \ldots, | f_n y' - f_n y | < \varepsilon/2\}.$$

Then for any $x' \in V_x$, $y' \in V_y$ and $z' = x' + y'$ we shall have

$$| f_k z' - f_k z | = | f_k x' + f_k y' - f_k x - f_k y |$$

$$\leq | f_k x' - f_k x | + | f_k y' - f_k y | < \varepsilon/2 + \varepsilon/2 = \varepsilon \quad (k = 1, \ldots, n),$$

i.e. $z' \in \{z' : | f_1 z' - f_1 z | < \varepsilon, \ldots, | f_n z' - f_n z | < \varepsilon\} \subset V_z$, and consequently, $V_x + V_y \subset V_z$.

Let x, λ be any vector and any number, V_u be any vicinity of the point $u = \lambda x$. It contains some base vicinity of this point u,

$$V_u \supset \{u' : | f_1 u' - f_1 u | < \varepsilon, \ldots, | f_n u' - f_n u | < \varepsilon\}.$$

We determine the vicinities of the points x and λ:

$$V_x = \{x' : | f_1 x' - f_1 x | < \varepsilon/2\lambda_1, \ldots, | f_n x' - f_n x | < \varepsilon/2\lambda_1\},$$

$$V_\lambda = \{\lambda' : | \lambda' - \lambda | < \varepsilon/2 [\max_{1 \leq m \leq n} | f_m x | + \varepsilon/2\lambda_1]\},$$

where $\lambda_1 = \max(1, | \lambda |)$. Then we shall have for any $x' \in V_x$, $\lambda' \in V_\lambda$, $u' = \lambda' x'$

$$| f_k u' - f_k u | \leq | \lambda | | f_k x' - f_k x | + | \lambda' - \lambda | | f_k x' |$$

$$< \varepsilon/2 + \varepsilon | f_k x' | /2 [\max_{1 \leq m \leq n} | f_m x | + \varepsilon/2\lambda_1] \leq \varepsilon/2 + \varepsilon [| f_k x |$$

$$+ \varepsilon/2\lambda_1]/2 [\max_{1 \leq m \leq n} | f_m x | + \varepsilon/2\lambda_1] < \varepsilon \quad (k = 1, \ldots, n),$$

i.e. $u' \in \{u' : | f_1 u' - f_1 u | < \varepsilon, \ldots, | f_n u' - f_n u | < \varepsilon\} \subset V_u$ and $\lambda' x' = u' \in V_u$ at any $x' \in V_x$, $\lambda' \in V_\lambda$. ◁

Theorem 4.51. *In order a topological linear space X with the topology τ_F be a T_2-space it is necessary and sufficient that for any two different points x, y, $x \neq y$, there appears such a functional f in F that $fx \neq fy$.*

▷ If V_x, V_y are nonintersecting vicinities of the points x, y, $x \neq y$ then there exists a base vicinity of the point x which is contained in V_x

$$\{x' : | f_1 x' - f_1 x | < \varepsilon, \ldots, | f_n x' - f_n x | < \varepsilon\} \subset V_x.$$

This vicinity does not also intersect with V_y, and consequently, does not contain the point y, i.e. $| f_k y - f_k x | > \varepsilon$ and $f_k x \neq f_k y$ $(k = 1, \ldots, n)$.

If for any points x, y, $x \neq y$ there exists such a functional $f \in F$ that $fx \neq fy$ then the vicinities

$$V_x = \{x' : | fx' - fx | < \varepsilon\}, \quad V_y = \{y' : | fy' - fy | < \varepsilon\}$$

of the points x, y do not intersect at any $\varepsilon < | fx - fy | / 2$. ◁

On the basis of this theorem the set of the linear functionals F satisfying the condition of the theorem is called the *set separating the points of the space X*.

Theorem 4.52. *In order the set of linear functionals F should separate the points of the space X it is necessary and sufficient that from $fx = 0$ at all $f \in F$ should follow $x = 0$ (i.e. the intersections of the kernels of all the functionals $f \in F$ should represent a singleton $\{0\}$).*

▷ For proving it is sufficient to notice that if $x \neq 0$ and $fx = 0$ for all $f \in F$ then for two different points x_1 and $x_2 = x_1 + x$, $fx_1 = fx_2$ for all $f \in F$, i.e. the condition is necessary. If $fx = 0$ all $f \in F$ only at $x = 0$ then at any x_1, x_2, $x_1 \neq x_2$ will occur such a functional $f \in F$ that $f(x_1 - x_2) \neq 0$, and consequently, $fx_1 \neq fx_2$; i.e. the condition is sufficient. ◁

Later on we shall always suppose that the set F of linear functionals separates the points of the space and besides is a linear space itself.

Theorem 4.53. *The linear topological space X with the topology τ_F is locally convex.*

▷ It is sufficient to prove that all the base zero vicinities are convex. Then it will follow from the definition of the base of the vicinities of the point that any zero vicinity contains a convex vicinity of zero. Let x and y, $x \neq y$ be the points of base vicinity of zero

$$V = \{z : | f_1 z | < \varepsilon, \ldots, | f_n z | < \varepsilon\}.$$

From $| f_k x | < \varepsilon$, $| f_k y | < \varepsilon$ $(k = 1, \ldots, n)$ follows at any $\alpha \in (0, 1)$

$$| f_k(\alpha x + (1 - \alpha)y) | = | \alpha f_k x + (1 - \alpha)f_k y | \leq \alpha | f_k x |$$

$$+(1 - \alpha) | f_k y | < \alpha \varepsilon + (1 - \alpha)\varepsilon = \varepsilon \quad (k = 1, \ldots, n),$$

i.e. $\alpha x + (1 - \alpha)y \in V$. ◁

Theorem 4.54. *The sequence $\{x_n\}$ converges to x in τ_F if and only if the numeric sequence $\{fx_n\}$ converges to fx at any $f \in F$.*

▷ From the convergence of $\{x_n\}$ follows that any vicinity $V_x \in \tau_F$ of point x contains all the points x_n beginning with someone. In particular, at any $\varepsilon > 0$, $f \in F$ the vicinity $V_x = \{x' : | fx' - fx | < \varepsilon\}$ at the point x contains all the points x_n beginning with someone, i.e. $| fx_n - fx | < \varepsilon$ at all sufficiently large n. It proves the convergence of the sequence $\{fx_n\}$ to fx at any $f \in F$. Consequently, the condition is necessary.

For proving the sufficiency we notice that any vicinity $V_x \in \tau_F$ of the point x contains some base vicinity

$$V_x \supset \{x' : | f_1 x' - f_1 x | < \varepsilon, \ldots, | f_N x' - f_N x | < \varepsilon\}.$$

From the convergence of the sequence $\{fx_n\}$ to fx at any $f \in F$ follows the convergence of all the sequences $\{f_k x_n\}$ $(k = 1, \ldots, N)$, i.e. we get the inequality $| f_k x_n - f_k x | < \varepsilon$ at all sufficiently large n. Consequently, V_x contains all the points x_n beginning with someone. This proves the sufficiency of the condition. ◁

4.6.5. Continuous Linear Functions in Topological Linear Spaces

Let us study peculiar properties of continuous linear functions in topological linear spaces.

Theorem 4.55. *The linear function $y = f(x)$ mapping a topological linear space X into another topological linear space Y is continuous everywhere if it is continuous at least at one point.*

▷ Suppose that the linear function $y = f(x)$ is continuous at the point x_0. Let x be any point of the space X, V_y be any vicinity of the point $y = f(x)$. The set $V_{y_0} = V_y + y_0 - y$ represents the vicinity of the point $y_0 = f(x_0)$. Due to the continuity of f at the point x_0 there exists the vicinity V_{x_0} of the point x_0 whose image is contained in V_{y_0}, $f(V_{x_0}) \subset V_{y_0}$. But then $V_x = V_{x_0} + x - x_0$ will be the vicinity of the point x whose image is contained in V_y:

$$f(V_x) = f(V_{x_0}) + f(x) - f(x_0) \subset V_{y_0} + y - y_0 = V_y . \quad \triangleleft$$

4.6.6. Bounded Linear Functions

The linear function $y = f(x)$ mapping a topological linear space X into a normed linear space Y is called a *bounded* if it is bounded in some zero vicinity V_0:

$$\| f(x) \| < c \quad \forall x \in V_0 . \tag{4.16}$$

Theorem 4.56. *In order the linear function $y = f(x)$ mapping a topological linear space X into a normed linear space Y be continuous it is necessary and sufficient that it should be bounded.*

$\triangleright$ If the function f is continuous then for any $\varepsilon > 0$ there exists such zero vicinity V_ε that

$$\| f(x) \| < \varepsilon \quad \forall x \in V_\varepsilon ,$$

i.e. the function f is bounded in V_ε. On the contrary, if the function f is bounded then there exist such number $c > 0$ and such zero vicinity V_0 that (4.16) is valid. But then at any $\varepsilon > 0$ and $x \in V_0$, $z = (\varepsilon/c)x \in (\varepsilon/c)V_0 = V_\varepsilon$ and $\| f(z) \| = (\varepsilon/c) \| f(x) \| < \varepsilon$. Consequently, at any $\varepsilon > 0$ there exists such zero vicinity V_ε that $\| f(z) \| < \varepsilon$ at all $z \in V_\varepsilon$ what proves the continuity of the function f in zero. By Theorem 4.55 it is continuous everywhere. $\triangleleft$

R e m a r k. Thus the linear function with the values in a normed linear space is continuous if and only if it is bounded.

4.6.7. Bounded Linear Functions in Normed Spaces

Bounded (continuous) linear functions in a normed linear space possess the peculiar properties.

Theorem 4.57. *The linear function $y = f(x)$ mapping a normed linear space X into a normed linear space Y is bounded if and only if there exists such number $C > 0$ that*

$$\| f(x) \| < C \| x \| \quad \forall x \in X . \tag{4.17}$$

▷ From the definition of the boundedness of the linear function follows that in the case of a normed linear space X

$$\| f(x) \| < c \quad \forall x, \quad \| x \| < r, \tag{4.18}$$

where $c > 0$, $r > 0$ are some numbers. Really, in this case any zero vicinity in X contains some ball base vicinity of zero. As at any $x \in X$ and any $\varepsilon > 0$ the norm of the vector $y = (r - \varepsilon)x/ \| x \|$ is less than r, $\| y \| = r - \varepsilon < r$ then from inequality (4.18) follows

$$\| f(y) \| = \frac{r - \varepsilon}{\| x \|} \| f(x) \| < c, \quad \| f(x) \| < C \| x \|, \quad C = c/(r - \varepsilon)$$

at all x. Thus condition (4.17) is necessary.

For proving the sufficiency of condition (4.17) notice that from (4.17) follows

$$\| f(x) \| < Cr = c \quad \forall x, \quad \| x \| < r,$$

i.e. we get (4.18) what proves the boundedness of the function f. ◁

4.6.8. Norm of a Linear Function

It is clear that if condition (4.17) is fulfilled at some $C > 0$ then it is fulfilled also at all large C. From condition (4.17) follows

$$\sup_x \frac{\| f(x) \|}{\| x \|} \leq C.$$

The variable

$$\| f \| = \sup_x \frac{\| f(x) \|}{\| x \|} = \sup_{\|x\|=1} \| f(x) \| \tag{4.19}$$

is called *a norm* of the continuous function f.

In particular, if the function $f(x)$ represents continuous linear functional $f(x) = fx$ formula (4.19) determines a norm of a continuous linear functional:

$$\| f \| = \sup_x \frac{| fx |}{\| x \|} = \sup_{\|x\|=1} | fx | . \tag{4.20}$$

If the function $f(x)$ represents a continuous linear operator $f(x) = Ax$ formula (4.19) determines a norm of continuous linear operator:

$$\| A \| = \sup_x \frac{\| Ax \|}{\| x \|} = \sup_{\|x\|=1} \| Ax \| . \tag{4.21}$$

It is easy to see that the norm of the linear function $y = f(x)$ represents a greatest lower bound of the numbers $C > 0$ for which inequality (4.17) is valid. Really, according to the definition of the supremum from (4.19) follows that for any $\varepsilon > 0$ there exists such x for which

$$\frac{\|\, f(x)\, \|}{\|\, x\, \|} > \|\, f\, \| - \varepsilon, \quad \text{i.e.} \quad \|\, f(x)\, \| > (\|\, f\, \| - \varepsilon) \|\, x\, \| \ .$$

Hence it is clear that at C as close as one likes to $\|\, f\, \|$ but smaller than $\|\, f\, \|$ inequality (4.17) is violated.

4.6.9. Hahn–Banach Theorem for Normed Linear Spaces

A norm of a vector in a normed linear space $\|\, x\, \|$ represents a convex functional (Subsection 1.4.6). The product of a norm $\|\, x\, \|$ by any number $a > 0$ also represents a convex functional. Therefore inequality (4.17) in the case of the linear functional f represents the condition of the subjection $|\, f(x)\, | \le p(x)$ at $p(x) = \|\, f\, \| \|\, x\, \|$. From general Hahn–Banach theorem about the extension of a linear functional (Subsection 1.4.7) follows the following statement.

Theorem 4.58. *Any continuous linear functional given on the subspace of a normed linear space may be extended on the whole space with the retention of the norm* (Hahn–Banach theorem for the normed spaces).

Corollary. *In a normed linear space there exists a continuous linear functional which has a given value y_0 in any given point x_0 whose norm is equal to $|\, y_0\, | \, / \, \|\, x_0\, \|$.*
▷ By Theorem 1.15 there exists a linear functional

$$f(x) = \alpha y_0 \, ,$$

determined on the one–dimensional subspace $L = \{x\, :\, x = \alpha x_0\}$ which takes the given value y_0 at the point $x = x_0$. From $\|\, x\, \| = |\, \alpha\, | \|\, x_0\, \|$, $|\, f(x)\, | = |\, \alpha\, | \, |\, y_0\, |$ follows

$$|\, f(x)\, | = \frac{|\, y_0\, |}{\|\, x_0\, \|} \|\, x\, \| \quad \forall x \in L \, .$$

Hence it is clear that $\|\, f\, \| = |\, y_0\, | \, / \, \|\, x_0\, \|$. It remains to notice that according to Hahn–Banach theorem the functional f may be extended on the whole space with the retention of a norm. ◁

R e m a r k. Notice that in a topological linear space a kernel of a continuous functional different from zero cannot contain a set which is dense in X. So, Corollary 2 of Theorem 1.18 should be correspondingly changed.

Corollary. *For any closed subspace L which does not coincide with the whole space X there exists a continuous linear functional whose kernel contains L.*

E x a m p l e 4.8. Let us consider on the space $C([-1, 1])$ the linear functional $fx = \int\limits_{-1}^{1} x(t)dt - cx(0), c > 0$. Determine whether it is continuous or not. If it is continuous find its norm.

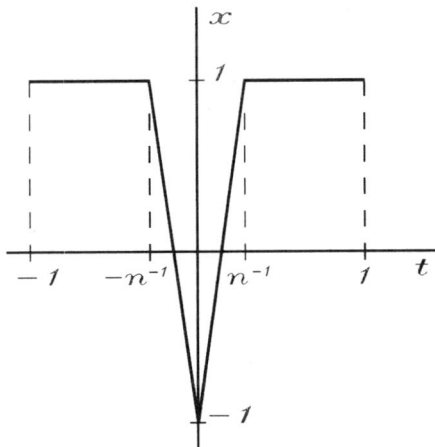

Fig. 4.3

In order to answer this question it is sufficient to find $\sup |fx|$ at $\|x\| = \sup |x(t)| = 1$. It is clear that $|fx| \le 2 + c$ at all $x(t)$, $\|x\| = 1$. Consequently, the functional fx is bounded and its norm is not more than $2 + c$. For finding $\|f\|$ consider the sequence of continuous functions

$$x_n(t) = \begin{cases} 1 & \text{at } t \le -n^{-1} \text{ и } t \ge n^{-1}, \\ -2nt - 1 & \text{at } t \in [-1, -n^{-1}), \\ 2nt - 1 & \text{at } t \in (n^{-1}, 1]. \end{cases}$$

On Fig.4.3 the graph of the function $x_n(t)$ is shown. For this sequence $fx_n = 2(1 - n^{-1}) + c$. Hence and from $\|f\| \le 2 + c$ follows

$$\|f\| = \sup_{\|x\|=1} |fx| = 2 + c.$$

Notice that the way applied for solving this problem is a standard way of constructing the sequence which gives the approximation with any degree of the accuracy to $\sup |fx| / \|x\|$.

E x a m p l e 4.9. Solve the same problem (Example 4.8) for the linear functional

$$fx = \int_0^1 x(t)\frac{dt}{t}$$

on the space $C_0([0,\ 1])$ of continuous functions which have zero value at $t = 0$.

Consider a sequence of continuous functions

$$x_n(t) = \begin{cases} nt & \text{at } t \in [0,\ n^{-1}], \\ 1 & \text{at } t \in (n^{-1},\ 1]. \end{cases}$$

For these functions $fx_n = 1+\ln n$. Consequently, the functional fx is not bounded. Thus we have an example of a discontinuous linear functional.

For more detailed treatment of the topologies in a linear space refer to (Schaefer 1966).

P r o b l e m s

4.6.1. Prove that any topological linear space satisfies the axiom of the separability T_3.

I n s t r u c t i o n. Use the result of Problem 4.2.8 and take into account that owing to the continuity of the operation of the vectors subtraction for any zero vicinity V there exists such zero vicinity W that $W - W \subset V$. And as $W - W$ represents an union of zero vicinities $W - x$ which correspond to all $x \in W$ then $W - W$ contains the closure $[W]$ of zero vicinity W.

4.6.2. Prove that the functional

$$fx = \int_a^b f(t)\,x(t)\,dt$$

on the space $C([a,\ b])$, $f(t) \in C([a,\ b])$ is a contionuous linear functional and its norm is equal to

$$\| f \| = \int_a^b |f(t)|\,dt\,.$$

4.6.3. The same (Problem 4.6.2) for the functional

$$fx = \int_a^b f(t)\,x(t)\,dt - x(a) - x(b)\,.$$

4.6.4. Determine whether the linear functional

$$fx = \int\limits_0^1 x(t)\,\frac{dt}{t^s}$$

is continuous or not on the space $C_0([0, 1])$ and if it is continuous find its norm. Consider the cases $s \in (0, 1)$ and $s > 1$.

4.6.5. Prove that a linear functional

$$fx = \int\limits_a^b f(t)\,x(t)\,dt$$

on the space $L_2([a, b])$ is continuous if and only if $f(t) \in L_2([a, b])$ and find its norm.

I n s t r u c t i o n. For proving the sufficiency of the condition and finding the norm use inequality (1.17). For proving the necessity of the condition suppose that it is not fulfilled and after taking the sequence of the functions

$$x_n(t) = \begin{cases} \overline{f(t)} & \text{at } |\,f(t)\,| < n, \\ 0 & \text{at } |\,f(t)\,| \geq n \end{cases}$$

and applying Theorem 3.27 about monotone convergence prove that $|\,fx_n\,| \,/\, \|\,x_n\,\| \rightarrow \infty$ at $n \rightarrow \infty$.

4.6.6. Generalize the results of Problem 4.6.5 on a linear functional

$$fx = \int\limits_T f(t)\,x(t)\,\mu(dt)$$

in the space of the scalar functions $L_2(T, \mathcal{A}, \mu)$ with nonnegative finite measure $\mu(A)$, $A \in \mathcal{A}$. After taking, in particular, $T = (-\infty, \infty)$,

$$\mu(A) = \int\limits_A e^{-t^2}\,dt\,,$$

give an example of the function $f(t)$ for which the functional f is not bounded.

4.6.7. Prove that if a separable topological linear space X is locally convex then the set of continuous linear functionals on X separates the points of X.

S o l u t i o n. Let V be a convex vicinity of zero satisfying the condition $\lambda x \in V \Rightarrow |\,\lambda\,|\,x \in V$. Such vicinity of zero always exists as for an arbitrary convex vicinity of zero U

$$V = \bigcup_{|\mu|=1} \mu U$$

is such vicinity of zero. We establish the correspondence between the set V and the functional

$$p_V(x) = \inf(\lambda \; : \; \lambda > 0, \, x \in \lambda V).$$

This functional is called *Minkowskii functional* correspondent to the set V. This functional is convex as

$$p_V(\alpha x) = \inf(\lambda \; : \; \lambda > 0, \, |\alpha| x \in \lambda V)$$

$$= \inf(|\alpha| \mu \; : \; \mu > 0, \, x \in \mu V) = |\alpha| \, p_V(x),$$

and for any x_1, x_2, $x_1 \in \lambda_1 V$, $x_2 \in \lambda_2 V$ at $\lambda_1 = p_V(x_1)$, $\lambda_2 = p_V(x_2)$. As a result $x_1 + x_2 \in (\lambda_1 + \lambda_2)V$ and $p_V(x_1 + x_2) \leq \lambda_1 + \lambda_2 = p_V(x_1) + p_V(x_2)$. It is evident that $p_V(x) \leq 1$ for any $x \in V$ and $p_V(x) \geq 1$ at $x \notin V$. Determine the linear functional $f_0 x = \alpha$ on one–dimensional subspace $L_0 = \{x \; : \; x = \alpha x_0\}$ where x_0 is any vector which does not belong to V. Then we shall have $f_0 x_0 = 1 \leq p_V(x_0)$ and $|f_0 x| = |\alpha| \leq |\alpha| \, p_V(x_0) = p_V(x)$. Thus the functional f_0 is subjected to the condition $|f_0 x| \leq p_V(x)$ on L_0. According to Hahn–Banach Theorem 1.17 we may extend the functional on the whole space X with the retention of this condition. The obtained functional f is bounded as $|fx| \leq p_V(x) \leq 1$ for any $x \in V$. By Theorem 4.56 it is continuous. As any point $x_0 \neq 0$ is not contained in some vicinity of zero V then for any x_0 there exists such continuous linear functional f that $fx_0 \neq 0$. By Theorem 4.52 the set of continuous linear functionals separates the points of the space X.

The proved statement establishes a close link between local convexity of a topological linear space and the existence of the set of continuous linear functionals on it which separates its points. In particular, from this statement follows that in any normed linear space the set of continuous linear functionals separates its points. If the space is not locally convex then there may be not anyone continuous linear functional besides zero one.

4.6.8. Prove that at continuous linear mapping of one topological linear space into another one the image of a convex set is a convex set.

4.6.9. Suppose that in the linear space X the metric is given. It is invariant relative to the shifts $d(x + a, \, y + a) = d(x, \, y) \; \forall x, \, y, \, a$. Such metric is completely determined by the distances of all the points from zero $d(x, \, y) = d(x - y, \, 0)$. Putting $d(x, \, 0) = |x|$ from the metric axioms we get the following properties of the function $|x| : |x| \geq 0$ and $|x| = 0$ only at $x = 0$, $|x + y| \leq |x| + |y|$, $|-x| = |x|$. Thus the function $|x|$ possesses all the properties of the norm except the second property (Subsection 1.3.6). The complete linear space X with such a metric satisfying the additional condition of the continuity of the vector multiplication by a number is called a F–*space*. It is clear that a F–space is a metric one but not normed linear space. A F–space with the topology induced by a metric represents a topological linear space (the addition of the vectors is continuous by virtue of the inequality of the triangular). May we state that a closed ball in a F-space is a convex set?

4.7. Weak Topologies

4.7.1. Definition of a Weak Topology

Let (X, τ) be a topological linear space, and F be a set of all continuous linear functionals on it. According to Theorem 4.48 the topology τ_F determined by a set of linear functionals F is the weakest topology in which all the functionals from F are continuous. Consequently, $\tau_F \subset \tau$. It gives the opportunity to introduce a notion of a weak topology in a topological linear space.

A *weak topology* in the topological linear space (X, τ) is called the topology τ_F determined by the set F of all continuous (in the topology τ) linear functionals of this space by means of the vicinities of zero

$$\{x \ : | f_1 x | < \varepsilon, \ldots, | f_n x | < \varepsilon\} \ \forall \varepsilon > 0, \forall n, f_1, \ldots, f_n \in F. \quad (4.22)$$

Otherwise if none third topology is determined in X then the topology τ is called *a strong topology* of the space X.

In particular, in a normed linear space X with the topology τ induced by the norm the topology τ_F determined by the set F of all continuous functionals of the space X will be a weak topology.

Unlike the topological notions connected with the topology τ all the topological notions connected with the weak topology τ_F are introduced with the addition of the word "weak".

So, the sequence of the points $\{x_n\}$ of a linear space X convergent to x (fundamental) in the weak topology τ_F is called *weakly convergent to x (weakly fundamental)*. In accordance with this fact the point x is called *a weak limit* of weakly convergent to x of the sequence $\{x_n\}$, $x = \text{w} \lim x_n, x_n \xrightarrow{\text{w}} x$.

A space X is called *weakly complete* if any weakly fundamental sequence has a weak limit in it.

The set A of a topological linear space X (or the whole space X) is called *weakly compact (weakly countable compact)* if any (any countable) its covering by open sets in weak topology contains a finite covering. For weak countable compactness of the set A in the topological linear space according to general Theorem 4.31 it is necessary and sufficient that any infinite subset of this set has the limit points in the weak topology.

The set A of a topological linear space X (the whole space X) is called *weakly sequential compact* if any sequence of its points contains weakly convergent subsequence.

The function $y = \varphi(x)$ mapping a topological space X (not obligatory linear) in a topological linear space Y is called *weakly continuous* if it is continuous in weak topology of a space Y.

The function $y = \varphi(x)$ mapping the measurable space $(X, \mathcal{A})$ into the topological linear space Y with the σ-algebra $\mathcal{B}_F$ induced by the weak topology τ_F is called *weakly measurable* if it is $(\mathcal{A}, \mathcal{B}_F)$-measurable, i.e. if the inverse image of any set from the σ-algebra $\mathcal{B}_F$ determined by the function φ belongs to the σ-algebra $\mathcal{A}$ (Subsection 3.1.1):

$$\varphi^{-1}(B) \in \mathcal{A} \quad \forall B \in \mathcal{B}_F.$$

In other words, if the inverse image of the σ-algebra $\mathcal{B}_F$ is contained in $\mathcal{A}$, then $\varphi^{-1}(\mathcal{B}_F) \subset \mathcal{A}$.

4.7.2. Weak Convergence

The notion of the weak convergence extends the notion of the convergence (Subsection 4.3.1). It is determined by the following theorem.

Theorem 4.59. *Any convergent sequence in a topological linear space weakly converges to the same limit.*

▷ If the sequence $\{x_n\}$ converges to x then any vicinity V_x of the point x contains all the points x_n beginning with someone. As any vicinity of the point x in the weak topology τ_F is at the same time the vicinity of the point x in the strong topology τ then any vicinity of the point x in the weak topology τ_F contains all the points x_n beginning with someone. This proves the weak convergence of the sequence $\{x_n\}$ to x. ◁

From this theorem due to the fact that the fundamental property of the sequence $\{x_n\}$ represents the convergence to zero of the sequence $\{x_n - x_m\}$ follows the statement.

Theorem 4.60. *Any fundamental sequence in a topological linear space is weakly fundamental.*

Thus in the topological linear space any convergent (fundamental) sequence weakly converges to the same limit (is weakly fundamental).

R e m a r k. But the inverse in the general case is not true. A weakly convergent sequence may be not convergent. Later on we shall give an example of such a sequence (Problem 4.7.3). A weakly fundamental sequence may be not fundamental.

Theorem 4.61. *For weak convergence of the sequence $\{x_n\}$ to x in a topological linear space X it is necessary and sufficient the convergence of the numeric sequence $\{fx_n\}$ to fx for any functional $f \in F$.*

This theorem is the direct Corollary of Theorem 4.54 about the convergence of the sequence in the topology τ_F.

From this theorem by virtue of the fundamental property of the sequence $\{x_n\}$ represents the convergence of the sequence $\{x_n - x_m\}$ to zero. So, we get the following two statements.

Theorem 4.62. *The sequence $\{x_n\}$ in a topological linear space is weakly fundamental if and only if the numeric sequence $\{fx_n\}$ is fundamental at any $f \in F$.*

Theorem 4.63. *Any weakly convergent sequence is weakly fundamental.*

▷ By Theorem 4.61 from weak convergence of the sequence $\{x_n\}$ follows the convergence of the numeric sequence $\{fx_n\}$ at any $f \in F$. But any convergent numeric sequence is fundamental and from this fact by virtue of Theorem 4.62 follows that the sequence $\{fx_n\}$ is weakly fundamental. ◁

Thus in a weak complete topological linear space for weak convergence of the sequence it is necessary and sufficient its property to be weak fundamental.

4.7.3. Weakly Continuous Functions

We shall consider the functions mapping the topological space (X, σ) in the topological linear space (Y, τ) in which the weak topology τ_F is determined by means of the set of all continuous linear functionals F. So, we have.

Theorem 4.64. *The function $y = \varphi(x)$ is weakly continuous if and only if the numeric function $z = f\varphi(x)$ is continuous at any $f \in F$.*

▷ For proving notice that the set

$$A = \{x : | f\varphi(x) - f\varphi(x') | < \varepsilon\}$$

at any $\varepsilon > 0$, $f \in F$ and $x' \in X$ represents simultaneously the inverse image of open set $\{y : | fy - fy' | < \varepsilon\}$, $y' = \varphi(x')$ of the space Y

determined by the function $y = \varphi(x)$ and the inverse image of open set $\{z : \mid z - z' \mid < \varepsilon\}$, $z' = f\varphi(x')$ of the scalars field K determined by the function $z = f\varphi(x)$. If the function $y = \varphi(x)$ is weakly continuous then by Theorem 4.19 the set A is open at all $\varepsilon > 0$, $f \in F$, $x' \in X$. But then the inverse image for any open set H of the scalars field K determined by the function $z = f\varphi(x)$ will be also open set of the space X, $(f\varphi)^{-1}(H) \in \sigma$ as the sets $\{z : \mid z - z' \mid < \varepsilon\}$ correspondent to all $\varepsilon > 0$, $z' \in K$ form the base of the topology of the scalars field K. Consequently, by the same Theorem 4.19 the function $z = f\varphi(x)$ is continuous at any $f \in F$ what proves the necessity of the condition.

If the function $z = f\varphi(x)$ is continuous at any $f \in F$ then the set A is open at all $\varepsilon > 0$, $f \in F$, $x' \in X$. And as the finite intersections of the sets $\{y : \mid fy - fy' \mid < \varepsilon\}$ correspondent to all $\varepsilon > 0$, $y' \in Y$, $f \in F$ form the base of the weak topology τ_F of the space Y then the inverse image of any open set G of the weak topology τ_F determined by the function $y = \varphi(x)$ is also the open set of the space X, $\varphi^{-1}(G) \in \sigma$. Consequently, by Theorem 4.19 the function $y = \varphi(x)$ is continuous what proves the sufficiency of the condition. ◁

Theorem 4.65. *Any continuous function is weakly continuous.*

▷ If the function $y = \varphi(x)$ is continuous then by Theorem 4.18 about the continuity of a composite function the function $z = f\varphi(x)$ is continuous for any continuous functional f, in particular, for $f \in F$. Hence by virtue of Theorem 4.64 follows the weak continuity of the function $y = \varphi(x)$. ◁

R e m a r k. Thus the notion of the weak continuity of the function extends the notion of the continuity. But weakly continuous function may be not continuous.

4.7.4. Weakly Measurable Functions

Let $\mathcal{B}$ and $\mathcal{B}_F$ be the σ–algebras in a linear space Y induced by the topology τ and the weak topology τ_F. We shall consider the functions mapping the measurable space $(X, \mathcal{A})$ into Y. The function $y = \varphi(x)$ is measurable if it is $(\mathcal{A}, \mathcal{B})$–measurable, i.e. if $\varphi^{-1}(\mathcal{B}) \subset \mathcal{A}$. The function $y = \varphi(x)$ is weakly measurable if it is $(\mathcal{A}, \mathcal{B}_F)$–measurable, i.e. if $\varphi^{-1}(\mathcal{B}_F) \subset \mathcal{A}$. It is evident that any measurable function is weakly measurable as from the relations $\mathcal{B}_F \subset \mathcal{B}$, $\varphi^{-1}(\mathcal{B}) \subset \mathcal{A}$ follows $\varphi^{-1}(\mathcal{B}_F) \subset \varphi^{-1}(\mathcal{B}) \subset \mathcal{A}$. But weakly measurable function may be not measurable.

Theorem 4.66. *The function* $y = \varphi(x)$ *is weakly measurable if and only if the numeric function* $z = f\varphi(x)$ *is measurable at any* $f \in F$.

▷ Let τ_K and $\mathcal{C}$ be the topology and (generated by it) the σ–algebra of Borel sets in the scalars field K. From the definition of the weak topology τ_F in the space Y follows that it is the weakest topology which contains all the inverse images $f^{-1}\tau_K$ of the topology τ_K determined by the functionals $f \in F$. Hence and from the definition of the σ–algebras $\mathcal{C}$ and $\mathcal{B}_F$ follows that $\mathcal{B}_F$ is the minimal σ–algebra which contains all the inverse images $f^{-1}\mathcal{C}$ the σ–algebra $\mathcal{C}$ determined by the functionals $f \in F$, $f^{-1}\mathcal{C} \subset \mathcal{B}_F \ \forall f \in F$. But in such case all the functionals $f \in F$ are not only $(\mathcal{B}, \mathcal{C})$–measurable but $(\mathcal{B}_F, \mathcal{C})$–measurable, $f^{-1}\mathcal{C} \subset \mathcal{B}_F \subset \mathcal{B}$.

From the $(\mathcal{A}, \mathcal{B}_F)$–measurability of the function $y = \varphi(x)$ and the $(\mathcal{B}_F, \mathcal{C})$–measurability of the functional $f \in F$ by virtue of Theorem 3.1 about the measurability of a composite function follows the $(\mathcal{A}, \mathcal{C})$–measurability of the function $z = f\varphi(x)$ at any $f \in F$ what proves the necessity of the condition.

If the function $z = f\varphi(x)$ is measurable at any $f \in F$ then

$$(f\varphi)^{-1}(\mathcal{C}) = \varphi^{-1}(f^{-1}\mathcal{C}) \subset \mathcal{A} \,.$$

As $\mathcal{B}_F$ is the minimal σ–algebra which contains all $f^{-1}\mathcal{C}$, $f \in F$ then $\varphi^{-1}(\mathcal{B}_F)$ by virtue of the properties of the inverse mappings is the minimal σ–algebra which contains all the inverse images $\varphi^{-1}(f^{-1}\mathcal{C})$ of the σ–algebra $\mathcal{C}$, $f \in F$. Consequently,

$$\varphi^{-1}(\mathcal{B}_F) \subset \mathcal{A} \,,$$

i.e. the function $y = \varphi(x)$ is $(\mathcal{A}, \mathcal{B}_F)$–measurable (weakly measurable) what proves the sufficiency of the condition. ◁

Theorem 4.67. *The weak limit of a weakly convergent sequence of weakly measurable functions represents a weakly measurable function.*

▷ Let the function $\varphi(x)$ represent the weak limit of weakly convergent sequence $\{\varphi_n(x)\}$ of weakly measurable functions. By Theorem 4.61 from the weak convergency of the sequence $\{\varphi_n(x)\}$ to $\varphi(x)$ follows the convergency of the sequence of the scalar functions $\{f\varphi_n(x)\}$ to $f\varphi(x)$ at any $f \in F$. From the weak measurability of the functions $\varphi_n(x)$ by Theorem 4.66 follows the measurability of the scalar functions $f\varphi_n(x)$ at all $f \in F$. By Theorem 3.3 the function $f\varphi(x)$ is measurable at any $f \in F$. And at last by Theorem 4.66 it follows the weak measurability of the function $\varphi(x)$. ◁

On the basis of Theorem 4.66 and Corollaries 1–3 of Theorem 3.5 the sum of weakly measurable functions, the product of weakly measurable function on the measurable numeric function and at last the quotient of the division of weakly measurable function by the measurable numeric function which nowhere vanishes are weakly measurable.

Similarly, on the basis of Theorem 4.66 and Corollary 4 of Theorem 3.5 we prove that the function

$$\varphi(x) = \sum_{k=1}^{n} \varphi_k(x) \mathbf{1}_{E_k}(x)$$

is weakly measurable if the functions $\varphi_1(x), \ldots, \varphi_n(x)$ are weakly measurable and the sets $E_1, \ldots, E_n$ are measurable $E_1, \ldots, E_n \in \mathcal{A}$.

4.7.5. Coincidence of Measurability and Weak Measurability of Functions with Values in a Separable B–Space

In the general case weakly measurable function may be not measurable. But for the function with the values in a separable B–space the notions of the measurability and weak measurability coincide.

Theorem 4.68. *Any weakly measurable function with the values in a separable B–space Y is measurable.*

▷ For proving it is sufficient to show that in this case the σ–algebras $\mathcal{B}$ and $\mathcal{B}_F$ induced in the space Y by strong and weak topologies coincide. As always $\mathcal{B}_F \subset \mathcal{B}$ and any open set in a separable B–space may be received by no more than a countable union of the balls then it is sufficient to show that any ball $S_r(0) = \{y : \| y \| < r\}$ belongs to the σ–algebra $\mathcal{B}_F$ induced by the weak topology. Then it will follow that any open set from $\mathcal{B}$ belongs also to $\mathcal{B}_F$. As $\mathcal{B}$ is the minimal σ–algebra which contains all the open sets of the strong topology then $\mathcal{B} \subset \mathcal{B}_F$, i.e. $\mathcal{B} = \mathcal{B}_F$.

Let $S = S_r(0) = \{y : \| y \| < r\}$ and $\{y_n\}$ be a dense countable set in $\bar{S} = Y \backslash S$. We establish the correspondence between each point y_n and such a functional f_n with the norm equal to unit that $| f_n y_n | = \| y_n \|$. According to Corollary of Theorem 4.58 such a functional exists. We set the sequence of positive numbers $\{\varepsilon_m\}$, $\varepsilon_m > 0$, $\varepsilon_m \to 0$ at $m \to \infty$ convergent to zero and form the sets

$$B_m^n = \{y : | f_n y | < r - \varepsilon_m\}, \quad C_m = \bigcap_{n=1}^{\infty} B_m^n,$$

$$C = \lim C_m = \bigcup_{m=1}^{\infty} C_m . \qquad (4.23)$$

Prove that $S = C$. Let y be an arbitrary point of the ball S, $y \in S$. Then $\| y \| < r$ and there exists such $\eta > 0$ that $\| y \| < r - \eta$. As the norms of all linear functionals f_n are equal to 1 then $| f_n y | \leq \| y \| < r - \eta < r - \varepsilon_m$ for all $\varepsilon_m < \eta$. And it means that the point y belongs to all the sets B_m^n for which $\varepsilon_m < \eta$, and consequently, and also to the sets C_m for which $\varepsilon_m < \eta$, i.e. $y \in C$. Thus any point of the ball S belongs to the set C, and consequently, $S \subset C$. Let y be any point of the set C, $y \in C$. Suppose that it does not belong to the set S, $y \notin S$. Then $y \in \bar{S}$ and at any m will occur such point y_{k_m} that $\| y_{k_m} - y \| < \varepsilon_m$. For this point y_{k_m} $| f_{k_m} y_{k_m} | = \| y_{k_m} \| \geq r$ and $| f_{k_m} (y_{k_m} - y) | \leq \| f_{k_m} \| \| y_{k_m} - y \| < \varepsilon_m$, and consequently,

$$| f_{k_m} y | = | f_{k_m} y_{k_m} - f_{k_m} (y_{k_m} - y) |$$

$$\geq | f_{k_m} y_{k_m} | - | f_{k_m} (y_{k_m} - y) | > r - \varepsilon_m .$$

Hence and from (4.18) it follows that $y \notin B_m^{k_m}$, $y \notin C_m$, $y \notin C$. The obtained contradiction proves that $C \subset S$. From this inclusion and the inclusion obtained before $S \subset C$ follows that $S = C$. Now it remains only to notice that all the sets B_m^n belong to the σ–algebra $\mathcal{B}_F$ induced by the weak topology τ_F. As a result the ball $S = S_r(0) = C$ which is obtained from the sets B_m^n by the countable intersections and unions belongs to the σ–algebra $\mathcal{B}_F$. ◁

R e m a r k. So, we proved that for the measurability of the function $\varphi(x)$ with the values in a separable B–space it is necessary and sufficient that it should be weakly measurable. In other words, the class of weakly measurable functions with the values in a separable B–space coincides with the class of measurable such functions.

The detailed treatment of the weak topologies is in (Schaefer 1966).

P r o b l e m s

4.7.1. Determine on the space of the continuous functions $C([a, b])$ the set of linear functionals F by the formula

$$fx = \int\limits_a^b f(t)\, x(t)\, dt ,$$

where $f(t)$ is a continuous function. Determine zero vicinities which induce the weak topology τ_F in $C([a, b])$. What represents the weak convergence in this space?

4.7.2. Under the conditions of the previous problem at $a = -\pi$, $b = \pi$ show that sequences of the functions $\{\sin nt\}$ and $\{\cos nt\}$ weakly converge. Find their weak limits. This problem gives the examples of nonconvergent but weakly convergent sequences.

I n s t r u c t i o n. Represent an integral for each function $x_n(t) = \sin nt$ or $\cos nt$ as the sum of the integrals on the semiperiods on which of them $x_n(t)$ does not change the sign. Apply to each integral the theorem about the mean and account the continuity of the function $f(t)$.

4.7.3. Whether the sequence of the functions be convergent or weakly convergent in $C([-\pi, \pi])$

$$x_n(t) = \begin{cases} (1 + nt)n & \text{at} \quad t \in (-n^{-1}, 0), \\ (1 - nt)n & \text{at} \quad t \in (0, n^{-1}), \\ 0 & \text{at} \quad t \in [-\pi, -n^{-1}] \quad \text{and} \quad t \in [n^{-1}, \pi]? \end{cases}$$

4.7.4. Consider the function

$$x(t) = \begin{cases} \varphi_1(s)\,\psi(t) & \text{at} \quad t \le 0, \\ \varphi_2(s)\,\psi(t) & \text{at} \quad t > 0, \end{cases}$$

where $\varphi_1(s) = \varphi_2(s) = \varphi(s)$ at $s \ne 0$ and $\varphi_1(0) = a$, $\varphi_2(0) = b \ne a$, $\varphi(s)$ is the continuous function s at the interval $[-1, 1]$ and $\psi(t)$ is the continuous function t at the interval $[-1, 1]$, $\psi(0) \ne 0$. It is clear that $x(t)$ is the bounded function s at each t. Consequently, $x(t)$ maps the interval $[-1, 1]$ into the space of the bounded functions $B([-1, 1])$ with the norm $\| y \| = \sup | y(s) |$. Determine the set of the linear functions F on $B([-1, 1])$ by the formula

$$fy = \int\limits_{-1}^{1} f(s)\,y(s)\,ds\,, \quad f(s) \in C([-1, 1])\,.$$

Is the function $x(t)$ continuous or not? If not find its points of the discontinuity and the jumps. Is the function $x(t)$ weakly continuous?

4.7.5. Determine on the B–space of the continuous scalar functions $C(T)$ where T is the bounded closed set on R^n the weak topology τ_F by the set of linear functionals

$$fx = \sum_{i=1}^{N} l_i\, x(t_i)\,, \qquad (*)$$

correspondent to all finite $t_1, \ldots, t_N \in T$ and the numbers $l_1, \ldots, l_N$. Whether the space $C(T)$ with the topology τ_F be separable, locally convex? Whether the operations of the vectors addition and the vector multiplication by a number be continuous in the topology τ_F? Prove that the sequence of the functions $\{x_k(t)\}$ weakly converges if and only if it converges at each t.

Generalize the obtained results on the space $C(T)$ of continuous vector functions. In this case $x(t_i)$ in $(*)$ represents a column–matrix and l_i represents a row–matrix with the same number of the elements.

CHAPTER 5
SPACES OF OPERATORS AND FUNCTIONALS

Chapter 5 is devoted to spaces of operators and spaces of functionals. In Section 5.1 the notions of spaces of linear operators and spaces of bounded linear operators are introduced. The topologies in the space of bounded linear operators are defined. The definition of the dual space for a given topological linear space is outlined and properties of dual spaces are studied. Section 5.2 contains the theory of weak integrals (Pettis integrals). Section 5.3 is devoted to the theory of generalized functions. The notions of the space of test functions and the space of generalized functions are introduced, operations on generalized functions are defined and their main properties are studied for important applications. In last Section 5.4 dual spaces are studied for the main types of functions spaces, i.e. for the space of bounded functions, space of continuous functions, spaces of differentiable functions and Lebesgue spaces.

5.1. General Theory

5.1.1. Operations on Operators and Functionals

Let A be an operator mapping some space X into a linear space Y. The equality $y = \alpha(Ax)$ where α is a complex number determines some operator C which also maps X into Y, $y = Cx$. This operator is called *a product of the operator A by the number α: $C = \alpha A$.*

If A and B are two operators mapping a space X into a linear space Y then the equality $y = Ax + Bx$ determines some operator C which also maps X into Y, $y = Cx$. This operator is called *a sum of the operators A and B: $C = A + B$.*

From these two definitions follows the definition of a *linear combination of the operators*:

$$A = \sum_{k=1}^{n} \alpha_k A_k,$$

if

$$Ax = \sum_{k=1}^{n} \alpha_k A_k x.$$

In the special case when Y is a field of the scalars (the real axis or the complex plane) the previous definitions give a product of the functional by a number, a sum and linear combination of the functionals definitions.

If the operator A maps a space X into Y, and the operator B maps a space Y into Z and the domain D_B of the operator B intersects with the range R_A of the operator A then the equality $z = B(Ax)$ determines the operator C mapping X into Z: $z = Cx$. This operator is called a *product of the operator A by the operator B*: $C = BA$. As the domain of the operator $C = BA$ serves the set $D_C = \{x : x \in D_A, Ax \in D_B\}$, and as its range serves the set $R_C = \{z : z = By, y \in D_B R_A\}$. From this definition follows the definition of a product of any number of the operators.

If Z is a field of the scalars then from this definition follows the definition of a *product of the operator A by the functional f* which evidently represents the functional $g = fA$.

In the special case when $Y = Z = X$ besides a product BA of the operator A by B a product AB of the operator B by A may be determined. It is evident that in the general case $AB \neq BA$. The domain of the operator AB serves $D_B \bigcap \{x : Bx \in D_A\}$. And the domain of the operator BA serves $D_A \bigcap \{x : Ax \in D_B\}$.

If $Y = Z = X$, $D_B \bigcap \{x : Bx \in D_A\} = D_A \bigcap \{x : Ax \in D_B\}$ and $ABx = BAx$ at any x in the domain of the operators AB and BA the operators A and B are called *commutative* or *permutable*.

From the preceding definitions follows the definition of any entire positive degree of the operator A mapping X into X and also the definition of a polynomial of the operator A which maps a linear space X into X.

If the operator A establishes the one–to–one correspondence between the points of the domain $D_A \subset X$ and the range $R_A \subset Y$ then there exists *the inverse operator A^{-1}* mapping R_A on D_A, $x \in A^{-1}y$, $y \in R_A$. The domain of the inverse operator A^{-1} serves the range R_A of the operator A and the range A^{-1} serves the domain D_A of the operator A, $D_{A^{-1}} = R_A$, $R_{A^{-1}} = D_A$.

It is evident that the operator A is the inverse one for A^{-1}, $A = (A^{-1})^{-1}$ and that a product of the operator by the inverse operator represents *an identity operator I* which leaves the elements of a space invariable. Really, from $y = Ax$ and $x = A^{-1}y$ follows $A^{-1}Ax = x$ and $AA^{-1}y = y$, whence it is clear that $A^{-1}A = I$ represents an identity operator in a space X, and $AA^{-1} = I$ is an identity operator in the space Y.

It is not difficult to ensure that a sum, a linear combination, a product and entire degrees of the linear operators, and also the operator inverse relative to the linear operator represents linear operators.

Let A be an operator one–to–one mapping a space X into Y, and B be an operator one–to–one mapping a space Y into Z and $R_A D_B \neq \emptyset$, and $C = BA$. Then there exists the inverse operators $C^{-1} = A^{-1} B^{-1}$. Really, from $y = Ax$, $z = By$ follows $y = B^{-1} z$, $x = A^{-1} y = (A^{-1} B^{-1})z$. Consequently, $C^{-1} = A^{-1} B^{-1}$. Thus the operator inverse relative to a product of the operators is equal to a product of the inverse operators which are taken in the reverse order.

E x a m p l e 5.1. The linear operator A mapping the finite–dimensional Euclidean space Y is determined by a matrix of the correspondent transformation. In this case the previous definitions give the known from linear algebra notions of a product of a matrix by a number, a sum, a linear combination and matrices product.

E x a m p l e 5.2. If the matrix determinant of a linear transformation is different from zero then as it is known from linear algebra there exists an inverse linear transformation which determines the correspondent inverse operator. Thus if X and Y are finite–dimensional Euclidean spaces of the same dimension then interinverse linear operators are determined by the correspondent interinverse matrices.

E x a m p l e 5.3. The relation between the input $x(t)$ and the output $y(t)$ for linear control system (Example 1.8) is determined in general case by the linear operator A of the system, $y(t) = Ax(t)$. The operators of some typical linear systems with lumped parameters described by ordinary differential equations with corresponding initial conditions are presented in Table A.1.1 (Appendix 1).

A system is called *stationary* if its output $y(t)$ to a given input $X(t)$ is independent of the instant of its application and depends only on the amount of time which has passed since that instant, i.e. $Ax(t - t_0) = y(t - t_0)$, where t_0 is an arbitrary initial instant. In *nonstationary* systems the output depends on the instant when the input is applied as well as on its shape, i.e. $Ax(t - t_0) = y(t, t_0)$. All systems except the 6-th presented in Table A.1.1 (Appendix 1) are stationary.

In control theory for stationary linear systems a *system operator* is usually determined by a *system transfer function* (D'Azzo and Houpis 1975, Pugachev and Sinitsyn 1987):

$$\Phi(s) = \frac{Y(s)}{X(s)} .$$

Here s being a complex variable, $X(s)$ and $Y(s)$ are *Laplace transforms* of input

$x(t)$ and output $Y(t)$ correspondingly (Appendix 3)

$$Y(s) = \mathcal{L}[y(t)] = \int\limits_0^\infty y(t)e^{-st}dt \, ,$$

$$X(s) = \mathcal{L}[x(t)] = \int\limits_0^\infty x(t)e^{-st}dt \, .$$

The transfer function of typical stationary linear systems described by ordinary differential equations (Table A.1.1) presents a rational function of the complex variable s. Some stationary linear systems with distributed parameters are described by transcendental or irrational functions of the complex variable (see Appendix 1, Table A.1.2, Systems $1, \ldots, 7$).

E x a m p l e 5.4. Let us consider the basic connections of stationary linear control systems (Example 5.3) and their operators.

The basic types of the connections in composite systems are the connections *in series, in parallel,* or *by a feedback loop. A series connection* of the systems is called such a connection when the output of each system is connected with the input of the following system, i.e. when the output variable of each system serves as the input variable for the following system. Here it is supposed that the connected systems possess *the directed action.* It means that while connecting the output of one system with the input of the other system the characterictics of the first system do not change, i.e. that at a given input $x(t)$ of a system its output variable represents one and the same function $y(t)$ no matter whether the output of this system is connected with the input of other system or not (Fig.5.1).

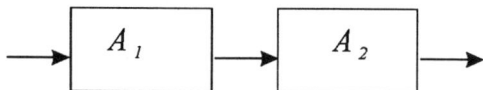

Fig. 5.1

The transfer function $\Phi(s)$ of a series connection of two linear stationary systems with the transfer functions $\Phi_1(s)$ and $\Phi_2(s)$ is equal to the product of transfer function,

$$\Phi(s) = \Phi_1(s)\Phi_2(s). \tag{$*$}$$

This formula is proved by taking the input signal for first system equal to e^{st}. Its output $\Phi_1(s)e^{st}$ will be the input of the second system. So, finally, we get on the output $\Phi_1(s)\Phi_2(s)e^{st}$.

A *parallel connection* of the systems is called such a connection at which the input variable is supplied simultaneously on several systems, and their output variables are summarized (Fig. 5.2). The transfer function $\Phi(s)$ of two linear stationary systems with transfer functions $\Phi_1(s)$ and $\Phi_2(s)$ is equal to the sum of transfer functions

$$\Phi(s) = \Phi_1(s) + \Phi_2(s). \qquad (**)$$

The proof of $(**)$ is analogous $(*)$.

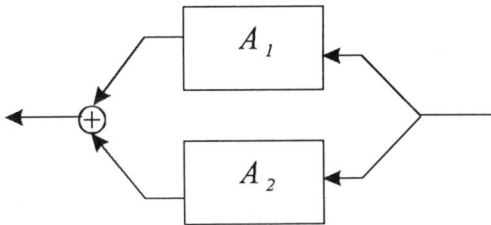

Fig. 5.2

A *feedback connection* is called a connection of the system output with its input (Fig. 5.3). If the output variable of the system is directly set on its input without any transformation then the feedback is called a *rigid*. A rigid feedback may be *positive* or *negative* in accordance with the fact whether the output variable of the system is summarized with its input variable or is subtracted from it. If some system is included into the chain of the feedback transforming the output variable of the basis system then such feedback is called *flexible* (Fig. 5.4).

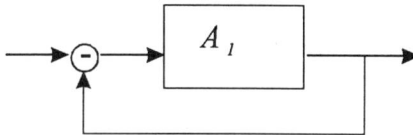

Fig. 5.3

The transfer function $\Phi(s)$ for linear stationary systems in Fig.5.4 is equal to

$$\Phi(s) = \frac{\Phi_1(s)}{1 + \Phi_1(s)\Phi_2(s)} \qquad (***)$$

Two systems are called *mutually inverse* or *reciprocal* if as a result of their sequential connection we obtain an ideal tracking system, i.e. if at any input of the

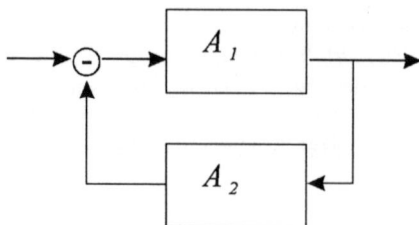

Fig. 5.4

first system the output of the second system at each instant is equal to the input of the first system. It is easy to understand that two reciprocal systems remain reciprocal while changing the order of their connection (Fig.5.5.). The transfer functions $\Phi_1(s)$ and $\Phi_2(s)$ of two reciprocal linear stationary systems yield the condition: $\Phi_1(s)\Phi_2(s) = 1$.

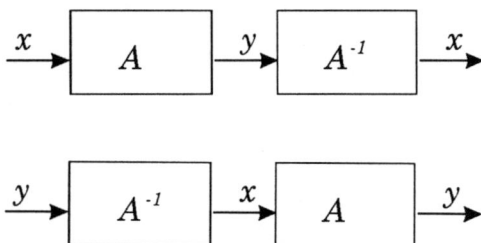

Fig. 5.5

5.1.2. Spaces of Linear Operators and Functionals

From the previous definitions follows that the operation of an operator addition and the multiplication of an operator by a number possess all the properties of the operations of the vectors addition and vector multiplication by a number (Subsection 1.3.1). Therefore the set of all the operators mapping a space X into a linear space Y both linear and nonlinear may be considered as a linear space. The spaces of linear operators and functionals are very important for the functional analysis.

We shall denote a space of all linear operators mapping a linear space X into a linear space Y in terms of $\mathcal{L}(X, Y)$. In particular, a space of all linear operators mapping a space X into X we shall denote by $\mathcal{L}(X)$, $\mathcal{L}(X) = \mathcal{L}(X, X)$. A space of all continuous linear operators mapping a topological linear space X into a topological linear space Y we shall denote by $\mathcal{B}(X, Y)$. In particular, a space of all continuous linear operators mapping X into X we shall denote by $\mathcal{B}(X)$, $\mathcal{B}(X) = \mathcal{B}(X, X)$. It is clear that if all the operators of spaces $\mathcal{L}(X)$ and $\mathcal{B}(X)$ are invertible then by virtue of the fact that the operation of the operators multiplication is determined in $\mathcal{L}(X)$ and $\mathcal{B}(X)$ these spaces represent an algebra with the identity (Subsection 1.3.10). If X is a normed space then $\mathcal{B}(X)$ is a normed algebra.

In the special case when X and Y are normed linear spaces a space of the continuous linear operators $\mathcal{B}(X, Y)$ is itself a normed linear space in which the norm is determined as a norm of a linear operator. In particular, a space $\mathcal{B}(X)$ in the case of a normed space X represents a normed algebra.

Theorem 5.1. *A space of the continuous (bounded) linear operators $\mathcal{B}(X, Y)$ mapping a normed linear space X into a B–space Y is complete (i.e. is a B–space).*

▷ Let $\{T_n\}$ be the fundamental sequence in $\mathcal{B}(X, Y)$. Then for any $\varepsilon > 0 \parallel T_n - T_m \parallel < \varepsilon$ at all $n, m > N_\varepsilon$ where N_ε is some number (dependent on ε) and from $\parallel T_n x - T_m x \parallel \le \parallel T_n - T_m \parallel \parallel x \parallel$ follows

$$\parallel T_n x - T_m x \parallel < \varepsilon \parallel x \parallel . \tag{5.1}$$

Hence it is clear that the sequence $\{T_n x\}$ is fundamental at any $x \in X$. As a space Y of the values of the some operator T is a B–space, i.e. is complete then at each $x \in X$ there exists the limit

$$T x = \lim_{n \to \infty} T_n x.$$

It is evident that the operator T is linear as

$$T \sum_{k=1}^{p} \alpha_k x_k = \lim_{n \to \infty} T_n \sum_{k=1}^{p} \alpha_k x_k = \lim_{n \to \infty} \sum_{k=1}^{p} \alpha_k T_n x_k = \sum_{k=1}^{p} \alpha_k T x_k.$$

In order to prove that it is continuous, i.e. $T \in \mathcal{B}(X, Y)$) we pass in (5.1) to the limit at $m \to \infty$ substituting first ε by the number $\varepsilon_1 < \varepsilon$. Then we obtain

$$\parallel T_n x - T x \parallel \le \varepsilon_1 \parallel x \parallel < \varepsilon \parallel x \parallel .$$

It is clear that the operator $T_n - T$ is bounded and its norm is smaller than ε. Consequently, the operator $T = T_n - (T_n - T)$ is also bounded, and $\| T_n - T \| \to 0$ at $n \to \infty$. Thus any fundamental sequence of the operators from $\mathcal{B}(X, Y)$ has in $\mathcal{B}(X, Y)$ the limit T. This fact proves the completeness of the space $\mathcal{B}(X, Y)$. ◁

Pay attention to the fact that $\mathcal{B}(X, Y)$ is complete no matter whether X is complete or not, only Y should be complete. $\mathcal{B}(X, Y)$ is a B–space if Y is a B–space.

5.1.3. Dual Spaces

A space of all continuous linear functionals on a topological linear space X is called *a space dual with* X and is denoted by X^*.

If a topology is given in the dual space X^* then we may determine *the second space* X^{**} *dual with* X as a space of continuous linear functionals on X^* dual with X^*. If X is a normed linear space then the dual space X^* will be also a normed linear space. Really, any continuous linear functional in a normed linear space is bounded and its norm may be assumed as a norm in the dual space X^*.

It is evident that a norm of a linear functional satisfies all the axioms of a norm (Subsection 1.3.6). The second dual space X^{**} will be also a normed linear space in this case.

▷ Notice now that the equality $z = fx$ at any fixed x determines a map of the space X^* into the field of the scalar K, i.e. the functional on X^*. Thus at fixed x fx represents a linear functional on X^*. This functional is continuous as at all f

$$| fx | \leq \| f \| \| x \|, \tag{5.2}$$

and consequently, determines some element φ_x of the second dual space X^{**}:

$$\varphi_x f = fx. \tag{5.3}$$

From (5.2) and (5.3) follows that a norm of the functional φ_x does not exceed $\| x \|$, $\| \varphi_x \| \leq \| x \|$. On the other hand, according to the Corollary of Hahn–Banach theorem in a normed linear space (4.58) at any given $x \in X$ in X^* there exists such functional f_0 that

$$| \varphi_x f_0 | = | f_0 x | = \| f_0 \| \| x \| .$$

Hence and from the inequality

$$| \varphi_x f | \leq \| \varphi_x \| \| f \|$$

at all f follows $\| x \| \leq \| \varphi_x \|$. Two obtained opposite inequalities give $\| \varphi_x \| = \| x \|$.

Thus equality (5.3) establishes the correspondence between the element φ_x of the second dual space X^{**} and each element x of the space X. This correspondence is the one–to–one mapping as if to one and the same functional $\varphi_x \in X^{**}$ correspond two elements x, x' of a space X then from $\varphi_x f = fx = fx'$ follows $f(x - x') = 0$ at all $f \in X^*$, i.e. $x' = x$. ◁

The outlined shows that equality (5.3) establishes the one–to–one mapping of the whole space X into the second dual space X^{**} with the retention of a norm.

In the special case when not only each $x \in X$ corresponds to the functional $\varphi_x \in X^{**}$ but also vice versa to each $\varphi \in X^{**}$ corresponds some element $x \in X$ (i.e. when (5.3) determines the one–to–one mapping of X on all X^{**}) we may identify the second dual space X^{**} with X, $X^{**} = X$. In this case the space X is called *a reflexive one*.

On the basis of correspondence (5.3) between the elements of the spaces X and X^{**} a weak topology in the dual space X^* may be given by two ways: by zero vicinities

$$\{ f : | \varphi_1 f | < \varepsilon, \ldots, | \varphi_n f | < \varepsilon \}, \tag{5.4}$$

correspondent to all n, $\varepsilon > 0$ и $\varphi_1, \ldots, \varphi_n \in X^{**}$ (an ordinary way of determining a weak topology in the normed linear space when $F = X^{**}$), or by zero vicinities

$$\{ f : | fx_1 | < \varepsilon, \ldots, | fx_n | < \varepsilon \}, \tag{5.5}$$

correspondent to all n, $\varepsilon > 0$ и $x_1, \ldots, x_n \in X$ (when as the set of the linear functionals F is assumed the space $X \subset X^{**}$). The second topology in the general case is weaker then the first one. For distinguishing these two weak topologies the second topology is called $*$–*weak* topology while the first one is simply called *a weak* topology. In the special case of the reflexive space X $*$–weak topology in X^* coincides with a weak. From Theorem 5.1 follows immediately.

Theorem 5.2. *The space X^* dual with a normed linear space X is complete, i.e. a B–space (no matter whether X is complete or not).*

▷ For proving it is sufficient to notice that in this case Y represents a field of the scalars K which is a complete normed linear space, i.e. a B–space. ◁

Now we show how the obtained results are generalized on the case of a topological linear space X. In this case a strong topology in the dual space X^* is determined by zero vicinities

$$\{\, f : |\, fx\, | < \varepsilon\ \forall x \in A\, \}, \tag{5.6}$$

correspondent to all $\varepsilon > 0$ and to all bounded sets A [a]. It is clear that in the case when X is a normed linear space the topology determined by zero vicinities (5.6) coincides with the topology determined by a norm. A weak and $*$–weak topology in X^* are determined in this case similarly as in the case of a normed space X.

Equality (5.3) analogously as in the case of a normed space X determines the one–to–one mapping X into X^{**}. In the special case when it is mapping X on all X^{**} a space X is called *semi–reflexive*. If the mapping X on X^{**} determined by formula (5.3) is continuous in the strong topologies of the spaces X and X^{**} [b] a semi–reflexive space X is called *reflexive*. It is clear that in the case of a normed space X the notions of the semi–reflexivity and the reflexivity coincide as in this case the mapping of X into X^{**} determined by equality (5.3) is always continuous as it retains a norm. The norm of this linear mapping is always equal to 1.

Analogously in the case of a complex linear space X we may determine the dual space $\bar{X}^*$ of dually linear functionals on X.

E x a m p l e 5.5. Let X be n–dimensional real Euclidean space. As it is easy to see a linear functional in X is determined by formula $fx = f_1 x_1 + \cdots + f_n x_n$ where $f_1, \ldots, f_n$ are arbitrary real numbers. Here any linear functional f in the finite–dimensional space is bounded and its norm $\|\, f\, \|$ is determined by formula

$$\|\, f\, \| = \sqrt{f_1^2 + \cdots + f_n^2}\,.$$

Thus a linear functional in n–dimensional Euclidean space is completely determined by n–dimensional vector f with the coordinates $f_1, \ldots, f_n$. To each linear functional corresponds the unique vector f and vice versa, to each vector f corresponds the unique linear functional and the norm of the functional coincides with the norm of the correspondent vector. This one–to–one correspondence between

---

[a] The set A of a topological linear space is called *bounded* if for any vicinity of zero V there exists such number $c > 0$ that $A \subset \lambda V$ at all λ, $|\lambda| \geq c$ (Subsection 4.2.1)

[b] A strong topology in X^{**} is determined similarly as in X^* by means of zero vicinities of form (5.6).

elements of the space X and the linear functionals with the retention of the norm gives the opportunity to identify the dual space X^* with the space X itself. Then the result of the action of any linear functional f on the vector $x \in X$ may be represented as a scalar product of the vector x of fixed vector f, $fx = (x, f)$.

E x a m p l e 5.6. If X is a complex n–dimensional Euclidean space then after writing a linear functional in the form $fx = \bar{f}_1 x_1 + \cdots + \bar{f}_n x_n$ we can represent it again as a scalar product $fx = (x, f)$ and identify the dual space X^* with the space X. Here the norm of the functional also coincides with the norm of the correspondent vector f:

$$\| f \| = \sqrt{|f_1|^2 + \cdots + |f_n|^2} \, .$$

In this case we may also assume $X^* = X$ [a].

E x a m p l e 5.7. The results of the previous Examples are easily generalized on the case $n = \infty$. In this case as the elements of the space X are all infinite sequences $x = \{x_k\}$ for which the series $\sum |x_k|^2$ converges, and the norm of the element $x = \{x_k\}$ is determined by formula

$$\| x \| = \sqrt{\sum_{k=1}^{\infty} |x_k|^2} \, .$$

This space is usually called the space l_2. In this case also to any linear functional corresponds the vector $f \in X$ and

$$fx = (x, f) = \sum_{k=1}^{\infty} \bar{f}_k x_k.$$

Thus in this case we may also assume $X^* = X$

E x a m p l e 5.8. On the basis of Theorem 4.49 as a space dual with the topological linear space X with the topology τ_F determined by zero vicinities serves F: $X^* = F$.

5.1.4. Topologies in a Space of Bounded Linear Operators

In a space $\mathcal{B}(X, Y)$ of bounded linear operators mapping a normed linear space X into a normed linear space Y a natural topology is determined by a norm. But this topology is sometimes too

---

[a] It should be remembered that in the case of a complex space X the scalar product (x, f) at fixed f represents a linear functional of x, and at fixed x represents a dually linear functional of f.

strong. Therefore two more topologies are usually determined in a space $\mathcal{B}(X,Y)$. One topology is determined by zero vicinities

$$\{T: \| Tx_1 \| < \varepsilon, \ldots, \| Tx_n \| < \varepsilon \}, \qquad (5.7)$$

correspondent to all $\varepsilon > 0$, n and $x_1, \ldots, x_n \in X$. The second topology is determined by a set of linear functionals of the form $fT = gTx$ where g is a continuous linear functional on a space Y, i.e. an element of the space Y^* dual with Y. Due to the continuity of the functional g and the operator T

$$| fT | \leq \| g \| \| Tx \| \leq \| g \| \| x \| \| T \| \ .$$

Consequently, the functional f is a continuous linear functional of a normed linear space $\mathcal{B}(X,Y)$ and its norm does not exceed $\| g \| \| x \|$. In the correspondence with the general way of determining a topology by means of a set of linear functionals the second topology in $\mathcal{B}(X,Y)$ is determined by zero vicinities

$$\{T: | g_1 Tx_1 | < \varepsilon, \ldots, | g_n Tx_n | < \varepsilon \}, \qquad (5.8)$$

correspondent $\varepsilon > 0$, n, $x_1, \ldots, x_n \in X$, $g_1, \ldots, g_n \in Y^*$.

Let us establish the relations between three topologies in $\mathcal{B}(X,Y)$.

Each vector $x \in X$ determines the map φ_x of a space $\mathcal{B}(X,Y)$ in $Y: \varphi_x(T) = Tx$. This mapping is linear and continuous as

$$\| \varphi_x(T) \| = \| Tx \| \leq \| T \| \| x \| \ .$$

The vicinity of zero (5.7) at $n = 1$, $x_1 = x$ represents the inverse image in $\mathcal{B}(X,Y)$ of zero vicinity $\{y: \| y \| < \varepsilon\}$ in Y:

$$\{T: \| Tx \| < \varepsilon\} = \varphi_x^{-1}(\{y: \| y \| < \varepsilon\}).$$

Therefore by virtue of the continuity of φ_x the set $\{T: \| Tx \| < \varepsilon\}$ and together with it all the vicinities of zero (5.7) according to Theorem 4.19 are the open sets in the topology of the space $\mathcal{B}(X,Y)$ generated by a norm. Consequently, the topology $\mathcal{B}(X,Y)$ generated by the vicinities of zero (5.7) is not stronger than the topology determined by a norm.

By Theorem 4.49 the topology in Y determined by the set of linear functionals Y^* is not stronger than the topology generated by a norm. Hence it follows that the topology in $\mathcal{B}(X,Y)$ determined by the vicinities of zero (5.8) is not stronger than the topology determined by the vicinities of zero (5.7).

In accordance with the stated relations between three topologies in a space of the operators $\mathcal{B}(X, Y)$ a topology determined by a norm is called *an uniform* topology, a topology determined by the vicinities of zero (5.7) is called *a strong* topology, and a topology determined by the vicinities of zero (5.8) is called *a weak* topology.

For detailed treatment of spaces of linear operators and functionals refer to (Dunford and Schwarz 1958).

P r o b l e m s

5.1.1. Prove that as the space dual with the space of the sequence l_p with the norm $\| x \| = \left(\sum_{n=1}^{\infty} |x_n|^p \right)^{1/p}$ at $p > 1$ serves l_q, $q^{-1} = 1 - p^{-1}$. Is l_p reflexive or not?

I n s t r u c t i o n. Use inequality (3.89) which is also valid for the spaces l_p, l_q, l_r (prove) at $r = 1$.

5.1.2. Prove that as the space dual with l_1 serves the space of the bounded sequences m with a norm $\| x \| = \sup_k |x_k|$.

5.1.3. Find the space dual with the space $C([a, b])$ with the topology τ_F of Problem 4.3.2. Will $C([a, b])$ with the topology τ_F be reflexive?

5.1.4. Find the space dual with the space $C(T)$ with the topology τ_F of Problem 4.3.6.

5.1.5. Find the space dual with a space $\mathcal{B}(X, Y)$ of the bounded linear operators mapping a normed linear space X into a normed linear space Y.

5.1.6. The linear operators $A_1, \ldots, A_n$ mapping the space X of the finite–dimensional vector functions of the variable t into the space Y of the finite–dimensional vector functions of the variable s are determined by the formulae

$$A_p x = \int g_p(s, t) x(t) \, dt \quad (p = 1, \ldots, n), \qquad (*)$$

where $g_1(s, t), \ldots, g_n(s, t)$ are some matrix functions, in general case rectangular. Find the sum of the operators $A_1, \ldots, A_n$.

5.1.7. The linear operator A mapping the space X of the functions of the variable t into the space Y of the functions of the variable s is determined by the formula of the form $(*)$, and the linear operator B mapping Y into the space Z of the functions of the variable u is determined by the same formula

$$By = \int h(u, s) y(s) \, ds.$$

Find the product of the operators A and B and express it by the formula of the form $(*)$ (find a kernel of the correspondent integral operator).

5.1.8. Prove that the linear operator T has the inverse operator T^{-1} if and only if its kernel $\{x : Tx = 0\}$ represents a singleton $\{0\}$.

5.2. Weak Integrals

5.2.1. Definition of a Weak Integral

Let us consider a space with the measure $(X, \mathcal{A}, \mu)$, $\mu(A) \geq 0 \; \forall A \in \mathcal{A}$ and the topological linear space Y with the topology τ_G generated by the set of linear functionals G (Subsection 4.2.4). This topology is determined by zero vicinities

$$\{ y : |g_1 y| < \varepsilon, \dots, |g_n y| < \varepsilon \}, \tag{5.9}$$

correspondent to all n, $\varepsilon > 0$, $g_1, \dots, g_n \in G$. Relative to the set of linear functionals G we shall suppose similarly as in Section 4.2 that it represents a linear space and divides the points of the space Y (Subsection 4.2.4). By virtue of Theorem 4.49 the set of linear functionals G coincides with the space Y^* dual with Y : $G = Y^*$. Let $\mathcal{B}_G$ be the σ–algebra of the sets of the space Y generated by the topology τ_G.

The $(\mathcal{A}, \mathcal{B}_G)$–measurable function $y = f(x)$ (Subsection 3.1.1) mapping the space X into the topological space Y is called *weak μ–integrable* if

(i) the numeric function $gf(x)$ is μ–integrable at any $g \in Y^*$;

(ii) there exists such vector y_f of the space Y that

$$g y_f = \int gf(x)\,\mu(dx) = \int gf\,d\mu \quad \forall g \in Y^*. \tag{5.10}$$

In this case a vector y_f is called *a weak integral (the Pettis integral)* of the function f over the measure μ:

$$y_f = \int f(x)\mu(dx) = \int f\,d\mu. \tag{5.11}$$

Theorem 5.3. *If the space Y is complete and there exists such a sequence of the simple functions $\{f^n(x)\}$ convergent almost everywhere (in the topology τ_G) to the function $f(x)$ that*

$$\int |gf^n - gf^m|\,d\mu \to 0 \quad \text{at} \quad n, m \to \infty$$

at any $g \in Y^*$ then the function $f(x)$ is weakly μ–integrable and

$$\int f\,d\mu = \lim_{n \to \infty} \int f^n\,d\mu. \tag{5.12}$$

▷ In this case all numeric functions gf, $g \in Y^*$ are μ–integrable due to Theorem 4.54 $gf^n \xrightarrow{\text{a.e.}} gf$ at any $g \in Y^*$ and

$$\int gf^n \, d\mu \to \int gf \, d\mu \quad \forall g \in Y^*. \tag{5.13}$$

But in property (V) of an integral of a simple function

$$\int gf^n \, d\mu = g \int f^n \, d\mu \quad (n = 1, 2, \ldots). \tag{5.14}$$

From (5.13) and (5.14) follows the convergence of the sequence $\{g \int f^n \, d\mu\}$ and consequently, its property to be fundamental. Thus the sequence of the integrals $\{\int f^n \, d\mu\}$ is fundamental (in the topology τ_G of the space Y). By virtue of the completeness of the space Y it converges to some limit

$$\int f^n \, d\mu \to y_f.$$

By Theorem 4.54 it follows

$$g \int f^n \, d\mu \to g y_f \quad \forall g \in Y^*. \tag{5.15}$$

From (5.13), (5.14) and (5.15) follows (5.10) and (5.11). Thus the limit of the sequence of the integrals $\{\int f^n \, d\mu\}$ of the simple functions represents a weak integral of the function f. So, it proves formula (5.12). ◁

If in the space Y the topology τ in which all the functionals are continuous $g \in G$ is determined then the topology τ_G will be a weak topology in the space Y (Subsection 4.3.1). Under the conditions of Theorem 5.3 the sequence of simple functions $\{f^n\}$ will be almost everywhere weakly convergent to the function f, and integral (5.11) of the function f will be a weak limit of the sequence of the integrals of simple functions f^n. Therefore the Pettis integral is called a weak integral.

A weak integral of the function $y = f(x)$ with the values in the topological linear space Y with the topology τ_G over any set $A \in \mathcal{A}$ is determined as a weak integral of the function $y = f(x)1_A(x)$:

$$\int_A f(x)\mu(dx) = \int f(x)1_A(x)\mu(dx). \tag{5.16}$$

It is evident that under the conditions of Theorem 5.3

$$\int_A f^n \, d\mu \to \int_A f \, d\mu$$

is uniform relative to $A \in \mathcal{A}$.

Theorem 5.3 gives a sufficient condition of the existence of an integral. But this condition is far from necessary. Integral (5.11) may also exist in the case when there is no sequence of the simple functions $\{f^n\}$ convergent almost everywhere to f. The condition of the existence of a weak integral which gives Theorem 5.3 is too bounding. A more general condition of the existence of a weak integral gives the following theorem.

Theorem 5.4. *If all the functions $gf(x)$, $g \in Y^*$ are μ–integrable and Y^* is a space with the first axiom of the countability then a weak integral exists as an element of the second dual space Y^{**}.*

▷ It is clear that the integral $\int gf\, d\mu$ represents the value of some linear functional φ at the point g of the space Y^*. To make sure that this functional belongs to Y^{**} it is sufficient to prove that it is continuous at zero (Theorem 4.55). For this purpose we take any sequence of the functionals $\{g_n\}$, $g_n \in Y^*$ convergent to zero. For this sequence $g_n y \to 0$ $\forall y \in Y$. On the the basis of Fatou lemma (Subsection 3.5.3)

$$\overline{\lim} \int |g_n f|\, d\mu \leq \int \overline{\lim} |g_n f|\, d\mu = \int \lim |g_n f|\, d\mu = 0.$$

Consequently, there exists

$$\lim \int g_n f\, d\mu = 0,$$

what proves the continuity of the functional $\varphi = \int f\, d\mu$. ◁

Corollary. *If the space Y is reflexive and Y^* is a space with the first axiom of the countability then for the existence of the weak integral (5.11) it is necessary and sufficient that all the functions $gf(x)$, $g \in Y^*$ be μ–integrable.*

5.2.2. Properties of a Weak Integral

From the linearity of the functionals $g \in Y^*$ and formula (5.10) immediately follows that the Pettis integral possesses properties (I)–(VII) of the Bochner integral (Subsecions 3.3.5).

From property (X) of the Bochner integral follows that for any $\varepsilon > 0$ and any functionals $g_1, \ldots, g_n \in Y^*$ there exists such $\delta > 0$ that

$$\left| \int_A g_1 f\, d\mu \right| < \varepsilon, \ldots, \left| \int_A g_n f\, d\mu \right| < \varepsilon$$

at all $A \in \mathcal{A}$, $\mu(A) < \delta$ for the given function $y = f(x)$. On the basis of (5.10) these inequalities may be rewritten in the form

$$\left| g_1 \int_A f \, d\mu \right| < \varepsilon, \ldots, \left| g_n \int_A f \, d\mu \right| < \varepsilon.$$

Thus for any base vicinity of zero

$$V_0 = \{ y : |g_1 y| < \varepsilon, \ldots, |g_n y| < \varepsilon \}$$

there exists such $\delta > 0$ that

$$\int_A f \, d\mu \in V_0 \quad \text{for all} \quad A \in \mathcal{A}, \quad \mu(A) < \delta. \tag{X}$$

This inclusion represents the extension of property (X) over weak integrals. Analogously property (XV) is extended over the weak integrals. On the basis of property (XV) of the Bochner integral for any functionals $g_1, \ldots, g_n \in Y^*$ and any $\varepsilon > 0$ there exists such set $A_\varepsilon^n \in \mathcal{A}$ that $\mu(A_\varepsilon^n) < \infty$ and

$$\left| \int_{A_\varepsilon^n} g_1 f \, d\mu \right| < \varepsilon, \ldots, \left| \int_{A_\varepsilon^n} g_n f \, d\mu \right| < \varepsilon.$$

On the basis of (5.10) it means that for any base vicinity of zero

$$V_0 = \{ y : |g_1 y| < \varepsilon, \ldots, |g_n y| < \varepsilon \}$$

there exists such set $A_\varepsilon^n \in \mathcal{A}$ that $\mu(A_\varepsilon^n) < \infty$ and

$$\int_{A_\varepsilon^n} f \, d\mu \in V_0. \tag{XV}$$

E x a m p l e 5.9. Let Y be a space of all scalar or finite–dimensional vector functions of the variable t (scalar or finite–dimensional vector) with the domain T, $G = Y^*$ be the set of linear functionals on Y determined by formula

$$gy(t) = \sum_{k=1}^{N} g_k^T y(t_k) \tag{5.17}$$

at all natural N, at all finite collections $t_1, \ldots, t_N \in T$ of the values of the argument t and all the vectors (in general case complex) $g_1, \ldots, g_N$. In this formula the vectors g_k and $y(t_k)$ are presented in the form of the matrices–columns. Let us consider the function $f(x)$ with the values in the space Y, $y = f(x) = f(x,t)$, $t \in T$. According to (5.10) a weak integral of the function f exists if and only if the function $f(x,t)$ considered as the function of two variables x and t is μ–integrable at any $t \in T$. For proving this it is sufficient to take functional (5.17) at $N = 1$, $g_1^T = [1 \ldots 1]$. In this case the integral

$$\int f(x,t)\mu(dx) = y_f(t), \quad t \in T,$$

represents a function of the variable t.

5.2.3. Relation between a Weak Integral and a Strong One

If Y is a separable B–space then a weak topology in it is determined in the ordinary way. In this case we may determine both the strong integrals (Bochner integrals) of the functions with the values in Y and the weak integrals (Pettis integrals). The question arises in what way these two types of the integrals are connected with each other. The answer to this question is given in the following theorem.

Theorem 5.5. *If there exists the Bochner integral of the functions $f(x)$ then there also exists the Pettis integral coinciding with the Bochner integral.*

▷ If there exists the Bochner integral of the function $f(x)$ then there exists the sequence of the simple functions which determines it (Subsection 3.3.5). Therefore for any functional $g \in Y^*$ by virtue of its continuity

$$|gf^n(x) - gf(x)| \leq \|g\| \, \| f^n(x) - f(x) \| \xrightarrow{\text{a.e.}} 0$$

and

$$\int |gf^n - gf^m|\,d\mu \leq \|g\| \int \| f^n - f^m \| \, d\mu \to 0.$$

Thus the sequence of the simple functions $\{gf^n(x)\}$ detemines the function $gf(x)$, and consequently, the function $gf(x)$ is μ–integrable. Further on the basis of property (V) of the integral of the simple function

$$g \int f^n \, d\mu = \int gf^n \, d\mu \to \int gf \, d\mu \quad \forall g \in Y^*.$$

On the other hand, by Theorem 4.59 from the convergence of the integrals $\{\int f^n \, d\mu\}$ to the Bochner integral $\int f \, d\mu$ follows the weak convergence of this sequence to the Bochner integral:

$$g \int f^n \, d\mu \to g \int f \, d\mu \quad \forall g \in Y^*.$$

Consequently,

$$\int gf \, d\mu = g \int f \, d\mu.$$

Hence on the basis of (5.10) and (5.11) we conclude that there exists the Pettis integral which coincides with the Bochner integral. ◁

R e m a r k. The proved theorem establishes the existence of the Pettis integral in those cases when the Bochner integral exists. But the Pettis integral of the function with values in a separable B–space may also exist when the Bochner integral is absent.

E x a m p l e 5.10. Let $X = (0, 1)$, μ be the Lebesgue measure. The function $f(x) = \sin(t/x)$ maps $(0, 1)$ into the space of the differentiable functions $C^1([0, 1])$ of the variable t. It is evident that the function $f(x)$ is integrable at any $t \in [0, 1]$ as it is modulus majorized by the integrable function $f_0(x) \equiv 1$ and

$$\left| \int\limits_0^1 \sin \frac{t}{x} \, dx \right| < 1.$$

But its norm is equal to

$$\| f(x) \| = \sup_{t \in [0,1]} \left| \sin \frac{t}{x} \right| + \sup_{t \in [0,1]} \left| \frac{1}{x} \cos \frac{t}{x} \right| > \frac{1}{x}$$

and consequently, is nonintegrable. Thus the function $f(x)$ may serve as an example of a weak integrable but non–integrable function with the values in a B–space $C^1([0, 1])$ with a weak topology determined by the set of linear functions (5.17).

E x a m p l e 5.11. In probability theory there are random variables with the values in the linear spaces. Such random variables are, for instance, *random functions* – random variables with the values in the functional spaces. Let $(\Omega, \mathcal{S}, P)$ be a probability space, i.e. the space of elementary events Ω with the σ–algebra of the sets $\mathcal{S}$ and the probability P which is determined on it (representing a nonnegative measure). As it is known a random variable with the values in the measurable space $(X, \mathcal{A})$ is called $(\mathcal{S}, \mathcal{A})$–measurable function $x(\omega)$ of elementary event ω with the domain $\Omega \backslash \Omega_0$ where Ω_0 is a set of elementary events with zero probability, $\Omega \in \mathcal{S}$,

$P(E) = 0$. In order to determine the expectation (mean value) of the random variable $x(\omega)$ with the values in the linear space X we introduce in this space the topology τ_F by means of the set of the linear functionals F. This topology is a weak topology if X is already endowed with some topology τ. In this case as F a set of all continuous in the topology τ linear functionals is assumed, $F = X^*$ (Subsection 4.3.1). Then a weak integral (the Pettis integral) of the function $x(\omega)$ over the probability measure P is called an expectation of the random variable $x(\omega)$

$$m_x = Ex(\omega) = \int x(\omega)P(d\omega) = \int x \, dP,$$

if it exists. So, if the function $fx(\omega)$ is P–integrable at any $f \in F$ and there exists such element m_x of the space X that

$$fm_x = \int fx(\omega)P(d\omega) \quad \forall f \in F.$$

In particular, if a random variable represents a real scalar random function of the variable $t \in T$, $x(\omega)$ represents the function ω with the values in the space of all the functions of the variable $t \in T$, i.e. in fact a function of two variables t and ω, $x(\omega) = x(t, \omega)$. If to each $t \in T$ corresponds the real axis R_t – a space of the values of a random function at given t then the space of all real scalar functions of the variable $t \in T$ will represent a direct product of all the spaces R_t (Subsection 2.1.9)

$$R^T = \prod_{t \in T} R_t.$$

For determining the topology on R^T we usually take the same set of linear functionals (5.17) as in Example 5.9. Then in accordance with the result of Example 5.9 an expectation of the random function $x(t, \omega)$ will be determined by formula

$$m_x(t) = Ex(t, \omega) = \int x(t, \omega)P(d\omega).$$

Here the integral represents an ordinary Lebesgue integral which exists if and only if the function $|x(t, \omega)|$ is P–integrable at each t (Subsection 3.4.1). To such a definition of an expectation of a random function leads a general definition of an expectation as a weak integral.

Notice that if we try, for instance, to determine an expectation of the random function $x(t, \omega)$ with the values in a separable B–space of the continuous functions $C(T)$ as a strong integral (the Bocher integral) then it will be not sufficient the P–integrability of the function $|x(t, \omega)|$ for an expectation existence. In accordance with the condition of the Bochner integral existence we had to require the P–integrability of the norm $\sup_{t \in T} |x(t, \omega)|$ of the function $x(t, \omega)$ what is significantly more rigorous requirement (Example 5.10). Thus the existence of the Bochner integral is not sufficient for determining an expectation of random variable with the values in a B-space.

5.2.4. Passage to the Limit under a Weak Integral Sign

The possibility of the limit passage under the integral sign is determined by the following theorem analogous to Theorem 3.30.

Theorem 5.6. *If the space* Y *is complete, the sequence of weakly* μ-*integralbe functions* $\{f_n(x)\}$ *with the values in* Y *converges almost everywhere* (*in the topology* τ_G) *to the function* $f(x)$ *and all the sequences* $\{gf_n(x)\}$, $g \in Y^*$ *are modulus–bounded by the* μ-*integrable functions* $v_g(x)$, $|gf_n(x)| \le v_g(x)$ *then the function* $f(x)$ *is weakly integrable and*

$$\lim_{n \to \infty} \int f_n \, d\mu = \int f \, d\mu. \tag{5.18}$$

▷ By virtue of Theorem 4.54 from the convergence of the sequence $\{f_n\}$ almost everywhere to f follows the almost everywhere convergence of the sequence $\{gf_n\}$ to gf at any $g \in Y^*$. Consequently, by Lebesgue Theorem 3.30 the function gf is μ-integrable at $g \in Y^*$ and

$$\lim_{n \to \infty} \int gf_n \, d\mu = \int gf \, d\mu \quad \forall g \in Y^*. \tag{5.19}$$

Thus the first condition of the weak μ-integrability of the function f is fulfilled. From the weak μ-integrability of the function f_n at any n we get the equality

$$\int gf_n \, d\mu = g \int f_n \, d\mu \, . \ \forall g \in Y^*. \tag{5.20}$$

From (5.19) and (5.20) follows that the sequence of the integrals $\{\int f_n \, d\mu\}$ is fundamental (in the topology τ_G). Owing to the completeness of the space Y this sequence has the limit y_f:

$$\lim_{n \to \infty} g \int f_n \, d\mu = g y_f \quad \forall g \in Y^*. \tag{5.21}$$

Finally, from (5.19)–(5.21) follows that the second condition of the weak μ-integrability of the function f is also fulfilled. Thus the function f is weakly μ-integrable and

$$\int f \, d\mu = y_f = \lim_{n \to \infty} \int f_n \, d\mu. \ \triangleleft$$

Corollary. *If the series*

$$\sum_{k=1}^{\infty} f_k(x), \qquad (5.22)$$

where $\{f_k(x)\}$ is the sequence of the weakly μ–integrable functions converges (in the topology τ_G) almost everywhere to some function $f(x)$ and at any $g \in Y^*$ all the numeric functions

$$gs_k(x) = \sum_{k=1}^{n} gf_k(x), \quad (n = 1, 2, \ldots)$$

are modulus–bounded by the μ–integrable function $v_g(x)$ (certainly dependent on $g \in Y^*$) then series (5.22) may be termwise integrable:

$$\int f\, d\mu = \int \sum_{k=1}^{\infty} f_k\, d\mu = \sum_{k=1}^{\infty} \int f_k\, d\mu. \qquad (5.23)$$

5.2.5. Fubini Theorem

Let us extend Fubini theorem 3.44 on Pettis integrals. Similarly as in Subsection 3.8.1 let us consider two measurable spaces $(X, \mathcal{A})$, $(Y, \mathcal{B})$ and their product $(Z, \mathcal{C}) = (X \times Y, \mathcal{A} \times \mathcal{B})$. Let λ_x be a collection of such nonnegative measures in the space $(Y, \mathcal{B})$ that at any $B \in \mathcal{B}$ the function $\lambda_x(B)$ of the variable x is measurable, μ – a nonnegative measure in the space $(X, \mathcal{A})$. We determine in the same way as in Subsection 3.8.1 the nonnegative measure ν in the product of the spaces $(Z, \mathcal{C})$. By Theorem 3.43 it is determined by formula (3.105):

$$\nu(C) = \int \mu(dx) \int \mathbf{1}_C(x, y)\lambda_x(dy), \quad C \in \mathcal{C}.$$

Consider the function $f(x, y)$ with the values in the topological linear space U with the topology τ_G determined simularly as in Subsection 5.2.1 by the set of the linear functionals $G = U^*$ and the σ–algebra $\mathcal{D}_G$ generated by the topology τ_G.

Theorem 5.7. *If the function $f(z) = f(x, y)$ is weakly ν–integrable, its section $f_x(y) = f(x, y)$ almost at all x relative to the measure μ represents a weakly λ_x–integrable function y and a weak integral*

$$w(x) = \int f_x(y)\lambda_x(dy) = \int f\, d\lambda_x$$

is a weakly μ–integrable function x then

$$\int f \, d\nu = \int w \, d\mu = \int d\mu \int f \, d\lambda_x. \qquad (5.24)$$

▷ From the weak ν–integrability of $f(x, y)$ follows ν–integrability of $gf(x, y)$ at any $g \in U^*$. By Theorem 3.44 and definition (5.10) we have

$$\int gf \, d\nu = \int d\mu \int gf \, d\lambda_x , \qquad (5.25)$$

$$\int gf \, d\lambda_x = g \int f \, d\lambda_x = gw(x), \quad \int gf \, d\nu = \int gw \, d\mu ,$$

for any $g \in Y^*$. Finally, by virtue of the weak ν–integrability of the function $f(x, y)$ and the weak μ–integrability of the function $w(x)$

$$\int gf \, d\nu = g \int f \, d\nu \quad \forall g \in U^* ,$$

$$\int gw \, d\mu = g \int w \, d\mu \quad \forall g \in U^* .$$

Consequently,

$$g \int f \, d\nu = g \int w \, d\mu = g \int d\mu \int f \, d\lambda_x \quad \forall g \in U^* .$$

This proves the validity of formula (5.24) for weak integrals of the functions $f(x, y)$ satisfying the conditions of the theorem. ◁

The theory of weak integrals is given in (Dunford and Schwarz 1958, Edwards 1965). For probability applications refer to (Cramer and Leadbetter 1967, Kolmogorov 1933, Parthasarathy 1980, Pugachev and Sinitsyn 1987).

Problems

5.2.1. All the integrals by the Wiener measure calculated in Examples 3.4. and 3.6 and in Problems 3.3.1–3.3.4 and 3.5.1–3.5.3 exist as Bochner integrals if the integrands have the values in the space $C(S)$ of the continuous functions. But they may exist only as weak integrals if the integrands have the values in another B–space or in the topological linear spaces. On the basis of the result of Example 5.10 give the examples where the integrals of Example 3.6 and Problems 3.5.1–3.5.3 exist only as weak integrals.

5.2.2. Prove that if X is a compact T_2-space with the nonnegative normed measure μ, $\mu(X) = 1$, $f(x)$ is a continuous function mapping X into the topological linear space Y with the topology τ_G generated by the set of linear functionals G and the convex shell H of the set $f(X)$ (i.e. the minimal convex set which contains $f(X)$) is precompact in Y (Subsection 4.4.4) then the weak integral y_f of the function $f(x)$ exists and $y_f \in [H]$.

I n s t r u c t i o n. Consider all kinds of the collection of the functionals $g_\alpha = \{g_1, \ldots, g_n\}$ $\forall g_1, \ldots, g_n \in Y^*$ and determine the correspondent sets

$$E_\alpha = \{ y : y \in [H], \ g_1 y = \textstyle\int g_1 f \, d\mu, \ldots, g_n y = \int g_n f \, d\mu \}.$$

These sets are closed as the inverse images in Y of the singletons of the space K^n ($K = R$ or C) at continuous mapping $g_\alpha : Y \to K^n$. None of them is empty as at any x $g_\alpha f(x) \in g_\alpha f(X) \subset g_\alpha[H]$ and by virtue of the convexity of the set $g_\alpha[H]$ and the equality $\mu(X) = 1$ at $\int g_\alpha f \, d\mu \in g_\alpha[H]$, hence if follows the existence of such $y \in [H]$ that $g_\alpha y = \int g_\alpha f \, d\mu$. All finite intersections of the sets E_α are also nonempty as the sets of the same type E_α. Thus the aggregate of all the sets E_α represents a centred system of closed subsets of the compact set $[H]$ (Subsection 4.4.2). It remains to use Theorem 4.25.

5.2.3. Under the conditions of Problem 5.2.2 suppose that Y is a locally convex topological linear space with the topology τ. In this case the topology τ_G generated by a set of continuous linear functionals $G = Y^*$ will be a weak topology and the topology τ – a strong topology in the space Y. Prove that in this case for any zero vicinity $V \in \tau$ of the space Y there exists such a finite partition of the compact X into the sets which are pairwise nonintersecting sets, $X = \bigcup E_i$ that

$$\int f \, d\mu - \sum_{i=1}^{n} f(x_i)\mu(E_i) \in V \quad \forall x_i \in E_i,$$

i.e. a weak integral represents in this case the strong limit of the sequence of Riemann integral sums.

I n s t r u c t i o n. Supposing that the vicinity of zero V is convex (what by virtue of the definition of locally convex space does not lead to the loss of generality) single out the covering of the compact $f(X)$ by open sets $V_x = f(x)+V$ the finite covering $\bigcup V_k \supset f(X)$ and construct from the inverse images of the sets $V_1, \ldots, V_n$ pairwise nonintersecting sets $E_1, \ldots, E_n$. On these sets will be $f(x) - f(x_i) \in V$ $\forall x \in E_i$ and by virtue of the convexity of V and the result of Problem 1.4.5 we shall have

$$\int_{E_i} [f(x) - f(x_i)]\frac{\mu(dx)}{\mu(E_i)} \in V \,,$$

$$\int f\, d\mu - \sum_{i=1}^{n} f(x_i)\mu(E_i) = \sum_{i=1}^{n} \mu(E_i) \int_{E_i} [f(x) - f(x_i)]\frac{\mu(dx)}{\mu(E_i)} \in V.$$

5.2.4. At any weakly μ–integrable function $f(x)$ the equality $z = gf(x)$ determines a linear operator mapping Y^* into the space $\bar{L}_1 = L_1(X, \mathcal{A}, \mu)$. What represents a norm of the element $z \in L_1$?

5.3. Generalized Functions

5.3.1. Notion of a Generalized Function

For mathematical description of such a physical phenomena as a mass or electrical charges, which are concentrated at the points, on the lines or at the surfaces, impulse forces etc. English physicist Dirac introduced the following notion of the *impulse delta–function (the δ–function)*:

$$\delta(0) = \infty, \quad \delta(t) = 0 \quad \forall t \neq 0,$$

$$\int_A \delta(t)\, dt = 1 \quad \forall A \ni 0.$$

It is clear that an ordinary function which possesses such properties does not exist. As we know an integral of an ordinary function as an absolutely continuous function of the set relative to the measure over which the integration is performed (in a given case the Lebesgue measure) tends to zero while the measure of the set A over which the integration is extended tends to zero. Therefore an integral of an ordinary function cannot remain equal to 1 for all the sets which contain the origin. Thus the δ–function differs in its properties from ordinary functions. In consequence of this fact the necessity to generalize the notion of a function and extend it on such functions as the δ–function and its derivatives appears.

5.3.2. Two Approaches to a Definition of Generalized Functions

Dirac δ–function is only a mathematical model of the concentrated and impulse phenomena. In nature actually the masses, the charges and the forces may be distributed in some regions whose volumes may be assumed negligibly small. Analogously it is commonly supposed that the phenomena which happen during negligibly small time intervals are instantaneous. Such picture of the real phenomena indicates the first approach to the definition of generalized functions: the limit passage from ordinary functions which describe the distribution of masses, charges and so on in small volumes and the phenomena happening during small time intervals at unbounded decrease of the volumes and time intervals.

The second approach to the definition of the generalized functions is based on the fact that as a result of all the operations on the δ–functions and their derivatives in applications they are always a part of the final formulae in the form of the multipliers under the integral sign. As the integral of the product of two functions always represents a linear functional of any of them then the δ–function and its derivatives may be identified with the definite linear functionals. Such approach leads to the definition of a generalized function as a linear functional which acts on some functional space.

The second approach is natural for functional analysis while the first approach is more typical for mathematical analysis. From the mathematical point of view both approaches are equivalent.

5.3.3. Space of Test Functions

In modern theory of the generalized functions the different type of functional spaces of these functions are considered.

A function $\varphi(t)$ of the scalar or finite–dimensional vector argument $t \in R^n$ is called *finite* if it differs from zero only on some bounded set. A set closure on which the function $\varphi(t)$ is different from zero is called *a carrier* or *support* of the function $\varphi(t)$ and is denoted by carr φ or supp φ.

A set Φ of all scalar finite functions which are continuous with their derivatives of all orders is called *a space of test functions*. The topology τ in this space is usually determined by zero vicinities

$$\{ \varphi : |\varphi(t)| < \gamma_0(t), \ldots, |\varphi^{(m)}(t)| < \gamma_m(t) \}, \tag{5.26}$$

correspondent to all nonnegative entire m and all the collections $\gamma_0(t)$, $\gamma_1(t), \ldots, \gamma_m(t)$ of continuous positive functions [a]. Endowed with such topology τ the space of test functions Φ represents a topological linear space. Let us study the topology τ.

First of all notice that the operations of the vectors addition and vector multiplication by a number are continuous in the topology τ and that the space Φ with the topology τ is locally convex. These statements are proved in the same way as they were proved in Subsection 4.6.4 for the topology τ_F. The answer to the question what represents the convergence in the topology τ gives the following statement.

Theorem 5.8. *In order the sequence of the functions* $\{\varphi_n\} \subset \Phi$ *should converge to the function* $\varphi(t) \in \Phi$ *it is necessary and sufficient that*

(i) *the carriers of all the functions* $\varphi_n(t)$ *should be contained in any compact* K;

(ii) *the sequences* $\{\varphi_n^{(k)}(t)\}$ *should uniformly converge to the correspondent functions* $\varphi^{(k)}(t)$ $(k = 0, 1, 2, \ldots)$.

▷ Without loss of generality we may suppose that $\varphi(t) \equiv 0$ and $\varphi_n(t) \to 0$. If the carriers of all the functions $\varphi_n(t)$ are not contained in any compact then there exists the subsequence $\{\varphi_{n_k}(t)\}$ and unboundedly modulus–increasing sequence of the values $\{t_k\}$ of the argument t that $\varphi_{n_k}(t_k) \neq 0$. It is clear that zero vicinity (5.26) correspondent to $m = 0$ and any positive function $\gamma_0(t)$ satisfying the condition $\gamma_0(t_k) < |\varphi_{n_k}(t_k)|/2$ $(k = 1, 2, \ldots)$ does not contain any function $\varphi_{n_k}(t)$. Consequently, the subsequence $\{\varphi_{n_k}(t)\}$ and also with it the whole sequence $\{\varphi_n(t)\}$ cannot converge to zero. The received contradiction proves the necessity of the first condition. If this condition is fulfilled then from the convergence of the sequence $\{\varphi_n\}$ to zero follows that at any positive functions $\gamma_0(t)$, $\gamma_1(t)$, $\ldots$

$$|\varphi_n^{(k)}(t)| < \gamma_k(t), \ \forall t \in K \quad (k = 0, 1, 2, \ldots)$$

for all sufficiently large n. But by Theorem 4.43 any continuous in the compact K the real function achieves on it its largest and smallest values. Therefore at all the same n

$$|\varphi_n^{(k)}(t)| < \varepsilon_k \ \forall t \quad (k = 0, 1, 2, \ldots),$$

---

[a] In the case of the vector argument t each derivative $\varphi^{(k)}(t)$ is substituted by the set of all the derivatives of the order k of the function $\varphi(t)$.

where ε_k is the greatest value of the function $\gamma_k(t)$ on K. Hence by virtue of the arbitrariness of the functions $\gamma_k(t)$ in (5.26), and consequently, the numbers $\varepsilon_k > 0$ follows the uniform convergence to zero of all the sequences $\{\varphi_n^{(k)}(t)\}$ $(k = 0, 1, 2, \ldots)$ what proves the necessity of the second condition.

For proving the sufficiency of the conditions notice that for any zero vicinity (5.26) we have

$$\{\varphi : |\varphi^{(k)}(t)| < \gamma_k(t) \ \forall t \quad (k = 0, 1, \ldots, m)\}$$

$$\supset \{\varphi : |\varphi^{(k)}(t)| < \gamma_k(t) \ \forall t \in K, \ \text{carr}\, \varphi \subset K \ (k = 0, 1, \ldots, m)\}$$

$$\supset \{\varphi : |\varphi^{(k)}(t)| < \varepsilon_k \ \forall t, \ \text{carr}\, \varphi \subset K \ (k = 0, 1, \ldots, m)\},$$

where $\varepsilon_0, \varepsilon_1, \ldots, \varepsilon_m$ are the smallest values of the functions $\gamma_0(t)$, $\gamma_1(t), \ldots, \gamma_m(t)$ on the compact K. From these inclusions and from uniform convergence to zero of all sequences $\{\varphi_n^{(k)}(t)\}$ follows that any zero vicinity in the space Φ contains all the functions $\varphi_n(t)$ beginning with someone. Thus $\varphi_n \to 0$ in the topology τ. $\triangleleft$

It is clear that the set of all the functions $\varphi \in \Phi$ whose carriers are contained in one and the same compact K represents the subspace Φ_K of the space Φ. Theorem 5.8 shows that any convergent sequence of the functions from the subspace Φ is contained in the subspace Φ_K correspondent to some compact K. The uniform convergence of the sequences $\{\varphi_n^{(k)}(t)\}$ corresponds to the topology τ_K in Φ_K determined by zero vicinities

$$\left\{\varphi : \max_{\substack{t \in K \\ 0 \leq k \leq m}} |\varphi^{(k)}(t)| < \varepsilon\right\}, \tag{5.27}$$

correspondent to all nonnegative entire m and all $\varepsilon > 0$.

Theorem 5.9. *The topology τ in Φ induces in Φ_K the topology τ_K determined by the base of zero vicinities (5.27).*

$\triangleright$ The topology τ in Φ induces in Φ_K the topology τ_K determined by the base of zero vicinities

$$\{\varphi : |\varphi(t)| < \gamma_0(t), \ldots, |\varphi^{(m)}(t)| < \gamma_m(t) \ \forall t\} \bigcap \Phi_K$$

$$= \{\varphi : |\varphi(t)| < \gamma_0(t), \ldots, |\varphi^{(m)}(t)| < \gamma_m(t) \ \forall t \in K, \ \text{carr}\, \varphi \subset K\}.$$

This base of zero vicinities contains all the vicinities of zero (5.27) in Φ_K. To make sure in this fact it is sufficient to take the functions $\gamma_0(t)$,

$\gamma_1(t), \ldots, \gamma_m(t)$ satisfying the condition $\gamma_0(t) = \gamma_1(t) = \ldots = \gamma_m(t) = \varepsilon$ at $t \in K$. On the other hand, at any m, $\gamma_0(t), \ldots, \gamma_m(t)$

$$\{\varphi : |\varphi(t)| < \gamma_0(t), \ldots, |\varphi^{(m)}(t)| < \gamma_m(t) \; \forall t \in K, \; \text{carr}\, \varphi \subset K \}$$

$$\supset \Big\{ \varphi : \max_{\substack{t \in K \\ 0 \leq k \leq m}} |\varphi^{(k)}(t)| < \varepsilon, \; \text{carr}\, \varphi \subset K \Big\},$$

where

$$\varepsilon = \min_{\substack{t \in K \\ 0 \leq k \leq m}} \gamma_k(t).$$

Consequently, the vicinities of zero (5.27) form the base of zero vicinities of the topology τ_K. ◁

Theorem 5.10. *The space of test functions* Φ *with the topology* τ *is complete.*

▷ Let $\{\varphi_n\}$ be an arbitrary fundamental sequence in Φ. In exactly the same way as at proving the necessity of the first condition in Theorem 5.8 it is proved that all the functions φ_n belong to one and the same subspace Φ_K correspondent to some compact K and that the sequence $\{\varphi_n\}$ is fundamental in the topology τ_K of the subspace Φ_K. But the subspace Φ_K is a complete space (it is proved analogously as Theorem 4.44 about the completeness of the space of continuous functions on the compact). Consequently, the sequence $\{\varphi_n\}$ has the limit in $\Phi_K \subset \Phi$. ◁

R e m a r k. In some cases we may reduce the requirement of infinite differentiability of the test functions and restrict ourselves to the requirement of their continuity or the continuity and the existence of their continuous derivatives till some order N. In such cases the topology in the space of the test functions is determined by zero vicinities (5.26) correspondent to $m = 0, 1, \ldots, N$.

E x a m p l e 5.12. As an example of the test function belonging to the space Φ, i.e. a finite function which is continuous together with its derivatives of all orders may serve the function

$$\varphi(t) = \alpha \exp\{[(t-a)^T C(t-a) - k^2]^{-1}\} 1_A(t),$$

where α is some number, a is vector, C is a positive determined matrix, k is a real number, $1_A(t)$ is an indicator of the set

$$A = \{ t : (t^T - a^T)C(t-a) < k^2 \},$$

i.e. the set of interior points of the ellipsoid,

$$(t^T - a^T)C(t - a) = k^2.$$

It is easy to verify that on the boundary of this ellipsoid all the derivatives of the function $\varphi(t)$ vanish. Really, all the derivatives of this functoin represent the linear combinations of the degrees of the variable $[(t^T - a^T)C(t - a) - k^2]^{-1}$ multiplied by $\exp\{[(t^T - a^T)C(t - a) - k^2]^{-1}\}$. At $(t^T - a^T)C(t - a) \to k^2$ all these products tend to zero as the exponential function e^{-u} decreases at $u \to \infty$ more fast than any degree of the variable u increases.

5.3.4. Definition of a Generalized Function

A *generalized function* or *a distribution* is called a continuous linear functional on the space of the test functions. In accordance with this definition the set Φ^* of all continuous linear functionals on the space of the test functions Φ, i.e. the space dual with Φ is called *a space of generalized functions*.

Theorem 5.11. *The functional f on the space of the test functions Φ with the domain $D_f = \Phi$ is continuous if and only if for any sequence $\{\varphi_n\} \subset \Phi$ convergent to some function $\varphi \in \Phi$ the sequence $\{f\varphi_n\}$ converges to $f\varphi$.*

▷ The necessity of the condition follows from the general Theorem 4.20. For proving the sufficiency we notice that by virtue of the fact that any convergent sequence $\{\varphi_n\}$ belongs to some subspace Φ_K and that Φ_K is the space with the first axiom of the countability (as the base of zero vicinities of the topology τ_K in Φ_K may serve the set of zero vicinities (5.27) with rational $\varepsilon > 0$) the contraction of the functional f on any subspace Φ_K by Theorem 4.21 represents a continuous functional. Therefore an inverse image of any open set G of the scalars field in Φ_K equal to $(f^{-1}G) \bigcap \Phi_K$ is an open set. But in the topology τ_K induced in Φ_K by the topology τ of the space Φ the set $U \bigcap \Phi_K$ is open if and only if the set U is open. Consequently, the set $f^{-1}G$ is open. According to Theorem 4.19 follows the continuity of the functional f. ◁

E x a m p l e 5.13. Let $f(t)$ be an integrable function over the Lebesgue measure on any bounded set. Such functions are called *locally integrable*. This function determines a continuous linear functional on Φ by formula

$$f\varphi = \int f(t)\varphi(t)\,dt, \quad \varphi \in \Phi, \tag{5.28}$$

where as usual an integral without specifying the integration region is appreciated as an integral over the whole space R^n. To make certain in the continuity of the functional f it is sufficient to show that it is continuous in zero (Theorem 4.55). Let $\{\varphi_n\}$ be the sequence of the functions from Φ convergent to zero. Due to the boundedness of the convergent sequence there exists such number $M > 0$ that $|\varphi_n(t)| < M$ at all n at all the points of the union of the carriers of the function $\varphi_n(t)$. Therefore all the functions $f(t)\varphi_n(t)$ are modulus–bounded on the union of their carriers of the integrable function $M|f(t)|$. By Theorem 3.30 the limit passage under the integral sign is possible in this case what gives $\lim\limits_{n\to\infty} f\varphi_n = 0$. This proves our statement. Thus any locally integrable function $f(t)$, in particular, any test function $\varphi(t) \in \Phi$ generates a generalized function $f \in \Phi^*$.

E x a m p l e 5.14. The generalized δ–function concentrated at the point s is determined by the equality

$$\delta_s\varphi = \varphi(s), \quad \varphi \in \Phi. \tag{5.29}$$

It is clear that it is a continuous linear functional which establishes the correspondence between each test function $\varphi \in \Phi$ and its value at $t = s$.

The p–th order derivative of the generalized δ–function concentrated at the point s is determined by the equality

$$\delta_s^{(p)}\varphi = (-1)^p\varphi^{(p)}(s), \quad \varphi \in \Phi. \tag{5.30}$$

The partial derivatives of the δ–function of a vector argument are determined by analogous formula.

It is clear that there is no locally integrable function $f(t)$ for which integral (5.28) is equal to $\varphi(s)$ or $(-1)^p\varphi^{(p)}(s)$ for all the functions $\varphi \in \Phi$. Nevertheless by the applications convention equalities (5.29) and (5.30) are written in the form

$$\delta_s\varphi = \int \delta(t-s)\varphi(t)\,dt = \varphi(s),$$

$$\delta_s^{(p)}\varphi = \int \delta^{(p)}(t-s)\varphi(t)\,dt = (-1)^p\varphi^{(p)}(s), \tag{5.31}$$

where $\delta(t-s)$ and $\delta^{(p)}(t-s)$ are recognized as the functions generating the functional δ_s and its derivatives $\delta_s^{(p)}$. These functions $\delta(t-s)$ and $\delta^{(p)}(t-s)$ are also called the generalized functions, i.e. are identified with the functionals δ_s and $\delta_s^{(p)}$. Notice that the second formula of (5.31) may be received by formal differentiation with respect to s of the first formula (5.31). This is the cause of the definition of the δ–function derivatives by formula (5.30).

For avoiding the multiplier $(-1)^p$ in the second formula of (5.31) the δ–function is usually assumed as even one $\delta(t-s) = \delta(s-t)$ and formulae (5.31) are written in the form

$$\int \delta(s-t)\varphi(t)\,dt = \varphi(s), \quad \int \delta^{(p)}(s-t)\varphi(t)\,dt = \varphi^{(p)}(s). \tag{5.32}$$

The second formula is also here received from the first one by formal differentiation with respect to s. In this case similarly as for an ordinary function the δ–function derivatives are odd, and the derivatives of even orders are even functions.

E x a m p l e 5.15. As an example of a generalized function may serve the functional $f\varphi = \int \varphi(t)\mu(dt)$, where μ is any σ–finite measure. It is apparent that the δ–function δ_s serves as a special case of this generalized function when μ is equal to 1 and is completely concentrated at the point s.

5.3.5. Space of Generalized Functions

Let us study the space of the generalized functions Φ^*. For introducing the topology in Φ^* notice that each function $\varphi \in \Phi$ generates the linear functional $f\varphi$ on Φ^* (at fixed $f \in \Phi^*$ $f\varphi$ represents the space map of the test functions Φ into the field of the scalars K, and at fixed function $\varphi \in \Phi$ $f\varphi$ represents the space map of the generalized functions Φ^* into the field of the scalars K). In accordance with this we define the topology in Φ^* by zero vicinities

$$\{f : |f\varphi_1| < \varepsilon, \ldots, |f\varphi_N| < \varepsilon\}, \tag{5.33}$$

correspondent to all $\varepsilon > 0$ and all $\varphi_1, \ldots, \varphi_N \in \Phi$. In other words, as the topology in Φ^* is assumed the $*$–weak topology (Subsection 5.1.3). This topology is the weakest in which all the functionals on Φ^* determined by the functions $\varphi \in \Phi$ are continuous (Theorem 4.49). According to Theorem 4.54 the sequence convergence of the generalized functions $\{f_n\}$ in this topology represents the convergence of numeric sequences $\{f_n\varphi\}$ at all $\varphi \in \Phi$. The space Φ^* of the generalized functions with the topology determined by zero vicinities (5.33) represents a topological linear space.

In modern functional analysis the space of the test functions Φ is often denoted by $\mathcal{D}(R^n)$, and the space of the generalized functions by $\mathcal{D}'(R^n)$.

E x a m p l e 5.16. Let $\{f_n\}$ be any sequence of nonnegative integrable finite functions whose carriers form decreasing sequence of the sets which has as the limit one point 0, and whose integrals are equal to 1:

$$\int f_n(t)\,dt = 1 \quad \forall n.$$

By integral mean theorem for any test function $\varphi \in \Phi$

$$f_n\varphi = \int f_n(t)\varphi(t)\,dt = \varphi(\tau_n),$$

where τ_n is some point belonging to the carrier of the function $f_n(t)$, $\tau_n \in \operatorname{carr} f_n$. As $\bigcap\limits_{n} \operatorname{carr} f_n = \{0\}$ then $\tau_n \to 0$ at $n \to \infty$, and consequently, $\varphi(\tau_n) \to \varphi(0)$. Therefore

$$\lim_{n \to \infty} f_n \varphi = \varphi(0) = \delta_0 \varphi.$$

Thus the sequence of the generalized functions induced by the functions f_n converges to the δ-function concentrated in zero δ_0. It is evident that as the functions f_n we may take, in particular, the test functions with required properties. By the shift of the argument t to s (i.e. by the change of the functions $f_n(t)$ by the functions $f_n(t - s)$) we get the δ-function concentrated at the point s as the sequence limit of the generalized functions induced by the functions $f_n(t - s)$.

E x a m p l e   5.17. Let $\{f_n\}$ be the sequence of the test functions of the scalar variable t with the same properties as in Example 5.16. Then the functionals may be determined

$$f_n' \varphi = \int f_n'(t)\varphi(t)\, dt.$$

After integration by parts and accounting that $\varphi(-\infty) = \varphi(+\infty) = 0$ we obtain

$$f_n' \varphi = - \int f_n(t)\varphi'(t)\, dt = -\varphi'(\tau_n'),$$

where $\tau_n' \in \operatorname{carr} f_n$. In this case by means of the limit passage we get

$$\lim_{n \to \infty} f_n' \varphi = -\varphi'(0) = -\delta_0 \varphi'$$

and by the induction

$$\lim_{n \to \infty} f_n^{(p)} \varphi = (-1)^p \varphi^{(p)}(0) = (-1)^p \delta_0 \varphi^{(p)}.$$

Thus the δ-function derivatives may be also determined as the limits of the sequences of the generalized functions induced by the correspondent derivatives of the functions f_n.

Analogously we may detemine the δ-function partial derivatives of a vector argument as the limits of the sequences of the generalized functions induced by the correspondent test functions possessing the required properties.

It is easy to see that for determining the δ-function it is sufficient to take as the space of the test functions Φ the space of continuous finite functions. For determining the δ-function derivatives till the order N inclusively it is sufficient to take as Φ the space of the continuous finite functions which have the derivatives till the order N inclusively.

5.3.6. Regular and Singular Generalized Functions

It was shown in Example 5.13 that any locally integrable function induces the correspondent generalized function. But not any generalized function may be induced by some locally integrable function. As the examples of such generalized functions may serve the δ–functions and their derivatives. In this connection the generalized functions are divided into two classes – regular and singular.

A generalized function is called *regular* if it is induced by some locally integrable function. The set of all regular generalized functions represents a subspace of the generalized functions space. A generalized function is called *singular* if there is no locally integrable functions which induces it.

R e m a r k. Examples 5.16 and 5.17 showed that some singular generalized functions may be represented as the limits of the sequences of the regular generalized functions induced by the test functions. It can be shown that any singular generalized function represents the sequence limit of the regular generalized functions induced by the test functions (Antosik et al. 1973).

5.3.7. Local Properties of Generalized Functions

It is apparent that for any function $\varphi \in \Phi$ whose carrier does not contain the point s,

$$\delta_s \varphi = \varphi(s) = 0, \quad \delta_s^{(p)} \varphi = (-1)^p \varphi^{(p)}(s) = 0.$$

This property of the δ–function and its derivatives suggests to consider the δ–function concentrated at the point s and all its derivatives equal to zero on any set which does not contain the point s. This corresponds to the definition of the δ–function given in Subsection 5.3.1 adopted in the applications. The singleton $\{s\}$ in this case represents a carrier of the δ–function and its derivatives. In accordance with this fact a carrier for any generalized function is determined.

A *carrier* of the generalized function f is called a minimal closed set F for which $f\varphi = 0$ for any function $\varphi \in \Phi$ whose carrier does not intersect with F. It is clear that such set F is not empty as for any generalized function besides $f = 0$ there exists such a function $\varphi \in \Phi$ that $f\varphi \neq 0$. The carrier of this function will obligatory intersect with F as in the opposite case it should be $f\varphi = 0$. On the basis of this definition the generalized function is accepted equal to zero on any set which does not intersect with its carrier. The generalized functions with bounded carrier represent an important class of the generalized functions.

E x a m p l e 5.18. From the definition of the δ–functions and their derivatives (Example 5.14) follows that the carrier of all the functions δ_s, $\delta_s^{(p)}$ represents the singleton $\{s\}$. A carrier of any regular generalized function coincides with the closure of the domain D_f correspondent to locally integrable function. If this function is determined on the whole space R^n then as a carrier of the correspondent regular generalized function serves the whole space R^n.

E x a m p l e 5.19. For giving an example of a singular generalized function whose carrier represents the whole space let us consider an integral

$$f\varphi = \int\limits_{-\infty}^{+\infty} \frac{\varphi(t)}{t}\, dt.$$

The function $1/t$ is not locally integrable and therefore the integral exists only as the main value in Cauchy sense. Really, if carr $\varphi(t) = [a, b]$, $a < 0, b > 0$ then

$$f\varphi = \int\limits_{a}^{b} \frac{\varphi(t)-\varphi(0)}{t}\, dt + \varphi(0) \int\limits_{a}^{b} \frac{dt}{t}.$$

The first integral represents here a regular generalized function, and the second integral exists only as the main value of the integral

$$\int\limits_{a}^{b} \frac{dt}{t} = \lim_{\varepsilon \to 0}\left(\int\limits_{a}^{-\varepsilon} \frac{dt}{t} + \int\limits_{\varepsilon}^{b} \frac{dt}{t}\right) = \ln\left|\frac{b}{a}\right|.$$

E x a m p l e 5.20. The Dirichlet integral

$$\int\limits_{a}^{b} \frac{\sin m(t-s)}{\pi(t-s)}\varphi(t)\, dt$$

for any function $\varphi \in \Phi$ converges at $m \to \infty$ to $\varphi(s)$ if $s \in (a, b)$ and to zero if $s \notin (a, b)$:

$$\lim_{m \to \infty} \int\limits_{a}^{b} \frac{\sin m(t-s)}{\pi(t-s)}\varphi(t)\, dt = \begin{cases} \varphi(s) & \text{at } s \in (a, b), \\ 0 & \text{at } s \notin (a, b). \end{cases} \tag{5.34}$$

Consequently, the sequence limit of the regular generalized functions induced by the functions $\sin m(t-s)/\pi(t-s)$ $(m = 1, 2, \ldots)$ represents the δ–function δ_s what may be written in the form

$$\delta(t - s) = \lim_{m \to \infty} \frac{\sin m(t-s)}{\pi(t-s)}. \tag{5.35}$$

The limit in the right–hand side does not exist at any t. It should be considered as a weak limit, i.e. as the sequence limit of the correspondent generalized function in the topology of the space Φ^*. It illustrates the fact that the generalized function is not a point function, it may have no concrete value at any value of the argument. Its equality to zero outside of its carrier should be considered only in the sense of the definition of a generalized function given above.

E x a m p l e   5.21.   In the same way we may represent the δ–function for n–dimensional vector argument as the limit

$$\delta(t-s) = \lim_{m \to \infty} \prod_{k=1}^{n} \frac{\sin m(t_k - s_k)}{\pi(t_k - s_k)}, \ \ t = \{t_1, \ldots, t_n\}, \ \ s = \{s_1, \ldots, s_n\}.$$

E x a m p l e   5.22.   Let $f(t)$ be locally integrable function $t \in R^m$. Find the limit of the sequence of regular generalized functions $f_n = n^m f(n(t-s))$ $(n = 1, 2, \ldots)$. We have

$$f_n \varphi = n^m \int f(n(t-s))\varphi(t)\,dt = \int f(u)\varphi(s + u/n)\,du \to \varphi(s) \int f(u)\,du.$$

Consequently, $\lim f_n = a\delta_s$, $a = \int f(u)\,du$ or

$$\lim_{n \to \infty} n^m f(n(t-s)) = a\delta(t-s).$$

This formula may be written in the form

$$\lim_{\varepsilon \to 0} \varepsilon^{-m} f(\varepsilon^{-1}(t-s)) = a\delta(t-s).$$

Pay attention to the fact that the carriers of all the functions f_n in Examples 5.20–5.22 coincide with the whole space while the carrier of the limit function represents the singleton $\{s\}$.

5.3.8. Change of Variables in Generalized Functions

As for any locally integrable function $f(u)$ and any one–to–one continuous and differentiable mapping $u = k(t)$ of the space R^n into R^n

$$\int f(k(t))\varphi(t)\,dt = \int f(u)\varphi(k^{-1}(u)) \, |\, J(u) \,|\, du, \qquad (5.36)$$

where $J(u)$ is an Jacobian of the coordinates of the vector function $k^{-1}(u)$ of the coordinates of the vector u then for generating function $f(k(t))$ of the regular generalized function the formula is valid

$$f(k(t))\varphi = f(\varphi(k^{-1}(u)) \, |\, J(u) \,|). \qquad (5.37)$$

The generalized function $f(k(t))$ for any given generalized function f $= f(t)$ is usually determined by this formula. In particular, putting $f = \delta_s$ we obtain the formula

$$\delta_s(k(t))\varphi = \delta_s(\varphi(k^{-1}(u)))\,|\,J(u)\,| = \varphi(k^{-1}(s))\,|\,J(s)\,|. \qquad (5.38)$$

After rewriting this formula in an integral form

$$\int \delta(k(t) - s)\varphi(t)\,dt = \varphi(k^{-1}(s))\,|\,J(s)\,|,$$

we come to the conclusion that for calculating the integral containing the δ–functions whose arguments serve any functions the change of variables should be performed by ordinary rules of an argument of the δ–function which depends on the integration variable.

In the case when $k^{-1}(t)$ is multifunction the integration region in (5.36) should be divided into the parts correspondent to the branches of the inverse function $k^{-1}(t)$. In this case the expressions in the right–hand sides of formulae (5.37) and (5.38) are substituted by the sums of the same expressions correspondent to all branches of the inverse function $k^{-1}(t)$. The necessity of the change of variables in the argument of the δ–function arises in many problems of probability theory (Pugachev and Sinitsyn 1987).

5.3.9. Differentiation of Generalized Functions

Let $f(t)$ be a differentiable locally integrable with its first derivative function. Consider a regular generalized function induced by its derivative $f'(t)$ [a]:

$$f'\varphi = \int f'(t)\varphi(t)\,dt.$$

Integrating by parts and taking into account that $\varphi(t)$ is a finite function we get

$$f'\varphi = -\int f(t)\varphi'(t)\,dt = -f\varphi'.$$

This formula gives the grounds to determine a *derivative* of any generalized function f by the formula

$$f'\varphi = -f\varphi'. \qquad (5.39)$$

---

[a] In the case of the vector argument t the prime means the differentiation with respect to any coordinate of a vector.

By the induction we obtain the formula for a derivative of any order p of the generalized function f:

$$f^{(p)}\varphi = (-1)^p f\varphi^{(p)} \quad (p = 1, 2, \ldots). \tag{5.40}$$

Analogously from (5.39) we get the formula for the derivatives of the generalized function f of the vector argument $t = \{t_1, \ldots, t_n\}$

$$\left[\frac{\partial^{|p|} f}{\partial t_1^{p_1} \cdots \partial t_n^{p_n}}\right]\varphi = (-1)^{|p|} f \frac{\partial^{|p|}\varphi}{\partial t_1^{p_1} \cdots \partial t_n^{p_n}}, \tag{5.41}$$

where $|p| = p_1 + \cdots + p_n$.

Formulae (5.40) and (5.41) show that any generalized function, in particular, any regular generalized function has the derivatives of all the orders which represent the generalized functions. It also follows from (5.40) and (5.41) that in the context of the generalized functions theory any ordinary locally integrable function has the derivatives of all the orders which may be singular generalized functions.

E x a m p l e 5.23. Find a derivative of an *unit step function* $\mathbf{1}(t - s)$,

$$\mathbf{1}(t) = \begin{cases} 1 & \text{at } t > 0, \\ 0 & \text{at } t < 0. \end{cases} \tag{5.42}$$

After denoting the correspondent regular generalized function in terms of Δ_s we shall have

$$\Delta_s\varphi = \int \mathbf{1}(t - s)\varphi(t)\, dt = \int_s^\infty \varphi(t)\, dt.$$

By formula (5.39) we find

$$\Delta_s'\varphi = -\Delta_s\varphi' = -\int_s^\infty \varphi'(t)\, dt = \varphi(s) = \delta_s\varphi.$$

Thus the derivative of the function $\mathbf{1}(t - s)$ represents the δ-function $\delta(t - s)$.

E x a m p l e 5.24. Let $f(t)$ be a piecewise continuous and piecewise differentiable function of a scalar variable with the discontinuities of the first kind at the points $t_1, \ldots, t_N$, which has at the points $t_1, \ldots, t_N$ the left and the right derivatives. We denote by $a_1, \ldots, a_N$ the jumps of the function $f(t)$ at the points $t_1, \ldots, t_N$. It is obvious that the function

$$g(t) = f(t) - \sum_{k=1}^N a_k \mathbf{1}(t - t_k)$$

is continuous and has a piecewise continuous derivative $g'(t)$ with the discontinuities of the first order at the points $t_1, \ldots, t_N$. On the basis of the results of the previous Example the derivative of the function $f(t)$ is expressed by the formula

$$f'(t) = g'(t) + \sum_{k=1}^{N} a_k \delta(t - t_k).$$

E x a m p l e 5.25. Analogously, if the function $f(t)$ is piecewise continuous and has piecewise continuous derivatives till the p-th order inclusively with the discontinuities of the first order at the points $t_1, \ldots, t_N$ then its derivative of the p-th order is determined by formula

$$f^{(p)}(t) = g^{(p)}(t) + \sum_{l=0}^{p-1} \sum_{k=1}^{N} a_k^{(l)} \delta^{(p-l)}(t - t_k),$$

where $a_1^{(l)}, \ldots, a_N^{(l)}$ are the jumps of the derivative $f^{(l)}(t)$ at the points $t_1, \ldots, t_N$ $(l = 0, 1, \ldots, p-1)$ and $g(t)$ is a continuous function with the continuous derivatives $g'(t), \ldots, g^{(p-1)}(t)$ and the piecewise continuous derivative $g^{(p)}(t)$ determined by formula

$$g(t) = f(t) - \sum_{l=0}^{p-1} \sum_{k=1}^{N} a_k^{(l)} \psi_l(t - t_k),$$

where

$$\psi_0(t-t_k) = \mathbf{1}(t-t_k), \quad \psi_l(t-t_k) = \int_{-\infty}^{t} \psi_{l-1}(\tau-t_k)\, d\tau \quad (l = 1, \ldots, p-1).$$

5.3.10. Series of Generalized Functions

Series whose terms are the generalized functions

$$\sum_{k=1}^{\infty} \alpha_k f_k \tag{5.43}$$

converges (in the topology of the space of the generalized functions Φ^*) to the generalized function f if the numeric series

$$\sum_{k=1}^{\infty} \alpha_k f_k \varphi \tag{5.44}$$

converges to $f\varphi$ for any test function $\varphi \in \Phi$. Here even in the case when the terms of the series represent the regular generalized functions the sum of the series may be a singular generalized function.

On the basis of the definition of the generalized functions derivatives (5.39)–(5.41) *the convergent series of the generalized functions may be differentiated termwise arbitrarily many times.* And as a result the convergent series of the generalized functions are always obtained.

E x a m p l e 5.26. Let $\{\psi_k\}$ be a complete orthonormal system of the functions in the space $L_2(R^n)$ of the scalar functions with the integrable square of the modulus. In this case any function with the integrable square of the modulus, in particular, any test function $\varphi \in \Phi$ may be represented by generalized Fourier series

$$\varphi(t) = \sum_{k=1}^{\infty} c_k \psi_k(t),$$

whose coefficients are determined by the formula

$$c_k = \int \varphi(\tau)\overline{\psi_k(\tau)}\, d\tau = \overline{\psi_k}\varphi.$$

After substituting this expression into the previous formula we get

$$\varphi(t) = \sum_{k=1}^{\infty} \psi_k(t) \int \varphi(\tau)\overline{\psi_k(\tau)}\, d\tau$$

$$= \lim_{N\to\infty} \int \sum_{k=1}^{N} \psi_k(t)\overline{\psi_k(\tau)}\varphi(\tau)\, d\tau. \tag{5.45}$$

Hence it is clear that the series (5.45) composed of the regular generalized functions of the variable τ generated by the functions $\overline{\psi_k(\tau)}$ with the coefficients $\psi_k(t)$ converges to the singular generalized function δ_t:

$$\sum_{k=1}^{\infty} \psi_k(t)\overline{\psi_k(\tau)} = \delta(\tau - t). \tag{5.46}$$

E x a m p l e 5.27. Quite in the same way as in Example 5.26 representing the derivative $\varphi^{(p)}(t)$ by the generalized Fourier series

$$\varphi^{(p)}(t) = \sum_{k=1}^{\infty} \psi_k(t)\overline{\psi_k}\varphi^{(p)} = (-1)^p \sum_{k=1}^{\infty} \psi_k(t)\overline{\psi_k}^{(p)}\varphi$$

$$= (-1)^p \int \sum_{k=1}^{\infty} \psi_k(t)\overline{\psi_k^{(p)}(\tau)}\varphi(\tau)\, d\tau$$

$$= (-1)^p \lim_{N \to \infty} \int \sum_{k=1}^{N} \psi_k(t) \overline{\psi_k^{(p)}(\tau)} \varphi(\tau) \, d\tau, \qquad (5.47)$$

we see that series (5.47) composed of the generalized functions $\overline{\psi_k^{(p)}}$ of the variable τ (not obligatory regular) with the coefficients $\psi_k(t)$ converges to the p-th derivative of the δ-function $\delta_s^{(p)}$:

$$\sum_{k=1}^{\infty} \psi_k(t) \overline{\psi_k^{(p)}(\tau)} = \delta^{(p)}(\tau - t). \qquad (5.48)$$

E x a m p l e 5.28. The periodic function $f(t)$ determined on the interval $(-\pi, \pi)$ by the formula

$$f(t) = t/2 \quad \text{at} \quad t \in (-\pi, \pi),$$

as it is known may be represented by Fourier series

$$f(t) = \sum_{n=1}^{\infty} (-1)^{n+1} \frac{\sin nt}{n}. \qquad (5.49)$$

This series converges at all t to the function

$$f(t) = t/2 - \pi \sum_{k=0}^{\infty} [\mathbf{1}(t - (2k+1)\pi) - \mathbf{1}(-t - (2k+1)\pi)],$$

if we assume as it is usually done $\mathbf{1}(0) = 1/2$. Thus series (5.49) composed of the regular generalized functions induced by the functions $\sin nt$ converges to the regular generalized function $f(t)$.

Differentiating formula (5.49) and accounting that the evenness of the δ-function we receive

$$f'(t) = 1/2 - \pi \sum_{k=0}^{\infty} [\delta(t - (2k+1)\pi) + \delta(-t - (2k+1)n)]$$

$$= 1/2 - \pi \sum_{k=-\infty}^{\infty} \delta(t - (2k+1)\pi) = \sum_{n=1}^{\infty} (-1)^{n+1} \cos nt. \qquad (5.50)$$

The function $f'(t)$ is singular and the series in the right-hand side diverges at any t. But series (5.50) composed of the regular generalized functions induced by the functions $\cos nt$ converges to the singular generalized function

$$f'(t) = 1/2 - \pi \sum_{k=-\infty}^{\infty} \delta(t - (2k+1)\pi).$$

5.3.11. Product of a Generalized Function and a Test Function

Let $f(t)$ be a locally integrable function, $\psi(t) \in \Phi$ be a test function. Consider a regular generalized function induced by the product $f(t)\psi(t)$:

$$(f\psi)\varphi = \int f(t)\psi(t)\varphi(t)\, dt = f(\psi\varphi).$$

Generalizing this result we shall call *a product of the generalized function f by the test function ψ* the generalized function $f\psi$ determined by the formula

$$(f\psi)\varphi = f(\psi\varphi). \tag{5.51}$$

Find the derivatives of the product of the generalized function f by the test function ψ. On the basis of formulae (5.40), (5.41) and (5.51)

$$(f\psi)^{(p)}\varphi = (-1)^p(f\psi)\varphi^{(p)} = (-1)^p f(\psi\varphi^{(p)}) \tag{5.52}$$

in the case of the scalar argument t and

$$\left[\frac{\partial^{|p|} f\psi}{\partial t_1^{p_1}\cdots\partial t_n^{p_n}}\right]\varphi = (-1)^{|p|} f\left[\psi\frac{\partial^{|p|}\varphi}{\partial t_1^{p_1}\cdots\partial t_n^{p_n}}\right], \tag{5.53}$$

in the case of the vector argument t.

In particular, the derivatives of the product of the δ-function δ_s by the test function ψ are determined by the formulae

$$(\delta_s\psi)^{(p)}\varphi = (-1)^p\delta_s(\psi\varphi^{(p)}) = (-1)^p\psi(s)\varphi^{(p)}(s) = \psi(s)\delta_s^{(p)}\varphi, \tag{5.54}$$

$$\left[\frac{\partial^{|p|}\delta_s\psi}{\partial t_1^{p_1}\cdots\partial t_n^{p_n}}\right]\varphi = (-1)^{|p|}\psi(s)\frac{\partial^{|p|}\varphi(s)}{\partial s_1^{p_1}\cdots\partial s_n^{p_n}} = \psi(s)\frac{\partial^{|p|}\delta_s}{\partial t_1^{p_1}\cdots\partial t_n^{p_n}}\varphi,$$

$$|p| = p_1 + p_2 + \cdots + p_n. \tag{5.55}$$

These formulae show that at differentiating the product of the δ-function by the test function *the multiplier at the δ-function is not differentiated* but is replaced by its value at the point s in which the δ-function is concentrated. This result may be foreseen because as the carrier of the functions $\delta_s^{(p)}$ ($p = 0, 1, 2, \ldots$) serves the singleton $\{s\}$ as a result of which the values of the function $\varphi(t)$ at $t \neq s$ influence in no way on the result of the operations over the δ-functions. The differentiation rule of the product of the δ-function by the test function (5.54), (5.55) repeatedly simplifies the differentiation of the composite expressions which contain the δ-functions and their derivatives and therefore are always used in applications.

R e m a r k. Notice that from the derived differentiation rule of the product $\delta_s \psi$ there is an important exception. If the δ-function substitutes in any formula an ordinary differentiable function then the products $\delta_s^{(p)} \psi$ must be obligatory differentiated by an ordinary formula of the differentiation of the functions product.

5.3.12. Primitive of a Generalized Function

Let f be a generalized function of the scalar variable t. According to the general definiton a *primitive* of the generalized function f is called such generalized function F whose derivative is the generalized function f. In the correspondence with (5.39) F is determined by the formula

$$F\varphi' = -f\varphi. \tag{5.56}$$

But this formula determines F as a continuous linear functional not on the whole space of the test functions Φ but only in the subspace $\Phi' \subset \Phi$ of the derivatives of the test functions. To make sure that Φ' do not coincide with Φ it is sufficient to notice that by virtue of the finiteness of the functions $\varphi \in \Phi$

$$\int_{-\infty}^{\infty} \varphi'(t)\,dt = \varphi(\infty) - \varphi(-\infty) = 0,$$

i.e. the integral of the derivative of any test function is equal to zero.

For extending the functional F over the whole Φ let us take an arbitrary test function φ_0 whose integral is equal to 1:

$$\int_{-\infty}^{\infty} \varphi_0(t)\,dt = 1. \tag{5.57}$$

It is evident that φ_0 does not belong to the subspace Φ', and consequently, the functional F is not determined for the function φ_0. Let $\varphi \in \Phi$ be an arbitrary test function. We shall show that the function

$$\psi(t) = \varphi(t) - \varphi_0(t) \int_{-\infty}^{\infty} \varphi(\tau)\,d\tau \tag{5.58}$$

belongs to the subspace Φ'. Obviously, that by virtue of equality (5.57) we have

$$\int\limits_{-\infty}^{\infty} \psi(\tau)\,d\tau = \int\limits_{-\infty}^{\infty} \varphi(\tau)\,d\tau - \int\limits_{-\infty}^{\infty} \varphi_0(\tau)\,d\tau \int\limits_{-\infty}^{\infty} \varphi(\tau)\,d\tau = 0.$$

On the basis of this formula a function

$$\omega(t) = \int\limits_{-\infty}^{t} \psi(\tau)\,d\tau \qquad (5.59)$$

is a finite unboundedly differentiable function, i.e. the test function $\omega \in \Phi$ whose derivative is equal $\omega'(t) = \psi(t)$. Consequently, $\psi \in \Phi'$ and due to (5.56) we get

$$F\psi = -f\omega. \qquad (5.60)$$

Thus any test function $\varphi \in \Phi$ may be presented by the formula

$$\varphi(t) = \psi(t) + \varphi_0(t) \int\limits_{-\infty}^{\infty} \varphi(\tau)\,d\tau, \qquad (5.61)$$

where $\psi(t)$ is a derivative of some test function $\omega(t)$. In order to extend the functional F on the whole space of the test functions it is sufficient now to prescribe arbitrarily its value at the point $\varphi_0 \in \Phi$. Putting $F\varphi_0 = C$ we get for any test function $\varphi \in \Phi$

$$F\varphi = -f\omega + C \int\limits_{-\infty}^{\infty} \varphi(\tau)\,d\tau. \qquad (5.62)$$

E x a m p l e 5.29. Find the primitive of the δ-function δ_s. After denoting it by Δ_s according to formula (5.62) we find

$$\Delta_s\varphi = -\delta_s\omega + C \int\limits_{-\infty}^{\infty} \varphi(\tau)\,d\tau = -\omega(s) + C \int\limits_{-\infty}^{\infty} \varphi(\tau)\,d\tau.$$

Taking into account (5.58) and (5.59) we get

$$\Delta_s\varphi = - \int\limits_{-\infty}^{s} \psi(\tau)\,d\tau + C \int\limits_{-\infty}^{\infty} \varphi(\tau)\,d\tau =$$

$$= - \int\limits_{-\infty}^{s} \varphi(\tau) \, d\tau + \left(\int\limits_{-\infty}^{s} \varphi_0(\tau) \, d\tau + C \right) \int\limits_{-\infty}^{\infty} \varphi(\tau) \, d\tau$$

$$= \int\limits_{s}^{\infty} \varphi(\tau) \, d\tau + \left(\int\limits_{-\infty}^{s} \varphi_0(\tau) \, d\tau + C - 1 \right) \int\limits_{-\infty}^{\infty} \varphi(\tau) \, d\tau.$$

But due to the arbitrariness of C the variable

$$\int\limits_{-\infty}^{s} \varphi_0(\tau) \, d\tau + C - 1$$

at fixed s is an arbitrary constant. Denoting it by C_1 we may rewrite the previous formula in the form

$$\Delta_s \varphi = \int\limits_{-\infty}^{\infty} [1(\tau - s) + C_1] \varphi(\tau) \, d\tau.$$

Hence it is obvious that the primitive of δ–function δ_s represents a regular generalized function equal to the sum of the unit step function and an arbitrary constant.

Formula (5.62) and the cited example show that similarly as an ordinary integrable function the generalized function has an infinite set of the primitives which differ one from another by an arbitrary constant.

5.3.13. Representation of the δ–Function by Fourier Integral

On the basis of the results of Examples 5.19 and 5.20 and accounting that

$$\frac{\sin m(t - s)}{\pi(t - s)} = \frac{1}{2\pi} \int\limits_{-m}^{m} e^{i\lambda(t-s)} \, d\lambda = \frac{1}{2\pi} \int\limits_{-m}^{m} e^{i\lambda(s-t)} \, d\lambda,$$

we come to the conclusion that the δ–function δ_s may be represented by the Fourier integral

$$\delta(t - s) = \frac{1}{(2\pi)^n} \lim_{m \to \infty} \int\limits_{-m}^{m} e^{i\lambda^T(t-s)} \, d\lambda = \frac{1}{(2\pi)^n} \lim_{m \to \infty} \int\limits_{-m}^{m} e^{i\lambda^T(s-t)} \, d\lambda,$$

$$(5.63)$$

where the integral is assumed as an integral in the limits from $-m$ till m taken in all the coordinates of the vector λ and all the vectors are represented in the form of the matrices–columns.

Formula (5.63) is usually written shortly in the form

$$\delta(t - s) = \frac{1}{(2\pi)^n} \int_{-\infty}^{\infty} e^{i\lambda^T(t-s)} \, d\lambda = \frac{1}{(2\pi)^n} \int_{-\infty}^{\infty} e^{i\lambda^T(s-t)} \, d\lambda. \qquad (5.64)$$

The integral in this formula naturally diverges. The limit in formula (5.63) does not exist in the ordinary sense as any convergent integral of the function dependent on the parameter represents an ordinary function of this parameter, and consequently, cannot be the δ–function. But this integral converges and the limit in (5.63) exists in the sense of the convergence of the generalized functions, i.e. in the topology of the space of the generalized functions Φ^*. Really, by Dirichlet theorem about an integral for any test function $\varphi \in \Phi$ we have

$$\lim_{m \to \infty} \int_{-\infty}^{\infty} \frac{\sin m(t - s)}{\pi(t - s)} \varphi(t) \, dt$$

$$= \lim_{m \to \infty} \int_{-\infty}^{\infty} \left[\frac{1}{2\pi} \int_{-m}^{m} e^{i\lambda(t-s)} \, d\lambda \right] \varphi(t) \, dt = \varphi(s). \qquad (5.65)$$

Analogous formula is also valid in the case of the vector argument s.

By virtue of the fact that the limit of an inner integral in the middle part of formula (5.65) in the context of standart mathematical analysis does not exist before the passage to the limit at $m \to \infty$ we change the integration order using Fubini theorem 3.44. As a result we get the representation of the function $\varphi(t)$ by Fourier integral

$$\varphi(s) = \frac{1}{(2\pi)^n} \lim_{m \to \infty} \int_{-m}^{m} \int_{-\infty}^{\infty} e^{i\lambda^T(t-s)} \varphi(t) \, dt \, d\lambda.$$

Later on to avoid confusion it is convenient to change the denotion of the integration variable t by τ, and the variable s replace by the variable t. Then we reduce the previous formula to the form

$$\varphi(t) = \frac{1}{(2\pi)^n} \lim_{m \to \infty} \int_{-m}^{m} \int_{-\infty}^{\infty} e^{i\lambda^T(\tau-t)} \varphi(\tau) \, d\tau \, d\lambda. \qquad (5.66)$$

The function

$$\hat{\varphi}(\lambda) = \mathcal{F}[\varphi(t)] = \int\limits_{-\infty}^{\infty} e^{i\lambda^T \tau} \varphi(\tau)\, d\tau \qquad (5.67)$$

is called the *Fourier transform* of function $\varphi(t)$ (Appendix 3). Here formula (5.66) determines the *inverse Fourier transform* of the function $\hat{\varphi}(\lambda)$:

$$\varphi(t) = \frac{1}{(2\pi)^n} \lim_{m\to\infty} \int\limits_{-m}^{m} e^{-i\lambda^T t}\hat{\varphi}(\lambda)\, d\lambda. \qquad (5.68)$$

This formula is usually written shortly in the form

$$\varphi(t) = \mathcal{F}^{-1}[\hat{\varphi}(\lambda)] = \frac{1}{(2\pi)^n} \int\limits_{-\infty}^{\infty} e^{-i\lambda^T t}\hat{\varphi}(\lambda)\, d\lambda, \qquad (5.69)$$

bearing that in the case when the function $\hat{\varphi}(\lambda)$ is not absolutely integrable the integral should be considered as the main value of the integral in Cauchy sense.

After comparing the last part of formula (5.64) with (5.69) we conclude that Fourier transform of the δ–function δ_s may be determined by the following formula:

$$\hat{\delta}_s(\lambda) = e^{i\lambda^T s}. \qquad (5.70)$$

5.3.14. Representation of Derivatives of the δ–Functions by the Fourier Integral

As any generalized function has the derivatives of all orders then formulae (5.63) and (5.64) may be differentiated as many times as desired. So, we receive the representation of the derivatives of the δ–functions by the Fourier integral

$$\delta^{(p)}(t-s) = \frac{1}{2\pi} \lim_{m\to\infty} \int\limits_{-m}^{m} (i\lambda)^p e^{i\lambda(t-s)}\, d\lambda$$

$$= \frac{(-1)^p}{2\pi} \lim_{m\to\infty} \int\limits_{-m}^{m} (i\lambda)^p e^{i\lambda(s-t)}\, d\lambda \qquad (5.71)$$

in the case of the scalar argument t and

$$\frac{\partial^{|p|}\delta(t-s)}{\partial t_1^{p_1}\cdots\partial t_n^{p_n}} = \frac{1}{(2\pi)^n}\lim_{m\to\infty}\int_{-m}^{m}(i\lambda_1)^{p_1}\cdots(i\lambda_n)^{p_n}e^{i\lambda^T(t-s)}\,d\lambda$$

$$= \frac{(-1)^{|p|}}{(2\pi)^n}\lim_{m\to\infty}\int_{-m}^{m}(i\lambda)^{p_1}\cdots(i\lambda_n)^{p_n}e^{i\lambda^T(s-t)}\,d\lambda \qquad (5.72)$$

in the case of the n–dimensional vector argument t, $|p|=p_1+\cdots+p_n$. These formulae may be written in more compact form

$$\delta^{(p)}(t-s) = \frac{1}{2\pi}\int_{-\infty}^{\infty}(i\lambda)^p e^{i\lambda(t-s)}\,d\lambda = \frac{(-1)^p}{2\pi}\int_{-\infty}^{\infty}(i\lambda)^p e^{i\lambda(s-t)}\,d\lambda \quad (5.73)$$

in the case of the scalar argument t and

$$\frac{\partial^{|p|}\delta(t-s)}{\partial t_1^{p_1}\cdots\partial t_n^{p_n}} = \frac{1}{(2\pi)^n}\int_{-\infty}^{\infty}(i\lambda_1)^{p_1}\cdots(i\lambda_n)^{p_n}e^{i\lambda^T(t-s)}\,d\lambda$$

$$= \frac{(-1)^{|p|}}{(2\pi)^n}\int_{-\infty}^{\infty}(i\lambda)^{p_1}\cdots(i\lambda_n)^{p_n}e^{i\lambda^T(s-t)}\,d\lambda \qquad (5.74)$$

in the case of the n–dimensional vector argument t, $|p|=p_1+\cdots+p_n$.

After comparing the last part of formula (5.73) and (5.74) with (5.69) we come to the conclusion that Fourier transforms of the derivatives of the δ–function should be determined by formula

$$\hat{\delta}_s^{(p)}(\lambda) = (-1)^p(i\lambda)^p e^{i\lambda s} \qquad (5.75)$$

in the case of the scalar argument t and

$$\left(\widehat{\frac{\partial^{|p|}\delta_s}{\partial t_1^{p_1}\cdots\partial t_n^{p_n}}}\right)(\lambda) = (-1)^{|p|}(i\lambda_1)^{p_1}\cdots(i\lambda_n)^{p_n}e^{i\lambda^T s} \qquad (5.76)$$

in the case of the vector argument t.

R e m a r k. Notice that formulae (5.70), (5.75) and (5.76) do not follow from (5.67) as they determine Fourier transforms of the generalized functions while formula (5.67) determines Fourier transform of the test function. Nevertheless formulae (5.70), (5.75) and (5.76) may be derived by formal use of formula (5.67).

5.3.15. Products of the δ–Functions and their Derivatives

If we consider the result of the functional $\delta_s^{(p)}$ action on the test function $\varphi \in \Phi$ as the function of the variable s then it will be also a test function. Therefore the result of the action of the δ–function $\delta_s^{(p)}$ derivative on any test function $\varphi \in \Phi$ may be again transformed by any functional f from the space of the generalized functions Φ^*. As a result we obtain the definition of the *product of the δ–function $\delta_s^{(p)}$ derivative by any generalized function f*:

$$(f\delta_s^{(p)})\varphi = f(\delta_s^{(p)}\varphi) = (-1)^p f\varphi^{(p)} = f^{(p)}\varphi. \qquad (5.77)$$

So, $f\delta_s^{(p)} = f^{(p)}$. In particular, at $f = \delta_u^{(q)}$ we get

$$(\delta_u^{(q)}\delta_s^{(p)})\varphi = \delta_u^{(q)}(\delta_s^{(p)}\varphi) = (-1)^p \delta_u^{(q)}\varphi^{(p)} = (-1)^{p+q}\varphi^{(p+q)}(u).$$

Thus we have

$$(\delta_u^{(q)}\delta_s^{(p)})\varphi = \delta_u^{(p+q)}\varphi \qquad (5.78)$$

and $\delta_u^{(q)}\delta_s^{(p)} = \delta_u^{(p+q)}$.

After writing formula (5.77) in the integral form we receive

$$(f\delta_s^{(p)})\varphi = \int f(s)\,ds \int \delta^{(p)}(t-s)\varphi(t)\,dt$$

$$= \int \left(\int f(s)\delta^{(p)}(t-s)\,ds \right)\varphi(t)\,dt = \int f^{(p)}(t)\varphi(t)\,dt.$$

Hence it follows

$$\int f(s)\delta^{(p)}(t-s)\,ds = (-1)^p \int f(s)\delta^{(p)}(s-t)\,ds = f^{(p)}(t). \qquad (5.79)$$

After comparing this formula with (5.31) we see that formula (5.31) for the integral of the product of the δ–function or its derivative by the test function is also true for the integral of the product of the δ–function or its derivative by any generalized function.

In particular, at $f(s) = \delta^{(q)}(s - u)$ formula (5.79) gives

$$\int \delta^{(p)}(t - s)\delta^{(q)}(s - u)\,ds = \delta^{(p+q)}(t - u). \qquad (5.80)$$

R e m a r k. Notice that formulae (5.71) and (5.72) may be derived without immediate differentiation of formulae (5.63) and (5.64). Namely, after writting formula (5.65) for the function $\varphi^{(p)}(t)$ and implementing to the integral under the limit sign the integrations by parts p times we get formula (5.71). Similarly formula (5.72) is derived.

For more detailed treatment of the generalized functions theory refer to (Antosik et al. 1973). Generalized random functions and fields are widely used in probability theory and its applications. See e.g. (Balakrishnan 1976, Cramer and Leadbetter 1967, Jazwinski 1990, McShane 1974, Pugachev and Sinitsyn 1987).

P r o b l e m s

5.3.1. At what $s > 0$ the functionals

$$f_1\varphi = \int\limits_{-\infty}^{\infty} \tfrac{f(t)}{t^s}\varphi(t)\,dt, \quad f_2\varphi = \int\limits_{-\infty}^{\infty} \tfrac{f(t)}{t^s}[\varphi(t) - \varphi(0)]\,dt,$$

$$f_3\varphi = \int\limits_{-\infty}^{\infty} \tfrac{f(t)}{t^s}\left[\varphi(t) - \sum_{k=1}^{N} \tfrac{\varphi^{(k)}(0)}{k!}t^k\right]dt,$$

where $f(t)$ is a locally integrable function, $f(0) \neq 0$ are regular generalized functions, singular generalized functions? Determine their carriers. In the case when they are singular represent them in the form of the sum of a regular generalized function and an item proportional to the δ–function or its derivative.

R e m a i n d e r. The spaces Φ and Φ^* are the complex linear spaces. Therefore both the test functions and the multipliers at them under the integral sign may be complex. In particular, $t^s = |t|^s e^{\pi i s}$ at $t < 0$.

5.3.2. Find the sequences limits of the generalized functions received from the generalized functions of Problem 5.3.1 by the change $f(t)$ by $f(nt)$.

5.3.3. Prove that the set of the regular generalized functions is dense in the space of all the generalized functions Φ^*. This statement establishes the equivalency of two approaches to the construction of the generalized functions theory which were mentioned in Subsection 5.3.2.

I n s t r u c t i o n. It is sufficient to prove that any base vicinity (5.33) of any generalized function f

$$\{g : |g\varphi_1 - f\varphi_1| < \varepsilon, \ldots, |g\varphi_N - f\varphi_N| < \varepsilon\} \qquad (*)$$

contains regular generalized functions. It is sufficient for this purpose to take any N of the locally integrable functions $g_1(t), \ldots, g_N(t)$, to put $g = c_1 g_1 + \cdots + c_N g_N$ and determine $c_1, \ldots, c_N$ from the equations $g\varphi_k = f\varphi_k + a_k$ $(k = 1, \ldots, N)$ where $|a_1|, \ldots, |a_N| < \varepsilon$. The received regular generalized function is determined in the form of an integral on the whole space Φ and is contained in the vicinity $(*)$ of the generalized function f.

5.3.4. Prove that the δ–function of the scalar argument may be presented as the limit at $\varepsilon \to 0$ of any cited regular generalized functions:

$$\delta(t - s) = \lim_{\varepsilon \to 0} \frac{1}{2\varepsilon} e^{-|s-t|/\varepsilon},$$

$$\delta(t - s) = \lim_{\varepsilon \to 0} \frac{1}{\pi} \frac{1}{\varepsilon^2 + (s-t)^2},$$

$$\delta(t - s) = \lim_{\varepsilon \to 0} \frac{1}{\sqrt{2\pi\varepsilon}} e^{-(t-s)^2/2\varepsilon}.$$

5.3.5. Prove that the δ–function of the m–dimensional vector argument may be presented by the formula

$$\delta(t - s) = (2\pi)^{-m/2} |K|^{-1/2} \lim_{\varepsilon \to 0} \varepsilon^{-m} \exp\{-(t^T - s^T)K^{-1}(t - s)/2\varepsilon^2\},$$

where K is any positively determined symmetric matrix, $|K|$ is its determinant.

I n s t r u c t i o n. For calculating the integral use the formula given in Appendix 2.

5.3.6. Find the derivatives of different orders of the generalized functions $|t|$, sgn$t$, $[t]$, $[t^r]$, $[t]^r$ (r is natural), $|\sin t|$, $|\cos t|$, sign$(\sin t)$, sgn$(\cos t)$ and all kinds of their products.

5.3.7. Examine that for any test function $\varphi(t)$ the following formula is valid:

$$\left(\sum_{m=0}^{p} C_p^m \psi^{(p-m)} \delta_s^{(m)} \right) \varphi = \psi(s) \delta_s^{(p)} \varphi,$$

where $\psi(t)$ is a given test function.

5.3.8. Calculate the integrals

$$f_1 \varphi = \int_{-\infty}^{\infty} \delta(t^3 - s)\varphi(t)\, dt,$$

$$f_2 \varphi = \int_{-\infty}^{\infty} \delta(t^2 - s)\varphi(t)\, dt,$$

$$f_3 \varphi = \int_{-\infty}^{\infty} \delta(\sin t - s)\varphi(t)\, dt,$$

$$f_4\varphi = \int\limits_{-\infty}^{\infty} \delta(\operatorname{tg} t - s)\varphi(t)\,dt.$$

I n s t r u c t i o n. In the case of multifunction $k^{-1}(t)$ (Subsection 5.3.8) partition the real axis into the intervals in which the function $k^{-1}(t)$ is single–valued.

5.3.9. Prove that in the general case when the function $k^{-1}(t)$ (Subsection 5.3.8) is a multifunction formulae (5.37) and (5.38) are replaced by the formulae

$$f(k(t))\varphi = \sum_\nu f(\varphi(k_\nu^{-1}(u)))\,|J_\nu(u)|,$$

$$\delta_s(k(t))\varphi = \sum_\nu \varphi(k_\nu^{-1}(s))\,|J_\nu(s)|,$$

where $J_\nu(u)$ is the Jacobian of the coordinates of the ν–th branch of the $k_\nu^{-1}(u)$ function along the coordinates of the vector u.

5.3.10. Derive formula (5.70) for Fourier transform of the δ–function δ_s by the limit passage at $\varepsilon \to 0$ in the formula received from the formula of Problem 5.3.5 by Fourier transform.

5.3.11. Let

$$L = \sum_{k=0}^{n} \frac{d^k}{dt^k}[a_k(t)\cdot]$$

be a linear differential operator, $v(t)$ be a function continuous at the point $t = \alpha$ together with its derivatives $v'(t), \ldots, v^{(m-1)}(t)$, the derivatives $v^{(m)}(t)$, $\ldots, v^{(n-1)}(t)$ have at the point α the discontinuities of the first kind with the jumps $\Delta_\alpha v^{(k)}$ $(k = m, \ldots, n-1)$. The variable Lv owing to the discontinuities of the derivatives $v^{(m)}, \ldots, v^{(n-1)}$ contains a linear combination of the generalized functions $\delta(t-\alpha), \delta'(t-\alpha), \ldots, \delta^{(n-m-1)}(t-\alpha)$. Derive the formula

$$Lv = Lu + \sum_{k=0}^{n-m-1} A_k \delta^{(k)}(t-\alpha),$$

where $u(t)$ is a function continuous together with its derivatives $u'(t)$, $\ldots, u^{(n-1)}(t)$ which is obtained from $v(t)$ by the subtraction of the function generated by the jumps of its derivatives $v^{(m)}(t), \ldots, v^{(n-1)}(t)$, and

$$A_k = \sum_{h=m}^{n-k-1} \sum_{l=m}^{h} (-1)^{k+h+1} C_h^l a_{k+h+1}^{(h-l)}(\alpha)\Delta_\alpha v^{(l)}.$$

I n s t r u c t i o n. At first deduce the formula for the case when only one derivative $v^{(l)}$ has the discontinuity at the point $t = \alpha$ introducing the derivatives of the δ–function of nonnegative orders

$$\delta^{(-1)}(t-\alpha) = \int\limits_{-\infty}^{t} \delta(s-\alpha)\,ds = 1(t-\tau),$$

$$\delta^{(-l)}(t - \alpha) = \int_{-\infty}^{t} \delta^{(-l+1)}(s - \alpha)\, ds \quad (l = 2, 3, \ldots)$$

and after calculating a part of the function $v(t)$ generated by the discontinuity of the derivative $v^{(l)}$:

$$v_l(t) = \delta^{(-l-1)}(t - \alpha)\Delta_\alpha v^{(l)},$$

we shall have

$$v(t) = u(t) + \delta^{(-l-1)}(t - \alpha)\Delta_\alpha v^{(l)},$$

Here $u(t)$ is the function continuous together with its derivatives $u'(t)$, $\ldots, u^{(n-1)}(t)$. At differentiating the products $a_k \delta^{(k-l-1)}(t - \alpha)$ remind that at $k - l - 1 \geq 0$ the multiplier at the δ–function or its derivative is not differentiated.

5.3.12. Prove that the linear differential equation with the δ–function in the right–hand side

$$\sum_{k=0}^{n} a_k(t)y^{(k)} = \delta(t - \tau)$$

is equivalent to the Cauchy problem

$$\sum_{k=0}^{n} a_k(t)y^{(k)} = 0,$$

$$y(\tau) = y'(\tau) = \cdots = y^{(n-2)}(\tau) = 0, \quad y^{(n-1)}(\tau) = 1/a_n(\tau).$$

The solution of this Cauchy problem is $y(t) = 0$ at all $t < \tau$.

5.3.13. Prove that the linear differential equation

$$\sum_{k=0}^{n} a_k(t)y^{(k)} = \sum_{k=0}^{m} b_k(t)\delta^{(k)}(t - \tau) \qquad (*)$$

is equivalent to the Cauchy problem

$$\sum_{k=0}^{n} a_k(t)y^{(k)} = 0, \quad m < n,$$

$$y(\tau) = y'(\tau) = \cdots = y^{(n-m-2)}(\tau) = 0,$$

$$y^{(n-m-1)}(\tau) = \alpha_1, \ldots, y^{(n-1)}(\tau) = \alpha_{m+1},$$

where $\alpha_1, \ldots, \alpha_{m+1}$ are expressed in terms of the values of the coefficients a_k, b_k at $t = \tau$. The solution of this Cauchy problem is $y(t) = 0$ at all $t < \tau$.

I n s t r u c t i o n. It is clear from $(*)$ that $y^{(n)}$ contains a linear combination of the generalized functions $\delta(t - \tau), \ldots, \delta^{(m)}(t - \tau)$ which corresponds to the jumpwise change $y^{(n-m-1)}, \ldots, y^{(n-1)}$ (or what is the same to the initial values of these derivatives at $t = \tau$). The same discontinuities cause the appearance of the linear combination of the functions $\delta(t - \tau), \ldots, \delta^{(k-n+m)}(t - \tau)$ in the expression of the derivative $y^{(k)}$ $(k = n - m, \ldots, n - 1)$. After adding to $y^{(n-m)}, \ldots, y^{(n)}$ in the equation $(*)$ these linear combinations of the δ–functions and its derivatives and comparing the coefficients at the identical derivatives of the δ–function in the left– and right–hand sides of the obtained equality we come to the system of linear algebraic equations with the triangular matrix for the initial values α_1, $\ldots, \alpha_{m+1}$ of the derivatives $y^{(n-m-1)}, \ldots, y^{(n-1)}$ at $t = \tau$.

5.3.14. A vector white noise $V(t)$ is determined as a generalized vector random function with a matrix of correlation and cross–correlation function, $K_V(t, t')$ $= \nu(t)\delta(t - t')$, $V(t)$ being an intensity matrix. Find $K_{V(p)}(t, t')$, $p > 1$.

5.4. Dual Spaces for Some Functions Spaces

5.4.1. Dual Space of the Space of Continuous Functions

Let us find a space dual with the space of the bounded continuous functions $C(T)$ (Subsection 1.3.11). For this purpose we establish a general form of a continuous linear functional on $C(T)$. But it is difficult to solve this problem for $C(T)$. Therefore we shall consider $C(T)$ as the subspace of the space $B(T)$. According to Hahn–Banach theorem 4.58 any continuous linear functional on $C(T)$ may be extended with retention of a norm over the whole space $B(T)$. After establishing a general form of continuous linear functional on $B(T)$ and narrowing it on $C(T)$ we may find a general form of a continuous linear functional on $C(T)$.

Theorem 5.12. *Any continuous linear functional on the space of the bounded functions $B(T)$ is determined by the formula*

$$fx = \int x(t)\varphi(dt) = \int x\, d\varphi, \qquad (5.81)$$

where $\varphi(A)$ is a finite additive measure.

▷ Let f be a continuous linear functional on $B(T)$, $f \in B^*(T)$. At first consider the values of the functional f on the indicators of the sets of the space T. It is clear that $f\mathbf{1}_A$ represents a finite additive function of the set

$$\varphi(A) = f\mathbf{1}_A, \qquad (5.82)$$

as for any pairwise nonintersecting sets $A_1, \ldots, A_n$, $A_k A_h = \emptyset$ at $k \neq h$,

$$\varphi\left(\bigcup_{k=1}^{n} A_k\right) = f \sum_{k=1}^{n} \mathbf{1}_{A_k} = \sum_{k=1}^{n} f \mathbf{1}_{A_k} = \sum_{k=1}^{n} \varphi(A_k)$$

and for any set A the variable $\varphi(A) = f \mathbf{1}_A$ is finite. Determine now the values of the functional f on the set of the simple functions

$$x(t) = \sum_{k=1}^{n} x_k \mathbf{1}_{E_k}(t), \quad E_k E_h = \emptyset \quad \text{at} \quad k \neq h \ ^a.$$

By virtue of the linearity of the functional f

$$fx = \sum_{k=1}^{n} x_k f \mathbf{1}_{E_k} = \sum_{k=1}^{n} x_k \varphi(E_k).$$

The latter part of this formula represents an integral of the simple function $x(t)$ with respect to the additive measure φ. Therefore

$$fx = \int x(t)\varphi(dt) = \int x \, d\varphi.$$

Thus any continuous linear functional on $B(T)$ is determined on the set of simple functions by formula (5.81). In order to extend this formula over all the functions from $B(T)$ it is sufficient to show that a set of simple functions is dense in the space $B(T)$. Let $x(t)$ be any function from $B(T)$. Its range R_x is completely placed inside the circle of the radius $\|x\|$ with the centre at the origin of the complex plane. Therefore at any $\varepsilon > 0$ the range R_x may be covered by a finite number of the circles $A_1, \ldots, A_N$ of the radius $\varepsilon/2$. Let us determine the sets

$$B_1 = A_1, \quad B_k = A_k \bigcap_{h=1}^{k-1} \bar{A}_h \quad (k = 2, \ldots, N)$$

and denote their inverse images in terms of $E_1, \ldots, E_N$, $E_k = x^{-1}(B_k)$. In each set B_k we shall take an arbitrary point x_k and determine the following simple function:

$$x'(t) = \sum_{k=1}^{N} x_k \mathbf{1}_{E_k}(t).$$

---

^{a} Any finite-valued function is measurable relative to the σ-algebra of all the sets of the space T, and consequently, may be considered as a simple function.

It is evident that $|x(t) - x'(t)| < \varepsilon$ at all t, and consequently, $\|x - x'\| < \varepsilon$. Thus any vicinity of any element from $B(T)$ contains simple functions. It means that a set of the simple functions is dense in $B(T)$ and any function from $B(T)$ may be presented as the limit of uniformly convergent sequence of the simple functions (i.e. convergent in the metric of the space $B(T)$). From this fact and the continuity of the functional f follows that formula (5.81) is valid for any function $x(t)$ from $B(T)$. ◁

Let us establish the relation between a norm of the functional f and complete variation of the additive measure φ.

Theorem 5.13. *The functional norm determined by formula (5.81) is equal to the value of complete variation of the additive measure φ on the whole space T.*

▷ Introduce a normed space of the finite additive measures φ a norm equal to the value of the complete variation $|\varphi|$ of the measure φ on the whole space T (Subsection 2.2.6):

$$\|\varphi\| = |\varphi|(T). \tag{5.83}$$

This space is often denoted by $ba(T)$. Then we shall have

$$|fx| \leq \int |x|\, d|\varphi| \leq \|x\|\, |\varphi|(T) = \|x\|\, \|\varphi\|.$$

Hence it follows that a norm of the functional f does not exceed the norm of the measure φ, $\|f\| \leq \|\varphi\|$. On the other hand, by the definition of a complete variation of a measure at any $\varepsilon > 0$ there exists such finite collection of the sets $A_1, \ldots, A_n$, $A_k A_h \neq \varnothing$ at $k \neq h$ that

$$\sum_{k=1}^{n} |\varphi(A_k)| > |\varphi|(T) - \varepsilon.$$

After determining a simple function

$$x(t) = \sum_{k=1}^{n} e^{-i\arg\varphi(A_k)} \mathbf{1}_{A_k}(t),$$

we get

$$fx = \sum_{k=1}^{n} e^{-i\arg\varphi(A_k)} \varphi(A_k) = \sum_{k=1}^{n} |\varphi(A_k)| > \|\varphi\| - \varepsilon.$$

Bearing in mind that $\|x\| = 1$ owing to the arbitrariness of $\varepsilon > 0$ we receive $\|f\| \geq \|\varphi\|$. Comparing this with the previous inequality we come to the conclusion that $\|f\| = \|\varphi\| = |\varphi|(T)$. ◁

Thus we proved that to each continuous linear functional f on $B(T)$ corresponds the unique finite additive measure $\varphi(A)$ and vice versa, to each such measure corresponds the unique continuous linear functional f on $B(T)$ determined by formula (5.81). The norm of the functional f is equal to complete variation of the correspondent additive measure φ (i.e. to its norm in the space $ba(T)$). It gives the opportunity to identify the dual space $B^*(T)$ of continuous linear functionals with the space $B(T)$ of finite additive measures determined on the σ–algebra of all the sets of the space T.

As $C(T)$ is a subspace of the space $B(T)$ and any continuous linear functional on $C(T)$ is extended on $B(T)$ with retention of a norm then from the obtained results follows that any continuous linear functional on $C(T)$ is determined by formula (5.81). But as a result of the ambiguity of the functional extension (Theorem 1.17) while narrowing the functionals from $B(T)$ to $C(T)$ the one–to–one correspondence between continuous linear functionals and the finite additive measures is violated. To each additive measure $\varphi \in ba(T)$ corresponds one continuous linear functional f on $C(T)$. But to each continuous linear functional f on $C(T)$ corresponds some set of different finite additive measures. It turns out that taking any additive measure φ of this set we may determine the unique such measure which possesses one important property. For formulating the correspondent theorem it is necessary to introduce one definition.

A measure or an additive measure μ determined on some class of the sets $\mathcal{A}$ of a topological space which contains the topology is called *regular* if for any $\varepsilon > 0$ and any set $A \in \mathcal{A}$ there exists such a set G and a closed set F that $F \subset A \subset G$ and

$$|\mu(G) - \mu(A)| < \varepsilon, \quad |\mu(A) - \mu(F)| < \varepsilon.$$

Theorem 5.14. *To each continuous linear functional f on the space of the bounded continuous functions $C(T)$ determined on the metric space T corresponds the unique regular additive measure μ_f on the algebra of the sets $\mathcal{B}$ induced by the topology of the space T in terms of which f is expressed by the integral*

$$fx = \int x(t)\mu_f(dt) = \int x\,d\mu_f \quad \forall x \in C(T), \tag{5.84}$$

and vice versa, to each regular finite additive measure μ_f on the algebra
B conrresponds the unique continuous linear functional f determined by
formula (5.84), and the norm of the functional f coincides with the value
of the complete variation of the additive measure μ_f on T:

$$\| f \| = \| \mu_f \| = | \mu_f | (T) \tag{5.85}$$

(Generalized Riesz theorem).

The proof of Theorem 5.14 is rather difficult. Therefore a great patience is required from the reader.

▷ As any numeric additive measure φ may be presented in the form $\varphi = \varphi_1 - \varphi_2 + i(\varphi_3 - \varphi_4)$ where φ_1, φ_2, φ_3, φ_4 are nonnegative additive measures (Corollary of Theorem 2.22) then it is sufficient to consider the case of the nonnegative φ.

Let us settle for a while to denote by the letter G with different indices the open sets of the space T and by the letter F with different indices the closed sets of the space T. Determine the function of the set $\psi(A)$ on all the sets of the space T by the formulae

$$\psi(F) = \inf_{G \supset F} \varphi(G), \quad \psi(A) = \sup_{F \subset A} \psi(F). \tag{5.86}$$

Let us study this function. First of all it is clear that $\psi(\varnothing) = 0$, $\psi(T) = \varphi(T)$ and $\psi(A) \le \psi(B)$ if $A \subset B$. By virtue of the nonnegativity of the additive measure φ for any two sets, in particular, for open sets G and G_1, we have $\varphi(G \bigcup G_1) \le \varphi(G) + \varphi(G_1)$. If F_1 is any closed set and $G \supset F_1 \bar{G}_1$, then $G \bigcup G_1 \supset F_1$ and from formulae (5.86) follows that $\psi(F_1) \le \varphi(G \bigcup G_1)$, and consequently,

$$\psi(F_1) \le \varphi(G) + \varphi(G_1).$$

As this inequality is valid for any $G \supset F_1 \bar{G}_1$ then by virtue of formulae (5.86) we have

$$\psi(F_1) \le \psi(F_1 \bar{G}_1) + \varphi(G_1)$$

for any closed set F_1 and open set G_1. Let us take an arbitrary closed set F. Then for any $G_1 \supset F_1 F$ we shall have $\bar{G}_1 \subset \bar{F}_1 \bigcup \bar{F}$, $F_1 \bar{G}_1 \subset F_1(\bar{F}_1 \bigcup \bar{F}) = F_1 \bar{F}$ and $\psi(F_1 \bar{G}_1) \le \psi(F_1 \bar{F})$. Therefore the previous inequality is amplified by the change $\psi(F_1 \bar{G}_1)$ by $\psi(F_1 \bar{F})$:

$$\psi(F_1) \le \psi(F_1 \bar{F}) + \varphi(G_1).$$

As this inequality is valid for any $G_1 \supset F_1 F$ then on the basis of definitions (5.86)

$$\psi(F_1) \le \psi(F_1 \bar{F}) + \psi(F_1 F)$$

for any closed sets F and F_1. Let now D be an arbitrary set. Then for any $F_1 \subset D$ $\psi(F_1 F) \le \psi(DF)$, $\psi(F_1 \bar{F}) \le \psi(D\bar{F})$ the previous inequality gives

$$\psi(F_1) \le \psi(DF) + \psi(D\bar{F}).$$

As it is valid for any $F_1 \subset D$ then formulae (5.86) give

$$\psi(D) \le \psi(DF) + \psi(D\bar{F}) \tag{5.87}$$

for any set D and any closed set F. On the other hand, taking arbitrary nonintersecting closed sets F_1, F_2 and nonintersecting open sets $G_1 \supset F_1$, $G_2 \supset F_2$, we get for any $G \supset F_1 \bigcup F_2$ $\varphi(G) \ge \varphi(GG_1) + \varphi(GG_2)$. As a result of the fact that $GG_1 \supset F_1$, $GG_2 \supset F_2$ on the basis of (5.86) follows $\varphi(G) \ge \psi(F_1) + \psi(F_2)$. As this is valid for any $G \supset F_1 \bigcup F_2$ then by virtue of formulae (5.86)

$$\psi(F_1 \bigcup F_2) \ge \psi(F_1) + \psi(F_2)$$

for any nonintersecting sets F_1 and F_2. After taking an arbitrary set D and an arbitrary closed set F we receive for any $F_1 \subset DF$, $F_2 \subset D\bar{F}$ the following inequalities:

$$\psi(D) \ge \psi(F_1 \bigcup F_2) \ge \psi(F_1) + \psi(F_2).$$

Hence by virtue of the arbitrariness of $F_1 \subset DF$, $F_2 \subset D\bar{F}$ by the definition (5.86) of the function ψ follows

$$\psi(D) \ge \psi(DF) + \psi(D\bar{F}).$$

Comparing this inequality with (5.87) we come to the conclusion that for any set D and any closed set F

$$\psi(D) = \psi(DF) + \psi(D\bar{F}).$$

It means that all the closed sets are Lebesgue measurable relative to the function ψ (Subsection 2.3.4) determined on all the sets of the space T, i.e. belong to the class of the sets

$$\mathcal{C} = \{A : \psi(D) = \psi(DA) + \psi(D\bar{A}), \ \forall D \subset T\}.$$

After repeating the proof of Theorem 2.31 we ensure that the class of the sets $\mathcal{C}$ represents an algebra which contains all the open and closed sets, and the function ψ is additive on $\mathcal{C}$.

Thus to each continuous linear functional f on $C(T)$ corresponds the finite additive measure μ_f on the algebra of the sets $\mathcal{B} \subset \mathcal{C}$ of the space T which represents the narrowing on $\mathcal{B}$ of the function ψ determined by formulae (5.86).

Let us prove that the additive measure μ_f is regular. From definition (5.86) of the functions ψ and μ_f follows that for any set $A \in \mathcal{B}$ and any $\varepsilon > 0$ there exist such closed sets $F \subset A$ and $F' \subset \bar{A}$ that

$$\mu_f(A) - \mu_f(F) < \varepsilon, \quad \mu_f(\bar{A}) - \mu_f(F') < \varepsilon.$$

But by virtue of the additivity of μ_f

$$\mu_f(\bar{A}) - \mu_f(F') = \mu_f(\bar{A}\bar{F}') = \mu_f(\bar{F}') - \mu_f(A).$$

Hence and from the previous inequalities follows that for any set $A \in \mathcal{B}$ and any $\varepsilon > 0$ there exist such closed set $F \subset A$ and open set $G = \bar{F}' \supset A$ that

$$\mu_f(A) - \mu_f(F) < \varepsilon, \quad \mu_f(G) - \mu_f(A) < \varepsilon. \tag{5.88}$$

These inequalities prove the regularity of μ_f.

So, by means of given continuous linear functional f on $C(T)$ we determined the correspondent regular finite additive measure $\mu_f(A)$. Prove that

$$\int x \, d\varphi = \int x \, d\mu_f \tag{5.89}$$

for any bounded continuous function $x(t)$.

As any numeric function $x(t)$ may be presented in the form $x = x_1 - x_2 + i(x_3 - x_4)$ where x_1, x_2, x_3, x_4 are nonnegative functions and $fx = \| x \| f(x/ \| x \|)$ then we may restrict ourselves by the proof of equality (5.89) for the case $0 \leq x(t) \leq 1$ and nonnegative additive measure μ_f. We take $\varepsilon > 0$ and approximate the function $x(t)$ from below by such a simple function

$$x'(t) = \sum_{k=1}^{n} x_k \mathbf{1}_{A_k}(t), \quad x_k = \inf_{t \in A_k} x(t),$$

that

$$x(t) - x'(t) < \varepsilon/3\mu_f(T).$$

Then shall have

$$\int x\, d\mu_f < \int x'\, d\mu_f + \varepsilon/3 = \sum_{k=1}^{n} x_k \mu_f(A_k) + \varepsilon/3.$$

By virtue of the regularity of μ_f there exist such closed sets $F_1, \ldots, F_n$, $F_k \subset A_k$ that $\mu_f(A_k) < \mu_f(F_k) + \varepsilon/3n$. On the basis of these inequalities and a result of the fact that $x_k \leq 1$,

$$\int x\, d\mu_f < \sum_{k=1}^{n} x_k \mu_f(F_k) + 2\varepsilon/3. \tag{5.90}$$

Due to the continuity of $x(t)$ there exist such nonintersecting open sets $G_1, \ldots, G_n$ that

$$F_k \subset G_k \text{ and } z_k = \inf_{t \in G_k} x(t) > x_k - \varepsilon/3\mu_f(T) \quad (k = 1, \ldots, n).$$

As a result of this and taking into account that $\mu_f(F_k) \leq \mu_f(G_k)$ inequality (5.94) may be replaced by the inequality

$$\int x\, d\mu_f < \sum_{k=1}^{n} z_k \mu_f(G_k) + \varepsilon.$$

Further from definition (5.86) of the function ψ follows that for any open set G

$$\mu_f(G) = \sup_{F \subset G} \mu_f(F) \leq \varphi(G).$$

Therefore we get

$$\int x\, d\mu_f < \sum_{k=1}^{n} z_k \varphi(G_k) + \varepsilon \leq \int x\, d\varphi + \varepsilon.$$

Hence due to the arbitrariness of $\varepsilon > 0$ follows

$$\int x\, d\mu_f \leq \int x\, d\varphi.$$

But this inequality is also valid for the function $1 - x(t)$:

$$\int (1 - x)\, d\mu_f \leq \int (1 - x)\, d\varphi,$$

whence

$$\int x \, d\mu_f \geq \int x \, d\varphi.$$

The obtained opposite inequalities prove the validity of formula (5.89) for any bounded continuous function $x(t)$.

It remained to prove the uniqueness of such additive measure μ_f and show that a norm of the functional f on $C(T)$ is equal to complete variation of this measure $|\mu_f|(T) = \|\mu_f\|$.

Suppose that to one continuous linear functional f on $C(T)$ correspond two regular finite additive measures μ_f and μ'_f and assume $\sigma = \mu_f - \mu_f$. Then we have

$$\int x \, d\sigma = 0$$

for any continuous function $x(t)$. Let A be an arbitrary set of the algebra $\mathcal{B}$. Because of the regularity of σ at any $\varepsilon > 0$ there are such open and closed sets G and F that $F \subset A \subset G$,

$$|\sigma(A) - \sigma(F)| < \varepsilon/3, \quad |\sigma(G) - \sigma(A)| < \varepsilon/3.$$

Let $x_1(t)$ be a continuous function $0 \leq x_1(t) \leq 1$, $x_1(t) = 0$ at $t \in \bar{G}$, $x_1(t) = 1$ at $t \in F^a$. Then taking

$$x(t) = e^{-i \arg \sigma(F)} x_1(t),$$

we shall have

$$\left| \int x \, d\sigma \right| = \left| \int_F x \, d\sigma + \int_{G \setminus F} x \, d\sigma \right| = \left| |\sigma(F)| + \int_{G \setminus F} x \, d\sigma \right|$$

$$\geq |\sigma(F)| - |\sigma(G \setminus F)| > |\sigma(A)| - \varepsilon$$

or by virtue of the arbitrariness of $\varepsilon > 0$,

$$0 = \left| \int x \, d\sigma \right| \geq |\sigma(A)|.$$

---

[a] In a metric space such function always exists. It may be determined, for instance, by formula

$$x_1(t) = \frac{d(t,\bar{G})}{d(t,\bar{G}) + d(t,F)},$$

where $d(t, A) = \inf_{s \in A} d(t, s)$ is the distance from the point t till the set A.

This is possible only at $\sigma(A) = 0$. Thus $\mu'_f(A) = \mu_f(A)$ at any $A \in \mathcal{B}$ what proves the uniqueness of the function μ_f.

Let now $A_1, \ldots, A_n$ be such pairwise nonintersecting sets $A_k \in \mathcal{B}$, $A_k A_h = \emptyset$ at $k \neq h$ that

$$\sum_{k=1}^{n} |\mu_f(A_k)| > |\mu_f|(T) - \varepsilon/4,$$

F_k and G_k $(k = 1, \ldots, n)$ are such closed and open sets that $F_k \subset A_k \subset G_k$, $|\mu_f(A_k) - \mu_f(F_k)| < \varepsilon/4n$, $|\mu_f(G_k) - \mu_f(A_k)| < \varepsilon/4n$, $x_1(t), \ldots, x_n(t)$ are some continuous functions, $0 \leq x_k(t) \leq 1$, $x_k(t) = 0$ at $t \in \bar{G}_k$, $x_k(t) = 1$ at $t \in F_k$. We shall take the function

$$x(t) = \sum_{k=1}^{n} e^{-i \arg \mu_f(F_k)} x_k(t).$$

This function is continuous and

$$|fx| = \left| \int x \, d\mu_f \right| = \left| \sum_{k=1}^{n} \left(|\mu_f(F_k)| + \int_{G_k \backslash F_k} x \, d\mu_f \right) \right|$$

$$\geq \sum_{k=1}^{n} |\mu_f(F_k)| - \sum_{k=1}^{n} |\mu_f(G_k \backslash F_k)| > \sum_{k=1}^{n} |\mu_f(A_k)| - 3\varepsilon/4$$

$$> |\mu_f|(T) - \varepsilon.$$

Hence by virtue of the arbitrariness $\varepsilon > 0$ and the fact that $\|x\| = 1$ we get $\|f\| \geq |\mu_f|(T)$. This in combination with apparent inequality $\|f\| \leq |\mu_f|(T)$ proves the equality $\|f\| = |\mu_f|(T)$. ◁

Thus the space $C^*(T)$ dual with the space of the bounded continuous functions $C(T)$ given on the metric space T may be identified with the space of the regular finite additive measures given on the algebra of the sets $\mathcal{B}$ of the space T induced by the topology. The space in which as a norm serves a complete variation of the correspondent additive measure is often denoted by $rba(T)$.

In the special case when T represents a compact of some metric space the measure μ_f is not only additive but is also σ–additive.

Theorem 5.15. *To each continuous linear functional f on the space of continuous functions $C(T)$ determined on the compact T corresponds the unique regular finite measure μ_f on the σ–algebra of the* *sets $\mathcal{A}$ induced by the topology of the compact T in terms of which f is expressed by formula (5.84), and vice versa, to each regular finite measure μ_f on the σ-algebra $\mathcal{A}$ corresponds the unique continuous linear functional f determined by formula (5.84), and the norm of the functional f is equal to the value of a complete variation of the measure μ_f on T (Riesz theorem).*

▷ On the basis of Theorem 2.18 for proving the σ-additivity of the measure μ_f it is sufficient to prove its semi–additivity. It is obvious that it is sufficient to prove it for the nonnegative μ_f.

Let $\{A_n\}$ be an arbitrary sequence of pairwise nonintersecting sets of the algebra $\mathcal{B}$ such that $A = \bigcup A_n \in \mathcal{B}$. Due to the regularity of μ_f on $\mathcal{B}$ at $\varepsilon > 0$ for each A_n exists such an open set $G_n \supset A_n$ that $\mu_f(G_n) - \mu_f(A) < \varepsilon/2^{n+1}$, and such closed set $F \subset A = \bigcup A_n$ that $\mu_f(F) > \mu_f(A) - \varepsilon/2$. On the other hand, owing to the compactness of T and the fact that

$$F \subset \bigcup_{n=1}^{\infty} G_n,$$

follows

$$F \subset \bigcup_{n=1}^{m} G_n$$

at some finite m. Consequently,

$$\sum_{n=1}^{\infty} \mu_f(A_n) > \sum_{n=1}^{\infty} \mu_f(G_n) - \varepsilon/2 \geq \sum_{n=1}^{m} \mu_f(G_n) - \varepsilon/2$$

$$\geq \mu_f\left(\bigcup_{n=1}^{m} G_n\right) - \varepsilon/2 \geq \mu_f(F) - \varepsilon/2 > \mu_f(A) - \varepsilon.$$

As it is valid at any $\varepsilon > 0$ then

$$\sum_{n=1}^{\infty} \mu_f(A_n) \geq \mu_f(A),$$

what proves the semi–additivity of μ_f on $\mathcal{B}$. By Theorem 2.18 from the additivity and semi–additivity of μ_f on $\mathcal{B}$ follows its σ-additivity on $\mathcal{B}$. According to Theorem 2.38 the measure μ_f may be uniquely extended over the σ-algebra of the sets of the compact T induced by the topology. Here at this extension of the measure μ_f the integral in formula (5.84) does not change (Property XVII of the integral, Subsection 3.3.5). ◁

So, the space $C^*(T)$ dual with the space of the continuous functions $C(T)$ determined on the compact T may be identified with the space of the regular measures which is often denoted by $rca(T)$.

On the basis of Riesz theorem 5.15 a finite measure is sometimes determined as a continuous linear functional on the space of continuous functions (the Radon measure).

5.4.2. The σ–Function as a Continuous Linear Functional on the Space of Continuous Functions

Consider a special case of the finite–dimensional space $T = R^n$ and the additive measure μ_f concentrated at one point s of the space R^n, $\mu_f(\{s\}) = 1$. In this case formula (5.84) will determine a continuous linear functional on $C(R^n)$ of the form

$$fx = \int x(t)\mu_f(dt) = x(s).$$

According to the definition of the δ–function (5.29) this functional represents the δ–function concentrated at the point s:

$$\delta_s x = \int\limits_{-\infty}^{\infty} x(t)\delta(t-s)\,dt = x(s).$$

Thus the δ–function represents a continuous linear functional on the space of the continuous functions $C(R^n)$ (the extension of δ_s from the space Φ on $C(R^n)$) as a result of which all the operations with the δ–functions may be extended over any continuous functions. This fact is widely used in applications.

5.4.3. Dual Spaces of the Spaces of Differentiable Functions

Let us find the space $C^{n*}(T)$ dual with the space of n times differentiable functions $C^n(T)$.

▷ For this purpose it is necessary to determine a general form of a continuous linear functional on $C^n(T)$. We shall restrict our consideration to the case when T is a closed interval of the real axis $T = [a, b]$. Notice that any function $x(t)$ from $C^n(T)$ represents a solution of the following differential equation:

$$x^{(n)} = z(t), \tag{5.91}$$

where $z(t)$ is a continuous function of T. For complete definiton of the function $x(t)$ for Eq. (5.91) it is also necessary to set the initial conditions: the values of $x(t)$ and its derivatives $x'(t), \ldots, x^{(n-1)}(t)$ at any point, for instance, at $t = a$, $x(a) = x_0$, $x'(a) = x_0', \ldots, x^{(n-1)}(a) = x_0^{(n-1)}$. Thus Eq. (5.91) establishes the one–to–one correspondence between the elements of the space $C^n(T)$ from one hand and the pairs "a continuous function and an n–dimensional vector" from another hand. In other words, Eq. (5.91) determines the one–to–one mapping of the space $C^n(T)$ onto the direct product $K^n \times C(T)$ of the n–dimensional space K^n and the space $C(T)$. This mapping is evidently linear.

Determine in the product of the spaces $K^n \times C(T)$ a norm of the element $u = \{x_0, x_0', \ldots, x_0^{(n-1)}, z(t)\}$ by formula

$$\|u\| = \sum_{p=0}^{n-1} |x_0^{(p)}| + \|z\| \, . \tag{5.92}$$

Prove that the operator A which performs the mapping $K^n \times C(T)$ on $C^n(T)$ and the inverse operator A^{-1} are bounded. For this purpose we notice that from Eq. (5.91) for various k follows the formulae

$$x^{(k)} = \sum_{p=k}^{n-1} x_0^{(p)} \frac{(t-a)^{p-k}}{(p-k)!}$$

$$+ \frac{1}{(n-k-1)!} \int_a^t (t-\tau)^{n-k-1} z(\tau) \, d\tau \quad (k = 0, 1, \ldots, n-1), \tag{5.93}$$

from which we get

$$\|x\| = \sum_{k=0}^{n} \sup_{t \in T} |x^{(k)}(t)| < c \left\{ \sum_{p=0}^{n-1} |x_0^{(p)}| + \sup_{t \in T} |z(t)| \right\} = c \, \|u\|,$$

where

$$c = \sum_{q=0}^{n} \frac{(b-a)^q}{q!} \, .$$

Consequently, the operator A is bounded and its norm is smaller than the number c. The boundedness of the inverse operator A^{-1} is evident as $\|u\| \le \|x\|$. ◁

Thus Eq. (6.91) *determines the one–to–one and mutually continuous linear mapping of the space* $C^n(T)$ *onto the product of the spaces*
$K^n \times C(T)$ *in which a norm is determined by formula* (5.92).

Let us pass to establishing a general form of a continuous linear
functional on $C^n(T)$.

▷ So, let f be a continuous linear functional on $C^n(T)$. Owing to the
continuity of the operator A the product fA represents the continuous
linear functional g on the product of the spaces $K^n \times C(T)$, $fA = g$.
Consequently,

$$fx = fAu = gu = g\{x_0, x_0', \ldots, x_0^{(n-1)}, 0\} + g\{0, 0, \ldots, 0, z(t)\}.$$

Thus any continuous linear functional on $C^n(T)$ represents the sum of
the continuous linear functional f_1 in the n–dimensional space K^n with
the elements $v = \{x_0, x_0', \ldots, x_0^{(n-1)}\}$ and the continuous linear functional f_2 on $C(T)$:

$$fx = f_1 v + f_2 z.$$

But as a result of the fact that $x_0^{(p)} = x^{(p)}(a)$ $(p = 0, 1, \ldots, n-1)$,

$$f_1 v = \sum_{p=0}^{n-1} \alpha_p x^{(p)}(a),$$

where $\alpha_0, \alpha_1, \ldots, \alpha_{n-1}$ are some complex numbers in a general case. By
Riesz theorem 5.15 we have

$$f_2 z = \int_a^b z(t)\mu_f(dt) = \int_a^b z \, d\mu_f,$$

here μ_f is some finite regular measure given on the σ–algebra $\mathcal{A}$ of Borel
sets of the interval $[a, b]$. Consequently,

$$fx = \sum_{p=0}^{n-1} \alpha_p x^{(p)}(a) + \int_a^b x^{(n)}(\tau)\mu_f(d\tau). \lhd \qquad (5.94)$$

Thus the following theorem is proved.

Theorem 5.16. *Any continuous linear functional on the space*
$C^n(T)$ *of the continuous functions together with their derivatives till
the n–th order inclusively is expressed by formula* (5.94), *therewith the*

correspondence between continuous linear functionals on $C^n(T)$ from one hand, and the sets of the numbers $\alpha_0, \alpha_1, \ldots, \alpha_{n-1}$ and regular finite measures μ_f from another hand, is reciprocal.

5.4.4. The δ-Function and its Derivatives as Continuous Linear Functionals on the Spaces of the Differentiable Functions

Consider a special case of the linear continuous functional f on the space of the differentialble functions $C^n(T)$ when $\alpha_0 = \alpha_1 = \ldots = \alpha_{k-1} = 0$, $\alpha_p = (s-a)^{p-k}/(p-k)!$, $(p = k, \ldots, n-1)$ and the measure μ_f is completely concentrated on the interval $[a, s]$ and is determined at $a \leq t_1 < t_2 \leq s \leq b$ by formula

$$\mu_f([t_1, t_2]) = \mu_f((t_1, t_2]) = \mu_f([t_1, t_2))$$

$$= \mu_f((a, b)) = \int_{t_1}^{t_2} \frac{(s-t)^{n-k-1}}{(n-k-1)!}\, dt.$$

In this case formula (5.94) takes the form

$$fx = \sum_{p=k}^{n-1} \frac{(s-a)^{p-k}}{(p-k)!} x^{(p)}(a) + \int_a^s \frac{(s-t)^{n-k-1}}{(n-k-1)!} x^{(n)}(t)\, dt.$$

After comparing this expression with (5.93) and taking into account Eq. (5.91) we get

$$fx = x^{(k)}(s) \quad (k = 0, 1, \ldots, n-1).$$

In the other special case when $\alpha_0 = \alpha_1 = \ldots = \alpha_{n-1} = 0$ and the measure μ_f is concentrated at the point s and $\mu(\{s\}) = 1$ formula (5.94) gives

$$fx = x^{(n)}(s).$$

After comparing the obtained formulae with definitions (5.29) and (5.30) of the δ-function and its derivatives we see that the δ-function and its derivatives till the n-th order inclusively represent the continuous linear functionals on the space of the differentiable functions $C^n(T)$ (the extensions of the functionals $\delta_s, \delta'_s, \ldots, \delta_s^{(n)}$ from Φ on $C^n(T)$):

$$\delta_s^{(k)} x = \int_a^b x(t)\delta^{(k)}(t-s)\, dt = (-1)^k x^{(k)}(s) \quad (k = 0, 1, \ldots, n).$$

This gives the opportunity to extend the operations with the δ–functions and their derivatives on the all functions which are continuous together with their derivatives till the correspondent order what is usually made in the applications.

E x a m p l e 5.30. Consider the linear control system (Example 1.8) whose input and output represent the continuous time functions (systems with lumped parameters). Let us introduce some definitions (Pugachev and Sinitsyn 1987, Zadeh and Desoer 1963).

A linear system is called *non–anticipative* if the value of its output $y(t)$ at every instant t does not depend on the values of its input $x(t)$ at $\tau > t$. Thus the value of the output of a non–anticipative system $y(t)$ at every instant t is a functional of the input $x(\tau)$ in the time interval $t_0 \leq \tau \leq t$. For the stability of the linear system in a given regime (i.e. deviation $\Delta y(t)$ of its output in this regime remains arbitrary small for any sufficiently small deviation $\Delta x(t)$ of its input) it is necessary that this functional should be continuous. By Riesz theorem 5.14 an output of the system at each time instant t is determined by formula

$$y(t) = \int x(\tau)\mu_t(d\tau) \tag{5.95}$$

where μ_t is some regular finite additive measure dependent on t (to different t correspond different continuous linear functionals). By Theorem 3.32 any finite or σ–finite measure in the finite–dimensional space may be expanded into absolutely continuous relative to the Lebesgue measure and singular components, therewith an absolutely continuous component represents an integral over the Lebesgue measure of some function. After applying Theorem 3.32 to μ_f we can write[a]

$$\mu_t(A) = \int\limits_{A} w(t, \tau)\, d\tau + \mu_t^s(A).$$

The singular part is always concentrated on some finite set of the points $\{t_1, \ldots, t_N\}$:

$$\mu_t^s(A) = \sum_{\{k: t_k \in A\}} \mu_t(\{t_k\}).$$

---

[a] Lebesgue theorem 3.32 was proved for the σ–additive measures. But accounting that in the applications we meet only the bounded time intervals and bounded space regions we may use Theorem 5.15 instead of Theorem 5.14 and consider μ_t as the σ–additive measure.

Therefore the measure μ_t in formula (5.95) may be expressed by formula

$$\mu_t(A) = \int\limits_A \left[w(t,\tau) + \sum_{k=1}^{N} \mu_t(\{t_k\})\delta(\tau - t_k) \right] d\tau.$$

As a result of this formula (5.95) takes the form

$$y(t) = \int g(t,\tau)x(\tau)\,d\tau, \qquad (5.96)$$

where

$$g(t,\tau) = w(t,\tau) + \sum_{k=1}^{N} \mu_t(t_k)\delta(\tau - t_k). \qquad (5.97)$$

Thus an output of the linear system $y(t)$ is expressed in terms of its input $x(t)$ by formula (5.96). The function $g(t,\tau)$ is assumed as one of the basic characteristics of a linear system and is called its *weighting function* or *impulse response function*. Riesz theorem gives a rigorous mathematical view of a weighting function.

A weighting function of a linear system has a simple physical sense. It represents a response of a system at the instant t on an instantaneous unit impulse which acts on the input of a system at the instant τ. To make sure it is sufficient to change in formula (5.96) the denotion of the variable integration from τ to σ and put $x(\sigma) = \delta(\sigma - \tau)$.

A weighting function of any non–anticipative linear system is equal to zero at the value of the second argument larger than the value of the first argument

$$g(t,\tau) = 0 \quad \text{at} \quad \tau > t. \qquad (5.98)$$

This property shows that none physical system may response at a given moment on the input which will act on it later. For such a system in formula (5.96) the integrand is equal to zero at $\tau < t_0$ and at $\tau > t$. Really, the input $x(t)$ does not act on a system till the instant t_0 what is equal to the supposition $x(\tau) = 0$ at $\tau < t_0$. At $\tau > t$ the weighting function $g(t,\tau)$ is equal to zero due to condition (5.98). Consequently, for a linear non–anticipative system relation (5.96) between the input and the output variables are rewritten in such a way:

$$y(t) = \int\limits_{t_0}^{t} g(t,\tau)x(\tau)d\tau. \qquad (5.99)$$

A weighting function of a stationary linear system (Example 5.3) depends only on the difference of the arguments $t - \tau$:

$$g(t,\tau) = w(t - \tau).$$

In this case formula (5.99) gives the following expression for the response $y(t)$ of a non–anticipative stationary linear system on any input $x(t)$:

$$y(t) = \int\limits_{-\infty}^{t} w(t - \tau)x(\tau)d\tau .$$

For the generality we consider the lower limit of the integration equal to $-\infty$ in order to cover the case of unboundedly long time acting disturbances. In the special case when the input begins to act on the system at the instant t_0 it should be taken in the latter formula $x(\tau) = 0$ at $\tau < t_0$.

After changing the variables in formula (5.101) $\xi = t - \tau$ we reduce it to the form of convolution

$$y(t) = \int\limits_{0}^{\infty} w(\xi)x(t - \xi)d\xi . \tag{5.100}$$

Applying to Eq. (5.100) the Laplace transform (Appendix 3) we get the following formulae for the transfer function $\Phi(s)$ in terms of the weighting function $w(t - \tau)$ and vice versa:

$$\Phi(s) = \int\limits_{0}^{\infty} w(\xi)e^{-s\xi} d\xi , \tag{5.101}$$

$$w(t - \tau) = \frac{1}{2\pi} \int\limits_{0}^{\infty} \Phi(i\omega)e^{i\omega(t-\tau)} d\omega . \tag{5.102}$$

In Table A.1.1 (Appendix 1) for some typical linear systems the correspondent weighting functions are given.

E x a m p l e 5.31. In the problems of control theory we have also consider the systems with lumped parameters possessing differentiating properties. The outputs of such systems may contain the derivatives of the input. It is clear that the relation between the input and the output for linear systems of such kind cannot be described by formula (5.96) even in that case when the weighting function $g(t, \tau)$ contains the δ–functions.

For extending formula (5.96) over such linear systems it is necessary to consider the input as an element of the correspondent space $C^n(T)$. Then the output of the system $y(t)$ will be expressed by formula

$$y(t) = \sum_{p=0}^{n-1} \alpha_p(t)x^{(p)}(a) + \int\limits_{a}^{b} x^{(n)}(\tau)\mu_t(d\tau) \tag{5.103}$$

On the same basis as in Example 5.30 the measure μ_t may be expanded into absolutely continuous and singular parts relative to the Lebesgue measure and suppose that the singular part is concentrated on the finite set of the points. Then formula (5.103) will be rewritten in the form

$$y(t) = \sum_{p=0}^{n-1} \alpha_p(t) x^{(p)}(a) + \int_a^b h(t, \tau) x^{(n)}(\tau) \, d\tau. \qquad (5.104)$$

Here we may always suppose that the function $h(t, \tau)$ is piecewise continuous and has piecewise continuous derivatives till n order inclusively. It always takes place if $h(t, \tau)$ is obtained theoretically on the basis of the adopted mathematical model of the studied system. If it is determined within the experimental error then it may be always approximated with sufficient accuracy by such function. Then we may perform n–multiple integration by parts after previous partitioning the integration region into the parts by the points of the discontinuity of the function $h(t, \tau)$ and its derivatives. Then we shall receive formula (5.96) where the weighting function $g(t, \tau)$ contains a linear combination of the δ–functions and their derivatives till n order inclusively. It goes without saying that in $g(t, \tau)$ the following item should be included:

$$\sum_{p=0}^{n-1} (-1)^p \alpha_p(t) \delta^{(p)}(\tau - a)).$$

The given arguments explain the use of formula (5.96) in control theory in the case of differentiable inputs.

In Table A.1.1 (Appendix 1) for some typical differential systems the correspondent weighting functions are given.

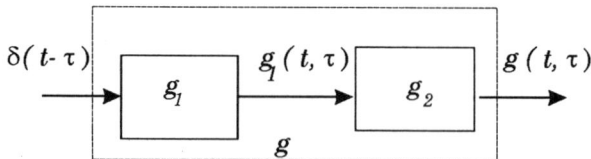

Fig. 5.6

E x a m p l e 5.32. Under the conditions of Example 5.4 let us find the weighting functions of the typical connections of linear systems. Consider at first a series connection of two linear systems whose weighting functions $g_1(t, \tau)$ and $g_2(t, \tau)$ are given (Fig. 5.6). Find the weighting functions $g(t, \tau)$ of a composite system. For this purpose we find its response on the unit impulse. As a result of the action of the unit inpulse on the input of the first system its output will represent its weighting function $g_1(t, \tau)$. Consequently, the input of the second system is expressed by formula $x_2(t) = g_1(t, \tau)$. The output of the second system accordingly to general formula (5.96) is expressed by the formula

$$y_2(t) = \int_{-\infty}^{\infty} g_2(t, \sigma) x_2(\sigma) d\sigma = \int_{-\infty}^{\infty} g_2(t, \sigma) g_1(\sigma, \tau) d\sigma. \qquad (5.105)$$

But the output of the second system in given case represents also the output of the whole series connection of considered systems, i.e. $y_2(t) = g(t, \tau)$. Comparing this formula with (5.105) we obtain the following formula for the weighting function of the series connection of two linear systems:

$$g(t, \tau) = \int_{-\infty}^{\infty} g_2(t, \sigma) g_1(\sigma, \tau) d\sigma. \qquad (5.106)$$

In the special case of the non-anticipative systems $g_1(\sigma, \tau) = 0$ at $\sigma < \tau$, $g_2(t, \sigma) = 0$ at $\sigma > t$ formula (5.106) takes the form

$$g(t, \tau) = \int_{\tau}^{t} g_2(t, \sigma) g_1(\sigma, \tau) d\sigma. \qquad (5.107)$$

This formula is valid at $t \geq \tau$. At $t < \tau$ the integrand in (5.106) is equal to zero at all σ. Thus we come to the conclusion that the series connection of non–anticipative systems always gives a non–anticipative system.

Using the obtained formulae (5.106) and (5.107) we can find the weighting function of a system received as a result of the series connection of any finite number of linear systems. Notice that the result of the series connection of the linear systems in the general case depends on the order of their connection. Really, changing the places of connecting systems in Fig 5.6 we should change the places of the weighting functions g_1 and g_2 in formula (5.106). As a result a weighting function of the sequential connection of two systems will be

$$g'(t, \tau) = \int_{-\infty}^{\infty} g_1(t, \sigma) g_2(\sigma, \tau) d\sigma. \qquad (5.108)$$

This expression in the general case does not coincide with formula (5.106).

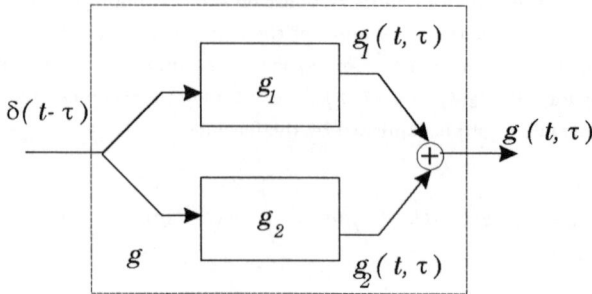

Fig. 5.7

Let us consider now the parallel connection of the linear systems which have the known weighting functions $g_1(t, \tau)$ and $g_2(t, \tau)$ (Fig. 5.7). For determining the weighting function $g(t, \tau)$ of a composite system we shall find its response on the unit impulse $\delta(t - \tau)$. Accounting that while delivering the unit impulse $\delta(t - \tau)$ on the inputs of the connected systems their outputs are equal to the correspondent weighting functions $g_1(t, \tau)$ and $g_2(t, \tau)$ we get

$$g(t, \tau) = g_1(t, \tau) + g_2(t, \tau). \tag{5.109}$$

Thus at the parallel connection of the linear systems their weighting functions are added (Example 5.3).

For studying the systems with the feedback we use the notion of reciprocal systems. Consider a series connection of two linear reciprocal systems. A weighting function of a linear system which is inverse relative to the system with the weghting function $g(t, \tau)$ we shall denote by the symbol $g^-(t, \tau)$. The weighting function of the series connection of the reciprocal systems is equal to $\delta(t - \tau)$. On the other hand, the weighting function of the series connection may be calculated using formula (5.106). As a result we obtain

$$\int_{-\infty}^{\infty} g^-(t, \sigma) g(\sigma, \tau) d\sigma = \delta(t - \tau). \tag{5.110}$$

This relation represents the main property of the weighting functions of the reciprocal linear systems. Changing the order of the connection of the systems we get another relation

$$\int\limits_{-\infty}^{\infty} g(t,\sigma)g^-(\sigma,\tau)d\sigma = \delta(t-\tau). \qquad (5.111)$$

Thus the weighting functions of two reciprocal linear systems satisfy two integral realtions (5.110) and (5.111) which in the general case do not coincide.

For non–anticipative reciprocal linear sytems formulae (5.110) and (5.111) may be rewritten in the form

$$\int\limits_{\tau}^{t} g^-(t,\sigma)g(\sigma,\tau)d\sigma = \delta(t-\tau), \qquad (5.112)$$

$$\int\limits_{\tau}^{t} g(t,\sigma)g^-(\sigma,\tau)d\sigma = \delta(t-\tau). \qquad (5.113)$$

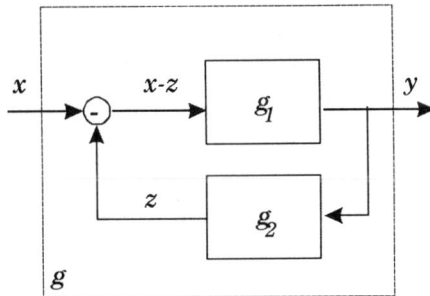

Fig. 5.8

A weighting function $g(t,\tau)$ of the system with a feedback loop illustrated in Fig. 5.8 may be determined by the same standard way that in the previous cases after observing the passage of the unit impulse through the system. But this way leads to the integral equation relative to the sought weighting function $g(t,\tau)$. It is rather easy to find a weighting function of an inverse system. It is evident that a system inverse relative to a system with a feedback is a parallel connection of a system inverse relative to a system which stands in the direct chain and a system which stands in the chain of a feedback (Fig. 5.9). Really, denoting the input of

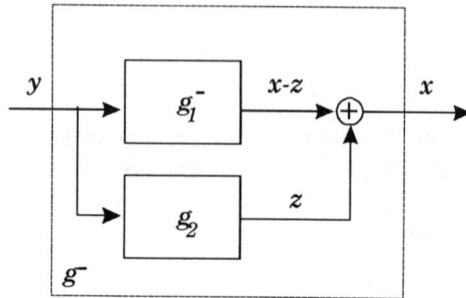

Fig. 5.9

the considered system by x, the output by y, and the output of a system in the chain of the feedback by z we see that the input in the direct chain will be $x - z$ and its output will be y. While giving on the input of the parallel connection of the systems with the weighting functions g_1 and g_2 the signal y we get on the output of the first system the signal $x - z$, and on the output of the second system in the main system the signal z. At adding these variables will give on the output of the parallel connection of the systems the signal x. Thus a series connection of the systems shown in Fig. 5.8 and Fig. 5.9 gives an ideal tracking system. Consequently, these systems are reciprocal. Applying formula (5.109) for a weighting function of the parallel connection of the linear systems we find

$$g^-(t, \tau) = g_1^-(t, \tau) + g_2(t, \tau). \qquad (5.114)$$

This formula determines a weighting function of a linear system inverse relative to a system with a feedback shown in Fig. 5.8.

E x a m p l e 5.33. Consider the linear control system with distrubuted parameters (Example 1.9) whose inputs and outputs represent the functions of the time t and the coordinates ξ, η, ζ of the point space. It is evident that formula (5.96) is also valid in the case when t is the 4–dimensional vector with components t, ξ, η, ζ. In applications the weighting function $g(t, \tau)$ in formula (5.96) is usually called the *Green function*. It depends on two point space vector and two time scalar arguments,

$$g = g(\xi, \eta, \zeta, \xi_1, \eta_1, \zeta_1, t, t_1).$$

For stationary system g depends on two point space vector arguments ξ, η, ζ and ξ_1, η_1, ζ_1 and one time scalar argument $t - t_1$,

$$g = g(\xi, \eta, \zeta, \xi_1, \eta_1, \zeta_1, t - t_1).$$

For stationary system the transfer function presents the Laplace transform of the Green function (Appendix 3):

$$\Phi(\xi, \eta, \zeta, \xi_1, \eta_1, \zeta_1, s) = \mathcal{L}[g(\xi, \eta, \zeta, \xi_1, \eta_1, \zeta_1, t - t_1)].$$

For stationary and space invariant system g depends on one point space vector argument $\{\xi - \xi_1, \eta - \eta_1, \zeta - \zeta_1\}$ and one time scalar argument $t - t_1$,

$$g = g(\xi - \xi_1, \eta - \eta_1, \zeta - \zeta_1, t - t_1).$$

Green functions and transfer functions for some typical linear systems with distributed parameters are presented in Table A.1.2 (Appendix 1).

E x a m p l e 5.34. In stationary optical processing system (Example 1.10) the operator A is often taken in the following integral form:

$$y(\xi, \eta) = Ax(\xi_1, \eta_1) = \int\limits_{-\infty}^{\infty} \int\limits_{-\infty}^{\infty} g(\xi, \eta, \xi_1, \eta_1)x(\xi_1, \eta_1)d\xi_1, d\eta_1.$$

For space invariant optical system (Pratt 1978)

$$g = g(\xi - \xi_1, \eta - \eta_1),$$

and the integral operator A presents space convolution

$$y(\xi, \eta) = \int\limits_{-\infty}^{\infty} \int\limits_{-\infty}^{\infty} g(\xi - \xi_1, \eta - \eta_1)x(\xi_1, \eta_1)d\xi_1, d\eta_1$$

or shortly

$$y = g * x.$$

Here asterisk means integral operation of space convolution. Convolution is always symmetric,

$$y(\xi, \eta) = \int\limits_{-\infty}^{\infty} \int\limits_{-\infty}^{\infty} g(\xi_1, \eta_1)x(\xi - \xi_1, \eta - \eta_1)d\xi_1, d\eta_1.$$

Notice that distinct from a stationary system (Example 5.30) a space invariant system may be an anticipative one, i.e. space coordinates may be negative relative to some origin (see also Table A.1.3 in Appendix 1).

5.4.5. Dual Spaces of Lebesgue Spaces

Determine the spaces dual with Lebesgue spaces L_p studied in Section 3.7. According to the proved in Subsection 3.7.4 all the spaces L_p,

$1 \leq p < \infty$ are the B–spaces therewith the norm of the element x of the space $L_p(T, \mathcal{B}, \mu)$ (i.e. the norm of the function $x(t)$ from $L_p(T, \mathcal{B}, \mu)$) is determined by formula

$$\| x \|_p = \left\{ \int \| x(t) \|^p \, \mu(dt) \right\}^{1/p} = \left\{ \int \| x \|^p \, d\mu \right\}^{1/p} . \qquad (5.115)$$

By Theorem 5.2 the space L_p^* dual with L_p also represents a B–space.

For finding a general form of a continuous linear functional on the space $L_p(T, \mathcal{B}, \mu)$ of the numeric functions at $1 < p < \infty$ notice that accordingly to Lemma (Subsection 3.7.2) for any numeric functions $x(t) \in L_p$ and $f(t) \in L_q$, $p^{-1} + q^{-1} = 1$ the composite function $f(t)x(t)$ is μ–integrable.

Theorem 5.17. *For any function $f(t) \in L_q$ formula*

$$fx = \int f(t)x(t)\mu(dt) = \int fx \, d\mu \qquad (5.116)$$

determines a continuous linear functional on L_p whose norm is equal to the norm of the function $f(t)$ in the space L_q.

▷ By virtue of Holder inequality (3.92)

$$| fx | \leq \| f \|_q \| x \|_p .$$

It means that the functional f is bounded and

$$\| f \| \leq \| f \|_q . \qquad (5.117)$$

On the other hand, putting

$$x(t) = | f(t) |^{q-1} e^{-i \arg f(t)}$$

and taking into consideration that $x(t) \in L_p$ as

$$| x(t) |^p = | f(t) |^{p(q-1)} = | f(t) |^q ,$$

and consequently, $\| x \|_p^p = \| f \|_q^q$ we obtain from (5.116)

$$| fx | = \| f \|_q^q \leq \| f \| \| x \|_p = \| f \| \| f \|_q^{q/p} = \| f \| \| f \|_q^{q-1} .$$

Hence it is clear that $\| f \|_q \leq \| f \|$. Together with inequality (5.117) it gives $\| f \| = \| f \|_q$.

Thus to any function $f(t) \in L_q$ corresponds a continuous linear functional f on the space L_p determined by formula (5.106) whose norm $\| f \|$ is equal to the norm $\| f \|_q$ of the function $f(t)$ in the space L_q. ◁

Therefore the theorem gives the opportunity to suppose that any continuous linear functional on L_p is determined by formula (5.106). We shall prove it now.

Theorem 5.18. *Any continuous linear functional on the space $L_p(T, \mathcal{B}, \mu)$, $1 < p < \infty$ is expressed by formula (5.116), where $f(t)$ is some function from the space $L_q(T, \mathcal{B}, \mu)$, $p^{-1} + q^{-1} = 1$, and the norm of this functional is equal to the norm $\| f \|_q$ of the function $f(t)$ in the space $L_q(T, \mathcal{B}, \mu)$.*

▷ Suppose at first that the measure μ is finite, $\mu(T) < \infty$, and consider the values of the functional f on the indicators of the measurable sets [a]. It is evident that $f 1_A$ represents an additive function of the set $\varphi(A) = f 1_A$. For any monotone sequence of the sets $\{A_n\}$, $A = \lim A_n$ we have $\| 1_{A_n} - 1_A \|_p \to 0$ and $\varphi(A_n) \to \varphi(A)$ by virtue of the continuity of the functional f. Consequently, the additive function $\varphi(A)$ is continuous. By Theorem 2.14 it is σ–additive. As $\| 1_A \|_p = 0$ at $\mu(A) = 0$ then $\varphi(A) = 0$ at $\mu(A) = 0$, i.e. the function $\varphi(A)$ is μ–continuous (Subsection 3.6.1). By Radon–Nikodym theorem 3.33 there exists such μ–integrable function $f(x)$ that

$$\varphi(A) = f 1_A = \int\limits_A f \, d\mu = \int f(t) 1_A(t) \mu(dt).$$

Hence it follows that for any μ–integrable simple function $x(t)$ the value of the functional f is expressed by formula (5.116). As the set of μ–integrable simple functions is dense in L_p (Theorem 3.40) and the functional f is continuous then formula (5.116) is also valid for all the functions $x(t) \in L_p$.

It remains to prove that $f(t) \in L_q$ and $\| f \|_q = \| f \|$. Let $\{v^n(t)\}$ be a nondecreasing sequence of nonnegative simple functions which determines $| f(t) |$ (Subsection 3.4.1). Then

$$[v^n(t)]^q \leq [v^n(t)]^{q-1} | f(t) | = [v^n(t)]^{q-1} f(t) e^{-i \arg f(t)}. \tag{5.118}$$

The function

$$z(t) = [v^n(t)]^{q-1} e^{-i \arg f(t)} \tag{5.119}$$

---

[a] The finiteness of the measure μ is necessary that the indicators of all measurable sets including $1_T(t)$ belong to L_p.

belongs to L_p and as all the functions v^n belong to L_q:

$$|z|^p = (v^n)^{p(q-1)} = (v^n)^q,$$

and consequently, $\| z \|_p^p = \| v^n \|_q^q$. Therefore by virtue of inequality (5.118) and formula (5.119) we get

$$\| v^n \|_q^q \le \int fz\, d\mu = fz \le \| f \| \| z \|_p = \| f \| \| v^n \|_q^{q/p} = \| f \| \| v^n \|_q^{q-1},$$

whence $\| v^n \|_q \le \| f \|$ at all n. As by Theorem 3.27 we have

$$\int |f|^q\, d\mu = \lim \int (v^n)^q\, d\mu = \lim \| v^n \|_q^q,$$

then $f(t) \in L_q$ and $\| f \|_q \le \| f \|$. Correlating this inequality with inequality (5.117) we get $\| f \|_q = \| f \|$.

If the measure μ is σ–finite then the space T may be presented as the limit of increasing sequence of the spaces $\{T_n\}$. The measure μ is finite on each of them, $\mu(T_n) < \infty$. In this case for any n the narrowing f_n of the functional f on the subspace $L_p^n = L_p(T_n, \mathcal{B}_n, \mu)$ of the functions from L_p equal to zero outside T_n is expressed by formula (5.116). The norm $\| f \|_q^{(n)}$ of the function $f(t)$ in L_q^n is equal to $\| f_n \| \le \| f \|$. By virtue of the uniqueness of the function $f(t)$ in each L_q^n the function $f(t)$ in L_q^m at $m > n$ is the extension of the function $f(t)$ in L_q^n from T_n to $T_m \supset T_n$. Thus the function $f(t)$ is determined on the whole space T. Prove that it belongs to the space L_q. For this purpose we determine the functions

$$f_n(t) = \begin{cases} f(t) & \text{at} \quad t \in T_n, \\ 0 & \text{at} \quad t \notin T_n. \end{cases}$$

It is obvious that the functions $|f_n(t)|$ form nondecreasing sequence convergent almost everywhere to $|f(t)|$. Consequently, on the basis of Theorem 3.27

$$\int |f|^q\, d\mu = \lim \int |f_n|^q\, d\mu = \lim(\| f \|_q^{(n)})^q.$$

As $\| f \|_q^{(n)} \le \| f \|$ at all n then the integral in the left–hand side of this equality converges, and consequently, $f(t) \in L_q$ and $\| f \|_q \le \| f \|$. After comparing this equality with (5.117) we come to the conclusion that $\| f \|_q = \| f \|$. Consequently, formula (5.116) determines the functional f

on the whole space L_p and the theorem is true in the general case of the σ-finite measure μ. ◁

Thus we proved that between continuous linear functionals on L_p and the elements of the space L_q, $p^{-1} + q^{-1} = 1$, $1 < p$, $q < \infty$ there exists the one-to-one correspondence retaining the norm. It gives the opportunity to identify the space L_p^* dual with L_p, $1 < p < \infty$ with the space L_q.

Corollary *The space L_p, $1 < p < \infty$ is reflexive because the second dual space L_p^{**} as dual with $L_p^* = L_q$, $p^{-1} + q^{-1} = 1$ coincides with L_p, $L_p^{**} = L_p$.*

R e m a r k. Theorem 5.18 is also true for the space L_1. The space L_1 is not reflexive. But for proving it is necessary in this case to study the space L_∞ ($q = \infty$ at $p = 1$). Theorems 5.17 and 5.18 are also true in more general case when $L_p = L_p(T, \mathcal{B}, \mu, S)$ represents Lebesgue space of the functions with the values in the separable B-space S and $L_q = L_q(T, \mathcal{B}, \mu, S^*)$ is Lebesgue space of the functions with the values in the separable space S^* dual with S. In this case the product $f(t)x(t)$, $x \in L_p$, $f \in L_q$ represents the numeric function.

The detailed theory of the dual spaces is given in (Dunford and Schwarz 1958, Edwards 1965). For applications in control theory refer to (Pugachev and Sinitsyn 1987, Zadeh and Desoer 1963), in signal and image processing to (Oppenheim and Schafer 1975, Papoulis 1968, Pratt 1978).

P r o b l e m s

5.4.1. Present in the form of (5.84) the following linear functionals on the space $C([-1, 1])$:

$$f_1 x = \sum_{k=1}^{N} a_k x(t_k), \quad t_1, \ldots, t_N \in [-1, 1],$$

$$f_2 x = \int\limits_{-1}^{1} f(t)x(t)\,dt, \quad f_3 x = \int\limits_{-1}^{1} f(t)[x(t) - x(0)]\,dt,$$

$$f_4 x = \int\limits_{-1}^{1} f(t)x(t)\,dt + \sum_{k=1}^{N} c_k x(t_k),$$

$$f_5 x = \int\limits_{-1}^{1} [f(t)x(t) + \sum_{k=1}^{N} g_k(t)x(t_k)]\,dt,$$

where $f(t), g_1(t), \ldots, g_N(t) \in C([-1, 1])$ and find their norms as the complete variations of the correspondent measures μ_f.

5.4.2. Is it possible to present the functional

$$fx = \int\limits_{-1}^{1} \frac{x(t) - x(0)}{t^s} dt$$

in the form of (5.84) for the cases $s \in (0, 1)$, $s = 1$ and $s > 1$?

5.4.3. Prove that the norm of the functional (5.94) on the space $C^n([a, b])$ is equal to $\| f \| = M = \max(| \alpha_0 |, \ldots, | \alpha_{n-1} |, | \mu_f |([a, b]))$.

I n s t r u c t i o n. Firstly prove the inequality $\| f \| \leq M$ and then consider the case $M = | \alpha_p |$ and $M = | \mu_f |([a, b])$. In the first case determine the function $x(t) \in C^n([a, b])$, $\| x \| = 1$ for which $fx = | \alpha_p |$. In the second case construct the sequence of the functions $\{x_n(t)\}$, $x_n(t) \in C^n([a, b])$, $\| x_n \| = 1$ for which $\sup fx_n = | \mu_f |([a, b])$.

5.4.4. Reduce to the form of (5.94) the following functionals on the correspondent spaces of the differentiable functions on the interval $[-1, 1]$:

$$f_1 x = x(0) + cx'(1) + \int\limits_{-1}^{1} (a + bt)x''(t) \, dt,$$

$$f_2 x = \int\limits_{-1}^{1} x(t) \, dt, \quad x(t) \in C^n([-1, 1]), \quad f_3 x = \sum_{k=1}^{N} [a_k x(t_k) + b_k x'(t_k)],$$

$$f_4 x = x(-1) + x(1) - 3x'(0) + \int\limits_{-1}^{1} t^2 x''(t) \, dt,$$

$$f_5 x = \int\limits_{-1}^{1} x'(t) \sin t \, dt + \sum_{k=1}^{N} [a_k x(t_k) + b_k x'(t_k) + c_k x''(t_k) + d_k x'''(t_k)],$$

$$t_1, \ldots, t_N \in [-1, 1], \quad x(t) \in C^3([-1, 1])$$

and find their norms.

5.4.5. Let a space X be a B–space. Prove that if X^* is separable then X is also separable. Is the inverse statement true?

5.4.6. Consider the space l_p $(p > 1)$ of the sequences $x = \{x_k\}_{k=1}^{\infty}$, $x_k \in R$, $\sum\limits_{k=1}^{\infty} | x_k |^p < \infty$, with the norm

$$\| x \| = \left[\sum_{k=1}^{\infty} | x_k |^p \right]^{1/p} .$$

Prove that $l_p^* = l_q$, $1/p + 1/q = 1$, i.e. any continuous linear functional in l_p may be presented in the form

$$fx = \sum_{k=1}^{\infty} x_k y_k ,$$

where $x = \{x_k\}_{k=1}^{\infty} \in l_p$, $y = \{y_k\}_{k=1}^{\infty} \in l_q$ and $\| f \| = \| y \|_{l_q}$.

5.4.7. Consider the space c_0 of the sequences $x = \{x_k\}_{k=1}^{\infty}$, $x_k \in R$, convergent to zero with the norm $\| x \| = \max_k |x_k|$. Prove that $c_0^* = l_1$, i.e. any continuous linear functional in c_0 may be presented in the form

$$fx = \sum_{k=1}^{\infty} x_k y_k \,,$$

where $x = \{x_k\}_{k=1}^{\infty} \in c_0$, $y = (y_1, y_2, \ldots) \in l_1$ and $\| f \| = \| y \|_{l_1^*}$.

5.4.8. Consider the space m of bounded sequences $x = \{x_k\}_{k=1}^{\infty}$, $x_k \in R$, with the norm $\| x \| = \sup_k |x_k|$. Is it true that $m^* \neq l_1$. Prove that $l_1^* = m$, i.e. any continuous linear functional in l_1 may be presented in the form

$$fx = \sum_{k=1}^{\infty} x_k y_k \,,$$

where $x = \{x_k\}_{k=1}^{\infty} \in l_1$, $y = \{y_k\}_{k=1}^{\infty} \in m$ and $\| f \| = \| y \|_m$.

5.4.9. Consider a non–anticipative linear system discribed by differential equations:

$$\dot{z} = az + a_1 x + a_0 \,, \quad y = bz + a_0 \,, \quad z_0 = z(t_0) \,, \tag{5.120}$$

where x is the input, y is the output, a is square matrix of order p, a_1 a $p \times n$–matrix, a_0 a p–dimensional vector, b a $m \times p$–matrix, b_0 m–dimensional vector. Prove the following formulae for weighting functions:

a) $g(t, \tau) = b(t) u(t, \tau) a_1(\tau) 1(t - \tau),$ \hfill (5.121)

where $1(t - \tau)$ is the unit step function, $u = u(t, \tau)$ is the matrix solution of the homogeneous equation $\dot{u} = au$, satisfying initial condition $u(\tau, \tau) = I$ (fundamental matrix);

b) $g(t, \tau) = w(t - \tau) = b \begin{vmatrix} \alpha_{11} e^{s_1 (t-\tau)} & \ldots & \alpha_{1p} e^{s_p (t-\tau)} \\ \ldots & \ldots & \ldots \\ \alpha_{p1} e^{s_1 (t-\tau)} & \ldots & \alpha_{pp} e^{s_p (t-\tau)} \end{vmatrix} \alpha^{-1} 1(t - \tau),$

$$\tag{5.122}$$

$s_1, \ldots, s_p$ are different roots of the characteristic equation $| a - sI | = 0$, the elements of the matrix α are determined by the sets of linear algebraic equations:

$$(a - s_k I)[\alpha_{1k} \ldots \alpha_{pk}]^T = 0 \quad (k = 1, \ldots, p). \tag{5.123}$$

5.4.10 Prove that the weighting function of the nonstationary linear system

$$\ddot{z} + t^{-1} \dot{z} + (1 - n^2 t^{-2}) z = x \,, \quad y = z$$

is equal to $g(t, \tau) = \frac{\pi \tau}{2}[J_n(\tau)N_n(t) - N_n(\tau)J_n(t)]$ where $J_n(t)$ and $N_n(t)$ are Bessel functions of the first and second kind respectively (Appendix 2).

5.4.11. Find matrix weighting functions $g(t, \tau)$ for the following systems:

a) $\dot{z}_1 = z_2$, $\dot{z}_2 = -t^{-2}z_1 - t^{-1}z_2 + x$, $t > 0$, $\qquad\qquad\qquad$ (5.124)

$$u(t, \tau) = \{u_{ij}(t, \tau)\}, \quad (i, j = 1, 2),$$

$$u_{11} = \cos \ln \frac{t}{\tau}, \quad u_{12} = \tau \ln \frac{t}{\tau}, \quad u_{21} = -\frac{1}{t} \ln \frac{t}{\tau}, \quad u_{22} = \cos \ln \frac{t}{\tau},$$

b) $\dot{z}_1 = z_2$, $\dot{z}_2 = -ct^{-2}z_1 + x$, $t > 0$, $\qquad\qquad\qquad\qquad$ (5.125)

$$u(t, \tau) = \{u_{ij}(t, \tau)\}, \quad i, j = 1, 2, 2\gamma = \sqrt{|1 - 4c|}, \quad c < 1/4,$$

$$u_{11}(t, \tau) = \frac{1}{4\gamma}\sqrt{\frac{t}{\tau}}\left[(1 + 2\gamma)\left(\frac{\tau}{t}\right)^\gamma - (1 - 2\gamma)\left(\frac{t}{\tau}\right)^\gamma\right],$$

$$u_{12}(t, \tau) = \frac{\sqrt{\tau t}}{2\gamma}\left[\left(\frac{t}{\tau}\right)^\gamma - \left(\frac{\tau}{t}\right)^\gamma\right],$$

$$u_{21}(t, \tau) = \frac{1 - 4\gamma^2}{8\gamma\sqrt{\tau t}}\left[\left(\frac{\tau}{t}\right)^\gamma - \left(\frac{t}{\tau}\right)^\gamma\right],$$

$$u_{22}(t, \tau) = \frac{1}{4\gamma}\sqrt{\frac{\tau}{t}}\left[(1 + 2\gamma)\left(\frac{t}{\tau}\right)^\gamma - (1 - 2\gamma)\left(\frac{\tau}{t}\right)^\gamma\right].$$

5.4.12. Using Appendix 3 derive transfer functions for weighting functions given in Appendix 1 and vice versa, derive weighting functions for transfer functions given in Table A.1.1 (Appendix 1).

5.4.13. Using Appendix 3 derive transfer functions for weighting functions given in Appendix 1 and vice versa, derive weighting functions for transfer functions given in Table A.1.2 (Appendix 1).

5.4.14. Prove that under conditions of Problem 5.3.14 a primitive $U(t)$ of $V(t)$ is an ordinary random function and

$$K_W(t, t') = \int_{t_0}^{\min\{t,t'\}} \nu(\tau)d\tau, \quad W(t) = U(t) - U(t_0).$$

Prove that an integral over the Wiener measure may be presented as an ordinary integral with the white noise in the integrand

$$\int_A X(t)Z(dt) = \int_A X(t)V(t)dt, \quad Z(A) = \int_A V(\tau)d\tau.$$

CHAPTER 6
LINEAR OPERATORS

Chapter 6 contains the general theory of linear operators and operator equations. In Section 6.1 the definition of closed operators is given and their main properties are studied. The definition of an adjoint operator is introduced and the conditions for its existence are established. The main theorems concerning bounded operators are proved. The definitions of isometric and unitary operators are given and the properties of isometric, unitary and unitarly equivalent operators are studied. Section 6.2 stands somewhat apart from the basic topic of Chapter 6, namely from the theory of linear operators. In Section 6.2 linear and nonlinear operator equations are considered. The method of contraction images is presented as one of the main methods for proving the existence and uniqueness of solutions of various operator equations. This method is necessary, in particular, for further exposition of the theory of linear operators. The last Section 6.3 is devoted to the notion of spectrum of an operator. After introducing the main definitions concerning the spectra of operators the operator equations connected with the notion of spectrum are considered. The general properties of the spectra of linear operators are outlined, in particular, the spectra of unitary operators are studied.

6.1. Basic Notions and Theorems

6.1.1. Closed Operators

We studied in Section 4.6 the general properties of continuous linear functionals and operators. In particular, we learned that a linear functional or an operator mapping a topological linear space into a normed space is continuous if and only if it is bounded. Unbounded linear operators along with bounded ones play important role in functional analysis.

Let T be the operator mapping a topological linear space X into a topological linear space Y. The set of pairs $\{\{x, y\} : y = Tx\}$ is called the *plot of the operator* T and denoted by $\mathrm{Gr}(T)$. Let the topology be determined in the product space $Z = X \times Y$ in the usual way in which any vicinity of zero W represents the product of some vicinity of zero U in the space X and some vicinity of zero V in the space Y:

$$W = U \times V = \{\{x, y\} : x \in U, y \in V\} \tag{6.1}$$

(the topology of Tychonoff product space, Section 4.1.6). The plot of the operator T is contained in some subspace of the space $Z = X \times Y$ since $\{x_k, y_k\} \in \mathrm{Gr}(T)$ $(k = 1, \dots, n)$ implies $\{c_1 x_1 + \dots + c_n x_n, c_1 y_1 + \dots + c_n y_n\} \in \mathrm{Gr}(T)$.

The operator T is called a *closed operator* if its plot $\mathrm{Gr}(T)$ represents a closed set in the product space $Z = X \times Y$.

Theorem 6.1. *If X and Y represent T_1-spaces with the first countability axiom then the operator T is closed if and only if the convergence of sequences $\{x_n\} \subset X$ and $\{Tx_n\} \subset Y$ implies $x = \lim x_n \in D_T$ and $y = \lim Tx_n = Tx$.*

▷ If X and Y represent T_1-spaces with the first countability axiom then $Z = X \times Y$ is also the T_1-space with the first countability axiom. By Theorem 4.17 the point $\{x, y\}$ is the limit point of $\mathrm{Gr}(T)$ if and only if $\mathrm{Gr}(T)$ contains the sequence $\{\{x_n, Tx_n\}\}$ converging to the point $\{x, y\}$: $x_n \to x$, $Tx_n \to y$. Moreover $\{x, y\} \in \mathrm{Gr}(T)$ if and only if $x \in D_T$, $y = Tx$. ◁

Theorem 6.1 is valid, in particular, for normed spaces X, Y. If the operator T is closed and the inverse operator T^{-1} exists then it is also closed since its plot coincides with the plot of T.

Theorem 6.2. *A continuous operator is closed if and only if its domain is closed.*

▷ If the continuous operator T is closed then its domain D_T is closed as the inverse image of the closed set $\mathrm{Gr}(T)$ at the continuous mapping $z = \{x, Tx\}$ of the space X into $Z = X \times Y$ (Corollary of Theorem 4.19). If D_T is closed then for any limit point $\{x, y\}$ of the plot $\mathrm{Gr}(T)$ of the operator T the point x is in D_T as the limit point of D_T. Hence $y = Tx$ and $\mathrm{Gr}(T)$ is closed. ◁

Corollary. *A continuous operator defined on the whole space is closed.*

If X and Y are B-spaces then the inverse proposition is valid: a *closed operator* defined on the whole space is continuous.

If the operator T is not closed, and each point of the closure of its plot $[\mathrm{Gr}(T)]$ is determined uniquely by the first element of the pair $\{x, y\}$, i.e. when $\{x_1, y_1\}, \{x_2, y_2\} \in [\mathrm{Gr}(T)]$, $x_1 = x_2$ implies $y_1 = y_2$ then the operator T may be extended by means of adding to its plot all its limit points which do not belong to it. The operator $\bar{T}$ obtained in this way is called the *closure of the operator T*.

6.1.2. Commutativity of an Integral with a Linear Operator

We now show that property (V) of an integral of a simple function (commutativity of an integral with a linear operator) may be extended to integrals of functions with values in a separable B–space or in a topological linear space. Let T be the operator mapping a linear space Y into a linear space Z. We first prove property (V) for a continuous operator T and after that we shall show that it is valid for any closed operator under some additional conditions.

Theorem 6.3. *If T is the continuous linear operator defined on the whole space Y and the function $f(x)$ is μ–integrable then the function $\varphi(x) = Tf(x)$ is μ–integrable and*

$$\int_A \varphi d\mu = \int_A Tf d\mu = T \int_A f d\mu . \tag{6.2}$$

We first prove the theorem for the Bochner integral (Section 3.3).
▷ If Y and Z are separable B–spaces then the continuity of T implies

$$\| Tf^n(x) - Tf(x) \| \leq \| T \| \| f^n(x) - f(x) \| ,$$

where $\| T \|$ is a norm of the operator T. This inequality and the almost everywhere convergence of $\{f^n(x)\}$ to $f(x)$ imply the almost everywhere convergence of the sequence $\{Tf^n(x)\}$ to $Tf(x)$, and

$$\int \| Tf^n - Tf^m \| d\mu \leq \| T \| \int \| f^n - f^m \| d\mu .$$

Hence the sequence of the simple functions $\{Tf^n(x)\}$ defines $Tf(x)$. Finally, equality (V) for integrals of the simple functions implies

$$\int_A Tf d\mu = \lim \int_A Tf^n d\mu = \lim T \int_A f^n d\mu .$$

But the operator T is continuous. Hence

$$\lim T \int_A f^n d\mu = T \lim \int_A f^n d\mu = T \int_A f d\mu ,$$

what proves property (V). ◁

Now we prove the theorem for the Pettis integral (Section 5.2).

▷ If Y and Z are topological linear spaces with the topologies τ_G and τ_H determined by the sets of linear functionals G and H respectively (Subsection 4.6.6) then the continuity of the operator T in the topologies τ_G, τ_H implies the continuity of hT for any $h \in H = Z^*$ (Theorem 4.18). Hence $hT \in G = Y^*$. Definitions (5.10) and (5.11) give for $g = hT$

$$\int_A hTf d\mu = hT \int_A f d\mu.$$

This formula shows that the Pettis integral of the function $\varphi(x) = Tf(x)$ exists and

$$\int_A Tf d\mu = T \int_A f d\mu,$$

proving the theorem. ◁

Theorem 6.4. *If T is a closed operator with the domain D_T mapping a separable B–space Y into a separable B–space Z, the function $f(x)$ has the values in D_T and the functions $f(x)$ and $\varphi(x) = Tf(x)$ are μ–integrable then*

$$\int_A f d\mu \in D_T \quad \text{and} \quad \int_A \varphi d\mu = \int_A Tf d\mu = T \int_A f d\mu. \qquad (6.3)$$

▷ Let $V = Y \times Z$ be the product space with the norm $\| u \| = \| y \| + \| z \|$ of the element $u = \{y, z\}$. The plot of the operator T, $U = \text{Gr}(T) = \{\{y, Ty\} : y \in D_T\}$, represents a subspace of the space V since it is a normed linear space by virtue of linearity of T. By the completeness of T any fundamental sequence $\{y_n, Ty_n\}$ in U has the limit $\{y, Ty\}$ in U. Hence U represents a B–space and is separable as the subspace of the separable space V.

Now we define for each $x \in X$ the point $u(x) = \{f(x), Tf(x)\}$ in the space $U \subset V$. We have thus determined the function $u(x)$ with values in V. Let k be a linear functional on V. Then $ku = k\{f, Tf\} = k\{f, 0\} + k\{0, Tf\}$. But $k\{y, 0\} = gy$ represents evidently the value of some linear functional g at the point $y \in Y$ and $k\{0, z\} = hz$ be the value of some other linear functional h at the point $z \in Z$. Hence any continuous linear functional k on V has the form

$$k\{y, z\} = gy + hz$$

and
$$ku = k\{f, Tf\} = gf + hTf,$$

where g and h are continuous linear functionals on Y and Z respectively. By the condition of the theorem the functions $f(x)$ and $Tf(x)$ are μ–integrable, and consequently, measurable. But any measurable function is weakly measurable (Subsection 4.7.4). Hence the numerical functions $gf(x)$ and $hTf(x)$ and their sum

$$ku(x) = gf(x) + hTf(x)$$

are measurable for any continuous linear functional k on V. Thus the function $u(x)$ is weakly measurable. But any weakly measurable functions with values in a separable B–space are measurable (Subsection 4.7.5). Hence the function $u(x)$ is measurable.

The μ–integrability of the functions $f(x)$ and $Tf(x)$ implies the μ–integrability of the norm of the function $u(x)$:

$$\| u(x) \| = \| f(x) \| + \| Tf(x) \| .$$

But by Corollary 1 of Theorem 3.20 the measurability of the function $u(x)$ and the μ–integrability of its norm implies the μ–integrability of $u(x)$ and

$$\int_A u d\mu = \left\{ \int_A f d\mu, \quad \int_A Tf d\mu \right\} .$$

Finally, $u(x) \in U = \mathrm{Gr}(T)$ implies $\int_A u d\mu \in U$ giving (6.3). ◁

6.1.3. Adjoint Operators

Let T be the linear operator mapping a topological linear space X into a topological linear space Y, X^* and Y^* respective dual spaces. The product of the operator T and a functional $g \in Y^*$ represents a functional $f = gT$ on the space X. This functional is linear since T and g are linear:

$$f(\alpha_1 x_1 + \alpha_2 x_2) = g(\alpha_1 T x_1 + \alpha_2 T x_2) = \alpha_1 f x_1 + \alpha_2 f x_2 .$$

Its domain coincides with the domain D_T of the operator T. If there exists the unique functional $f \in X^*$ for any functional g from some set $G \subset Y^*$ such that $f = gT$ then the operator T determines the mapping of G into X^* along with the usual mapping $y = Tx$ of D_T into Y. In other words, the linear operator T mapping X into Y generates in this case some operator T^* mapping the dual space Y^* of Y into the dual space X^* of X

$$y = Tx, \quad f = T^*g \quad D_{T^*} = G.$$

This operator T^* is called the *adjoint or conjugate operator* of T. The domain D_{T^*} of T^* represents the set G of all the functionals $g \in Y^*$ for which the functional gT is continuous, i.e. belongs to X^*, and X^* contains the unique functional f coinciding with gT. Hence the operator T has the adjoint operator T^* if and only if there exist functionals $g \in Y^*$ for each of which X^* contains the unique functional f coinciding with gT.

Thus if the linear operator T has the adjoint operator T^* then it is determined by

$$(T^*g)x = g(Tx) \quad \text{for all} \quad x \in D_T \quad \text{and} \quad g \in D_{T^*}. \tag{6.4}$$

This equality may be written shortly as $T^*g = gT$.

The adjoint operator is linear since $g = \alpha_1 g_1 + \alpha_2 g_2$ implies

$$T^*g = gT = \alpha_1 g_1 T + \alpha_2 g_2 T = \alpha_1 T^* g_1 + \alpha_2 T^* g_2.$$

Formula (6.4) determines the adjoint operator T^* as the function of the operator T mapping the space $\mathcal{L}(X, Y)$ into the space $\mathcal{L}(X^*, Y^*)$: $T^* = \varphi(T)$. This function is linear since $T = \alpha_1 T_1 + \alpha_2 T_2$ implies

$$T^*g = g(c_1 T_1 + c_2 T_2) = (c_1 T_1^* + c_2 T_2^*)g.$$

Theorem 6.5. *If the linear operator T is mapping a topological linear space X into a topological linear space Y and the linear operator S is mapping Y into a topological linear space Z and the adjoint operators T^* and S^* exist and $R_T \subset D_S$, $D_{T^*} R_{S^*} \neq \emptyset$ then the operator ST mapping X into Z exists and has the adjoint operator which is given by*

$$(ST)^* = T^* S^*. \tag{6.5}$$

▷ Let X^*, Y^*, Z^* be dual spaces of X, Y, Z respectively. For any $h \in D_{S^*}$ there exists in Y^* the unique functional $g = S^* h$ satisfying

$$gy = h(Sy) \quad \forall y \in D_S.$$

If, in particular, $y \in R_T \subset D_S$ then $y = Tx$ for some x implying

$$g(Tx) = h(STx) \quad \forall x \in D_T .$$

On the other hand, for any $g \in D_{T^*}$ there exists the unique functional $f = T^* g \in X^*$, for which

$$fx = g(Tx) \quad \forall x \in D_T .$$

Thus for any $h \in D_{S^*}$ satisfying $g = S^* h \in D_{T^*}$ there exists the unique functional $f \in X^*$ for which

$$fx = h(STx) \quad \forall x \in D_T .$$

This proves the existence of the adjoint operator $(ST)^*$ of ST

$$f = (ST)^* h$$

with the domain $\{h : h \in D_{S^*}, S^* h \in D_{T^*}\}$. Finally, $f = T^* g$, $g = S^* h$ implying

$$f = T^* S^* h .$$

Comparing two expressions for f we get (6.5). ◁

E x a m p l e 6.1. Let X be n–dimensional Euclidean space, T the linear operator mapping X into m–dimensional Euclidean space Y. This operator is completely determined by the matrix A of the corresponding linear transform $y = Ax$. For typical signal and image processing operators refer to Table A.1.3 (Appendix 1). Since the dual spaces X^* and Y^* coincide in this case with X and Y respectively (Examples 5.5 and 5.6) the adjoint operator T^* maps Y into X and is given by the matrix of the corresponding linear transform $x = A^* y$. Eq.(6.4) determining the adjoint operator is now $(Ax, g) = (x, A^* g)$ or in a scalar form

$$\sum_{r=1}^{m} (Ax)_r \bar{g}_r = \sum_{r=1}^{m} \left(\sum_{s=1}^{n} a_{rs} x_s \right) ,$$

$$\bar{g}_r = \sum_{s=1}^{n} x_s \left(\overline{\sum_{r=1}^{m} \bar{a}_{rs} g_r} \right) = \sum_{s=1}^{n} x_s \overline{(A^* g)}_s .$$

This equality shows that the adjoint operator is determined in this case by the matrix obtained from A by transposition and replacement of all the elements by their complex conjugate numbers: $A^* = \bar{A}^T$. In the special case of $X = R^n$,

$Y = R^m$ the matrix A^* of the adjoint operator represents the transposed matrix A of the operator T: $A^* = A^T$.

E x a m p l e 6.2. Let T be the operator mapping the space $X = C(K)$ of continuous functions into the space $Y = C(L)$ of continuous functions:

$$Tx = \int w(s,\, t)\, x(t)\, dt\,, \quad s \in L\,,$$

$K \subset R^n$, $L \subset R^m$ being compacts, $w(s,\, t)$ a continuous function. Let G be the set of linear functionals on $Y = C(L)$ representable in the form

$$gy = \int g(s)\, y(s)\, ds\,. \tag{$*$}$$

By Riesz theorem 5.15 the set G of such linear functionals is the subspace of the dual space $Y^* = C^*(L)$ of $Y = C(L)$. For all such functionals

$$\begin{aligned} g(Tx) &= \int g(s)\, ds \int w(s,\, t)\, x(t)\, dt \\ &= \int \left\{ \int w(s,\, t)\, g(s)\, ds \right\} x(t)\, dt\,. \end{aligned}$$

Comparing this relation with Eq. (6.4) we see that the adjoint operator T^* of T is determined in this case by

$$T^* g = \int w(s,t) g(s)\, ds\,.$$

This operator maps G into the set $F \subset X^*$ of the functionals f of the form $(*)$:

$$fx = \int f(t)\, x(t)\, dt\,.$$

Thus the adjoint operator T^* of the linear integral operator T represents the integral operator whose kernel $w^*(t,\, s)$ differs from the kernel $w(t,\, s)$ of T by changing the roles of arguments:

$$w^*(t,\, s) = w(t,\, s)\,.$$

The same result is obtained in the case of the linear integral operator mapping the space $C^N(K)$ of continuous functions of t with continuous derivatives of orders up to N into the space $C^M(L)$ of continuous functions of s with continuous derivatives of orders up to M. In this case F and G represent the spaces of functions each of which is equal to the sum of a piecewise continuous function and a linear combination of the δ–functions and their derivatives of respective orders.

E x a m p l e 6.3. The result obtained in Example 6.2 allows to determine a weighting function of a linear control system by the so–called the *adjoint systems method* (Laning and Battin 1956, Pugachev and Sinitsyn 1987). It consists in the following.

While giving on an input of a system (Example 1.8, Fig.1.8) the identity impulse $\delta(t - \tau)$ or while simulating the act of this impulse on the output we get on the output a weighting function of a system in the dependence of the first argument t at a given fixed value of the moment of the supply of the identity impulse τ. For complete determining in such a way a weighting function as a function of two variables t and τ it is necessary to determine the responses of the system on the identity impulses which act at different moments τ. Hence we determine the surface which represents a weighting function by its sections parallel to the axis t. For calculating the response of a linear system on an arbitrary disturbance at the moment t it is necessary to know one section of the surface which represents a weighting function of a system parallel to the axis τ. In order to avoid repeatedly simulation of a given system while determining its weighting function it is naturally to replace a given system by another one which has the same weighting function but with changed roles of the arguments. In other words, to replace a given system by an adjoint system for which a weighting function is represented by the same surface as a weighting function of a given system but the coordinates axes change places: the axis t takes a position of the axis τ and vice versa. Then the first argument will play the role of the instant of the impulse supply, and the second one – the role of the current instant. This fact will give us the opportunity to determine the necessary surface section representing a weighting function of a given system parallel to the axes τ by means of one simulation of the identity impulse action on an adjoint system.

Thus if the given linear control system has the weighting function $g(t, \tau)$ then the weighting function of the adjoint system will be $g^*(t, \tau)$. In this denotion the first argument as usual represents the current time, and the second argument represents the instant of the identity impulse action. According to the defintion of an adjoint system (Example 6.2) we have: $g^*(t, \tau) = g(t, \tau)$. This equality expresses the fact that the weighting functions of two interadjoint systems are represented by one and the same surface in the coordinates which differ from one another by the denotion of the axis t and τ.

It is evident that an adjoint non–anticipative system is an anticipative system. At first glance it involves some difficulties. But we may easily overcome them. Really, by virtue of condition (5.98) of the given system an adjoint system at each given instant t must response only on such distrubances which follows the given instant t and cannot response on the previous distrubances. If the direction of time duration is reversed then the notions "the previous" and the "subsequent" will change the places and the adjoint system will become non–anticipative. Consequently, the adjoint system may be simulated after changing the direction of time reading, i.e. while simulating the negative time by the natural physical time. While simulating the adjoint system in the negative time we may obtain just such section of a weighting function of a given system which is necessary for the calculations by formula (5.96).

E x a m p l e 6.4. Consider the space $X = L_p(\Delta, \mathcal{C}, \mu, Y)$ of the functions with values in a B–space Y. The linear operator

$$Tx = \int \alpha(t)\, x(t)\, \mu(dt)\,,$$

$\alpha(t)$ being the numerical function from the space $L_q' = L_q(\Delta, \mathcal{C}, \mu, K)$, $p^{-1} + q^{-1} = 1$, is mapping L_p into Y. Let g be a continuous linear functional on Y, $g \in Y^*$. Then due to property $(\vee)$ of the integral (commutativity of an integral with a continuous linear operator or functional, Theorem 6.3)

$$g(Tx) = \int \alpha(t)\, gx(t)\, \mu(dt) = (T^*g)x\,.$$

Hence the adjoint operator T^* is determined by

$$T^*g = \int \alpha(t) g(\cdot)\, \mu(dt)\,,$$

where the point in brackets indicates the place of the argument x of the functional T^*g.

By the last Remark in Subsection 5.4.5 the dual space of L_p is the space $L_q'' = L_q(\Delta, \mathcal{C}, \mu, Y^*)$ of functions with values in Y^*. If the spaces Y and Y^* are separable then the operator T^* is mapping any functional $g \in Y^*$ onto the function $\alpha(t)\, g \in L_q''$. Consequently, the adjoint operator T^* of T represents in this case the operator of multiplication by the numerical function $\alpha(t) \in L_q'$.

E x a m p l e 6.5. Consider the operator

$$Tx = \int \varphi(s,\, t)\, x(t)\, \mu(dt)$$

in the space $L_p(\Delta, \mathcal{C}, \mu)$, $1 < p < \infty$, of numerical functions of t supposing that the numerical function $\varphi(s,\, t)$ possesses the following properties:

(i) $\varphi(s, t) \in L_q(\Delta, \mathcal{C}, \mu)$, $p^{-1} + q^{-1} = 1$, at any s and its norm $\| \varphi \|_q(s)$ is an element of the space $L_h(S, \mathcal{S}, \nu)$, $1 < h < \infty$, of functions of s;

(ii) $\varphi(s, t) \in L_h(S, \mathcal{S}, \nu)$ at any t and its norm $\| \varphi \|_h(t)$ is and element of the space $L_q(\Delta, \mathcal{C}, \mu)$.

The operator T is mapping $L_p(\Delta, \mathcal{C}, \mu)$ into $L_h(S, \mathcal{S}, \nu)$. By Theorems 5.17 and 3.44 we have for any linear functional g on $L_h(S, \mathcal{S}, \nu)$

$$
\begin{aligned}
g(Tx) &= \int g(s)\, \nu(ds) \int \varphi(s,\, t)\, x(t)\, \mu(dt)\\
&= \int x(t)\, \mu(dt) \int \varphi(s,\, t)\, g(s)\, \nu(ds)\,,
\end{aligned}
$$

where $g(s) \in L_k(S, \mathcal{C}, \nu)$, $h^{-1} + k^{-1} = 1$. Under our assumptions the function

$$f(t) = \int \varphi(s, t) \, g(s) \, \nu(ds)$$

belongs to the space $L_q(\Delta, \mathcal{C}, \mu)$ and consequently, determines the continuous linear functional on $L_p(\Delta, \mathcal{C}, \mu)$. Thus the adjoint operator T^* is given by

$$T^* g = \int \varphi(s, t) \, g(s) \, \nu(ds) \,.$$

It is mapping the dual space $L_k(S, \mathcal{J}, \nu)$ of $L_h(S, \mathcal{S}, \nu)$, $h^{-1} + k^{-1} = 1$ into the dual space $L_q(\Delta, \mathcal{C}, \mu)$, $p^{-1} + q^{-1} = 1$ of $L_p(\Delta, \mathcal{C}, \mu)$.

E x a m p l e   6.6. Differentiation operator D in the space $C^1([a, b])$ of differentiable functions in the interval $[a, b]$ is mapping $C^1([a, b])$ into the space $C([a, b])$ of continuous functions. This operator is bounded and its norm does not exceed unity. In fact it transforms the element $x(t) \in C^1([a, b])$ with the norm $\| x \| = \sup | x(t) | + \sup | x'(t) |$ to the element $x'(t)$ with the norm $\| x' \|$ $= \sup | x'(t) | \leq \| x \|$.

By Riesz theorem 5.15 the dual space of $C([a, b])$ represents the space of finite regular measures and any continuous linear functional g on $C([a, b])$ is expressed in terms of this measure by (5.84). Hence

$$g(Dx) = \int\limits_a^b x'(t) \, \mu(dt) \,,$$

μ being a finite regular measure. But this formula determines the continuous linear functional on $C^1([a, b])$ by virtue of (5.94). Taking into account that the dual space of $C^1([a, b])$ represents the space of pairs $\{\alpha_0, \mu\}$, α_0 being a complex number and μ a finite regular measure (Theorem 5.16), we see that the adjoint operator D^* of D in $C^1([a, b])$ transforms μ to the pair $\{0, \mu\}$. The domain of T^* is the whole dual space of $C([a, b])$.

Now we restrict D^* to the class of measures μ whose absolutely continuous part has the piecewise differentiable Radon–Nikodym derivative with respect to the Lebesgue measure and singular part is absent. Then the preceding formula may be written in the form

$$g(Dx) = \int\limits_a^b w(t) \, x'(t) \, dt \,.$$

Integrating by parts we get

$$g(Dx) = w(b - 0) \, x(b) - w(a + 0) \, x(a) - \int\limits_a^b w'(t) \, x(t) \, dt$$

$$= - \int\limits_a^b [\, w(a+0)\, \delta(t-a) - w(b-0)\delta(t-b) + w'(t)\,]\, x(t)\, dt\,,$$

$w'(t)$ being the generalized derivative of $w(t)$ without account of discontinuities of $w(t)$ at the points a and b. Now we assume $w(a) = w(b) = 0$ for all the measures μ of the class considered. This is possible because the Radon–Nikodym derivative with respect to the Lebesgue measure may be arbitrarity changed on any set of points of zero Lebesgue measure. Then the expression in square brackets becomes the derivative $w'(t)$ of $w(t)$ with the accont of all discontinuities of $w(t)$, and the preceding formula takes the form

$$g(Dx) = \int\limits_a^b w(t)\, x'(t)\, dt = - \int\limits_a^b w'(t)\, x(t)\, dt\,. \qquad (6.6)$$

Hence the considered restriction of the operator D^* transforms $w(t)$ into $-w'(t)$:

$$D^* w(t) = -w'(t)\,.$$

Thus the considered restriction of the adjoint operator D^* of the differentiation operator D with respect to t in the space $C^1([\,a,\,b\,])$ represents the differentiation operator whith respect to $-t$.

Let us now consider the space $L_1(X,\,\mathcal{A},\,\nu)$ of functions $f(x)$ with values in the space $C^1([\,a,\,b\,])$, i.e. the space of functions of two variables $f(t,\,x)$, $a \leq t \leq b$, belonging to $C^1([\,a,\,b\,])$ at any x (the integrals of such functions by the measure ν also belong to $C^1([\,a,\,b\,])$ by the properties of the Bochner integral). Since the differentiation operator $D = d\,/\,dt$ is continuous in $C^1([\,a,\,b\,])$ it is commutative with the integral operator by Theorem 6.3:

$$D \int f(x)\, \nu(dx) = \int Df(x)\, \nu(dx)\,,$$

or

$$\frac{\partial}{\partial t} \int f(t,\,x)\, \nu(dx) = \int \frac{\partial f(t,\,x)}{\partial t}\, \nu(dx)\,.$$

Thus the differentiation operator in $C^1([\,a,\,b\,])$ is commutative with the Bochner integral with values in $C^1([\,a,\,b\,])$. In other words, the differentiation of the Lebesgue integral with respect to a parameter may be fullfiled under the integral sign [a].

---

[a] The Bochner integral represents in this case the Lebesgue integral of the numerical function $f(t,\,x)$ depending on the parameter t.

E x a m p l e 6.7. The differentiation operator D in the space of continuous functions $C([a, b])$ transforms the element $x(t) = C([a, b])$ with the norm $\| x \| = \sup | x(t) |$ to the element $x'(t) \in C([a, b])$ with the norm $\| x' \| = \sup | x'(t) |$. The domain of D is the subspace of functions of $C([a, b])$ which have the continuous derivative. Evidently the domain of D is dense in $C([a, b])$ since by known Weierstrass theorem any continuous function may be uniformly approximated by a polynomial with any degree of accuracy.

It is clear that the differentiation operator D in $C([a, b])$ is unbounded since the quantity $\| x' \| / \| x \| = \sup | x'(t) | / \sup | x(t) |$ may be arbitrarily large. For instance, the function $x(t) = a \sin \omega t$ is continuous at any $a > 0$ and $\omega > 0$, and for this function $\| x' \| / \| x \| = \omega$ may be equal to any positive number.

Since

$$g(Dx) = \int Dx(t)\,\mu(dt) = \int x'(t)\,\mu(dt)\,,$$

by (5.84) the domain of the adjoint operator D^* is the set of all finite regular measures μ for which

$$\int x'(t)\,\mu(dt) = \int x(t)\,\varphi(dt)\,, \tag{6.7}$$

φ being a finite regular measure

$$\varphi = D^* \mu\,.$$

It is clear that functions μ for which relation (6.7) is valid exist. In fact, formula (6.6) of Example 6.6 shows that the class of functions μ to which the operator D^* was restricted in Example 6.6 satisfies (6.7). Thus to any function μ of the form

$$\mu(A) = \int_A w(t)dt\,,$$

where the function $w(t)$ is piecewise continuous and differentiable corresponds the unique function

$$\varphi(A) = - \int_A w'(t)\,dt\,,$$

for which (6.7) is valid.

If D^* is restricted to the measures μ for which there exists the unique measure φ satisfying (6.7) then the operator D^* becomes the differentiation operator with respect to $-t$ as in Example 6.6.

The differentiation operator in $C([a, b])$ is closed. In fact the space $C^1([a, b])$ is complete. Hence $x_n(t) \to x(t)$, $x'_n(t) \to z(t)$ implies the differentiability of $x(t)$ and the equality $z(t) = x'(t)$. Therefore Theorem 6.4 legitimates the permutation of the differentiation operator and the integral operator once more.

E x a m p l e 6.8. The differentiation operator D in the space of test functions Φ (Subsection 5.3.3) transforms any test function $\varphi \in \Phi$ to its derivative φ'. In accordance with (6.4) the adjoint operator D^* is determined by

$$f(D\varphi) = (D^*f)\varphi\,,$$

where f is a generalized function. But

$$f(D\varphi) = f\varphi' = -f'\varphi\,.$$

by (5.39). Comparing this relation with the preceding one we get

$$D^*f = -f'\,.$$

Thus the adjoint operator D^* is defined in this case on the whole space of generalized functions Φ^* and represents the differentiation operator with respect to $-t$ as in Example 6.6.

6.1.4. Existence of Adjoint Operator

The considered examples show the existence of operators having adjoint operators. Now we shall find the general conditions of existence of a adjoint operator.

Theorem 6.6. *Continuous linear operator T has the adjoint operator T^* if and only if it is defined on the whole space X, $D_T = X$. The adjoint operator T^* is in this case defined on the whole space Y^*.*

▷ If the operator T is continuous then the functional gT is continuous for any $g \in Y^*$ by Theorem 4.18 about the continuity of a composite function. But if $D_T \neq X$ then its extension to the whole space X may be not unique. In other words, X^* may contain a set of functionals f for which

$$fx = g(Tx) \quad \text{if} \quad x \in D_T\,. \tag{6.8}$$

for a given $g \in Y^*$. The adjoint operator T^* does not exist in this case. In fact if $D_T \neq X$ then we may assume that D_T is not dense in X, $[D_T] \neq X$, since the operator T may be extended to the whole X by the continuity if $[D_T] = X$. Taking any functional $f_1 \in X^*$, $f_1 \neq 0$, for which $D_T \subset \ker f_1$ we get $f_1 x = 0$ and $(f + f_1)x = fx$ for all $x \in D_T$ [a].

---

[a] If $[D_T] \neq X$, then the functionals $f_1 \in X^*$ exist for which $D_T \subset \ker f_1$ since D_T is a subspace of X (Corollary 2 of Theorem 1.18).

Hence the condition $D_T = X$ is necessary. It is also sufficient since $fx = f'x$ at any x implies $f' = f$. ◁

Theorem 6.7. *If T is not continuous then the adjoint operator T^* exists if and only if the domain D_T of T is dense in X and therefore the closure $[D_T]$ coincides with X, $[D_T] = X$.*

▷ If the operator T is not continuous then the functional gT may be not continuous. If in this case there exists a set $G \subset Y^*$ of functionals g for which gT is continuous then it is necessary for the existence of the adjoint operator T^* that X^* will contain the single functional f coinciding with gT on D_T for any $g \in G$. If $[D_T] \neq X$ then $f_1 x = 0$ for any functional $f_1 \in X^*$ whose kernel contains $[D_T]$ and $(f + f_1)x = fx \ \forall x \in D_T$. Consequently, X^* contains the infinite set of functionals f coinciding with gT on D_T. Thus the condition $[D_T] = X$ is necessary.

To prove the sufficiency of this condition we notice first of all that the coincidence of functionals $f, f' \in X^*$ on D_T implies their coincidence on $[D_T]$ by continuity. Therefore if $[D_T] = X$ then X^* contains the single functional f coinciding with gT on D_T for each $g \in G$.

It remains to prove that the set G of functionals g for each of which gT is continuous is not empty. To prove this we consider the plot $\mathrm{Gr}(T)$ of the operator T in the product space $Z = X \times Y$. As

$$hz = h\{x, y\} = h\{x, 0\} + h\{0, y\} = fx + gy \qquad (6.9)$$

for any continuous linear functional h on $Z = X \times Y$, $h \in Z^*$, f and g being continuous linear functionals on X and Y respectively, $f \in X^*$, $g \in Y^*$, we have $Z^* = X^* \times Y^*$ implying $h = \{f, g\}$, $f \in X^*$, $g \in Y^*$. Thus the dual space of the product space $Z = X \times Y$ is the corresponding product space $Z^* = X^* \times Y^*$. Rewriting (6.8) in the form

$$fx - g(Tx) = 0$$

and comparing with (6.9) we see that all the points $\{x, -Tx\}$, $x \in D_T$, belong to the kernel of the functional $h = \{f, g\}$ for each pair $\{f, g\}$ satisfying (6.8). But the set of points $\{x, -Tx\} \in Z$, $x \in D_T$ represents the plot of the operator T transformed by the operator U mapping each point $\{x, y\} \in Z$ onto the point $\{x, -y\} : U\{x, y\} = \{x, -y\}$. Thus to any continuous linear functional $h \in Z^*$ whose kernel contains the plot of the operator T transformed by the operator U,

$$U\mathrm{Gr}(T) \subset \ker h, \qquad (6.10)$$

corresponds the pair $\{f, g\}$ satisfying (6.8). Now we recall that the linear functionals whose kernels contain any given subspace exist always (Corollary 2 of Theorem 1.18). Hence the continuous linear functionals h whose kernels contain the plot of the operator T transformed by the operator U exist always. For each such functional the pair $\{f, g\}$ satisfies (6.8). If $[D_T] = X$ then for each pair $\{f, g\}$ satisfying (6.8) f is the single element of X^* for which $gT = f$ as was proved earlier. This proves the existence of the adjoint operator T^*,

$$f = T^* g . \tag{6.11}$$

The domain D_{T^*} of the operator T^* represents the set G of all the functionals $g \in Y^*$ to each of which corresponds such $f \in X^*$ that the kernel of the functional $h = \{f, g\}$ contains the plot of the operator T transformed by the operator U. Thus the condition $[D_T] = X$ is sufficient for the existence of the adjoint operator T^*. ◁

R e m a r k. The theorem proved implies that the domain D_{T^*} of the adjoint operator cannot be widen while the operator T is extended, since the set of functionals h satisfying (6.8) can only be restricted while extending the domain D_T.

Theorem 6.8. *The adjoint operator is closed.*

▷ First of all we notice that at any $x \in D_T$ the equality

$$\varphi_x h = h\{x, -Tx\} = fx - g(Tx)$$

determines the continuous linear functional φ_x on the space $Z^* = X^* \times Y^*$ (Subsection 5.1.3). The kernel of this functional

$$\ker \varphi_x = \{\{f, g\} : fx - g(Tx) = 0\} \tag{6.12}$$

is closed as the inverse image of the closed set $\{0\}$. The plot of the adjoint operator T^* represents the intersection of all the sets (6.12) corresponding to all $x \in D_T$:

$$\mathrm{Gr}\,(T^*) = \bigcap_{x \in D_T} \{\{f, g\} : fx - g(Tx) = 0\} . \tag{6.13}$$

But the intersection of closed sets is always closed. Hence the plot of T^* represents a closed set, i.e. the operator T^* is closed. ◁

Theorem 6.9. *If the operator T admits the closure $[D_T]$ and $[D_T] = X$ then its closure $\bar{T}$ has the adjoint operator $\bar{T}^*$ coinciding with T^*.*

▷ If the operator T admits the closure then the plot of its closure $\bar{T}$ represents the closure of the plot of the operator T, $\mathrm{Gr}\,(\bar{T}) = [\,\mathrm{Gr}(T)\,]$ by definition. Taking into account that the kernel of any continuous linear functional is closed as the inverse image of the singleton $\{0\}$ and that any closed set contains the closures of all its subsets we see that if the kernel of the functional $h \in Z^*$ contains the plot of the operator T transformed by the operator U then it contains also its closure $[\,U\,\mathrm{Gr}(T)\,] = U[\,\mathrm{Gr}(T)\,]$ $= U\mathrm{Gr}(\bar{T})$, and vice versa. Hence the set of pairs $\{f,\,g\}$ satisfying (6.8) for the operator $\bar{T}$ coincides with that for the operator T. In other words, the plots of the operator $\bar{T}^*$ and T^* coincide, i.e. $\bar{T}^* = T^*$. ◁

6.1.5. Positive Operators

Let X be a topological linear space, X^* its dual space. The linear operator T mapping the space X into the dual space X^* is called a *positive operator* if

$$(Tx)x \geq 0 \quad \forall x \in X\,. \tag{6.14}$$

Similarly the linear operator T mapping the dual space X^* into X is called a *positive operator* if for any functional $f \in X^*$

$$fTf = f(Tf) \geq 0\,. \tag{6.15}$$

In the same way the *negative operator* mapping X into X^* or X^* into X is defined. The operator T is negative if and only if $-T$ is positive.

The notion of a positive operator enables one to order partially the set of all linear operators mapping X into X^* or X^* into X and write inequality signs between some operators. So, for instance, $T \geq 0$ means that the operator T is positive (exceeds "zero" operator mapping X or X^* into zero element of the space X^* of X respectively). The inequality $T_2 \geq T_1$ means that $T_2 - T_1 \geq 0$. The equality sign is rejected if X (respectively X^*) does not contain an element for which the equality $T_2x = Tx_1$ (respectively $T_2f = Tf_1$) holds.

The linear operator T mapping a normed linear space X into the dual space X^* is called *positive definite operator* if

$$(Tx)x > c \parallel x \parallel \quad \forall x \in X$$

for some $c > 0$.

E x a m p l e 6.9. The linear operator T with Hermite nonnegative definite matrix A mapping n–dimensional Euclidean space X into itself is positive since

$$(Tx)x = (x,\,Ax) = \sum_{k=1}^{n} x_k \sum_{l=1}^{n} \bar{a}_{kl}\bar{x}_l = \sum_{k,l=1}^{n} \bar{a}_{kl}x_k\bar{x}_l \geq 0$$

at all x_1 , ... , x_n by virtue of nonnegative definiteness of the matrix A. See also Table A.1.3 (Appendix 1).

E x a m p l e 6.10. The linear operator T of Example 6.5 at $S = \Delta, \nu = \mu,$ $h = q$ is mapping the space $L_p(\Delta, \mathcal{C}, \mu)$ into the dual space $L_q(\Delta, \mathcal{C}, \mu)$. It is positive if

$$(Tx)x = \int x(s)\,\mu(ds)\int \varphi(s,\,t)\,x(t)\,\mu(dt)$$

$$= \int \int \varphi(s,\,t)\,x(t)\,x(s)\,\mu(dt)\,\mu(ds) \geq 0 \qquad (6.16)$$

for any function $x(t) \in L_p(\Delta, \mathcal{C}, \mu)$.

E x a m p l e 6.11. Consider the linear operators T_1 and T_2 of the same type as in Example 6.10. The inequality $T_1 \leq T_2$ holds if the difference of the kernels $\varphi(s,\,t) = \varphi_2(s,\,t) - \varphi_1(s,\,t)$ of T_2 and T_1 satisfies (6.16).

6.1.6. Isometric Operators

The linear operator V mapping the whole B–space X onto the whole B–space Y and preserving the norm,

$$\| Vx \| = \| x \|, \quad \forall x \in X, \quad R_V = Y,$$

is called an *isometric operator*. Obviously an isometric operator is bounded and its norm is equal to unity: $\| V \| = 1$.

Theorem 6.10. *An isometric operator V is always invertible and its inverse operator V^{-1} is also isometric.*

▷ If $Vx = Vx' = y$ then $0 = \| Vx - Vx' \| = \| V(x - x') \| = \| x - x' \|$, i.e. $x' = x$. Thus the isometric operator V is the one–to–one mapping of X onto Y preserving the norm. Hence the inverse operator V^{-1} exists and is evidently linear. Also $D_{V^{-1}} = R_V = Y$, $R_{V^{-1}} = D_V = X$. Hence it is isometric. ◁

Theorem 6.11. *An isometric operator V has the adjoint operator V^* which is also isometric and $(V^*)^{-1} = (V^{-1})^*$.*

▷ The adjoint operator V^* exists by Theorem 6.1.6 since $D_V = X$ and $D_{V^*} = Y^*$. Eq. (6.4) determining V^* is

$$(V^*g)x = g(Vx). \qquad (6.17)$$

But $\| Vx \| = \| x \|$. Consequently,

$$| (V^*g)x | = | g(Vx) | \leq \| g \| \| x \|, \qquad (6.18)$$

By definition of the norm of a functional for any $\varepsilon > 0$ there exists such $x_0 \in X$ that

$$| (V^*g)x_0 | = | g(Vx_0) | > (\| g \| - \varepsilon) \| x_0 \| . \tag{6.19}$$

On the other hand, we have

$$| (V^*g)x | \leq \| V^*g \| \| x \| \quad \forall x \in X . \tag{6.20}$$

Comparing (6.19) with (6.20) yields $\| V^*g \| \geq \| g \|$. But from (6.18) we find $\| V^*g \| \leq \| g \|$. These inequalities prove that $\| V^*g \| = \| g \|$ $\forall g \in Y^*$. Now let f be any continuous linear functional on X, $f \in X^*$. Since the isometric operator V^{-1} exists by Theorem 6.10 it has the adjoint operator $(V^{-1})^*$ defined on the whole space X^*. Hence

$$fx = f[V^{-1}(Vx)] = [(V^{-1})^*f](Vx) = (V^*g)x , \quad \forall x \in X ,$$

where $g = (V^{-1})^*f$. Consequently, to any $f \in X^*$ corresponds such a functional $g \in Y^*$ that $f = V^*g$. This means that V^* is mapping the whole space Y^* onto the whole space X^*. The properties of the operator V^* established show that it is isometric. Finally, the relations $f = V^*g$ and $g = (V^{-1})^*f$ imply $(V^*)^{-1} = (V^{-1})^*$. ◁

E x a m p l e 6.12. The arguments of Subsection 5.1.3 show that the mapping of the second dual space X^{**} onto X defined by $\varphi_x f = fx$, $x \in X$, $f \in X^*$, $\varphi_x \in X^{**}$ represents the isometry if and only if X is a reflexive B–space.

E x a m p l e 6.13. Operator V transforming a continuous linear functional f on L_p, $1 \leq p < \infty$, to the element $g(x)$ of the space L_q, $p^{-1} + q^{-1} = 1$, is isometric because it establishes the one–to–one correspondence between the set of all continuous linear functionals on L_p and the set of all the elements of the space L_q retaining the norm.

6.1.7. Unitary Operators

The linear operator U mapping the whole B–space X onto itself and preserving the norm, $D_U = R_U = X$, $\| Ux \| = \| x \|$ $\forall x$, is called an *unitary operator*. In other words, the unitary operator represents the isometric operator mapping a B–space X onto itself.

From the general properties of isometric operators follows that an unitary operator is bounded, its norm being equal to unity, $\| U \| = 1$, and has the unitary inverse operator U^{-1} and the unitary adjoint operator U^* satisfying $(U^*)^{-1} = (U^{-1})^*$.

E x a m p l e 6.14. The operator U with an orthogonal matrix A in R^n is unitary since

$$\| Ax \|^2 = \sum_{p=1}^{n} (Ax)_p^2 = \sum_{p=1}^{n} \left(\sum_{q=1}^{n} a_{pq} x_q \right)^2$$

$$= \sum_{p,q,r=1}^{n} a_{pq} a_{pr} x_q x_r = \sum_{q,r=1}^{n} \left(\sum_{p=1}^{n} a_{pq} a_{pr} \right) x_q x_r$$

$$= \sum_{q,r=1}^{n} \delta_{qr} x_q x_r = \sum_{q=1}^{n} x_q^2 = \| x \|^2 .$$

See also Table A.1.3 (Appendix 1).

6.1.8. Unitarly Equivalent Operators

Let T_1 be the operator mapping a B–space X into X and T_2 the operator mapping a B–space Y into Y. The operators T_1 and T_2 are called *unitarly equivalent operators* if there exists such an isometric operator V mapping X onto Y that $x \in D_{T_1}$, $y \in D_{T_2}$ and $y = Vx$ imply $T_2 y = VT_1 x$. From this definition follows that the operators T_1 and T_2 are unitarly equivalent if and only if there exists such an isometric operator V mapping X onto Y that

$$T_2 = VT_1V^{-1}, \quad T_1 = V^{-1}T_2V, \quad VT_1 = T_2V . \qquad (6.21)$$

6.1.9. Banach–Steinhaus Theorem

Let us consider the set of continuous (and consequently, bounded by Theorem 4.56) operators $\{T_\alpha\}$ mapping a B–space X into a B–space Y defined on the whole space X.

Theorem 6.12. *If the set* $\{\| T_\alpha x \|\}$ *is bounded at any* $x \in X$ *then the set of norms* $\{\| T_\alpha \|\}$ *of operators* T_α *is bounded* (Banach–Steinhaus theorem).

▷ Consider the functional

$$\varphi(x) = \sup_\alpha \| T_\alpha x \| . \qquad (6.22)$$

From the definition of a supremum follows that for any $x \in X$, $\varepsilon > 0$ such an operator T_α may be found that

$$\varphi(x) - \| T_\alpha x \| < \varepsilon/2 . \qquad (6.23)$$

From the definition of a supremum follows that for any $x \in X$, $\varepsilon > 0$ such an operator T_α may be found that

$$\varphi(x) - \| T_\alpha x \| < \varepsilon/2 . \tag{6.23}$$

By continuity of T_α to any $\varepsilon > 0$ corresponds such $\delta > 0$ that

$$| \| T_\alpha x' \| - \| T_\alpha x \| | \leq \| T_\alpha x' - T_\alpha x \| < \varepsilon/2 \tag{6.24}$$

at all x', $\| x' - x \| < \delta$. Inequalities (6.23) and (6.24) and the definition (6.22) of $\varphi(x)$ imply

$$\varphi(x) - \varphi(x') \leq \varphi(x) - \| T_\alpha x \| + \| T_\alpha x \| - \| T_\alpha x' \| < \varepsilon ,$$

i.e.

$$\varphi(x') > \varphi(x) - \varepsilon \quad \forall x', \| x' - x \| < \delta . \tag{6.25}$$

This property of $\varphi(x)$ means that the *functional* $\varphi(x)$ *is semi–continuous from below*. Furthermore from the properties of the norm and the supremum follows that $\varphi(x)$ is a convex functional (Subsection 1.4.4) since

$$\varphi(x_1 + x_2) \leq \varphi(x_1) + \varphi(x_2) , \tag{6.26}$$

$$\varphi(\gamma x) = | \gamma | \varphi(x) \tag{6.27}$$

at any x_1, x_2, x, γ. Thus $\varphi(x)$ represents a convex semi–continuous from bellow functional.

To prove the theorem it is sufficient to show that the functional $\varphi(x)$ is bounded in the unit ball $S_1(0)$ (the ball whose centre is 0 and radius is equal to 1). In fact, if

$$\varphi(x) < c \quad \forall x, \| x \| < 1, \tag{6.28}$$

for some $c > 0$ then (6.27) yields at any $x \in X$, $\varepsilon > 0$

$$\varphi(x) = (\| x \| + \varepsilon) \, \varphi \left(\frac{x}{\| x \| + \varepsilon} \right) < c \, (\| x \| + \varepsilon) .$$

Hence

$$\varphi(x) = \sup_\alpha \| T_\alpha x \| < c_1 \| x \|, \quad c_1 > c ,$$

as $\varepsilon > 0$ is arbitrary. Thus $\| T_\alpha \| < c_1$ at all α and the theorem will be proved.

$\varphi(x) < a$ at any x, $\| x - x_0 \| < r$, and putting $z = (x - x_0)/r$ we get by (6.26) and (6.27)

$$\varphi(z) = \varphi\left(\frac{x - x_0}{r}\right) = \frac{1}{r}\varphi(x - x_0) \leq \frac{\varphi(x) + \varphi(x_0)}{r} < \frac{2a}{r}.$$

This means that $\varphi(z) < c = 2a/r$ at any z $\| z \| < 1$, i.e. $\varphi(x)$ is bounded in $S_1(0)$ if it is bounded in some ball $S_r(x_0)$.

Now we suppose that $\varphi(x)$ is unbounded in $S_1(0)$. Let $\{\eta_n\}$ be any increasing sequence of numbers, $\eta_n > 0$, $\eta_n \to \infty$ as $n \to \infty$, and x_1 such a point in $S_1(0)$ that $\varphi(x_1) > \eta_1$. By the semi–continuity of $\varphi(x)$ from below there exists for any $\varepsilon > 0$ such a vicinity $S_{\delta_1}(x_1)$ of the point x_1 that $\varphi(x) > \eta_1 - \varepsilon$ at all $x \in S_{\delta_1}(x_1)$. Taking an arbitrary $\delta > 0$ we may choose δ_1 in such a way that $\delta_1 < \delta/2$, $S_{\delta_1}(x_1) \subset S_1(0)$. Since $\varphi(x)$ cannot be bounded in $S_{\delta_1}(x_1)$ there exists such a point $x_2 \in S_{\delta_1}(x_1)$ that $\varphi(x_2) > \eta_2$ and such δ_2 that $\varphi(x) > \eta_2 - \varepsilon$ at all x in the ball $S_{\delta_2}(x_2)$. Here δ_2 may be chosen in such a way that $\delta_2 < \delta/2^2$, $S_{\delta_2}(x_2) \subset S_{\delta_1}(x_1)$. Continuing this procedure we obtain the sequence of points $\{x_n\} \subset S_1(0)$, $\varphi(x_n) > \eta_n$, and the sequence of balls $\{S_{\delta_n}(x_n)\}$ imbedded into one another, $S_{\delta_n}(x_n) \subset S_{\delta_{n-1}}(x_{n-1})$, $\delta_n < \delta/2^n$. The sequence $\{x_n\}$ is obviously a fundamental sequence because

$$\| x_n - x_m \| < \delta_m < \delta/2^m$$

at all m and $n > m$ since $x_n \in S_{\delta_m}(x_m)$. By completeness of the B–space X the sequence $\{x_n\}$ is convergent to some point x_0 and $x_0 \in S_{\delta_n}(x_n)$ at any n. Hence $\varphi(x_0) > \eta_n - \varepsilon$ $\forall n$, and $\varphi(x_0) = \infty$ because $\eta_n \to \infty$ as $n \to \infty$. But this conclusion is in contradiction with the assumption of the theorem that the sequence $\{T_\alpha x\}$ is bounded at any $x \in X$. This contradiction proves that $\varphi(x)$ is bounded in $S_1(0)$ completing thus the proof of the theorem. ◁

Corollary 1. *If the sequence $\{T_n x\}$ converges at any $x \in X$, $T_n x \to T x$ then the limiting operator T is a bounded linear operator.*

▷ The linearity of T follows immediately from the linearity of all the operators T_n. The convergence of the sequence $\{T_n x\}$ at any x implies the boundedness of this sequence. By the theorem proved the sequence of norms $\{\| T_n \|\}$ is bounded. Hence

$$\frac{\| T x \|}{\| x \|} = \lim \frac{\| T_n x \|}{\| x \|} \leq \sup_n \| T_n \| < c, \quad \text{i.e.} \quad \| T \| < c$$

for some $c > 0$. ◁

for some $c > 0$. ◁

Corollary 2. *Any weakly convergent sequence in a normed linear space is bounded.*

▷ To prove this statement it is sufficient to notice that any $x_n \in X$ generates the continuous linear functional $f x_n$ on the dual space X^* which represents a B–space by Theorem 5.2. The norm of this functional is equal to $\| x_n \|$ (Subsection 5.1.3). Therefore $\{f x_n\}$ represents the sequence of values of the functionals x_n at the point $f \in X^*$. By the weak convergence of $\{x_n\}$ to x the sequence $\{f x_n\}$ converges to $f x$ at all $f \in X^*$ and therefore is bounded at any $f \in X^*$. By the proved theorem the sequence of norms $\{\| x_n \|\}$ is bounded. ◁

Corollary 3. *The dual space X^* of a reflexive B–space X is weakly complete.*

▷ Let the sequence $\{f_n\} \subset X^*$ be a weak fundamental sequence. By Theorem 4.62 the sequence of numbers $\{f_n x\}$ is a fundamental sequence at any $x \in X$ and consequently, converges to some limit (depending on x) $f x = \lim f_n x$. By Corollary 1 f is a bounded linear functional, $f \in X^*$. Thus X^* contains such a functional f that $f_n x \to f x$ at any $x \in X$. By Theorem 4.61 this means that any weak fundamental sequence $\{f_n\} \subset X^*$ weakly converges to some limit $f \in X^*$ proving that the space X^* is weakly complete. ◁

6.1.10. Bounded Linear Operators in a Normed Space

Theorem 6.6 establishes the existence of the adjoint operator for any continuous operator T defined on a whole topological linear space X. Now we continue to study the operator T^* in the case of the normed linear spaces X, Y and a continuous operator T.

Theorem 6.13. *If X and Y are the normed linear spaces then the adjoint operator T^* of T is bounded and $\| T^* \| = \| T \|$.*

▷ Operator T being bounded we may write

$$|(T^* g) x| = |g T x| \leq \| g \| \| T \| \| x \| \ .$$

Hence

$$\| T^* g \| \leq \| g \| \| T \| \quad \text{and} \quad \| T^* \| \leq \| T \| \ . \tag{6.29}$$

On the other hand, Corollary of Theorem 4.58 implies the existence in Y^* of a functional g with the unit norm whose value is equal to $\| y_0 \|$

at a given $y_0 : \| g \| = 1$, $gy_0 = \| y_0 \|$. Putting $y_0 = Tx$ we obtain $gTx = \| Tx \|$. But $gTx = (T^*g)x$. Hence

$$\| Tx \| \leq \| T^* \| \| g \| \| x \| = \| T^* \| \| x \| .$$

yielding $\| T \| \leq \| T^* \|$. Together with (6.29) this inequality gives $\| T \| = \| T^* \|$. ◁

Theorem 6.14. *If X is a reflexive B-space and the domain of a bounded linear operator T does not coincide with X then T may be extended to the whole space X retaining its norm.*

▷ If $D_T \neq X$ then from the proof of Theorem 6.6 follows the existence in X^* of an infinite set of functionals satisfying (6.8) at a given $g \in T^*$:

$$fx = g(Tx) \quad \forall x \in D_T .$$

Nevertheless all these functionals coincide on D_T. Therefore taking someone of these functionals f at each $g \in Y^*$ we may define the adjoint operator T^* also in this case assuming $T^*g = f$. This operator is defined on the whole space Y^*. It is bounded by virtue of (6.29) and $\| T^* \| \leq \| T \|$. Consequently, there exists the adjoint operator $T^{**} = (T^*)^*$ of T^* defined on the whole space $X = X^{**}$. By Theorem 6.13 T^{**} is bounded and $\| T^{**} \| = \| T^* \|$. Formula (6.4) determining the adjoint operator takes in this case the form

$$g(T^{**}x) = (T^*g)x \quad \forall x \in X ,$$

because any continuous linear functionals φ on X^* and ψ on Y^* are defined in this case by $\varphi f = fx$, $\psi g = gy$ for some $x \in X$ и $y \in Y$. Comparing this formula with (6.4) we see that the operator T^{**} represents the extension of the operator T to the whole space X. Consequently, $\| T^{**} \| \geq \| T \|$. This inequality together with the inequality $\| T^{**} \| = \| T^* \| \leq \| T \|$ obtained before yields $\| T^{**} \| = \| T \|$. ◁

Thus the domain of a bounded linear operator in a reflexive B-space may always be assumed coinciding with the whole space.

Theorem 6.15. *If X and Y are the normed linear spaces, the operator T is defined on the whole space X and its adjoint operator T^* is defined on the whole space Y^* then the operator T is bounded.*

▷ Suppose that T is unbounded. Then such a sequence $\{x_n\}$ exists in X that $\| Tx_n \| > n \| x_n \| \forall n$. Putting $z_n = x_n / \| x_n \|$ we get $\| Tz_n \| > n$. On the other hand, $g(Tz_n) = (T^*g)z_n$, and consequently,

$| g(Tz_n) | \leq \| T^*g \| \| z_n \| = \| T^*g \| \; \forall g \in Y^*$. Thus the sequence of continuous linear functionals $\{Tz_n\}$ on Y^* is such that the sequence $\{| g(Tz_n) |\}$ is bounded at any $g \in Y^*$. By Theorem 5.2 the space Y^* is a B–space. Therefore by Banach–Steinhaus theorem 6.12 the sequence of norms $\{\| Tz_n \|\}$ of functionals Tz_n is bounded, i.e. $\| Tz_n \| < c$ for some $c < \infty$ at all n. The contradiction obtained proves the boundedness of the operator T. ◁

6.1.11. Banach Theorem about Inverse Operator

Banach theorem about the boundedness of inverse operator plays an important role in functional analysis. The proof of this theorem is based on some other theorems which will now be proved.

Theorem 6.16. *If a B–space represents a countable union of closed sets then at least one of these closed sets contains an open nonempty set (Baire theorem).*

▷ Suppose that the B–space X represents a countable union of the closed sets $\{F_n\}$:

$$X = \bigcup_{n=1}^{\infty} F_n , \qquad (6.30)$$

and none of the sets F_n contain a nonempty open set. If F_1 does not contain a nonempty open set then $F_1 \neq X$. Hence the open set $\bar{F}_1$ is nonempty. Let x_1 be any point of $\bar{F}_1$, $S_{r_1}(x_1)$ a ball vicinity of x_1 contained in $\bar{F}_1$, $S_1 = S_{r_1}(x_1) \subset \bar{F}_1$. If F_2 does not contain a nonempty open set then it cannot contain the ball $S_{r_1/2}(x_1)$. Hence the intersection $\bar{F}_2 \cap S_{r_1/2}(x_1)$ represents a nonempty open set. Let x_2 be any point of $\bar{F}_2 \cap S_{r_1/2}(x_1)$, $S_2 = S_{r_2}(x_2)$ be a ball vicinity of x_2 contained in $\bar{F}_2 \cap S_{r_1/2}(x_1)$. The radius r_2 may always be taken smaller than $r_1/2$, $r_2 < r_1/2$. Continuing in this way we get after $n - 1$ steps the points $x_1 , \ldots , x_{n-1}$ and the balls imbedded into one another $S_1 , \ldots , S_{n-1}$, $S_k \subset S_{k-1}$ with the centers $x_1 , \ldots , x_{n-1}$ and the radii $r_1 , \ldots , r_{n-1}$, $r_k < r_{k-1}/2$. As F_n does not contain a nonempty open set then it cannot contain the ball $S_{r_{n-1}/2}(x_{n-1})$. Therefore the intersection $\bar{F}_n \cap S_{r_{n-1}/2}(x_{n-1})$ represents a nonempty open set. Taking any point x_n of this set together with its ball vicinity $S_n = S_{r_n}(x_n) \subset \bar{F}_n \cap S_{r_{n-1}/2}(x_{n-1})$, $r_n < r_{n-1}/2$, we obtain n points $x_1 , \ldots , x_n$ and n balls imbedded into one another $S_1 , \ldots , S_n$. Continuing this procedure we get the sequence of points $\{x_n\}$ and the sequence of balls imbedded into one another $\{S_n\}$, $\| x_n - x_{n-1} \| < r_{n-1}/2$, at any natural p. Hence $\{x_n\}$ is a fundamental sequence and by completeness of the B–space X has the limit $x = \lim x_n$. Since

$$\| x - x_n \| \le \| x - x_m \| + \| x_m - x_n \| < \| x - x_m \| + r_n ,$$

at any n and $m > n$, passing to the limit as $m \to \infty$ yields $\| x - x_n \| < r_n$. Therefore $x \in S_n$ i.e. the limit point x is contained in all the balls S_n ($n = 1, 2, \ldots$). But $S_n \subset \bar{F}_n$ at any n. Hence the point x cannot belong to F_n at any n and (6.30) shows that $x \notin X$, what is impossible. This contradiction proves the theorem. ◁

Theorem 6.17. *If T is the bounded linear operator mapping the whole B–space X onto the whole B–space Y, $TX = Y$ then the image of any ball vicinity of zero in X contains a ball vicinity of zero in Y.*

▷ Let $S = S_r(0)$ be a ball vicinity of zero in the space X, $A = S/2 = S_{r/2}(0)$. Obviously, $x_1 - x_2 \in S$ for any two points $x_1, x_2 \in A$ since $\| x_1 - x_2 \| \le \| x_1 \| + \| x_2 \| < r/2 + r/2 = r$. Therefore $y_1 - y_2 \in TS$ for any two points $y_1, y_2 \in TA$.

Now we notice that any point $x \in X$ belongs to all the sets nA, corresponding to $n > 2 \| x \| /r$. Hence any point $y \in Y$ belongs to all the sets nTA at sufficiently large n, $y \in nTA \subset [nTA] = n[TA]$. Consequently,

$$Y = \bigcup_{n=1}^{\infty} [nTA] , \qquad (6.31)$$

i.e. the B–space Y represents a countable union of closed sets. By Theorem 6.16 at least one of the sets $[nTA] = n[TA]$ contains a non-empty open set. The mapping $z = y/n$ is the one–to–one and both mappings $z = y/n$ and $y = nz$ are continuous (i.e. the mapping $z = y/n$ represents a homeomorphism, Subsection 4.4.5). Hence the set $[TA]$, and consequently, the set TA contains a nonempty open set G. Let y be any point of the set G and $S_\rho(y)$ its ball vicinity contained in G. As the points y and $z + y$ belong to $S_\rho(y) \subset G \subset TA$ for any $z \in S_\rho(0)$, the point $z = (z + y) - y$ belongs to TS by proved above. But z is any point of $S_\rho(0)$. Consequently, $S_\rho(0) \subset TS$ and the theorem is proved. ◁

Theorem 6.18. *If T is a continuous linear mapping of the whole B–space X onto the B–space Y then the images of open sets are open* (Principle of a mapping openness).

▷ Let T be the bounded linear operator mapping the whole B–space X onto the whole B–space Y, G any open set in X, y any point of the image TG of the set G, $y \in TG$, $x \in G$ the point of G for which $y = Tx$,

▷ Let T be the bounded linear operator mapping the whole B-space X onto the whole B-space Y, G any open set in X, y any point of the image TG of the set G, $y \in TG$, $x \in G$ the point of G for which $y = Tx$, $S_r(x)$ a ball vicinity of the point x contained in G, $S_r(x) \subset G$. By Theorem 6.17 the image of the ball vicinity of zero $S_r(0)$ contains some ball vicinity of zero $S_\rho(0)$ of the space Y, $TS_r(0) \supset S_\rho(0)$. Hence taking into account that $S_r(x) = S_r(0) + x$, $S_\rho(y) = S_\rho(0) + y$, $y = Tx$, we obtain $S_\rho(y) \subset TS_r(x) \subset TG$. Thus any point of the set TG is contained in TG together with some its vicinity. This proves that TG is the open set. ◁

Theorem 6.19. *If T is the bounded linear operator mapping the whole B-space X onto the whole B-space Y and the inverse operator T^{-1} exists then this operator is bounded* (Banach theorem).

▷ By Theorem 6.18 the images of all open sets represent open sets. But the image B of the set A, $B = TA$ for the mapping T is the inverse image of the set A for the inverse mapping T^{-1}. Consequently, the inverse images of all open sets are open for the mapping T^{-1}. By Theorem 4.19 the mapping T^{-1} is continuous proving the boundedness of the operator T^{-1}. ◁

For detailed treatment of linear operators basic notions and theorems refer to (Dunford and Schwarz 1958, Edwards 1965, Yosida 1965). Some applications in discrete linear signal and image processing are given in (Bose 1988, Elliot 1987, Openheim and Schafer 1975, Papoulis 1968, Pratt 1978, Rabiner and Gold 1975).

Problems

6.1.1. Prove the theorem: the closed operator T mapping the whole B-space X into a B-space Y is continuous (Subsection 6.1.1).

I n s t r u c t i o n. The plot of the operator T represents a closed subspace V of the B-space $Z = X \times Y$, i.e. represents itself a B-space. There is the one-to-one mapping between the spaces X and V. The projecting operator P_x of V on X is continuous (Problem 4.3.9) (this is easily proved directly: the norm of the element $v = \{x, Tx\} \in V$ is equal to $\| x \| + \| Tx \|$ and the norm of its image $x = P_x v$ is equal to $\| x \|$). It remains to use Theorem 6.19.

6.1.2. Prove the existence of the adjoint operator A^* for the operator

$$Ax = \int\limits_T K(s, t)\, x(t)\, dx \,,$$

mapping $L_2(T)$ into itself, $T \subseteq R^n$, and find A^*.

in the space $L_2([\alpha, \beta])$ and find L^* $(\alpha > -\infty, \beta < \infty)$.

I n s t r u c t i o n. Recall Theorem 3.41 and Weirestrass theorem about approximation of a continuous function by a polynomial.

6.1.4. The same problem (Problem 6.1.3) for the operator

$$L = \sum_{k=1}^{n} a_k(z_1, \dots, z_n) \frac{\partial}{\partial z_k} + \frac{1}{2} \sum_{k,l=1}^{n} b_{kl}(z_1, \dots, z_n) \frac{\partial^2}{\partial z_k \partial z_l}$$

in the space $L_2(Z)$, Z being a compact in the space R^n.

6.1.5. Generalize the result of Problem 6.1.4 to the operator

$$L = a_0(z) + \sum_{k=1}^{N} \sum_{|r|=k} a_r(z) \frac{\partial^{|r|}}{\partial z_1^{r_1} \dots \partial z_n^{r_n}},$$

$z \in Z$ and r being the n-dimensional vectors $z = [z_1 \dots z_n]^T$, $r = [r_1 \dots r_n]^T$, $|r| = r_1 + \dots + r_n$.

6.1.6. Prove Banach–Steinhaus theorem 6.12 using Baire theorem 6.16.

I n s t r u c t i o n. Define the sets $F_n = \{x : \| T_\alpha x \| \le n \, \forall \alpha\}$. These sets are closed owing to the continuity of the operators T_α and $\bigcup_{n=1}^{\infty} F_n = X$. By Theorem 6.16 at least one of the sets F_n contains a nonempty open set G. Take any point $x_0 \in G$ and its ball vicinity $S_r(x_0) \subset G$ and prove that $\| T_\alpha x \| \le 2n \| x \| \, \forall \alpha$ for any point $x \in X$ yielding $\| T_\alpha \| < c$ for any $c > 2n$ for all α. This proof is much simpler than the proof in Subsection 6.1.9. The advantage of the proof given in Subsection 6.1.9 consists in its independence of Baire theorem.

6.1.7. Find the weighting and transfer functions for the adjoint linear systems given in Tables A.1.1 and A.1.2 (Appendix 1). Describe these functions.

6.1.8. Find the weighting and transfer functions for the adjoint stationary linear systems of Example 5.30.

6.1.9. Find the weighting and transfer functions for the adjoint stationary systems of Example 5.31.

6.1.10. Find the weighting and transfer functions for the adjoint systems presenting typical connections of stationary linear systems.

6.1.11. Find the weighting and transfer functions for the adjoint systems of Example 5.33.

6.1.12. Find the weighting and transfer functions of the adjoint system of Example 5.34.

6.2. Operator Equations

6.2.1. General Form of an Operator Equation

It is necessary for further exposition of the theory of linear operators to know how to prove the existence and uniqueness of solutions of some operator equations. We study here a general method for proving such theorems, i.e. the *method of contraction mappings*, which is often successfully used. This method is also called a *principle of contraction mappings*. The method is applicable to operator equations of the general form either linear or nonlinear ones.

Many equations encountered in a lot of mathematical problems may be written in the form

$$Ax = y \qquad (6.32)$$

or

$$Ax = x, \qquad (6.33)$$

A being some operator, nonlinear in general, mapping some space X into another space Y in the case of Eq. (6.32) and into the same space X in the case of Eq. (6.33).

But Eq. (6.32) may always be reduced to Eq. (6.33). For this purpose it is sufficient to take any operator B mapping one–to–one the whole space Y onto some linear space Z and any operator C mapping one–to–one Z onto X, put $x = Cz$ and write Eq. (6.32) as

$$BACz + z - By = z.$$

This is the equation of the form of Eq. (6.33) with the operator $Dz = BACz + z - By$ (depending on y). Therefore we shall study in the sequel only Eq. (6.33).

At first we consider some important problems leading to Eq. (6.33). To do this we need to define the derivatives of B–space–valued functions with respect to a real variable.

6.2.2. Derivatives of B–Space–Valued Functions with Respect to a Real Variable

The derivatives of B–space–valued functions of a real scalar variable may be defined in the usual way. Let $y = f(t)$ be the function of a real scalar variable t with values in the B–space Y. Its derivatives with respect to t are defined by the usual formulae

$$y' = f'(t) = \frac{dy}{dt} = \lim_{\Delta t \to 0} \frac{f(t + \Delta t) - f(t)}{\Delta t},$$

$$y^{(p)} = f^{(p)}(t) = \frac{d^p y}{dt^p} = \frac{d}{dt} f^{(p-1)}(t) \quad (p = 2, 3, \ldots). \tag{6.34}$$

Owing to the completeness of the B–space X the derivative $f'(t_0)$ of $f(t)$ at the point t_0 exists if and only if the sequence

$$f_n(t_0) = \frac{f(t_0 + h_n) - f(t_0)}{h_n} \quad (n = 1, 2, \ldots) \tag{6.35}$$

is a fundamental sequence for any sequence of real numbers $\{h_n\}$ convergent to zero. Certainly the convergence represents here the convergence in the metric of the space Y induced by the norm:

$$\left\| \frac{f(t_0 + \Delta t) - f(t_0)}{\Delta t} - f'(t_0) \right\| \to 0 \quad \text{as} \quad \Delta t \to 0, \tag{6.36}$$

$$\|f_n(t_0) - f_m(t_0)\| \to 0 \quad \text{as} \quad n, m \to \infty. \tag{6.37}$$

The function $f(t)$ is differentiable in an interval if it has the derivative $f'(t)$ at any point t of this interval.

It is evident that the derivatives of B–space–valued functions possess all the properties of derivatives of numerical functions:

(i) $\frac{d}{dt}[f(t) + c] = f'(t) \; \forall c \in Y$ independent of t;

(ii) $\frac{d}{dt}[f_1(t) + f_2(t)] = f_1'(t) + f_2'(t)$;

(iii) $\frac{d}{dt}[\alpha(t)f(t)] = \alpha'(t)f(t) + \alpha(t)f'(t)$ for any scalar function $\alpha(t)$;

(iv) $\frac{d}{dt}f(\varphi(t)) = f'(\varphi(t))\,\varphi'(t)$ for any real function $\varphi(t)$.

It is clear that the derivative of a B–space–valued function is an element of the same B–space.

E x a m p l e 6.15. Let $f(t)$ be the function of a real variable $t \in T$ with values in the space $C(X)$ of continuous functions of x defined on the set X. This means that $f(x)$ is a continuous function of x at any fixed $t \in T$:

$$f(t) = u(x, t), \quad x \in X.$$

The derivative of the function $f(t)$ with respect to t is defined by

$$f'(t) = \lim_{h_n \to 0} \frac{u(x, t + h_n) - u(x, t)}{h_n},$$

$\{h_n\}$ being the sequence of real numbers convergent to 0. In accordance with (6.36) the convergence represents the convergence in the metric of the space $C(X)$:

$$\sup_{x \in X} \left| \frac{u(x, t + h_n) - u(x, t)}{h_n} - f'(t) \right| \to 0,$$

i.e. as the uniform convergence with respect to x.

E x a m p l e 6.16. Let $f(t)$ be the function of a real variable $t \in T$ with values in the space $L_p(X, \mathcal{A}, \mu)$. The value of $f(t)$ at each t represents a function of x from $L_p(X, \mathcal{A}, \mu)$:

$$f(t) = u(x, t) \quad u(x, t) \in L_p(X, \mathcal{A}, \mu) \quad \forall t \in T.$$

The convergence in (6.36) and the fundamental property (6.37) represent in this case the convergence in the metric of the space L_p induced by the norm:

$$\left\| \frac{u(x, t + h_n) - u(x, t)}{h_n} - f'(t) \right\|_p$$

$$= \left\{ \int \left\| \frac{u(x, t + h_n) - u(x, t)}{h_n} - f'(t) \right\|^p \mu(dx) \right\}^{1/p} \to 0.$$

6.2.3. Integrals of B–Space–Valued Functions by a Real Variable

The Bochner and the Riemann integrals of B–space–valued functions by the Lebesgue measure were considered in Subsection 3.4.6. Consider the Riemann integral by the Lebesgue measure on the real axis R with the variable upper limit:

$$F(t) = \int_a^t f(\tau) \, d\tau. \tag{6.38}$$

Theorem 6.20 *If the function $f(\tau)$ is continuous at the point t then $f(t)$ represents the derivative of $F(t)$ with respect to t.*

▷ Let $\{h_n\}$ be any sequence of real numbers convergent to 0, $I_n = [t, t + h_n)$ at $h_n > 0$ and $I_n = (t + h_n, t]$ at $h_n < 0$. We have

$$\frac{F(t + h_n) - F(t)}{h_n} - f(t) = \frac{1}{|h_n|} \int_{I_n} f(\tau) \, d(\tau) - f(t)$$

$$= \frac{1}{|h_n|} \int_{I_n} [f(\tau) - f(t)] \, d\tau .$$

By continuity of $f(\tau)$ at the point t there exists for any $\varepsilon > 0$ such $\delta = \delta_\varepsilon > 0$ that

$$\| f(\tau) - f(t) \| < \varepsilon \quad \forall \tau, \quad | \tau - t | < \delta .$$

Therefore

$$\left\| \frac{F(t + h_n) - F(t)}{h_n} - f(t) \right\| < \varepsilon \quad \forall h_n, |h_n| < \delta .$$

This implies

$$\lim_{n \to \infty} \left\| \frac{F(t + h_n) - F(t)}{h_n} - f(t) \right\| = 0 ,$$

i.e.

$$f(t) = \lim_{n \to \infty} \frac{F(t + h_n) - F(t)}{h_n} = F'(t) . \quad ◁$$

6.2.4. Differential Equations in B–Spaces

A lot of problems of modern mathematics and its applications lead to the equations of the form

$$y' = f(t, y) , \tag{6.39}$$

where y is an element of some B–space Y and $f(t, y)$ the function with values in the B–space Y with the domain D_f in the product space $R \times Y$.

As the derivative y' is invariant while adding to the function $y = \varphi(t)$ an arbitrary constant element of the space Y then Eq. (6.39) cannot have the unique solution. To obtain the unique solution of Eq. (6.39) it is necessary to assign the initial value y_0 of y at $t = t_0$, $\{t_0, y_0\} \in D_f$. As a result we obtain the Cauchy problem

$$y' = f(t, y) , \quad y = y_0 \quad \text{at} \quad t = t_0 \tag{6.40}$$

in the same way as in the theory of ordinary differential equations in the finite–dimensional spaces.

We call the *solution of differential equation* (6.40) satisfying the given initial condition the function $y = \varphi(t)$, $\varphi : R \to Y$ possessing the following properties:

(i) $\varphi(t)$ is a continuous function of t, $\varphi(t) \in D_f(t)$ at any $t \in T$ [a];

(ii) $\varphi(t)$ has the derivative $\varphi'(t)$ at all $t \in T$;

(iii) $\varphi'(t) = f(t, \varphi(t))$ for each $t \in T$;

(iv) $\varphi(t_0) = y_0$.

From Eq. (6.40) the integral equation

$$y(t) = y_0 + \int_{t_0}^{t} f(\tau, y(\tau))\, d\tau. \tag{6.41}$$

follows by integration. This equation has the form of Eq. (6.33). Here $x = y(t)$ is a continuous function with the values in the B–space Y. The right–hand side of Eq. (6.41) represents the image of x determined by the respective operator.

Since the function $f(t, y)$ maps the B–space Y into itself at any fixed t it represents an operator in the general case. If this operator is linear $f(t, y) = A(t)y$ then Eq. (6.40) becomes

$$\frac{dy}{dt} = A(t)y, \quad y = y_0 \quad \text{at} \quad t = t_0. \tag{6.42}$$

This is the *linear differential equation* in the B–space Y.

E x a m p l e 6.17. If $Y = K^n$ ($K = R$ or C as always) then Eq. (6.40) represents the set of ordinary differential equations

$$y_k' = f_k(t, y_1, \ldots, y_n), \quad y_k = y_{k0} \quad \text{at} \quad t = t_0 \quad (k = 1, \ldots, n). \tag{6.43}$$

It is well known that the differential equation of the n–th order solved with respect to the n–th derivative

$$y^{(n)} = f(t, y, y', \ldots, y^{(n-1)}),$$

$$y = y_0, \quad y' = y_0', \ldots, y^{(n-1)} = y_0^{(n-1)} \quad \text{at} \quad t = t_0 \tag{6.44}$$

---

[a] By $D_f(t)$ we denote the section of the domain of the function f at the point t. Here T represents the set of all t for which $D_f(t)$ is nonempty.

may also be reduced to the set of equations (6.43).

E x a m p l e 6.18. Consider the *heat conduction equation*

$$\frac{\partial y}{\partial t} = k^2 \frac{\partial^2 y}{\partial x^2} . \tag{6.45}$$

Here $y = \varphi(t, x)$ represents a continuous function of x at any t, i.e. an element of the B-space $C(X)$, X being the interval $[a, b] \in R$. Eq. (6.45) may be written as

$$y' = Ay , \tag{6.46}$$

A being the operator of double differentation with respect to the variable x,

$$A = k^2 \frac{\partial^2}{\partial x^2} .$$

The domain of A represents the set of all the functions on $X = [a, b]$ continuous with their derivatives of the first and second orders and satisfying given boundary conditions, say $y(t, a) = \psi_1(t)$, $y(t, b) = \psi_2(t)$. The initial condition is

$$y = y_0(x) \quad \text{at} \quad t = t_0 .$$

E x a m p l e 6.19. The *heat conduction equation* in R^3

$$\frac{\partial y}{\partial t} = k^2 \left(\frac{\partial^2 y}{\partial x_1^2} + \frac{\partial^2 y}{\partial x_2^2} + \frac{\partial^2 y}{\partial x_3^2} \right)$$

may also be written in the form of (6.46), A being the operator

$$A = k^2 \left(\frac{\partial^2}{\partial x_1^2} + \frac{\partial^2}{\partial x_2^2} + \frac{\partial^2}{\partial x_3^2} \right) .$$

The variable y represents at each t a continuous function of the vector $x = \{x_1, x_2, x_3\}$ with the domain X, i.e. the element of the B-space $C(X)$. The domain X is defined by the inequality of the form $\Phi(x) \geq 0$. The domain of the operator A represents the set of all the functions of the vector x continuous with their first and second derivatives defined at $\Phi(x) \geq 0$ and satisfying some boundary condition, for instance,

$$y = \psi(t) \quad \text{at} \quad \Phi(x) = \Phi(x_1, x_2, x_3) = 0 .$$

The initial condition has the form

$$y = y_0(x) = y_0(x_1, x_2, x_3) \quad \text{at} \quad t = t_0 .$$

E x a m p l e   6.20. In the same way the *wave equation*

$$\frac{\partial^2 u}{\partial t^2} = k^2 \left(\frac{\partial^2 u}{\partial x_1^2} + \frac{\partial^2 u}{\partial x_2^2} + \frac{\partial^2 u}{\partial x_3^2} \right) \tag{6.47}$$

may be written in the form of Eq. (6.46). Putting in R^3

$$y = [\, u_1 \, u_2 \,]^T \,, \quad u_1 = u \,, \quad u_2 = \frac{\partial u}{\partial t} \,,$$

we see that Eq. (6.47) is replaced by the set of two equations

$$u_1' = u_2 \,, \quad u_2' = A u_1 \,,$$

where A is the same operator as in Example 6.19. This set of equations may be considered as the equation of the form of (6.46):

$$y' = \begin{bmatrix} 0 & I \\ A & 0 \end{bmatrix} y \,.$$

E x a m p l e   6.21. The *integro–differential equation*

$$\frac{\partial y}{\partial t} = f(t, x, y) + \int\limits_X g(t, x, \xi, y(t, \xi)) \, d\xi \,, \tag{6.48}$$

where $x,\, \xi \in R^n$, $y \in R^m$, $f(t, x, y) \in R^m$ $\forall t, x, y$, $g(t, x, \xi, y) \in R^m$ $\forall t, x, \xi, y$, may also be written as Eq. (6.46), y being a continuous m–dimensional vector function of n–dimensional vector variable x with the domain $X \subset R^n$ at each t.

E x a m p l e   6.22. In probability theory and its numerous applications differential equations with random functions play an important role. A *random function* is by definition such a function whose value at any value of the argument represents a random variable, i.e. a measurable function of the elementary event ω, $X(t) = x(t, \omega)$, $\omega \in \Omega$. This function of the variable ω is often an element of the H–space $L_2(\Omega, \mathcal{S}, P, R^n)$ of n–dimensional vector (scalar at $n = 1$) functions with integrable by measure P square of moduli

$$\int | x(t, \omega) |^2 \, P(d\omega) < \infty \,.$$

The measure P represents the probability defined on the σ–algebra $\mathcal{S}$ of sets of the space of elementary events Ω.

The differential equation determining the random function $Y(t) = y(t, \omega)$, $y(t, \omega) \in R^n$ at fixed t, ω, has generally the form

$$Y' = F(t, Y), \quad Y = Y_0 \quad \text{at} \quad t = t_0 \,. \tag{6.49}$$

$F(t, y) = f(t, y, \omega)$ being a random function of the variables t, y, F : R $\times L_2(\Omega, \mathcal{S}, P, R^n) \rightarrow L_2(\Omega, \mathcal{S}, P, R^n)$. At fixed t the function $F(t, y)$ represents the operator mapping the space $L_2(\Omega, \mathcal{S}, P, R^n)$ into itself. Eq. (6.49) represents obviously an equation of the form of (6.40).

The examples considered show that Eq. (6.40) and its special case (6.42) cover a variety of equations encountered in mathematics and its applications. Therefore the theory of differential equations in abstract spaces (in particular, in B–spaces) play an important role in modern mathematics. The elements of this theory will be outlined in the sequel (Section 9.3).

6.2.5. Integral Equations

Many problems of mathematics and its applications lead to integral equations. The integral equations in a B–space have in the general case the form

$$\int_B f(x, \xi, y(\xi)) \, \mu(d\xi) + \varphi(x) = y(x). \qquad (6.50)$$

The variable x here may have the values in any space X, $y(x)$, $\varphi(x)$ are the functions with the values in a B–space Y, $f(x, \xi, y)$ the function mapping $X \times X \times Y$ into Y, μ any σ–finite numerical measure and the integral represents the Bochner integral. It is clear that Eq. (6.50) is an equation of the form of (6.33).

Eq. (6.41) which is equivalent to Cauchy problem (6.40) represents the special case of Eq. (6.50) where x is a real variable t, the function $f(t, \tau, y)$ is equal to zero at all $t < \tau$ and at $\tau < t_0$ and is independent of t at $t > \tau$, the function $\varphi(t)$ is a constant and μ is the Lebesgue measure on the real axis R.

E x a m p l e 6.23. The *linear Fredholm integral equation*

$$\int_B K(t, s) \, x(s) \, ds - \lambda x(t) = \varphi(t) \qquad (6.51)$$

where $x(t)$ and $\varphi(t)$ are scalar or finite–dimensional vector functions of the scalar or finite–dimensional vector variable, $K(t, s)$ the scalar or matrix–valued function of two variables respectively, λ a complex parameter. The functions x, φ, K may be either real or complex (we call matrix, in particular, a matrix–column a real matrix if all its elements are real numbers, and a complex matrix if some of its elements are complex numbers). The functions x, φ, K are assumed to be given.

E x a m p l e 6.24. The *linear Volterra integral equation*

$$\int_0^t K(t,s)\, x(s)\, ds - \lambda x(t) = \varphi(t) \tag{6.52}$$

represents a special case of Fredholm equation (6.51) where t (and accordingly s) is a real variable and $K(t,s) = 0$ at $s > t$.

E x a m p l e 6.25. The *nonlinear Hammerstein integral equation*

$$\int_B K(t,s)\, f(s, x(s))\, ds + \varphi(t) = x(t) \tag{6.53}$$

where $f(s,x)$ being a given function is also an equation of the form of (6.33). The integral operator on the left–hand side of Eq. (6.53) is called a *Hammerstein operator*.

6.2.6. Contraction Mappings

Let X be a complete metric space (in particular, a B–space), A being the operator mapping the space X into itself, $A : X \to X$. The operator A is called a *contraction operator* (a *contraction mapping*) if

$$d(Ax, Ay) < \alpha\, d(x, y) \tag{6.54}$$

for any elements x, y of the space X, α being a positive number smaller than unity, $\alpha \in (0,1)$. It is clear that the contraction operator is always continuous.

The point x is called a *fixed point* of the mapping A if $Ax = x$. The fixed point of the mapping A represents evidently the solution of the equation

$$x = Ax. \tag{6.55}$$

The *powers of the operator* A are defined as usual by the recursive formula

$$A^p x = A(A^{p-1}x) \quad (p = 2, 3, \ldots). \tag{6.56}$$

6.2.7. Existence of the Unique Fixed Point of an Operator

The existence and uniqueness of the solution of Eq. (6.55) is established by the following theorem.

Theorem 6.21. *If the operator A^p is a contraction operator at some natural p then the operator A has the unique fixed point.*

▷ Let x_0 be any point of the space X. Define the sequence of points $\{x_n\}$:

$$x_{n+1} = A^p x_n \quad (n = 0, 1, 2, \ldots). \tag{6.57}$$

As A^p is a contaction operator

$$d(A^p x, A^p y) < \alpha d(x, y), \quad \alpha \in (0, 1),$$

at all $x, y \in X$. Putting $x = x_{k-1}$, $y = x_k$ we have

$$d(x_k, x_{k+1}) < \alpha\, d(x_{k-1}, x_k) \quad (k = 1, 2, \ldots).$$

Hence

$$d(x_k, x_{k+1}) < \alpha^k\, d(x_0, x_1)$$

and

$$d(x_n, x_m) \le \sum_{k=n}^{m-1} d(x_k, x_{k+1}) < \sum_{k=n}^{m-1} \alpha^k\, d(x_0, x_1)$$

$$< \sum_{k=n}^{\infty} \alpha^k\, d(x_0, x_1) = \frac{\alpha^n}{1-\alpha} d(x_0, x_1)$$

at all n and $m > n$. Thus

$$d(x_n, x_m) \to 0 \quad \text{at} \quad n, m \to \infty,$$

proving that $\{x_n\}$ is a fundamental sequence. Since the space X is complete there exists the limit $x = \lim_{n \to \infty} x_n \in X$ and by continuity of A^p we have $A^p x = \lim_{n \to \infty} A^p x_n$. Consequently,

$$d(x, A^p x) = \lim_{n \to \infty} d(x_n, A^p x_n) = \lim_{n \to \infty} d(x_n, x_{n+1}) = 0.$$

Thus x is the fixed point of the operator A^p. Hence

$$d(x, Ax) = d(A^p x, A^{p+1} x) < \alpha d(x, Ax).$$

Taking into account that $\alpha < 1$ this gives $d(x, Ax) = 0$ proving that x is the fixed point of the operator A.

To prove the uniqueness of the fixed point of the operator A we notice that any fixed point of the operator A is also the fixed point of the operator A^p. Hence the assumption that two fixed points x, y of the operator A exist leads to contradiction $d(A^p x, A^p y) = d(x, y)$. ◁

The proved theorem underlies the method of contraction mappings representing a powerful tool for proving the existence of solutions for variety of equations.

Theorem 6.21 is also valid if the operator A^p is a contraction operator only on some closed set $D \neq X$ if the operator A^p maps D into D, $A^p D \subset D$.

The process of successive approximations to the fixed point used while proving Theorem 6.21 may in some cases be applied to find the approximate solutions of equations in problems of practice.

Theorem 6.21 establishes the existence of the fixed point of the operator A in the case of a contraction operator A^p. However the fixed point may exist also in the case where A^p is at none p a contraction operator. If, for instance, the operator A^p is a *dilatation operator* at some p

$$d(A^p x, A^p y) > \beta d(x, y), \quad \beta > 1, \tag{6.58}$$

and the inverse operator A^{-1} exists then the operator A^{-p} will be a contraction operator since putting $u = A^p x$, $v = A^p y$ we obtain from (6.58)

$$d(A^{-p} u, A^{-p} v) < d(u, v)/\beta.$$

In the same way as in Theorem 6.21 the fixed point of the operator A exists and is unique in this case if A^p is a dilatation operator only on some set C contained in the closure of its image: $C \subset [A^p C]$. However the process of successive approximations of Theorem 6.21 is divergent in the case of a dilatation operator A^p (the fixed point is unstable). Using the inverse operator gives the convergent process of successive approximations. This approach is widely used in practice (certainly if the inverse operator may easily be found). In this case the existence of the inverse operator A^{-1} is sufficient only in some region $C \subset X$ in which A^{-p} is a contraction operator and $C \subset [A^p C]$.

E x a m p l e   6.26. Consider the equation

$$x = f(x), \tag{6.59}$$

where x is a real scalar variable. The solution of Eq. (6.59) is evidently the absciss of the intersection point of the plot $y = f(x)$ of the function f and the bisector of the coordinate angle $y = x$ (Fig. 6.1 and Fig. 6.2). It is intuitively clear that the sequence $\{x_n\}$ $x_1 = f(x_0), \ldots, x_n = f(x_{n-1}), \ldots$ will converge to the point x if

$$| f(\xi) - f(\eta) | < \alpha | \xi - \eta |, \tag{6.60}$$

at any ξ, η where $\alpha \in (0, 1)$. Approximation of x_n to the fixed point x of the operator $Ax = f(x)$ may be illustrated as the mouvement along the "staircase" on Fig.6.1, where the case of an increasing function f is shown, or along the rectangular spiral on Fig.6.2 where the case of a decreasing function f is shown.

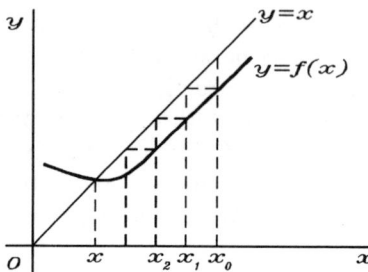   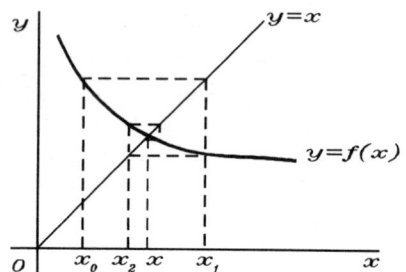

Fig. 6.1   Fig. 6.2

The equation $F(x) = 0$ may be reduced to Eq. (6.59) by putting $f(x) = x + kF(x)$. Condition (6.60) under which the operator $Ax = f(x)$ is a contraction operator becomes in this case

$$| \xi + kF(\xi) - \eta - kF(\eta) | < \alpha | \xi - \eta |$$

or

$$\left| 1 + k \frac{F(\xi) - F(\eta)}{\xi - \eta} \right| < \alpha$$

at all ξ, η in some sufficiently large vicinity of the fixed point x.

The results of this example may easily be generalized to the case of a finite-dimensional vector variable x.

The principle of contraction mapping gives in this case not only the proof of existence of solutions of equations $x = f(x)$ and $F(x) = 0$ but also the practical method for finding approximate solutions of these equations. This method is usually called a *successive approximations method* or an *iterations method*. The iterations process is finished in practical problems when two successive approximations x_n and x_{n+1} coincide within the adopted accuracy of calculations (after that all further iterations will coincide with x_n).

E x a m p l e 6.27. The results of Example 6.26 may be used to prove the existence of the function $x = \varphi(t)$ defined implicitly by the equation

$$F(t, x) = 0$$

at $t \in [t_1, t_2]$, $x \in [a, b]$. Using the iterations method we prove the existence of the solution x at each t, representing the function $x = \varphi(t)$ with the domain $D_\varphi = [t_1, t_2]$ and the range $R_\varphi \subset [a, b]$.

6.2.8. Existence of the Unique Solution of an Integral Equation

Now we apply the contraction mappings method for proving the existence and uniqueness of the solution of the integral equation (6.50) in the B–space Y in the case where $x = t$ is a real variable, μ the Lebesgue measure on the real axis R and the region of integration B represents the interval $[t_0, t]$

$$y(t) = \varphi(t) + \int_{t_0}^{t} f(t, \tau, y(\tau)) d\tau . \qquad (6.61)$$

Here $\varphi(t)$ is the function of the real variable t with values in the B–space Y, $f(t, \tau, y)$ the function mapping the product space $R \times R \times Y$ into Y, i.e. the operator in the space Y depending on two real variables t, τ.

Theorem 6.22. *If there exist such positive numbers a, b, c, M, $aM < b$ that*
(i) $\| f(t, \tau, y) \| < M$ *at* $| t - t_0 | < a$, $| \tau - t_0 | < a$, $\| y - \varphi(t) \| < b$,
(ii) $\| f(t, \tau, y_1) - f(t, \tau, y_2) \| < c \| y_1 - y_2 \|$ *at* $| t - t_0 | < a$,
$| \tau - t_0 | < a$, $\| y_1 - \varphi(t) \| < b$, $\| y_2 - \varphi(t) \| < b$,
then Eq. (6.61) has the unique solution $y(t)$ representing the continuous function of t in the interval $| t - t_0 | < a$, satisfying the condition

$$\| y(t) - \varphi(t) \| < b \quad at \quad | t - t_0 | < a .$$

▷ Define the operator

$$Au(t) = \varphi(t) + \int_{t_0}^{t} f(t, \tau, u(\tau)) d\tau , \qquad (6.62)$$

mapping the space $C([t_0 - a, t_0 + a], Y)$ of continuous functions $u(t)$ of the variable t in the interval $| t - t_0 | \le a$ with values in the B–space Y into itself.

We shall prove that the operator A^p is a contraction operator at sufficiently large natural p. For this purpose we notice that condition (i) and inequalities $\| u_1(t) - \varphi(t) \| < b$, $\| u_2(t) - \varphi(t) \| < b$ imply

$$\| f(t, \tau, u_1(\tau)) \| , \ \| f(t, \tau, u_2(\tau)) \| < M$$

yielding

$$\| Au_i(t) - \varphi(t) \| < M|t - t_0| < Ma < b \quad (i = 1, 2),$$

and by condition (ii)

$$\| f(t, \tau, Au_1(\tau)) - f(t, \tau, Au_2(\tau)) \| < c \| Au_1(\tau) - Au_2(\tau) \| . \quad (6.63)$$

Thus the iterative action of the operator A does not lead out of the region where conditions (i) and (ii) are fulfilled. From condition (ii) and Eq. (6.62) follows

$$\| Au_1(t) - Au_2(t) \| < c \left| \int_{t_0}^{t} \| u_1(\tau) - u_2(\tau) \| \, d\tau \right|$$

$$\leq c \sup_{|t-t_0| \leq a} \| u_1(t) - u_2(t) \| |t - t_0| = c \| u_1 - u_2 \|_C |t - t_0|, \quad (6.64)$$

where $\| x \|_C$ is the norm of $x(t)$ in the space $C([t_0 - a, t_0 + a], Y)$. Then we find from (6.62)–(6.64)

$$\| A^2 u_1(t) - A^2 u_2(t) \| < c \left| \int_{t_0}^{t} \| Au_1(\tau) - Au_2(\tau) \| \, d\tau \right|$$

$$< \| u_1 - u_2 \|_C \left| \int_{t_0}^{t} |\tau - t_0| \, d\tau \right| = c^2 \| u_1 - u_2 \|_C |t - t_0|^2 / 2$$

at all t, $|t - t_0| \leq a$. Continuing in such a way we come to the inequality

$$\| A^p u_1(t) - A^p u_2(t) \| < c^p \| u_1 - u_2 \|_C |t - t_0|^p / p!,$$

giving

$$\| A^p u_1 - A^p u_2 \|_C < c^p \| u_1 - u_2 \|_C a^p / p!. \quad (6.65)$$

Taking a sufficiently large p we shall have $c^p a^p / p! < \alpha < 1$ and

$$\| A^p u_1 - A^p u_2 \|_C < \alpha \| u_1 - u_2 \|_C . \quad (6.66)$$

The operator A^p is thus a contraction operator. By Theorem 6.21 the equation $y = Ay$ has the unique solution representing a continuous function of t in the interval $|t - t_0| < a$ with values in the B–space Y, satisfying the condition $\| y(t) - \varphi(t) \| < b$ at all t, $|t - t_0| < a$. ◁

If condition (ii) of Theorem 6.22 which is usually called the *Lipschitz condition* is fulfilled at all y_1, $y_2 \in Y$ then condition (i) may be rejected. Thus we come to the theorem.

Theorem 6.23. *If there exist such positive numbers a and c that*

$$\| f(t, \tau, y_1) - f(t, \tau, y_2) \| < c \, \| y_1 - y_2 \| \quad at \quad |t - t_0|, \ |\tau - t_0| < a$$

and all y_1, $y_2 \in Y$ then Eq. (6.61) has the unique solution $y(t)$ representing a continuous function of t in the interval $|t - t_0| < a$.

▷ From the proof of Theorem 6.22 follows that the operator A^p represents in this case a contraction operator at any u_1 and u_2. ◁

6.2.9. Existence of the Unique Solution of a Differential Equation

The differential equation in a B–space with the initial condition Cauchy problem (6.40) is equivalent to the integral equation (6.41). Hence Theorems 6.22 and 6.23 are applicable to Eq. (6.40). The corresponding theorems for Eq. (6.40) are as follows.

Theorem 6.24. *If there exist such positive numbers a, b, c, M, $aM < b$ that*
(i) $\| f(t, y) \| < M$ *at* $|t - t_0| < a$, $\| y - y_0 \| < b$,
(ii) $\| f(t, y_1) - f(t, y_2) \| < c \, \| y_1 - y_2 \|$ *at* $|t - t_0| < a$, $\| y_1 - y_0 \| < b$, $\| y_2 - y_0 \| < b$,
then Eq. (6.41), and consequently, the Cauchy problem has the unique solution $y(t)$ representing a continuous function of t in the interval $|t - t_0| < a$ satisfying the condition $\| y(t) - y_0 \| < b$.

Theorem 6.25. *If there exist such positive numbers a and c that*

$$\| f(t, y_1) - f(t, y_2) \| < c \, \| y_1 - y_2 \|$$

at $|t - t_0| < a$ at all y_1, $y_2 \in Y$, then Eq. (6.41), and consequently, Cauchy problem (6.40) has the unique solution $y(t)$ representing a continuous function of t in the interval $|t - t_0| < a$.

E x a m p l e 6.28. Theorems 6.24 and 6.25 establish the sufficient conditions of existence of the unique solution of the ordinary differential equations of Example 6.17.

E x a m p l e 6.29. By Theorem 6.24 the integro–differential equation of Example 6.21 has the unique solution representing a continuous function of t in the interval $[t_0, t_0 + a]$ with values in the space $C(X)$ of m–dimensional vector functions of x if the following conditions are fulfilled:

1_1) $\displaystyle\sup_{x \in X} \left| f(t, x, y(x)) + \int_X g(t, x, \xi, y(\xi))d\xi \right| < M$ at $0 < t - t_0 < a$,

$\displaystyle\sup_{x \in X} | y(x) - y_0(x) | < b$,

2_1) $\displaystyle\sup_{x \in X} \left| f(t, x, y_1(x)) - f(t, x, y_2(x)) + \int_X [g(t, x, \xi, y_1(\xi)) \right.$

$\left. - g(t, x, \xi, y_2(\xi))] d\xi \right| < c \sup_{x \in X} | y_1(x) - y_2(x) |$ at $| t - t_0 | < a$,

$\displaystyle\sup_{x \in X} | y_1(x) - y_0(x) |$, $\sup_{x \in X} | y_2(x) - y_0(x) | < b$.

By Theorem 6.25 the integro–differential equation of Example 6.21 has the unique solution if condition 2_1 is fulfilled at $0 < t - t_0 < a$ at all $y_1(x)$ and $y_2(x)$.

E x a m p l e 6.30. The equation with random functions of Example 6.22 has the unique solution representing the continuous function of t with values in the space $L_2(\Omega, \mathcal{S}, P, R^n)$ if the following conditions are fulfilled:

1_2) $\int \| f(t, y(\omega), \omega) \|^2 P(d\omega) < M^2$ at $0 < t - t_0 < a$,

$\int \| y(\omega) - y_0(\omega) \|^2 P(d\omega) < b^2$, $aM < b$,

2_2) $\int \| f(t, y_1(\omega), \omega) - f(t, y_2(\omega), \omega) \|^2 P(d\omega)$

$< c^2 \int \| y_1(\omega) - y_2(\omega) \|^2 P(d\omega)$ at $0 < t - t_0 < a$,

$\int \| y_1(\omega) - y_0(\omega) \|^2 P(d\omega) < b^2$, $\int \| y_2(\omega) - y_0(\omega) \|^2 P(d\omega) < b^2$.

Condition 1_2 may be rejected if condition 2_2 is fulfilled at $0 < t - t_0 < a$ at all $y_1(\omega)$ and $y_2(\omega)$.

R e m a r k. Unlike Examples 6.17, 6.21 and 6.22 Theorems 6.24 and 6.25 do not establish the existence and uniqueness of solutions of the partial differential equations of Examples 6.18, 6.19 and 6.20. The reason is that the differential operator A is unbounded and does not satisfy the Lipschitz condition under which Theorems 6.24 and 6.25 were proved. To prove the existence and uniqueness of solutions theorems for a wider class of functions $f(t, y)$ other methods are used in the theory of differential equations in finite–dimensional and abstract spaces. Such methods are outlined in Section 9.3 for equations in B–spaces.

6.2.10. Existence of the Unique Solution of a Volterra Linear Integral Equation

Linear integral equations play an important role in modern mathematics, mathematical physics and engineering sciences.

The linear integral equation with the variable upper limit:

$$\int_{t_0}^{t} K(t,\tau)\, y(\tau)\, d\tau - \lambda\, y(t) = \varphi(t) \tag{6.67}$$

is called a *Volterra equation* (Example 6.24). The function $K(t,\tau)$ is called the *kernel* of the integral operator on the left–hand side of Eq. (6.67).

Eq. (6.67) takes the form of Eq. (6.61) with $f(t,\tau,y) = K(t,\tau)y/\lambda$ after dividing by λ. It is clear that the condition of Theorem 6.23 is satisfied in this case at any $\lambda \neq 0$ if the kernel $K(t,\tau)$ is bounded on the square $|\,t - t_0\,|,\,|\,\tau - t_0\,| < a,\,|\,K(t,\tau)\,| < C$ since

$$|\,K(t,\tau)\,y_1(\tau)/\lambda - K(t,\tau)\,y_2(\tau)/\lambda\,| < C\,|\,y_1(\tau) - y_2(\tau)\,|\,/\,|\,\lambda\,|\,.$$

Thus the following theorem is valid.

Theorem 6.26. *Volterra linear integral equation* (6.67) *has the unique solution at any value of the parameter* $\lambda \neq 0$ *if the function* $\varphi(t)$ *is continuous and* $|\,K(t,\tau)\,| < C$ *at* $0 < t - t_0,\,\tau - t_0 < a$.

R e m a r k. This theorem is easily generalized to the case of functions $y(t)$ and $\varphi(t)$ with values in R^n. The condition $|\,K(t,\tau)\,| < C$ is replaced in this case by similar condition $|\,K_{pq}(t,\tau)\,| < C$ ($p, q = 1, \ldots, n$).

6.2.11. Existence of the Unique Solution of a Fredholm Linear Integral Equation

Another important class of integral equations encountered in mathematical physics and engineering is a Fredholm linear integral equation (Example 6.23)

$$\int_{X} K(x,\xi)\, y(\xi)\, d\xi - \lambda y(x) = \varphi(x)\,, \quad x \in X\,, \tag{6.68}$$

where x, ξ are vector variables x, $\xi \in R^q$ and X some regions in the space R^q. Volterra equation (6.67) may certainly be considered as a special case of Fredholm equation (6.68) where $q = 1$, $X = [t_0, t_0 + a]$ and $K(t, \tau) = 0$ at $\tau > t$. But unlike the Volterra equation the Fredholm equation has a solution not at all values of the parameter λ.

Theorem 6.27. *Linear Fredholm integral equation* (6.68) *has the unique solution* $y(x)$ *representing a continuous function in the region* X *if the kernel* $K(x, \xi)$ *is bounded,* $\mid K(x, \xi) \mid < C$, *the function* $\varphi(x)$ *is continuous in* X *and* $\mid \lambda \mid > Cv(X)/\alpha$ *where* $v(X)$ *is the volume of the region* X,

$$v(X) = \int\limits_X dx\,,$$

and $\alpha \in (0, 1)$.

▷ It is sufficient to show that the operator

$$Au(x) = \frac{1}{\lambda}\varphi(x) + \frac{1}{\lambda}\int\limits_X K(x, \xi)\,u(\xi)\,d\xi \qquad (6.69)$$

is a contraction operator under the conditions of the theorem. We have using Eq. (6.69)

$$\mid Au_1(x) - Au_2(x) \mid < \frac{C}{\mid \lambda \mid}\int\limits_X \mid u_1(\xi) - u_2(\xi) \mid d\xi$$

$$\leq C \sup_{x \in X} \mid u_1(x) - u_2(x) \mid v(X)/\mid \lambda \mid$$

and

$$\| Au_1 - Au_2 \|_C = \sup_{x \in X} \mid Au_1(x) - Au_2(x) \mid$$

$$\leq C \| u_1 - u_2 \|_C\, v(X)/\mid \lambda \mid\,.$$

We see that the operator A is a contraction operator since $\mid \lambda \mid > Cv(X)/\alpha$ implies $Cv(X)/\mid \lambda \mid < \alpha < 1$. By Theorem 6.21 the equation $y = Ay$ has in this case the unique solution $y(x)$ representing a continuous function of x in the region X. ◁

R e m a r k. The proved theorem may easily be extended to the case of vector functions $y(x)$, $\varphi(x)$ with values in R^n and $(n \times n)$–matrix function $K(x, \xi)$. The condition $\mid K(x, \xi) \mid < C$ is replaced in this case by the condition $\mid K_{pq}(x, \xi) \mid < C$ $(p, q = 1, \ldots, n)$. The condition

$|\lambda| > Cv(X)/\alpha$ is sufficient for the existence of the solution of Fredholm Eq. (6.68). Nevertheless Eq. (6.68) may have solutions at smaller values of the parameter λ.

Theory of linear and nonlinear operator equations will be considered in Chapters 9 and 10. For detailed study of the method of contraction mapping and similar methods refer to (Edwards 1965, Schwartz 1969). Some applications for stochastic equations are given in (Balakrishnan 1976, Da Prato and Zabczyk 1992, Laning and Battin 1956, Pugachev 1965, Pugachev and Sinitsyn 1987).

Problems

6.2.1. Consider the space E^n of columns $x = \{x_k\}_{k=1}^m$, $x_k \in R$, with the norm $\| x \| = \left[\sum_{k=1}^m | x_k |^2 \right]^{1/2}$. Prove that the linear mapping $A : E^n \to E^n$ with matrix $\| a_{ij} \|$ $(i, j = 1, 2, \ldots, n)$ will be a contraction mapping if

$$\sum_{i=1}^n \sum_{j=1}^n | a_{ij}^2 | < 1.$$

6.2.2. Consider the space c^m of columns $x = \{x_k\}_{k=1}^m$, $x_k \in R$, with the norm $\| x \| = \max_{1 \le k \le m} | x_k |$. Prove that the linear mapping $A : c^n \to c^n$ with matrix $\| a_{ij} \|$ $(i, j = 1, 2, \ldots, n)$ will be a contraction mapping if

$$\max_{1 \le i \le n} \sum_{j=1}^n | a_{ij} | < 1.$$

6.2.3. Consider the space l^n of columns $x = \{x_k\}_{k=1}^m$, $x_k \in R$, with the norm $\| x \| = \sum_{k=1}^m | x_k |$. Prove that the linear mapping $A : l^n \to l^n$ with matrix $\| a_{ij} \|$ $(i, j = 1, 2, \ldots, n)$ will be a contraction mapping if

$$\max_{1 \le j \le n} \sum_{i=1}^n | a_{ij} | < 1.$$

6.2.4. The function $f(x)$ is differentiable and $| f'(x) | < \alpha < 1$ in the interval $[a, b]$. Is the function $y = f(x)$ a contraction mapping? Under what conditions the equation $x = f(x)$ has a solution in the interval $[a, b]$? The same questions for the case where $| f'(x) | > \beta > 1$. Illustrate graphically the divergence of successive approximations in the second case.

6.2.5. Under what restrictions imposed to the continuous function $f(s, t)$, s, $t \in [0, 1]$ the operators

$$A_1 x = \int_0^1 f(s, t)\, x(t)\, dt\,,$$

$$A_2 x = \int_0^1 f(s, t)\, x^n(t)\, dt \quad (n - \text{natural})\,,$$

$$A_3 x = \int_0^1 f(s, t)\, \sin x(t)\, dt$$

are contraction operators? Construct the iteration process of Theorem 6.21 for solving the equation $x = A_i x$ $(i = 1, 2, 3)$ in the case of a contraction operator A_i.

6.2.6. Prove that the number k in the case of the equation $F(x) = 0$ in Example 6.26 may always be chosen in such a way that the operator $Ax = f(x)$ be a contraction operator if the function $F(x)$ is monotone in a sufficiently large vicinity of the fixed point.

6.2.7. Prove the theorem similar to Theorem 6.21 for the set of two equations $Ax = y$, $By = x$, A being the operator mapping a complete metric space X into a complete metric space Y and B the operator mapping Y into X. Construct the respective iteration process. Should both operators A and B be contraction operators? Should both spaces X and Y be complete?

6.2.8. Apply the successive approximations of Problem 6.2.7 to the case of the set of scalar equations $f(x) = y$, $g(y) = x$. Illustrate graphically the approximation to the solution of this set of equations and find the conditions under which the approximations converge. Give the graphical illustration of the divergent iteration process in the case where the conditions of the theorem in Problem 6.2.7 are not fulfilled.

6.2.9. Prove that the mapping $\Psi : R \to R$, $\Psi(x) = x + \pi/2 - \text{arctg}\, x$ satisfy the condition $\| \Psi(x) - \Psi(y) \| < \| x - y \|$, $x \neq y$, $x, y \in R$ but it has no fixed point.

6.2.10. Consider the operator $\Psi(x)$ mapping convex set $D \subset X$ into itself and being continiously differentiable on D. Prove that the operator $\Psi(x)$ will be contractive if and only if

$$\sup_{x \in D} \| \Psi'(x) \| < 1\,.$$

6.3. Spectrum of an Operator

6.3.1. Eigenvalues

The equation

$$Tx - \lambda x = y \tag{6.70}$$

and the respective homogeneous equation

$$Tx = \lambda x \tag{6.71}$$

play a great role in the theory of linear operators. Here T is the linear operator mapping a topological linear space X into itself, λ is a complex parameter, y is the given element of the space X, x is the unknown element of X. Eq. (6.71) has a trivial solution $x = 0$ at any λ. Nevertheless it may have a solution different from zero at some values of the parameter λ. Such values of λ play an exceptional role in the linear operators theory.

The values of the parameter λ at which Eq. (6.71) has a solution different from zero $x \neq 0$ are called *eigenvalues of the operator T*. The solutions of Eq. (6.71) corresponding to an eigenvalue are called *eigenvectors of the operator T*. If a set of eigenvectors $\{x_\alpha\}$ corresponds to the eigenvalue λ then the linear span of these vectors (i.e. the set of all their finite linear combinations, Subsection 1.3.3) is called an *eigensubspace of the operator T* corresponding to the eigenvalue λ. In the special case where only one eigenvector (determined up to a numerical factor) corresponds to the eigenvalue λ the respective eigensubspace is the one–dimensional subspace of X. If the set $\{x_\alpha\}$ is infinite then the eigensubspace is the infinite–dimensional subspace of X. It is evident that any element of the eigensubspace represents an eigenvector corresponding to a given eigenvalue λ.

It is clear that the eigensubspaces G_1 and G_2 corresponding to two different eigenvalues λ_1 and λ_2 of the operator T have only one common point 0. Really, if G_1, G_2 contains a point $x \neq 0$ then two relations $Tx = \lambda_1 x$ and $Tx = \lambda_2 x$ are valid what is impossible.

A subspace $X' \subset X$ is called an *invariant subspace of the operator T* if $Tx \in X'$ at any $x \in X'$.

R e m a r k. Obviously any eigensubspace of the operator T is its invariant subspace. But an invariant subspace may be not an eigensubspace. Moreover an invariant subspace may contain none of eigenvectors.

Theorem 6.28. *If a parameter λ is an eigenvalue of the operator T then* Eq. (6.70) *cannot have the unique solution.*

▷ Assume that Eq. (6.70) has a solution x_0. Let x be the eigenvector corresponding to the eigenvalue λ. Then $x_1 = x_0 + cx$ is a solution of Eq. (6.70) at any value of c. Thus if λ is an eigenvalue then Eq. (6.70) either has no a solution or has an infinite set of solutions. In other words, the operator $T - \lambda I$ (I being the identity operator) has not the inverse operator $(T - \lambda I)^{-1}$ if λ is an eigenvalue of operator T. ◁

6.3.2. Resolvent and Spectrum of an Operator

If the inverse operator

$$R_\lambda = (T - \lambda I)^{-1},$$

exists at a given value of λ it is called the *resolvent of the operator T*.

The domain of the resolvent R_λ represents the range $R_T(\lambda)$ of the operator $T - \lambda I$.

The values of λ at which the resolvent R_λ exists, continuous and defined on the whole space X are called the *regular points of the operator T*. The set of regular points of the operator T is callled its *resolvent set* and is denoted by $\rho(T)$.

The set of values of λ for which the resolvent does not exist, or exists but is not continuous or is defined not on the whole space X is called the *spectrum of the operator T* and is denoted by $\sigma(T)$. All the values of λ belonging to the spectrum are called *spectrum points*.

It is evident that the resolvent set and the spectrum of an operator represent the complements of one another. From these definitions follows directly that λ can be a regular point only when $R_T(\lambda) = X$. But it is not sufficient in the general case. It is necessary that the resolvent R_λ be continuous. Nevertheless there is an important case where the condition $R_T(\lambda) = X$ is necessary and sufficient for the regularity of the point λ. It is the case of a closed operator T in a B–space X. The resolvent R_λ represents a closed operator in this case as the inverse operator of the closed operator $T - \lambda I$. And the closed operator defined on the whole space is continuous (Problem 6.1.1). It is clear that all the eigenvalues of the operator T belong to its spectrum.

It is usual to distinguish three types of the spectrum points of the closed operator in a B–space. Accordingly the spectrum is divided into three parts.

A set of eigenvalues of the operator T is called a *point spectrum of the operator* T and is denoted by $\sigma_p(T)$. The resolvent does not exist for $\lambda \in \sigma_p(T)$.

A set of of spectrum points of the operator T for which the resolvent exists and its domain is dense in X, $R_T(\lambda) \neq X$, $[R_T(\lambda)] = X$ is called a *continuous spectrum of the operator* T and is denoted by $\sigma_c(T)$.

A set of spectrum points of the operator T for which the resolvent exists and its domain is not dense in X, $[R_T(\lambda)] \neq X$ is called a *residual spectrum of the operator* T and is denoted by $\sigma_r(T)$.

Thus $\sigma(T) = \sigma_p(T) \bigcup \sigma_c(T) \bigcup \sigma_r(T)$.

E x a m p l e 6.31. It is known from linear algebra that the set of homogeneous algebraic equations

$$Ax = \lambda x$$

in n–dimensional space has a solution different from zero if and only if the determinant of the matrix $A - \lambda I$ is equal to zero:

$$| A - \lambda I | = \begin{vmatrix} a_{11} - \lambda & a_{12} & \cdots & a_{1n} \\ a_{21} & a_{22} - \lambda & \cdots & a_{2n} \\ \vdots & \vdots & \ddots & \vdots \\ a_{n1} & a_{n2} & \cdots & a_{nn} - \lambda \end{vmatrix} = 0. \qquad (6.72)$$

This algebraic equation of the n–th degree has always n solutions $\lambda_1, \ldots, \lambda_n$. Thus any linear operator in n–dimensional space has n eigenvalues some of which may coincide (the case of multiple roots of Eq. (6.72)). All the other values of λ are the regular points, since the nonhomogeneous equation

$$Ax - \lambda x = y$$

has always the unique solution at any y if $\lambda \neq \lambda_1, \ldots, \lambda_n$. Thus the spectrum of any linear operator in n–dimensional space consists only of n eigenvalues $\lambda_1, \ldots, \lambda_n$ The resolvent represents in this case the linear operator with the matrix $(A - \lambda I)^{-1}$.

E x a m p l e 6.32. Consider the operator of multiplication by the independent variable in the space of continuous functions $C([a, b])$:

$$Qx = tx(t).$$

This operator has not eigenvalues because there is no a function $x(t)$ satisfying the equation

$$tx(t) = \lambda x(t) \quad (\text{at all} \quad t \in [a, b])$$

at some λ. On the other hand, all the values of $\lambda \in [a, b]$ are the points of the residual spectrum $\sigma_r(Q)$ since the equation

$$tx(t) - \lambda x(t) = y(t)$$

has the solution at $a \leq \lambda \leq b$

$$x(t) = \frac{y(t)}{t - \lambda}$$

not for all the functions $y(t)$ but only for those which are representable in the form $y(t) = (t - \lambda)z(t)$, $z(t) \in C([a, b])$, and the set of functions with zero at $t = \lambda$ is not dense in $C([a, b])$. All the values of λ not belonging to the interval $[a, b]$ are regular points. The resolvent represents in this case the multiplication operator by the function $(t - \lambda)^{-1} \in C([a, b])$.

E x a m p l e 6.33. Consider the differentiation operator D in the space of continuous functions $C([a, b])$. All the values of λ are the eigenvalues of the operator D since the differential equation

$$x'(t) = \lambda x(t)$$

has the solution $x(t) = ce^{\lambda t}$ in $C([a, b])$ at any λ, c being an arbitrary constant. Consequently, the differentiation operator in $C([a, b])$ has no regular points. The whole complex plane C represents its spectrum. In accordance with general Theorem 6.28 the nonhomogeneous equation

$$x'(t) - \lambda x(t) = y(t),$$

has the infinite set of solutions determined by the well known formula

$$x(t) = ce^{\lambda t} + e^{\lambda t} \int_a^t y(\tau)e^{-\lambda \tau} d\tau.$$

at any λ and $y(t) \in C([a, b])$.

6.3.3. Properties of Resolvents

Let λ and μ be two regular points of the operator T. By the definition of the resolvent (Subsection 6.3.2) we have

$$R_\lambda = R_\mu(T - \mu I)R_\lambda, \quad R_\mu = R_\mu(T - \lambda I)R_\lambda.$$

Taking into account that any linear operator is commutative with the identity operator I we obtain

$$R_\lambda - R_\mu = (\lambda - \mu)R_\mu R_\lambda. \tag{6.73}$$

This relation between the resolvents corresponding to two different regular points λ, μ is generally called *Hilbert formula*.

Writing (6.73) as

$$R_\mu R_\lambda = \frac{R_\lambda - R_\mu}{\lambda - \mu} \tag{6.74}$$

and interchanging λ and μ we come to the conclusion that the resolvents R_λ and R_μ corresponding to different regular points are commutative,

$$R_\mu R_\lambda = R_\lambda R_\mu.$$

Theorem 6.29. *Any operator is commutative with its resolvent at any regular point.*

▷ This is the immediate consequence of the following equalities:

$$TR_\lambda = I + \lambda R_\lambda, \quad R_\lambda T = I + \lambda R_\lambda. \quad ◁$$

Theorem 6.30. *An operator S is a commutative with an operator T if and only if it is commutative with its resolvent R_λ at any regular point.*

▷ If $ST = TS$ then

$$(T - \lambda I)SR_\lambda = S(T - \lambda I)R_\lambda = S.$$

Multiplhing this equality from the left by R_λ we get $SR_\lambda = R_\lambda S$. If $SR_\lambda = R_\lambda S$ then

$$(T - \lambda I)SR_\lambda = (T - \lambda I)R_\lambda S = S.$$

Multiplying this equality from the right by $T - \lambda I$ we get

$$(T - \lambda I)S = S(T - \lambda I),$$

i.e. $TS = ST$. $\triangleleft$

Corollary. *Operators T and S are commutative if and only if their resolvents are commutative at any regular points.*

$\triangleright$ If $ST = TS$ then $SR_\lambda = R_\lambda S$ by the proved theorem. But the operator R_λ commutative with the operator S is also commutative with its resolvent Q_μ, i.e. $Q_\mu R_\lambda = R_\lambda Q_\mu$. The same reasoning in the inverse order make sure that $Q_\mu R_\lambda = R_\lambda Q_\mu$ imply $ST = TS$. $\triangleleft$

6.3.4. Properties of the Spectrum of a Linear Operator in a B–Space

Now we continue the study of the spectrum of linear operators restricting ourselves to operators in a B–space. We prove first an auxiliary proposition.

Lemma. *If the linear operator A mapping a B–space X onto the whole B–space Y $(R_A = Y)$ has the bounded inverse operator A^{-1} then for any bounded linear operator B mapping X into Y whose norm is smaller than $\|A^{-1}\|^{-1}$, $\|B\| < \|A^{-1}\|^{-1}$, the operator $A + B$ has the bounded inverse operator $(A + B)^{-1}$ defined on the whole space Y and $\|(A + B)^{-1}\| \le (\|A^{-1}\|^{-1} - \|B\|)^{-1}$.*

$\triangleright$ Consider the equation

$$(A + B)x = y. \tag{6.75}$$

As the linear operator A has the inverse operator A^{-1} defined on the whole space Y Eq. (6.75) may be represented in the form

$$x = A^{-1}(y - Bx).$$

We shall prove that the operator C, $Cx = A^{-1}(y - Bx)$ is a contraction operator (Subsection 6.2.6). Taking any $x_1, x_2 \in X$, $x_1 \neq x_2$ we have

$$Cx_1 - Cx_2 = A^{-1}B(x_2 - x_1)$$

and

$$\|Cx_1 - Cx_2\| \le \|A^{-1}\| \|B\| \|x_1 - x_2\|. \tag{6.76}$$

By the condition of Theorem 6.21 $\| B \| < \| A^{-1} \|^{-1}$ whence $\| A^{-1} \|$ $\times \| B \| < 1$ and (6.76) takes the form

$$\| Cx_1 - Cx_2 \| < \alpha \| x_1 - x_2 \|, \quad \| A^{-1} \| \| B \| < \alpha < 1 .$$

Hence the operator C is a contraction operator. Consequently, Eq. (6.75) has the unique solution at any y. It remains to prove the boundedness of the operator $(A + B)^{-1}$ and to estimate its norm. Taking into account that $x = A^{-1} Ax$ at any $x \in X$, and consequently, $\| x \| \le \| A^{-1} \| \| Ax \|$ and $\| Ax \| \ge \| A^{-1} \|^{-1} \| x \|$ we find

$$\| y \| = \| (A + B)x \| \ge \| Ax \| - \| Bx \|$$

$$\ge \| A^{-1} \|^{-1} \| x \| - \| B \| \| x \| = (\| A^{-1} \|^{-1} - \| B \|) \| x \| .$$

Hence

$$\| x \| = \| (A + B)^{-1} y \| \le (\| A^{-1} \|^{-1} - \| B \|)^{-1} \| y \|$$

at any $y \in Y$. Thus the operator $(A + B)^{-1}$ is bounded and

$$\| (A + B)^{-1} \| \le (\| A^{-1} \|^{-1} - \| B \|)^{-1} . \quad \triangleleft$$

R e m a r k. The operator A may be unbounded in the proved Lemma.

Theorem 6.31. *The spectrum of a bounded linear operator T defined on the whole B–space X is contained in the closed circle of the radius $\| T \|$ with the centre at 0.*

▷ To prove this theorem we use the proved Lemma. Putting $| \lambda | > \| T \|$, $A = -\lambda I$, $B = T$ and taking into account that $A^{-1} = -\lambda^{-1} I$, $\| A^{-1} \| = | \lambda |^{-1}$, $\| B \| = \| T \|$ we come to the conclusion that Eq. (6.70) has the unique solution $x = R_\lambda y$ defined on the whole space X at $| \lambda | < \| T \|$ and $\| R_\lambda \| \le (| \lambda | - \| T \|)^{-1}$. Thus any exterior point λ of the circle $| \lambda | \le \| T \|$ is regular. ◁

Theorem 6.32. *The spectrum of a linear operator T represents a closed set.*

▷ Let λ_0 be a regular point of an operator T. Using Lemma in the case $A = T - \lambda_0 I$, $B = (\lambda_0 - \lambda)I$ and taking into account that $A^{-1} = R_{\lambda_0}$, $\| B \| = | \lambda - \lambda_0 |$ we come to the conclusion that the bounded resolvent R_λ exists defined on the whole space X at any λ in the circle

$|\lambda - \lambda_0| < \|R_{\lambda_0}\|^{-1}$. Consequently, any regular point has the vicinity whose all the points are regular. This means that the resolvent set $\rho(T)$ is an open set and the spectrum is a closed set as the complement of $\rho(T)$. ◁

Due to the openness of the resolvent set Hilbert formula (6.74) is valid at any μ in some vicinity of λ. Therefore we may pass to the limit as $\mu \to 0$ yielding

$$\frac{dR_\lambda}{d\lambda} = R_\lambda^2 . \tag{6.77}$$

Thus the resolvent R_λ considered as a function of λ is differentiable at any regular point.

6.3.5. Spectrum of an Unitary Operator

Theorem 6.33. *The moduli of all the eigenvalues of an unitary operator are equal to unity.*

▷ If λ is an eigenvalue of U and x is the respective eigenvector then

$$Ux = \lambda x , \tag{6.78}$$

But $\|Ux\| = \|x\|$, $\|\lambda x\| = |\lambda| \|x\|$. Hence $|\lambda| = 1$. ◁

Theorem 6.34. *If λ and x are an eigenvalue and the corresponding eigenvector of an unitary operator U then $\bar{\lambda} = \lambda^{-1}$ and x represent an eigenvalue and the corresponding eigenvector of the inverse operator U^{-1}.*

▷ Dividing Eq. (6.78) by λ and taking into account that $\lambda^{-1} = \bar{\lambda}$ we rewrite Eq. (6.78) in the form

$$U^{-1}x = \bar{\lambda}x . \quad ◁$$

Theorem 6.35. *If $\lambda \neq 0$ is a regular point of an unitary operator U then λ^{-1} is a regular point of the inverse operator U^{-1}.*

▷ If $\lambda \neq 0$ is a regular point of the unitary operator U then the equation

$$Ux - \lambda x = y \tag{6.79}$$

has the unique solution at any $y \in X$ and $\|x\| < c\|y\|$ at some $c > 0$. Putting $z = Ux$, $u = -\lambda^{-1}y$ we may rewrite Eq. (6.79) as

$$U^{-1}z - \lambda^{-1}z = u . \tag{6.80}$$

The vector $z = Ux$ is the unique solution of this equation at a given $u = -\lambda^{-1}y$ and $\|z\| = \|x\| < c\|y\| = c|\lambda|\|u\|$. But u is an arbitrary since y is arbitrary. Hence λ^{-1} is a regular point of the operator U^{-1}. ◁

Theorem 6.36. *The spectrum of an unitary operator U is disposed on the circumference $|\lambda| = 1$.*

▷ By Theorem 6.31 all the points λ exterior of the circle $|\lambda| \leq 1$ are regular points of the unitary operator U since $\|U\| = 1$. Let $\lambda \neq 0$ be any interior point of the circle $|\lambda| \leq 1$. The inverse operator U^{-1} is an unitary operator by Theorem 6.10. Hence λ^{-1} is its regular point as $|\lambda^{-1}| \geq 1$. By Theorem 6.35 the point λ is a regular point of the operator $U = (U^{-1})^{-1}$. Consequently, only the points of the circumference $|\lambda| = 1$ can belong to the spectrum of U [a]. ◁

Spectral theory of linear operators will be considered in Chapters 7 and 8. For detailed study of basic notions refer to (Dowson 1978, Dunford and Schwarz 1958, Edwards 1965).

Problems

6.3.1. Find the spectrum of the multiplication operator by the continuous function $\varphi(t)$ in the space $C([a, b])$: $Ax = \varphi(t)x(t)$. What are the point spectrum, continuous spectrum and residual spectrum in this case?

6.3.2. Find the spectrum of the multiplication operator by an elementary function in the space of bounded functions $B([a, b])$. What are the point spectrum, continuous spectrum and residual spectrum in this case?

6.3.3. Find the spectrum of the multiplication operator M by "Cantor ladder" (the function $F(x)$ of Example 3.9) in the space of bounded functions $B([a, b])$. Describe $\sigma_p(M)$, $\sigma_c(M)$, $\sigma_r(M)$ in this case.

6.3.4. Find the spectrum of the multiplication operator B by the function $\varphi(t) \in B(T)$ in the space of bounded functions $B(T)$, $T \subset R^n$. What must be a function $\varphi(t)$ for which the point spectrum $\sigma_p(B)$ is not empty?

6.3.5. Prove that the multiplication operator by the continuous function $\varphi(t)$, $\varphi(0) \neq 0$ in the space $C_0([0, 1])$ of continuous functions $x(t)$ satisfying the condition $x(0) = 0$ has nonempty continuous spectrum. Describe this spectrum.

6.3.6. Find the spectra of the operators

$$Ax = \int_{-1}^{1} ts x(s)\, ds, \quad x(t) \in C([-1, 1]),$$

---

[a] The point $\lambda = 0$ is evidently regular since $R_U(0) = X$.

$$Bx = \int\limits_{-\pi}^{\pi} \sin \omega(t - s)\, x(s)\, ds\,, \quad x(t) \in C([-\pi, \pi])\,,$$

$$Tx = \int\limits_{T_1} \sum_{n=1}^{N} \varphi_n(t)\, \psi_n(s)\, x(s)\, ds\,, \quad x(t) \in C(T_1)\,, \quad T_1 \subset R^n\,.$$

6.3.7. Find the spectra of the following operators:

$$A_1 x = [\, x(t) + x(-t)\,]/2\,,$$

$$A_2 x = [\, x(t) - x(-t)\,]/2\,.$$

6.3.8. Let T be the bounded linear operator acting in a B–space X, $T \in \mathcal{B}(\mathcal{X})$. Prove that the spectrum of the adjoint operator T^* coincides with that of T.

I n s t r u c t i o n. Prove that the existence of the bounded resolvent R_λ of the operator T defined on the whole space X, $R_\lambda \in \mathcal{B}(\mathcal{X})$ implies the existence of the resolvent of the operator T^* representing the operator $R_\lambda^* \in \mathcal{B}(\mathcal{X}^*)$ adjoint of R_λ. In its turn from the existence of the bounded resolvent of the operator T^* defined on the whole space X^*, $(T^* - \lambda I^*)^{-1} \in \mathcal{B}(X^*)$ (I^* is the identity operator in X^*) follows by above proved the existence of the resolvent of the operator T^{**} belonging to the space $\mathcal{B}(X^{**})$. But T^{**} represents the extension of T from $X \subset X^{**}$ to X^{**}. Therefore the bounded resolvent R_λ of the operator T exists. But T is a closed operator as continuous operator defined on the whole space (Corollary of Theorem 6.2). Hence the domain of its resolvent is closed $[R_T(\lambda)] = R_T(\lambda)$ (Theorem 6.2). If $R_T(\lambda) \neq X$ then by Corollary 2 of Theorem 1.18 there exists the functional $f \in X^*$, $f \neq 0$, whose kernel contains the subspace $R_T(\lambda)$ i.e. $[(T^* - \lambda I^*)f]x = f(T - \lambda I)x = 0\ \forall x$. This implies $(T^* - \lambda I^*)f = 0$ what is in contradiction with the existence of the resolvent of the operator T^*.

6.3.9. Let $T \in \mathcal{B}(X)$. Prove the following inclusions:

$$\sigma_p(T) \subset \sigma_p(T^*)\bigcup\sigma_r(T^*)\,, \quad \sigma_r(T) \subset \sigma_p(T^*) \subset \sigma_p(T)\bigcup\sigma_r(T)$$

I n s t r u c t i o n. The proof of the first inclusion is based on the existence for any $x \neq 0$ of such a functional $f \in X^*$ that $fx \neq 0$. The proof of the second inclusion is based on Corollary 2 of Theorem 1.18. This Corollary together with Remark at the end of Subsection 4.6.11 asserts that for any closed subspace $L \subset X$, $L \neq X$ a functional $f \in X^*$ exists and whose kernel contains L. To prove the third inclusion recall that the kernel of the functional $f \neq 0$ is a closed set which cannot coincide with the whole space X.

6.3.10. Formula (6.77) shows that the resolvent R_λ of the operator T is differentiable at all regular points, i.e. represents an analytic function of the complex variable λ on the resolvent set $\rho(T)$ with values in the B–space X. Show that all the points of the spectrum $\sigma(T)$ of the operator T represent singularities of R_λ.

I n s t r u c t i o n. Let $d(\lambda)$ be the distance of the point λ from the spectrum $\sigma(T)$ of T, $d(\lambda) = \inf(|\lambda - \mu|, \mu \in \sigma(T))$. While proving Theorem 6.32 it was shown that for any point $\lambda \in \rho(T)$ all the points inside the circle $|\lambda' - \lambda| < \|R_\lambda\|^{-1}$ are regular. Hence $d(\lambda) \geq \|R_\lambda\|^{-1}$.

6.3.11. Prove that the spectrum of a bounded operator T in a B–space X is not empty.

S o l u t i o n. Consider the analytical function $f(z)$ with regular point $z = 0$. It is representable by the series $f(z) = \sum a_\nu z^\nu$ with some radius of convergence r. Then for any operator T, $\|T\| < r$ the series $\sum a_\nu T^\nu$ converges in the uniform topology of the space $\mathcal{B}(X)$ (Subsection 5.1.4) since $\|T^\nu\| \leq \|T\|^\nu$ and

$$\left\| \sum_{\nu=n}^{n+p} a_\nu T^\nu \right\| \leq \sum_{\nu=n}^{n+p} |a_\nu| \|T\|^\nu \to 0 \quad \text{as} \quad n \to \infty.$$

at any n, $p > 0$. In particular, the convergence of the series $\sum z^\nu \lambda^{-\nu-1} = -(z - \lambda)^{-1}$ at $|z| < \lambda$ implies the convergence of the series $\sum T^\nu \lambda^{-\nu-1}$ at $\|T\| < \lambda$ in the uniform topology of the space $\mathcal{B}(X)$. Multiplying the sum S of this series from the left and from the right by $T - \lambda I$ we get $(T - \lambda I)S = S(T - \lambda I) = -I$ whence $S = -R_\lambda$. Thus the resolvent R_λ of the operator T is representable by the series

$$R_\lambda = - \sum_{\nu=0}^{\infty} T^\nu \lambda^{-\nu-1}$$

convergent in the uniform topology at all λ, $|\lambda| > \|T\|$. But the convergence in the uniform topology implies the convergence in the weak topology of the space $\mathcal{B}(X)$ (Subsection 5.1.4). Consequently, the function $fR_\lambda x$ of the complex variable λ is representable by the Loran series

$$fR_\lambda x = - \sum_{\nu=0}^{\infty} fT^\nu x \lambda^{-\nu-1} = -\frac{1}{\lambda} \sum_{\nu=0}^{\infty} fT^\nu x \lambda^{-\nu}$$

at any $x \in X$, $f \in X^*$ converging outside the circle $|\lambda| \leq \|T\|$. If the spectrum of the operator T is empty then its resolvent R_λ has no singularities. Hence $fR_\lambda x$ represents an entire function equal to zero at the infinity. By Liouville theorem such a function is identical to zero. Thus $fR_\lambda x = 0$ at any x, λ implying $R_\lambda = 0$ what is impossible.

6.3.12. Prove that the spectrum of the differentiation operator $D = d/dt$ in the space $C_0([0, 1])$ of continuous functions $x(t)$ satisfying the condition $x(0) = 0$ is empty. This result shows that contrary to bounded operators (Problem 6.3.11) the spectrum of an unbounded operator may be empty.

6.3.13. Find the spectrum of the differentiation operator in the space of continuous functions on $[0, 1]$ satisfying the condition $x(0) = x(1)$.

6.3.14. Consider the invertible operator $U \in \mathcal{B}(X)$ possessing the property $\| U^\nu \| < c$ at all integers ν for some $c > 0$. Show that the spectrum of the operator U lies on the circumference $| \lambda | = 1$. Does this imply that U is an unitary operator?

I n s t r u c t i o n. Expanding the resolvent of the operator U in positive powers of λ show that all λ, $| \lambda | < 1$ are regular. In the same way expanding the resolvent in negative powers of λ show that all the point λ, $| \lambda | > 1$ are regular.

6.3.15. Prove that the spectra of the unitarly equivalent operators T_1 and T_2 coincide completely:

$$\sigma_p(T_1) = \sigma_p(T_2), \quad \sigma_c(T_1) = \sigma_c(T_2), \quad \sigma_r(T_1) = \sigma_r(T_2).$$

6.3.16. Find the adjoint operator of $T = c_1 T_1 + \ldots + c_n T_n$ if each of the operators T_k have the adjoint operator T_k^*.

CHAPTER 7
LINEAR OPERATORS IN HILBERT SPACES

Chapter 7 contains the theory of linear operators in H-spaces. In Section 7.1 the definitions of orthogonal subspaces, their orthogonal sums and orthogonal complements are given. Section 7.2 is devoted to the general form of a continuous linear functional on a H-space. Bounded and unbounded linear operators and unitary operators are studied. As a special case of an unitary operator Fourier – Plancherel operator is considered. Symmetric and self–adjoint operators and their spectra are studied in Section 7.3. The general theory of orthogonal projection operators (orthoprojectors) is outlined in Section 7.4. In Section 7.5 orthogonal, orthonormal, biorthogonal and biorthonormal systems of vectors are studied. The notion of a basis is introduced and the condition of existence of a basis in a H-space is established. Then the expansions of any vectors in terms of a basis are derived. As special cases the expansions of functions are studied in terms of systems of orthonormal functions forming bases in the respective function spaces representing H-spaces. In the last Section 7.6 the general theory of compact linear operators is outlined and spectra of compact operators on H-spaces are studied. Hilbert — Schmidt operators and trace type operators are considered as special cases. Fredholm linear integral equations are also considered and the theory of such equations are formulated following from the general theory of compact linear operators.

7.1. Hilbert Spaces

7.1.1. Orthogonal Subspaces

In the theory of the H–spaces only *closed subspaces* are considered. In this case any subspace G of the H–space X contains the limits of all fundamental sequences of their elements. These limits exist and belong to X by virtue of the completeness of X. Thus all the subspaces of the H–space are the H–spaces.

The subspaces G_1 and G_2 of the H–space X are called *orthogonal* if any vector $x_1 \in G_1$ is orthogonal to any vector $x_2 \in G_2$, $(x_1, x_2) = 0$.

Orthogonal sum of the orthogonal subspaces G_1, G_2 is called a set of the vectors $x \in X$ which may be presented in the form $x = x_1 + x_2$, $x_1 \in G_1$, $x_2 \in G_2$. Orthogonal sum of the subspaces G_1 and G_2 is denoted by $G_1 \oplus G_2$. The orthogonal sum of the subspaces evidently also represents a subspace. Each of the subspaces G_1 and G_2 is called *an orthogonal difference* of the subspace $G = G_1 \oplus G_2$ and G_2 or G_1 correspondingly, $G_1 = G \ominus G_2$, $G_2 = G \ominus G_1$.

Orthogonal supplement of the subspace G in the H–space X is called a set of all the vectors $x \in X$ orthogonal to G, $(x, y) = 0$ for all $y \in G$. Orthogonal supplement G in X is denoted by $G^\perp$ or $X \ominus G$. It is obvious that orthogonal supplement of the subspace is also a subspace.

Let G be a subspace of the H–space X and $x \in X$ be any vector which does not belong to G. The variable

$$\rho = \inf_{y \in G} \|y - x\| \tag{7.1}$$

represents a distance of the point x from the subspace G.

Theorem 7.1. *For any point x which does not belong to the subspace G of the H–space X there exists in G the point y_0 the distance of which from x is equal to the distance x from G:*

$$\rho = \|y_0 - x\| . \tag{7.2}$$

▷ For proving we take any sequence $\{y_n\}$ of the vectors of the subspace G for which

$$\rho = \lim \|y_n - x\| .$$

Then at any $\varepsilon > 0$ and any sufficiently large n we shall have $\| y_n - x \| < \rho + \varepsilon$. On the other hand, from the identity of the parallelogram (1.19)

$$\|a + b\|^2 + \|a - b\|^2 = 2 \|a\|^2 + 2 \|b\|^2 ,$$

which is valid for any elements of the H–space follows

$$\|y_n - y_m\|^2 + \|y_n + y_m - 2x\|^2 = 2 \|y_n - x\|^2 + 2 \|y_m - x\|^2 .$$

Hence taking into account that in the consequence of the definition of a subspace

$$\frac{y_n + y_m}{2} \in G ,$$

$$\left\| \frac{y_n + y_m}{2} - x \right\| \geq \rho ,$$

we get at all sufficiently large n, m

$$\|y_n - y_m\|^2 < 4(\rho + \varepsilon)^2 - 4\rho^2 = 8\rho\varepsilon + 4\varepsilon^2 .$$

It means that by virtue of the arbitariness of $\varepsilon > 0$ the sequence $\{y_n\}$ is fundamental. As the subspace G is complete then there exist in it the limit $y_0 = \lim y_n$ and $\lim \| y_n - x \| = \| y_0 - x \|$. Thus we proved the existence of the point $y_0 \in G$ for which equality (7.2) is valid. ◁

Theorem 7.2. *Under the conditions of Theorem 7.1 the vector* $z = x - y_0$ *is orthogonal to the subspace* G:

$$(x - y_0, y) = 0 \quad \text{for all} \quad y \in G. \tag{7.3}$$

▷ Suppose that there exists the vector $y_1 \in G$ for which

$$(x - y_0, y_1) = \alpha \neq 0.$$

Then after taking the vector

$$y = y_0 + c\alpha y_1 \in G, \quad c > 0$$

and using the formula

$$\| y - x \|^2 = \| y_0 - x \|^2 + \| y - y_0 \|^2 + (y_0 - x, y - y_0) + (y - y_0, y_0 - x), \tag{7.4}$$

which is valid for all y we obtain

$$\| y - x \|^2 = \| y_0 - x \|^2 - c \, | \alpha |^2 (2 - c \, \| y_1 \|^2).$$

Hence it is clear that $\| y - x \| < \| y_0 - x \| = \rho$ at any $c \in (0, 2/ \| y_1 \|^2)$ what is impossible by virtue of the definition of ρ.

Thus $(x - y_0, y) = 0$ for any vector $y \in G$ and formula (7.4) gives

$$\| y - x \|^2 = \| y_0 - x \|^2 + \| y - y_0 \|^2, \quad y \in G. \tag{7.5}$$

This formula shows that the point $y_0 \in G$ whose distance from x is equal to the distance x from G is unique. ◁

Corollary 1. *Any vector of the H-space X may be represented uniquely in the form of the sum $x = y + z$ where y is the projection of the vector x on any subspace G of the space X, $y \in G$ and $z \in G^\perp$.*

▷ To be certain it is sufficient to take $y = y_0$, $z = x - y_0$ and take into consideration condition (7.3) and the uniqueness of y_0. ◁

Corollary 2. *Any H-space X is an orthogonal sum of any its subspace $G \subset X$ and the orthogonal supplement of this subspace $X = G \oplus G^\perp$.*

▷ This follows from the previous Corollary and the definition of the orthogonal sum of the subspaces. ◁

7.1.2. Linear Functionals

A general form of a continuous linear functional in the H–space the following theorem establishes.

Theorem 7.3. *To any continuous linear functional $f \in X^*$ in the H–space X corresponds the unique vector $x_f \in D_f$ such one that $fx = (x, x_f)$ at all x and $\| x_f \| = \| f \|$ (Riesz theorem).*

▷ Let f be a continuous linear functional in the H–space X, $f \in X^*$, G be its kernel, $G = \ker f$, D_f be its domain [a]. It is evident that G is a subspace as it is a closed set as an inverse image of a closed singleton $\{0\}$ (Corollary of Theorem 4.19). The domain D_f of the functional f may be also considered as a subspace. Really, if D_f is a nonclosed subspace then the functional f may be extended due to the continuity and the domain D_f may be expanded in such a way till a closed subspace.

Let us take some point $x_0 \in G^\perp \cap D_f$. Such point always exists if the functional f is not zero. At any $x \in D_f$ the vector $y = x f x_0 - x_0 f x$ belongs to G as

$$fy = f(x f x_0 - x_0 f x) = f x_0 \cdot f x - f x \cdot f x_0 = 0 \,.$$

Consequently, $(y, x_0) = 0$, i.e. $(x f x_0 - x_0 f x, x_0) = 0$. Hence by the properties of the scalar product follows

$$(f x_0)(x, x_0) - (f x)(x_0, x_0) = 0 \,,$$

$$fx = \frac{f x_0}{\| x_0 \|^2}(x, x_0) = \left(x, \frac{\overline{f x_0}}{\| x_0 \|^2} x_0 \right) \,.$$

Introducing the vector

$$x_f = \frac{\overline{f x_0}}{\| x_0 \|^2} x_0 \,,$$

we rewrite the obtained equality in the form

$$fx = (x, x_f) \,. \tag{7.6}$$

[a] We know that according to Hahn–Banach theorem 4.58 for the normed spaces any continuous linear functional may be extended with retention of the norm from the subspace over the whole space. Here we receive an explicit expression of this extention for the H–space. Therefore we suppose that $D_f \neq X$.

Thus any continuous linear functional f in the H–space X represents a scalar product of the vector $x \in D_f$ by some vector $x_f \in D_f$.

Supposing that in D_f there exist two vectors x_f and x'_f for which

$$fx = (x, x_f) = (x, x'_f),$$

we receive $(x, x_f - x'_f) = 0$ at all $x \in D_f$. As D_f is a subspace then from x_f, $x'_f \in D_f$ follows $x_f - x'_f \in D_f$. Putting $x = x_f - x'_f$ we obtain $\| x_f - x'_f \| = 0$, i.e. $x'_f = x_f$.

Finally, as at any $x \in D_f$ accordingly to inequality (1.17)

$$|fx| = |(x, x_f)| \le \| x_f \| \, \| x \| \tag{7.7}$$

and $|f x_f| = \| x_f \|^2$ then $\| f \| = \| x_f \|$. Thus to each continuous linear functional f in the H–space X corresponds the unique vector $x_f \in D_f$ for which equality (7.6) is valid and here $\| f \| = \| x_f \|$. And vice versa to each vector $x_f \in X$ corresponds continuous linear functional f determined by formula (7.6) on the whole space X and here $\| f \| = \| x_f \|$. ◁

7.1.3. Extension of a Linear Functional

Now notice that the right–hand side of equality (7.6) is determined at all $x \in X$. Consequently, formula (7.6) determines the extension of the functional f given on the subspace D_f over the whole space X with retention of the norm as inequality (7.7) is valid at all $x \in X$.

If we give up the requirement $x_f \in D_f$ then the element x_f at which the functional f given on the subspace $D_f \ne X$ is expressed by formula (7.6) will be not unique as from $(x, x_f - x'_f) = 0$ at all $x \in D_f$ follows in this case only $x_f - x'_f \in D_f^\perp$. Therefore for any $y \in D_f^\perp$ putting $z = x_f + y$, $x_f \in D_f \bigcap G^\perp$ along with formula (7.6) we shall have

$$fx = (x, z).$$

This formula also gives the extension of the functional f on the whole space X. But the norm of f in this case increases as from the fact that $x_f \in D_f$ and $(x, x_f - z) = 0$ at all $x \in D_f$ follows $(x_f, x_f - z) = 0$ and

$$\| z \|^2 = \| x_f \|^2 + \| z - x_f \|^2 - (x_f, x_f - z) - (x_f - z, x_f)$$

$$= \| x_f \|^2 + \| z - x_f \|^2,$$

i.e. $\parallel z \parallel > \parallel x_f \parallel$ if $z \neq x_f$. Thus formula (7.6) gives the unique extension of f over the whole space X with retention of the norm.

E x a m p l e 7.1. In Examples 5.5 and 5.6 we saw that any linear functional in the finite–dimensional Euclidean space may be presented in the form of a scalar product. This is the special case of the proved general theorem.

E x a m p l e 7.2.   According to Theorem 5.18 any continuous linear functional in the space L_2 is expressed by formula (5.116) where $f(x) \in L_2$. After writting it in the form

$$ fx = \int x(t)\,\overline{f(t)}\,\mu(dt)\,, $$

where $f(x) \in L_2$ we present the functional f in L_2 in the form of a scalar product of the elements $x(t)$ and $f(t)$ of the space L_2.

R e m a r k. The established one–to–one correspondence between the elements of the H–space X and its dual space X^* with retention of the norm gives the opportunity to identify the spaces dual with the H–spaces with the correspondent H–spaces. Therefore further on we shall consider $X^* = X$ and write in formula (7.6) f instead of x_f. This agreement completely corresponds to the general definition of a dual space given in Section 5.1.3 in the case of the real H–space. Really, in this case a scalar product is linear both on the first multiplier and the second one. But for the complex H–space a scalar product is linearly adjoint on the second multiplier and this agreement represents a deviation from the general theory which leads to another different from general definition of a adjoint operator in the theory of the H–spaces. To remain in the context of the general theory it is necessary to substitute formula (7.6) by formula $fx = (x, Vf)$ where V is a adjoint linear isometric operator which establishes the one–to–one correspondence between the spaces X^* and X.

From the fact that $X^* = X$ follows that the second dual space X^{**} also coincides with X. Thus any H–space is reflexive. This statement is also valid when it is not assumed that $X^* = X$ and in accordance with Theorem 7.3 it is assumed that X^* and X are connected by the one–to–one linear isometric mapping. In this case the norm in X^* determines a scalar product and on the basis of the same Theorem 7.3 the spaces X^{**} and X^* are connected by the one–to–one adjoint linear isometric mapping. The composition of these two mappings represents the one–to–one linear isometric mapping of X^{**} on X what proves the reflexivity of the space X. On the basis of Corollary 3 of Banach–Steinhaus theorem 6.12 and Riesz theorem 7.3 any H–space is weakly complete, i.e. any weakly fundamental sequence has a weak limit in it.

For the detailed study of H–spaces basic notions refer to (Balakrishnan 1976, Dowson 1978, Dunford and Schwarz 1958).

P r o b l e m s

7.1.1. In Subsection 7.1.3 it is proved that formula (7.6) gives the unique extension of a continuous linear functional on the whole H–space X while general Hahn–Banach theorem 1.17 and Hahn–Banach theorem 4.58 for the normed linear spaces establish in the general case the ambiguity of linear functional extension. Prove that the inequalities of Theorem 1.17 for the value of a functional at each new point at its extension in the case of the H–space X turn into the rigorous equalities.

I n s t r u c t i o n. Account that in the case of the normed space X $p(x)$ $=\| f \| \cdot \| x \|^2$ (Theorem 4.58).

7.1.2. Accounting the Remark at the end of Subsection 5.4.5 and using Riesz theorem 7.3 find the general form of a continuous linear functional on the space L_2 of the finite–dimentional vector functions.

7.1.3. Prove that for any set A in a H–space the set $A^\perp$ is the subspace and $A \subset (A^\perp)^\perp$.

7.1.4. Find the orthogonal supplements in H–space $L_1 (0, 1)$ of the following sets: 1) polynomials of x; 2) polynomials of x^2; 3) polynomials with zero constant term.

7.2. Linear Operator

7.2.1. Bounded Linear Operators

Let T be a continuous (and consequently, bounded by Theorem 4.56) linear operator mapping the H–space X into the H–space Y. The correspondent dual spaces X^* and Y^* may be considered coinciding with X and Y correspondingly. By Theorem 6.6 the operator T has adjoint operator T^* if and only if $D_T = X$. Equality (6.4) which determines the adjoint operator by virtue of Riesz theorem 7.3 takes in this case the form

$$(x, T^*g) = (Tx, g). \tag{7.8}$$

The domain of the adjoint operator T^* according to Theorem 6.6 coincides with the whole space Y, $D_{T^*} = Y$.

In the case of the complex H–space X the definition of adjoint operator (7.8) differs from general definiton (6.4). Formula (6.4) determines the adjoint operator T^* as a linear function of the operator T, and formula (7.8) as the adjoint linear function of operator T. In

accordance with the Remark at the end of Subsection 7.1.3 the operator T_1^* adjoint with the operator T in the sense of definition (6.4) should
be determined by the formula

$$(x, V T_1^* g_1) = (Tx, W g_1),$$

where V is an adjoint linear isometric operator mapping X^* on X, and
W is an adjoint linear isometric operator mapping Y^* on Y. To underline the difference of the operator $T^* = V T_1^* W^{-1}$ which is determined by formula (7.8) from the adjoint operator T_1^* which is given
by (6.4) in the general theory the operator T^* is called *Hilbert adjoint operator* with the operator T. Taking into account that in the
H–spaces theory only Hilbert adjoint operators are considered they are
called for brevity simply *adjoint operator*.

Notice that the left–hand side of equality (7.8) represents a scalar
product in the H–space X while the right–hand side is a scalar product
in the H–space Y. This difference is not marked in any way for avoiding
the complexity in the denotions. It cannot lead to the confusion because
it is always clear from the character of the multipliers to what space a
scalar product concerns.

If $D_T \neq X$ then at each $g \in Y$ there exists an infinite set of the
vectors $f \in X$ satisfying condition (6.8) which has in the given case the
form

$$(x, f) = (Tx, g). \tag{7.9}$$

But by virtue of Riesz theorem 7.3 in the domain D_T of the operator
T, and consequently, the functional (Tx, g) of x there always exists the
unique vector f, $f \in D_T$ satisfying condition (7.9). Taking this unique
vector f we may determine on the whole space Y the adjoint operator
T^* putting $T^* g = f$, $g \in Y$, $f \in D_T$. After this in the correspondence
with Theorem 6.14 and acconting the reflexivity of the H–space we may
determine on the whole space X the second adjoint operator T^{**} which
serves as the extension of the operator T on the whole space X with the
retention of the norm. Therefore later on we shall always assume that
the bounded linear operators are determined on the whole H–space.

On the basis of Theorem 6.14 the operators T and T^* are mutually
adjoint and their norms coincide, $\|T\| = \|T^*\|$.

E x a m p l e 7.3. The operator T of Example 6.5 at $p = h = 2$ maps
the H–space $L_2(\Delta, \mathcal{C}, \mu)$ into the H–space $L_2(S, \mathcal{L}, \nu)$. It is bounded as in the
consequence of inequality (1.17)

$$\|Tx\|_2^2 = \int |\int \varphi(s,t)\, x(t)\, \mu(dt)|^2\, \nu(ds) \leq \|x\|_2^2 \int \{\|\varphi\|_2\, (s)\}^2\, \nu(ds),$$

and the function $\|\varphi\|_2(s)$ by the condition belongs to $L_2(S, \mathcal{L}, \nu)$. As

$$\int \{\|\varphi\|_2(s)\}^2 \nu(ds) = \int \nu(ds) \int |\varphi(s,t)|^2 \mu(dt)$$
$$= \int \int |\varphi(s,t)|^2 \mu(dt)\nu(ds),$$

then the condition $\varphi(x,s) \in L_2(\Delta, \mathcal{C}, \mu), \|\varphi\|_2(s) \in L_2(S, \mathcal{L}, \nu)$ is equivalent to the condition of belonging the function $\varphi(s,t)$ of two variables s and t to the space $L_2(\Delta \times S, \mathcal{C} \times \mathcal{L}, \mu \times \nu)$. The adjoint operator T^* for this case is determined accordingly (7.8) by the formula

$$T^*g = \int \overline{\varphi(s,t)}\, g(s)\, \nu(ds).$$

E x a m p l e 7.4. The operator of the multiplication by independent variable Q_μ in the space $L_2([a,b],\mu)$ of the functions of real variable t on the interval $[a,b]$ with the finite measure μ is bounded at finite a, b as for any function $x(t) \in L_2([a,b],\mu)$ we have

$$\|Q_\mu x\|^2 = \int\limits_a^b t^2\, |x(t)|^2\, \mu(dt) \le c^2\, \|x\|^2,$$

where $c = \max\{|a|, |b|\}$. As

$$(Q_\mu x, z) = \int\limits_a^b t\, x(t)\, \overline{z(t)}\, \mu(dt) = (x, Q_\mu z),$$

then the adjoint operator Q_μ^* coincides with Q_μ.

E x a m p l e 7.5. The correspondence between the function $x(t)$ from the space $L_2([-\pi, \pi])$ and its Fourier coefficients

$$u_\nu = \frac{1}{2\pi} \int\limits_{-\pi}^{\pi} x(t)\, e^{-i\nu t} dt \quad (\nu = 0, \pm 1, \pm 2, \ldots)$$

represents the linear operator T mapping the H–space $L_2([-\pi, \pi])$ into the H–space l_2 of the vectors u with the components u_ν, $\nu = (0, \pm 1, \pm 2, \ldots)$, $u = Tx$. As

$$\|Tx\|^2 = \|u\|^2 = \sum_{\nu=-\infty}^{\infty} |u_\nu|^2 = \frac{1}{2\pi} \int\limits_{-\pi}^{\pi} |x(t)|^2\, dt = \frac{1}{2\pi}\, \|x\|^2$$

for all $x(t) \in L_2([-\pi, \pi])$ then the operator T is bounded and its norm is equal to $1/\sqrt{2\pi}$. The presentation of the function $x(t)$ by its Fourier series

$$x(t) = \sum_{\nu=-\infty}^{\infty} u_\nu e^{i\nu t}$$

determines the inverse operator $x = T^{-1}u$. This operator is also bounded and its norm is equal to $\sqrt{2\pi}$. Finally, from the relations

$$(Tx, z) = (u, z) = \sum_{\nu=-\infty}^{\infty} u_\nu \bar{z}_\nu = \frac{1}{2\pi} \int_{-\pi}^{\pi} x(t) \sum_{\nu=-\infty}^{\infty} \bar{z}_\nu e^{-i\nu t} dt = (x, v),$$

$$v(t) = \frac{1}{2\pi} \sum_{\nu=-\infty}^{\infty} z_\nu e^{i\nu t},$$

follows that the adjoint operator T^* is determined by formula

$$v = T^* z = \frac{1}{2\pi} T^{-1} z.$$

Thus in a given case $T^* = T^{-1}/2\pi$. It is easy to see that the operator $U = \sqrt{2\pi}T$ is isometric (Subsection 6.1.6).

7.2.2. Isometric and Unitary Operators in H–Spaces

Besides general properties studied in Subsections 6.1.6, 6.1.7 and 6.3.5 isometric and unitary operators in the H–spaces possess some other specific properties. Really, from the identity

$$(x, y) = \frac{1}{4}\{\|x + y\|^2 - \|x - y\|^2 + i\|x + iy\|^2 - i\|x - iy\|^2\},$$

which is valid for any vectors x, y of the H–space it follows directly that a linear operator mapping the H–space X into the H–space Y with the retention of the norm also retains the scalar product

$$(Vx, Vy) = (x, y) \quad \text{at all} \quad x, y \in X. \tag{7.10}$$

Consequently, the isometric and the unitary operators in the H–spaces retain the scalar products.

Remember that the isometric operator V always has the inverse operator V^{-1} (determined on the whole space Y) and take an arbitrary vector $u \in Y$ and put in equality (7.10) $y = V^{-1}u$. Then we obtain

$$(Vx, u) = (x, V^{-1}u) \quad \text{for all} \quad x \in X, \quad u \in Y.$$

Hence it is clear that the operator V^* adjoint with the isometric operator V coincides with the innverse operator $V^* = V^{-1}$.

Theorem 7.4. *If the linear operator V is determined on the whole H-space X and has the adjoint operator determined on the whole H-space Y and $(Vx_1, Vx_2) = (x_1, x_2)$ for all x_1, $x_2 \in X$, $(V^*y_1, V^*y_2) = (y_1, y_2)$ for all y_1, $y_2 \in Y$ then the operators V and V^* are isometric (and consequently, reciprocal).*

▷ If is sufficient to prove that the operator V maps X on the whole space Y. By definition of the adjoint operator $(Vx, y) = (x, V^*y)$ at any $x \in X$ and $y \in Y$. After taking an arbitrary vector $z \in X$ and putting $y = Vz$ we get $(Vx, Vz) = (x, V^*Vz)$ for all x, $z \in X$. On the other hand, $(Vx, Vz) = (x, z)$. Consequently, $V^*V = I$. Similarly after taking an arbitrary vector $u \in Y$ and putting $x = V^*u$ we receive $(VV^*u, y) = (V^*u, V^*y)$. On the other hand, $(V^*u, V^*y) = (u, y)$. Consequently, $VV^* = I$. Thus the operators V and V^* are reciprocal, and consequently, $R_V = D_{V^*} = Y$, $R_{V^*} = D_V = X$ which proves the statement. ◁

Let us pass to the unitary operators. We showed in Subsection 6.3.5 that the spectrum of an unitary operator is completely placed on the circle $| \lambda | = 1$ and, in particular, all eigenvalues are modulus equal to 1.

Theorem 7.5. *In the H-space the eigenvectors of the unitary operator U correspondent to any two different eigenvalues λ_1, λ_2 are orthogonal.*

▷ Really, from $Ux_1 = \lambda_1 x_1$, $Ux_2 = \lambda_2 x_2$ and the unitary property of U follows $(x_1, x_2) = (Ux_1, Ux_2) = \lambda_1 \bar{\lambda}_2 (x_1, x_2)$, and by virtue of $\lambda_2 \neq \lambda_1$ it is possible only at $(x_1, x_2) = 0$. ◁

E x a m p l e 7.6. It follows from the results of Example 7.5 that the operator $V = \sqrt{2\pi T}$ mapping $L_2([-\pi, \pi])$ on l_2 is isometric.

7.2.3. Fourier–Plancherel Operators

One of the most important for the applications isometric operators is the *Fourier–Plancherel operator*

$$y(\omega) = Fx = \frac{1}{\sqrt{2\pi}} \frac{d}{d\omega} \int_{-\infty}^{\infty} \frac{e^{i\omega t} - 1}{it} x(t)\, dt, \qquad (7.11)$$

which maps the space $L_2(R)$ of the functions of the variable t on the space $L_2(R)$ of the functions of the variable ω with the retention of the norm.

Theorem 7.6. *The Fourier–Plancherel operator represents an isometric operator* [a].

▷ For proving we shall take an integrable simple function

$$x(t) = \sum_{k=1}^{n} x_k 1_{[a_{k-1}, a_k)}(t).$$

Then we shall get

$$y(\omega) = Fx = \frac{1}{\sqrt{2\pi}} \sum_{k=1}^{n} x_k \frac{e^{ia_k\omega} - e^{ia_{k-1}\omega}}{i\omega}.$$

As the functions

$$\varphi_k(\omega) = \frac{e^{ia_k\omega} - e^{ia_{k-1}\omega}}{i\omega} \quad (k = 1, \ldots, n)$$

belong to $L_2(R)$, are orthogonal and

$$\int_{-\infty}^{\infty} |\varphi_k(\omega)|^2 \, d\omega = 4 \int_{0}^{\infty} \frac{1 - \cos(a_k - a_{k-1})\omega}{\omega^2} \, d\omega = 2\pi(a_k - a_{k-1}),$$

then the function $y(\omega)$ belongs to $L_2(R)$ and

$$\int_{-\infty}^{\infty} |y(\omega)|^2 \, d\omega = \sum_{k=1}^{n} |x_k|^2 \, (a_k - a_{k-1}) = \int_{-\infty}^{\infty} |x(t)|^2 \, dt.$$

Remember now that the set of integrable simple functions is dense in L_2 (Subsection 3.7.5). Consequently, for any function $x(t) \in L_2(R)$ we may find the sequence of integrable simple functions $\{x^n(t)\}$ convergent in mean square (i.e. on the norm of the space L_2) to $x(t)$, $x^n(t) \xrightarrow{2} x(t)$.

---

[a] From the mathematical point of view the variables t and ω may be considered as different denotions of one and the same variable. In this case the Fourier–Plancherel operator maps the space $L_2(R)$ on the same space $L_2(R)$ with the retention of the norm, and consequently, is an unitary one. But in applications the variables t and ω have different physical sense (for instance, time and frequency). So, it is expedient to consider the space of the functions t and the space of the functions ω as two different spaces.

Let $y_n(\omega) = Fx^n$. As at any n, m the function $x^n(t) - x^m(t)$ is an integrable simple function then according to the proved

$$\int_{-\infty}^{\infty} |y_n(\omega) - y_m(\omega)|^2 \, d\omega = \int_{-\infty}^{\infty} |x^n(t) - x^m(t)|^2 \, dt \,.$$

Then it follows that the sequence $\{y_n(\omega)\}$ is fundamental in L_2 together with $\{x^n(t)\}$. By virtue of the completeness of L_2 there exists the limit function $y(\omega) \in L_2(R)$, $y_n(\omega) \xrightarrow{\;2\;} y(\omega)$ and

$$\left| \int_0^\omega y_n(\sigma)d\sigma - \int_0^\omega y(\sigma)d\sigma \right|^2 \leq \int_0^\omega d\sigma \cdot \int_0^\omega |y_n(\sigma) - y(\sigma)|^2 d\sigma \leq \omega \, \|y_n - y\|^2 \,.$$

Consequently,

$$\int_0^\omega y(\sigma)d\sigma = \lim \int_0^\omega y_n(\sigma)d\sigma = \lim \frac{1}{\sqrt{2\pi}} \int_{-\infty}^{\infty} \frac{e^{i\omega t} - 1}{it} x^n(t) \, dt \,.$$

As

$$z_\omega(t) = \frac{e^{i\omega t} - 1}{-\sqrt{2\pi}it} = \frac{1 - e^{i\omega t}}{\sqrt{2\pi}it} \in L_2(R), \qquad (7.12)$$

then the latter integral represents a scalar product of the elements $x^n(t)$ and $z_\omega(t)$ of the space $L_2(R)$. Thus

$$\int_0^\omega y(\sigma)d\sigma = \lim(x^n, z_\omega) = (x, z_\omega) = \frac{1}{\sqrt{2\pi}} \int_{-\infty}^{\infty} \frac{e^{i\omega t} - 1}{it} x(t) \, dt \,.$$

Hence it follows that $y(\omega) = Fx$.

So, we proved that to any function $x(t) \in L_2(R)$ the operator F sets up the correspondence of the function $y(\omega) \in L_2(R)$. Thus the whole space $L_2(R)$ is the domain D_F of the operator F.

Finally, from the fact that the operator F retains the norm in L_2 at the transformation of a simple function and two evident equalities $\|x\| = \lim \|x^n\|$, $\|y\| = \lim \|y_n\|$ follows

$$\|y\|^2 = \int_{-\infty}^{\infty} |y(\omega)|^2 d\omega = \lim \int_{-\infty}^{\infty} |y_n(\omega)|^2 d\omega$$

$$= \lim \int\limits_{-\infty}^{\infty} |x^n(t)|^2 dt = \int\limits_{-\infty}^{\infty} |x(t)|^2 dt = \| x \|^2 .$$

Thus the operator F maps the space $L_2(R)$ of the functions of the variable t into the space $L_2(R)$ of the function of the variable ω with the retention of the norm. It remains to prove that the range R_F of the operator F represents the whole space $L_2(R)$.

As any linear operator in the H–space retaining the norm retains also a scalar product then for any functions $x_1(t)$, $x_2(t) \in L_2(R)$ and $y_1(\omega) = Fx_1$, $y_2(\omega) = Fx_2$ we have

$$\int\limits_{-\infty}^{\infty} x_1(\tau)\overline{x_2(\tau)}\, d\tau = \int\limits_{-\infty}^{\infty} y_1(\omega)\overline{y_2(\omega)}\, d\omega .$$

After taking $x_1(\tau) = x(\tau)$, $x_2(\tau) = 1_{[0,t]}(\tau)$ and correspondingly

$$y_1(\omega) = y(\omega) = Fx , \quad y_2(\omega) = \frac{e^{it\omega} - 1}{\sqrt{2\pi}\, it} ,$$

we receive

$$\int\limits_{0}^{t} x(\tau)d\tau = \frac{1}{\sqrt{2\pi}} \int\limits_{-\infty}^{\infty} \frac{1 - e^{-it\omega}}{it} y(\omega)\, d\omega .$$

Hence it follows that almost at all t

$$x(t) = \frac{1}{\sqrt{2\pi}} \frac{d}{dt} \int\limits_{-\infty}^{\infty} \frac{1 - e^{-it\omega}}{it} y(\omega)\, d\omega .$$

This formula determines the inverse operator F^{-1} which evidently possesses the same properties that the operator F. In particular, the whole space $L_2(R)$ is the domain $D_{F^{-1}}$ of the inverse operator F^{-1}. As the domain of the inverse operator always coincides with the range of the given operator then $R_F = D_{F^{-1}} = L_2(R)$. This proves that the Fourier–Plancherel operator F maps the whole space $L_2(R)$ on the whole space $L_2(R)$. ◁

If $x(t)$ belongs to $L_1(R)$ (i.e. $x(t)$ is an integrable in Lebesgue sense) then the function

$$z(t) = x(t)\frac{e^{it\omega} - 1}{it}$$

belongs to the space $L_1(R)$ of the functions t with the values in the space $C^1(R)$ of the differentiable functions of the variable ω. Therefore definiton (7.11) we differentiate with respect to ω under the integral sign (Example 7.6). As a result we get

$$y(\omega) = \frac{1}{\sqrt{2\pi}} \int\limits_{-\infty}^{\infty} e^{i\omega t} x(t)\, dt\,.$$

This formula shows that the Fourier–Plancherel operator coincides with the Fourier operator on the subspace of the integrable functions from $L_2(R)$ (i.e. on $L_1(R) \cap L_2(R)$). As $L_1(R)$ contains, in particular, all integrable simple functions then the domain of the Fourier operator is dense in $L_2(R)$. Thus the Fourier–Plancherel operator is the extension of the Fourier operator by the continuity from the space $L_1(R) \cap L_2(R)$ on the all space $L_2(R)$.

Let us now consider t and ω as different denotions of one and the same independent variable. Then the Fourier–Plancherel operator F will map the space $L_2(R)$ on the same space $L_2(R)$, i.e. will be unitary. From the formula (Appendix 2)

$$\int\limits_{-\infty}^{\infty} e^{it\tau + \tau^2/2} \frac{d^n e^{-\tau^2}}{d\tau^n}\, d\tau = i^n \sqrt{2\pi} e^{t^2/2} \frac{d^n e^{-t^2}}{dt^n}$$

follows that the Hermite functions

$$\varphi_n(t) = (-1)^n e^{t^2/2} \frac{d^n e^{-t^2}}{dt^n} = H_n(t) e^{-t^2/2} \quad (n = 0, 1, 2, \dots)$$

satisfy the equation

$$F\varphi_n = \frac{1}{\sqrt{2\pi}} \int\limits_{-\infty}^{\infty} e^{it\tau} \varphi_n(\tau)\, d\tau = i^n \varphi_n(t)\,.$$

Hence taking into account that $\varphi_n(t) \in L_1(R)$, we see that the numbers $1,\ i,\ -1,\ -i$ are the eigenvalues of the Fourier–Plancherel operator, and the functions $\varphi_{4n}(t)$, $\varphi_{4n+1}(t)$, $\varphi_{4n+2}(t)$, $\varphi_{4n+3}(t)$ $(n = 0, 1, 2, \dots)$ are the correspondent eigenfunctions. Thus the Fourier–Plancherel operator has the following four eigenvalues of the infinite multiplicity: $1,\ i,\ -1,\ -i$. The subspace formed by all linear combinations of the Hermite functions coincides with the whole space $L_2(R)$. From this fact on the basis of Theorem 7.5 follows that the Fourier–Placherel operator has no other eigenvalues.

In the same way for $z_\omega(t)$ determined by formula (7.12) it is proved that the Fourier–Plancherel operator

$$y(\omega_1, \ldots, \omega_n) = Fx = \frac{\partial^n}{\partial\omega_1 \ldots \partial\omega_n} \int\limits_{-\infty}^{\infty} \cdots \int\limits_{-\infty}^{\infty} \overline{z_{\omega_1}(t_1)}$$

$$\ldots \overline{z_{\omega_n}(t_n)}\, x(t_1, \ldots, t_n)\, dt_1 \ldots t_n\,, \qquad (7.13)$$

is the isometric operator which maps the space $L_2(R^n)$ of the functions of n–dimensional vector t on the space $L_2(R^n)$ of the functions of n–dimensional vector ω.

R e m a r k. If we consider t and ω as different denotions of one and the same vector variable then F will be an unitary operator whose eigenvalues are only the numbers $1, i, -1, -i$. As the correspondent eigenvectors serve the products of the Hermite functions of the variable $t_1, \ldots, t_n$ (Appendix 2) whose indexes sums are the four modulus–comparable with 0, 1, 2, 3 correspodingly.

7.2.4. Unbounded Linear Operators

On the basis of Theorem 6.7 the operator T^* adjoint with unbounded linear operator T exists if and only if the domain of the operator T is dense in the space X, $[D_T] = X$. In the case of the H–spaces X and Y, $TX \subset Y$, the adjoint operator T^* analogously as in the case of the bounded operator T is determined by equality (7.8)

$$(x, T^*g) = (Tx, g), \quad x \in D_T\,. \qquad (7.14)$$

The domain of the operator T^* in the space Y (with which in the case of the H–space Y coincides Y^*) is the set G of the vectors $g \in Y$ (Theorem 6.7). For each of vector g there exists such a vector $f \in X$ that the kernel of the linear functional $h = \{f, g\}$ on the product of spaces $Z = X \times Y$ contains the plot $\mathrm{Gr}(T)$ of the operator T transformed by the operator U, $U\{x, y\} = \{x, -y\}$ and therewith $T^*g = f$ (Subsection 6.1.4). It easy to see that in the case of the H–spaces X and Y this set represents the set of all those vector g for which the pair $\{f, g\} = \{T^*g, g\} \in Z$ is orthogonal to the pair $\{x, -Tx\} \in Z$, i.e. belongs to the orthogonal supplement of the closure of the operator T plot transformed by the operator U in the product of the spaces $Z = X \times Y$, $\{f, g\} \in [U\mathrm{Gr}(T)]^\perp$. Thus the subspace $[U\mathrm{Gr}(T)]^\perp$ serves as a plot of the adjoint operator T^*.

The operator U in the case of the B–spaces (in the special case of the H–spaces) X and Y is unitary as for any $x \in X$, $y \in Y$, $\{x, y\} \in X \times Y = Z$,

$$\| \{x, y\} \| = \| x \| + \| y \| = \| \{x, -y\} \| \,.$$

If there exists the second adjoint operator T^{**} (i.e. if D_{T^*} is dense in Y) then it is the extension of T, $D_T \subset D_{T^{**}}$, $T^{**}x = Tx$ at all $x \in D_T$. It is clear from equality (7.8) and analogous equality which determines the adjoint operator T^{**}:

$$(x, T^*g) = (T^{**}x, g)\,, \quad g \in D_{T^*}\,.$$

But by Theorem 6.8 the operator T^{**} is closed as it is adjoint with T^*. And the closed operator T^{**} may be the extension of T only in that case when T admits the closure.

Theorem 7.7. *If T is a closed operator then the second adjoint operator T^{**} exists and coincides with T, $T^{**} = T$.*

▷ In order to prove theorem we notice that in this case the plot $\mathrm{Gr}(T)$ of the operator T is closed and by virtue of the continuity of the unitary operator U^{-1} the set $U\mathrm{Gr}(T)$ is also closed, and consequently, $[U\mathrm{Gr}(T)] = U\mathrm{Gr}(T)$, $\mathrm{Gr}(T^*) = U\mathrm{Gr}(T)^\perp$. Thus in the case of the closed operator T the space Z is an orthogonal sum of the plot of the adjoint operator T^* and the plot of the operator T transformed by the operator U:

$$Z = \mathrm{Gr}(T^*) \oplus U\mathrm{Gr}(T)\,. \tag{7.15}$$

Notice now that the operator U as an unitary one and maps the space Z on the whole space Z (it simply changes the sign of the second element of each pair $\{x, y\} \in Z$). Operator U transforms the set $\mathrm{Gr}(T^*)$ into $U\mathrm{Gr}(T^*)$, and the set $U\mathrm{Gr}(T)$ – into $\mathrm{Gr}(T)$ as $U^2 = I$. The orthogonal sum of the transformed subspaces $U\mathrm{Gr}(T^*)$ and $\mathrm{Gr}(T)$ coincides with the whole space Z. Therefore at transforming the space Z by the operator U equality (7.15) will become

$$Z = U\mathrm{Gr}(T^*) \oplus \mathrm{Gr}(T)\,. \tag{7.16}$$

The second adjoint operator T^{**} exists if and only if the subspace $U\mathrm{Gr}(T^*)^\perp$ may serve as a plot of a closed operator, i.e. when each pair

$\{x, y\} \in U\mathrm{Gr}(T^*)^\perp$ is completely determined by the vector x. In a given case this condition is fulfilled as from equality (7.16) follows that $U\mathrm{Gr}(T^*)^\perp = \mathrm{Gr}(T)$, i.e. is the plot of the closed operator T. Thus in a given case the operator T^{**} exists and its plot $U\mathrm{Gr}(T^*)^\perp$ coincides with the plot $\mathrm{Gr}(T)$ of the operator T. Consequently, $T^{**} = T$. ◁

Corollary. *If the operator T admits the closure $\bar{T}$ then the second adjoint operator T^{**} exists and coincides with $\bar{T}$.*

▷ Really, it follows from the proved theorem that in a given case there exists $\bar{T}^{**} = \bar{T}$. But by Theorem 6.9 $\bar{T}^* = T^*$. Consequently, the operator T^* has the adjoint operator T^{**} which coincides with $\bar{T}^{**}$, $T^{**} = \bar{T}^{**} = \bar{T}$. ◁

Thus for the existence of the second adjoint operator T^{**} it is necessary and sufficient the existence of the closure of $\bar{T}$ of the operator T and therewith $T^{**} = \bar{T}$.

It follows from definition (7.8) that if there exists T^* then for any vector $g \in [R_T]^\perp$

$$(x, T^*g) = (Tx, g) = 0$$

at all $x \in D_T$. As D_T is dense in X then it is possible only at $T^*g = 0$. And vice versa if $T^*g = 0$ then $(Tx, g) = 0$, i.e. $g \in [R_T]^\perp$. Consequently, the subspace $[R_T]^\perp$ is a kernel of the operator T^*. In other words, the range R_T of the operator T is orthogonal to the kernel of the adjoint operator T^*.

Theorem 7.8. *If the operator T has a dense in X the domain D_T and dense in Y the range R_T and there exists the inverse operator T^{-1} then the adjoint operator T^* has the inverse operator $(T^*)^{-1}$, $(T^*)^{-1} = (T^{-1})^*$.*

▷ In this case by virtue of the density of the domains of the operators T and T^{-1} there exist the adjoint operators T^* and $(T^{-1})^*$ and at any $x \in D_T$, $y = Tx$, $g \in D_{T^*}$, $f = T^*g$ we have $(x, f) = (y, g)$. Therewith by virtue of the density of D_T to each $g \in D_{T^*}$ corresponds the unique vector $f \in R_{T^*}$. And as this equality is valid for any $y \in D_{T^{-1}} = R_T$ and $x = T^{-1}y$ then by virtue of the density of R_T to each $f \in R_{T^*}$ corresponds the unique vector $g \in D_{T^*}$. Consequently, the adjoint operator T^* has the inverse operator $(T^*)^{-1}$. After writing equality (7.14) in the form

$$(Tx, g) = (T^{-1}Tx, T^*g) = (Tx, (T^{-1})^*T^*g) \quad \text{at all} \quad g \in D_{T^*},$$

by virtue of the density of R_T we get $(T^{-1})^*T^* = I$. On the other hand, after writing equality (7.14) by means of $x = T^{-1}y$ in the form

$$(T^{-1}y, f) = (TT^{-1}y, (T^{-1})^*f) = (T^{-1}y, T^*(T^{-1})^*f) \quad \text{at all} \quad f \in R_{T^*},$$

owing to the density of $R_{T^{-1}}$ we receive $T^*(T^{-1})^* = I$. Two received equalities show that $(T^*)^{-1} = (T^{-1})^*$. ◁

Thus if the operator T has the inverse T^{-1} and there exist the adjoint operators T^* and $(T^{-1})^*$ then the operator T^* has also the inverse $(T^*)^{-1}$ and $(T^*)^{-1} = (T^{-1})^*$.

E x a m p l e 7.7. Consider the operator Q_μ of the multiplication by the independent variable t in the space $L_2(R, \mu)$ in the case when the measure μ is finite on any finite interval of the real axis R, and for any infinite interval measure μ is infinite $\mu((-\infty, b)) = \mu((a, \infty)) = \infty$ at all a and b. The domain of the operator Q_μ evidently contains all finite functions from L_2, i.e. the functions with bounded carrier. In particular, D_{Q_μ} contains all the μ–integralbe simple functions. As by Theorem 3.40 the set of the μ–integrable simple is dense in L_2 then the domain of D_{Q_μ} of the operator Q_μ is dense in L_2. The operator Q_μ is unbounded as at any $c > 0$ in D_{Q_μ} there will be the finite function $x(t)$ equal to zero at $\mid t \mid < c$ and for this function

$$\| Q_\mu x \|^2 = \int t^2 \mid x(t) \mid^2 \mu(dt) \geq c^2 \int \mid x(t) \mid^2 \mu(dt) = c^2 \| x \|^2 \ .$$

It follows from the density of D_{Q_μ} in $L_2(R, \mu)$ that there exists the adjoint operator Q_μ^*. Equality (7.14) which determines Q_μ^* in a given case has the form

$$(Q_\mu x, g) = \int t x(t) \, \overline{g(t)} \, \mu(dt) = (x, Q_\mu^* g) \ .$$

Here the integral represents the scalar product of $x(t)$ by some fucntion from $L_2(R, \mu)$ only in that case when $tg(t) \in L_2(R, \mu)$, and consequently, $g(t) \in D_{Q_\mu}$. Thus the domain Q_μ^* coincides with the domain Q_μ and $Q_\mu^* g = tg(t)$ at all $g(t) \in D_{Q_\mu}$. Consequently, $Q_\mu^* = Q_\mu$, i.e. the operator of the multiplication by an independent variable in $L_2(R, \mu)$ coincides with its adjoint operator.

E x a m p l e 7.8. Consider the operator of the differentiation D in $L_2(R)$. Its domain contains the Hermite functions (Subsection 7.2.3 and Appendix 2), and consequently, all their finite linear combinations. We shall know in Subsection 7.5.5 that the set of finite linear combinations of Hermite functions is dense in $L_2(R)$. Consequently, the domain of the operator of the differentiation is dense in $L_2(R)$. It means that there exists the adjoint operator D^*. Eq. (7.14) which determines D^* has the form

$$(Dx, g) = \int\limits_{-\infty}^{\infty} x'(t) \, \overline{g(t)} \, dt = (x, D^* g) \ .$$

Here the integral may be presented as a scalar product of $x(t)$ by some function from $L_2(R)$ only in that case when $g(t)$ is differentiable and $g'(t) \in L_2(R)$, i.e. when $g(t)$ belongs to the domain of the operator D. In this case while integrating by parts and taking into account that any differentiable function $z(t) \in L_2(R)$ satisfies the conditions $z(-\infty) = z(\infty) = 0$ we get

$$(Dx, g) = \int\limits_{-\infty}^{\infty} x'(t)\, \overline{g(t)}\, dt = -\int\limits_{-\infty}^{\infty} x(t)\, \overline{g'(t)}\, dt\,.$$

After comparing this formula with the previous one we come to the conclusion that the domain of the operator D^* coincides with the domain of the operator D and that $D^* g = -g'(t)$, i.e. $D^* = -D$. Thus the operator D^* adjoint with the operator of the differentiation with respect to the variable t in $L_2(R)$ represents the operator of the differentiation with respect to the correspondent negative variable (i.e. with respect to $-t$).

The detailed treatment of bounded and unbounded linear operators, isometric and unitary operators in H–spaces is given in (Balakrishnan 1976, Dowson 1978, Dunford and Schwarz 1958).

P r o b l e m s

7.2.1. Under the conditions of Example 7.4 find an operator adjoint with the operator of the multiplication by a bounded function.

7.2.2. Find an operator adjoint with the operator

$$\Delta\{x_1, x_2, \ldots\} = \{\underbrace{0 \ldots 0}_{n-1}, x_1, x_2, \ldots\}$$

in the space l_2.

7.2.3. Find an operator adjoint with the operator

$$A\{x_1, x_2, \ldots\} = \{y_1, y_2, \ldots\}, \quad y_n = \sum_{m=1}^{\infty} a_{nm}\, x_m\,,$$

in l_2 under the condition

$$\sum_{m=1}^{\infty} |a_{nm}|^2 < \infty \quad \forall n\,.$$

7.2.4. At what conditions the operator of Problems 7.2.3 will be unitary?

7.2.5. Under the conditions of Example 7.7 find the operator of the multipication by a bounded function from $L_2(R, \mu)$.

7.2.6. Find the operator adjoint with $T = c_1 T_1 + \cdots + c_n T_n$ if each of the operators T_k has the adjoint operator T_k^*. Explain the discrepancy between the obtained result with the result of Problem 6.3.18.

7.2.7. Prove that $|\,A^* A\,| = |\,A A^*\,| = \|\,A\,\|^2 = \|\,A^*\,\|^2$ for any bounded operator in a H–space.

7.3. Self–Adjoint Operators

7.3.1. Self–Adjoint and Symmetric Operators

A linear operator A mapping the H–space X into X is called *self–adjoint* or *self–conjugate* if it coincides with its adjoint $A = A^*$. It follows from this definition that the operator A is self–adjoint if and only if

(i) there exists the adjoint operator A^*;

(ii) the domain of the adjoint operator A^* coincides with the domain of the operator A, $D_{A^*} = D_A$;

(iii) $A^*x = Ax$ at all $x \in D_A$.

If only the first and the third of these conditions are fulfilled then the operator A is called *symmetric*. Thus a self–adjoint operator is a special case of a symmetric operator.

It follows from definition (7.14) that the symmetric operator A satisfies the condition

$$(Ax, y) = (x, Ay) \quad \text{at all} \quad x, y \in D_A. \tag{7.17}$$

The domain of the operator A^* adjoint with the symmetric operator A cannot be narrower than the domain of the operator A, $D_{A^*} \supset D_A$. If follows from the third condition in accordance of which the operator A^* is determined and coincides with A everywhere in D_A. Thus the adjoint operator A^* serves as the extension of the symmetric operator A. But in the general case this extension may be not symmetric. If A^* is a symmetric operator then it is self–adjoint as according to Remark after Theorem 6.7 for any (also including a symmetric one) extension B of the operator A from $D_B \supset D_A$ follows $D_{B^*} \subset D_{A^*}$. In this case the symmetric operator A may be extended till the self–adjoint operator and as its extension serves A^*.

As by Theorem 6.8 the adjoint operator of any linear operator is closed then the self–adjoint operator is always closed.

If the self–adjoint operator A is bounded then it is determined on the whole space $D_A = X$ (Theorem 6.6). And vice versa, if the self–adjoint operator A is determined on the whole space, $D_A = X$, then on the basis of Theorem 6.15 it is bounded.

The domain of the unbounded self–adjoint operator on the basis of Theorem 6.7 is dense in X.

Theorem 7.9. *If the self–adjoint operator A has the inverse operator A^{-1} then the operator A^{-1} is also the self–adjoint operator.*

▷ At first we show that the domain $D_{A^{-1}} = R_A$ of the inverse operator is dense in X. If R_A is not dense then there exists in X the vector $z \neq 0$ orthogonal to R_A, $(Ax, z) = 0$ at all $x \in D_A$. It is clear that the functional of x (Ax, z) is bounded on D_A. Consequently, z belongs to the domain of the adjoint operator A^* coinciding in a given case with A (Subsection 6.1.3), $z \in D_{A^*} = D_A$ and $(Ax, z) = (x, Az) = 0$ at all $x \in D_A$. As D_A is dense in X then $Az = 0$. But it is impossible at $z \neq 0$ as there exists the inverse operator A^{-1}. Thus there is no vector $z \neq 0$ in X orthogonal to R_A what proves the density fo R_A in X. But then the inverse operator has the adjoint operator $(A^{-1})^*$ which by Theorem 7.8 coincides with the operator $(A^*)^{-1} = A^{-1}$. This proves the self–adjointness of the operator A^{-1}. ◁

Theorem 7.10. *If the range R_A of the symmetric operator A coincides with the whole space X then the operator A is a self-adjoint operator.*

▷ Really, for any $x \in D_A$, $y \in D_{A^*}$ we have

$$(Ax, y) = (x, A^*y).$$

As $R_A = X$ then there exists such vector $z \in D_A$ that $Az = A^*y$ and by virtue of the symmetry of A we get

$$(Ax, y) = (x, Az) = (Ax, z).$$

But due to the fact that $R_A = X$ it is possible only at $y = z$. Consequently, $y \in D_A$, i.e. $D_{A^*} \subset D_A$ what in the connection with the inclusion $D_{A^*} \supset D_A$ which follows from the definition of a symmetric operator gives $D_{A^*} = D_A$. ◁

7.3.2. Formula for the Norm of a Self–Adjoint Operator

For study the spectrum of the self–adjoint operator the following theorem is useful.

Theorem 7.11. *If the self-adjoint operator A is bounded then its norm is determined by formula*

$$\| A \| = \sup \frac{|(Ax, x)|}{\| x \|^2}. \tag{7.18}$$

▷ As $|(Ax, x)| \leq \| Ax \| \| x \| \leq \| A \| \| x \|^2$ then

$$c = \sup \frac{|(Ax, x)|}{\| x \|^2} \leq \| A \|. \tag{7.19}$$

On the other hand, from the identity

$$(Au, v) + (Av, u) = \frac{1}{2}\{(Au + Av, u + v) - (Au - Av, u - v)\},$$

which is valid for any $u, v \in X$ and for any operator A mapping X into X and from the fact that $|(Az, z)| \leq c \, \|z\|^2$ for any $z \in X$ follows

$$|(Au, v) + (Av, u)| \leq \frac{c}{2}(\|u + v\|^2 + \|u - v\|^2) = c(\|u\|^2 + \|v\|^2). \tag{7.20}$$

Let x, y be arbitrary vectors $x, y \in X$. After presenting the complex number (Ax, y) in the exponential form $(Ax, y) = re^{i\alpha}$, $r > 0$ and putting

$$u = \frac{xe^{-i\alpha}}{\|x\|}, \quad v = \frac{y}{\|y\|},$$

we get

$$(Au, v) = (Av, u) = \frac{r}{\|x\| \|y\|}, \quad \|u\| = \|v\| = 1.$$

Substituting these expressions into inequality (7.20) we shall have

$$\frac{2r}{\|x\| \|y\|} \leq 2c,$$

whence $r = |(Ax, y)| \leq c \, \|x\| \, \|y\|$. At $y = Ax$ hence it follows $\|Ax\|^2 \leq c \, \|x\| \, \|Ax\|$ and $\|Ax\| \leq c \, \|x\|$ at all x. Consequently, $\|A\| \leq c$. Hence and from inequality (7.19) it follows that $\|A\| = c$, i.e. inequality (7.18). ◁

7.3.3. Spectrum of a Self-Adjoint Operator

Theorem 7.12. *All eigenvalues of a self-adjoint operator are real.*

▷ Suppose that λ is an eigenvalue and x is the correspondent eigenvector. Then $Ax = \lambda x$ and $(Ax, x) = \lambda \, \|x\|^2$ whence

$$\lambda = \frac{(Ax, x)}{\|x\|^2}.$$

Whence by virtue of condition (7.17) follows that λ is a real number. ◁

Corollary. *If the adjoint operator A is a positive one, i.e. $(Ax, x) \geq 0$ at all $x \in D_A$ then all eigenvalues of the operator A are nonnegative.*

Theorem 7.13. *The eigenvectors of the self-adjoint operator correspondent to different eigenvalues are orthogonal.*

▷ Let λ_1 and λ_2 be the different eigenvalues of the self-adjoint operator A, $\lambda_1 \neq \lambda_2$ and x_1 and x_2 be the correspondent eigenvectors. Then $Ax_1 = \lambda_1 x_1$, $Ax_2 = \lambda_2 x_2$ and $(Ax_1, x_2) = \lambda_1(x_1, x_2)$, $(x_1, Ax_2) = \lambda_2(x_1, x_2)$. And as $(Ax_1, x_2) = (x_1, Ax_2)$ then $(\lambda_1 - \lambda_2)(x_1, x_2) = 0$ whence $(x_1, x_2) = 0$. ◁

R e m a r k. This theorem is also true in that case when A is a symmetric operator.

Theorem 7.14. *In order that λ be an eigenvalue of the self-adjoint operator A it is necessary and sufficient that the closure of the range $R_A(\lambda)$ of the operator $A - \lambda I$ do not coincide with the whole space X, $[R_A(\lambda)] \neq X$.*

▷ If λ is an eigenvalue, and x is the correspondent eigenvector then as λ is a real then $0 = (Ax - \lambda x, y) = (x, Ay - \lambda y)$ at all $y \in D_A$. Hence it follows that the eigenvector x is orthogonal to the range $R_A(\lambda)$ of the operator $A - \lambda I$, i.e. $x \in [R_A(\lambda)]^\perp$, and consequently, $[R_A(\lambda)] \neq X$. And vice versa if $[R_A(\lambda)] \neq X$ then for any x from the orthogonal supplement $[R_A(\lambda)]$, $x \in [R_A(\lambda)]^\perp$ $(x, Ay - \lambda y) = 0$ at any $y \in D_A$. After writing this equality in the form $(x, Ay) - (\bar{\lambda} x, y) = 0$ we see that the pair $\{y, -Ay\}$ belongs to the kernel of the functional $h = \{x, \bar{\lambda} x\}$. According to Theorem 6.7 by virtue of the self-adjointness of A follows $x \in D_{A^*} = D_A$ and $\bar{\lambda} x = A^* x = Ax$. This proves that $\bar{\lambda}$ is an eigenvalue of the operator A and x is the correspondent eigenvector. But then by Theorem 7.12 $\bar{\lambda}$ is a real number and $\bar{\lambda} = \lambda$. Consequently, λ is an eigenvalue of the operator A. ◁

Corollary 1. *The eigensubspace of the self-adjoint operator A correspondent to the eigenvalue λ represents an orthogonal supplement of the subspace $[R_A(\lambda)]$ in X.*

Corollary 2. *The residual spectrum of the self-adjoint operator is empty, $\sigma_r(A) = \varnothing$.*

▷ Really, it follows from the fact as for any $\lambda \notin \sigma_p(A)$ we have $[R_A(\lambda)] = X$. ◁

Theorem 7.15. *Any complex value of λ with an imaginary part different from zero is a regular point of the self-adjoint operator A.*

▷ Let us take any complex value of the parameter λ, $\lambda = \mu + i\nu$, $\nu \neq 0$. This value of λ cannot be an eigenvalue. Therefore $[R_A(\lambda)] = X$ and there exists the resolvent $R_\lambda = (A - \lambda I)^{-1}$. Let us take an arbitrary vector $x \in D_A$ and put $y = Ax - \lambda x = Ax - \mu x - i\nu x$. Then we shall have in consequence of definition (7.17)

$$\| y \|^2 = (y,y) = \| Ax - \mu x \|^2 - i\nu(x, Ax - \mu x) + i\nu(Ax - \mu x, x) + \nu^2 \| x \|^2$$

$$= \| Ax - \mu x \|^2 + \nu^2 \| x \|^2 \geq \nu^2 \| x \|^2,$$

whence $\| x \| \leq \| y \| / | \nu |$ at all $y \in R_A(\lambda)$. Consequently, the operator R_λ is bounded. Operators A and $A - \lambda I$ are closed and therefore the inverse operator R_λ is continuous and closed. But it takes place only in that case when its domain $R_A(\lambda)$ is closed (Theorem 6.2). Consequently, $R_A(\lambda) = [R_A(\lambda)] = X$, i.e. the bounded operator R_λ is determined on the whole space X. Hence it follows that λ is a regular point of the operator A. ◁

Corollary. *The spectrum of the self–adjoint operator is displaced entirely on the real axis.*

Theorem 7.16. *The spectrum of the positive self–adjoint operator A is displaced entirely on the positive part of the real axis.*

▷ It is sufficient to show that any real $\lambda < 0$ is a regular point. As by the Corollary of Theorem 7.12 $\lambda < 0$ cannot be an eigenvalue then $[R_A(\lambda)] = X$ and there exists the resolvent $R_\lambda = (A - \lambda I)^{-1}$. After taking an arbitrary $x \in D_A$ and putting $y = Ax - \lambda x$ we shall have

$$\| y \|^2 = \| Ax \|^2 - 2\lambda(Ax, x) + \lambda^2 \| x \|^2 \geq \lambda^2 \| x \|^2,$$

in consequence of that $(Ax, x) \geq 0$ at all $x \in D_A$, and $\lambda < 0$. So we have $\| x \| \leq \| y \| / | \lambda |$ at all $y \in R_A(\lambda)$, i.e. the resolvent R_λ is bounded on $R_A(\lambda)$. While proving Theorem 7.15 it was shown that R_λ is a closed operator. Consequently, $R_A(\lambda) = [R_A(\lambda)] = X$, i.e. the bounded resolvent R_λ is determined on the whole space X. ◁

Theorem 7.17. *If $R_A(\lambda) = X$ then λ is a regular point of the self–adjoint operator A.*

▷ As the self–adjoint operator is closed (Subsection 7.3.1) then the condition $R_A(\lambda) = X$ is necessary and sufficient for the regularity of λ (Subsection 6.3.2). ◁

R e m a r k. So, we proved that for the self–adjoint operator A all the values of λ for which $R_A(\lambda) = X$ are regular points, all the values of λ for which $[R_A(\lambda)] \neq X$ are the eigenvalues, and all the values of λ for which $R_A(\lambda) \neq X$ are the points of the spectrum different from the eigenvalues (the points of continuous spectrum).

E x a m p l e 7.9. It was shown in Examples 7.4 and 7.7 that the operator Q_μ of the multiplication by the independent variable t in the space $L_2(R, \mu)$ is self–adjoint. If the measure μ is completely concentrated on the finite interval $[a, b]$ then Q_μ is a bounded self–adjoint operator determined on the whole space $L_2(R, \mu)$. If the measure μ is finite on any finite interval and is infinite on any infinite interval then Q_μ is an unbounded self–adjoint operator with dense domain in $L_2(R, \mu)$.

In exactly the same way as in Example 6.32 we come to the conclusion that the operator Q_μ of the multiplication by an independent variable in the space $L_2(R, \mu)$ has no eigenvalue and its spectrum serves the whole real axis if the measure μ is not concentrated on any finite interval. According to Corollary 2 of Theorem 7.16 the whole spectrum of the operator Q_μ is continuous $\sigma(Q_\mu) = \sigma_c(Q_\mu)$.

E x a m p l e 7.10. From the results of Example 7.8 follows that the operator D_i of the differentiation with respect to the imaginary argument it in $L_2(R)$ represents a non–bounded self–adjoint operator. Its domain is dense in $L_2(R)$. Operator D_i in $L_2(R)$ is also unitary equivalent to the operator of the multiplication by the independent variable Q in $L_2(R)$ (Subsection 6.1.8).

For proving this we notice that any function $x(t) \in L_2(R)$ which belongs to the domain D_Q of the operator Q also belongs to $L_1(R)$. Really, distinguishing from the real axis the finite interval $A = (-a, a)$ we may present the function $x(t)$ in $L_2(A)$ in the form of the product of the unit functions $y(t) \equiv 1$ and $x(t)$ which belong to $L_2(A)$. Analoguosly in $L_2(R \backslash A)$ we get presentation in the form of the product of the functions $z(t) = 1/t$ and $tx(t)$ which belong to $L_2(R \backslash A)$. Then on the basis of Lemma of Subsection 3.7.2 we come to the conclusion that $x(t)$ is integrable both on the interval $A = (-a, a)$ and on the remained part of the real axis $R \backslash A$ what proves the integrability of $x(t)$ on the whole real axis R, i.e. its belonging to the space $L_1(R)$. From this fact and from the results of Subsection 7.2.3 follows that if $x(t) \in D_Q$ then

$$y(t) = Fx = \frac{1}{\sqrt{2\pi}} \frac{d}{dt} \int\limits_{-\infty}^{\infty} \frac{e^{it\tau} - 1}{i\tau} x(\tau)\, d\tau = \frac{1}{\sqrt{2\pi}} \int\limits_{-\infty}^{\infty} e^{it\tau} x(\tau)\, d\tau.$$

The integrand belongs here to the space $L_1(R)$ of the functions of the variable τ with the values in the space $C^1(R)$ of the differentiable functions of the variable t. In accordance with the proved in Example 6.7 in this case the integral belongs to $C^1(R)$ and the differentiation with respect to it may be performed under the integral sign. Then we obtain

$$D_i y = \frac{d}{i\, dt} y(t) = \frac{1}{\sqrt{2\pi}} \int\limits_{-\infty}^{\infty} e^{it\tau} \tau x(\tau)\, d\tau = FQx.$$

Thus the unitary operator F transforms any function $x(t) \in D_Q$ into $y(t) \in D_{D_i}$ and the function Qx transforms it into $D_i y$. It means that the operators Q and D_i in $L_2(R)$ are unitary equivalent. From the obtained formulae follows and D_i in $L_2(R)$ are unitary equivalent. From the obtained formulae follows that $D_i F = FQ$ and $D_i - \lambda I = F(Q - \lambda I)F^{-1}$. Then we may conclude (Problem 6.3.15) that the spectra of the operators Q and D_i in $L_2(R)$ completely coincide. The operator Q has no eigenvalues and all real values of λ are the points of its continuous spectrum. Consequently, the operator of the differentiation with respect to the imaginary argument D_i has no eigenvalues and all real values of λ belong to its continuous spectrum $\sigma(D_i) = \sigma_c(D_i)$.

For detailed study of self–adjoint operators refer to (Dowson 1978, Dunford and Schwarz 1963).

P r o b l e m s

7.3.1. At what conditions the differential operator

$$L = \sum_{k=0}^{n} a_k D^k, \quad D = d/dt$$

in the space $L_2(R)$ will be self–adjoint? Consider the case of the constant coefficients a_k separately and the case when they are the functions of t.

7.3.2. Prove that Sturm–Liouville operator S

$$Sx = D(p(t)Dx) - q(t)x, \quad D = d/dt, \quad p(t) \in C^1([a, b]), \quad q(t) \in C([a, b])$$

is symmetric in the space $L_2([a, b])$ if as its domain is considered the set of twice differentiable functions $x(t) \in L_2([a, b])$ satisfying the boundary conditions

$$c_{i1}x(a) + c_{i2}x(b) + c_{i3}x'(a) + c_{i4}x'(b) = 0 \quad (i = 1, 2).$$

May we extend it till the self–adjoint operator?

7.3.3. Prove that for any differential operator

$$L = a_2 D^2 + a_1 D + a_0, \quad D = d/dt$$

in $L_2(R)$ the product of this operator and the operator M of the multiplication by the function

$$\varphi(t) = c \exp\left\{\int_0^t \frac{a_1(\tau) - a_2'(\tau)}{a_2(\tau)} d\tau\right\},$$

ML represents a self–adjoint operator.

7.3.4. Prove that the Laplace differential operator

$$A = \sum_{i=1}^{n} \frac{\partial^2}{\partial x_i^2}$$

in $L_2(R^n)$ is a self–adjoint operator.

7.3.5. Let T be an arbitrary operator which maps the H–space X into the H–space Y, T^* be its adjoint operator. Show that the operators TT^* and T^*T are positive self–adjoint operators.

7.3.6. Let T be an arbitrary operator which maps the H–space X into the X, T^* be its adjoint operator. Show that the operators $T + T^*$ and $i(T - T^*)$ are self–adjoint.

7.3.7. On the basis of the results of Problem 7.3.6 present any operator T : $X \to X$ which has an adjoint operator T^* in the form of a linear combination of two self–adjoint operators.

7.3.8. Prove that operator A : $l_2 \to l_2$, $Ax = \{\lambda_k x_k\}_{k=1}^\infty$, $x = \{x_k\}_{k=1}^\infty$, $\lambda_k \in R$, $\sup_k | \lambda_k | < \infty$, is a self–adjoint one.

7.3.9. Let $h \in R$. Prove that the difference operator A : $L_2(-\infty, \infty)$ $\to L_2(-\infty, \infty)$

$$Ax(t) = \tfrac{t}{h} \left[x \left(t + \tfrac{h}{2} \right) - x \left(t - \tfrac{h}{2} \right) \right]$$

is a self–adjoint one.

7.3.10. Let $A \in \delta(H)$, $\delta(H)$ being the set of self–adjoint operators. Prove that $\| A^2 \| = \| A \|^2$.

7.4. Projectors

7.4.1. Projectors and their Properties

Let G be a subspace of the H–space X. According to Corollary 1 of Theorem 7.2 any vector $x \in X$ may be presented uniquely in the form of the sum $x = u + v$, $u \in G$, $v \in G^\perp$. The vector u represents a *projection* of the vector x on the subspace G.

The operator P which establishes the correspondence between each vector $x \in X$ and its projection u on G, $u = Px$ is called an *operator of the projection on G* or shortly *projector on G*. If we wish to underline that we are dealing with *the orthogonal* projection, i.e. the finding of such a vector $u \in G$ that $(u, x - u) = 0$ then the operator P is called *an orthoprojector on G*. Later on we shall consider essentially the orthoprojectors and therefore we may call them simply the projectors.

It is evident that the projector P represents a bounded linear operator and its norm is equal to 1. From the fact that $Px \in G$ for any $x \in X$ and $Py = y$ for any $y \in G$ follows that $P^2x = Px$ for any $x \in X$. It means that $P^2 = P$ and generally $P^n = P$ at any natural n.

The operators possessing such properties are called *idempotent operators*. Thus any projector represents an idempotent operator.

As $(Px, y - Py) = (x - Px, Py) = 0$ for any x, y then

$$(Px, y) = (Px, Py) + (Px, y - Py) = (Px, Py)$$

$$= (Px, Py) + (x - Px, Py) = (x, Py)$$

for all $x, y \in X$. Consequently, any projector is a self–adjoint operator. From these properties of the projector follows for any $x \in X$

$$\| Px \|^2 = (Px, Px) = (P^2 x, x) = (Px, x).$$

Hence it is clear that $(Px, x) \geq 0$ for all $x \in X$. It means that any projector is a positive operator (Subsection 6.1.5). Remember that in the case of the H–space X the dual space X^* according to the proved in Subsections 7.1.2 and 7.1.3 may be identified with X.

It is obvious that the subspace G on which the operator P projects is its range $G = R_P$, and the orthogonal supplement $G^\perp$ of the subspace G serves as the kernel of the projector P, $G^\perp = \ker P$ as $P(x - Px) = 0$ for any x.

Theorem 7.18. *Any idempotent self–adjoint linear operator P is a projector.*

▷ If $P^2 = P$ and $P^* = P$ then $\| Px \|^2 = (Px, Px) = (P^2 x, x)$ $= (Px, x) \leq \| Px \| \cdot \| x \|$, whence $\| Px \| \leq \| x \|$. Consequently, P is a bounded operator with the norm not exceeding 1. Let us consider the set G of all the vectors which the operator P leaves unchanged, $G = \{u : Pu = u\}$. Apparently that G represents a subspace as it is a linear space and for any convergent sequence $\{u_n\}$, $u_n \in G$, $u = \lim u_n$, $Pu_n = u_n \to u$. On the other hand, $Pu_n \to Pu$ in the consequence of the continuity of P, i.e. $Pu = u$ and $u \in G$. We denote by P_G the projector on G. Then we shall have for any $y \in G$

$$(Px - P_G x, y) = (Px, y) - (P_G x, y)$$

$$= (x, Py) - (x, P_G y) = (x, y) - (x, y) = 0.$$

It is evident that $P_G x \in G$ and $Px \in G$ for any x as $P(Px) = P^2 x$ $= Px$. Therefore $Px - P_G x \in G$ and the received equality is valid at $y = Px - P_G x$. Consequently, $\| Px - P_G x \| = 0$ and $Px = P_G x$ at all x, i.e. P coincides with P_G. ◁

In the theory of linear operators there often occur the sums, the products and the differences of the projectors. Therefore it is necessary to clarify the conditions at which the addition, the multiplication and the subtraction of the projectors give as a result the projectors. These conditions are determined by Theorems 7.19–7.21.

Theorem 7.19. *If P_1 and P_2 are the projectors on the subspaces G_1 and G_2 correspondingly then the product $P_1 P_2$ represents a projector if and only if P_1 and P_2 are commutative, $P_1 P_2 = P_2 P_1$. In this case $P_1 P_2$ is a projector on the subspace $G = G_1 \bigcap G_2$.*

▷ Really, if $P_1 P_2$ is a projector then in the consequence of its self–adjoint $(P_1 P_2)^* = P_1 P_2$. On the other hand, by property (6.3) of the adjoint operators (Theorem 6.15) $(P_1 P_2)^* = P_2^* P_1^* = P_2 P_1$. So the condition $P_1 P_2 = P_2 P_1$ is necessary. If $P_1 P_2 = P_2 P_1$ then $(P_1 P_2)^2 = P_1 P_2 P_1 P_2 = P_1^2 P_2^2 = P_1 P_2$ and $(P_1 P_2)^* = P_2^* P_1^* = P_2 P_1 = P_1 P_2$, i.e. the operator $P = P_1 P_2$ is an idempotent self–adjoint operator. Consequently, $P_1 P_2$ is a projector. We denote by G' the subspace on which projects operator $P = P_1 P_2$. As $Px = P_1 P_2 x \in G_1$, $Px = P_2 P_1 x \in G_2$, and consequently, $Px \in G = G_1 \bigcap G_2$ at all x then $G' \subset G$. On the other hand, for any $x \in G'^\perp$, $y \in G = G_1 \bigcap G_2$

$$(x, y) = (x, P_1 y) = (P_1 x, y) = (P_1 x, P_2 y) =$$
$$= (P_2 P_1 x, y) = (Px, y) = (0, y) = 0.$$

Consequently, $x \in G^\perp$, i.e. $G'^\perp \subset G^\perp$ and $G' \supset G$. From two opposite inclusions follows that $G' = G = G_1 \bigcap G_2$. ◁

Theorem 7.20. *If P_1 and P_2 are the projectors on the subspaces G_1 and G_2 correspondingly then the sum $P_1 + P_2$ is a projector if and only if the subspaces G_1 and G_2 are orthogonal. In this case $P = P_1 + P_2$ is a projector on the orthogonal sum $G = G_1 \oplus G_2$ of the subspaces G_1 and G_2.*

▷ Really, if $P = P_1 + P_2$ is a projector then $\| Px \|^2 = (Px, x) = (P_1 x, x) + (P_2 x, x) = \| P_1 x \|^2 + \| P_2 x \|^2$, and consequently, $\| x \|^2 \geq \| P_1 x \|^2 + \| P_2 x \|^2$ at any x. In particular, taking $x = P_1 y$ we get

$$\| P_1 y \|^2 \geq \| P_1 y \|^2 + \| P_2 P_1 y \|^2$$

at all y. This inequality is possible only at $P_2 P_1 y = 0$ for any y, i.e. at $P_2 P_1 = 0$. It means that the subspaces G_1 and G_2 are orthogonal and $P_1 P_2 = 0$. And vice versa if $P_2 P_1 = P_1 P_2 = 0$ then

$$(P_1 + P_2)^2 = (P_1 + P_2)P_1 + (P_1 + P_2)P_2 = P_1^2 + P_2^2 = P_1 + P_2,$$

$$(P_1 + P_2)^* = P_1^* + P_2^* = P_1 + P_2.$$

Consequently, in this case $P = P_1 + P_2$ is a projector. We denote by G' the subspace on which P projects. As $P_1 x \in G_1$, $P_2 x \in G_2$, and consequently, $(P_1 + P_2)x \in G_1 \oplus G_2 = G$ at all x then $G' \subset G$. On the other hand, for any $x \in G'^{\perp}$

$$0 = \| Px \|^2 = (Px, x) = (P_1 x, x) + (P_2 x, x) = \| P_1 x \|^2 + \| P_2 x \|^2 .$$

Therefore $P_1 x = P_2 x = 0$ for any $x \in G'^{\perp}$. It means that the vector x is orthogonal to the subspaces G_1 and G_2, i.e. $x \in (G_1 \oplus G_2)^{\perp} = G^{\perp}$. Consequently, $G'^{\perp} \subset G^{\perp}$, i.e. $G^{\perp} = G$. ◁

Corollary. *If $P_1, \dots, P_n$ are the projectors on the subspaces G_1, $\dots, G_n$ correspondingly then $P = \sum_{k=1}^{n} P_k$ is the projector if and only if $G_1, \dots, G_n$ are pairwise orthogonal. In this case P is a projector on the orthogonal sum $G = \bigoplus_{k=1}^{n} G_k$ of the subspaces $G_1, \dots, G_n$.*

Theorem 7.21. *If P_1 and P_2 are the projectors on the subspaces G_1 and G_2 correspondingly then the difference $P = P_1 - P_2$ is the projector if and only if $G_2 \subset G_1$. In this case P is a projector on the orthogonal supplement $G = G_1 \ominus G_2$ of the subspace G_2 in G_1.*

▷ Suppose that P is a projector on some subspace G'. As $P_1 = P_2 + P$ is a projector then by Theorem 7.20 the subspaces G_2 and G' are orthogonal and P_1 is a projector on $G_2 \oplus G'$. Consequently, $G_1 = G_2 \oplus G'$ whence $G_2 \subset G_1$ and $G' = G_1 \ominus G_2 = G$. For proving the sufficiency of the condition suppose that $G_2 \subset G_1$ and denote by P_G the projector on $G = G_1 \ominus G_2$. According to Theorem 7.20 we shall have $P_1 = P_2 + P_G$ whence $P = P_1 - P_2 = P_G$. ◁

Corollary 1. *If $G_2 \subset G_1$ then $\| P_2 x \| \leq \| P_1 x \|$.*

▷ Really, in this case $P_1 = P_2 + P$ and for any x we have $\| P_1 x \|^2 = (P_1 x, x) = (P_2 x, x) + (Px, x) = \| P_2 x \|^2 + \| Px \|^2$. Prove that also inversely from $\| P_2 x \| \leq \| P_1 x \|$ follows $G_2 \subset G_1$. So, if $x \in G_1^{\perp}$ then $P_1 x = 0$. As $\| P_2 x \| \leq \| P_1 x \|$ then also $P_2 x = 0$, i.e. $x \in G_2^{\perp}$. Consequently, $G_1^{\perp} \subset G_2^{\perp}$ whence $G_2 \subset G_1$. ◁

Corollary 2. *If $G_2 \subset G_1$ then $P_1 P_2 = P_2 P_1 = P_2$.*

▷ Really, $P_2 x \in G_1$ whence it follows that $P_1 P_2 x = P_2 x$ for all x. On the other hand, $Px = P_1 x - P_2 x \in G_1 \ominus G_2$, in consequence of which $P_2 Px = 0$ and $P_2 P_1 x = P_2^2 x = P_2 x$ for all x. ◁

E x a m p l e 7.11. The studied properties are well illustrated on the example projectors in 3–dimensional Euclidean space. If G_1 and G_2 are the planes then $P_1 P_2$ transform any point x into some point of the plane G_1, and $P_2 P_1$ transform into some point of the plane G_2. These points in the general case do not coincide. And only in the case when the planes G_1 and G_2 are orthogonal $P_1 P_2$ and $P_2 P_1$ transform x into one and the same point on the line of the intersection G_1 and G_2, i.e. the projection of x on this line. If G_1 and G_2 are two orthogonal lines (which pass through the origin) then $P_1 + P_2$ is an operator of the projection on the plane $G_1 \oplus G_2$. If G_1 and G_2 are nonorthogonal then $P_1 x + P_2 x$ is not a projection of the point x. The difference $P_1 - P_2$ is a projector only if G_2 is a line and G_1 contains its plane or the whole space (or if G_2 is a plane and G_1 is the whole space).

E x a m p l e 7.12. The operator P_A in the space $L_2(X)$ which establishes the correspondence between any function $y(x) \in L_2(X)$ and the function

$$y_A(x) = y(x)\mathbf{1}_A(x) = \begin{cases} y(x) & \text{at } x \in A, \\ 0 & \text{at } x \in \bar{A}, \end{cases}$$

represents a projector on the subspace $L_2(A)$. Here A is any measurable set. For any measurable sets $A, B \subset X$ we have $P_A P_B = P_B P_A = P_{AB}$, $P_A + P_B = P_{A \cup B}$ if and only if AB is a set of zero measure. In other words, when any function from $L_2(A)$ is orthogonal to any function from $L_2(B)$. The difference $P_A - P_B$ is the projector $P_{A \setminus B}$ if and only if $B \subset A$, and consequently, $L_2(B) \subset L_2(A)$ [a].

7.4.2. Convergence of Sequences of Operators

In the linear operators theory we often meet with the sequences of the operators. Consider the space $\mathcal{B}(X)$ of the bounded linear operators mapping the H–space X into X. Any bounded linear operator represents a point of this space which evidently is a normed linear space (namely, the B–space in accordance with general Theorem 5.1). In order to study the convergence of the sequences of the operators in the space $\mathcal{B}(X)$ serve three topologies (Subsection 5.1.4): an uniform topology induced by a norm, strong topology determined by zero vicinities

$$\{T : \|T x_1\| < \varepsilon, \ldots, \|T x_n\| < \varepsilon\}, \qquad (7.21)$$

correspondent to all n, $\varepsilon > 0$ and $x_1, \ldots, x_n \in X$ and a weak topology determined by zero vicinities

$$\{T : |(T x_1, y_1)| < \varepsilon, \ldots, |(T x_n, y_n)| < \varepsilon\}, \qquad (7.22)$$

---

[a] As $L_2(X)$ we may assume here the space L_2 of the scalar functions which are determined on the arbitrary space X with a nonnegative measure.

correspondent to all n, $\varepsilon > 0$ and all pairs $x_1, y_1; \ldots; x_n, y_n \in X \times X$.

These three topologies determine three types of the convergence of the sequences of the operators $\{T_n\}$: the uniform convergence $\| T_n - T \| \to 0$, the strong convergence or simply the convergence $\| T_n x - Tx \| \to 0$ $\forall x$ and the weak convergence $(T_n x - Tx, y) \to 0$ $\forall x, y$.

R e m a r k. It was shown in Subsection 5.1.4 that from the uniform convergence of the sequence of the operators follow its convergence and a weak convergence, and from the convergence follows a weak convergence. The inverse statements in the general case are not true. In particular, from a weak convergence of the sequence of the operators in the general case its convergence does not follow. But in the special case for the projectors this statement as we shall see later on is sometimes true.

7.4.3. Sequences of the Projectors

Let $\{G_n\}$ be a monotone sequence of the subspaces of the H–space X, $\{P_n\}$ be a sequence of the correspondent projectors.

Theorem 7.22. *If the sequence $\{P_n\}$ weakly converges to the projector P then it converges to P.*

▷ If the sequence $\{P_n\}$ weakly converges to some projector P then by virtue of the relation $\| P'x \|^2 = (P'x, x)$ which is valid for any projector P' and the fact that or $P - P_n$ or $P_n - P$ by Theorem 7.21 is a projector, $\{P_n\}$ converges to P. Thus from a weak convergence of the monotone sequence of the projectors $\{P_n\}$ to the projector P follows the convergence of $\{P_n\}$ to P. ◁

Theorem 7.23. *Any monotone sequence of the projectors $\{P_n\}$ converges to the projector P on the subspace $G = \lim G_n$ ($\bigcup G_n$ in the case of increasing $\{G_n\}$ and $\bigcap G_n$ in the case of decreasing $\{G_n\}$).*

▷ As $P_m - P_n$ is a projector at any m, n, $m > n$ in the case of increasing $\{G_n\}$, and $m < n$ in the case of decreasing $\{G_n\}$ then

$$\| P_m x - P_n x \|^2 = (P_m x - P_n x, x) = (P_m x, x) - (P_n x, x).$$

The right–hand side is an infinitesimal at all sufficiently large m and n as a monotone sequence of positive numbers $\{(P_n x, x)\}$ upper bounded by the number $\| x \|^2$ converges. Consequently, the sequence $\{P_n x\}$ is fundamental and due to the completeness of the H–space converges to some limit (dependent on x), $\lim P_n x = Px$. Operator P is linear and as $(P_n x, y) \to (Px, y)$, $(x, P_n y) \to (x, Py)$, $(P_n x, P_n y) \to (Px, Py)$ and $(P_n x, y) = (x, P_n y) = (P_n x, P_n y)$ at all x, y then $(Px, y) = (x, Py)$ $= (Px, Py)$ at all x, y. From this fact follows that P is a self–adjoint idempotent operator $P^* = P^2 = P$. Therefore P is a projector. We denote by G' the subspace on which P projects. If $G' \subset G$, $G \ominus G' \neq \{0\}$ then $G' \subset G_n$, $G_n \ominus G' \neq \{0\}$ for all n in the case of decreasing $\{G_n\}$ and for all n no smaller than some N in the case of increasing $\{G_n\}$. After taking in the first case $x \in G \ominus G'$ and in the second case $x \in G_N \ominus G'$ we get $Px = 0$, $P_n x = x$ for all $n \geq N$ what contradicts to the proved convergence $\{P_n x\}$ to Px at all x. If $G \subset G'$, $G' \ominus G \neq \{0\}$ then $G_n \subset G'$, $G' \ominus G_n \neq \{0\}$ for all n in the case of increasing $\{G_n\}$ and for all n no smaller than some N in the case of decreasing $\{G_n\}$. After taking in the first case $x \in G' \ominus G$, and in the second case $x \in G' \ominus G_N$ we receive $Px = x$, $P_n x = 0$ for all $n > N$ what also contradicts to the convergence $\{P_n x\}$ to Px at all x. The obtained contradictions prove that $G' = G$ and thus conclude the proof of the theorem. ◁

7.4.4. General Definition of a Projector

The orthogonal projection has the sense only for the H–spaces. And even for the H–spaces we have sometimes to consider nonorthogonal projection. We shall meet with this fact in Subsections 7.5.7 and 7.5.8. Therefore the necessity of studying the operators of the projection in any spaces and the operators of nonorthogonal projection in the H–spaces arises. We have already met with the notion of the point projection in any products of the spaces (Subsections 2.1.8 and 2.1.9). The map of a point of the set spaces product at anyone from these spaces or on the product of some subset of these spaces is a projection operator or a projector. From the definiton of the point projection in Subsections 2.1.8 and 2.1.9 follows that any projector represents an idempotent operator. This leads us to the following general definition of a projector.

An operator of the projection or *a projector* in any linear space X is called any not zero idempotent linear operator.

The range of the projector P cannot be empty and serves as a set on which the points of the space are projected. Really, putting $u = Px$ for any x we obtain $Pu = P^2 x = Px = u$, and vice versa from $Pu = u$ follows $u \in R_P$. Thus $Pu = u$ if and only if $u \in R_P$. On the other hand, $P(x - Px) = Px - P^2 x = Px - Px = 0$ for any x. Consequently, and projector P gives a presentation of any vector x in the form of the sum $x = u + v$ where $u \in R_P$ and $v \in \ker P$. Correspondingly the space X is represented in the form of the sum of two nonintersecting subspaces

$$X = G_1 + G_2, \quad G_1 = R_P, \quad G_2 = \ker P.$$

Here the sum is assumed as the sum of the sets (Subsection 1.4.1), and as nonintersecting subspaces are considered the subspaces whose the unique point of the intersection is 0. This corresponds to the orthogonal subspaces in the H–space.

For detailed study of projectors refer to (Balakrishnan 1976, Dowson 1978, Dunford and Schwarz 1963).

P r o b l e m s

7.4.1. Generalize Example 7.12 on the case of the space of the scalar functions $L_2(X, \mathcal{A}, \mu)$.

7.4.2. Let $x_1, \ldots, x_n$ be the vectors in the H–space X satisfying the conditions $(x_k, x_l) = \delta_{kl}$. Will the operator be a projector

$$Px = \sum_{k=1}^{n} (x, x_k) x_k?$$

If so then at what subspace it projects the vectors $x \in X$?

7.4.3. Generalize Theorems 7.19–7.21 on the projectors in any linear space.

7.4.4. Let $x_1, \ldots, x_n$ be linear independent vectors of the linear space X. Show that there exist the functionals $f_1, \ldots, f_n$ satisfying the conditions $f_k x_l = \delta_{kl}$. May the functionals $f_1, \ldots, f_n$ be linear independent? Will the operator

$$Px = \sum_{k=1}^{n} (f_k x) x_k$$

be a projector if so then at what subspace it projects the vectors $x \in X$?

7.4.5. Show that the operator of the multiplication by an indicator of the set A in the space of the scalar functions $L_p(X, \mathcal{A}, \mu)$ is a projector on the subspace $L_p(A, \mathcal{A}, \mu)$.

7.4.6. Let $A \in \mathcal{L}(H)$, $\mathcal{L}(H)$ being the space of bounded linear operators in the Hilbert space H. Prove the equivalence of the following statements.

1) If $x_n, x, y_n, y \in H$ and $x_n \to x$ (weakly), $y_n \to y$ (weakly) at $n \to \infty$ then $(Ax_n, y_n) \to Ax, y)$.

2) If $x_n, x \in H$, $x_n \to x$ (weakly) at $n \to \infty$ then $Ax_n \to Ax$.

7.5. Sequences of Vectors and Bases

7.5.1. Sequences of Vectors

Let $\{x_k\}$ be any sequence of the vectors in the H–space X. Without loss of the generality we may consider these vectors linear independent as we may always achieve it by rejecting from the sequence each vector which represents a linear combination of the previous ones. Let us denote by G_n a subspace formed by the first n vectors of our sequence $x_1, \ldots, x_n$. According to Corollary 1 of Theorem 7.2 any vector $x \in X$ may be uniquely presented in the form of the sum $x = u_n + z_n$ where u_n is a projection of x on G_n, and $z_n \in G_n^\perp$. Thus any vector $x \in X$ may be presented in the form

$$x = \sum_{k=1}^{n} \alpha_k^{(n)} x_k + z_n \, . \tag{7.23}$$

For determining the coefficients $\alpha_1^{(n)}, \ldots, \alpha_n^{(n)}$ we scalar multiply (7.23) by x_p. Then taking into account that the vector z_n is orthogonal to all the vectors $x_1, \ldots, x_n$ we get a set of the equations:

$$(x, x_p) = \sum_{k=1}^{n} \alpha_k^{(n)} (x_k, x_p) \quad (p = 1, \ldots, n) \, . \tag{7.24}$$

This set of equations has always the unique solution as by Theorem 1.12 the determinant of the Gram matrix of linearly independent vectors is strictly positive. We shall not solve here Eqs. (7.24) in the general case but restrict ourselves by the remark that for any sequence of the vectors $\{x_n\}$, any vector x of the H–space X may be presented by the sequence of expansions (7.23) in each of them the vector z_n is orthogonal to the vectors $x_1, \ldots, x_n$.

7.5.2. Orthogonal and Orthonormal Sequences

A sequence of the vectors $\{x_k\}$ is called *orthogonal* if all the vectors of this sequence are pairwise orthogohal $(x_p, x_q) = 0$ at $p \neq q$. A sequence of vectors $\{x_k\}$ is called *orthonormal* if it is orthogonal and the norms of all its vectors are equal to 1, $(x_p, x_q) = \delta_{pq}$.

Theorem 7.24. *Any sequence of linearly independent vectors $\{u_k\}$ may be orthonormalized, namely substituted by such orthonormal sequence $\{x_k\}$ that each vector u_n will be a linear combination of the vectors $x_1, \ldots, x_n$.*

▷ We put $x_1 = u_1/\|u_1\|$, $x_2 = \gamma_2(\alpha_{21}x_1 + u_2)$ and choose α_{21} from the condition $(x_2, x_1) = 0$, and γ_2 from the condition $\|x_2\| = 1$. As a result we obtain $\alpha_{21} = -(u_2, x_1)$, $\gamma_2 = \|\alpha_{21}x_1 + u_2\|^{-1}$. After this we put $x_n = \gamma_n(\alpha_{n1}x_1 + \ldots + \alpha_{n,n-1}x_{n-1} + u_n)$ at any $n > 2$ and determine $\alpha_{n1}, \ldots, \alpha_{n,n-1}$ sequentially from the conditions $(x_n, x_k) = 0$ ($k = 1, \ldots, n-1$), and γ_n from the condition $\|x_n\| = 1$. Then we receive $\alpha_{np} = -(u_n, x_p)$, $(p = 1, \ldots, n-1)$, $\gamma_n = \|\alpha_{n1}x_1 + \ldots + \alpha_{n,n-1}x_{n-1} + u_n\|^{-1}$ ($n = 3, 4, \ldots$). ◁

If we assume in the previous formulae $x_1 = u_1$, $\gamma_2 = \gamma_3 = \ldots = 1$ then the stated way will give an orthogonal sequence of the vectors. These sequence in the general case will be not orthonormal.

Thus after taking any sequence of linearly independent vectros $\{u_k\}$ we may construct such orthonormal sequence $\{x_n\}$ that

$$u_1 = \|u_1\| x_1, \quad u_n = \sum_{k=1}^{n}(u_n, x_k)x_k \quad (n = 2, 3, \ldots). \tag{7.25}$$

Hence it follows that at the stated way of the orthonormalization of the sequence of the vectors the subspace formed by the vectors $x_1, \ldots, x_n$ coincides with the subspace G_n formed by the vectors $u_1, \ldots, u_n$. Therefore without loss of the generality we may consider the sequence of the vectors $\{x_k\}$ in (7.23) orthonormal.

7.5.3. Expansion of a Vector in Terms of Orthonormal Vectors

In the case of the orthonormal sequence of the vectors $\{x_k\}$ Eqs. (7.24) give

$$\alpha_p^{(n)} = (x, x_p) \quad (p = 1, \ldots, n), \tag{7.26}$$

and formula (7.23) takes the form

$$x = \sum_{k=1}^{n}(x, x_k)x_k + z_n. \tag{7.27}$$

After scalar multiplying this equality by x we find

$$\|x\|^2 = \sum_{k=1}^{n}|(x, x_k)|^2 + (z_n, x).$$

Substituting in the last item expression (7.27) of the vector x we obtain by virtue of orthogonality of the vector z_n to $x_1, \ldots, x_n$. So, $(z_n, x) = \|z_n\|^2$ and the previous equality will take the form

$$\|x\|^2 = \sum_{k=1}^{n} |(x, x_k)|^2 + \|z_n\|^2 . \qquad (7.28)$$

Hence we obtain the inequality which is valid at all n:

$$\sum_{k=1}^{n} |(x, x_k)|^2 \leq \|x\|^2 . \qquad (7.29)$$

This inequality is usually called the *Bessel inequality*.

The vector z_n in formula (7.27) represents a residual term of the approximation of the vector x by its projection of the subspace G_n. Formula (7.28) may serve for the estimate of a norm of the residual term z_n in expansion (7.27).

From Bessel inequality (7.29) follows that the series

$$\sum_{k=1}^{\infty} |(x, x_k)|^2 \qquad (7.30)$$

converges for any vector x. As

$$\left\| \sum_{k=m}^{m+s} (x, x_k) x_k \right\|^2 = \sum_{k=m}^{m+s} |(x, x_k)|^2$$

at all natural m and s then from the convergence of series (7.30) follows the convergence of the series

$$\sum_{k=1}^{\infty} (x, x_k) x_k . \qquad (7.31)$$

Consequently, and vector $x \in X$ is uniquely expressed in terms of the expansion

$$x = \sum_{k=1}^{\infty} (x, x_k) x_k + z_\infty , \qquad (7.32)$$

where $z_\infty = \lim z_n$, and consequently, $(z_\infty, x_k) = 0$ at all k, i.e. the vector z_∞ is orthogonal to all the vectors of the sequence $\{x_k\}$.

From the convergence of series (7.30) follows that $(x, x_n) \to 0$ at $n \to \infty$ and at all x. On the basis of Riesz theorem 7.1 it means that any orthonormal sequence of the vectors $\{x_n\}$ in the H–space weakly converges to zero (Subsection 4.7.2). Thus we have an example of weakly convergent but not convergent sequence.

Let P_n be a projector on the subspace G_n. The sequence of the spaces $\{G_n\}$ is monotone increasing. Consequently, by Theorem 7.23 the sequence of the projectors $\{P_n\}$ converges to the projector P on the subspace $G = \lim G_n = \bigcup G_n$. Therefore the series in expansion (7.32) converges to the projection Px of the vector x on the subspace G, and z_∞ belongs to the orthogonal supplement $G^\perp$ of the subspace G.

7.5.4. Complete Sequences of Vectors and Bases

The sequence of the vectors $\{x_k\}$ is called *complete* if in X there is no vector which is orthogonal to all the vectors of this sequence. If the orthonormal sequence of the vectors $\{x_k\}$ is complete then the orthogonal supplement $G^\perp$ of the subspace G formed by the sequence $\{x_k\}$ consists of one zero and $z_\infty = 0$ in expansion (7.32). In this case any vector x of the H–space X is expreseed in terms of the expansion

$$x = \sum_{k=1}^{\infty} (x, x_k) x_k \,, \tag{7.33}$$

and formula (7.28) in the limit at $n \to \infty$ gives

$$\| x \|^2 = \sum_{k=1}^{\infty} |(x, x_k)|^2 \,. \tag{7.34}$$

This equality is usually called the *Parseval equality*.

The received results we shall formulate in the form of the theorem.

Theorem 7.25. *Let $\{x_k\}$ be a complete orthonormal sequence of the vectors in the H–space X then any vector $x \in X$ may be presented by expansion* (7.33) *and therewith Parseval equality* (7.34) *is valid.*

Let x, y be two vectors of the H–space X, $\{x_n\}$ be a complete orthonormal sequence of the vectors. After presenting the vectors x, y by expansion (7.33) we obtain the following formula for a scalar product of the vectors x, y:

$$(x, y) = \sum_{k=1}^{\infty} (x, x_k)(x_k, y) \,. \tag{7.35}$$

A sequence of the vectors $\{x_k\}$ of a topological linear space X is called *a basis* if any vector $x \in X$ is uniquely expressed by the expansion

$$x = \sum_{k=1}^{\infty} \alpha_k x_k$$

convergent in the topology of this space.

Thus any complete orthogonal sequence of the vectors in the H–space X represents a basis. Further we shall meet with nonorthogonal bases (Subsection 7.5.8).

7.5.5. Representation of Functions by Series

We shall apply the stated theory for studying the functions expansions by series. Let us consider the H–space $L_2(T)$ of the functions with the integrable modulus square on Lebesgue measure over the region T. Let $\{x_n(t)\}$ be a complete orthonormal sequence of the functions in $L_2(T)$,

$$\int_T x_n(t)\,\overline{x_m(t)}\,dt = \delta_{nm} \,.$$

Such sequence may be received by the orthonormalization of any complete sequence of the function $\{u_n(t)\}$ [a]. If the region T is finite we may assume as the functions $u_n(t)$, for example, the degrees of independent variable $u_n(t) = t^{n-1}$ $(n = 0, 1, 2, \ldots)$ in the case of the scalar variable t and all power monomials relatively to the coordinates of the vector t in the case of the vector variable t. If the region T is infinite as the functions $u_n(t)$ are usually assumed the power monomials multiplied by some sufficiently quick decreasing function t when even only one of the coordinates of the vector t unboundedly increases in modulus.

If follows from the general theory of Subsection 7.5.4 that any function $x(t) \in L_2(T)$ may be presented in terms of series (7.33):

$$x(t) = \sum_{k=1}^{\infty} \alpha_k x_k(t) \,, \tag{7.36}$$

$$\alpha_k = (x, x_k) = \int_T x(t)\,\overline{x_k(t)}\,dt \quad (k = 1, 2, \ldots) \,. \tag{7.37}$$

---

[a] The complete sequences of the functions $L_2(T)$ exist in consequence of the separability of the space $L_2(T)$ (Subsections 3.7.6 and 7.5.6).

Series (7.36) converges to $x(t)$ in the norm of the space $L_2(T)$ (i.e. in the mean square):

$$\int_T \left| x(t) - \sum_{k=1}^n \alpha_k x_k(t) \right|^2 dt \to 0 \quad \text{at } n \to \infty.$$

The norm of the residual term of series (7.36) due to formula (7.28) may be estimated by

$$\| z_n \|^2 = \int_T \left| x(t) - \sum_{k=1}^n \alpha_k x_k(t) \right|^2 dt = \int_T |x(t)|^2 \, dt - \sum_{k=1}^n |\alpha_k|^2 .$$

(7.38)

We may use formula (7.38) for the estimate of the approximation accuracy of the function $x(t)$ by a segment of series (7.36).

R e m a r k. The most difficulties at applying the expansions of the functions in terms of the orthogonal bases presents the proof of the completeness of the chosen sequence of the functions. For proving the completeness of the functions sequence in the spaces L_2 of the functions of a vector variable it is useful to apply the following statement. If $\{x_{pk}(t_p)\}$ are complete orthonormal sequences in the spaces $L_2(T_p)$ $(p = 1, \ldots, n)$ then all possible products $x_{1k_1}(t_1) \ldots x_{nk_n}(t_n)$ form a complete orthonormal sequence in the space $L_2(T_1 \times \ldots \times T_n)$ of the functions of the n–dimensional vector variable $t = \{t_1, \ldots, t_n\}$. This statement in many cases gives the opportunity to restrict ourselves to the proof of the completeness of the systems of the functions of a scalar variable.

For establishing the completeness of the systems of the functions in the spaces L_2 the following theorem is useful.

Theorem 7.26. *If the function $\varphi(t)$ which is not equal to zero almost everywhere and increases at $|t| \to \infty$ faster than the exponential function $e^{-\alpha|t|}$, $\alpha > 0$ then the sequence of the functions $t^k \varphi(t)$ ($k = 0, 1, 2, \ldots$) is complete in $L_2(R)$.*

▷ For proving we suppose that there exists such a function $z(t) \in L_2(R)$ which is orthogonal to all the functions $t^k \varphi(t)$:

$$\int_{-\infty}^{\infty} t^k \varphi(t) z(t) \, dt = 0 \quad (k = 0, 1, 2, \ldots).$$

(7.39)

As $\varphi(t)\,z(t) \in L_1(R)$ by virtue of inequality (3.89) at $p = q = 2$ then the function

$$u(\omega) = \frac{1}{\sqrt{2\pi}} \int\limits_{-\infty}^{\infty} \varphi(t)\,z(t)\,e^{i\omega t}\,dt$$

is an analytical function of the complex variable ω at $|\mathrm{Im}\,\omega| < \alpha$. As all its derivatives are equal to zero at $\omega = 0$ by virtue of condition (7.39) then $u(\omega) \equiv 0$. Hence owing to the unitary property of Fourier operator (Subsection 7.2.3) follows that $\varphi(t)\,z(t) = 0$ almost everywhere. As $\varphi(t)$ is not equal to zero almost everywhere then $z(t) = 0$ almost everywhere. Thus there is no function different from zero in $L_2(R)$ which is orthogonal to all the functions $t^k \varphi(t)$ what proves the completeness of the sequence of the functions $t^k \varphi(t)$ $(k = 0, 1, 2, \ldots)$. ◁

R e m a r k. Theorem 7.26 is also true for the spaces $L_2(R^n)$ at any natural n; at $n > 1$ t^k should be considered as $t_1^{k_1} \ldots t_n^{k_n}$ $(k_1, \ldots, k_n = 0, 1, 2, \ldots)$, and ωt as the scalar product $\omega^T t$ of the vectors $\omega = [\omega_1 \ldots \omega_n]^T$ and $t = [t_1 \ldots t_n]^T$. Putting, in particular, $\varphi(t) = 0$ everywhere outside the finite region T and $\varphi(t) = 1$ at $t \in T$ owing to Theorem 7.26 we come to the conclusion that the system of the functions t^k $(k = 0, 1, 2, \ldots)$ is complete in the space $L_2(T)$ for any finite region $T \subset R^n$. If the function $\varphi(t)$ satisfies the conditions of Theorem 7.26 and is equal to zero at $t < 0$ then follows the completeness of the functions $t^k \varphi(t)$ in the space $L_2([0, \infty))$.

E x a m p l e 7.13. By the orthogonalization of the functions system t^k $(k = 0, 1, 2, \ldots)$ on the interval $[a, b]$ we receive the sequence of the *Legendre polynomials*:

$$P_k(t) = c_k \frac{d^k}{dt^k}\left[(t - a)^k (t - b)^k\right] \quad (k = 0, 1, 2, \ldots).$$

By Theorem 7.26 the sequence of the Legendre polynomials is complete, and consequently, is the basis in $L_2([a, b])$. Therefore any function $x(t) \in L_2([a, b])$ may be presented by the expansion in terms of the Legendre polynomials

$$x(t) = \sum_{k=0}^{\infty} \alpha_k P_k(t),$$

convergent in the mean square. The coefficients of this expansion according to (7.37) are determined by the following formula:

$$\alpha_k = \int\limits_a^b x(t)\,P_k(t)\,dt \Big/ \int\limits_a^b P_k^2(t)\,dt \quad (k = 0, 1, 2, \ldots).$$

The approximation accuracy of the function $x(t)$ by a segment of the expansion in terms of the Legendre polynomials may be estimated by formula (7.38).

Legendre polynomials are usually used on the interval $[-1, 1]$. In this case the coefficient c_k is taken equal to $2^k k!$ (Appendix 2).

E x a m p l e 7.14. The orthonormal sequence of trigonometrical functions $e^{ikt}/\sqrt{2\pi}$ $(k = 0, \pm 1, \pm 2, \ldots)$ is complete, and consequently, is the base in $L_2([-\pi, \pi])$. Really, supposing that there is the function $z(t)$ which is orthogonal to all the functions $e^{ikt}/\sqrt{2\pi}$ we get

$$\int\limits_{-\pi}^{\pi} e^{ikt} z(t)\, dt = 0 \quad (k = 0, \pm 1, \pm 2, \ldots). \tag{7.40}$$

Hence by the integration by parts at $k \neq 0$ we find

$$\int\limits_{-\pi}^{\pi} e^{ikt} u(t)\, dt = 0 \quad (k = \pm 1, \pm 2, \ldots), \tag{7.41}$$

where

$$u(t) = \int\limits_{-\pi}^{t} z(\tau)\, d\tau.$$

According to condition (7.41) the function $u(t)$ is orthogonal to all the vectors $e^{ikt}/\sqrt{2\pi}$ besides $1/\sqrt{2\pi}$. In order to receive the function which is also orthogonal to $1/\sqrt{2\pi}$ we put

$$v(t) = u(t) - \frac{1}{2\pi} \int\limits_{-\pi}^{\pi} u(\tau)\, d\tau.$$

Then by virtue of condition (7.41) we shall have

$$\int\limits_{-\pi}^{\pi} e^{ikt} v(t)\, dt = 0 \quad (k = 0, \pm 1, \pm 2, \ldots). \tag{7.42}$$

As the function $v(t)$ is continuous and due to condition (7.40) at $k = 0$ $u(\pi) = u(-\pi) = 0$ as a result of which $v(\pi) = v(-\pi)$. Then by the Weierstrass theorem $v(t)$ may be uniquely approximated with any degree of accuracy by a trigonometrical polynomial on the interval $[-\pi, \pi]$. Let $p(t)$ be such polynomial that $|v(t) - p(t)| < \varepsilon$ at all $t \in [-\pi, \pi]$. Then due to condition (7.42) and inequality (1.17)

$$\int\limits_{-\pi}^{\pi} |v(t)|^2\, dt = \int\limits_{-\pi}^{\pi} \overline{v(t)}\, [v(t) - p(t)]\, dt$$

$$< \varepsilon \int\limits_{-\pi}^{\pi} |v(t)|\, dt \leq \varepsilon \sqrt{2\pi \int\limits_{-\pi}^{\pi} |v(t)|^2\, dt}$$

or

$$\| v \| = \sqrt{\int\limits_{-\pi}^{\pi} |v(t)|^2 \, dt} < \varepsilon\sqrt{2\pi} \,.$$

Hence in consequence of the arbitrariness of $\varepsilon > 0$ we conclude that $\| v \| = 0$, and consequently, $v(t) = 0$ almost everywhere. From the definition of the function $v(t)$ follows that also $z(t) = 0$ almost everywhere, i.e. $\| z \| = 0$ what prove the completeness of the sequence of the functions e^{ikt} $(k = 0, \pm 1, \pm 2, \ldots)$.

Thus the sequence $e^{ikt}/\sqrt{2\pi}$ $(k = 0, \pm 1, \pm 2, \ldots)$ represents a basis and any function $x(t) \in L_2([-\pi, \pi])$ may be presented by Fourier series

$$x(t) = \frac{1}{\sqrt{2\pi}} \sum_{k=-\infty}^{\infty} \alpha_k \, e^{ikt} \,,$$

where in the correspondence with general formula (7.37)

$$\alpha_k = \frac{1}{\sqrt{2\pi}} \int_{-\pi}^{\pi} x(t) \, e^{-ikt} dt \quad (k = 0, \pm 1, \pm 2, \ldots) \,.$$

Therewith this series converges to $x(t)$ in mean square.

E x a m p l e 7.15. Accounting the Remark before Theorem 7.26 the sequence of the functions $(2\pi)^{-n/2} e^{ik^T t}$ $(k_1, \ldots, k_n = 0, \pm 1, \pm 2, \ldots)$ of n–dimensional vector variable t represents an orthonormal base in the space $L_2([-\pi, \pi]^n)$ and any function $x(t) \in L_2([-\pi, \pi]^n)$ may be presented by Fourier series convergent in mean square

$$x(t) = (2\pi)^{-n/2} \sum_{p=-\infty}^{\infty} \sum_{|k|=p} \alpha_k \, e^{ik^T t} \,,$$

whose coefficients according to expansion (7.37) are determined by the formula

$$\alpha_k = (2\pi)^{-n/2} \int\limits_{[-\pi,\pi]^n} x(t) \, e^{-ik^T t} dt \,.$$

E x a m p l e 7.16. By the orthogonalization of the sequence of the functions $t^k e^{-t^2/2}$ $(k = 0, 1, 2, \ldots)$ in the space $L_2(R)$ an orthogonal sequence of the *Hermite functions* is obtained

$$\varphi_k(t) = (-1)^k \, e^{t^2/2} \frac{d^k}{dt^k} e^{-t^2} = e^{-t^2/2} H_k(t) \quad (k = 0, 1, 2, \ldots) \,,$$

where

$$H_k(t) = 2^k t^k + \sum_{p=1}^{[k/2]} (-1)^p (2p-1)!! \, C_k^{2p} 2^{k-p} t^{k-2p}$$

are the *Hermite polynomials*

$$\int\limits_{-\infty}^{\infty} \varphi_m(t)\,\varphi_n(t)\,dt = \int\limits_{-\infty}^{\infty} e^{-t^2} H_m(t)\,H_n(t)\,dt = 0 \quad \text{at} \quad m \neq n \,,$$

$$\int\limits_{-\infty}^{\infty} \varphi_n^2(t)\,dt = \int\limits_{-\infty}^{\infty} e^{-t^2} H_n^2(t)\,dt = 2^n\, n!\sqrt{\pi}\,.$$

On the basis of the proved theorems the sequence of the Hermite functions represents a basis in $L_2(R)$ and any function $x(t) \in L_2(R)$ may be represented by a series convergent in mean square

$$x(t) = \sum_{k=0}^{\infty} \alpha_k\, \varphi_k(t) = \sum_{k=0}^{\infty} \alpha_k\, e^{-t^2/2} H_k(t)\,,$$

where in the correspondence with general formula (7.37) and Appendix 2

$$\alpha_k = \int\limits_{-\infty}^{\infty} x(t)\,\varphi_k(t)\,dt \Big/ \int\limits_{-\infty}^{\infty} \varphi_k^2(t)\,dt = \frac{1}{2^k\,k!\sqrt{\pi}} \int\limits_{-\infty}^{\infty} x(t)\,\varphi_k(t)\,dt$$

$$= \frac{1}{2^k\,k!\sqrt{\pi}} \int\limits_{-\infty}^{\infty} e^{-t^2/2} x(t)\,H_k(t)\,dt\,.$$

For the estimate of the approximation accuracy we may use formula (7.38).

E x a m p l e 7.17. By the orthogonalization of the functions $t^{k+\nu/2} e^{-t/2}$ $(k = 0, 1, 2, \ldots)$ at a given real $\nu > -1$ in the space $L_2([0, \infty))$ the sequence of the *Laguerre functions* is received

$$\psi_k^{(\nu)}(t) = \frac{1}{k!} t^{-\nu/2} e^{t/2} \frac{d^k}{dt^k} (t^{\nu+k} e^{-t}) = t^{\nu/2} e^{-t/2} L_k^{(\nu)}(t)\,,$$

where

$$L_k^{(\nu)}(t) = \sum_{l=0}^{k} \frac{(-1)^l \Gamma(k+\nu+1)}{l!\,(k-l)!\,\Gamma(l+\nu+1)}\, t^l$$

are the *Laguerre polynomials*

$$\int\limits_{0}^{\infty} \psi_m^{(\nu)}(t)\,\psi_n^{(\nu)}(t)\,dt = \int\limits_{0}^{\infty} t^\nu e^{-t} L_m^{(\nu)}(t)\,L_n^{(\nu)}(t)\,dt = 0\,,$$

$$\int\limits_{0}^{\infty} [\psi_n^{(\nu)}(t)]^2\,dt = \int\limits_{0}^{\infty} t^\nu e^{-t} [L_n^{(\nu)}(t)]^2\,dt = \frac{1}{n!}\Gamma(n+\nu+1)\,.$$

By Theorems 7.25 and 7.26 the sequence of the Laguerre functions represents a base
in $L_2([0, \infty])$ and any function $x(t) \in L_2([0, \infty))$ may be presented by a series

$$x(t) = \sum_{k=0}^{\infty} \alpha_k \, \psi_k^{(\nu)}(t) = \sum_{k=0}^{\infty} \alpha_k t^{\nu/2} e^{-t/2} L_k^{(\nu)}(t) \,,$$

where (see Appendix 2)

$$\alpha_k = \frac{k!}{\Gamma(k + \nu + 1)} \int_0^{\infty} x(t) \, \psi_k^{(\nu)}(t) \, dt$$

$$= \frac{k!}{\Gamma(k + \nu + 1)} \int_0^{\infty} t^{\nu/2} e^{-t/2} x(t) \, L_k^{(\nu)}(t) \, dt \,.$$

Therewith the series converges to the function $x(t)$ in mean square. For the estimate
of the approximation accuracy of the function $x(t)$ by a segment of the expansion
in terms of the Laguerre functions we may calculate the norm of a residual term by
formula (7.38). Usually we apply the Laguerre polynomials correspondent to $\nu = 0$
which are denoted simply by $L_k(t)$.

E x a m p l e 7.18. In signal processing the orthonormal sequence of the
Walsh functions $\{w_n(t)\}_0^{\infty}$ in the space $L_2([0, 1])$ is very convenient. These functions are determined in the following way. The first function $w_0(t)$ is assumed equal
to 1. The second function $w_1(t)$ is assumed equal to 1 on the interval $[0, 1/2]$
and to -1 on the interval $[1/2, 1]$. Further the interval $[0, 1]$ is divided into four
parts and on each of them the function is assumed equal to $+1$ or -1 so that the
product of each newly introduced function by each from the previous ones be equal
to $+1$ on the intervals of the total length $1/2$ and -1 on the remainder intervals of
the summary length $1/2$. This condition is necessary for the case when each newly
introduced function will be orthogonal to all previous ones. Hence the square of any
function $w_n(t)$ is identically equal to 1 in consequence of which the norms of all the
functions $w_n(t)$ in $L_2([0, 1])$ are equal 1.

In order to give a general formula which determines the Walsh functions let
introduce the sequence of the functions $\{r_n(t)\}_{n=0}^{\infty}$ whose plots represent "combs"
with rectangular cog–wheels of width $1/2^k$ ($k = 1, 2, \ldots$) (Fig.7.1):

$$r_n(t) = \begin{cases} +1 & \text{at} \quad \frac{2k}{2^n} \leq t < \frac{2k+1}{2^n}, \\[2mm] -1 & \text{at} \quad \frac{2k+1}{2^n} \leq t < \frac{2k+2}{2^n} \end{cases}$$

$$(k = 0, 1, \ldots, 2^{n-1} - 1; \quad n = 1, 2, \ldots) \,.$$

This is the so–called *Rademacher functions*. Evidently these functions form the
orthonormal sequence. But this sequence is not complete.

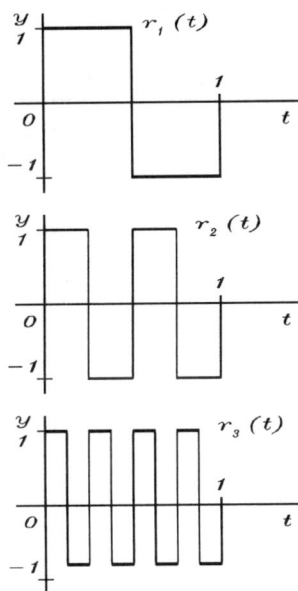

Fig. 7.1

Walsh function $w_n(t)$ at $n \geq 1$ is determined in the following way. The number n is presented in binary form:

$$n = 2^0 \nu_1 + 2^1 \nu_2 + \ldots + 2^{p-1} \nu_p , \qquad (7.43)$$

where each of the numbers $\nu_1 , \ldots , \nu_p$ represents the correspondent binary digit of the number n, i.e. is equal to 0 or 1. After that the function $w_n(t)$ is determined as the product of the Rademacher functions whose numbers coincide with the numbers of those ν_k which are equal to 1, i.e.

$$w_n(t) = \prod_{k=1}^{p} r_k^{\nu_k}(t) . \qquad (7.44)$$

Using this formula and binary presentation of the numbers of the functions we write

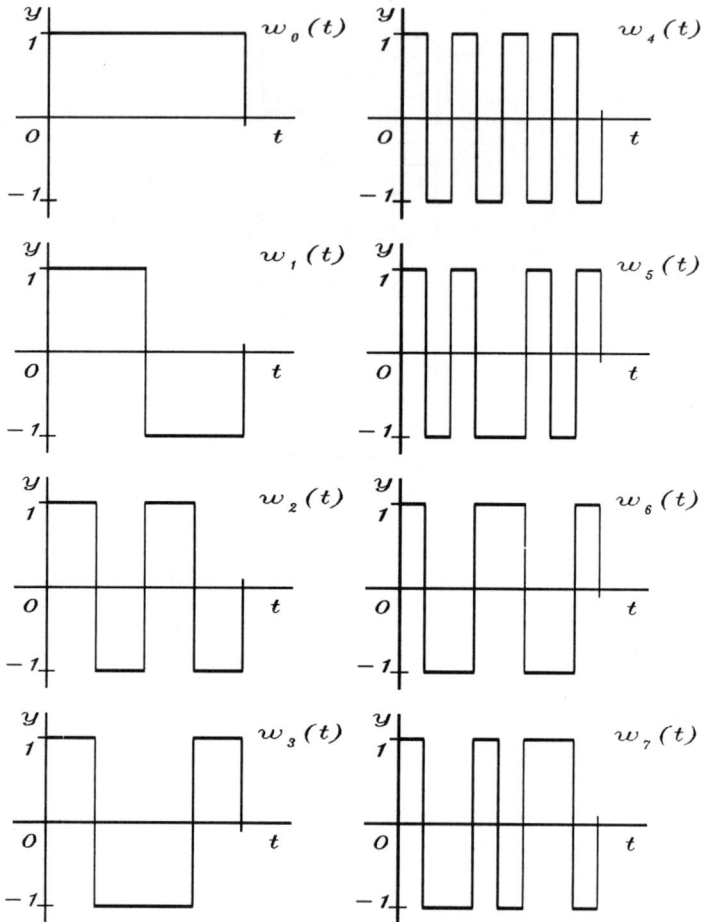

Fig. 7.2

out the first eight Walsh functions:

$$w_0(t) = 1\,,$$

$$w_1(t) = w_{001}(t) = r_1(t) \quad (= r_1^1(t)\, r_2^0(t)\, r_3^0(t))\,,$$

$$w_2(t) = w_{010}(t) = r_2(t) \quad (= r_1^0(t)\, r_2^1(t)\, r_3^0(t))\,,$$

$$w_3(t) = w_{011}(t) = r_1(t)\, r_2(t) \quad (= r_1^1(t)\, r_2^1(t)\, r_3^0(t))\,,$$

$$w_4(t) = w_{100}(t) = r_3(t)\,,$$

$$w_5(t) = w_{101}(t) = r_1(t)\, r_3(t)\,,$$

$$w_6(t) = w_{110}(t) = r_2(t)\, r_3(t)\,,$$

$$w_7(t) = w_{111}(t) = r_1(t)\, r_2(t)\, r_3(t)\,.$$

Fig.7.2 illustrates these functions. We see that for determining $w_n(t)$ to each unit of binary number n the correspondence should be established of the multiplier $r_k(t)$ whose number k represents the number of such digit of binary integer n in which this unit exists: if the unit is in the first digit then the multiplier $r_1(t)$ is taken, if the unit is in the second digit then the multiplier $r_2(t)$ is taken and in general, if the unit is in the k-th digit then the multiplier $r_k(t)$ is taken.

For proving the completeness of the Walsh functions system we subdivide the interval $[0,1]$ into 2^N equal intervals Δ_k $(k = 1, \ldots, 2^N)$ of length 2^{-N}. The number of the Walsh functions each of them retains the constant value on each intervals Δ_k is equal to 2^N. The set X_N of all step functions which retain the constant values on the intervals Δ_k $(k = 1, \ldots, 2^N)$ is evidently the 2^N-dimensional linear space. By virtue of linear independence of the Walsh functions $w_0(t)\, w_1(t)\,, \ldots,\, w_{2^N-1}(t)$ they form the basis in X_N. Therefore each step function of the set X_N represents a linear combination of the functions $w_0(t)$, $w_1(t)\,, \ldots,\, w_{2^N-1}(t)$. By the Corollary of Theorem 3.40 the set of linear combinations of the indicators of the intervals with binary rational ends is dense in $L_2([0,1])$. But this set coincides with the set $\bigcup\limits_{N=1}^{\infty} X_N$, i.e. with the set of all finite linear combinations of the Walsh functions. Consequently, the set of all finite linear combinations of the Walsh functions is dense in $L_2([0,1])$. This fact proves the completeness of the sequence of the Walsh functions $\{w_n(t)\}_0^{\infty}$. So, the orthonormal sequence of the Walsh functions represents the base in $L_2([0,1])$. Therefore any function $x(t) \in L_2([0,1])$ may be presented by the convergent in mean square expansion in terms of the Walsh functions

$$x(t) = \sum_{k=0}^{\infty} \alpha_k w_k(t)\,, \qquad (7.45)$$

$$\alpha_k = \int_0^1 x(t)\, w_k(t)\, dt\,. \qquad (7.46)$$

It is obvious that the integral in formula (7.46) represents the sum of the integrals of the function $x(t)$ with respect to those intervals on which $w_k(t) = 1$ minus the sum of the integrals with respect to those intervals on which $w_k(t) = -1$. It is clear that any segment of series (7.45) approximates the function $x(t)$ by step curve.

Walsh functions and the expansion with respect to them are widely used in signal and image processing (Table A.1.3, Appendix 1). As any finite interval $[a, b]$ may be reduced by the change of variables $s = (t - a)/(b - a)$ to the interval $[0, 1]$ then we may use Walsh functions expansion for any functions with integrable modulus square over any finite interval.

E x a m p l e 7.19. In signal processing instead of the Walsh functions it is sometimes convenient to use the *Haar functions* which are determined on the interval $[0, 1]$ by the formulae

$$h_{00}(t) = 1\,,$$

$$h_{np}(t) = \begin{cases} \sqrt{2^n} & \text{at } t \in \left[\dfrac{2p-2}{2^{n+1}}, \dfrac{2p-1}{2^{n+1}}\right], \\[2mm] -\sqrt{2^n} & \text{at } t \in \left[\dfrac{2p-1}{2^{n+1}}, \dfrac{2p}{2^{n+1}}\right], \\[2mm] 0 & \text{in other points } t \end{cases}$$

$$\left(p = 1, 2, \ldots, 2^n\,; \quad n = 0, 1, 2, \ldots\right).$$

In Fig.7.3 the plot of $h_{23}(t)$ is given.

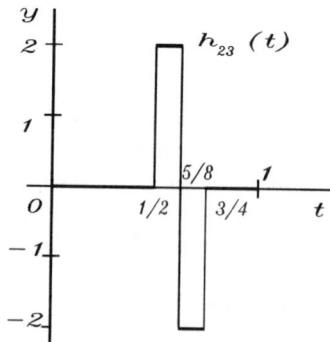

Fig. 7.3

It is evident that the Rademacher functions, and consequently, also the Walsh functions may be expressed in terms of the Haar functions

$$r_n(t) = \frac{1}{\sqrt{2^n}} \sum_{p=1}^{2^{n-1}} h_{n-1,p}(t) \quad (n = 1, 2, \ldots).$$

In exactly the same way as in Example 7.18 we may prove that the sequence of the Haar functions $h_{00}(t)$, $h_{np}(t)$ ($p = 1, \ldots, 2^n$; $n = 0, 1, 2, \ldots$) is orthonormal and complete in $L_2([0, 1])$. Therefore any function $x(t) \in L_2([0, 1])$ may be presented by the expansion in terms of the Haar functions

$$x(t) = \alpha_{00} + \sum_{n=0}^{\infty} \sum_{p=1}^{2^n} \alpha_{np} h_{np}(t),$$

$$\alpha_{00} = \int_0^1 x(t)\, dt,$$

$$\alpha_{np} = \int_0^1 x(t)\, h_{np}(t)\, dt = \left[\int_{\frac{2p-2}{2^{n+1}}}^{\frac{2p-1}{2^{n+1}}} x(t)\, dt - \int_{\frac{2p-1}{2^{n+1}}}^{\frac{2p}{2^{n+1}}} x(t)\, dt \right] \sqrt{2^n}.$$

It is clear that the segment of the series

$$x(t) \approx \alpha_{00} + \sum_{k=0}^{n} \sum_{p=1}^{2^k} \alpha_{kp} h_{kp}(t),$$

as the correspondent segment of the Walsh functions expansion approximates the function $x(t)$ by a step curve with the lengths of the steps $1/2^{n+1}$ (the same which gives the correspondent segment of the expansion in terms of the Walsh functions).

7.5.6. Conditions of Existence of a Basis in a H–Space

Let us establish now a necessary and sufficient condition of the existence of a basis in a H–space.

Theorem 7.27. *A basis in the H–space exists if and only if it is separable.*

▷ If $\{x_k\}$ is an orthonormal basis then for any vector x at any $\varepsilon > 0$ there exists such natural number n that

$$\left\| x - \sum_{k=1}^{n} \alpha_k x_k \right\| < \frac{\varepsilon}{2}, \qquad \alpha_k = (x, x_k).$$

On the other hand, there are such rational numbers r_k that

$$|\alpha_k - r_k| < \varepsilon/2n \quad (k = 1, \ldots, n).$$

Hence

$$\left\| \sum_{k=1}^{n} \alpha_k x_k - \sum_{k=1}^{n} r_k x_k \right\| < \frac{\varepsilon}{2},$$

and consequently,

$$\left\| x - \sum_{k=1}^{n} r_k x_k \right\| < \varepsilon.$$

Thus a countable set of the vectors

$$\left\{ \sum_{k=1}^{n} r_k x_k \right\},$$

correspondent to all natural n and all rational $r_1, \ldots, r_n$ is dense in X. So, X is separable. And vice versa if $\{\xi_k\}$ is a dense countable set then after rejection from it each vector which represents a linear combination of the previous vectors we obtain a complete sequence of the vectors $\{x_k\}$ as in this case $G = \bigcup G_n$ is a subspace which is dense in X, and consequently, coinciding with X. Orthonormalizing $\{x_k\}$ we receive an orthonormal basis. ◁

7.5.7. Biorthogonal and Biorthonormal Sequences

A sequence of the vectors pairs $\{x_k, y_k\}$ is called *biorthogonal* if $(x_k, y_l) = 0$ at all k, l, $k \neq l$. A sequence of the vectors pairs $\{x_k, y_k\}$ is called *biorthonormal* if $(x_k, y_l) = \delta_{kl}$ at all k, l.

Theorem 7.28. *For any sequence of linearly independent vectors $\{x_k\}$ there exists such sequence of linearly independent vectors $\{y_k\}$ that the sequence of the pairs $\{x_k, y_k\}$ is biorthonormal.*

▷ Let us denote by L_p a subspace formed by the vectors x_1, $\ldots, x_{p-1}, x_{p+1}, \ldots, x_n, \ldots$ ($p = 1, 2, \ldots$). Due to the linear independence of the vectors $\{x_k\}$ the subspace L_p does not contain x_p, and consequently, does not coincide with the whole space X. By virtue of Corollary 2 of Theorem 1.18 there exists a continuous linear functional whose kernel contains L_p. But according to the Riesz theorem (Subsection 7.1.3) any continuous linear functional on X represents the scalar product x by some vector of the space X. Therefore in X there exists such vector z_p that $(x, z_p) = 0$ at all $x \in L_p$, and consequently, $(x_k, z_p) = 0$ at all $k \neq p$. After putting $y_p = z_p/(x_p, z_p)$ we shall have $(x_k, y_p) = \delta_{kp}$. For proving the linear independence of the vectors $\{y_k\}$ it is sufficient to notice that if even one of them, for instance, y_p is expressed in terms of the previous ones then by virtue of the fact that $(x_p, y_q) = 0$ at all $q \neq p$, $(x_p, y_p) = 0$ what contradicts to the equality $(x_p, y_p) = 1$. ◁

Corollary. *If the sequence $\{x_k\}$ is complete then the sequence $\{y_k\}$ is unique.*

▷ In this case the subspace formed by the subspace L_k and the vector x_k coincides with X. Therefore the orthogonal supplement $L_k^\perp$ of the subspace L_k is one–dimensional. ◁

Theorem 7.28 establishes the existence of the vectors sequence which supplies the given sequence of linearly independent vectors till the biorthogonal one and till the biorthonormal sequence of the vectors pairs but does not give a constructive way of finding such a sequence. The following two theorems give effective ways of constructing the biorthogonal and biorthonormal sequences.

Theorem 7.29. *Any two sequences of linearly independent vectors $\{u_k\}$, $\{v_k\}$ satisfying the condition: for each vector u_k will be found the vector v_l for which $(u_k, v_l) \neq 0$ may be biorthonormalized, i.e. to substitute by such biorthonormal sequence of the pairs of the vectors $\{x_k, y_k\}$ that each vector u_n will be a linear combination of the vectors $x_1, \ldots, x_n$ and each vector v_n be a linear combination of the vectors $y_1, \ldots, y_n$.*

▷ Let $\{u_k\}$, $\{v_k\}$ be arbitrary sequences of linearly independent vectors satisfying the condition of the theorem. Without loss of the generality we may suppose that $(u_1, v_1) \neq 0$. We put $x_1 = \mu_1 u_1$, $y_1 = \nu_1 v_1$ and determine μ_1 and ν_1 from the condition $(x_1, y_1) = 1$. We receive $\mu_1 \nu_1 = 1/(u_1, v_1)$. After that we put

$$x_2 = \mu_2(\alpha_{21} x_1 + u_2), \quad y_2 = \nu_2(\beta_{21} y_1 + v_2)$$

and determine α_{21}, β_{21}, μ_2, ν_2 from the conditions: $(x_2, y_1) = (x_1, y_2) = 0$, $(x_2, y_2) = 1$. We get

$$\alpha_{21} = -(u_2, y_1), \quad \beta_{21} = -(v_2, x_1), \quad \mu_2 \nu_2 = [(u_2, v_2) - \alpha_{21} \bar{\beta}_{21}]^{-1}.$$

Suppose that continuing in such a way we found $n - 1$ pairs of the vectors $\{x_k, y_k\}$ satisfying the condition of the biorthonormality of $(x_k, y_l) = \delta_{kl}$ $(k, l = 1, \ldots, n - 1)$. We put

$$x_n = \mu_n(\alpha_{n1} x_1 + \ldots + \alpha_{n,n-1} x_{n-1} + u_n),$$

$$y_n = \nu_n(\beta_{n1} y_1 + \ldots + \beta_{n,n-1} y_{n-1} + v_n).$$

After scalar multiplying the first equation from these equations by y_k and the second one by x_k ($k < n$) and equalizing the left–hand sides of the obtained equalities to zero we find $\alpha_{nk} = -(u_n, y_k)$, $\beta_{nk} = -(v_n, x_k)$ ($k = 1, \ldots, n - 1$). After this from the condition $(x_n, y_n) = 1$ we have

$$\mu_n \nu_n = \left[(u_n, v_n) - \sum_{k=1}^{n-1} \alpha_{nk} \bar{\beta}_{nk}\right]^{-1}.$$

Continuing this process we get the biorthonormal sequence of the vectors pairs $\{x_k, y_k\}$ therewith each vector u_n will represent a linear combination of the vectors $x_1, \ldots, x_n$, and each vector v_n be a linear combination of the vectors $y_1, \ldots, y_n$. It is obvious that from two numbers μ_n and ν_n one may be chosen arbitrarily. In particular, μ_n and ν_n may be chosen in such a way that the norms of all the vectors x_n or y_n will be equal to 1. ◁

Theorem 7.30. *If A is a positive self–adjoint operator which maps one–to–one the whole H–space X on the whole space X, and $(Ax_k, x_k) \to 0$ only at $x_k \to 0$ then from any vector sequence we may receive such a vector sequence $\{x_k\}$ that the sequence of the pairs $\{x_k, y_k\}$, $y_k = Ax_k$ will be orthonormal.*

▷ Let us introduce in the space X the second scalar product

$$(x, y)_A = (x, Ay) = (Ax, y).$$

As by Theorem 6.15 the operator A is bounded then from the condition of theorem follows that the sequence $\{z_k\}$ converges (is fundamental) in the space X with the scalar product $(x, y)_A$ if and only if it converges (is fundamental) in X with the scalar product (x, y). Consequently, the space X with the scalar product $(x, y)_A$ also represents the H–space. Therefore any sequence of linearly independent vectors $\{u_k\}$ by Theorem 7.24 may be orthonormalized relatively to the scalar product $(x, y)_A$. The received sequence of the vectors $\{x_k\}$ will satisfy the condition $(x_p, x_q)_A = (x_p, Ax_q) = \delta_{pq}$. Hence the sequence of the vectors pairs $\{x_k, y_k\}$, $y_k = Ax_k$ will be biorthonormal. ◁

The notions of the biorthogonality and biorthonormality of the sequence of the vectors may be also generalized on the topological and, in particular, on the normed linear spaces. Let X be a topological linear space, X^* be its dual space of continuous linear functionals on X.

The vector x of the space X is called *orthogonal* to the vector f of the space X^* if $fx = 0$. A sequence of the pairs vectors $\{x_k, f_k\}$, $x_k \in X$, $f_k \in X^*$ is called *biorthogonal* if $f_k x_l = 0$ at $k \neq l$ and *biorthonormal* if $f_k x_l = \delta_{kl}$. Theorems 7.28–7.30 are also valid in this case. The only change in their proof will be the substitution of all scalar products (x, y) by the correspondent expressions fx. Here the sequence of the vectors $\{x_k\}$ is called *complete* if there is no functional in X^* which is different from zero and orthogonal to all the vectors x_n.

R e m a r k. Biorthonormal sequences in the normed linear spaces are also of a great importance for solving different problems connected with the B–spaces as the orthonormal sequences in H–spaces for solving the correspondent problems.

7.5.8. Expansion of a Vector in Terms of a Basis Generated by a Biorthonormal Sequence

Theory of Subsection 7.5.3 is easily generalized on the biorthonormal bases. Let $\{x_k, y_k\}$ be an orthonormal sequence of the pairs vectors.

Theorem 7.31. *If there exists the operator A satisfying the conditions of Theorem 7.30 and $y_k = Ax_k$ $(k = 1, 2, \ldots)$ then the sequence $\{x_k\}$ is complete in the space X with the scalar product $(x, y)_A$ if and only if it is complete in the space X with the scalar product (x, y). Hence the sequence of the vectors $\{y_k\}$, $y_k = Ax_k$, is also complete.*

▷ As $R_A = X$ then for any $z \in X$ there exists such $y \in X$ that $z = Ay$. Therefore $(x, z) = (x, Ay) = (x, y)_A$ and $(x_k, z) = 0$ at all x_k if and only if $(x_k, y)_A = 0$ at all x_k. As $z = Ay = 0$ only at $y = 0$ then from the completeness of the sequence $\{x_k\}$ in X with scalar product (x, y) follows its completeness in X with $(x, y)_A$ and vice versa. Finally, from the equalities $(x_k, z) = (x_k, Ay) = (Ax_k, y) = (y_k, y)$ follows that $\{x_k\}$ is complete if and only if $\{y_k\}$ is complete. ◁

Theorem 7.32. *If there exists such operator A satisfying the conditions of Theorem 7.30 that $y_k = Ax_k$ $(k = 1, 2, \ldots)$ and the sequence of the vectors $\{x_k\}$ is complete then any vector x of the H–space X may be presented by the expansion*

$$x = \sum_{k=1}^{\infty} (x, y_k) x_k \qquad (7.47)$$

and hence

$$Ax = \sum_{k=1}^{\infty} (x, y_k) y_k, \quad A^{-1}y = \sum_{k=1}^{\infty} (y, x_k) x_k . \qquad (7.48)$$

▷ Accordingly to formula (7.27) any vector $x \in X$ may be presented by the expansion

$$x = \sum_{k=1}^{n} (x, x_k)_A \, x_k + \zeta_n . \tag{7.49}$$

The norm of the residual term ζ_n in (7.49) will be equal to

$$\| \zeta \|_A^2 = \| x \|_A^2 - \sum_{k=1}^{n} |(x, x_k)_A|^2 . \tag{7.50}$$

If the sequence $\{x_k\}$ is complete then $\| \zeta_\infty \|_A = \lim \| \zeta_n \|_A = 0$ and by virtue of the conditions of Theorem 7.30 we have $\| \zeta_\infty \| = \lim \| \zeta_n \| = 0$. Hence it follows that any vector $x \in X$ may be presented by the following expansion:

$$x = \sum_{k=1}^{\infty} (x, x_k)_A x_k = \sum_{k=1}^{\infty} (x, A x_k) x_k \tag{7.51}$$

and therewith

$$A x = \sum_{k=1}^{\infty} (x, A x_k) y_k . \tag{7.52}$$

Putting in expansion (7.51) $x = A^{-1} y$ and accounting that $(x, A x_k) = (A x, x_k) = (y, x_k)$ we get

$$x = A^{-1} y = \sum_{k=1}^{\infty} (y, x_k) x_k . \tag{7.53}$$

As $A x_k = y_k$ then expansion (7.51) coincides with (7.47) and (7.52), expansion (7.53) coincide with (7.48). ◁

E x a m p l e 7.20. Consider the sequence of the pairs *of the Hermite functions of n-dimensional vector argument t*, $\{\varphi_\nu(t), \psi_\nu(t)\}$

$$\varphi_\nu(t) = e^{-t^T K^{-1} t/4} H_\nu(t) = (-1)^{|\nu|} e^{t^T K^{-1} t/4} \frac{\partial^{|\nu|}}{\partial t_1^{\nu_1} \dots \partial t_n^{\nu_n}} e^{-t^T K^{-1} t/2} ,$$

$$\psi_\nu(t) = e^{-t^T K^{-1} t/4} G_\nu(t)$$

$$= (-1)^{|\nu|} e^{t^T K^{-1} t/4} \left[\frac{\partial^{|\nu|}}{\partial s_1^{\nu_1} \dots \partial s_n^{\nu_n}} e^{-s^T K s/2} \right]_{s = K^{-1} t} ,$$

where ν is n–dimensional vector index, $\nu = \{\nu_1, \ldots, \nu_n\}$, $|\nu| = \nu_1 + \ldots + \nu_n$, K is an arbitrary invertible positively determined matrix, and $H_\nu(t)$ and $G_\nu(t)$ are the *Hermite polynomials of n–dimensional vector argument* t. The pair $\{\varphi_\nu(t), \psi_\nu(t)\}$ is biorthogonal in the space $L_2(R^n)$ of the functions of n–dimensional vector argument. By Theorem 7.26 the sequence $\{\varphi_\nu(t)\}$ is complete in $L_2(R^n)$ as it is connected with the sequence of the functions $t_1^{k_1}$ $\ldots t_n^{k_n} e^{-t^T K^{-1} t/4}$ ($k = 0, 1, 2, \ldots$) by the one–to–one relation. Therefore any function $x(t) \in L_2(R^n)$ may be presented by the convergent in mean square series

$$x(t) = \sum_{p=0}^{\infty} \sum_{|\nu|=p} a_\nu \varphi_\nu(t), \tag{7.54}$$

whose coefficients are determined by the formula

$$a_\nu = \int\limits_{R^n} x(t)\, \psi_\nu(t)\, dt \Big/ \int\limits_{R^n} \varphi_\nu(t)\, \psi_\nu(t)\, dt$$

$$= \int\limits_{R^n} x(t)\, \psi_\nu(t)\, dt \Big/ \left(\sqrt{(2\pi)^n\, |K|}\, \nu_1! \ldots \nu_n! \right)$$

($|K|$ is a determinant of the matrix K). The approximation accuracy may be estimated after calculating the norm of the residual term ζ_n in series (7.54):

$$\| \zeta_N \|^2 = \left\| x - \sum_{k=1}^{N} \alpha_k x_k \right\|^2 = \int\limits_{R^n} \left| x(t) - \sum_{p=0}^{M} \sum_{|\nu|=p} a_\nu \varphi_\nu(t) \right|^2 dt,$$

where $N = C_{n+M}^n - 1 = (n+M)!/n!\, M! - 1$.

For detailed study of sequences of vectors and bases refer to (Balakrishnan 1976, Dowson 1978). Properties of special functions and polynomials are given in (Bateman and Erdelyi 1955, Korn and Korn 1968).

Problems

7.5.1. Derive a formula

$$\| z_n \|^2 = \frac{\begin{vmatrix} (x, x) & (x, x_1) & \ldots & (x, x_n) \\ (x_1, x) & (x_1, x_1) & \ldots & (x_1, x_n) \\ \vdots & \vdots & \ddots & \vdots \\ (x_n, x) & (x_n, x_1) & \ldots & (x_n, x_n) \end{vmatrix}}{\begin{vmatrix} (x_1, x_1) & (x_1, x_2) & \ldots & (x_1, x_n) \\ (x_2, x_1) & (x_2, x_2) & \ldots & (x_2, x_n) \\ \vdots & \vdots & \ddots & \vdots \\ (x_n, x_1) & (x_n, x_2) & \ldots & (x_n, x_n) \end{vmatrix}} \tag{$*$}$$

for the norm of the residual term of formula (7.23).

I n s t r u c t i o n. Multiply scalar formula (7.23) by $x, x_1, \ldots, x_n$ and accounting that $(z_n, x) = (x, z_n) = \| z_n \|^2$ get a system of $n + 1$ equations relatively to the variables $1, \alpha_1^{(n)}, \ldots, \alpha_n^{(n)}$ which has the unique solution because of the linear independence of the vectors $x, x_1, \ldots, x_n$ (the Gram determinants in $(*)$ in this case are strictly positive). After expressing from these equations 1 we obtain the formula $(*)$.

7.5.2. Prove for the real H–space that the Gram determinant of linearly independent vectors is equal to square of the volume of the parallelpiped (in general case a multi–dimensional one) constructed on these vectors.

I n s t r u c t i o n. Derive it at first for the parallegram constructed on two vectors.   Then   by   induction   calculate   the   Gram   determinant   taking into consideration that the variable $\| z_n \|$ in $(*)$ represents the height of the $(n + 1)$–dimensional parallelepiped constructed on the vectors $x, x_1, \ldots, x_n$ whose base serves n–dimensional parallelpiped constructed on the vectors $x_1, \ldots, x_n$.

7.5.3. Expand in terms of the Legendre polynomials in $L_2([-1, 1])$ (Example 7.13) the functions: $e^{\alpha t}$, $\sin \pi t$, $\cos \pi t$, $\text{sh}t$, $\text{ch}t$, $\ln t$. Find the norms of the residual terms of these expansions.

7.5.4. Expand in the Fourier series (complex) the functions: $1_{[-\pi, \pi]}(t)$, t, t^2, t^3 in $L_2([-\pi, \pi])$ and estimate the residual terms.

7.5.5. Expand in the Fourier series (complex) the functions: $t_1 t_2$, $t_1^2 t_2^3$, $t_1^2 + t_2^3$ in $L_2([-\pi, \pi]^2)$ and estimate the residual terms.

7.5.6. Prove that the Hermite function (Example 7.16) $\varphi_n(t) = e^{-t^2/2} H_n(t)$ in $L_2(R)$ is orthogonal to the function $e^{-t^2/2} t^m$ at $m < n$.

7.5.7. Expand in series in terms of the Hermite functions in $L_2(R)$ the functions $e^{-t^2/2} t^m$ ($m = 0, 1, 2, \ldots$), $1_{[-\pi, \pi]}(t)$ and estimate the residual terms.

7.5.8. Prove that the Laguerre function (Example 7.17) $\psi_n^{(\nu)}(t) = t^{\nu/2} \times e^{-t/2} L_n^{(\nu)}(t)$ in $L_2([0, \infty))$ is orthogonal to the function $t^{\nu/2+m} e^{-t/2}$ at $m < n$.

7.5.9. Expand in series in terms of the Laguerre functions in $L_2([0, \infty))$ the functions $t^{\nu/2+m} e^{-t/2}$ ($m = 0, 1, 2, \ldots$) and estimate the residual terms.

7.5.10. Find the first 8 terms of the expansion in series in terms of the Walsh function (Example 7.18) in $L_2([0, 1])$ of the functions: t^n, $\sin \pi t$, $\cos \pi t$, $\text{sh}t$, $\text{ch}t$, $e^{\alpha t}$, $\ln t$ and estimate the residual terms.

7.5.11. Prove that any separable H–space is isomorphic to the space l_2 and this isomorphism is isometric.

I n s t r u c t i o n. Use formula (7.33), Parseval equality (7.34) and the Riesz–Fisher theorem (Problem 7.5.16).

7.5.12. Under the conditions of Example 7.20 prove that the Hermite function $\varphi_\nu(t) = e^{-t^T K^{-1} t/4} H_\nu(t)$ is orthogonal to the functions $t_1^{\mu_1} \ldots t_n^{\mu_n}$

$\times\, e^{-t^T K^{-1} t/4}$ at $|\mu|<|\nu|$.

I n s t r u c t i o n. Represent $t_1^{\mu_1}\ldots t_n^{\mu_n}$ as a linear combination of the Hermite polynomials $G_\lambda(t)$, $|\lambda|\leq|\mu|$.

7.5.13. Under the conditions of Example 7.20 prove that the Hermite function $\psi_\nu(t) = e^{-t^T K^{-1} t/4} G_\nu(t)$ is orthogonal to the functions $t_1^{\mu_1}\ldots t_n^{\mu_n} e^{-t^T K^{-1} t/4}$ at $|\mu|<|\nu|$.

7.5.14. Under the conditions of Example 7.20 find the expansion of any function $x(t) \in L_2(R^n)$ in terms of the Hermite functions $\psi_\nu(t) = e^{-t^T K^{-1} t/4} G_\nu(t)$ and write the formula for the norm of the residual term.

7.5.15. Prove that under the conditions of Example 7.20 the operator A of Theorem 7.30 is determined by formula

$$Ax = \int\limits_{R^n} a(t,s)\, x(s)\, ds\,,$$

where

$$\begin{aligned}
a(t,s) = {}& [(2\pi)^n\, |K|\,|I - K^2|]^{-1/2} \exp\{-t^T K^{-1} t/4 \\
& - s^T K^{-1} s/4 - (t^T + s^T)(I + K)^{-1}(t + s)/4 \\
& + (t^T - s^T)(I - K)^{-1}(t - s)/4\}.
\end{aligned}$$

I n s t r u c t i o n. Use the first formula of (7.48) and formula

$$\sum_{p=0}^{\infty}\sum_{|\nu|=p} \frac{G_\nu(t)\, G_\nu(s)}{\nu_1!\ldots\nu_n!} = |I - K^2|^{-1/2}\exp\{-(t^T + s^T)(I + K)^{-1}(t + s)/4$$

$$+ (t^T - s^T)(I - K)^{-1}(t - s)/4\}\,,$$

which is valid under the condition that the matrix $I - K$ is positively determined.

7.5.16. Prove the Riesz–Fisher theorem *if the series $\sum |c_n|^2$ converges then in the separable H-space for any orthonormal basis $\{x_n\}$ such vector x will be found for which $c_n = (x, x_n)$.*

I n s t r u c t i o n. Take the sequence of the vectors

$$u_m = \sum_{n=1}^{m} c_n x_n\,,$$

and account that $(u_m, x_n) = c_n$ at $n \leq m$ and that the series $\sum c_n x_n$ converges to some vector x by virtue of the convergence of the series $\sum |c_n|^2$. So, as result we have $\|u_m - x\|\to 0$ at $m \to \infty$.

7.5.17. Theorem 7.27 establishes the existence of the basis in any separable H-space. It is easy to understand that the proof of the separability necessity of the space for the basis existence is also transferred without any changes on the case of the

B–space. Thus for the basis existence in the B–space its separability is necessary. But this condition is not sufficient. As it is known in any separable B–space X there exist complete systems of the vectors (see the latter part of Subsection 7.5.7). Prove that any countable dense set of the vectors represents a complete system. But in order to receive a basis if it is possible it is not sufficient to have a complete system. Besides it is necessary that a complete system will be *the minimal*, i.e. that none of the vectors of this system belongs to closed linear spans of other vectors. Prove that a closed linear span of a complete system of the vectors coincides with the whole B–space X (remember Corollary 2 of Theorem 1.18).

7.5.18. Prove that for any complete minimal system of the vectors $\{x_n\}$ in the separable B–space X there exists the unique *adjoint* system of the functionals $\{f_n\} \subset X^*$ which forms together with $\{x_n\}$ the biorthonormal system $f_k x_l = \delta_{kl}$.

7.5.19. Let X be a separable B–space, $\{x_n\}$ be a complete minimal system of the vectors in X, $\{f_n\}$ be the correspondent adjoint system of the functionals in X^*. Prove that the operator which acts in X

$$P_n x = \sum_{k=1}^{n} (f_k x) x_k$$

represents a projector on the subspace L_n formed by the vectors $x_1, \ldots, x_n$.

I n s t r u c t i o n. It is sufficient to prove that $P_n x = x$, $\forall x \in L_n$ and $P_n x = 0$ at x which does not belong to L_n (Subsection 7.4.4).

7.5.20. Prove the theorem: *in order the system of the vectors $\{x_n\}$ of Problem 7.5.19 will be the basis in X it is necessary and sufficient that the projectors P_n be uniformly bounded* $\|P_n\| < C < \infty$.

I n s t r u c t i o n. For proving the necessity of the condition use Banach–Steinhaus theorem 6.12. To prove the sufficiency of the condition use the fact that a linear span of a system of the vectors $\{x_n\}$ is dense in consequence of which for any vector $x \in X$ and any $\varepsilon > 0$ there exist such numbers $c_1, \ldots, c_n$ that

$$\left\| x - \sum_{k=1}^{n} c_k x_k \right\| < \varepsilon.$$

As a result we get

$$\left\| P_n x - \sum_{k=1}^{n} c_k x_k \right\| < C\varepsilon.$$

7.5.21. Prove that if the B–space X in Problems 7.5.19 and 7.5.20 is reflexive and $\{x_n\}$ is a basis in X then a closed linear span of a system of the vectors $\{f_n\}$ coincide with the space X^* and the system $\{f_n\}$ represents a basis in X^*.

7.5.22. Let $\{x_n\}$ be the minimal complete system of the vectors in B–space X. Find its adjoint system of the functionals $\{f_n\}$, $f_n \in X^*$ satisfying the conditions $f_n x_m = \delta_{nm}$.

Instruction. Take an arbitrary sytem of linearly independent linear functionals $\{g_n\}$ which satisfies the condition $g_n x_n \neq 0$ and determine the following functionals:

$$h_n = \sum_{k=1}^{n-1} a_{nk} h_k + a_{nn} g_n, \qquad (**)$$

by choosing $a_{n1}, \ldots, a_{n,n-1}$ from the conditions $h_n x_m = 0$ $(m = 1, \ldots, n-1)$, and a_{nn} from the condition $h_n x_n \neq 0$. After that determine the functionals

$$f_n = \sum_{i=n}^{\infty} b_{nl} h_l, \qquad (***)$$

satisfying the conditions $f_n x_m = \delta_{nm}$. By virtue of the definition of the functionals h_l these conditions automatically are fulfilled at $m < n$. And at $m \geq n$ we have the following equations for b_{nl}:

$$\sum_{l=n}^{m} b_{nl} h_l x_m = \delta_{nm} \quad (m = n, n+1, \ldots; \ n = 1, 2, \ldots).$$

Write the explicit formulae for the coefficients a_{nk} in $(**)$ and b_{nl} in $(***)$.

7.5.23. Construct in the space $C([0,1])$ a system of the functions $x_1(t) = 1$, $x_2(t) = t$, $x_3(t) = 1 - |1 - 2t|$ the other functions $x_n(t)$ are illustrated graphically by isosceles triangles with the height 1 on each from 2^{r-1} intervals of the length 2^{-r+1} on which the interval $[0,1]$ is subdivided and are equal to zero outside of these intervals. These functions beginning with $x_3(t)$ are determined by the formula

$$x_n(t) = \max(0, 1 - |2k - 1 - 2^r t|), \quad k = n - 1 - 2^{r-1}$$

$$(n = 2 + 2^{r-1}, \ldots, 1 + 2^r; \ r = 1, 2, \ldots).$$

Prove that this system of the functions represents a basis in the space $C([0,1])$. Find an adjoint system of the functionals $\{f_n\}$ representing the expansion of the function $x(t) \in C([0,1])$ in the form

$$x(t) = \sum_{n=1}^{\infty} (f_n x) x_n(t).$$

7.5.24. Extend the results of Problem 7.5.23 on the space $C(T)$ for any interval $T \subset R$ including the infinite intervals.

Instruction. Use any continuous strictly monotone mapping of the interval T on the interval $[0,1]$.

7.5.25. Prove that a system of the Haar functions (Example 7.18) represents a basis in any space $L_p([0,1])$, $p > 1$, therewith the same system of the Haar functions in the dual space $L_q([0,1])$, $p^{-1} + q^{-1} = 1$ represents an adjoint basis in $\{f_n\}$.

Analogously as in Problem 7.5.24 determine the basis and the adjoint basis of the linear functionals for the space $L_p(T)$, $p > 1$ in the case of the arbitrary interval $T \subset R$ (including the infinite intervals).

Using the Remark preceding Theorem 7.26 construct a basis and an adjoint basis of the linear functionals in the space $L_p(T_1 \times \cdots \times T_N)$, $T_1, \ldots, T_N \subset R$, $p > 1$ of the functions of the N–dimenstional vector argument.

7.6. Compact Operators

7.6.1. Definition of a Compact Operator

The operator T mapping the topological space X into the topological space Y is called *compact* if its image TM of any bounded set M is precompact. A continuous compact operator is called *completely continuous*.

In the special case when Y is the T_1–space with the first countability axiom the operator T is compact if and only if its image $\{Tx_n\}$ of any bounded sequence $\{x_n\}$ contains a convergent subsequence (Theorems 4.17 and 4.31)[a].

Any linear operator in a finite–dimensional space is compact as the image of any bounded set is bounded, and a bounded set in the finite-dimensional space is usually precompact.

In the infinite–dimensional spaces not each even bounded linear operator is compact as not each bounded set in such a space is precompact.

7.6.2. Properties of Compact Operators

Theorem 7.33. *A compact linear operator which maps the B–space X into the B–space Y is bounded, and consequently, is also continuous.*

▷ If the operator T is unbounded then there exists such sequence $\{x_n\}$ that $\|Tx_n\| > n\|x_n\|$. Putting $z_n = x_n / \|x_n\|$ we obtain such bounded set $\{z_n\}$ that $\|Tz_n\| > n$ at all n. The sequence $\{Tz_n\}$ evidently does not contain any convergent subsequence. Consequently, the image $\{Tz_n\}$ of the bounded set $\{z_n\}$ is not precompact and the operator T cannot be compact. ◁

Corollary 1. *If X and Y are the B–spaces then any compact linear operator is completely continuous.*

---

[a] If a set is precompact then the limit of the sequence which is contained in it may not belong to this set.

R e m a r k. Thus for the linear operators in the B–spaces the notions of the compactness and the complete compactness coincide.

Corollary 1. *Unbounded linear operator cannot be compact.*

Corollary 3. *A compact linear operator T which maps the reflexive B–space X into B–space Y is determined on the whole space X, has the bounded adjoint operator T^* determined on the whole space X^* and $\|T^*\| = \|T\|$* (Subsection 6.1.10).

Corolary 4. *A compact linear operator is closed as a bounded linear operator determined on the whole space* (Corollary of Theorem 6.2).

Theorem 7.34. *If the linear operator T is compact then its image of any weakly convergent sequence represents a convergent sequence.*

▷ Really, if $fx_n \to fx$ at all $f \in X^*$ then by virtue of Corollary 2 of Banach–Steinhaus theorem 6.18 the set $\{x_n\}$ is bounded. In consequence of the compactness of the operator T the set $\{Tx_n\}$ is precompact. After choosing from it the convergent subsequence $\{Tx_{n_k}\}$ we get

$$g(Tx_{n_k}) = (T^*g)x_{n_k} \to (T^*g)x = g(Tx)$$

at any $g \in Y^*$. Hence it follows that $\lim Tx_{n_k} = Tx$. It remains to prove that the whole subsequence $\{Tx_n\}$ also converges to Tx. Supposing the opposite we may choose such $\varepsilon > 0$ and such unbounded increasing subsequence of the natural numbers $\{m_l\}$ that $\|Tx_{m_l} - Tx\| > \varepsilon$ at all l. The sequence $\{Tx_{m_l}\}$ does not contain any convergent subsequence as according to the proved any convergent subsequence from $\{Tx_n\}$ converges to Tx. This fact contradicts to the compactness of the operator T. Consequently, $Tx_n \to Tx$. ◁

The inverse statement is also true. *If the operator T (not obligatory linear) maps any weakly convergent sequence into convergent sequence then it is compact.*

First we shall prove an auxiliary statement restricting our consideration to the case of the H–space X.

Lemma. *Any bounded set in the H–space is weakly precompact, i.e. we may choose from it a weakly convergent sequence.*

▷ Let M be a bounded set in the H–space X. Take any sequence $\{x_n\} \subset M$ and denote by G a subspace formed by this sequence. Choose from $\{x_n\}$ such subsequence $\{x_{1n}\}$ that the numeric sequence $\{(x_1, x_{1n})\}$ will be convergent. From $\{x_{1n}\}$ we choose such subsequence $\{x_{2n}\}$ that the sequence $\{(x_2, x_{2n})\}$ will be convergent. Continuing this process we choose from sequence $\{x_{r-1,n}\}$ such subsequence $\{x_{rn}\}$ that the sequence $\{(x_r, x_{rn})\}$ will be convergent $(r = 3, 4, \ldots)$. It is evident that the diagonal sequence $\{x_{nn}\}$ possesses such property that all the sequences $\{(x_r, x_{nn})\}$ $(r = 1, 2, \ldots)$ converge.

As the sequence $\{x_n\}$ is complete in G according to the definition of G then any vector $y \in G$ my be presented by formula (7.23) as the limit of the sequence $\{y_N\}$ of the finite linear combinations of the vector x_r. The norms of the residual term $z_N = y - y_N$ will tend to zero at $N \to \infty$. For proving the convergence of the sequence $\{(y, x_{nn})\}$ at any $y \in G$ it is sufficient to prove its fundamental property. For this purpose we take an arbitrary $\varepsilon > 0$ and choose the finite linear combination y_N of the vectors x_r from the sequence $\{y_N\}$, $y_N \to y$ in such a way that in the equality

$$(y, x_{nn} - x_{mm}) = (y_N, x_{nn} - x_{mm}) + (y - y_N, x_{nn} - x_{mm})$$

will be $|(y - y_N, x_{nn} - x_{mm})| < \varepsilon/2$. It is possible as $|(y - y_N, x_{nn} - x_{mm})| \leq \|y - y_N\| \sup_{n,m} \|x_{nn} - x_{mm}\| \to 0$ at $N \to \infty$. After this taking into account that the sequence $\{(y_N, x_{nn})\}$ converges, and consequently, is also fundamental at any y_N we choose n_ε in such a way that $|(y_N, x_{nn} - y_{mm})| < \varepsilon/2$ at all $n, m > n_\varepsilon$. Then we shall have $|(y, x_{nn} - x_{mm})| < \varepsilon$ at all $n, m > n_\varepsilon$. This fact proves the fundamental property of the sequence $\{(y, x_{nn})\}$ at any $y \in G$. Consequently, for any vector $y \in G$ the sequence $\{(y, x_{nn})\}$ converges. On the other hand, $(z, x_{nn}) = 0$ for any $z \in G^\perp$. As any vector $x \in X$ may be uniquely presented in the form of the sum $x = y + z$, $y \in G$, $z \in G^\perp$ then the sequence $\{(x, x_{nn})\}$ converges at any $x \in X$. Thus the sequence separated from M is weakly fundamental. Because of the weak completeness of the H–space (Corollary 3 of Banach–Steinhaus theorem 6.12) the sequence $\{x_{nn}\}$ weakly converges to some vector $x \in X$ what proves the weak sequential precompactness of the set M. ◁

Theorem 7.35. *If the operator T maps the H–space X into the B–space Y and its image of any weakly convergent sequence represents a convergent sequence then the operator T is compact.*

▷ Let $\{x_n\}$ be any bounded sequence in the space X. According to the proved Lemma it contains the weakly convergent subsequence $\{x_{n_k}\}$. By the theorem condition the correspondent subsequence $\{T x_{n_k}\}$ of the sequence $\{T x_n\}$ converges. Thus the image of any bounded sequence contains a convergent subsequence. Consequently, the operator T is compact. ◁

Theorem 7.36. *If S is a continuous operator which maps the B–space X into the B–space Y and T is a compact operator which maps*

Y into the B–space Z then the operator TS which maps X into Z is compact.

▷ It follows from the fact that the image SM of the bounded set $M \subset X$ in Y is bounded, and consequently, the image TSM of the set M in Z is precompact. ◁

Theorem 7.37. *If T is a compact operator which maps the B–space X into the B–space Y, and S is a continuous operator which maps Y into the B–space Z then the operator ST which maps X into Z is compact.*

▷ It is sufficient to prove the precompactness in space Z of the image STM of any bounded set $M \subset X$. For this purpose we take an arbitrary sequence $\{z_n\} \subset STM$. Its inverse image $\{y_n\} \subset TM$ in space Y contains the convergent subsequence $\{y_{n_k}\}$ in consequence of the precompactness of TM. The image $\{z_{n_k}\}$ of this subsequence in $STM \subset Z$ represents a convergent sequence by virtue of the continuity of the operator S. Consequently, any sequence of the points of the set STM contains a convergent subsequence what proves the precompactness of the set STM and the compactness of the operator ST. ◁

It follows from Theorems 7.36 and 7.37 that an operator inverse relatively to a compact operator cannot be continuous as the unit operator I is not compact.

R e m a r k. Notice that Theorems 7.35 and 7.37 are also valid for the nonlinear operators.

Theorem 7.38. *If the linear operator T which maps the H–space into another H–space is bounded, and the operator T^*T is compact then the operator T is compact.*

▷ If the sequence $\{x_n\}$ weakly converges to x then the sequence $\{T^*Tx_n\}$ converges to T^*Tx, and consequently,

$$\|Tx_n - Tx\|^2 = (Tx_n - Tx, Tx_n - Tx) = (x_n - x, T^*Tx_n - T^*Tx)$$

$$\leq \|x_n - x\|\|T^*Tx_n - T^*Tx\| \to 0 \text{ at } n \to \infty$$

by virtue of the boundedness of $\{x_n\}$ (Corollary 2 of Theorem 6.12). Thus the operator T maps any weakly convergent sequence into convergent one, and consequently, is compact. ◁

Corollary. *A compact linear operator which maps one H–space into another H–space has a compact adjoint operator.*

▷ If T is a compact linear operator then by Corollary 3 of Theorem 7.33 it has a bounded adjoint operator T^*. Consequently, by Theorem 7.36 the operator $TT^* = T^{**}T^*$ is compact. By Theorem 7.38 it follows that the operator T^* is compact. ◁

R e m a r k. Notice one more obvious fact: *if the operators* $T_1, \ldots, T_n$ *are compact then the operator* $\alpha_1 T_1 + \cdots + \alpha_n T_n$ *at any complex* α_1, $\ldots, \alpha_n$ *is also compact.* Thus any finite linear combination of the compact operators represents a compact operator.

In order to establish the compactness of the operator T which maps the H–space into the B–space we may use the following theorem which gives the sufficient condition of the compactness.

Theorem 7.39. *If there exists such set of the compact operators* $\{T_\alpha\}$ *that for any* $\varepsilon > 0$ *in this set will be found the operator* T_{α_ε} *satisfying the condition* $\|T_{\alpha_\varepsilon} - T\| < \varepsilon$ *then the operator* T *is compact.*

▷ For proving let us take an arbitrary weakly convergent sequence $\{x_p\}$, $x_p \overset{w}{\to} x$ and prove that the operator T maps it into the convergent sequence $\{Tx_p\}$, $Tx_p \to Tx$. Let $\{\varepsilon_n\}$ be an arbitrary sequence of positive numbers convergent to zero $\varepsilon_n > 0$, $\varepsilon_n \to 0$ at $n \to \infty$. For each n we shall choose from $\{T_\alpha\}$ the operator T_{α_n} satisfying the condition $\|T_{\alpha_n} - T\| < \varepsilon_n$. By virtue of the boundedness of weakly convergent sequence we have $\|x_p\| < c$ at some $c > 0$ for all p and

$$\|Tx_p - Tx\| = \|(T - T_{\alpha_n})x_p + T_{\alpha_n}x_p - T_{\alpha_n}x + (T_{\alpha_n} - T)x\|$$

$$\leq \|T - T_{\alpha_n}\|\|x_p - x\| + \|T_{\alpha_n}x_p - T_{\alpha_n}x\|$$

$$< 2c\varepsilon_n + \|T_{\alpha_n}x_p - T_{\alpha_n}x\|.$$

We shall set now an arbitrary $\varepsilon > 0$ and choose n in such a way that $2c\varepsilon_n < \varepsilon/2$. After that we shall choose p_ε in such a way that $\|T_{\alpha_n}x_p - T_{\alpha_n}x\| < \varepsilon/2$ at all $p > p_\varepsilon$. It is possible as the sequence $\{T_{\alpha_n}x_p\}$ converges to $T_{\alpha_n}x$ by virtue of the compactness of all the operator T_{α_n}. Then we shall have

$$\|Tx_p - Tx\| < \varepsilon,$$

what proves the convergence of the sequence $\{Tx_p\}$ to Tx, and consequently, the compactness of the operator T. ◁

7.6.3. Spectrum of a Compact Operator

Let us study now the properties of the compact linear operators restricting ourselves to the operators in the H–spaces.

From the general properties of the spectra of the closed linear operators in the B–spaces (Subsection 6.3.2) follows.

Theorem 7.40. *If T is a compact operator then any value λ for which $R_T(\lambda) = X$ is regular, $\lambda \in \rho(T)$.*

Theorem 7.41. *A compact operator may have only finite number of the eigenvalues modulus exceeding the given number $\delta > 0$.*

▷ Let $\{\lambda_\alpha\}$ be a set of the eigenvalues of the compact operator T modulus exceeding δ. Supposing that this set is infinite we choose from it the sequence of the eigenvalues $\{\lambda_n\}$ placed in moduluses nonincreasing order and orthonormalize (Subsection 7.5.2) the correspondent sequence of linearly independent eigenvectors $\{x_n\}$, $Tx_n = \lambda_n x_n$. As a result we receive the orthonormal sequence $\{y_n\}$,

$$y_n = \sum_{k=1}^n a_{nk} x_k \quad (n = 1, 2, \ldots).$$

The shall have

$$Ty_n - \lambda_n y_n = \sum_{k=1}^n a_{nk}(\lambda_k - \lambda_n)x_k \quad (n = 2, 3, \ldots).$$

Substituting here the vectors $x_1, \ldots, x_{n-1}$ by their expressions in terms of $y_1, \ldots, y_{n-1}$ we get

$$Ty_n = \sum_{k=1}^{n-1} b_{nk} y_k + \lambda_n y_n \quad (n = 2, 3, \ldots). \tag{7.55}$$

Therewith accounting the orthonormality of the vectors system $\{y_n\}$ by formula (7.33) we find at any m and $n > m$

$$\|Ty_n - Ty_m\|^2 = \sum_{k=1}^{m-1} |b_{nk} - b_{mk}|^2 + |b_{nm} - \lambda_m|^2$$

$$+ \sum_{k=m+1}^{n-1} |b_{nk}|^2 + |\lambda_n|^2 \geq |\lambda_n|^2 > \delta^2.$$

This inequality shows that none of the subsequences of the sequence $\{Ty_n\}$ may be convergent what contradicts the compactness of the operator T. This proves that the set $\{\lambda_\alpha\}$ cannot be infinite. ◁

Corollary 1. *A set of the eigenvalues of a compact operator is finite or countable and in the second case has the unique limit point 0.*

Corollary 2. *To each eigenvalue of a compact operator different from zero may correspond only finite–dimensional proper subspace (finite number of linearly independent eigenvectors).*

▷ Supposing that to the eigenvalue $\lambda \neq 0$ corresponds infinite-dimensional proper subspace and choosing from it the sequence of linearly independent eigenvectors $\{x_n\}$ and assuming $\lambda_n = \lambda$ $(n = 1, 2, \ldots)$ we come to the conflict with the proved theorem. ◁

Theorem 7.42. *If $\lambda \neq 0$ is not eigenvalue of the operator T then the resolvent R_λ is the bounded operator on $R_T(\lambda)$.*

▷ Assuming the opposite we take such sequence $\{y_n\} \subset R_T(\lambda)$ that $\| R_\lambda y_n \| \,/\, \| y_n \| \to \infty$ at $n \to \infty$. Put $x_n = R_\lambda y_n$, $z_n = x_n / \| x_n \|$, $u_n = y_n / \| x_n \|$. Then in consequence of the fact that $Tx_n - \lambda x_n = y_n$ we shall have $Tz_n - \lambda z_n = u_n$. By virtue of the compactness of T we may choose from the sequence $\{Tz_n\}$ the convergent subsequence $\{Tz_{n_k}\}$, $\lim Tz_{n_k} = v$. As $u_n \to 0$ and $z_{n_k} = \lambda^{-1}(Tz_{n_k} - u_{n_k})$ then the subsequence $\{z_{n_k}\}$ converges and $\lim z_{n_k} = \lambda^{-1}v$. Hence it follows that $\lim Tz_{n_k} = \lambda^{-1}Tv$. After comparing two expressions of $\lim Tz_{n_k}$ and putting $\lim z_{n_k} = \lambda^{-1}v = z$ we shall have $\| z \| = 1$ and $Tz = \lambda z$. But this is impossible as by the condition λ is not an eigenvalue of the operator T. Consequently, there exists such number $c > 0$ that the solution x of the equation

$$Tx - \lambda x = y \qquad\qquad (7.56)$$

satisfies the condition $\| x \| < c \| y \|$ at all $y \in R_T(\lambda)$. This proves the boundedness of the resolvent on $R_T(\lambda)$. ◁

Theorem 7.43. *If $\lambda \neq 0$ is not eigenvalue of the compact operator T then the domain $R_T(\lambda)$ of its resolvent R_λ represents a closed set, and consequently, is a subspace.*

▷ By virtue of the completeness property of the operator T the operator $T - \lambda I$ is closed in consequence of which the inverse operator R_λ is also closed (Subsection 6.1.1). By Theorem 7.42 it is continuous, and by Theorem 6.2 the domain $R_T(\lambda)$ of a continuous closed operator is closed. ◁

The range $R_T(\lambda)$ of the operator $T - \lambda I$ in the case of the compact T is also closed when $\lambda \neq 0$ is an eigenvalue of the operator T even if the resolvent in this case does not exist. In order to prove it we use the following auxiliary statement.

Lemma. *If* $\lambda \neq 0$ *is not an eigenvalue of the compact operator* T *then there exists such number* c *and such solution* x *of Eq.(8.56) that* $\|x\| < c\,\|y\|$ *at all* $y \in R_T(\lambda)$.

▷ Let us take in proper subspace of the operator T which corresponds to the eigenvalue λ a complete system of linearly independent vectors $x^{(1)}, \ldots, x^{(m)}$. Then all the solutions of Eq. (7.56) will be determined by formula $x = x^{(0)} + c_1 x^{(1)} + \cdots + c_m x^{(m)}$ where $c_1, \ldots, c_m$ are arbitrary complex numbers, and $x^{(0)}$ is some solution of Eq. (7.56). Let us choose from these solutions at each $y \in R_T(\lambda)$ the solution x' with the minimal norm $\| x' \| = \min\limits_{c_k} \| x^{(0)} + c_1 x^{(1)} + \cdots + c_m x^{(m)} \|$. Then $z = (x^{(0)} + c_1 x^{(1)} + \cdots + c_m x^{(m)})/\|x'\|$ will be the solution of the equation

$$Tz - \lambda z = u, \quad u = y/\|x'\| \ . \tag{7.57}$$

And $z' = x'/\|x'\|$ will be a solution with the minimal norm $\|z'\| = 1$. Supposing that at any $c > 0$ will be found such vector $y \in R_T(\lambda)$ that $\|x'\| > c\|y\|$ (x' is a solution of Eq. (7.56) with the minimal norm) we take such sequence of the vectors $\{y_n\}$ and the correspondent sequence of the solutions $\{x_n\}$ with the minimal norm that $\|x_n\| \,/\, \|y_n\| \to \infty$. Then $z_n = x_n/\|x_n\|$ will be the solution of Eq. (7.57) at $u = u_n = y_n/\|x_n\|$ with the minimal norm $\|z_n\| = 1$. After this accounting that by the supposition $u_n \to 0$ at $n \to \infty$ similarly as while proving Theorem 7.42 we shall prove the existence on the convergent subsequence $\{z_{n_k}\}$ and $z = \lim z_{n_k}$, $\|z\| = 1$ is an eigenvector of the operator T correspondent to the eigenvalue λ. After taking arbitrary $\varepsilon > 0$ and such n_k that $\|z_{n_k} - z\| < \varepsilon$ we see that $z_{n_k} - z$ is the solution of Eq. (7.57) at $u = u_{n_k}$ with the minimal norm equal to 1. The obtained contradiction proves the Lemma. ◁

Theorem 7.44. *If* $\lambda \neq 0$ *is an eigenvalue of the compact operator* T *then the range* $R_T(\lambda)$ *of the operator* $T - \lambda I$ *is a closed set.*

▷ Let y be the limit point $R_T(\lambda)$, $\{y_n\} \subset R_T(\lambda)$ be the subsequence convergent to y, $y = \lim y_n$. Accordingly to the proved lemma at each y_n there exists such solution x_n of the equation $Tx_n - \lambda x_n = y_n$ that $\| x_n \| < c\| y_n \|$ at some $c < \infty$. After taking an arbitrary $\varepsilon > 0$ and sufficiently large n we get $\| x_n \| < c\,(\| y \| + \varepsilon)$. It means that the set $\{x_n\}$ is bounded and by virtue of the compactness of T there exists such subsequence $\{x_{n_k}\}$ that $\{Tx_{n_k}\}$ converges. As $x_{n_k} = \lambda^{-1}(Tx_{n_k} - y_{n_k})$ then the subsequence $\{x_{n_k}\}$ also converges. Therefore putting $x = \lim x_{n_k}$, $v = \lim Tx_{n_k}$ we shall have $x = \lambda^{-1}(v - y)$. But $\lim Tx_{n_k} = Tx$ by virtue of the continuity of T. Consequently, $Tx = v = \lambda x + y$

and $Tx - \lambda x = y$, i.e. $y \in R_T(\lambda)$. Thus the set $R_T(\lambda)$ contains all its limit points, i.e. is closed, $[R_T(\lambda)] = R_T(\lambda)$. It means that $R_T(\lambda)$ represents a subspace. ◁

R e m a r k. Notice that while proving Theorem 7.44 in passing we also proved Theorem 7.43 in another way.

Theorem 7.45. *If $\lambda \neq 0$ is an eigenvalue of the compact operator T then $\bar{\lambda}$ is an eigenvalue of the adjoint operator T^* and its correspondent proper subspace represents an orthogonal supplement of the subspace $R_T(\lambda)$.*

▷ If $\lambda \neq 0$ is an eigenvalue of the operator T then by Theorem 7.40 $R_T(\lambda) \neq X$. After taking any vector z from the orthogonal supplement $R_T(\lambda)$, $z \in R_T(\lambda)^\perp$ we obtain $(Tx - \lambda x, z) = 0$ for any $x \in X$. Supposing that $(x, T^*z - \bar{\lambda}z) = 0$ at all $x \in X$ what is possible only at $T^*z = \bar{\lambda}z$. Consequently, $\bar{\lambda}$ is the eigenvalue of the operator T^* and any vector $z \in R_T(\lambda)^\perp$ serves as the correspondent eigenvector. On the other hand, for any eigenvector z of the operator T^* correspondent to the eigenvalue $\bar{\lambda}$ and any vector $y \in R_T(\lambda)$ there exists such vector $x \in X$ that

$$(y, z) = (Tx - \lambda x, z) = (x, T^*z - \bar{\lambda}z) = 0.$$

Consequently, the vector z is orthogonal to $R_T(\lambda)$, i.e. $z \in R_T(\lambda)^\perp$. ◁

Corollary 1. *If $\lambda \neq 0$ is an eigenvalue of the compact operator T then its proper subspace correspondent to this eigenvalue represents an orthogonal supplement of the subspace $R_{T^*}(\bar{\lambda})$.*

Corollary 2. *If $\lambda \neq 0$ is an eigenvalue of the compact operator T then Eq. (7.56) has the solution if and only if the vector y is orthogonal to the proper subspace of the operator T^* correspondent to the eigenvalue $\bar{\lambda}$.*

Theorem 7.46. *If $\lambda \neq 0$ is not an eigenvalue of the compact operator T then the point λ is regular.*

▷ In accordance with Theorem 7.40 it is sufficient to prove that in this case $R_T(\lambda) = X$. Supposing that $R_T(\lambda) \neq X$ we shall take any vector $z \in R_T(\lambda)^\perp$. Then for any vector $x \in X$ we shall have $(Tx - \lambda x, z) = 0$ when it follows $(x, T^*z - \bar{\lambda}z) = 0$ at all x. It is possible only at $z = 0$ as $\bar{\lambda}$ in this case cannot be eigenvalue of the operator T^*. Consequently, $R_T(\lambda)^\perp = \{0\}$ and $R_T(\lambda) = X$. ◁

Corollary. *If $\lambda \neq 0$ is not an eigenvalue of the compact operator T then Eq. (7.56) has the solution at any $y \in X$.*

Theorem 7.47. *The proper subspaces* Y *and* Z *of the compact operators* T *and* T^* *correspondent to the eigenvalues* $\lambda \neq 0$ *and* $\bar{\lambda}$ *have one and the same finite dimension.*

▷ The finiteness of the dimensions of proper subspaces Y and Z follows from Corollary 2 of Theorem 7.41. Suppose that Y has the dimension m, and Z has the dimension $n > m$. Let $x_1, \ldots, x_m$ and $y_1, \ldots, y_n$ are orthonormal bases in Y and Z correspondingly. Determine the operator

$$Sx = Tx + \sum_{k=1}^{m} (x, x_k) y_k.$$

It is clear that it is a compact operator. We prove that λ is not an eigenvalue of the operator S. For this purpose we multiply the equation

$$Tx - \lambda x + \sum_{k=1}^{m} (x, x_k) y_k = 0$$

scalar by $Tx - \lambda x$, $y_1, \ldots, y_m$. Taking into consideration that $Tx - \lambda x \in R_T(\lambda)$, and $y_1, \ldots, y_m \in R_T(\lambda)^\perp$ we get the equations

$$Tx - \lambda x = 0, \quad (x, x_k) = 0 \quad (k = 1, \ldots, m).$$

The first from these equations shows that $x = c_1 x_1 + \cdots + c_m x_m$ at any $c_1, \ldots, c_m$, and the others give $c_1 = \cdots = c_m = 0$. Consequently, the equation $Sx = \lambda x$ has the unique solution $x = 0$, and λ cannot be an eigenvalue of the operator S. But then according to Corollary of Theorem 7.46 the equation $Sx - \lambda x = y$ has the solution at any y, in particular, at $y = y_l$, $l > m$. But this leads to the contradiction. Really, after multiplying the equation

$$Tx - \lambda x + \sum_{k=1}^{m} (x, x_k) y_k = y_l, \quad l > m,$$

scalar by y_l we receive at the left 0, and at the right 1. Thus $n \leq m$. And as the operators T and T^* are self–adjoint then $m \leq n$. Consequently, $m = n$. ◁

If follows from determined facts that the spectrum of the compact operator T represents a finite or countable set of the eigenvalues and the unique point of a spectrum which is not an eigenvalue may be only 0.

An eigenvalue is called *simple* if an one–dimensional proper subspace corresponds to it and an *m–multiple* if m–dimensional proper subspace corresponds to it.

The eigenvalues of a compact operator are enumerated in the order of nonincreasing of the moduluses and m–multiple eigenvalue is assumed as m coinciding eigenvalues.

Thus the pointwise spectrum $\sigma_p(T)$ of the compact operator T represents a finite or a countable set. This set contains the point $\lambda = 0$ if the inverse operator T^{-1} does not exist. If the inverse operator T^{-1} exists then it cannot be bounded (Subsection 7.6.2), and consequently, cannot be determined on the whole space X. Therefore $R_T(0) \neq X$. It means that the point $\lambda = 0$ always belongs to the spectrum of a compact operator no matter whether $\lambda = 0$ is the the limit point of the set of the eigenvalues or not. Therewith if $[R_T(0)] = X$ then 0 is the unique point of the continuous spectrum $\sigma_c(T)$ of the operator T and if $[R_T(0)] \neq X$ then 0 is the unique point of the residual spectrum $\sigma_r(T)$. Thus 0 may be the eigenvalue of the compact operator T (in this case $\sigma_c(T) = \sigma_r(T) = \varnothing$), the point of a continuous spectrum (in this case $\sigma_r(T) = \varnothing$) or the point of the residual spectrum (in this case $\sigma_c(T) = \varnothing$).

7.6.4. Normal Operators

The linear operator T which acts in the H–space is called *normal* if it is commutative with its adjoint operator, $TT^* = T^*T$. The self–adjoint and unitary operators represent the special types of the normal operators.

Theorem 7.48. *The normal operator T in the separable H–space and its adjoint operator T^* have one and the same orthonormal system of the eigenvectors.*

▷ Let λ be an eigenvalue, S be the correspondent proper subspace of the operator T. For any vector $x \in S$

$$Tx = \lambda x.$$

Hence in consequence of the commutativity of the operators T and T^* we receive

$$T^*Tx = TT^*x = \lambda T^*x.$$

Thus $T^*x \in S$. Let $\{u_n\}$ be an orthonormal basis in S. Using expansion (7.33) we have

$$x = \sum_k (x, x_k)x_k,$$

$$T^*x = \sum_k (T^*x, x_k)x_k = \sum_k (x, Tx_k)x_k = \bar{\lambda}x.$$

Consequently, any eigenvector of the operator T is an eigenvector of the adjoint operator T^*. Let now λ_1 and $\lambda_2 \neq \lambda_1$ be two eigenvalues of the operator T, x_1 and x_2 be the correspondent eigenvectors. According to the proved

$$Tx_1 = \lambda_1 x_1, \quad T^*x_2 = \bar{\lambda}_2 x_2.$$

After scalar multiplying the first from these equalities on the right by x_2, and the second one on the left by x_1 we get

$$(Tx_1, x_2) = \lambda_1(x_1, x_2), \quad (x_1, T^*x_2) = \lambda_2(x_1, x_2).$$

But $(Tx_1, x_2) = (x_1, T^*x_2)$. Consequently, $\lambda_1(x_1, x_2) = \lambda_2(x_1, x_2)$ what is possible only at $(x_1, x_2) = 0$. ◁

R e m a r k. The proved theorem is valid for any normal operators including unbounded operators. For compact operators this theorem establishes one more property of the spectrum typical of the spectra of the normal operators.

7.6.5. Operators with the Schmidt Norm

It is expedient besides an ordinary norm to introduce a notion of an absolute norm for the bounded linear operators in the separable H–space. Let T be a bounded linear operator which maps the separable H–space X into the separable H–space Y, $\{x_n\}$ and $\{y_n\}$ be orthonormal bases in X and Y correspondingly. The variable N_T determined by the following formula

$$N_T^2 = \sum_{p,q=1}^{\infty} |(Tx_p, y_q)|^2, \tag{7.58}$$

is called *the absolute norm* or *the Schmidt norm* of the operator T.

Theorem 7.49. *The Schmidt norm does not depend on the choice of the bases in X and Y and is completely determined by the operator T.*

▷ For proving we expand the vectors $Tx_p \in Y$ in terms of the basis $\{y_n\}$. By formula (7.33) we obtain

$$Tx_p = \sum_{q=1}^{\infty} (Tx_p, y_q) y_q .$$

The condition of the completeness of the sequence $\{y_q\}$ (7.34) gives

$$\|Tx_p\|^2 = \sum_{q=1}^{\infty} |(Tx_p, y_q)|^2 .$$

Substituting this expression of the sum of q in formula (7.58) we shall have

$$N_T^2 = \sum_{p=1}^{\infty} \|Tx_p\|^2 . \tag{7.59}$$

This formula make us sure that the Schmidt norm does not depend on the choice of the basis $\{y_n\}$. On the other hand, after presenting the vectors $T^* y_q \in X$ by the expansion in terms of the basis $\{x_n\}$ we obtain

$$T^* y_q = \sum_{p=1}^{\infty} (T^* y_q, x_p) x_p,$$

hence

$$\|T^* y_q\|^2 = \sum_{p=1}^{\infty} |(T^* y_q, x_p)|^2 = \sum_{p=1}^{\infty} |(Tx_p, y_q)|^2.$$

From this formula and (7.58) follows

$$N_T^2 = \sum_{q=1}^{\infty} \|T^* y_q\|^2 . \tag{7.60}$$

This formula shows that the norm N_T does not depend on the choice of the basis $\{x_n\}$. ◁

Theorem 7.50. *The Schmidt norm of an operator cannot be smaller than its norm,* $N_T \geq \|T\|$.

▷ Setting an arbitrary $\varepsilon > 0$ and choosing the basis $\{x_n\}$ in such a way that $\|Tx_1\| > (\|T\| - \varepsilon) \|x_1\| = \|T\| - \varepsilon$ we obtain from formula (7.59) the following inequalities:

$$N_T^2 = \sum_{p=1}^{\infty} \|Tx_p\|^2 \geq \|Tx_1\|^2 > (\|T\| - \varepsilon)^2.$$

Hence by virtue of the arbitrariness $\varepsilon > 0$ follows $N_T^2 \geq \|T\|^2$. ◁

Theorem 7.51. *The Schmidt norm possesses all properties of a norm:* $N_{\alpha T} = |\alpha| N_T$, $N_{T+S} \leq N_T + N_S$, $N_T \geq 0$ *with* $N_T = 0$ *only at* $T = 0$.

▷ It is evident that only the second statement needs the proof. As

$$N_{T+S}^2 = \sum_{p=1}^{\infty} \|Tx_p + Sx_p\|^2 \leq \sum_{p=1}^{\infty} (\|Tx_p\| + \|Sx_p\|)^2$$

$$= \sum_{p=1}^{\infty} \|Tx_p\|^2 + \sum_{p=1}^{\infty} \|Sx_p\|^2 + 2\sum_{p=1}^{\infty} \|Tx_p\| \|Sx_p\|$$

and

$$\sum_{p=1}^{\infty} \|Tx_p\| \|Sx_p\| \leq \sqrt{\sum_{p=1}^{\infty} \|Tx_p\|^2 \sum_{p=1}^{\infty} \|Sx_p\|^2} = N_T N_S$$

then

$$N_{T+S}^2 \leq N_T^2 + N_S^2 + 2N_T N_S = (N_T + N_S)^2. ◁$$

Theorem 7.52. *If the Schmidt norm of the operator T is finite then it is compact.*

▷ For proving we determine a set of the operators

$$T_n = \sum_{q=1}^{n} (T\cdot, y_q) y_q \quad (n = 1, 2, \ldots),$$

where the place of the vector on which act T_n is indicated by a point, and $\{y_q\}$ is any orthonormal basis in Y. All these operators are compact as their ranges R_{T_n} are finite–dimensional spaces. On the other hand, as

$$T = \sum_{q=1}^{\infty} (T\cdot, y_q) y_q$$

then at any $x \in X$

$$\|T_n x - Tx\|^2 = \left\| \sum_{q=n+1}^{\infty} (Tx, y_q) y_q \right\|^2 = \sum_{q=n+1}^{\infty} |(Tx, y_q)|^2$$

$$= \sum_{q=n+1}^{\infty} |(x, T^* y_q)|^2 \leq \|x\|^2 \sum_{q=n+1}^{\infty} \|T^* y_q\|^2 .$$

If $N_T < \infty$ then on the basis of (7.59) at any $\varepsilon > 0$ and sufficiently large n we shall have $\|T_n x - T x\| < \varepsilon \|x\|$ whence $\|T_n - T\| < \varepsilon$. From this fact by Theorem 7.39 follows that the operator T is compact. ◁

R e m a r k. The finiteness of the Schmidt norm is sufficient for the compactness of a linear operator. But not every compact linear operator T in the separable H–space has the Schmidt norm. Therefore the class of the operators with the Schmidt norm represents a part of the class of the compact linear operators.

E x a m p l e 7.21. Let us take the orthonormal basis $\{x_n\}$ in the separable H–space X and such sequence of the complex numbers $\{\lambda_n\}$ convergent to zero that the series $\sum |\lambda_n|^2$ diverges. Form an operator

$$T = \sum_{q=1}^{\infty} \lambda_q(\cdot, x_q) x_q.$$

As $T x_p = \lambda_p x_p$ then

$$N_T^2 = \sum_{p=1}^{\infty} |\lambda_p|^2 = \infty,$$

i.e. the Schmidt norm of the operator T is infinite. Beyond that point it is compact. Really, after determining a set of the compact operators

$$T_n = \sum_{q=1}^{n} \lambda_q(\cdot, x_q) x_q \quad (n = 1, 2, \ldots),$$

we obtain

$$\|T_n x - T x\|^2 = \sum_{q=n+1}^{\infty} |\lambda_q|^2 |(x, x_q)|^2 \leq |\lambda_{n+1}|^2 \|x\|^2$$

at all x. Hence it follows $\|T_n - T\| \leq |\lambda_{n+1}|$. Therefore $\|T_n - T\| < \varepsilon$ at all sufficiently large n whatever be $\varepsilon > 0$. By Theorem 7.39 follows the compactness of the operator T. Thus we have an example of a compact operator with an infinite absolute (Schmidt) norm.

7.6.6. Hilbert–Schmidt Operators

Let $X = L_2(\Delta, \mathcal{A}, \mu)$, $Y = L_2(S, \mathcal{B}, \nu)$ be two spaces of numeric functions given on the spaces with nonnegative measure $(\Delta, \mathcal{A}, \mu)$ and $(S, \mathcal{B}, \nu)$ correspondingly. Consider an integral operator

$$Tx = \int K(s, t) x(t) \mu(dt), \qquad (7.61)$$

which maps X into Y with the kernel $K(s,t)$ satisfying the condition

$$\iint |K(s,t)|^2 \mu(dt)\nu(ds) < \infty \qquad (7.62)$$

$(K(s,t) \in L_2(\Delta \times S, \mathcal{A} \times \mathcal{B}, \mu \times \nu))$ (Subsections 2.1.8 and 3.8.3). If the measures μ and ν possess the approximation property (Corollary of Theorem 3.40) then the spaces X and Y are separable. In this case the operator T is called the *Hilbert–Schmidt operator*.

Theorem 7.53. *The absolute norm of the Hilbert–Schmidt operator is determined by the formula*

$$N_T^2 = \iint |K(s,t)|^2 \mu(dt)\nu(ds). \qquad (7.63)$$

▷ For proving we take in spaces X and Y the orthonormal bases $\{x_n(t)\}$ and $\{y_n(s)\}$ correspondingly. From condition (7.62) and Fubini theorem 3.44 follows that almost at all $s \in S$

$$\int |K(s,t)|^2 \mu(dt) < \infty,$$

i.e. $K(s,t)$ considered as the function of t belongs to X almost at all s. Consequently, almost at all s it may be presented by the expansion

$$K(s,t) = \sum_{p=1}^{\infty} \alpha_p(s)\overline{x_p(t)},$$

where

$$\alpha_p(s) = \int K(s,t)x_p(t)\mu(dt) = Tx_p.$$

With

$$\int |K(s,t)|^2 \mu(dt) = \sum_{p=1}^{\infty} |\alpha_p(s)|^2 \quad \text{almost at all } s. \qquad (7.64)$$

Further

$$\int |\alpha_p(s)|^2 \nu(ds) = \int \left| \int K(s,t)x_p(t)\mu(dt) \right|^2 \nu(ds)$$

$$\leq \iint |K(s,t)|^2 \mu(dt)\nu(ds) < \infty,$$

as

$$\left| \int K(s,t)x_p(t)\mu(dt) \right|^2 \leq \int |K(s,t)|^2\mu(dt) \cdot \int |x_p(t)|^2\mu(dt),$$

and $\int |x_p(t)|^2\mu(dt) = 1$. Consequently, $\alpha_p(s) \in Y$ and may be presented by the expansion in terms of the basis $\{y_n\}$:

$$\alpha_p(s) = \sum_{q=1}^{\infty} \beta_q y_q(s),$$

where

$$\beta_q = \int \alpha_p(s)\overline{y_q(s)}\nu(ds) = \iint K(s,t)x_p(t)\overline{y_q(s)}\mu(dt)\nu(ds) = (Tx_p, y_q).$$

So,

$$\int |\alpha_p(s)|^2\nu(ds) = \sum_{q=1}^{\infty} |\beta|^2 = \sum_{q=1}^{\infty} |(Tx_p, y_q)|^2.$$

Hence and from expansion (7.64) follows [a]

$$N_T^2 = \sum_{p,q=1}^{\infty} |(Tx_p, y_q)|^2 = \int \sum_{p=1}^{\infty} |\alpha_p(s)|^2\nu(ds)$$

$$= \iint |K(s,t)|^2\mu(dt)\nu(ds). \triangleleft$$

Thus the absolute norm of the Hilbert–Schmidt operator is equal to the norm of its kernel considered as an element of the space $L_2(\Delta \times S, \mathcal{A} \times \mathcal{B}, \mu \times \nu)$. As this norm in consequence of (7.62) is finite then the Hilbert–Schmidt operator is compact.

Theorem 7.54. *Any operator T with a finite absolute norm which maps $X = L_2(\Delta, \mathcal{A}, \mu)$ into $Y = L_2(S, \mathcal{B}, \nu)$ is the Hilbert–Schmidt operator.*

---

[a] Here the termwise integration of the series is possible because the series converges almost at all s and its finite sums are upper majorized by the ν–integrable function $\int |K(s,t)|^2\mu(dt)$ (Lebesgue theorem 3.30).

▷ For proving we take arbitrary orthonormal bases $\{x_n(t)\}$ and $\{y_n(s)\}$ in spaces X and Y correspondingly and put

$$a_{pq} = (Tx_p, y_q), \quad K(s,t) = \sum_{p,q=1}^{\infty} a_{pq}\overline{x_p(t)}y_q(s).$$

Let K be a linear integral operator with the kernel $K(s,t)$. It is evident that

$$Kx_p = \sum_{q=1}^{\infty} a_{pq}y_q(s), \quad (Kx_p, y_q) = a_{pq}$$

and

$$N_K = \iint |K(s,t)|^2 \mu(dt)\nu(ds) = \sum_{p,q=1}^{\infty} |a_{pq}|^2 = N_T < \infty.$$

So, K is a Hilbert–Schmidt operator. Further for any vector $x \in X$

$$x = \sum_{r=1}^{\infty} (x, x_r)x_r ,$$

and consequently,

$$(Tx, y_q) = \sum_{p=1}^{\infty} (x, x_p)(Tx_p, y_q) = \sum_{p=1}^{\infty} a_{pq}(x, x_p),$$

$$(Kx, y_q) = \sum_{p=1}^{\infty} (x, x_p)(Kx_p, y_q) = \sum_{p=1}^{\infty} a_{pq}(x, x_p).$$

Therefore $(Tx - Kx, y_q) = 0$ at all q for any $x \in X$. By virtue of the completeness of the basis $\{y_n\}$ follows $T = K$. ◁

Thus in the separable spaces L_2 the class of the operators with the finite absolute (Schmidt) norm coincides with the class of the Hilbert–Schmidt operators. All stated here is extended with small changes over the Hilbert–Schmidt operators (7.61) in the space $L_2(\Delta, \mathcal{A}, \mu)$ of the finite dimensional functions.

Let us consider a special case of the Hilbert–Schmidt operator when the space Y coincides with the space $X = L_2(\Delta, \mathcal{A}, \mu)$. In this case the Hilbert–Schmidt operator as any compact operator has a spectrum which consists of no more than a countable set of eigenvalues and the point 0.

We shall consider a case of the normal Hilbert–Schmidt operator which has a countable set of the eigenvalues $\{\lambda_n\}$ and the correspondent set of the eigenvectors $\{x_n(t)\}$ representing an orthonormal basis. We present the function $\overline{K(s,t)}$ by expansion (7.33)

$$\overline{K(s,t)} = \sum_{p=1}^{\infty} \int \overline{K(s,\tau)}\,\overline{x_p(\tau)}\mu(d\tau)x_p(t).$$

But

$$\int K(s,\tau)x_p(\tau)\mu(d\tau) = \lambda_p x_p(s).$$

Consequently,

$$\overline{K(s,t)} = \sum_{p=1}^{\infty} \bar{\lambda}_p \overline{x_p(s)}x_p(t).$$

After mutiplying this equality by $K(s,t)$ and integration at first with respect to t and later with respect to s we obtain

$$\iint |K(s,t)|^2\mu(dt)\mu(ds) = \sum_{p=1}^{\infty}|\lambda_p|^2 \int |x_p(s)|^2\mu(ds) = \sum_{p=1}^{\infty}|\lambda_p|^2.$$

Comparing this equality with definition (7.63) we see that the absolute norm of the considered Hilbert–Schmidt operator is equal to the sum of moduluses squares of its eigenvalues. Thus the characteristic property of the Hilbert–Schmidt operators which act on the spaces L_2 is the convergence of the series whose moduluses squares of the eigenvalues serve as its numbers. This peculiarity of the Hilbert–Schmidt operators gives the ground for the generalization of the notion of the Hilbert–Schmidt operator. A compact operator which acts in a separable H–space is called the *Hilbert-Schmidt operator* if the series whose moduluses squares of its eigenvalues serve as its members converges.

7.6.7. Trace Type Operators

A compact operator T which acts in a separable H–space is called a *trace type* or a *nuclear* operator if the series whose eigenvalues serve as its members absolutely converges. The sum of this series is called *the trace* of the operator T and is denoted by

$$\operatorname{tr} T = \sum_{p=1}^{\infty} \lambda_p. \tag{7.65}$$

R e m a r k. This definition is quite conformed with the known theorem of linear algebra which states that a trace of a symmetric matrix in R^n is equal to the sum of its eigenvalues.

R e m a r k. In this Subsection we shall consider only normal trace type operators whose eigenvectors form any orthonormal basis in X.

Theorem 7.55. *A trace of a normal trace type operator T is determined by the formula*

$$\operatorname{tr} T = \sum_{n=1}^{\infty} (Tu_n, u_n), \tag{7.66}$$

where $\{u_n\}$ is an arbitrary orthonormal basis in X.

▷ Let $\{x_n\}$ be an orthonormal basis of the eigenvectors of the operator T. Then

$$\operatorname{tr} T = \sum_{p=1}^{\infty} \lambda_p (x_p, x_p) = \sum_{p=1}^{\infty} (Tx_p, x_p).$$

After expressing the vectors x_p by expansion (7.33) with respect to basis $\{u_n\}$ we receive

$$\operatorname{tr} T = \sum_{p=1}^{\infty} (Tx_p, x_p) = \sum_{m,n=1}^{\infty} (Tu_m, u_n) \sum_{p=1}^{\infty} (x_p, u_m)(u_n, x_p).$$

But according to (7.34) and (7.35) we have $\sum_p (u_n, x_p)(x_p, u_m)$ $= (u_n, n_m) = \delta_{nm}$. In consequence of this fact the previous formula takes the form (7.66). ◁

Formula (7.66) gives the opportunity to derive easily the following properties of the trace type operators and their traces:

(i) a trace is a linear function of an operator:

$$\operatorname{tr} \sum_{k=1}^{n} c_k T_k = \sum_{k=1}^{n} c_k \operatorname{tr} T_k$$

for any trace type operators $T_1, \ldots, T_n$ and for any numbers $c_1, \ldots, c_n$;

(ii) $\operatorname{tr} T^* = \overline{\operatorname{tr} T}$;

(iii) if T is a trace type operator, and A is a bounded operator which acts in X then AT and TA are trace type operators, and $\operatorname{tr} AT = \operatorname{tr} TA$, $|\operatorname{tr} AT| = |\operatorname{tr} TA| < \|A\| \, |\operatorname{tr} T|$.

Property (i) is obvious. Property (ii) follows directly from definition (7.65). To prove Property (iii) it is sufficient to take as the basis $\{u_n\}$ the basis of the eigenvectors $\{x_p\}$ of the operator T and account that it is also the basis of the eigenvectors of the adjoint operator T^*.

R e m a r k. Notice that the operators AT and TA may be the trace type in the case of the unbounded operator A. All depends on the quickness of the decreasing of the moduluses of the operator T eigenvalues.

E x a m p l e 7.22. Let $\{x_n\}$ be an orthonormal basis. Consider a trace type operator

$$Tx = \sum_{n=1}^{\infty} \lambda_n(x, x_n)x_n$$

and an unbounded operator

$$Ax = \sum_{n=1}^{\infty} \mu_n(x, x_n)x_n,$$

where $\{\mu_n\}$ is unboundedly increasing sequence of positive numbers. It is clear that x_n are eigenvectors of the operators T and A correspondent to the eigenvalues λ_n and μ_n in consequence of which

$$ATx = TAx = \sum_{n=1}^{\infty} \lambda_n\mu_n(x, x_n)x_n.$$

It is apparent from this fact that the operators AT and TA will be trace type if the series $\sum |\lambda_n| |\mu_n|$ converges.

Trace type operators play a great role in probability theory.

7.6.8. Fredholm Linear Integral Equations

If under conditions of Subsection 7.6.6 $S = \Delta$, $\nu = \mu$, and consequently, $Y = X$ then the equations

$$\int K(s,t)x(t)\mu(dt) = \lambda x(s), \tag{7.67}$$

$$\int K(s,t)x(t)\mu(dt) - \lambda x(s) = y(s), \tag{7.68}$$

whose kernel $K(s,t)$ satisfies condition (7.62) represent linear the *Fredholm integral equations of the second kind.*

In the integral equations theory Eqs. (7.67) and (7.68) are usually studied in the case when the measure μ represents the Lebesgue measure in the finite–dimensional space, and correspondingly, t and s are the finite–dimensional vector (in the special case scalar) variables.

Notice that when T is the Hilbert–Schmidt operator then Eqs. (7.68) and (7.67) represent the equation of Eq. (7.56) type.

From the general theorems of Subsection 7.6.3 follow as the special case the main theorems of the theory of the Fredholm linear integral equations (Fredholm theorems).

1) If $\lambda \neq 0$ is an eigenvalue of Eq. (7.67) then $\bar{\lambda}$ is an eigenvalue of the adjoint integral equation:

$$\int \overline{K(s,t)}y(s)\mu(ds) = \bar{\lambda}y(t) \tag{7.69}$$

and Eqs. (7.67) and (7.69) have one and the same finite number of linearly independent solutions.

2) Eq. (7.68) at $\lambda \neq 0$ has the solution which is the unique for any function $y(s) \in X$ if and only if λ is not an eigenvalue.

Thus either Eq. (7.68) has the unique solution at any function $y(s) \in L_2(\Delta, \mathcal{A}, \mu)$ or homogeneous Eq. (7.67) has non–trivial solution (Fredholm alternative).

3) It $\lambda \neq 0$ is an eigenvalue then Eq. (7.68) has the solution if and only if the function $y(s)$ is orthogonal to all eigenvectors of adjoint Eq. (7.69).

R e m a r k. Similarly from the general theorems of Subsection 7.6.3 follow the Fredholm theorems for the equations with the Hilbert–Schmidt operators in the space $L_2(\Delta, \mathcal{A}, \mu)$ of the finite–dimensional vector functions. Notice that the general theory of Subsection 7.6.3 is extended over the Fredholm integral equations in the H–spaces which play especially important role in modern mathematical physics. In the general theory of integral equations the integral equations in the spaces of the continuous functions are also studied. Sufficient conditions of the existence of the unique solution of the Volterra integral equation and Fredholm integral equation (7.68) in the space of continuous functions were given in Subsections 6.2.8 and 6.2.11. The condition to which λ was subjected is the condition of the regularity of the value λ.

E x a m p l e 7.23. Find eigenvalues and eigenvectors (functions) of the following integral homogeneous equation:

$$\int_{-\pi}^{\pi} e^{\alpha t} \cos(s-t)x(t)\,dt = \lambda x(s) \tag{*}$$

and solve the equation

$$\int_{-\pi}^{\pi} e^{\alpha t} \cos(s-t) x(t)\, dt - \lambda x(s) = y(s). \qquad (**)$$

After rewriting Eq. $(*)$ in the form

$$\cos s \int_{-\pi}^{\pi} e^{\alpha t} \cos t\, x(t)\, dt + \sin s \int_{-\pi}^{\pi} e^{\alpha t} \sin t\, x(t)\, dt = \lambda x(s),$$

we see that its solution may be only a linear combination $\cos t$ and $\sin t$. Therefore we put

$$x(t) = c_1 \cos t + c_2 \sin t. \qquad (***)$$

Then we shall have

$$\int_{-\pi}^{\pi} e^{\alpha t} \cos t\, x(t)\, dt = A(\alpha^2 + 2)c_1 - A\alpha c_2,$$

$$\int_{-\pi}^{\pi} e^{\alpha t} \sin t\, x(t)\, dt = -A\alpha c_1 + 2A c_2,$$

where $A = (e^{\alpha \pi} - e^{-\alpha \pi})/\alpha(\alpha^2 + 4)$. So, Eq. $(*)$ takes the form

$$A\{[(\alpha^2 + 2)c_1 - \alpha c_2]\cos s - (\alpha c_1 - 2c_2)\sin s\} = \lambda c_1 \cos s + \lambda c_2 \sin s.$$

Then we get the equations for c_1 and c_2 (recall that Eq. $(*)$ must be satisfied at all $s \in [-\pi, \pi]$):

$$[A(\alpha^2 + 2) - \lambda]c_1 - A\alpha c_2 = 0,$$

$$-A\alpha c_1 + (2A - \lambda)c_2 = 0.$$

After equalizing to zero the determinant of this set of equations and solving the received square equation we shall find two values of λ at which these equations have non–trivial solution. These values of λ will be the eigenvalues of Eq. $(*)$, and formula $(***)$ at the correspondent c_1 and c_2 will determine the eigenvectors. The value $\lambda = 0$ is also the eigenvalue, and all the functions $x(t) \in L_2([-\pi, \pi])$ orthogonal to the functions $e^{\alpha t} \cos t$ and $e^{\alpha t} \sin t$ are the correspondent eigenvectors.

The solution of Eq. $(**)$ evidently is expressed by the formula

$$x(t) = a_1 \cos t + a_2 \sin t - y(t)/\lambda.$$

In exactly the same way we shall obtain the equations for the coefficients a_1 and a_2

$$[A(\alpha^2 + 2) - \lambda]a_1 - A\alpha a_2 = \frac{1}{\lambda} \int_{-\pi}^{\pi} e^{\alpha t} \cos t\, y(t)\, dt,$$

$$-A\alpha a_1 + (2A - \lambda)a_2 = \tfrac{1}{\lambda} \int\limits_{-\pi}^{\pi} e^{\alpha t} \sin t\, y(t)\, dt.$$

This set of equations has the solution at any funciton $y(t) \in L_2([-\pi, \pi])$ if λ is not an eigenvalue.

The detailed theory of compact operators is given in (Dowson 1978, Dunford and Schwarz 1963, Edwards 1965).

Problems

7.6.1. The operator T is called *finite–dimensional* if its range R_T represents a finite–dimensional space. We shall call such operator n–*dimensional* operator if R_T is n–dimensional space. Prove that any n–dimensional linear operator which acts in the H–space has n eigenvalues different from zero (some of them may coincide), and zero is the eigenvalue of the infinite multiplicity and the eigenvectors correspondent to different eigenvalues are linearly independent.

7.6.2. Let $\lambda_1, \ldots, \lambda_n$ be the eigenvalues of n–dimensional linear operator T different from zero, $y_1, \ldots, y_n$ be the correspondent vectors, $\| y_k \| = 1$. Prove that $T = \lambda_1 P_1 + \cdots + \lambda_n P_n$, where P_k is a projector on the eigenvector y_k : $P_k x = c_k y_k$ for any x (P_k in the general case is not an orthoprojector, see Subsection 7.4.4).

7.6.3. Prove that any finite–dimensional operator is compact (and consequently, is also bounded).

7.6.4. Prove that if the sequence of the finite–dimensional operator $\{T_n\}$ converges to some operator T in the uniform topology of the space $\mathcal{B}(X, Y)$ (Subsection 5.1.4) then the operator T is compact.

7.6.5. Find the absolute (Schmidt) norm of a finite–dimensional operator.

7.6.6. Let T be n–dimensional operator in the H–space X. Solve the equation $Tx - \lambda x = y$. Find the condition of the existence of the unique solution at any $y \in X$.

7.6.7. Find the eigenvalues of the following integral equation:

$$\int\limits_{-1}^{1} (st^2 + s^2 t^3 + s^3 t)x(t)\, dt = \lambda x(s).$$

Solve the equation

$$\int\limits_{-1}^{1} (st^2 + s^2 t^3 + s^3 t)x(t)\, dt - \lambda x(s) = y(s).$$

Establish the conditions at which this equation has a solution at any function $y(t) \in L_2([-1, 1])$. Is the operator in this equation a finite–dimensional? Describe the range of this operator.

7.6.8. Find the eigenvalues and eigenvectors (functions) of the following integral equation:

$$\int_T \sum_{k=1}^{N} \varphi_k(s)\psi_k(t)x(t)\,dt = \lambda x(s)$$

and solve the equation

$$\int_T \sum_{k=1}^{N} \varphi_k(s)\psi_k(t)x(t)\,dt - \lambda x(s) = y(s).$$

Establish the conditions at which this equation has a solution at any function $y(t)$ $\in L_2(T)$. Describe the range of the operator in this equation.

7.6.9. Let $\{x_n\}$ be an orthonormal basis in the H–space X, and $\{\lambda_n\}$ be the sequence of complex numbers. Prove that the operator which acts in X

$$Tx = \sum_{p=1}^{\infty} \lambda_p(x, x_p)x_p$$

is normal. Find its eigenvalues and eigenvectors.

7.6.10. Prove that if eigenvectors of the linear operator T in the H–space form an orthonormal basis then the operator T is normal. Thus the normality of the operator is the necessary condition of the fact that the set of its eigenvectors will represent an orthonormal basis.

To what condition must satisfy the kernel $K(s,t)$ of the Hilbert–Schmidt integral operator which acts in the space L_2 in order it be normal?

7.6.11. Consider a trace type operator which acts in the space $L_2([-T,T])$,

$$Ax = \int_{-T}^{T} K(t,\tau)x(\tau)\,d\tau,$$

$$K(t,\tau) = \tfrac{1}{T}\Big\{1 + \sum_{n=1}^{\infty} \lambda_n[\cos n\omega t \cos n\omega\tau + \sin n\omega t \sin n\omega\tau]\Big\},$$

$\omega = \pi/T$. Find the spectrum of the operator A. To what conditions will satisfy the numbers λ_n in order the operator AD^2, $D = d/dt$ be trace type?

7.6.12. Let T be a trace type operator, A be the unbounded closed linear operator. At what conditions the operators AT and TA are trace type?

I n s t r u c t i o n. Use formula (7.66) taking as the basis $\{u_n\}$ the basis of the eigenvectors of the operator T.

CHAPTER 8
SPECTRAL THEORY OF LINEAR OPERATORS

Chapter 8 contains the spectral theory of linear operators in H-spaces. In Section 8.1 the theorem of existence of eigenvalues is proved for self-adjoint compact operators. The spectral decomposition of such an operator is derived. The application of this theory is given for the Fredholm integral equations with symmetric kernels. The method for solving integral equations of a sufficiently large class is outlined. In Section 8.2 the definition of the operator–valued measure is given whose values represent orthoprojectors. Such a measure represents the decomposition of the identity. Then the theory of integrals of numerical functions by operator–valued measures is outlined and the properties of operators are studied determined by the decomposition of the identity. Also the properties of spectra of such operators are studied. A wide class of functions of an unitary and self–adjoint operators is considered. This theory implies the expression for the decomposition of the identity, i.e. the spectral measure of a given self–adjoint operator. Special Section 8.3 is devoted to functions of operators. In Section 8.4 using this spectral measure the representation of the self–adjoint operator in the form of the integral by its spectral measure is derived (the spectral decomposition of a self–adjoint operator). Hence the spectral decompositions of an unitary operator, a group of unitary operators and a normal operator are derived.

8.1. Spectral Decomposition of a Compact Self–Adjoint Operator

8.1.1. Existence of Eigenvalues

It was shown in Subsection 7.6.3 that a spectrum of any compact operator consists of only eigenvalues and only 0 may belong to the spectrum deprived of eigenvalue. In this case the question of the existence of eigenvalues remains unsolved. This problem is positively solved for the self–adjoint compact operators.

Theorem 8.1. *Any compact self–adjoint operator has at least one eigenvalue.*

▷ Let A be a compact self–adjoint operator in the H–space X. To prove the existence of eigenvalue of A use formula (7.18) derived in Subsection 7.3.2:

$$\|A\| = \sup \frac{|(Ax, x)|}{\|x\|^2} = \sup_{\|x\|=1} |(Ax, x)|. \tag{8.1}$$

On the basis of the properties of the upper bound there exists such sequence $\{x_n\}$, $\|x_n\| = 1$ that

$$\lim |(Ax_n, x_n)| = \|A\|. \tag{8.2}$$

As in consequence of the self–adjointness of the operator A all the numbers (Ax_n, x_n) are real then the numeric sequence $\{(Ax_n, x_n)\}$ or converges to $\|A\|$ or to $-\|A\|$ or has two limit points $\|A\|$ and $-\|A\|$. In the latter case after dropping from it all negative (or all positive) members we receive the sequence which converges to $\|A\|$ (correspondingly $-\|A\|$). Thus the sequence $\{(Ax_n, x_n)\}$ may be always assumed as convergent. We put

$$\lambda = \lim(Ax_n, x_n) \tag{8.3}$$

and prove that λ is an eigenvalue of operator A. By virtue of the compactness of A and the boundedness of the sequence $\{x_n\}$ we may choose from it such subsequence $\{x_{n_k}\}$ that the sequence $\{Ax_{n_k}\}$ will converge to some limit z. Then taking into consideration that due to the self–adjointness of A the number λ is real and we shall have

$$\|Ax_{n_k} - \lambda x_{n_k}\|^2 = \|Ax_{n_k}\|^2 - 2\lambda(Ax_{n_k}, x_{n_k}) + \lambda^2$$

and on the basis of definition (8.3)

$$\lim \|Ax_{n_k} - \lambda x_{n_k}\|^2 = \|z\|^2 - \lambda^2. \tag{8.4}$$

But $\|z\| = \lim \|Ax_{n_k}\| \leq \|A\| = |\lambda|$. This inequality and formula (8.4) show that $\|z\| = |\lambda|$ and $y_{n_k} = Ax_{n_k} - \lambda x_{n_k} \to 0$ at $k \to \infty$. Hence it follows that the sequence $\{x_{n_k}\}$ converges and

$$x = \lim x_{n_k} = \lim(Ax_{n_k} - y_{n_k})/\lambda = z/\lambda.$$

From the convergence of $\{x_{n_k}\}$ to x follows the convergence of $\{Ax_{n_k}\}$ to Ax. Thus from one hand, $\lim Ax_{n_k} = z = \lambda x$, and from the other hand, $\lim Ax_{n_k} = Ax$. Consequently, $Ax = \lambda x$. This proves that λ

is an eigenvalue of the operator A, and $x = \lim x_{n_k}$, $\| x \| = 1$ is the corresponding eigenvector. ◁

Corollary 1. *The largest from the moduluses of the eigenvalues of the compact self–adjoint operator A is equal to its norm $|\lambda|_{\max} = \| A \|$.*

▷ By Theorem 6.29 the eigenvalues of the bounded operator cannot exceed by modulus its norm, and the modulus of the eigenvalue which is determined by formula (8.3) is equal to the norm of the operator A. ◁

Corollary 2. *The largest from the moduluses of the eigenvalues of the compact self–adjoint operator A is equal to the maximum of the functional $| (Ax, x) | / \| x \|^2$:*

$$|\lambda|_{\max} = \max \frac{|(Ax, x)|}{\| x \|^2},$$

and the vector x for which this maximum is achieved is the corresponding eigenvector.

R e m a r k. The extremal property of the eigenvalues suggests a practical way of finding the eigenvalues of the compact self–adjoint operators. By this way we may find sequentially the eigenvalues in the order of nonincreasing of their moduluses and the corresponding eigenvectors.

8.1.2. Spectral Decomposition

Let λ_1 be the largest eigenvalue by the modulus, and x_1, $\| x_1 \| = 1$ is the corresponding eigenvector. Let us denote by G_1 an one–dimensional subspace formed by the vector x_1 and put $H_1 = X$, $H_2 = G_1^\perp$. Let us contract the operator A on the subspace H_2. According to Theorem 8.1 the operator A on the subspace H_2 has the eigenvalues. Let λ_2 be the largest from them by modulus (evidently $|\lambda_2| \leq |\lambda_1|$). It may be found in the same way as it was used while proving Theorem 8.1. For this purpose accordingly to Corollary 2 of Theorem 8.1 it is necessary to find the conditional maximum of the functional $| (Ax, x) | / \| x \|^2$ at additional condition $(x, x_1) = 0$. As a result we shall find the eigenvalue λ_2 and the corresponding eigenvector x_2, $\| x_2 \| = 1$. We denote by G_2 the subspace formed by the vectors x_1, x_2 and contraction A of the subspace $H_3 = G_2^\perp$. Continuing this process we denote by G_{n-1} the subspace formed by the found first $n - 1$ orthonormal eigenvectors $x_1, \ldots, x_{n-1}$ and contract the operator A on the subspace $H_n = G_{n-1}^\perp$. After determining the conditional maximum of the functional $| (Ax, x) | / \| x \|^2$ at additional conditions $(x, x_1) = \cdots = (x, x_{n-1}) = 0$ we shall find the following eigenvalue by modulus λ_n and the corresponding eigenvector x_n, $\| x_n \| = 1$.

This process will be over after finite number n of the steps if

$$\max_{x \in H_{n+1}} \frac{|(Ax, x)|}{\|x\|^2} = 0. \qquad (8.5)$$

In this case $Ax = 0$ at all $x \in H_{n+1}$ as the left–hand side of equality (8.5) represents the norm of the operator A in H_{n+1}. Consequently, in this case 0 is an eigenvalue, and H_{n+1} is the corresponding proper subspace.

If equality (8.5) is not achieved at any n then the process will be continued unboundedly and will give as a result a countable set of the eigenvalues $\{\lambda_n\}$ and the corresponding orthonormal sequence of the eigenvectors $\{x_n\}$. Here on the basis of Corollary 1 of Theorem 8.1 at any $z \in H_{n+1}$

$$\|Az\| \le |\lambda_{n+1}| \|z\|. \qquad (8.6)$$

Consequently, taking any vector $x \in X$ we shall have

$$x = \sum_{k=1}^{n} (x, x_k)x_k + z_n, \quad z_n \in H_{n+1},$$

$$Ax = \sum_{k=1}^{n} \lambda_k(x, x_k)x_k + Az_n, \qquad (8.7)$$

therewith on the basis of inequality (8.6)

$$\left\| Ax - \sum_{k=1}^{n} \lambda_k(x, x_k)x_k \right\| \le |\lambda_{n+1}| \|z_n\| \le |\lambda_{n+1}| \|x\|.$$

As according to Theorem 7.41 $\lambda_{n+1} \to 0$ at $n \to \infty$ then the series $\sum \lambda_k(x, x_k)x_k$ converges to Ax at any $x \in X$:

$$Ax = \sum_{k=1}^{\infty} \lambda_k(x, x_k)x_k. \qquad (8.8)$$

Hence it follows that the sequence of the eigenvectors $\{x_n\}$ is complete in $R_A = G_\infty$. If in this case $G_\infty \ne X$ then $Ax = 0$ at all $x \in H_\infty = G_\infty^\perp$. Thus in the case of a countable set of the eigenvalues 0 serves as the eigenvalue, and H_∞ is the corresponding proper subspace, if $G_\infty \ne X$. In particular, it always takes place if X is not separable as the subspace G_∞ is always separable. And only in that case when X is separable and

$G_\infty = X$ the point 0 may be not an eigenvalue. But it belongs to the spectrum as the limit point of the set of the eigenvalues.

Passing in the first formula (8.7) to the limit at $n \to \infty$ we obtain the following decomposition:

$$x = x_0 + \sum_{k=1}^{\infty} (x, x_k)x_k, \quad x_0 \in H_\infty. \tag{8.9}$$

In the special case when X is the separable one and a system of the vectors $\{x_n\}$ is complete in X, $G_\infty = X$ and the subspace H_∞ consists of only one point 0, and consequently, $x_0 = 0$. Thus we have.

Theorem 8.2. *Any compact self–adjoint operator has nonempty finite or countable set of real eigenvalues and here for any $x \in X$ expansions (8.8) and (8.9) are valid.*

R e m a r k. In the sequence of the eigenvalues $\{\lambda_n\}$ received by the stated way each eigenvalue is repeated as many as linearly independent eigenvectors correspond to it. If to a given eigenvalue corresponds an one–dimensional proper subspace, i.e. one eigenvector then it is called *a simple* one. If the dimension of the proper subspace is more that 1 then the eigenvalue is called *a multiple* one, and the dimension of the proper subspace is called *its multiplicity*. Thus in the obtained sequence $\{\lambda_n\}$ each multiple eigenvalue is presented by the corresponding number of coinciding eigenvalues.

Formula (8.8) gives *a spectral decomposition* of a compact self–adjoint operator. It shows that such operator is completely determined by its spectrum and the corresponding set of proper subspaces. Later on we shall consider only the case when the H–space X is separable and the system of the eigenvector $\{x_n\}$ of the operator A is complete, i.e. represents a basis in X. In this case decomposition (8.9) of any vector $x \in X$ has the form

$$x = \sum_{k=1}^{\infty} (x, x_k)x_k. \tag{8.10}$$

Consider an operator equation

$$Ax - \lambda x = y. \tag{8.11}$$

Presenting y by decomposition (8.10):

$$y = \sum_{k=1}^{\infty} (y, x_k)x_k \tag{8.12}$$

and substituting expressions (8.8), (8.10) and Eq. (8.12) into Eq. (8.11) we get the following equation:

$$\sum_{k=1}^{\infty} (\lambda_k - \lambda)(x, x_k)x_k = \sum_{k=1}^{\infty} (y, x_k)x_k,$$

which gives $(x, x_k) = (y, x_k)/(\lambda_k - \lambda)$. Substituting this expression into formula (8.10) we obtain the solution of Eq. (8.11)

$$x = \sum_{k=1}^{\infty} \frac{(y, x_k)}{\lambda_k - \lambda} x_k. \tag{8.13}$$

The series converges here for any regular point λ as

$$\sum_{k=1}^{\infty} \left| \frac{(y, x_k)}{\lambda_k - \lambda} \right|^2 \leq \sup_k \left| \frac{1}{\lambda_k - \lambda} \right|^2 \sum_{k=1}^{\infty} |(y, x_k)|^2$$

and the series $\sum |(y, x_k)|^2$ converges by virtue of Parseval formula (7.34). From solution (8.13) follows a spectral decomposition of the resolvent R_λ of the operator A:

$$R_\lambda = \sum_{k=1}^{\infty} \frac{(\cdot, x_k)}{\lambda_k - \lambda} x_k. \tag{8.14}$$

The norm of the resolvent is determined by formula

$$\| R_\lambda \| = \sup_k | \lambda - \lambda_k |^{-1}. \tag{8.15}$$

Let us consider now *an operator equation of the first kind*

$$Ax = y. \tag{8.16}$$

This equation represents Eq. (8.11) which corresponds to the point of the spectrum $\lambda = 0$ (Eq. (8.11) at $\lambda \neq 0$ is called *an operator equation of the second kind*). It is clear that Eq. (8.16) may have the solution only in that case when $\lambda = 0$ is not an eigenvalue. In this case the resolvent $R_\lambda = R_0 = T^{-1}$ is unbounded (Subsection 7.6.3) and Eq. (8.16) cannot have the solutions at any y. But formula (8.13) at $\lambda = 0$ determines the solution of Eq. (8.16)

$$x = \sum_{k=1}^{\infty} (y, x_k)x_k/\lambda_k. \tag{8.17}$$

The series converges here only at the condition of the following series convergence:

$$\sum_{k=1}^{\infty} |(y, x_k)/\lambda_k|^2. \tag{8.18}$$

Consequently, Eq. (8.16) has the solution only for such y for which series (8.18) converges.

Let us make it clear how the absolute norm of a compact self–adjoint operator is connected with its spectrum. For this purpose we assume as the basis $\{x_n\}$ in X a basis formed by the orthonormal eigenvectors of the operator A including the eigenvectors corresponding to zero eigenvalue (if 0 is an eigenvalue). As $Ax_n = \lambda_n x_n$ for any eigenvector $x_n \in G_\infty$ and $Ax'_n = 0$ for any eigenvector $x'_n \in H_\infty$ then according to formula (7.59) the absolute (Schmidt) norm of the operator A is determined by

$$N_A^2 = \sum_{n=1}^{\infty} \|Ax_n\|^2 + \sum_{n=1}^{\infty} \|Ax'_n\|^2 = \sum_{n=1}^{\infty} \lambda_n^2 \|x_n\|^2 = \sum_{n=1}^{\infty} \lambda_n^2.$$

Thus the square of the absolute norm of the compact self–adjoint operator A is equal to the sum of the squares of all its eigenvalues

$$N_A^2 = \sum_{n=1}^{\infty} \lambda_n^2. \tag{8.19}$$

Hence it follows that the absolute (Schmidt) norm of a compact self–adjoint operator is finite if and only if the series of the squares of the moduluses of its eigenvalues converges, i.e. when it is the Hilbert–Schmidt operator (Subsection 7.6.6).

8.1.3. Linear Fredholm Integral Equations with Symmetric Kernel

The Hilbert–Schmidt integral operator which acts in the space L_2 will be self–adjoint if its kernel is symmetric $K(t, s) = \overline{K(s, t)}$ (see Examples 6.2 and 6.5). It was shown in Subsection 7.6.6 that the Hilbert–Schmidt operator is compact. Consequently, the Hilbert–Schmidt self–adjoint operator A has nonempty finite or countable set of the eigenvalues $\{\lambda_n\}$ and the corresponding orthonormal sequence of the eigenfunctions $\{x_n(t)\}$,

$$(x_p, x_q) = \int x_p(t)\overline{x_q(t)}\mu(dt) = \delta_{pq}.$$

Here for any function $x(t) \in L_2$ decompositions (8.9) and (8.8) are valid:

$$x(t) = x_0(t) + \sum_{\nu=1}^{\infty} (x, x_\nu) x_\nu(t), \qquad (8.20)$$

$$\int K(t,s) x(s) \mu(ds) = \sum_{\nu=1}^{\infty} \lambda_\nu (x, x_\nu) x_\nu(t), \qquad (8.21)$$

$x_0(t)$ is the function which belongs to the kernel of the operator A,

$$\int K(t,s) x_0(s) \mu(ds) = 0.$$

In consequence of the finiteness of the absolute norm of the Hilbert–Schmidt operator (Subsection 7.6.6) and formulae (7.63) and (8.19) it is clear that the series of the squares of the eigenvalues of the Hilbert–Schmidt self–adjoint integral operator converges and

$$\sum_{n=1}^{\infty} \lambda_n^2 = \iint |K(t,s)|^2 \mu(dt) \mu(ds). \qquad (8.22)$$

After rewriting decomposition (8.21) in the form

$$\int K(t,s) x(s) \mu(ds) = \int \sum_{\nu=1}^{\infty} \lambda_\nu x_\nu(t) \overline{x_\nu(s)} x(s) \mu(ds)$$

and accounting that this formula is valid for any function $x(t) \in L_2$ we come to the conclusion that

$$K(t,s) = \sum_{\nu=1}^{\infty} \lambda_\nu x_\nu(t) \overline{x_\nu(s)}. \qquad (8.23)$$

Formula (8.23) determines a spectral decomposition of the kernel of the Hilbert–Schmidt operator. In the theory of integral equations formula (8.23) is known as the *Mercer formula*.

But the cited derivation of formula (8.23) is not rigorous as it is based on the formal change of the order of the series summation and integration. For rigorous proof of formula (8.23) we calculate the norm of the residual member of series (8.23):

$$\iint \left| K(t,s) - \sum_{\nu=1}^{n} \lambda_\nu x_\nu(t) \overline{x_\nu(s)} \right|^2 \mu(ds) \mu(dt)$$

$$= \iint |K(t,s)|^2 \mu(dt)\mu(ds)$$

$$- \sum_{\nu=1}^{n} \lambda_\nu \int \overline{x_\nu(t)}\mu(dt) \int K(t,s)x_\nu(s)\mu(ds)$$

$$- \sum_{\nu=1}^{n} \lambda_\nu \int x_\nu(t)\mu(dt) \int \overline{K(t,s)}\,\overline{x_\nu(s)}\mu(ds) + \sum_{\nu=1}^{n} \lambda_\nu^2.$$

Hence taking into consideration that $\int K(t,s)x_\nu(s)\mu(ds) = \lambda_\nu x_\nu(t)$ we find

$$\iint \left| K(t,s) - \sum_{\nu=1}^{n} \lambda_\nu x_\nu(t)\overline{x_\nu(s)} \right|^2 \mu(ds)\mu(dt)$$

$$= \iint |K(t,s)|^2 \mu(dt)\mu(ds) - \sum_{\nu=1}^{n} \lambda_\nu^2.$$

In consequence of (8.22) the right–hand side of this equality tends to zero at $n \to \infty$ what proves formula (8.23).

Formula (8.13) determines the solution of the following integral equation of the second kind:

$$\int K(t,s)x(s)\mu(ds) - \lambda x(t) = y(t). \tag{8.24}$$

Accordingly to (8.13) the solution of Eq. (8.24) is determined by formula

$$x(t) = \sum_{k=1}^{\infty} \frac{x_k(t)}{\lambda_k - \lambda} \int y(s)\overline{x_k(s)}\mu(ds). \tag{8.25}$$

It is clear that the resolvent of the Hilbert–Schmidt operator with the symmetric kernel represents an integral operator with a kernel

$$R(t,s,\lambda) = \sum_{k=1}^{\infty} \frac{x_k(t)\overline{x_k(s)}}{\lambda_k - \lambda}. \tag{8.26}$$

The solution of the following integral equation of the first kind:

$$\int K(t,s)x(s)\mu(ds) = y(t) \tag{8.27}$$

accordingly to (8.17) is determined by the formula

$$x(t) = \sum_{k=1}^{\infty} \frac{x_k(t)}{\lambda_k} \int y(s)\overline{x_k(s)}\mu(ds). \qquad (8.28)$$

The solution of Eq. (8.27) exists if and only if the following series converges:

$$\sum_{k=1}^{\infty} \frac{1}{\lambda_k^2} \left[\int y(s)\overline{x_k(s)}\mu(ds) \right]^2. \qquad (8.29)$$

If this condition is not fulfilled then Eq. (8.27) has no solution in the space L_2 or in dense spaces in L_2 of continuous or differentiable functions (Subsection 3.7.7). But it may have a solution as we shall see in Subsection 8.1.4 in more wide space of the generalized functions.

Thus from the theorems about a spectrum of a compact operator follows the main theorems and the formulae of the theory of Fredholm linear integral equations with a symmetric kernel in the space L_2.

8.1.4. Method for Solving Integral Equations of Some Class

In applications the integral equations with real symmetric kernel of following type:

$$K(t,s) = \int_{\alpha}^{\beta} w(t,\tau)w(s,\tau)\,d\tau, \qquad (8.30)$$

where the functions $w(t,s)$ satisfy the ordinary linear differential equation

$$Fw(t,\tau) = H\delta(t-\tau), \qquad (8.31)$$

$$F = \sum_{k=0}^{n} a_k D^k, \quad H = \sum_{k=0}^{m} b_k D^k, \quad D = \frac{d}{dt}, \quad m < n \qquad (8.32)$$

(about the ordinary differential equations with δ–functions see Problems 5.3.12 and 5.3.13). In such cases the solution of the following integral equations of the first kind:

$$\int_{t_1}^{t_2} K(t,s)x(s)\rho(s)\,ds = y(t) \qquad (8.33)$$

and of the second kind:

$$\int_{t_1}^{t_2} K(t,s)x(s)\rho(s)\,ds = \lambda x(t) \qquad (8.34)$$

$$\int_{t_1}^{t_2} K(t,s)x(s)\rho(s)\,ds - \lambda x(t) = y(t) \qquad (8.35)$$

at $t_1 > \alpha$, $t_2 < \beta$ is reduced to the integration of the ordinary linear differential equations. In Eqs. (8.33), (8.34) and (8.35) the function $\rho(t)$ is an arbitrary strictly positive function. So, the measure μ in this case is expressed by formula

$$\mu(A) = \int_A \rho(t)\,dt.$$

At first we outline this method for the equation of the first kind (8.33). On the basis of (8.30) we have

$$\int_{t_1}^{t_2} K(t,s)x(s)\rho(s)\,ds = \int_{t_1}^{t_2} \left(\int_{\alpha}^{\beta} w(t,\tau)w(s,\tau)\,d\tau \right) x(s)\rho(s)\,ds$$

$$= \int_{\alpha}^{\beta} w(t,\tau)\,d\tau \int_{t_1}^{t_2} w(s,\tau)x(s)\rho(s)\,ds.$$

Putting

$$u(t) = \int_{t_1}^{t_2} w(s,t)x(s)\rho(s)\,ds, \qquad (8.36)$$

we reduce Eq. (8.33) to the form

$$\int_{\alpha}^{\beta} w(t,\tau)u(\tau)\,d\tau = y(t).$$

After acting on both parts of this equation by the operator F and accounting Eq. (8.31) we receive

$$\int\limits_{\alpha}^{\beta} H_t \delta(t - \tau) u(\tau) \, d\tau = F y(t)$$

or after calculating the integrals which contain the δ–functions

$$H u = F y. \tag{8.37}$$

This is an ordinary linear differential equation with unknown function $u(t)$ which is connected by formula (8.36) with the function $x(t)$. For expressing $x(t)$ in terms of $u(t)$ let us introduce the integral operator W with the kernel $w(t, s)$. From Eq. (8.31) accounting that $\delta(t - s)$ represents a kernel of an unit operator we obtain for the operator W the formula

$$W = F^{-1} H, \tag{8.38}$$

where F^{-1} is an inverse operator relative to F, i.e. an operator of the solution of the differential equation $F x = y$. Remembering that the kernel of the adjoint integral operator W^* represents the same function $w(t, s)$ with changed roles of the arguments we may rewrite formula (8.36) in the form

$$u = W^* x \rho.$$

But the operator adjoint with the product of the operators is equal to the product of the corresponging operators taken in the inverse order (Theorem 6.5). Therefore on the basis of (8.38) we receive for u the following formula:

$$u = H^* F^{*-1} x \rho.$$

From this formula putting

$$z = F^{*-1} x \rho,$$

we get the differential equation for the function $z(t)$

$$H^* z = u \tag{8.39}$$

and the formula for the solution of Eq. (8.33)

$$x = F^* z / \rho. \tag{8.40}$$

Thus the solution of Eq. (8.33) is reduced in a given case to the integration of linear differential equations (8.37) and (8.39) and to application of formula (8.40). But not everyone solution of Eqs. (8.37) and (8.39) gives the solution of Eq. (8.33).

Notice firstly that in order to find the equations at which the solution of Eqs. (8.37), (8.39) and formula (8.40) determine the solution Eq. (8.33) the function $u(t)$ which is determined by formula (8.36) must not contain the δ–functions and their derivatives although may have the discontinuities of the first kind. But it is so only in that case when the solution $x(t)$ of Eq. (8.33) contains only the derivatives of the δ–functions of the orders no higher than $n - m - 1$ as according to Eq. (8.31) the derivative $w_t^{(n-m)}(t, \tau)$ contains the δ–function, and consequently, the function $u(t)$ will contain the δ–function if $x(t)$ contains $\delta^{(n-m)}(t - \tau)$. Formula (8.40) shows that $x(t)$ will not contain the derivatives of the δ–function of higher order than $n - m - 1$ if and only if the function $z(t)$ is continuous together with its derivatives $z'(t), \ldots, z^{(m-1)}(t)$. It gives the conditions

$$z(t_2) = z'(t_2) = \cdots = z^{(m-1)}(t_2) = 0. \tag{8.41}$$

At $t < t_1$ and $t > t_2$ it is natural to assume the function $y(t)$ and the functions $u(t)$ and $z(t)$ which are determined by Eqs. (8.37) and (8.39) identically equal to zero.

Secondly in order to provide in Eq. (8.33) the equality of the integral within given boundaries of the right–hand side of this equation it is necessary to extend the function $y(t)$ to the interval (α, t_1) in such a way that the solution $x(t)$ Eq. (8.33) will be equal to zero at $t < t_1$. By solution (8.40) it is necessary for this purpose to determine the function $z(t)$ on the interval (α, t_1) by the equation

$$F^* z(t) = 0. \tag{8.42}$$

The general solution of this equation has the form

$$z(t) = \sum_{r=1}^{n} c_r z_r(t),$$

where $z_1(t), \ldots, z_n(t)$ are any linearly independent solutions of Eqs. (8.43). Substituting this expression $z(t)$ into Eq. (8.39) we find the function $u(t)$ on the interval (α, t_1):

$$u(t) = \sum_{r=1}^{n} c_r u_r(t), \quad u_r(t) = H^* z_r(t).$$

Substituting this expression of the function $u(t)$ into Eq. (8.37) and integrating this equation we shall find the extension of the function $y(t)$ on the interval (α, t_1):

$$y(t) = \sum_{r=1}^{n} c_r y_r(t),$$

where $y_r(t)$ is the solution of Eq. (8.37) at $u = u_r(t)$ $(r = 1, \ldots, n)$ identically equal to zero at $u_r(t) \equiv 0$.

In order to find the equations for determining the constants c_1, $\ldots, c_n$ and m constants of the integration in the solution of Eq. (8.37) relative to u we notice that conditions (8.41) provide the continuity of the function $z(t)$ and its derivatives $z'(t), \ldots, z^{(m-1)}(t)$ at the point t_2. Here the function $x(t)$ which is determined by formula (8.40) will contain the δ–function and its derivatives till the order $n - m - 1$ inclusively with the carrier at the point t_2. For the order of the derivatives of the δ–function be no higher than $n - m - 1$ at the point t_1 it is naturally to require the continuity at the point t_1 of the function $z(t)$ which is determined by Eq. (8.39) on the interval (t_1, t_2) and by Eq. (8.42) at $t < t_1$ and its derivatives till $m - 1$ order inclusively:

$$z^{(k)}(t_1) = \sum_{r=1}^{n} c_r z_r^{(k)}(t_1) \quad (k = 0, 1, \ldots, m - 1). \tag{8.43}$$

Formula for the solution of $z(t)$ of Eq. (8.42) evidently determines the function $z(t)$ on the interval (α, t_1) as the solution of Eq. (8.39). But this solution does not coincide with the solution at boundary conditions (8.41). So, for finding the solution of Eq. (8.39) on the interval (α, t_1) satisfying conditions (8.41) at founded function $u(t)$ on the interval (α, t_1) it is necessary to extend the functions $u_r(t)$ $(r = 1, \ldots, n)$ on the interval $[t_1, t_2]$ by means of the integration of Eq. (8.37) at $y = y_r(t)$ on the interval (α, t_1), $y = 0$ at $t \geq t_1$. While determining the initial values $u_r(t_1)$ it is necessary to account that the solution of Eq. (8.37) relatively to u contains a linear combination of the functions $y(t), y'(t), \ldots, y^{(n-m)}(t)$. This linear combination vanishes to zero by a jump at the transition of the point t_1. Therefore the corresponding linear combinations of the functions $y_r(t), y'_r(t), \ldots, y_r^{(n-m)}(t)$ determine the jump of the functions $u_r(t)$ $(r = 1, \ldots, n)$ at the point t_1, and consequently, the initial values $u_r(t_1)$ for the integration of the equation $H u_r = 0$ on the interval $[t_1, t_2]$. After finding the functions

$u_r(t)$ the integration of Eq. (8.39) gives

$$z(t) = \sum_{r=1}^{n} c_r \zeta_r(t) + z_0(t) .$$

Here $\zeta_r(t)$ $(r = 1, \ldots, n)$ are the solutions of the equations $H^* \zeta_r = u_r(t)$ satisfying conditions (8.41), $z_0(t)$ is the solution of Eq. (8.39) statisfying conditions (8.41) at $u = 0$ on the interval (α, t_1), $u = u_0(t)$ on the interval $[t_1, t_2]$, and $u_0(t)$ is the solution of Eq. (8.37) which identically vanishes at $y = 0$. The initial value $u_0(t_1)$ is determined similarly as the inital values $u_r(t_1)$. Thus we receive two solutions of Eq. (8.39) on the interval $(\alpha, t_1]$. It is necessary for their coincidence that these two solutions will coincide at the point t_1 together with their derivatives till $m - 1$ order inclusively. It gives equations

$$\sum_{r=1}^{n} c_r [z_r^{(k)}(t_1) - \zeta_r^{(k)}(t_1)] = z_0^{(k)}(t) \quad (k = 0, 1, \ldots, m - 1). \qquad (8.44)$$

Finally, from Eqs. (8.37) and (8.39) it is clear that for providing the continuity at the point t_1 of the functions $z(t), z'(t), \ldots, z^{(n-1)}(t)$ it is necessary to require that the function $y(t)$ given on the interval $[t_1, t_2]$ and determined by Eqs. (8.42), (8.39) and (8.37) at $t < t_1$ will be continuous together with its derivatives till $n - m - 1$ order inclusively at the point t_1:

$$y^{(k)}(t_1) = \sum_{r=1}^{n} c_r y_r^{(k)}(t_1) \quad (k = 0, 1, \ldots, n - m - 1). \qquad (8.45)$$

Eqs. (8.43)–(8.45) represent a set of $n + m$ linear algebraic equations which determines the constants $c_1, \ldots, c_r$ and m arbitrary constants which occur at the integration of Eq. (8.37) at $t \geq t_1$. The function $z(t)$ in Eq. (8.43) depends on these constants. The solution of Eq. (8.33) is concluded by the determination of all unknown constants by means of the solution of Eqs. (8.43)–(8.45).

The solution $x(t)$ of Eq. (8.33) which is determined by formula (8.40) contains a linear combination of the δ–functions and their derivatives

$$\sum_{k=0}^{n-m-1} [A_k \delta^{(k)}(t - t_1) + B_k \delta^{(k)}(t - t_2)]. \qquad (8.46)$$

For determining the coefficients A_k and B_k we may apply the formula obtained in Problem 5.3.11:

$$A_k = \sum_{h=m}^{n-k-1} \sum_{l=m}^{h} (-1)^{k+h+1} C_h^l a_{k+h+1}^{(h-l)}(t_1) \Delta z^{(l)}(t_1),$$

$$B_k = \sum_{h=m}^{n-k-1} \sum_{l=m}^{h} (-1)^{k+h+1} C_h^l a_{k+h+1}^{(h-l)}(t_2) \Delta z^{(l)}(t_2). \qquad (8.47)$$

E x a m p l e 8.1. Solve Eq. (8.33) for the case $\rho = 1$, $F = a_1 D + a_0$, $H = 1$. In this case the kernel of Eq. (8.33) is determined by formula

$$K(t, s) = \begin{cases} q_1(t)q_2(s) & \text{at } s < t, \\ q_1(s)q_2(t) & \text{at } s > t, \end{cases}$$

where

$$q_1(t) = \exp\left\{-\int_\alpha^t \frac{a_0(\tau)}{a_1(\tau)} \, d\tau\right\}, \quad q_2(t) = q_1(t) \int_\alpha^t \frac{d\tau}{a_1^2(\tau)q_1^2(\tau)}.$$

To make sure that it is sufficient to integrate Eq. (8.31) for a given case and calculate $K(t, s)$ by formula (8.30) accounting that $w(t, \tau) = 0$ at $\tau > t$. On the contrary setting the kernel of Eq. (8.33) by the previous formula we make sure that it is determined by formula (8.30) and Eq. (8.31) at

$$a_1(t) = (q_1 q_2' - q_1' q_2)^{-1/2}, \quad a_0(t) = -a_1(t)q_1'/q_1.$$

Eqs. (8.39) and (8.37) give

$$u(t) = z(t) = a_1 y'(t) + a_0 y(t).$$

So, Eq. (8.43) has the form

$$-a_1 z' + (a_0 - a_1')z = 0.$$

After integrating this equation we find $z(t) = c/a_1(t)q_1(t)$. Substituting this expression into Eq. (8.39) we obtain

$$a_1 y' + a_0 y = c/a_1(t)q_1(t).$$

After integrating this equation we find $y(t) = cq_2(t)$. For determining the only constant c we have one Eq. (8.45): $y(t_1) = cq_2(t_1)$. Hence we find c

$= y(t_1)/q_2(t_1)$. To determine the coefficients A_0 and B_0 at the δ–functions by formulae (8.47) we find the discontinuities of the function $z(t)$ at the points t_1 and t_2:

$$\Delta z(t_1) = a_1(t_1)y'(t_1) + a_0(t_1)y(t_1) - (y(t_1)/q_2(t_1))a_1(t_1)q_1(t_1)$$
$$= a_1(t_1)[y'(t_1) - q_2'(t_1)y(t_1)/q_2(t_1)],$$
$$\Delta z(t_2) = -a_1(t_2)y'(t_2) - a_0(t_2)y(t_2)$$
$$= -a_1(t_2)[y'(t_2) - q_1'(t_2)y(t_2)/q_1(t_2)].$$

After this formulae (8.40), (8.46) and (8.47) give the solution of Eq. (8.33):

$$x(t) = -a_1^2(t)y''(t) - 2a_1(t)a_1'(t)y'(t) + [a_0^2(t) - a_0(t)a_1'(t)$$
$$- a_0'(t)a_1(t)]y(t) - a_1^2(t_1)[y'(t_1) - q_2'(t_1)y(t_1)/q_2(t_1)]\delta(t - t_1)$$
$$+ a_1^2(t_2)[y'(t_2) - q_1'(t_2)y(t_2)/q_1(t_2)]\delta(t - t_2).$$

R e m a r k. The obtained results show that the equations of the first kind considered here have no solution in the space $L_2([a, b])$ at any $a, b > a$ but have the solution in the space of the generalized functions.

Let us use now the stated method for finding the eigenvalues and eigenfunctions of Eq. (8.34). Accounting that Eq. (8.34) is received from Eq. (8.33) by the replacement of the function $y(t)$ by the function $\lambda x(t)$ we substitute into Eq. (8.37) the expression of the function $u(t)$ from Eq. (8.39) and replace in it the function $y(t)$ by the function $\lambda x(t)$. Then we get for the function $z(t)$ the following linear differential equation:

$$F(F^* z/\rho(t)) - HH^* z/\lambda = 0. \tag{8.48}$$

The general solution of this equation contains $2n$ arbitrary constants. The solution of Eq. (8.43) which should be solved at finding the extension of $y(t)$ in the right–hand side of the integral equation on the interval (α, t_1) in addition contains n arbitrary constants. For determining all these constants we have the conditions of the absence of the δ–functions and their derivatives in expression (8.40) of the eigenfunctions. From these conditions follow, firstly, the conditions of the continuity of the function $z(t)$ and its derivatives till $n - 1$ order inclusively on the ends of the integration interval in Eq. (8.34), and secondly, the condition of the continuity agreed upon them at the point t_1 of the function equal to $\lambda x(t)$ on the interval (t_1, t_2) and its extension $y(t)$ in the correspondence with the stated method at $t < t_1$ and its derivatives till $n - 1$ order inclusively. These conditions give a set of homogeneous linear algebraic equations:

$$\sum_{h=1}^{2n} \gamma_h z_h^{(k)}(t_1, \lambda) - \sum_{r=1}^{n} c_r z_r^{(k)}(t_1) = 0,$$

$$\sum_{h=1}^{2n} \gamma_h z_h^{(k)}(t_2, \lambda) = 0, \tag{8.49}$$

$$\lambda \sum_{h=1}^{2n} \gamma_h x_h^{(k)}(t_1, \lambda) - \sum_{r=1}^{n} c_r y_r^{(k)}(t_1) = 0 \quad (k = 0, 1, \ldots, n-1),$$

where $z_1(t, \lambda), \ldots, z_{2n}(t, \lambda)$ are some linearly independent solutions of Eq. (8.48),

$$x_h(t, \lambda) = F^* z_h(t, \lambda)/\rho(t) \quad (h = 1, \ldots, 2n),$$

and $z_1(t), \ldots, z_n(t)$, $y_1(t), \ldots, y_n(t)$ are the same functions that in Eqs. (8.44) and (8.45). After equalizing to zero the determinant of the set of Eqs. (8.49) determine for each eigenvalue all the constants besides one. This constants will be determined from the condition of the equality to unit of the norm of the corresponding eigenfunction $x(t)$.

E x a m p l e 8.2. Find the eigenvalues and eigenfunctions of Eq. (8.34) with the kernel of Example 8.1.

Eq.(8.48) has in a given case the form

$$\frac{d}{dt}\left(a_1^2 \frac{dz}{dt}\right) - (a_0^2 + a_0' a_1 - a_0 a_1' - a_1 a_1'' - 1/\lambda)z = 0.$$

Its general solution is determined by the formula

$$z(t) = \gamma_1 z_1(t, \lambda) + \gamma_2 z_2(t, \lambda),$$

where $z_1(t, \lambda)$ and $z_2(t, \lambda)$ are some linearly independent partial solutions. Formula (8.40) gives the corresponding functions

$$x_h(t, \lambda) = -\frac{d}{dt}[a_1(t)z_h(t, \lambda)] + a_0(t)z_h(t, \lambda) \quad (h = 1, 2)$$

and determines the eigenfunction

$$x(t) = \gamma_1 x_1(t, \lambda) + \gamma_2 x_2(t, \lambda).$$

At $t < t_1$ the functions z and y are determined analogously as in Example 8.1 by the formulae

$$z(t) = c/a_1(t)q_1(t), \quad y(t) = cq_2(t).$$

Eqs. (8.49) for the determination of the constants γ_1, γ_2 and c have the form

$$\gamma_1 z_1(t_1, \lambda) + \gamma_2 z_2(t_1, \lambda) - c/a_1(t_1)q_1(t_1) = 0,$$
$$\gamma_1 z_1(t_2, \lambda) + \gamma_2 z_2(t_2, \lambda) = 0,$$
$$\lambda[\gamma_1 x_1(t_1, \lambda) + \gamma_2 x_2(t_1, \lambda)] - cq_2(t_1) = 0.$$

Equalizing to zero the determinant of this set of the equations we obtain an equation which determines the eigenvalues

$$\Delta(\lambda) = \begin{vmatrix} z_1(t_1, \lambda) & z_2(t_1, \lambda) & 1/a_1(t_1)q_1(t_1) \\ z_1(t_2, \lambda) & z_2(t_2, \lambda) & 0 \\ \lambda x_1(t_1, \lambda) & \lambda x_2(t_1, \lambda) & q_2(t_1) \end{vmatrix} = 0.$$

After determining the eigenvalues we shall express for each eigenvalue two constants γ_1 and γ_2 in terms of the third constant c. This constant c will be determined from the condition of the equality to 1 of the norm of the corresponding eigenfunction.

     E x a m p l e 8.3. In the special case of Example 8.2 when $a_0 = \sqrt{\alpha/2}$, $a_1 = 1/\sqrt{2\alpha}$, where α is some positive constant,

$$q_1(t) = 1/q_2(t) = e^{-\alpha t}$$

and the kernel of Eq. (8.34) is expressed by the formula

$$K(t, s) = e^{-\alpha|t-s|}.$$

Eq. (8.48) has the form

$$z'' + (2\alpha/\lambda - \alpha^2)z = 0.$$

Integrating this equation we find its solution

$$z(t) = \gamma_1 e^{i\omega t} + \gamma_2 e^{-i\omega t}.$$

After that the unknown function $x(t)$ will be determined by the formula

$$x(t) = -\frac{1}{\sqrt{2\alpha}}z'(t) + \sqrt{\frac{\alpha}{2}}z(t) = \gamma_1 \frac{\alpha - i\omega}{\sqrt{2\alpha}}e^{i\omega t} + \gamma_2 \frac{\alpha + i\omega}{\sqrt{2\alpha}}e^{-i\omega t}, \quad (*)$$

where $\omega = \sqrt{2\alpha/\lambda - \alpha^2}$. The equation for the eigenvalues at $t_1 = 0$, $t_2 = T$ has the form

$$\Delta(\lambda) = \begin{vmatrix} 1 & 1 & \sqrt{2\alpha} \\ e^{i\omega T} & e^{-i\omega T} & 0 \\ \lambda(\alpha - i\omega) & \lambda(\alpha + i\omega) & \sqrt{2\alpha} \end{vmatrix} = 0.$$

Opening the determinant after elementary transformations we receive the equation

$$\operatorname{tg}\omega T = -\frac{2\alpha\omega}{\alpha^2 - \omega^2}. \qquad (**)$$

This equation determines unbounded increasing sequence of the values $\omega = \omega_1, \omega_2, \ldots$ (Fig.8.1 at $\alpha = 7\pi/4T$). After finding the solution ω_ν of Eq. $(**)$ corresponding eigenvalue λ_ν will be determined by the formula

$$\lambda_\nu = \frac{2\alpha}{\alpha^2 + \omega_\nu^2} \quad (\nu = 1, 2, \ldots).$$

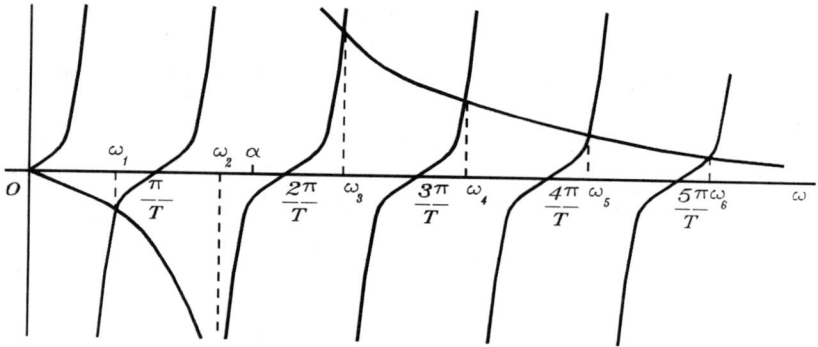

Fig. 8.1

This formula gives an unbounded decreasing sequence of the eigenvalues. The set of the equations for γ_1, γ_2 and c has the form

$$\gamma_1 + \gamma_2 - \sqrt{2\alpha}c = 0, \quad \gamma_1 e^{i\omega T} + \gamma_2 e^{-i\omega T} = 0,$$

$$\lambda[(\alpha - i\omega)\gamma_1 + (\alpha + i\omega)\gamma_2] - \sqrt{2\alpha}c = 0.$$

The second equation from these ones gives $\gamma_2 = -\gamma_1 e^{2i\omega T}$. Substituting this expression into anyone from two remained equations we may also express c in terms of γ_1. Substituting the obtained expression γ_2 into formula $(*)$, performing elementary transformations and determmining after that γ_1 from the condition of the normalization we shall find the eigenfuncitons:

$$x_\nu(t) = \sqrt{\frac{2}{T + \lambda_\nu}} \sin\left[\omega_\nu\left(t - \frac{T}{2}\right) + \frac{\nu\pi}{2}\right] \quad (\nu = 1, 2, \ldots).$$

It remained to show how the stated method is applied for the solution of the nonhomogeneous integral equation of the second kind (8.35). Taking into account that Eq. (8.35) is obtained from Eq. (8.33) by the change of the function $y(t)$ by the function $y(t) + \lambda x(t)$ in exactly the same way as in the case of homogeneous Eq. (8.34) we get for the function $z(t)$ the following linear differential equation:

$$F(F^* z / \rho(t)) - H H^* z / \lambda = -F y(t) / \lambda. \qquad (8.50)$$

The solution of this equation is determined by the formula

$$z(t) = \sum_{h=1}^{2n} \gamma_h z_h(t, \lambda) + v(t, \lambda), \qquad (8.51)$$

where in addition to the previous denotions $v(t, \lambda)$ is the particular solution of Eq. (8.50) which vanishes identically to zero at $y(t) = 0$. In order to find this solution it is sufficient to integrate Eq. (8.50) at the initial conditions: $z(t_1) = z'(t_1) = \cdots = z^{(2n-1)}(t_1) = 0$. Correspondingly Eqs. (8.49) will be replaced by the following set of equations

$$\sum_{h=1}^{2n} \gamma_h z_h^{(k)}(t_1, \lambda) - \sum_{r=1}^{n} c_r z_r^{(k)}(t_1) = 0,$$

$$\sum_{h=1}^{2n} \gamma_h z_h^{(k)}(t_2, \lambda) = -v^{(k)}(t_2, \lambda), \qquad (8.52)$$

$$\lambda \sum_{h=1}^{2n} \gamma_h x_h^{(k)}(t_1, \lambda) - \sum_{r=1}^{n} c_r y_r^{(k)}(t_1) = -\lambda w^{(k)}(t_1, \lambda)$$

$$(k = 0, 1, \ldots, n - 1),$$

where $w(t, \lambda) = F^* v(t, \lambda) / \rho(t)$. After the solution of this set of algebraic equations formulae (8.51) and (8.40) will completely determine the solution of Eq. (8.35). It goes without saying that the determinant of the set of Eqs. (8.52) should be different from zero, i.e. λ cannot be an eigenvalue.

E x a m p l e 8.4. Solve Eq. (8.35) with the kernel of Example 8.1.

Eq. (8.50) has in a given case the form

$$\frac{d}{dt}\left(a_1^2 \frac{dz}{dt}\right) - \left(a_0^2 + a_0' a_1 - a_0 a_1' - a_1 a_1'' - \frac{1}{\lambda}\right) z = -\frac{1}{\lambda}[a_1 y'(t) + a_0 y(t)].$$

Its solution is determined by the formula

$$z(t) = \gamma_1 z_1(t, \lambda) + \gamma_2 z_2(t, \lambda) + v(t, \lambda),$$

where

$$v(t, \lambda) = \frac{1}{\lambda} \int_{t_1}^{t} \frac{z_1(t, \lambda)z_2(\tau, \lambda) - z_2(t, \lambda)z_1(\tau, \lambda)}{z_1(\tau, \lambda)z_2'(\tau, \lambda) - z_1'(\tau, \lambda)z_2(\tau, \lambda)}$$

$$\times [a_1(\tau)y'(\tau) + a_0(\tau)y(\tau)]d\tau.$$

Eqs. (8.52) have the form

$$\gamma_1 z_1(t_1, \lambda) + \gamma_2 z_2(t_1, \lambda) - c/a_1(t_1)q_1(t_1) = 0,$$

$$\gamma_1 z_1(t_2, \lambda) + \gamma_2 z_2(t_2, \lambda) = -v(t_2, \lambda),$$

$$\lambda[\gamma_1 x_1(t_1, \lambda) + \gamma_2 x_2(t_1, \lambda)] - cq_2(t_1) = -\lambda w(t_1, \lambda),$$

where

$$w(t, \lambda) = -a_1(t)v'(t, \lambda) + [a_0(t) - a_1'(t)]v(t, \lambda).$$

After finding the coefficients γ_1, γ_2 and c formula (8.40) will determine the solution of Eq. (8.35)

$$x(t) = \gamma_1 x_1(t, \lambda) + \gamma_2 x_2(t, \lambda) + w(t, \lambda).$$

The stated method with non–essential changes is extended over the sets of the integral equations. In this case the functions $K(t, s)$, $w(t, \tau)$ and the coefficients a_k and b_k of the operator polynomials F and H represent the square matrices, and all constants of the integration and the functions $x(t)$, $y(t)$, $z(t)$ and $u(t)$ are the vectors.

R e m a r k. The most difficult fact in the implementation of the stated method is the establishment of the belonging the kernel of the equation to the class of the kernels which may be represented by formula (8.40) and Eq. (8.31). In the application problems formulae (8.30) and (8.31) are often follow directly from the sense of the problem. A widely spread class of the kernels of such kind is the class of the kernels which represent the even functions of the arguments difference $K(t, s) = k(\tau)$, $\tau = t - s$, whose Fourier transformations are the rational functions. All such kernels are expressed by formulae (8.30) and (8.31), and the coefficients a_k and b_k of the operators F and H are constant. Therefore the solution of the integral equations with such kernels by the stated method is reduced to the integration of the linear differential equations with the constant coefficients which gives the opportunity to get the analytical solution of the problem. We shall present two typical kernels for which Eq. (8.31) are known and are often encountered in applications.

To the kernel

$$k(\tau) = De^{-a|\tau|}(\cos \omega_0 \tau + \gamma \sin \omega_0 |\tau|), \quad a, \omega_0 > 0,$$

corresponds Eq. (8.31) of the form

$$w''(t,\tau) + 2aw'(t,\tau) + b^2 w(t,\tau) = \sqrt{2D(a - \gamma\omega_0)}[\delta'(t-\tau) + b_1\delta(t-\tau)],$$

where $b^2 = a^2 + \omega_0^2$, $b_1^2 = b^2(a + \gamma\omega_0)(a - \gamma\omega_0)^{-1}$.

To the kernel

$$k(\tau) = De^{-\alpha|\tau|}(1 + \beta|\tau|), \quad \alpha > 0,$$

corresponds Eq. (8.31) of the form

$$w''(t,\tau) + 2\alpha w'(t,\tau) + \alpha^2 w(t,\tau) = \sqrt{2D}[\sqrt{\alpha - \beta}\,\delta'(t-\tau)$$

$$+\alpha\sqrt{\alpha + \beta}\,\delta(t-\tau)].$$

8.1.5. Integral Representation of an Operator

Let us analyse the obtained spectral decomposition (8.8) of the operator A. It is clear that the vector $(x, x_k)x_k$ represents a projection of the vector x on the eigenvector x_k, i.e. on the one–dimensional subspace $G_k \ominus G_{k-1}$ formed by the vector x_k. Thus formula (8.8) represents the result of the operator A action on the vector x in the form of the sum of the products of the eigenvalues by the projections of the vector x by corresponding eigenvectors of the operator A. In the case of m–multiple eigenvalue $\lambda_n = \lambda_{n+1} = \cdots = \lambda_{n+m-1}$ the sum

$$\sum_{k=n}^{n+m-1} (x, x_k)x_k$$

represents the projection of the vector x on the corresponding m–dimensional proper subspace. Therefore if we change the numeration of the eigenvalues in such a way that multiple eigenvalues will be not repeated then formula (8.8) may be rewritten in the form

$$Ax = \sum_{k=0}^{\infty} \lambda_k P_k x . \tag{8.53}$$

Here P_k is the projector on the proper subspace corresponding to the eigenvalue λ_k, P_0 is the projector on the kernel $H_\infty = G_\infty^\perp = R_A^\perp$ of the operator A, and $\lambda_0 = 0$ may be presented by the decomposition over the set of the projectors $\{P_k\}$

$$A = \sum_{k=0}^{\infty} \lambda_k P_k. \tag{8.54}$$

It is natural to try to generalize this result on more wide class of the linear operators. For this purpose we notice that the set of the projectors $\{P_k\}$ determines on the real axis R the operator measure, i.e. the measure with the values in the space of the bounded linear operators $\mathcal{B}(X)$. Really, after determining the values of the operator function of the set E on any Borel set B by the formula

$$E(B) = \sum_{\{k:\lambda_k \in B\}} P_k, \tag{8.55}$$

on the basis of Theorems 7.19–7.21 and 7.23 about the orthoprojectors we assure that this function is additive and continuous (Subsection 2.2.2). By Theorem 2.12 this measure is also σ–additive on the σ–algebra of Borel sets of the real axis R. Its value on any set B represents a projector on the orthogonal sum of the proper subspaces corresponding to all eigenvalues which belong to B. In particular, $E(R)$ represents a projector on the whole space X, i.e. an identity operator $E(R) = I$, and $E(\emptyset)$ represents a projector into point 0, $E(\emptyset) = 0$. Thus formula (8.55) at $B = R$ determines the decomposition of the operator I on the projectors corresponding to all proper subspaces of the operator A. Therefore the operator measure E is usually called *a decomposition of the identity*. Taking into consideration that the measure E is completely concentrated on the discrete set of the points $\{\lambda_k\}$ (it is equal to zero on any set which contains none of these points) we may rewrite formula (8.54) in the form

$$A = \int_{-\infty}^{\infty} t E(dt). \tag{8.56}$$

This formula indicates the way of the generalization of the spectral theory of the linear operators. Considering more general operator measures we must reveal at first the class of the operators which may be presented in the form of the integrals with respect to the operator measure and then to solve the problem of finding the operator measure corresponding to the given operator.

8.1.6. Decomposition of the Identity

Let $(\Delta, \mathcal{A})$ be an arbitrary measurable space. Let us determine on the σ–algebra $\mathcal{A}$ the operator function of the set $E(A)$ possessing the following properties:

(i) the value $E(A)$ on any set $A \in \mathcal{A}$ represents a projector on some subspace G_A of the H–space X;

(ii) $E(\emptyset) = 0$, $E(\Delta) = I$;

(iii) if $A \subset B$, $A, B \in \mathcal{A}$ then $E(A) \le E(B)$ (i.e. $(E(A)x, x)$ $\le (E(B)x, x)$ at all x);

(iv) for any pairwise nonintersecting sets $A_1, \ldots, A_n \in \mathcal{A}$

$$E\left(\bigcup_{k=1}^{n} A_k\right) = \sum_{k=1}^{n} E(A_k),$$

i.e. the function $E(A)$ is additive;

(v) $E(A)E(B) = E(B)E(A) = E(AB)$ for any $A, B \in \mathcal{A}$.

The operator function of the set $E(A)$ possessing these properties is called *a decomposition of the identity*.

On the basis of Theorems 7.19–7.21 about the projectors the second property of the decomposition of the identity means that $E(\emptyset)$ represents a projector on the point 0, $G_\emptyset = 0$, and $E(\Delta)$ is a projector on the whole space X, $G_\Delta = X$. The third property means that $G_A \subset G_B$ for any sets $A, B \in \mathcal{A}$, $A \subset B$. The fourth property means that the subspaces $G_{A_1}, \ldots, G_{A_n}$ corresponding to pairwise nonintersecting sets $A_1, \ldots, A_n \in \mathcal{A}$ are orthogonal and the subspace G_A, $A = \bigcup_{k=1}^{n} A_k$ represents the orthogonal sum of the subspaces $G_{A_1}, \ldots, G_{A_n}$. In particular, hence it follows $G_A \oplus G_{\bar{A}} = X$ and $E(A) + E(\bar{A}) = I$ for any $A \in \mathcal{A}$. The fifth property means that $G_{AB} = G_A \cap G_B$ for any sets $A, B \in \mathcal{A}$.

From Theorem 7.23 about the monotone sequences of the projectors follows the continuity of the function $E(A)$ (Subsection 2.2.2). So, for any monotone sequence of the sets $\{A_n\} \subset \mathcal{A}$, $A = \lim A_n$,

$$\lim E(A_n)x = E(\lim A_n)x = E(A)x \text{ for all } x.$$

From the additivity and the continuity of the decomposition of the identity by virtue of Theorem 2.12 follows its σ–additivity.

Thus the decomposition of the identity $E(A)$ represents an operator measure, i.e. the measure with the values in the space of the bounded linear operators $\mathcal{B}(X)$.

The decomposition of the identity considered in Subsection 8.1.5 generated by the compact self–adjoint operator A is received in the special case when Δ represents the real axis R with the σ–algebra of Borel sets $\mathcal{A}$, and G_B is the orthogonal sum of the proper subspaces of the operator A corresponding to all eigenvalues which belong to the set B. This decomposition of the identity is called *a spectral measure* of the operator A.

It is natural to suppose that not only a compact but a self–adjoint operator generates some decomposition of the identity, i.e. its spectral measure in terms of which it is expressed by formula (8.56). The proof of this fact and the statements which follow from it is the essence of the spectral theory of the linear operators.

To the operator measure E we may correspond the set of the complex–valued measures

$$\mu_{x,y}(A) = (E(A)x, y), \tag{8.57}$$

which correspond to all $x, y \in X$. In particular, at $y = x$ we obtain the set of nonnegative measures

$$\mu_x(A) = (E(A)x, x) = \| E(A)x \|^2 . \tag{8.58}$$

All these measures are finite as by virtue of the second property of the decomposition of the identity $\mu_{x,y}(\Delta) = (x, y)$, $\mu_x(\Delta) = \| x \|^2$ and the identity

$$(Px, y) = \frac{1}{4}\{(P(x + y), x + y) - (P(x - y), x - y)$$

$$+ i(P(x + iy), x + iy) - i(P(x - iy), x - iy)\},$$

follows

$$\mu_{x,y}(A) = \frac{1}{4}\{\mu_{x+y}(A) - \mu_{x-y}(A) + i\mu_{x+iy}(A) - i\mu_{x-iy}(A)\}, \quad A \in \mathcal{A}.$$

For detailed study of spectral decomposition of a compact self–adjoint operator refer to (Dowson 1978, Dunford and Schwarz 1963). The corresponding theory of not self–adjoint operators is given in (Ringrose 1971). Applications in control theory are considered in (Balakrishnan 1976, Laning and Battin 1956, Pugachev 1965, Pugachev and Sinitsyn 1987).

P r o b l e m s

8.1.1. Prove that if $\lambda = 0$ is not an eigenvalue of the compact self–adjoint operator A then $\lambda = 0$ is the point of the continuous spectrum A.

I n s t r u c t i o n. It is sufficient using formula (8.17) to prove the density of the domain of the inverse operator A^{-1} in X.

8.1.2. Prove that any compact normal operator T (Subsection 7.6.4) has a finite or a countable set of the eigenvalues $\{\lambda_n\}$ and the corresponding orthonormal system of the eigenvectors $\{x_n\}$. Prove the following decompositin for operator T:

$$Tx = \sum_{k=1}^{\infty} \lambda_k(x, x_k)x_k.$$

I n s t r u c t i o n. Use the apparent equality

$$T = \frac{T + T^*}{2} + i\frac{T - T^*}{2i},$$

which expresses any operator acting in the H–space in terms of the self–adjoint operators $(T + T^*)/2$ and $(T - T^*)/2i$. From the properties of a normal operator it is clear that any eigenvector of the operator T is an eigenvector of the operators $(T + T^*)/2$ and $(T - T^*)/2i$. For proving the suggested statement it is sufficient to show that the operators $(T + T^*)/2$ and $(T - T^*)/2i$ have one and the same orthonormal system of the eigenvectors. Let μ be an eigenvalue of the operator $(T + T^*)/2$, S be the corresponding m–dimensional proper subspace. From the normality of the operator T follows that the operators $(T + T^*)/2$ and $(T - T^*)/2i$ are commutative. Hence it follows that $[(T - T^*)/2i]x \in S$ for any vector $x \in S$. Therefore the solution of equation $[(T - T^*)/2i]x = \nu x$ should be found in the form

$$x = \sum_{l=1}^{m} c_l y_l.$$

Here $\{y_1, \ldots, y_m\}$ is any orthonormal basis in S. As a result we shall find m eigenvalues $\nu_1, \ldots, \nu_m$ of the operator $(T - T^*)/2i$ and the corresponding eigenvectors $x_1, \ldots, x_m$ which will be the eigenvectors of the operator $(T + T^*)/2$.

8.1.3. Prove that for any compact normal operator T formulae (8.13)–(8.15) are valid for solving operator equation of the second kind

$$Tx - \lambda x = y,$$

for the resolvent and its norm. Prove that formula (8.17) gives the solution of equation of the first kind $Tx = y$.

8.1.4. Prove that if $\lambda = 0$ is not the eigenvalue of the compact normal operator T then the point $\lambda = 0$ is the point of continuous spectrum.

8.1.5. Prove that for any commutative self–adjoint operators A and B the operator $T = A + iB$ is a normal one.

8.1.6. Prove that any two commutative compact self–adjoint operators have the same orthonormal system of the eigenvectors.

8.1.7. Solve the integral Eqs. (8.33)–(8.35) for the following kernels:

$$K(t, s) = De^{-a|t-s|} \cos \omega_0(t - s), \ a > 0,$$

$$K(t, s) = De^{-a|t-s|}[\cos \omega_0(t - s) + \gamma \sin \omega_0|t - s|], \ a > 0,$$

$$K(t, s) = De^{-\alpha|t-s|}(1 + \alpha|t - s|), \ \alpha > 0,$$

$$K(t, s) = De^{-\alpha|t-s|}(1 - \alpha|t - s|), \ \alpha > 0,$$

$$K(t, s) = De^{-\alpha|t-s|}(1 + \beta|t - s|), \ \alpha > 0.$$

8.1.8. Solve integral Eqs. (8.33), (8.34) and (8.35) for the case $F = D^2 + 2aD + b^2$, $H = ke^{\gamma t}(D + b)$, $y(t) = e^{\alpha t}$, $0 < a < b$. Prove the following formula:

$$w(t, \tau) = \frac{k}{2i\omega_0}e^{-\alpha t} + a_1\tau[(b - a_1 + i\omega_0)e^{i\omega_0(t-\tau)}$$

$$-(b - a_1 - i\omega_0)e^{-i\omega_0(t-\tau)}], \ \omega_0 = \sqrt{b^2 - a^2}, \ a_1 = a + \gamma$$

and formula (8.30) at $\alpha = -\infty$

$$K(t, s) = D\exp\{-a|t - s| + 2\gamma\min(t, s)\}[\cos\omega_0(t-s) + \nu\sin\omega_0|t - s|],$$

where $D > 0$ and ν are expressed in terms of a, b, k, γ.

8.2. Operators Determined by Decomposition of the Identity

8.2.1. Integral of a Simple Function over an Operator–Valued Measure

Integral of a simple function over an operator measure is determined in exactly the same way as in Subsection 3.3.1 because in formulae (3.16) and (3.17) it does not matter whether the simple function f is a function with the values in the linear space, and the measure μ is a numeric one or vice versa, the function f is a numeric function and the measure μ has the values in the linear space. Let

$$\varphi(t) = \sum_{k=1}^{n} a_k 1_{A_k}(t), \qquad (8.59)$$

where $a_1, \ldots, a_n$ are any complex numbers, and $A_1, \ldots, A_n$ are any pairwise nonintersecting measurable sets.

In the correspondence with the general definition of Subsection 3.3.1 the integral of the simple function $\varphi(t)$ over the operator measure E with respect to $A \in \mathcal{A}$ is determined by formula

$$\int_A \varphi(t) E(dt) = \int_A \varphi\, dE = \sum_{k=1}^{n} a_k E(AA_k). \qquad (8.60)$$

In particular, the integral over the whole space Δ is obtained at $A = \Delta$ and is written without indicating the integration region

$$\int \varphi(t) E(dt) = \int \varphi\, dE = \sum_{k=1}^{n} a_k E(A_k). \qquad (8.61)$$

It is evident that integral (8.61) represents a linear operator

$$T = \int \varphi\, dE = \sum_{k=1}^{n} a_k E(A_k). \qquad (8.62)$$

In consequence of the projectors properties and the identity decomposition at any x

$$\|Tx\|^2 = \sum_{k,h=1}^{n} a_k \bar{a}_h (E(A_k)x, E(A_h)x)$$

$$= \sum_{k,h=1}^{n} a_k \bar{a}_h (E(A_k A_h)x, x) = \sum_{k=1}^{n} |a_k|^2 \mu_x(A_k)$$

or

$$\|Tx\|^2 = \int |\varphi|^2 \, d\mu_x. \qquad (8.63)$$

Hence putting $c = \max\{\varphi(t)\} = \max\{|a_k|\}$, we receive $\|Tx\|^2 \leq c^2 \|x\|^2$. Thus the operator T is bounded.

Further, as

$$(Tx, y) = \sum_{k=1}^{n} a_k (E(A_k)x, y) = \sum_{k=1}^{n} a_k (x, E(A_k)y) = (x, \sum_{k=1}^{n} \bar{a}_k E(A_k)y),$$

then the adjoint operator T^* is determined by formula

$$T^* = \sum_{k=1}^{n} \bar{a}_k E(A_k) = \int \bar{\varphi} \, dE. \qquad (8.64)$$

Thus the decomposition of the identity E establishes the correspondence between the simple function $\varphi(t)$ and the bounded linear operator T and the corresponding adjoint operator T^* which are determined by formulae (8.62) and (8.64).

8.2.2. Integral of a Measurable Function over an Operator–Valued Measure

Now let $\varphi(t)$ be an arbitrary measurable (relative to $\mathcal{A}$ and the σ–algebra of Borel sets of a complex plane) function. After modifying in corresponding way the method of Subsection 3.1.4 we shall construct the sequence of the simple functions $\{\varphi^n(t)\}$ which converges to $\varphi(t)$. For this purpose we introduce the sets $B_n = \{t : |\varphi(t)| \le n\}$ $(n = 1, 2, \ldots)$ and take an arbitrary sequence of positive numbers $\{\varepsilon_n\}$ convergent to zero. We cover the circle $C_n = \{z : |z| \le n\}$ of a complex plane by the circles of the radius ε_n and denote their number in terms of N_n, and their centers in terms of a_k^n. It is evident that all the points a_k^n may be taken inside the circle C_n or on its bound $|a_k^n| \le n$. After determining for each n the sets

$$F_k^n = \{t : |\varphi(t) - a_k^n| < \varepsilon_n\} \quad (k = 1, \ldots, N_n),$$

$$A_1^n = B_n F_1^n, \quad A_k^n = B_n F_k^n \bigcap_{m=1}^{k-1} \bar{F}_m^n \quad (k = 2, \ldots, N_n),$$

we construct the simple functions

$$\varphi^n(t) = \sum_{k=1}^{N_n} a_k^n \mathbf{1}_{A_k^n}(t).$$

It is obvious that

$$|\varphi^n(t) - \varphi(t)| < \varepsilon_n \text{ at all } t \in \bigcup_{k=1}^{N_n} A_k^n = B_n, \qquad (8.65)$$

and at any t at which $|\varphi(t)| < \infty$ the sequence $\{\varphi^n(t)\}$ converges to $\varphi(t)$ and this convergence is uniform of any set B_m. Besides $|\varphi^n(t)| \le n$ at all t and $\varphi^n(t) = 0$ at $t \in \bar{B}_n$.

Putting $T_n = \int \varphi^n \, dE$ we obtain the sequence of the bounded linear operators $\{T_n\}$. By Theorem 3.39 for any x for which

$$\int |\varphi|^2 d\mu_x < \infty, \qquad (8.66)$$

i.e. $\varphi(t) \in L_2(\Delta, \mathcal{A}, \mu_x)$, the sequence $\{\varphi^n(t)\}$ converges to $\varphi(t)$ and in the mean square

$$\|\varphi^n - \varphi\|_2^2 = \int |\varphi^n - \varphi|^2 d\mu_x \to 0 \text{ at } n \to \infty.$$

This is due to the fact that all the functions $\varphi^n(t)$ by virtue of inequality (8.65) are modulus–bounded by the function $\varphi(t) + \varepsilon_1$ which belongs to $L_2(\Delta, \mathcal{A}, \mu_x)$. Consequently, the sequence $\{\varphi^n(t)\}$ is fundamental in the mean square, and for any x which satisfies condition (8.66)

$$\|T_n x - T_m x\| = \left[\int |\varphi^n - \varphi^m|^2 d\mu_x \right]^{1/2} = \|\varphi^n - \varphi^m\|_2 \to 0$$

$$\text{at } n, m \to \infty.$$

Thus for any x which satisfies condition (8.66) the sequence $\{T_n x\}$ is fundamental. Due to the completeness of the H–space it converges to some limit (certainly if it depends on x): $\lim T_n x = T x$. Consequently, the sequence of the operators $\{T_n\}$ converges to some linear operator T. It is natural to call this operator *an integral* of the simple function $\varphi(t)$ over the operator measure E:

$$T = \int \varphi(t) E(dt) = \int \varphi \, dE. \qquad (8.67)$$

As the domain D_T of the operator T serves the set of the vectors x which satisfy condition (8.66):

$$D_T = \{x : \int |\varphi|^2 d\mu_x < \infty\}. \qquad (8.68)$$

Theorem 8.3. *An integral over an operator measure does not depend on the choice of the sequence of the simple functions which converges to $\varphi(t)$.*

▷ Really, if $\{\psi^n(t)\}$ is another sequence of the simple functions which converges to $\varphi(t)$, $T'_n = \int \psi^n \, dE$, $T' = \lim T'_n$ then by virtue of Theorem 3.39 $\{\psi^n(t)\}$ also converges to $\varphi(t)$ in $L_2(\Delta, \mathcal{A}, \mu_x)$ at any $x \in D_T$. Consequently, at any $x \in D_T$

$$\|T'x - Tx\| = \lim \|T'_n x - T_n x\| = \lim \|\psi^n - \varphi^n\|_2$$

$$\leq \lim\{\|\psi^n - \varphi\|_2 + \|\varphi^n - \varphi\|_2\} = 0. \; ◁$$

Notice that on the basis of formula (8.63)

$$\|Tx\|^2 = \lim \|T_n x\|^2 = \lim \int |\varphi^n|^2 d\mu_x = \int |\varphi|^2 d\mu_x.$$

Thus formula (8.63) is valid for any measurable function $\varphi(t)$.

Analogously we obtain for any $x \in D_T$ and any y

$$(Tx, y) = \lim(T_n x, y) = \lim \int \varphi^n(t)(E(dt)x, y)$$

$$= \int \varphi(t)(E(dt)x, y) = \int \varphi(t)\mu_{x,y}(dt). \tag{8.69}$$

Theorem 8.4. *The operator T determined by the decomposition of the identity E is bounded if and only if the function $\varphi(t)$ is bounded almost everywhere relatively to the measure E, i.e. when there exists such number $c > 0$ that $E(\{t : |\varphi(t)| > c\}) = 0$.*

▷ In the case of the function $\varphi(t)$, $|\varphi(t)| \leq c$ bounded almost everywhere relatively to E from (8.63) follows $\|Tx\| \leq c\|x\|$. Thus to the bounded function $\varphi(t)$ corresponds a bounded linear operator determined by formula (8.67).

And vice versa, if the operator T which is determined by formula (8.67) is bounded then the function $\varphi(t)$ is bounded (except may be the set of the points t of zero measure E). Really, if $\varphi(t)$ is unbounded then the set $\bar{B}_n = \{t : |\varphi(t)| > n\}$ has the measure E different from zero at any n, $E(\bar{B}_n) \neq 0$. After taking in each subspace $G_{\bar{B}_n}$ the point x_n

we get $E(\bar{B}_n)x_n = x_n$, and consequently, $\mu_{x_n}(A) = 0$ for any set $A \in \mathcal{A}$ which does not intersect with $\bar{B}_n$. Therefore

$$\|Tx_n\|^2 = \int |\varphi|^2 \, d\mu_{x_n} = \int_{\bar{B}_n} |\varphi|^2 \, d\mu_{x_n} > n^2 \, \|x_n\|^2 \, .$$

Consequently, the operator T cannot be bounded if the function $\varphi(t)$ is unbounded. ◁

Corollary. *If the operator T which is determined by formula (8.67) is bounded then it is determined on the whole space X.*

▷ It follows immediately from (8.68). ◁

Formula

$$Tx = \int \varphi(t)E(dt)x, \quad x \in D_T, \tag{8.70}$$

which follows from definiton (8.67) determines the element Tx of the H–space X as an integral of the numeric function $\varphi(t)$ over the measure $E(A)x$ with the values in the same H–space X. But this integral does not exist as an integral over a measure with the values in the B–space in the sense of Subsection 3.3.9 as the complete variation of the measure $E(A)x$ is infinite on any set $A \in \mathcal{A}$. Therefore we had to give the definition of an integral over an operator measure which is independent of the theory of Subsection 3.3.9.

The obtained results show that in the case of the function $\varphi(t)$ which is bounded almost everywhere relatively to E when the operator determined by formula (8.67) is bounded the integral in (8.67) exists as the limit of the integrals sequence of the simple functions in the space $\mathcal{B}(X)$. In the case of the unbounded function $\varphi(t)$ there exists only integral (8.70) at each $x \in D_T$ as the strong limit of the integrals sequence of the simple functions in the topology of the H–space X.

8.2.3. Adjoint Operator

Let us establish the condition of the existence of the integral operator T^* adjoint with the operator T which is determined by formula (8.67) and find it.

Theorem 8.5. *The adjoint operator T^* which is determined by the decomposition of the identity exists if and only if the function $\varphi(t)$ is almost everywhere finite relatively to the measure E,*

$$\lim_{n \to \infty} E(\{t : |\varphi(t)| > n\}) = 0. \tag{8.71}$$

▷ In order the operator T will have the adjoint T^* it is necessary and sufficient that its domain D_T will be dense in the space X (Theorem 6.7). It is necessary and sufficient for this purpose that the function $\varphi(t)$ will be almost everywhere finite relatively to the operator measure E. Naturally, if conditon (8.71) is fulfilled then putting $B = \bigcup B_n = \lim B_n$ we shall have $G_{\bar{B}} = 0$ and $G_B = X$. By virtue of the continuity of E at any x and any $\varepsilon > 0$ we may find such natural n that

$$\| x - E(B_n)x \| = \| E(B)x - E(B_n)x \| < \varepsilon.$$

Putting $y = E(B_n)x$ we obtain $\mu_y(A) = 0$ for any set $A \subset \bar{B}_n$ and

$$\int |\varphi|^2 \, d\mu_y = \int_{B_n} |\varphi|^2 d\mu_y \le n^2 \, \|y\|^2 < \infty.$$

Thus in any vicinity of any vector x there exist the vectors $y \in D_T$. Consequently, D_T is dense in X and condition (8.71) is sufficient.

If condition (8.71) is not fulfilled then $G_{\bar{B}} \ne \{0\}$ also for any x for which $E(\bar{B})x \ne 0$,

$$\int |\varphi|^2 \, d\mu_x \ge \int_{\bar{B}} |\varphi|^2 d\mu_x > n^2 \, \| E(\bar{B})x \|^2 \quad \text{at any } n,$$

i.e. $\int |\varphi|^2 \, d\mu_x = \infty$, and consequently, $x \notin D_T$. Therefore if $y \in D_T$ then $E(\bar{B})y = 0$ and $\| x - y \| \ge \| E(\bar{B})x \|$. So, D_T cannot be dense in X and the adjoint operator does not exist. This proves the necessity of condition (8.71). ◁

Theorem 8.6. *If the function $\varphi(t)$ is measurable and is almost everywhere finite relatively to E then there exists the adjoint operator T^* determined by formula*

$$T^* = \int \overline{\varphi(t)} E(dt) = \int \bar{\varphi} \, dE. \tag{8.72}$$

▷ The existence of the adjoint operator T^* follows from Theorem 8.5. For deriving formula (8.72) we notice that the sequence of the simple functions $\{\varphi^n(t)\}$ convergent to $\varphi(t)$ determines along with the sequence of the bounded linear operators $\{T_n\}$ the sequence of the adjoint operators $\{T_n^*\}$, $T_n^* = \int \bar{\varphi}^n \, dE$ convergent on D_T to the operator $S = \int \bar{\varphi} \, dE$, $Sx = \lim T_n^* x$ at $x \in D_T$. Hence for any $x, y \in D_T$

$$(Tx, y) = \lim(T_n x, y) = \lim(x, T_n^* y) = (x, Sy).$$

Consequently, $Sy = T^*y$ at $y \in D_T$. It means that $D_T \subset D_{T^*}$. In order to prove the equality $S = T^*$ it remains to show that $D_{T^*} = D_T$. We shall take for this purpose an arbitrary $z \in D_{T^*}$. Putting $x_n = T_n^* z$ we shall have

$$E(A)x_n = \sum_{k=1}^{N_n} \overline{a_k^n} E(A_k^n A)z = \int_A \overline{\varphi^n}\, dEz, \quad A \in \mathcal{A},$$

and

$$\mu_{x_n, y}(A) = (E(A)x_n, y) = \int_A \overline{\varphi^n}(dEz, y) = \int_A \overline{\varphi^n}\, d\mu_{z,y}. \qquad (8.73)$$

In particular, at $y = z$ we get

$$\mu_{x_n, z}(A) = (E(A)x_n, z) = \int_A \overline{\varphi^n}(dEz, z) = \int_A \overline{\varphi^n}\, d\mu_z \qquad (8.74)$$

and

$$\mu_{z, x_n}(A) = (E(A)z, x_n) = (z, E(A)x_n) = \int_A \varphi^n\, d\mu_z. \qquad (8.75)$$

Thus the functions $\mu_{x_n, z}$ and μ_{z, x_n} are continuous and their Radon–Nikodym derivatives with respect to the measure μ_z are equal to $\overline{\varphi^n}(t)$ and $\varphi^n(t)$ correspondingly (Subsections 3.6.1 and 3.6.3). Therefore putting in (8.73) $y = x_n$ on the basis of rule (3.82) of the measure change in the integral we receive

$$\mu_{x_n}(A) = (E(A)x_n, x_n) = \int_A \overline{\varphi^n}\, d\mu_{z,x_n}$$

$$= \int_A \overline{\varphi^n}\, \frac{d\mu_{z,x_n}}{d\mu_z}\, d\mu_z = \int_A |\varphi^n|^2\, d\mu_z$$

and

$$\int |\varphi|^2\, d\mu_{x_n} = \int |\varphi|^2 \frac{d\mu_{x_n}}{d\mu_z}\, d\mu_z = \int |\varphi \varphi^n|^2\, d\mu_z.$$

Hence accounting that $\varphi^n = 0$ on the set $\bar{B}_n$ and $|\varphi|, |\varphi^n| \leq n$ on B_n we obtain

$$\int |\varphi|^2 d\mu_{x_n} = \int_{B_n} |\varphi|^2 |\varphi^n|^2 d\mu_z \leq n^4 \|z\|^2 < \infty.$$

Thus $x_n \in D_T$ at all n. Therefore

$$(x_n, T^*z) = (Tx_n, z) = \int \varphi \, d\mu_{x_n, z}.$$

After replacing the measure $\mu_{x_n, z}$ by its expression (8.74) we shall have

$$(x_n, T^*z) = \int \varphi \overline{\varphi^n} \, d\mu_z = \int |\varphi^n|^2 d\mu_z + \int (\varphi - \varphi^n) \overline{\varphi^n} \, d\mu_z.$$

So using formula (8.63) and again accounting that $\varphi^n = 0$ on $\bar{B}_n$ we find

$$(x_n, T^*z) = \|x_n\|^2 + \int_{B_n} (\varphi - \varphi^n) \overline{\varphi^n} \, d\mu_z. \tag{8.76}$$

The functions φ and φ^n evidently belong to the space $L_2(B_n, B_n\mathcal{A}, \mu_z)$, and therewith the norms of the functions $\varphi - \varphi^n$ and φ^n in the space $L_2(B_n, B_n\mathcal{A}, \mu_z)$ on the basis of (8.65) and (8.63) satisfy the relations

$$\|\varphi - \varphi^n\|_2^2 = \int_{B_n} |\varphi - \varphi^n|^2 d\mu_z \leq \varepsilon_n^2 \|E(B_n)z\|^2 \leq \varepsilon_n^2 \|z\|^2,$$

$$\|\varphi^n\|_2^2 = \int_{B_n} |\varphi^n|^2 d\mu_z = \int |\varphi^n|^2 d\mu_z = \|T_n^*z\|^2 = \|x_n\|^2 .$$

Therefore from Holder inequality (3.92) follows

$$\left| \int_{B_n} (\varphi - \varphi^n) \overline{\varphi^n} \, d\mu_z \right| \leq \|\varphi - \varphi^n\|_2 \|\varphi^n\|_2 \leq \varepsilon_n \|z\| \|x_n\| .$$

Thus from (8.76) follows

$$\left| \|x_n\|^2 - |(x_n, T^*z)| \right| \leq \left| \|x_n\|^2 - (x_n, T^*z) \right| \leq \varepsilon_n \|z\| \|x_n\| .$$

Hence taking into consideration that $|(x_n, T^*z)| \leq \|x_n\| \|T^*z\|$, we find

$$\|x_n\|^2 \leq \|x_n\| \|T^*z\| + \varepsilon_n \|z\| \|x_n\|$$

or

$$\|x_n\| \leq \|T^*z\| + \varepsilon_n \|z\|.$$

Accounting that $\varepsilon_n \leq \varepsilon_1 < 2\varepsilon_1$ and putting $c = \|T^*z\| + 2\varepsilon_1 \|z\|$ we get $\|x_n\| < c$ and

$$\|x_n\|^2 = \|T_n^* z\|^2 = \int |\varphi^n| d\mu_z < c^2 \text{ at all } n.$$

Consequently,

$$\int |\varphi|^2 d\mu_z = \lim \int |\varphi^n|^2 d\mu_z \leq c^2 < \infty.$$

Thus any vector $z \in D_{T^*}$ satisfies condition (8.66), i.e. belongs to D_T. Therefore $D_{T^*} \subset D_T$. In connection with the inclusion $D_T \subset D_{T^*}$ received before it gives $D_{T^*} = D_T$ what proves the equality $S = \lim T_n^*$ $= T^*$ and along with it formula (8.72). ◁

From the proved theorem by the symmetry follows $S^* = T^{**} = T$. Hence on the basis on Theorem 6.8 we conclude that the operator T is closed. Thus all the operators with the dense domain generated by the decomposition of the identity are closed operators.

So, the decompostion of the identity E establishes the correspondence of the finite function $\varphi(t)$ measurable almost everywhere relatively to E between the closed linear operator T and the corresponding adjoint operator T^* which are determined by the formulae

$$T = \int \varphi(t) E(dt) = \int \varphi \, dE, \qquad T^* = \int \overline{\varphi(t)} \, E(dt) = \int \bar{\varphi} \, dE.$$

8.2.4. Commutativity of Operators

An integral over an operator measure represents a linear operator. Thus the decomposition of the identity determines the set of the linear operators in the H-space. Let us study the properties of these operators.

Theorem 8.7. *Any operator determined by the decomposition of the identity commutates with this decomposition of the identity.*

▷ If $\varphi(t)$ is a simple function and $T = \int \varphi \, dE$ then in consequence of the fifth property of the decomposition of the identity (Subsection 8.1.6) for any x we have

$$E(A)Tx = TE(A)x = \sum_{k=1}^{n} a_k E(AA_k)x = \int_A \varphi \, dEx.$$

For any measurable function $\varphi(t)$ after determining the sequence of the simple functions $\{\varphi^n(t)\}$ which converges to it and the corresponding operator $T_n = \int \varphi^n \, dE$ we receive in consequence of the boundedness of $E(A)$ for any $x \in D_T$ and for any $A \in \mathcal{A}$

$$E(A)Tx = E(A)\lim T_n x = \lim E(A)T_n x = \lim \int_A \varphi^n \, dEx = \int_A \varphi \, dEx,$$

$$TE(A)x = \lim T_n E(A)x = \lim \int_A \varphi^n \, dEx = \int_A \varphi \, dEx.$$

The domain of these operators set functions on the basis of (8.68) serves D_T. This proves the fact that

$$TE(A) = E(A)T = \int_A \varphi \, dE. \quad \triangleleft \qquad (8.77)$$

Now let us take two measurable functions $\varphi(t)$ and $\psi(t)$ and the operators which correspond to them

$$T = \int \varphi \, dE, \qquad S = \int \psi \, dE.$$

Theorem 8.8. *The operators T and S which are determined by one and the same decomposition of the identity E are commutative on the intersection of the domains of the operators ST and TS.*

▷ Let $\{\varphi^n(t)\}$ and $\{\psi^n(t)\}$ be the sequences of the simple functions convergent to $\varphi(t)$ and $\psi(t)$ correspondingly,

$$T_n = \int \varphi^n \, dE, \qquad S_n = \int \psi^n \, dE.$$

It is obvious that

$$T_n S_n = S_n T_n = \int \varphi^n \psi^n \, dE.$$

The sequence of these operators determines the operator

$$Q = \lim \int \varphi^n \psi^n \, dE = \int \varphi \psi \, dE$$

with the domain

$$D_Q = \{x : \int |\varphi \psi|^2 d\mu_x < \infty\}.$$

The product of the operators ST has the domain

$$D_{ST} = \{x : x \in D_T, \ Tx \in D_S\}$$
$$= \{x : \int |\varphi|^2 d\mu_x < \infty, \ \int |\psi|^2 d\mu_{Tx} < \infty\}.$$

It is easy to understand that $D_{ST} \subset D_Q$. Really, from (8.77) follows for any $A, B \in \mathcal{A}$

$$\mu_{Tx}(A) = \| E(A)Tx \|^2 = (E(A)Tx, Tx)$$

$$= \int_A \varphi(t)(E(dt)x, Tx) = \int_A \varphi \, d\mu_{x,Tx},$$

$$\mu_{x,Tx}(B) = (E(B)x, Tx) = \overline{(E(B)Tx, x)} = \int_B \bar{\varphi} \, d\mu_x.$$

After performing the change of variables in the first integral (Subsection 3.6.3) we obtain

$$\mu_{Tx}(A) = \int_A \varphi \frac{d\mu_{x,Tx}}{d\mu_x} \, d\mu_x = \int_A |\varphi|^2 \, d\mu_x. \qquad (8.78)$$

Consequently,

$$\int |\psi|^2 \, d\mu_{Tx} = \int |\psi|^2 \frac{d\mu_{Tx}}{d\mu_x} \, d\mu_x = \int |\varphi \psi|^2 \, d\mu_x. \qquad (8.79)$$

This formula shows that if $x \in D_{ST}$ then $x \in D_Q$, i.e. $D_{ST} \subset D_Q$.
    Further for any $x \in D_{ST}$ and any y

$$(STx, y) = \int \psi(dETx, y) = \int \psi \, d\mu_{Tx,y}.$$

But according to (8.77) we have

$$\mu_{Tx,y}(A) = (E(A)Tx, y) = \int_A \varphi(dEx, y) = \int_A \varphi \, d\mu_{x,y}.$$

Consequently,

$$(STx, y) = \int \psi \frac{d\mu_{Tx,y}}{d\mu_{x,y}} \, d\mu_{x,y} = \int \varphi\psi \, d\mu_{x,y} = (Qx, y)$$

or $(STx - Qx, y) = 0$ for any $x \in D_{ST}$ and any y. Thus $STx = Qx$ at
all $x \in D_{ST}$. It means that the operator ST is the contraction of Q on
D_{ST}. By the symmetry we conclude that $D_{TS} \subset D_Q$ and $TSx = Qx$
at all $x \in D_{TS}$, i.e. the operator TS is the contraction of Q on D_{TS}.
Consequently, on the intersection $D_{ST}D_{TS}$ the operators ST and TS
coincide with Q, $ST = TS = Q$. $\triangleleft$
    In the special case of the bounded function $\psi(t)$ the operator S is
bounded and $D_{ST} = D_T \subset D_Q$, $D_{TS} = D_T \subset D_Q$. Thus if even one of
the operators T and S is bounded then at all $x \in D_T$

$$STx = TSx = \int \varphi\psi \, dEx = \int \varphi(t)\psi(t)E(dt)x, \qquad (8.80)$$

i.e. the operators T and S are commutative.

8.2.5. Normal Operators

Let us consider now the case of the finite function $\varphi(t)$ almost
everywhere relatively to E and put $\psi(t) = \overline{\varphi(t)}$, and consequently,
$S = T^*$. By Theorem 8.8 the operators TT^* and T^*T serve the contraction of the operator

$$A = \int |\varphi|^2 \, dE,$$

which has the domain

$$D_A = \{x : \int |\varphi|^4 \, d\mu_x < \infty\}.$$

Theorem 8.9. *The operators T and T^* are commutative and*
$TT^* = T^*T = A$.

▷ It is sufficient to show that $D_{TT^*} = D_{T^*T} = D_A$. But it follows
directly from inequality (3.89) at $r = 2$, $p = q = 4$ in accordance of
which

$$\int |\varphi|^2 \, d\mu_x \leq \sqrt{\mu_x(\Delta) \int |\varphi|^4 \, d\mu_x} = \|x\| \sqrt{\sqrt{\int |\varphi|^4 \, d\mu_x}}.$$

From this inequality and formula (8.79) at $\psi(t) = \overline{\varphi(t)}$ follows that if
$x \in D_A$ then $x \in D_T$ and $Tx \in D_T = D_{T^*}$, i.e. $x \in D_{T^*T}$. It means
that $D_A \subset D_{T^*T}$, and consequently, $D_{T^*T} = D_A$. By the symmetry
$D_{TT^*} = D_A$. ◁

Thus any operator determined by the decomposition of the identity
is commutative with its adjoint one, i.e. is a normal operator (Subsection
7.6.4).

If follows from Theorem 8.9 that all the operators with dense domain
which are determined by the decomposition of the identity refer to the
class of the normal operators.

8.2.6. Inverse Operators

Let us establish the conditions at which the operator T which is
determined by the decomposition of the identity has the inverse operator
T^{-1}.

Theorem 8.10. *In order the operator T which is determined by
the decomposition of the identity has the inverse one it is necessary and
sufficient that the functions $\varphi(t)$ and $1/\varphi(t)$ be almost everewhere finite
relatively to the operator measure E. In this case*

$$T^{-1} = \int \frac{E(dt)}{\varphi(t)} = \int \frac{dE}{\varphi}.$$

▷ If T^{-1} exists then $Tx = 0$ only at $x = 0$. But from $Tx
= 0$ on the basis of formula (8.63) follows that $\int |\varphi|^2 \, d\mu_x = 0$, i.e.
$\mu_x(\{t : \varphi(t) \neq 0\}) = 0$ and $E(\{t : \varphi(t) \neq 0\})x = 0$. To this condition
satifies any vector $x \in G_{\{t:\varphi(t)=0\}}$. Consequently, if $Tx = 0$ only at $x = 0$
then $G_{\{t:\varphi(t)=0\}} = \{0\}$. Hence in consequence of the continuity of the
decomposition of the identity follows

$$E(\{t : \varphi(t) = 0\}) = \lim_{n \to \infty} E\left(\left\{t : \frac{1}{|\varphi(t)|} > n\right\}\right) = 0.$$

Thus the function $1/\varphi(t)$ is finite almost everywhere relatively to E, and consequently, there exists the operator

$$S = \int \frac{dE}{\varphi} = \int \frac{E(dt)}{\varphi(t)}$$

with the dense domain

$$D_S = \left\{ x : \int \frac{d\mu_x}{|\varphi|^2} < \infty \right\}.$$

We shall prove that $S = T^{-1}$. On the basis of Theorem 8.8

$$TSx = \int \varphi \frac{dEx}{\varphi} = \int dEx = x$$

for any $x \in D_{TS}$. Therefore it is sufficient to show that $D_S = R_T = D_{T^{-1}}$, $R_S = D_T = R_{T^{-1}}$. Take any vector $y \in R_T$. Such vector $x \in D_T$ corresponds to it that $y = Tx$. From (8.78) follows that

$$\mu_y(A) = \mu_{Tx}(A) = \int_A |\varphi|^2 \, d\mu_x.$$

Therefore

$$\int \frac{d\mu_y}{|\varphi|^2} = \int \frac{1}{|\varphi|^2} \frac{d\mu_y}{d\mu_x} \, d\mu_x = \int d\mu_x = \|x\|^2 < \infty,$$

i.e. $y \in D_S$. It means that $R_T \subset D_S$. And vice versa, for any vector $y \in D_S$ putting $x = Sy$ we have by virtue of (8.78)

$$\mu_x(A) = \mu_{Sy}(A) = \int_A \frac{d\mu_y}{|\varphi|^2}.$$

Consequently,

$$\int |\varphi|^2 \, d\mu_x = \int |\varphi|^2 \frac{d\mu_x}{d\mu_y} \, d\mu_y = \int d\mu_y = \|y\|^2 < \infty,$$

i.e. $x = Sy \in D_T$. It means that $y \in D_{TS}$ and $Tx = TSy = y$. Thus from $y \in D_S$ follows $y \in R_T$, i.e. $D_S \subset R_T$. From the obtained opposite inclusions follows that $D_S = R_T$. In passing it is proved that

$R_S \subset D_T$. As $y = Tx \in R_T = D_S$ for any $x \in D_T$ in consequence of which $Sy = STx = x$ then $D_T \subset R_S$, and consequently, $R_S = D_T$.

Applying the same arguments to the operator $S = T^{-1}$ we come to the conclusion that the function $\varphi(t)$ is finite almost everywhere relatively to E.

And vice versa if the functions $\varphi(t)$ and $1/\varphi(t)$ are almost everywhere finite relatively to E then from the previous statements follows that the operator T has the inverse one $T^{-1} = S$. ◁

Corollary. *The resolvent R_λ of the operator T with the dense domain which is determined by the decomposition of the identity is expressed by formula*

$$R_\lambda = (T - \lambda I)^{-1} = \int \frac{dE}{\varphi - \lambda} = \int \frac{E(dt)}{\varphi(t) - \lambda}. \qquad (8.81)$$

▷ Really, as

$$T - \lambda I = \int \varphi \, dE - \lambda I = \int (\varphi - \lambda) \, dE,$$

then from the proved theorem follows formula (8.81). ◁

8.2.7. Spectrum

Let us study now the spectrum of the operator determined by the decompostion of the identity.

Theorem 8.11. *The value λ is the eigenvalue of the operator T determined by the decomposition of the identity if and only if $G_{\{t:\varphi(t)=\lambda\}} \neq \{0\}$. In this case $G_{\{t:\varphi(t)=\lambda\}}$ is the corresponding proper subspace.*

▷ If the value λ is the eigenvalue of T and x is the corresponding eigenvector then $Tx = \lambda x$, and consequently,

$$Tx - \lambda x = \int \varphi \, dEx - \lambda x = \int (\varphi - \lambda) \, dEx = 0.$$

Formula (8.63) shows that in this case

$$\int |\varphi - \lambda|^2 \, d\mu_x = 0$$

and it means

$$\mu_x(\{t : \varphi(t) \neq \lambda\}) = \| E(\{t : \varphi(t) \neq \lambda\})x \|^2 = 0. \qquad (8.82)$$

Consequently, $x \in G_{\{t:\varphi(t)=\lambda\}}$ and $G_{\{t:\varphi(t)=\lambda\}} \neq \{0\}$, $E(\{t : \varphi(t) = \lambda\}) \neq 0$. And vice versa if $G_{\{t:\varphi(t)=\lambda\}} \neq \{0\}$ then for any vector $x \in G_{\{t:\varphi(t)=\lambda\}}$ formula (8.28) is valid and for none $x \notin G_{\{t:\varphi(t)=\lambda\}}$ it can be valid. ◁

Theorem 8.12. *The value λ is the regular point of the operator T determined by the decomposition of the identity if and only if there exists such $\varepsilon > 0$ that $G_{\{t:|\varphi(t)-\lambda|<\varepsilon\}} = \{0\}$.*

▷ If λ is a regular point of T then the resolvent R_λ is bounded and is determined on the whole space X. On the basis of formula (8.81) and Theorem 8.4 in this case the function $[\varphi(t) - \lambda]^{-1}$ is bounded almost everywhere relatively to E, i.e. there exists such $\varepsilon > 0$ that $E(\{t : |\varphi(t) - \lambda| < \varepsilon\}) = 0$, $G_{\{t:|\varphi(t)-\lambda|<\varepsilon\}} = \{0\}$. And vice versa, if there exists such $\varepsilon > 0$ that $G_{\{t:|\varphi(t)-\lambda|<\varepsilon\}} = \{0\}$ then the function $[\varphi(t)-\lambda]^{-1}$ is bounded and formula (8.81) determines on the whole space X the bounded operator $R_\lambda = (T - \lambda I)^{-1}$. Consequently, the value λ is the regular point. ◁

Corollary. *The value λ is the point of the spectrum of the operator T determined by the decomposition of the identity different from the eigenvalues if and only if $G_{\{t:\varphi(t)=\lambda\}} = \{0\}$ and $G_{\{t:|\varphi(t)-\lambda|<\varepsilon\}} \neq \{0\}$ at any $\varepsilon > 0$. Here if the function $[\varphi(t) - \lambda]^{-1}$ is almost everywhere finite relatively to E then accordingly to Theorem 8.5 and formula (8.81) the domain of the resolvent R_λ which coincides with the range $R_T(\lambda)$ of the operator $T - \lambda I$ is dense in X and λ is the point of a continuous spectrum. In the opposite case λ is the point of discrete spectrum.*

R e m a r k. At the given function $\varphi(t)$ the decomposition of the identity determines completely the spectrum of the operator T generated by this decomposition of the identity.

8.2.8. Canonical Forms of Operators

On the basis of Theorem 8.6 for any real measurable almost everywhere finite relatively to E the function $\varphi(t)$ the decomposition of the identity E determines the self–adjoint operator

$$A = \int \varphi(t)E(dt) = \int \varphi \, dE.$$

After performing the change of variables $s = \varphi(t)$ we determine on the σ–algebra $\mathcal{B}$ of Borel sets of the real axis R the decomposition of the identity $F(B) = E(\varphi^{-1}(B))$, $B \in \mathcal{B}$. Hence the formula which determines the operator A takes the form

$$A = \int_{-\infty}^{\infty} sF(ds). \qquad (8.83)$$

Thus any self–adjoint operator A which is determined by the decomposition of the identity is reduced to the form of (8.83). Therefore formula (8.83) gives *a canonical form* of a self–adjoint operator determined by the decomposition of the identity.

In the special case of the bounded operator A all the values λ for which $|\lambda| > \|A\|$ on the basis of Theorem 6.29 are the regular points. Therefore in consequence of Theorem 8.12 the decomposition of the identity F is different from zero only on the finite interval (a, b), $a = \inf \varphi(t)$, $b = \sup \varphi(t)$, $a \geq -\|A\|$, $b \leq \|A\|$ and formula (8.83) takes the form

$$A = \int_{a}^{b} sF(ds). \qquad (8.84)$$

If the operator determined by the decomposition of the identity is unitary $T = U$ then on the basis of Theorem 6.34 its spectrum entirely lies on the unit circumference of the complex plane. But as any regular point λ has the vicinity whose inverse image $\{t : |\varphi(t) - \lambda| < \varepsilon\}$ has zero operator measure E then $G_A = \{0\}$ and $E(A) = 0$ for any set $A \in \mathcal{A}$ which does not intersect with $\{t : |\varphi(t)| = 1\}$. Thus $|\varphi(t)| = 1$ almost everywhere relatively to E. In consequence of this fact we may put $\varphi(t) = e^{i\psi(t)}$ where $\psi(t)$ is a measurable real function. Then the unitary operator U determined by the decomposition of the identity will be expressed by the formula

$$U = \int e^{i\psi(t)} E(dt) = \int e^{i\psi} \, dE.$$

It is easy to see that also vice versa, any operator determined by this formula is unitary. Really, from (8.63) follows that for any x

$$\|Ux\|^2 = \int |e^{i\psi}|^2 \, d\mu_x = \mu_x(\Delta) = \|x\|^2 \ .$$

Further, as the functins $e^{i\psi(t)}$ and $e^{-i\psi(t)}$ are finite then there exists an inverse operator

$$U^{-1} = \int e^{-i\psi(t)} E(dt) = \int e^{-i\psi} \, dE.$$

As this operator is determined on the whole space X then $R_U = X$. Thus the operator U maps X on the whole space X with the retention of the norm. Consequently, it is unitary.

After performing the change of variables $s = \psi(t)$ we determine on the σ-algebra $\mathcal{B}$ of Borel sets of the real axis the decomposition of the identity $F(B) = E(\psi^{-1}(B))$, $B \in \mathcal{B}$. Hence the formula which determines the operator U will take the form

$$U = \int_{-\infty}^{\infty} e^{is} F(ds). \tag{8.85}$$

Now we notice that in consequence of the periodicity of the function e^{is} formula (8.85) may be transformed into the form

$$U = \int_{0}^{2\pi} e^{is} F_1(ds), \tag{8.86}$$

where $F_1(B)$ is the decomposition of the identity determined on the σ-algebra $\mathcal{B}_1$ of Borel sets of the interval $[0, 2\pi]$ by the formula

$$F_1(B) = \sum_{k=-\infty}^{\infty} F(B + 2k\pi), \quad B \in \mathcal{B}_1.$$

The series in this formula converges by virtue of Theorem 7.23 about the monotone sequences of the projectors.

Thus any unitary operator determined by the decomposition of the identity is reduced to the form of (8.85) and (8.86). Therefore formulae (8.85) and (8.86) give a *canonical* form of an unitary operator determined by the decomposition of the identity.

For any operator T generated by the decomposition of the identity E the change of variables $z = \varphi(t)$ determines on the σ-algebra $\mathcal{B}$ of Borel sets of a complex plane the decomposition of the identity $F(B) = E(\varphi^{-1}(B))$, $B \in \mathcal{B}$, and formula (8.67) takes the form

$$T = \int z F(ds) = \iint (u + iv) F(du\, dv), \tag{8.87}$$

where the integration is extended over the whole complex plane. In the special case when the function $\varphi(t)$ is bounded, i.e. R_φ represents a bounded set of the complex numbers, the operator measure F is equal to zero everywhere outside R_φ and integral (8.37) is reduced to the integral over the region R_φ.

So, any operator which is determined by the decomposition of the identity is reduced to the form of (8.67). By Theorem 8.9 this operator is normal. Therefore formula (8.87) gives *a canonical form* of a normal operator determined by the decomposition of the identity.

E x a m p l e 8.5. Let $E(B)$ be the decomposition of the identity determined by a compact self–adjoint operator constructed in Subsection 8.1.6. As for any point λ_n the subspace $G_{\lambda_n} = G_{\{t:t=\lambda_n\}}$ is different from $\{0\}$, $G_{\{t:t=\lambda_n\}} \neq \{0\}$ then all the points λ_n are the eigenvalues. As for any interval (a, b) which does not contain the eigenvalues $G_{(a,b)} = \{0\}$ then any real λ which is not the eigenvalue similary as any λ with the imaginary part different from zero is a regular point. Therefore the spectrum of the operator A consists of only the eigenvalues and the point 0. Thus all the results obtained in Subsection 8.1.6 follow from the more general theory just constructed.

E x a m p l e 8.6. Let $f(s)$ be a monotone mapping of the real axis R into itself, G_B be the subspace of the functions from $L_2(R)$ equal to zero outside the set $f^{-1}(B)$, $E(B)$ be the decomposition of the identity which represents a projector on G_B for any Borel set B. Then for any function $x = x(s) \in L_2(R)$

$$E(B)x = x(s)\mathbf{1}_{f^{-1}(B)}(s). \qquad (*)$$

Consequently, the operator $E(B)$ represents an operator of the multiplication by the indicator of the set $f^{-1}(B)$.

Let us consider a self–adjoint operator

$$A = \int\limits_{-\infty}^{\infty} tE(dt) = \int\limits_{-\infty}^{\infty} t\mathbf{1}_{f^{-1}([t,t+dt))}(s).$$

It is apparemt that for any function $x(s) \in D_A$

$$Ax = \int\limits_{-\infty}^{\infty} t\mathbf{1}_{f^{-1}([t,t+dt))}(s)x(s) = f(s)x(s),$$

as the function $\mathbf{1}_{f^{-1}([t,t+\varepsilon))}(s)$ is different from zero at the infinitesimal $\varepsilon > 0$ only at $s = f^{-1}(t)$, i.e. at $t = f(s)$. Thus the operator A represents an operator of the multiplication by the function $f(s)$. In particular, at $f(s) = s$ the operator A represents an operator of the multiplication by an independent variable.

If the function $f(s)$ is constant on the interval (a, b), $f(s) = c$ then $f^{-1}(c) = (a, b)$ and $E(\{c\}) = \mathbf{1}_{[a,b]}(s) \neq 0$. Consequently, $\lambda = c$ is the eigenvalue and any function $x(s)$ equal to zero outside the interval (a, b) is the corresponding eigenvector. By virtue of the separability of $L_2(R)$ (Subsection 3.7.5) the subspace $G_{\{c\}} = L_2((a, b))$ contains a countable set of linearly independent functions. Therefore the eigenvalue $\lambda = c$ has an infinite multiplicity.

If the function $f(s)$ has no values in the interval (α, β), for instance, has the discontinuity of the first kind at some point s_0 and $f(s_0 - 0) = \alpha$, $f(s_0 + 0) = \beta$ then all the points of the interval (α, β) are regular, as $f^{-1}((\alpha', \beta')) = \emptyset$ at any $\alpha' > \alpha$, $\beta' < \beta$. Consequently, for any $\lambda \in (\alpha, \beta)$ we may find such $\varepsilon > 0$ that

$$f^{-1}(\{t : |t - \lambda| < \varepsilon\}) = \emptyset \text{ and } E(\{t : |t - \lambda| < \varepsilon\}) = 0.$$

If the function $f(s)$ is continuous on the interval (a, b) and in none part of this interval is constant then at any $s \in (a, b)$ the point $\lambda = f(s)$ is not an eigenvalue. But it belongs to the spectrum of the operator A as

$$f^{-1}(\{t : |t - f(s)| < \varepsilon\}) \neq \emptyset \text{ и } E(\{t : |t - f(s)| < \varepsilon\}) \neq 0$$

at $s \in (a, b)$ for all $\varepsilon > 0$. In the correspondence with the general theory the point $\lambda = f(s)$ is the point of the continuous spectrum if the function $\psi(t) = [t - f(s)]^{-1}$ is almost everywhere finite relatively to the measure E determined by formula $(*)$ and by the point of a discrete spectrum in the opposite case.

For details concerning operators determined by decomposition of the identity address (Dowson 1978, Dunford and Schwarz 1963).

Problems

8.2.1. Prove that the inverse image of the resolvent set $\rho(T)$ of the operator T determined by formula (8.67) and which is given by the function $\varphi(t)$ has zero measure $E : E(\varphi^{-1}(\rho(T))) = 0$. Correspondingly, $E(\varphi^{-1}(\sigma(T))) = I$. In particular, if the space Δ represents the complex plane C and $\varphi(t) = t$, $E(\rho(T)) = 0$, $E(\sigma(T)) = I$, i.e. the operator measure E is completely concentrated on the spectrum of the operator T.

I n s t r u c t i o n. Theorem 8.12 establishes the existence of the vicinity of any regular point whose inverse image has zero measure E. The complex plane C as the separable metric space has a countable basis.

8.2.2. As $E(A)x = x$ for any $x \in G_A$ then the right–hand side of formula (8.77) determines the contraction of the operator T on the subspace $G_A = E(A)X$. Using this fact and accounting the latter result of Problem 8.2.1 prove that the spectrum of the contraction $T \mid G_A$ of the operator T on the subspace G_A satisfies the relation $\sigma(T \mid G_A) \subset [A]$.

8.2.3. The decomposition of the identity may be determined in any B–space X. Certainly, the norms of the projectors $E(A)$, $A \in \mathcal{A}$ in this case will not be equal to 1 and to the definition of the operator measure E in Subsection 8.1.6 we have to add the requirement of the uniform boundedness of the norms of the projectors $E(A) : \| E(A) \| < M \ \forall A \in \mathcal{A}$. Prove that in this case $E(A)^*$ represents the decomposition of the identity in the dual space X^* therewith $\| E(A)^* \| < M \ \forall A \in \mathcal{A}$. Analogously as in the case of the H–space X the operator measure E generates a set of the numeric measures $\mu_{x,f}(A) = fE(A)x, \ \forall x \in X, \ f \in X^*$. These measures are finite as $| \mu_{x,f}(A) | \leq M \| f \| \| x \|$. But $fE(A)x = (E(A)^* f)x$. Therefore the same measures are generated by the decomposition of the identity E^*.

8.2.4. Determine an integral of a simple function over an operator measure E in the B–space X and prove that it represents a bounded linear operator.

I n s t r u c t i o n. An integral of a simple function is determined by the same formulae (8.60) and (8.61) as in the case of the H–space X. But to prove the boundedness of the obtained operator in the case of the B–space X is more difficult. For this purpose we have to use the inequality

$$\left\| \sum_{k=1}^{n} a_k E(A_k) \right\| < 4M \sup_k | a_k |. \qquad (*)$$

For deriving this inequality we put $a_k = \alpha_k + i\beta_k$ and notice that the function

$$\varphi(u) = \left\| \sum_{k=1}^{n} u_k E(A_k) \right\|, \quad u_1, \ldots, u_n \in R,$$

is a continuous convex linear functional of the vector $u \in R^n$ (Subsection 1.4.6) as a result of which at any $t \in (0, 1)$ we have inequalities

$$\varphi(tu + (1 - t)v) \leq t\varphi(u) + (1 - t)\varphi(v) \leq \max(\varphi(u), \varphi(v)). \qquad (**)$$

Therefore $\varphi(u)$ achieves the maximum on n–dimensional cube $| u_k | \leq 1$ ($k = 1, \ldots, n$) at one of its tops $u_1, \ldots, u_n = \pm 1$. The supposition that the point of the maximum lies inside the cube or on its bound leads to the contradiction with inequality $(**)$. After combining all the items of the sum corresponding to u_k equal to $+1$ we get $\sum u_k E(A_k) = \sum E(A_k) = E(\bigcup A_k)$ and $\| \sum u_k E(A_k) \| \leq M$. And the same inequality we shall obtain for the part of the sum corresponding to u_k equal to -1. Consequently, $\varphi(u) < 2M$ at $| u_1 |, \ldots, | u_n | \leq 1$. Applying this inequality to the cases $u_k = \alpha_k / \max | a_l |$ and $u_k = \beta_k / \max | a_l |$ we receive inequality $(*)$.

8.2.5. Determine an integral over the operator measure E in the B–space X of the measurable function bounded almost everywhere relatively to the measure E and prove that it represents a bounded linear operator.

8.2.6. Let T be an operator determined by the decomposition of the identity in the B–space X:

$$T = \int \varphi(t)E(dt).$$

Prove that an adjoint operator is determined by the formula

$$T^* = \int \varphi(t) E(dt)^*.$$

Explain the discrepancy of this formula with formula (8.72) for the case of the H–space X.

8.3. Functions of Operators

8.3.1. Operator Polynomials and Series

The degrees of an operator in any linear space are determined naturally as the products $T^2 = TT$, $T^n = T^{n-1}T$ $(n = 2, 3, \ldots)$. The linear combinations of the degrees determine the following polynomials of the operator T:

$$p_n(T) = a_0 + a_1 T + \cdots + a_n T^n. \tag{8.88}$$

For the bounded linear operators in the normed linear space X we may determine the analytical functions of the operators. Let $f(z)$ be an analytical function of the complex variable z which has no singular points in some region containing the origin. Such function is representable in the vicinity of the origin by the power series

$$f(z) = \sum_{k=0}^{\infty} a_k z^k.$$

Then for any linear operator T whose norm is smaller than the radius of the series convergence we may determine the function

$$f(T) = \sum_{k=0}^{\infty} a_k T^k. \tag{8.89}$$

This series converges in the uniform topology of the space $\mathcal{B}(X)$ of the bounded linear operators acting in X (Subsection 5.1.4).

If the operator T is determined by the decomposition of the identity E then in the correspondence with the latter result of Subsection 8.2.4 formulae (8.88) and (8.89) take the form

$$p_n(T) = \int p_n(\varphi(t)) E(dt), \quad f(T) = \int f(\varphi(t)) E(dt).$$

In particular, for the bounded self–adjoint operator A formula (8.83) gives

$$f(A) = \int f(s)F(ds). \tag{8.90}$$

But for constructing the spectral theory of the linear operators it is necessary to extend the class of the functions of the operators.

8.3.2. Polynomials of an Unitary Operator

As any unitary operator U is determined on the whole H–space and has the inverse operator U^{-1} which is also determined on the whole H–space then the entire degrees of the unitary operator and the polynomials are naturally determined by

$$p(U) = \sum_{k=m}^{n} c_k U^k, \tag{8.91}$$

where m, n are any integers, and $c_m, \ldots, c_n$ are any complex numbers.

It is apparent that $p(U)$ represents a bounded linear operator and therefore there exists the adjoint operator $p(U)^*$. As $U^* = U^{-1}$ (Subsection 7.2.2) then at any x and y

$$(p(U)x, y) = \sum_{k=m}^{n} c_k(U^k x, y) = \sum_{k=m}^{n} c_k(x, U^{-k}y) = \left(x, \sum_{k=m}^{n} \bar{c}_k U^{-k}y\right).$$

Hence it is clear that the adjoint operator $p(U)^*$ is determined by formula

$$p(U)^* = \sum_{k=m}^{n} \bar{c}_k U^{-k}. \tag{8.92}$$

This result suggests that there exists a definite correspondence between the polynomials of an unitary operator U and the trigonometrical polynomials, i.e. the polynomials of e^{is}, $s \in [0, 2\pi]$. So, we have

$$p(e^{is}) = \sum_{k=m}^{n} c_k e^{iks}, \quad \overline{p(e^{is})} = \sum_{k=m}^{n} \bar{c}_k e^{-iks}.$$

This correspondence we shall use while defining the functions of the unitary operator U. At first we shall study this correspondence in detail.

It is evident that if the polynomial $p(e^{is})$ represents a sum or a product of the polynomials $p_1(e^{is})$ and $p_2(e^{is})$ then the operator polynomial $p(U)$ is the sum or the product of the operator polynomials $p_1(U)$ and $p_2(U)$:

$$p(e^{is}) = p_1(e^{is}) + p_2(e^{is}) \sim p(U) = p_1(U) + p_2(U), \qquad (8.93)$$

$$p(e^{is}) = p_1(e^{is})p_2(e^{is}) \sim p(U) = p_1(U)p_2(U). \qquad (8.94)$$

It is clear from (8.91) and (8.92) that the operator $p(U)$ is a self–adjoint operator $p(U)^* = p(U)$ if and only if $m = -n$, $\bar{c}_k = c_{-k}$, i.e. when the polynomial $p(e^{is})$ has the real values at all s, $\overline{p(e^{is})} = p(e^{is})$.

Now we suppose that the polynimial $p(e^{is})$ is nonnegative $p(e^{is}) \geq 0$. In this case there exists such polynomial $q(e^{is})$ that $p(e^{is}) = |q(e^{is})|^2 = q(e^{is})\overline{q(e^{is})}$ [a]. To this relation corresponds the opera-

---

[a] Really, as in this case for any complex z

$$p(\bar{z}^{-1}) = \sum_{k=-n}^{n} c_k \bar{z}^{-k} = \sum_{k=-n}^{n} \bar{c}_k z^k = \overline{p(z)},$$

then to each zero z_p of the polynomial $p(z)$ corresponds zero $\bar{z}_p^{-1}$ located symmetrically to z_p relatively to the unit circumference. Therefore the algebraic polynomials $z^n p(z)$ and

$$p_1(z) = \left(1 - \frac{z}{z_1}\right) \cdots \left(1 - \frac{z}{z_n}\right)\left(z - \frac{1}{\bar{z}_1}\right) \cdots \left(z - \frac{1}{\bar{z}_n}\right)$$

have the same zeros, and consequently, their ratio is constant. Hence follows the equality

$$p(z) = az^{-n}p_1(z) = a\left(1 - \frac{z}{z_1}\right) \cdots \left(1 - \frac{z}{z_n}\right)\left(1 - \frac{z^{-1}}{\bar{z}_1}\right) \cdots \left(1 - \frac{z^{-1}}{\bar{z}_n}\right).$$

From the inequality $p(e^{is}) \geq 0$ follows that the constant a is positive. Therefore putting

$$q(z) = \sqrt{a}\left(1 - \frac{z}{z_1}\right) \cdots \left(1 - \frac{z}{z_n}\right),$$

we shall obtained the required equality:

$$p(e^{is}) = q(e^{is})\overline{q(e^{is})} = |q(e^{is})|^2.$$

tor relation $p(U) = q(U)q(U)^* = q(U)^*q(U)$. Therefore for any x we have

$$(p(U)x, x) = (q(U)^*q(U)x, x) = (q(U)x, q(U)x) = \|q(U)x\|^2 \geq 0.$$

Consequently, to the nonnegative trigonometrical polynomial $p(e^{is})$ corresponds the nonnegative operator polynomial $p(U)$. Hence it follows that if $p_1(e^{is}) \leq p_2(e^{is})$ then $p_1(U) \leq p_2(U)$ [a]. In the special case at $p_1(e^{is}) = p(e^{is})$, $p_2(e^{is}) = c$ to the inequality $p(e^{is}) \leq c$ corresponds the operator inequality $p(U) \leq cI$. As the inequality $|p(e^{is})| \leq c$ is equivalent to $p(e^{is})\overline{p(e^{is})} = |p(e^{is})|^2 \leq c^2$ then from $|p(e^{is})| \leq c$ follows the operator inequality $p(U)^*p(U) \leq c^2 I$. Hence it follows that at any x

$$\|p(U)x\|^2 = (p(U)^*p(U)x, x) \leq (c^2 Ix, x) = c^2(x, x) = c^2\|x\|^2$$

and $\|p(U)\| \leq c$. Thus

$$|p(e^{is})| \leq c \sim \|p(U)\| \leq c. \tag{8.95}$$

8.3.3. Functions of an Unitary Operator

Let us consider the class C_0 of real bounded functions of the variable e^{is} each of them is continuous or represents the limit of the nonincreasing sequence of the continuous functions. Any function of this class may be presented as the limit of nonincreasing sequence of the trigonometric polynomials. Really, if $f(e^{is})$ is continuous then on the basis of the known Weierstrass theorem it may be uniformly approximated with any degree of the accuracy by a trigonometric polynomial on the interval $[0, 2\pi]$. Therefore after taking an arbitrary sequence of the positive numbers $\{\varepsilon_n\}$ convergent to zero and constructing for each ε_n the polynomial $p_n(e^{is})$ which approximates the function $f(e^{is}) + (\varepsilon_n + \varepsilon_{n+1})/2$ with the accuracy till $(\varepsilon_n - \varepsilon_{n+1})/2$ we receive at each n

$$\varepsilon_{n+1} \leq p_n(e^{is}) - f(e^{is}) \leq \varepsilon_n.$$

---

[a] According to the definition of Subsection 6.1.5 the inequality between the operators $A \leq B$ in the case of the H–space X should be assumed as the inequality $(Ax, x) \leq (Bx, x)$ at all x.

Hence and from the same inequality obtained by the substitution of n by $n-1$ follows

$$p_n(e^{is}) \le f(e^{is}) + \varepsilon_n \le p_{n-1}(e^{is}).$$

Thus we obtained the nonincreasing sequence of the trigonometric polynomials $\{p_n(e^{is})\}$ convergent to $f(e^{is})$ uniformly on the interval $[0, 2\pi]$. If $f(e^{is}) = \lim f_n(e^{is})$ where $\{f_n(e^{is})\}$ is nonincreasing sequence of the continuous functions then after constructing for each ε_n the polynomial $p_n(e^{is})$ which approximates the function $f_n(e^{is}) + (\varepsilon_n + \varepsilon_{n+1})/2$ with the accuracy till $(\varepsilon_n - \varepsilon_{n+1})/2$ we get at each n

$$\varepsilon_{n+1} \le p_n(e^{is}) - f_n(e^{is}) \le \varepsilon_n.$$

Hence and from the same inequality obtained by the substitution of n by $n-1$ follows

$$p_n(e^{is}) \le f_n(e^{is}) + \varepsilon_n \le f_{n-1}(e^{is}) + \varepsilon_n \le p_{n-1}(e^{is}).$$

Thus $\{p_n(e^{is})\}$ represents a nonincreasing sequence of the trigonometric polynomials convergent to $f(e^{is})$ at each point $s \in [0, 2\pi]$.

Let us consider now the corresponding decreasing sequence of the operator polynomials $\{p_n(U)\}$. As $p_n(e^{is}) \ge f(e^{is}) \ge \gamma = \inf f(e^{is})$ then $p_n(U) \ge \gamma I$ at any n. Thus $(p_n(U)x, x) \ge (\gamma I x, x) = \gamma \|x\|^2$ at any n, i.e. the nonincreasing sequence $\{(p_n(U)x, x)\}$ is lower bounded by the number $\gamma \|x\|^2$, and consequently, converges. As $p_n(U) - p_m(U) \ge 0$ at any n and $m > n$ then the function $(p_n(U)x - p_m(U)x, y)$ of the variables $x, y \in X$ possesses all the properties of the scalar product besides the property $(x, x) = 0$ only at $x = 0$. Therefore on the basis of inequality (1.17) we get

$$|(p_n(U)x - p_m(U)x, y)|^2 \le (p_n(U)x - p_m(U)x, x)(p_n(U)y$$

$$-p_m(U)y, y) \le (p_n(U)x - p_m(U)x, x)(p_1(U)y, y)$$

$$\le [(p_n(U)x, x) - (p_m(U)x, x)] \, \|p_1(U)\| \, \|y\|^2 \, .$$

By virtue of the convergence of $(p_n(U)x, x)$ $(p_n(U)x, x)$ $- (p_m(U)x, x) < \varepsilon^2 / \, \|p_1(U)\|$ at any $\varepsilon > 0$ at all sufficiently large n and $m > n$. Therefore

$$|(p_n(U)x - p_m(U)x, y)|^2 < \varepsilon^2 \, \|y\|^2$$

at sufficiently large n and all y and $m > n$. Putting here $y = p_n(U)x - p_m(U)x$ we shall obtain after the reduction $\| p_n(U)x - p_m(U)x \| < \varepsilon$. Thus the sequence $\{p_n(U)x\}$ is fundamental and due to the completeness of the H–space has the limit (certainly dependent on x) $\lim p_n(U)x = Tx$. Consequently, the sequence of the operator polynomials $\{p_n(U)\}$ converges to some operator T. This operator dependent on U naturally may be assumed as the function $f(U)$ of the unitary operator U,

$$f(U) = T = \lim p_n(U). \tag{8.96}$$

In order the definiton of the function $f(U)$ be correct it remains to show that the operator T does not depend on the choice of the sequence of the polynomials $\{p_n(e^{is})\}$ convergent to $f(e^{is})$. Taking another non–increasing sequence of the polynomials $\{q_n(e^{is})\}$ convergent to $f(e^{is})$ we put $S = \lim q_n(U)$. In consequence of the monotone convergence of the sequences $\{p_n(e^{is})\}$ and $\{q_n(e^{is})\}$ to one and the same function $f(e^{is})$ at any n and at any point $s \in [0, 2\pi]$ for sufficiently large m the inequality $q_m(e^{is}) < p_n(e^{is}) + \varepsilon_n$ will be fulfilled. By virtue of the continuity of the polynomials q_m and p_n this inequality will be also fulfilled in the vicinity of the point s. Owing to the compactness of the interval $[0, 2\pi]$ there exists a finite number of such vicinities (corresponding to different s) which cover the whole interval $[0, 2\pi]$. Therefore at all sufficiently large m the inequality $q_m(e^{is}) < p_n(e^{is}) + \varepsilon_n$ will be fulfilled at all $s \in [0, 2\pi]$ for any natural n. Similarly at any m and sufficiently large l the inequality $p_l(e^{is}) < q_m(e^{is}) + \varepsilon_m$ will be fulfilled at all $s \in [0, 2\pi]$. From these inequalities follows that at any n and for any sufficiently large m and l

$$p_l(e^{is}) - \varepsilon_m < q_m(e^{is}) < p_n(e^{is}) + \varepsilon_n,$$

i.e.

$$0 < p_n(e^{is}) - q_m(e^{is}) + \varepsilon_n < p_n(e^{is}) - p_l(e^{is}) + \varepsilon_n + \varepsilon_m$$

everywhere on the interval $[0, 2\pi]$. To these inequalities correspond the operator inequalities

$$0 < p_n(U) - q_m(U) + \varepsilon_n I < p_n(U) - p_l(U) + (\varepsilon_n + \varepsilon_m)I.$$

Hence in exactly the same way as we proved the fundamental property of the sequence $\{p_n(U)x\}$ we come to the conclusion that for any $\varepsilon > 0$

and any given vector $x \in X$ at all sufficiently large n and m will be $\|p_n(U)x - q_m(U)x\| < \varepsilon$. Noticing that at any n and m

$$\|Tx - Sx\| \leq \|Tx - p_n(U)x\| + \|p_n(U)x - q_m(U)x\| + \|q_m(U)x - Sx\|,$$

we may choose for any $\varepsilon > 0$ at first n and further m in such a way that

$$\|Tx - p_n(U)x\| < \varepsilon/3, \quad \|p_n(U)x - q_m(U)x\| < \varepsilon/3,$$

$$\|q_m(U)x - Sx\| < \varepsilon/3.$$

Then we get $\|Tx - Sx\| < \varepsilon$. Hence by virtue of the arbitrariness of $\varepsilon > 0$ follows that $T = S$ what proves the uniqueness of the limit in the definition of the function $f(U)$.

The stated definition of an operator function is easily extended over any linear combinations of the functions of the class C_0. It is sufficient for each linear combination

$$f(e^{is}) = \sum_{k=1}^{n} c_k f_k(e^{is})$$

of the functions $f_1, \ldots, f_n$ of the class C_0 to establish the correspondence with the operator function

$$f(U) = \sum_{k=1}^{n} c_k f_k(U).$$

In particular, the operator functions $f(U)$ corresponding to the complex $f(e^{is})$ are determined in such a way.

It is easy to see that for the operator functions and the corresponding functions of the variable e^{is} determined in such a way all the relations by which the polynomials of an unitary operator with the trigonometric polynomials are connected retain their validity. In particular, correspondences (8.93), (8.94) and (8.95) are valid:

$$f(e^{is}) = f_1(e^{is}) + f_2(e^{is}) \sim f(U) = f_1(U) + f_2(U), \qquad (8.97)$$

$$f(e^{is}) = f_1(e^{is})f_2(e^{is}) \sim f(U) = f_1(U)f_2(U), \qquad (8.98)$$

$$|f(e^{is})| \leq c \sim \|f(U)\| \leq c. \qquad (8.99)$$

It is also valid that to real function $f(e^{is})$ corresponds the self–adjoint operator $f(U)$.

Thus the functions of an unitary operator are determined for all the functions of the class C which represent a set of finite linear combinations of the functions of the class C_0. It is clear that the class C contains the limits of the nondecreasing sequences of the continuous functions.

It follows from relation (8.98) that any functions of one and the same unitary operator are commutative.

R e m a r k. Analogously the functions of two (or larger number) commutative unitary operators U_1, U_2 are determined. In this case to the trigonometric polynomial

$$p(e^{is}, e^{it}) = \sum_{k,l} a_{kl} e^{iks+ilt}$$

corresponds the operator polynomial

$$p(U_1, U_2) = \sum_{k,l} a_{kl} U_1^k U_2^l.$$

As any degrees of the commutative operators are commutative relations (8.93)–(8.95) and (8.97)–(8.99) together with them retain their validity. From relations (8.99) for two operators

$$f(e^{is}, e^{it}) = f_1(e^{is}, e^{it}) f_2(e^{is}, e^{it}) \sim f(U_1, U_2) = f_1(U_1, U_2) f_2(U_1, U_2),$$

in the special case when f_1 depends only on the first argument, and f_2 depends only on the second one follows that any functions $f_1(U_1)$ and $f_2(U_2)$ of the commutative unitary operators U_1 and U_2 are commutative.

E x a m p l e 8.7. The function $f(e^{is}) = \ln e^{is} = is$ is continuous. Therefore it determines the function $T = f(U) = \ln U$ of the unitary operator U. As $\overline{\ln e^{is}} = -is = -\ln e^{is}$ then the adjoint operator $(\ln U)^*$ is equal $-\ln U$, $T^* = -\ln U = -T$. So, as $|\ln e^{is}| \le 2\pi$ in the interval $[0, 2\pi]$ then $\ln U$ is a bounded linear operator.

E x a m p l e 8.8. If the unitary operator U is determined by some decomposition of the identity E,

$$U = \int_0^{2\pi} e^{it} E(dt),$$

then any function $f(U)$ of the considered class is expressed in terms of the corresponding function $f(e^{is})$ by formula

$$f(U) = \int_0^{2\pi} f(e^{it}) E(dt).$$

Really, this equality is apparent for the case when $f(e^{is})$ is a polynomial. For any function $f(e^{is})$ of the considered class C it is valid by virtue of Theorem 3.39 about the convergence of the sequences of the functions from L_2 majorized by some function from L_2 as any such function represents a linear combination of the limits of nonincreasing sequences of the polynomials.

8.3.4. Cayley Transform

Let A be an arbitrary self–adjoint operator. As the numbers i and $-i$ cannot be the eigenvalues of A then the ranges $R_A(i)$ and $R_A(-i)$ of the operators $A - iI$ and $A + iI$ coincide with the whole space X, $R_A(i) = R_A(-i) = X$ (Subsection 7.3.3). Therefore each of the formulae

$$y = (A + iI)x, \quad z = (A - iI)x \qquad (8.100)$$

gives the mapping D_A on the whole X. Owing to the existence of the bounded resolvent $R_{-i} = (A + iI)^{-1}$ determined on the whole X to any $y \in X$ also corresponds such $x \in D_A$ that

$$x = R_{-i}y = (A + iI)^{-1}y.$$

Consequently, formula

$$z = (A - iI)(A + iI)^{-1}y$$

determines the mapping of the space X on the whole space X. Hence formulae (8.100) give at any $x \in D_A$

$$\|y\|^2 = \|z\|^2 = \|Ax\|^2 + \|x\|^2.$$

Thus, the operator
$$U = (A - iI)(A + iI)^{-1} \qquad (8.101)$$

is linear, maps the space X on the whole space X and retains the norm. Consequently, U is an unitary operator.

The unitary operator U determined by formula (8.101) is called the *Cayley transform* of the self–adjoint operator A.

From (8.100) and the equality $z = Uy$ follows

$$x = \frac{y - z}{2i} = \frac{1}{2i}(I - U)y, \quad Ax = \frac{y + z}{2} = \frac{1}{2}(I + U)y. \qquad (8.102)$$

Comparing the first formula from these formulae with the first one of of (8.100) we see that the operators $A + iI$ and $(I - U)/2i$ are mutually inverse or reciprocal. In other words, the first equation of Eqs. (8.102) has the unique solution relatively to y which is determined by the first formula of (8.100). Consequently, 1 cannot be the eigenvalue of the Cayley transform U of the self-adjoint operator A and there exists the inverse operator $(I - U)^{-1}$ with the domain D_A. Hence from Eqs. (8.102) we receive $y = 2i(I - U)^{-1}x$ and $Ax = i(I + U)(I - U)^{-1}x$, i.e.

$$A = i(I + U)(I - U)^{-1}. \tag{8.103}$$

Formulae (8.101) and (8.103) establish the one–to–one correspondence between the operator A and its Cayley transform – the unitary operator U. To each self–adjoint operator A corresponds the unitary operator U determined by formula (8.101). And vice versa, to each unitary operator U for which 1 is not the eigenvalue corresponds the self–adjoint operator A determined by formula (8.103).

R e m a r k. As on the basis of Theorem 6.27 any operator is commutative with its resolvent at any regular point then the operator multipliers in formulae (8.101) and (8.103) may be replaced.

8.3.5. Functions of a Self-Adjoint Operator

Let $g(e^{is})$ be a function of the class $\mathcal{C}$, i.e. a continuous function or a finite linear combination of the functions which represent the limits of the nonincreasing sequences of the continuous functions. For such functions the corresponding functions of an unitary operator were determined in Subsection 8.3.3. Therefore for any unitary operator U the function $g(U)$ is determined. Let now A be an arbitrary self–adjoint operator, U be its Cayley transform $U = (A - iI)(A + iI)^{-1}$. Then formula

$$f(A) = g((A - iI)(A + iI)^{-1}) \tag{8.104}$$

will determine the function of the self–adjoint operator A. Thus any function $g(e^{is})$ of class $\mathcal{C}$ considered in Subsection 8.3.3 determines the corresponding function $f(A)$ of the self–adjoint operator A. The function of the real variable $t = i(1 + e^{is})/(1 - e^{is})$ determined by formula

$$f(t) = g((t - i)/(t + i)), \quad t \in R,$$

apparently, belongs to the same class $\mathcal{C}$. And vice versa, any function $f(t)$ of this class determines the function

$$g(e^{is}) = f(i(1 + e^{is})/(1 - e^{is})), \quad s \in [0, 2\pi]$$

of the same class. Thus the Cayley transform allows to determine a wide class of the functions of the self–adjoint operator. Here relations (8.97)–(8.99) are generalized on the functions of the self–adjoint operator:

$$f(t) = f_1(t) + f_2(t) \sim f(A) = f_1(A) + f_2(A), \qquad (8.105)$$

$$f(t) = f_1(t)f_2(t) \sim f(A) = f_1(A)f_2(A), \qquad (8.106)$$

$$|f(t)| \le c \sim \|f(A)\| \le c. \qquad (8.107)$$

It follows from (8.106) that any functions of one and the same operator are commutative.

R e m a r k. In exactlly the same way the functions of two (or larger number) of the commutative self–adjoint operators are determined (Remark in Subsection 8.3.3). So, it is sufficient to show that if two self–adjoint operators A_1, A_2 are commutative then their Cayley transforms

$$U_1 = (A_1 - i\,I)(A_1 + i\,I)^{-1}, \quad U_2 = (A_2 - i\,I)(A_2 + i\,I)^{-1}$$

are also commutative. But it follows directly from Theorem 6.28 and its Corollary in accordance with which each of two commutative operators is commutative with the resolvent of the other and their resolvents are also commutative. Really, after determining the functions of two (or larger number) commutative self–adjoint operators from relation (8.108) for the functions of two self–adjoint operators A_1, A_2

$$f(t_1, t_2) = f_1(t_1, t_2)f_2(t_1, t_2) \sim f(A_1 A_2) = f_1(A_1, A_2)f_2(A_1, A_2),$$

in the special case when f_1 depends only on the first argument, and f_2 depends only on the second one we may make the conclusion that any functions $f_1(A_1)$, $f_2(A_2)$ of the commutative operators A_1 and A_2 are commutative.

8.3.6. Spectral Measure of a Self–Adjoint Operator

The class of the functions $\mathcal{C}$ for which the functions of the self–adjoint operator are determined includes the indicators of Borel sets. Really, by virtue of the evident regularity of the Lebesgue measure l on R (for a definiton of a regular measure see Subsection 5.4.1) for any Borel set B and any $\varepsilon > 0$ there exists such decreasing sequence of the open sets $\{G_n\}$, $B \subset G_n$ that

$$l(G_n \backslash B) < \varepsilon/2^n.$$

Let us construct the sequence of the functions

$$f_n(t) = \frac{d(t, \bar{G}_n)}{d(t, \bar{G}_n) + d(t, B)} \quad (k = 1, 2, \ldots),$$

where

$$d(t, C) = \inf_{s \in C} |t - s|$$

is the distance from the point t till the set C. It is clear that the functions $f_n(t)$ are continuous, equal to 1 on the set B and equal to zero outside the corresponding sets G_n and form a decreasing sequence convergent to the indicator of the set B, $f_n(t) \to 1_B(t)$ at each t. Thus the indicators of Borel sets belong to the class $C_0 \subset C$. Consequently, for any Borel set B the function $1_B(A)$ of the self–adjoint operator A is determined.

Theorem 8.13. *For any self–adjoint operator A which maps the H–space X into X and any Borel set B the operator $1_B(A)$ represents a projector on some subspace of the space X.*

▷ As $1_B^2(t) = 1_B(t)$ then by virtue of relation (8.106) $1_B^2(A) = 1_B(A)$, i.e. $1_B(A)$ is an idempotent operator. And as the function $1_B(t)$ is real then $1_B(A)$ is a self–adjoint operator. Consequently, according to Theorem 7.18 it represents a projector on some subspace of the H–space X. ◁

Theorem 8.14. *For any self–adjoint operator A the operator set function $1_B(A)$ represents the decomposition of the identity.*

▷ From the relations

$$1_\emptyset(t) \equiv 0, \quad 1_R(t) = 1,$$

$$1_{B_1}(t) \leq 1_{B_2}(t), \quad \text{if} \quad B_1 \subset B_2,$$

$$1_{\bigcup B_k}(t) = \sum 1_{B_k}(t), \quad \text{if} \quad B_k B_h = \emptyset \text{ at } k \neq h,$$

$$1_{B_1}(t) 1_{B_2}(t) = 1_{B_2}(t) 1_{B_1}(t) = 1_{B_1 B_2}(t)$$

follows that the operator set function $1_B(A)$ at any fixed operator A possesses all properties of the decomposition of the identity

$$1_\emptyset(A) = 0, \quad 1_R(A) = I,$$

$$1_{B_1}(A) \leq 1_{B_2}(A), \quad \text{if} \quad B_1 \subset B_2,$$

$$\mathbf{1}_{\bigcup B_k}(A) = \sum \mathbf{1}_{B_k}(A), \quad \text{if} \quad B_k B_h = \emptyset \text{ at } k \neq h,$$

$$\mathbf{1}_{B_1}(A)\mathbf{1}_{B_2}(A) = \mathbf{1}_{B_2}(A)\mathbf{1}_{B_1}(A) = \mathbf{1}_{B_1 B_2}(A). \;\triangleleft$$

Thus any self–adjoint operator A determines on the σ–algebra $\mathcal{B}$ of Borel sets of the real axis R the decomposition of the identity

$$E_A(B) = \mathbf{1}_B(A), \quad B \in \mathcal{B}. \tag{8.108}$$

This decomposition of the identity is called *the spectral measure* of the operator A. The function

$$F_A(t) = E_A((-\infty, t)) \tag{8.109}$$

is called *the spectral function* of the operator A.

The detailed spectral theory of functions of linear operators is given in (Dowson 1978).

8.4. Spectral Decompositions of Linear Operators

8.4.1. Integral Representation of Operator Functions of Class $\mathcal{C}$

After determining the spectral measure of the self–adjoint operator we may present any function of this operator which belongs to the class $\mathcal{C}$ considered in Subsection 8.3.5 in the form of an integral over a spectral measure.

Theorem 8.15. *Let A be a self–adjoint operator in the H–space X, $f(t)$ be a function of the class $\mathcal{C}$. Then the operator function $f(A)$ is expressed by formula*

$$f(A) = \int\limits_{-\infty}^{\infty} f(t) E_A(dt), \tag{8.110}$$

where $E_A(B)$, $B \in \mathcal{B}$, is a spectral measure of the operator A determined by formula (8.108).

▷ Putting in formula (8.110) $f(t) = \mathbf{1}_B(t)$ then on the basis of formula (8.108) we get

$$\int\limits_{-\infty}^{\infty} \mathbf{1}_B(t) E_A(dt) = \int\limits_{B} E_A(dt) = E_A(B) = \mathbf{1}_B(A).$$

Thus formula (8.110) is valid for the indicators of Borel sets. Consequently, it is valid for all simple functions. But any nonnegative function $f(t)$ of the class C as a measurable function may be presented in the form of the limit of nondecreasing sequence of the simple functions $\{f^n(t)\}$ (Subsection 3.4.1). The corresponding sequence of the operator functions $\{f^n(A)\}$ converges to the operator function $f(A)$. In order to make sure in this fact it is sufficient to notice that owing to the boundedness of $f(t)$ the sequence $\{f^n(t)\}$ converges to $f(t)$ uniformly on the whole real axis R. Therefore for any $\varepsilon > 0$ at all t and at sufficiently large n the inequality $|f^n(t) - f(t)| < \varepsilon$ is valid. From this inequality by virtue of (8.107) follows the inequality $\|f^n(A) - f(A)\| < \varepsilon$ at all sufficiently large n. Consequently, formula (8.110) is valid for all nonnegative functions $f(t)$ of the class C. And as any function of the class C is expressed by the formula

$$f(t) = f_R^+(t) - f_R^-(t) + i[f_I^+(t) - f_I^-(t)],$$

where f_R^+, f_R^-, f_I^+, f_I^- are nonnegative functions of the class C then formula (8.110) is valid for all functions of the class C. ◁

8.4.2. Spectral Decomposition of a Self-Adjoint Operator

Theorem 8.16. *Any self-adjoint operator A admits an integral representation*:

$$A = \int_{-\infty}^{\infty} t E_A(dt). \tag{8.111}$$

▷ Let us consider Cayley transform (8.101) of the operator A,

$$U = (A - iI)(A + iI)^{-1}.$$

As the function $f(t) = (t - i)/(t + i)$ belongs to the class C then formula (8.110) is valid for it. Consequently,

$$U = (A - iI)(A + iI)^{-1} = \int_{-\infty}^{\infty} \frac{t - i}{t + i} E_A(dt).$$

Hence follow formulae

$$I + U = 2 \int_{-\infty}^{\infty} \frac{t}{t + i} E_A(dt), \quad I - U = 2i \int_{-\infty}^{\infty} \frac{1}{t + i} E_A(dt). \tag{8.112}$$

As there exists the inverse operator $(I - U)^{-1}$ then on the basis of Theorem 8.10 it is determined by formula

$$(I - U)^{-1} = \frac{1}{2i} \int_{-\infty}^{\infty} (t + i) E_A(dt). \qquad (8.113)$$

In consequence of the fact that the operator $I + U$ is bounded from formulae (8.112), (8.113), (8.80) and (8.103) we obtain

$$A = i(I + U)(I - U)^{-1} = \int_{-\infty}^{\infty} \frac{t}{t + i} (t + i) E_A(dt) = \int_{-\infty}^{\infty} t E_A(dt),$$

what proves the validity of formula (8.111). ◁

Formula (8.111) gives a *spectral decomposition (integral representation)* of the self–adjoint operator A.

Corollary 1. *The spectrum of a self–adjoint operator is completely determined by its spectral measure.*

▷ From formula (8.111) and Theorem 8.11 follows that λ is the eigenvalue of the operator A if and only if $G_{\{\lambda\}} \neq \{0\}$, i.e. when the value of the spectral measure of the operator A on the singleton $\{\lambda\}$ is different from zero operator $E_A(\{\lambda\}) \neq 0$. From formula (8.111) and Theorem 8.12 follows that λ is a regular point of the operator A if and only if there exists such $\varepsilon > 0$ that $G_{\{t:|t-\lambda|<\varepsilon\}} = \{0\}$, i.e. $E_A(\{t : |t - \lambda| < \varepsilon\}) = 0$. From these two facts follows that λ is a point of the spectrum of the operator A different from the eigenvalues if and only if at any $\varepsilon > 0$ $G_{\{t:|t-\lambda|<\varepsilon\}} \neq \{0\}$, i.e. $E_A(\{t : |t - \lambda| < \varepsilon\}) \neq 0$. ◁

Corollary 2. *If the operator A is bounded then $E_A(B) = 0$ for any Borel set B which does not intersect with the interval $(- \|A\|, \|A\|)$, and formula (8.111) takes the form*

$$A = \int_{-\|A\|}^{\|A\|} t E_A(dt). \qquad (8.114)$$

Corollary 3. *For any $x \in D_A$*

$$\int_{-\infty}^{\infty} t^2 \, \|E_A(dt)x\|^2 = \int_{-\infty}^{\infty} t^2 \mu_x(dt) < \infty. \qquad (8.115)$$

▷ This follows from formula (8.63) in accordance of which the latter integral is equal to $\|Ax\|^2$. ◁

Using the spectral function $F_A(t)$ which is determined by formula (8.109) we may rewrite integral representation (8.111) of the operator A in the form of the Lebesgue–Stieltjes integral

$$A = \int\limits_{-\infty}^{\infty} t\,dF_A(t). \tag{8.116}$$

As integral in (8.110) exists for any measurable function $f(t)$ then by means of formula (8.110) we may determine the operator function $f(A)$ for any measurable function $f(t)$ for which the set $\{x \; : \; f(t) \in L_2(R, \mathcal{B}, \mu_x)\}$ is not empty.

E x a m p l e 8.9. It was shown in Example 8.6 that to the operator of the multiplication by the function $f(s)$ in the space $L_2(R)$ (with Lebesgue measure) corresponds the decomposition of the identity $E(B)$ which represents an operator of the multiplication by the indicator of the set $f^{-1}(B)$. In particular, at $f(s) = s$ we get $f^{-1}(B) = B$ and the operator considered in Example 8.6 represents the operator Q of the multiplication by the independent variable s in the space $L_2(R)$. Thus the operator Q of the multiplication by the independent variable in $L_2(R)$ generates the decomposition of the identity E_Q which is determined by formula

$$E_Q(B)x = x(s)1_B(s), \quad x(s) \in L_2(R), \quad B \in \mathcal{B},$$

and here

$$Q = \int\limits_{-\infty}^{\infty} t E_Q(dt) = \int\limits_{-\infty}^{\infty} t 1_{[t,t+dt)}(s).$$

From the results of the analysis of a spectrum of the operator of the multiplication by the function $f(s)$ which was performed in Example 8.6 follows that the operator Q has no eigenvalues and any real λ is a point of its spectrum.

E x a m p l e 8.10. It is easy to see that the operator of the multiplication Q_μ by an independent variable in the space $L_2(R, \mathcal{B}, \mu)$ corresponding to the arbitrary nonnegative measure μ is determined by the same formula that Q:

$$Q_\mu = \int\limits_{-\infty}^{\infty} t E_Q(dt) = \int\limits_{-\infty}^{\infty} t 1_{[t,t+dt)}(s),$$

and consequently, corresponds to the same decomposition of the identity $E_Q(B)x = x(s)1_B(s)$. Let us study a spectrum of the operator Q_μ.

If s_0 is the point of the measure concentration μ, $\mu(s_0) > 0$, then the indicator $\mathbf{1}_{s_0}(s)$ of the singleton $\{s_0\}$ is different from zero in $L_2(R, \mathcal{B}, \mu)$ as its norm is equal $\sqrt{\mu(s_0)}$, $\| \mathbf{1}_{\{s_0\}}(s) \|^2 = \int \mathbf{1}^2_{\{s_0\}}(s)\mu(ds) = \mu(s_0)$, and consequently, $E_Q(\{t : t = s_0\}) = \mathbf{1}_{\{s_0\}}(s) \neq 0$. Therefore $\lambda = s_0$ is the eigenvalue of the operator Q_μ. The corresponding proper subspace $G_{\{s_0\}}$ represents a set of all the functions different from zero only at the point s_0, and consequently, is one-dimensional. Thus all the points of the measure μ concentration are the eigenvalues of the operator Q_μ and the proper subspaces which correspond to them are one-dimensional.

If real λ is not the point of the measure μ concentration but $\mu(\{t : |t - \lambda| < \varepsilon\}) \neq 0$ at any $\varepsilon > 0$ then this λ is the point of the spectrum of the operator Q_μ.

If there exists such $\varepsilon > 0$ that $\mu(\{t : |t - \lambda| < \varepsilon\}) = 0$ then the point λ is a regular point of the operator Q_μ.

E x a m p l e 8.11. It was shown in Example 7.10 that the operator of the differentiation with respect to the imaginary argument D_i in $L_2(R)$ is unitary equivalent to the operator of the multiplication by the independent variable Q:

$$D_i = FQF^{-1},$$

where F is the Fourier–Plancherel operator. Hence it follows that the operator D_i has the same spectrum that the operator Q. Thus the operator D_i in $L_2(R)$ has no eigenvalues and any real λ belongs to the spectrum D_i. All the complex λ are the regular points of D_i. For finding the decomposition of the identity E_{D_i} corresponding to the operator D_i we substitute into the previous formula the spectral decomposition Q from Example 8.9:

$$D_i = \int\limits_{-\infty}^{\infty} t F E_Q(dt) F^{-1}.$$

Hence we come to the conclusion that for any set $B \in \mathcal{B}$

$$E_{D_i}(B) = F E_Q(B) F^{-1}.$$

It is easy to verify that this operator function of the set possesses all the properties of the decomposition of the identity, and consequently, is the spectral measure of the operator D_i. To obtain the explicit expression $E_{D_i}(B)$ let us take any function $x = x(s) \in L_2(R)$ and transform it by the operator F^{-1}:

$$y(s) = F^{-1}x = \frac{1}{\sqrt{2\pi}} \frac{d}{ds} \int\limits_{-\infty}^{\infty} \frac{1 - e^{-is\tau}}{i\tau} x(\tau)\, d\tau.$$

Then taking into consideration that $E_Q(B)y = 1_B(s)y(s)$ we receive

$$F E_Q(B) F^{-1} x = \frac{1}{2\pi} \frac{d}{ds} \int_{-\infty}^{\infty} \frac{e^{is\sigma} - 1}{i\sigma}$$

$$\times \left\{ 1_B(\sigma) \frac{d}{d\sigma} \int_{-\infty}^{\infty} \frac{1 - e^{-i\sigma\tau}}{i\tau} x(\tau)\, d\tau \right\} d\sigma$$

$$= \frac{1}{2\pi} \frac{d}{ds} \int_B \frac{e^{is\sigma} - 1}{i\sigma} d \left(\int_{-\infty}^{\infty} \frac{1 - e^{-i\sigma\tau}}{i\tau} x(\tau)\, d\tau \right).$$

Thus the decomposition of the identity E_{D_i} corresponding to the operator of the differentiation with respect to the imaginary argument D_i is determined by the formula

$$E_{D_i}(B) x = \frac{1}{2\pi} \frac{d}{ds} \int_B \frac{e^{is\sigma} - 1}{i\sigma} d \left(\int_{-\infty}^{\infty} \frac{1 - e^{-i\sigma\tau}}{i\tau} x(\tau)\, d\tau \right).$$

8.4.3. Spectral Decomposition of an Unitary Operator

In exactly the same way (Subsection 8.4.2) we may determine a spectral measure of the unitary operator U and derive its integral representation (8.86). But we shall come to this fact in another way.

Theorem 8.17. *Any unitary operator U in the H–space X determines the bounded self-adjoint operator $A = -i \ln U$ in terms of which it is expressed by formula*

$$U = e^{iA}. \tag{8.117}$$

▷ To the function $-i \ln e^{is} = s$, $s \in [0, 2\pi]$ corresponds the operator function $-i \ln U$. From (8.99) follows that the operator $A = -i \ln U$ is bounded as $| - i \ln e^{is}| = s \le 2\pi$ at $s \in [0, 2\pi]$, and consequently, $\|A\| \le 2\pi$. As the function $-i \ln e^{is} = s$ is real then A is a self–adjoint operator. Finally, from $e^{is} = \exp\{i(-i \ln e^{is})\}$ follows that

$$U = \exp\{i(-i \ln U)\} = e^{iA}. \quad \triangleleft$$

Corollary. *Any unitary operator U admits the integral representation*

$$U = \int_{-\infty}^{\infty} e^{it} E_A(dt), \tag{8.118}$$

where $E_A(B)$ is a spectral measure of the self–adjoint operator A $= -i \ln U$.

▷ It follows immediately from formulae (8.110) and (8.117). ◁

Now recall that the operator A is bounded and $\|A\| \leq 2\pi$. Besides that the operator A is nonnegative by virtue of $-i \ln e^{is} = s \geq 0$ at $s \in [0, 2\pi]$ and the correspondence between the functions $f(e^{is})$ and the operators $f(U)$ established in Subsection 8.3.3. Consequently, on the basis of Theorem 6.29 about the boundedness of the spectrum of the bounded operator and Theorem 7.16 the spectrum of the operator A is completely concentrated on the interbal $[0, 2\pi]$. Hence by Theorem 8.12 follows that $E_A(B) = 0$ on any set $B \in \mathcal{B}$ nonintersecting with the interval $[0, 2\pi]$. Therefore formula (8.118) may be rewritten in the form.

$$U = \int_0^{2\pi} e^{it} E_A(dt). \tag{8.119}$$

But on the basis of decomposition formula (8.110) and the results of Subsection 8.2.8 formulae (8.117) and (8.118) determine the unitary operator U at any self–adjoint operator A. In this case for reducing formula (8.118) to the form of (8.119) one should apply the fact that the function e^{is} has the period 2π and perform partition of the integration region into the intervals of the length 2π. Then introducing the decomposition of the identity

$$E_U(B) = \sum_{k=-\infty}^{\infty} E_A(B + 2k\pi), \tag{8.120}$$

determined on the σ–algebra $\mathcal{B}_1$ of Borel sets of the interval $[0, 2\pi]$ we represent formula (8.118) in the form

$$U = \int_0^{2\pi} e^{it} E_U(dt). \tag{8.121}$$

The series in (8.120) converges by virtue of Theorem 7.23 about the monotone sequences of the projectors.

Formulae (8.118) and (8.121) give *a spectral decomposition (integral representation)* of the unitary operator U.

The decomposition of the identity $E_U(B)$ is called *a spectral measure* of the unitary operator U. The operator function

$$F_U(t) = E_U([0, t)) \tag{8.122}$$

is called *a spectral function* of the unitary operator U.

Using the spectral function we may rewrite formula (8.121) in the form

$$U = \int_0^{2\pi} e^{it} dF_U(t). \tag{8.123}$$

From formula (8.121) and the results of Subsection 8.2.7 follows that the spectral measure of the unitary operator completely determines its spectrum. Really, from (8.121) and Theorem 8.11 follows that λ is the eigenvalue of the operator U if and only if $G_{\{-i \ln \lambda\}} \neq \{0\}$, i.e. the value of the spectral measure of the operator U on the singleton $\{-i \ln \lambda\}$ is different from zero operator $E_U(\{-i \ln \lambda\}) \neq 0$. From (8.121) and Theorem 8.12 follows that λ is a regular point of the operator U if and only if there exists such $\varepsilon > 0$ that $G_{\{t:|e^{it}-\lambda|<\varepsilon\}} = \{0\}$, i.e. $E_U(\{t : |e^{it} - \lambda| < \varepsilon\}) = 0$. From these two facts follows that λ is a point of the spectrum U different from the eigenvalues if and only if at any $\varepsilon > 0$ $G_{\{t:|e^{it}-\lambda|<\varepsilon\}} \neq \{0\}$, i.e. $E_U(\{t : |e^{it} - \lambda| < \varepsilon\}) \neq 0$.

Now notice that on the basis of the results of Subsection 8.2.7 formula

$$A_1 = \int_0^{2\pi} t E_U(dt)$$

determines the bounded self–adjoint operator with the spectral measure $E_U(B)$ whose spectrum is completely concentrated on the interval $[0, 2\pi]$. On the basis of formulae (8.110) and (8.121) we have $U = e^{iA_1}$. On the other hand, from formulae (8.110) and (8.118) follows that $U = e^{iA}$ at any self–adjoint operator A with the spectral measure E_A. Thus the following theorem is proved.

Theorem 8.18. *The operator A in the representation of the unitary operator U (8.117) is non–uniquely determined. Formula (8.117) is valid for any self–adjoint operator A whose spectral measure $E_A(B)$ is connected with the spectral measure $E_U(B)$ of the operator U by the relation (8.120).*

In paricular, to this condition satisfy all the operators which are obtained from A by the shift of the spectrum on $2k\pi$ ($k = \pm 1, \pm 2, \ldots$) as the spectral measure $E_U(B)$ does not change at such shifts. So, as

$$A_k = \int_{-\infty}^{\infty} t E_A(dt + 2k\pi) = \int_{-\infty}^{\infty} (t + 2k\pi) E_A(dt + 2k\pi)$$

$$-2k\pi \int\limits_{-\infty}^{\infty} E_A(dt + 2k\pi) = \int\limits_{-\infty}^{\infty} sE_A(ds) - 2k\pi E_A(R)$$

$$= A - 2k\pi I,$$

then formula (8.117) remains valid at the addition to the operator A the operator $2k\pi I$.

As for any function $g(e^{it})$ of the class $\mathcal{C}$ the function $g(U) = g(e^{it})$ represents the function of the class $\mathcal{C}$ of the self–adjoint operator A then from formula (8.110) follows

$$g(U) = \int\limits_{-\infty}^{\infty} g(e^{it}) E_A(dt) = \int\limits_{0}^{2\pi} g(e^{it}) E_U(dt) \qquad (8.124)$$

for any function $g(e^{it})$ of the class $\mathcal{C}$.

The right–hand side of formula (8.124) has the sense for any measurable function $g(u)$. Therefore it is natural to determine the function $g(U)$ of the unitary operator for any measurable function $g(u)$ by formula (8.124).

E x a m p l e   8.12.   It was shown in Subsection 7.2.3 that the Fourier–Plancherel operator in $L_2(R)$ has the eigenvalues $1, i, -1, -i$ of the infinite multiplicity and the proper spaces G_1, G_i, G_{-1}, G_{-i} correspond to them. These subspaces are formed by the Hermite functions $\varphi_{4r}(t)$, $\varphi_{4r+1}(t)$, $\varphi_{4r+2}(t)$, $\varphi_{4r+3}(t)$ $(r = 0, 1, 2, \ldots)$ correspondingly. As the sequence of the functions $\{\varphi_k(t)\}$ according to the proved in Example 7.16 is complete in $L_2(R)$ then the orthogonal sum of the proper subspaces G_1, G_i, G_{-1}, G_{-i} coincide with $L_2(R)$, $G_1 \oplus G_i \oplus G_{-1} \oplus G_{-i} = L_2(R)$. Hence it follows that the decomposition of the identity corresponding to the Fourier–Plancherel operator is completely concentrated at the points $1, i, -1, -i$ and is determined by the projectors P_1, P_i, P_{-1}, P_{-i} on the subspaces G_1, G_i, G_{-1}, G_{-i} correspondingly. Therefore spectral decomposition (8.121) for the Fourier–Plancherel operator takes the form

$$F = P_1 + iP_i - P_{-1} - iP_{-i}.$$

Thus the spectrum of the Fourier–Plancherel operator consists of four points $1, i, -1, -i$. All other values of λ are its regular points. So, any function $x(s) \in L_2(R)$ is expressed by the decomposition of the Hermite functions (Example 7.16):

$$x(s) = \sum_{k=0}^{\infty} \alpha_k \varphi_k(s), \qquad \alpha_k = \frac{1}{2^k k! \sqrt{\pi}} \int\limits_{-\infty}^{\infty} x(s) \varphi_k(s) \, ds.$$

Its projections on the subspaces G_1, G_i, G_{-1}, G_{-i} are equal correspondingly

$$E_F(1)x = P_1 x(s) = \sum_{r=0}^{\infty} \alpha_{4r}\varphi_{4r}(s),$$

$$E_F(i)x = P_i x(s) = \sum_{r=0}^{\infty} \alpha_{4r+1}\varphi_{4r+1}(s),$$

$$E_F(-1)x = P_{-1} x(s) = \sum_{r=0}^{\infty} \alpha_{4r+2}\varphi_{4r+2}(s),$$

$$E_F(-i)x = P_{-i} x(s) = \sum_{r=0}^{\infty} \alpha_{4r+3}\varphi_{4r+3}(s).$$

The Fourier–Plancherel transform of the function $x(s)$ according to the obtained result is determined by formula

$$Fx(s) = \sum_{r=0}^{\infty} [\alpha_{4r}\varphi_{4r}(s) + i\alpha_{4r+1}\varphi_{4r+1}(s)$$

$$-\alpha_{4r+2}\varphi_{4r+2}(s) - i\alpha_{4r+3}\varphi_{4r+3}(s)].$$

8.4.4. Spectral Decomposition of the Group of Unitary Operators

Let us consider a set of the unitary operators $\{U_\tau\}$, $\tau \in R$ which possesses the following properties:

(i) $U_0 = I$;

(ii) $U_\tau U_\sigma = U_{\tau+\sigma}$;

(iii) the numeric function $(U_\tau x, y)$ of the variable τ is continuous at any $x, y \in X$.

This set of the operators represents an one–parametric continuous commutative (Abelian) group.

Theorem 8.19. *All the operators of the group $\{U_\tau\}$ represent the degrees of the operator $U = U_1$:*

$$U_\tau = U^\tau. \tag{8.125}$$

▷ Putting in property 2 that $\sigma = \tau$ we shall have $U_{2\tau} = U_\tau^2$. After this by the induction we get $U_{n\tau} = U_\tau^n$ at any natural n. Putting $n = q$, $\tau = 1/q$ we obtain $U_{1/q} = U_1^{1/q} = U^{1/q}$ for any natural q. Putting after this $n = p$, $\tau = 1/q$ we find $U_{p/q} = U_{1/q}^p = U^{p/q}$ for any natural p and q. Thus formula (8.125) is valid for all rational τ. Now notice that owing to the continuity of the function $(U_\tau x, y)$ for any sequence $\{\tau_n\}$ convergent to τ, $(U_{\tau_n} x, y) \to (U_\tau x, y)$ at any $x, y \in X$ and therefore

$$\| U_{\tau_n} x - U_\tau x \|^2 = \| U_{\tau_n} x \|^2 - (U_{\tau_n} x, U_\tau x) - (U_\tau x, U_{\tau_n} x)$$

$$+ \| U_\tau x \|^2 = 2 \| x \|^2 - (U_{\tau_n} x, U_\tau x) - (U_\tau x, U_{\tau_n} x) \to 0,$$

as $(U_{\tau_n} x, U_\tau x) \to \| U_\tau x \|^2 = \| x \|^2$. Consequently, $U_{\tau_n} x \to U_\tau x$ at any $x \in X$. After taking the sequence of the rational numbers $\{\tau_n\}$ convergent to τ we get $U_{\tau_n} = U^{\tau_n} \to U_\tau$ in the strong topology of the space $\mathcal{B}(X)$ of the bounded linear operators on X. On the other hand, by virtue of the continuity of the function $e^{i\tau s}$ and formula (8.124) $U_{\tau_n} = U^{\tau_n} \to U^\tau$. So, $U_\tau = U^\tau$ for any τ. ◁

Corollary. *Any continuous one–parametric commutative group of the unitary operators admits an integral representation*

$$U_\tau = \int\limits_{-\infty}^{\infty} e^{i\tau t} E_A(dt), \qquad (8.126)$$

where $E_A(B)$ is a spectral measure of the operator A in decomposition (8.117) of the operator $U = U_1$ (Stone theorem).

▷ As $U_\tau = U^\tau$ formula (8.126) follows directly from (8.124). ◁

Formula (8.126) gives *a spectral decomposition (integral representation) of one–parameter group of the unitary operators.*

Using the spectral function $F_A(t)$ determined by formula (8.109) we may rewrite formula (8.126) in the form

$$U_\tau = \int\limits_{-\infty}^{\infty} e^{i\tau t} dF_A(t). \qquad (8.127)$$

E x a m p l e 8.13. Consider the operator of the shift U_τ in the space $L_2(R)$ which associates to any function $x(s) \in L_2(R)$ the function $x(s + \tau)$. Obviously, U_τ is an unitary operator. The set of the operators $\{U_\tau\}$ corresponding to all the shifts of τ form the commutative group as $U_0 = I$, $U_{\tau+\sigma} = U_\tau U_\sigma$. Hence for any continuous function $x(s) \in L_2(R)$

$$\| U_\tau x - x \|^2 = \int\limits_{-\infty}^{\infty} | x(s + \tau) - x(s) |^2 ds \to 0 \text{ at } \tau \to 0.$$

Recalling that the set of the continuous functions is dense in $L_2(R)$ we come to the conclusion that for any $x(s) \in L_2(R)$ and any $\varepsilon > 0$ we may find such continuous function $y(s) \in L_2(R)$ that $\|x - y\| < \varepsilon/3$. Then due to the inequality

$$\|U_\tau x - x\| \leq \|U_\tau x - U_\tau y\| + \|U_\tau y - y\| + \|y - x\|,$$

we shall have $\|U_\tau x - x\| < \varepsilon$ at all τ at which $\|U_\tau y - y\| < \varepsilon/3$. Here for any functions $x(s), y(s) \in L_2(R)$ we have

$$|(U_{\tau'} x, y) - (U_\tau x, y)| = |(U_\tau (U_{\tau'-\tau} x - x), y)|$$

$$\leq \|U_{\tau'-\tau} x - x\| \|y\| < \varepsilon \|y\|$$

at all τ' sufficiently close to τ. Consequently, at any x, y the function $(U_\tau x, y)$ of the variable τ is continuous.

Thus the group of the shifts $\{U_\tau\}$ is continuous in one–parametric commutative group. Therefore all the operators of the shift U_τ may be represented by the spectral decomposition (8.126). For determining the decomposition of the identity E_A which corresponds to the group of the shifts $\{U_\tau\}$ we notice that for any unboundedly differentiable function $x(s)$ which belongs to the space $L_2(R)$ with all its derivatives

$$U_\tau x = x(s + \tau) = \sum_{n=0}^{\infty} \frac{\tau^n}{n!} x^{(n)}(s)$$

$$= \sum_{n=0}^{\infty} \frac{(i\tau)^n}{n!} D_i^n x(s) = e^{i\tau D_i} x.$$

Hence it is clear that the operator of the shift U_τ represents the function $e^{i\tau D_i}$ of the operator of the differentation D_i with respect to the imaginary argument. Therefore the decomposition of the identity E_A which corresponds to the group of the shifts coincides with the decomposition of the identity E_{D_i} of the operator D_i obtained in Example 8.11 and

$$U_\tau = e^{i\tau D_i} = e^{\tau D} = \int_{-\infty}^{\infty} e^{i\tau t} E_{D_i}(dt).$$

Here despite the fact that the equality $U_\tau x = e^{i\tau D_i} x$ is obtained not on the whole space $L_2(R)$ it is valid owing to the boundedness of the operator $e^{i\tau D_i}$ for all x and is extended by the continuity on the whole space $L_2(R)$.

The obtained expression of all the operators of the shift in terms of the operator of the differentiation is often used in the control theory and signal processing.

8.4.5. Spectral Decomposition of a Normal Operator

It remains to receive a spectral decomposition of an arbitrary normal operator. Let T be a normal operator. It is apparent that the operators

$$A_1 = \frac{T + T^*}{2}, \quad A_2 = \frac{T - T^*}{2i}$$

are self–adjoint operators and therefore may be represented by spectral decompositions (8.111):

$$A_1 = \int\limits_{-\infty}^{\infty} t E_1(dt), \quad A_2 = \int\limits_{-\infty}^{\infty} t E_2(dt),$$

where E_1 and E_2 are the corresponding spectral measures. Noticing that $T = A_1 + iA_2$ we get

$$T = \int\limits_{-\infty}^{\infty} t E_1(dt) + i \int\limits_{-\infty}^{\infty} t E_2(dt). \tag{8.128}$$

As the operators A_1 and A_2 are commutative then according to Remark at the end of Subsection 8.3.5 any functions of the operators A_1 and A_2 are commutative. In particular, their spectral measures $E_1(B_1)$ $= \mathbf{1}_{B_1}(A_1)$ and $E_2(B_2) = \mathbf{1}_{B_2}(A_2)$ are commutative. Consequently, taking into account that $E_1(R) = E_2(R) = I$ and changing the denotion of the integration variable in the second integral we may rewrite (8.128) in the form

$$T = \int\limits_{-\infty}^{\infty} t E_1(dt) E_2(R) + i \int\limits_{-\infty}^{\infty} s E_2(ds) E_1(R)$$

$$= \int\limits_{-\infty}^{\infty} \int\limits_{-\infty}^{\infty} t E_1(dt) E_2(ds) + i \int\limits_{-\infty}^{\infty} \int\limits_{-\infty}^{\infty} s E_1(ds) E_2(ds)$$

$$= \int\limits_{-\infty}^{\infty} \int\limits_{-\infty}^{\infty} (t + is) E_1(dt) E_2(ds). \tag{8.129}$$

By virtue of the commutativity of E_1 and E_2 the product $E_1(B)E_2(B)$ at any sets $B_1, B_2 \in \mathcal{B}$ represent a projector. It is easy to assure that the operator set function is determined by formula

$$E_T(B_1) = E_1(B_1) E_2(B_2) \tag{8.130}$$

on the semi–algebra of the rectangles of the complex plane C also possesses all other properties of the decomposition of the identity. It may be extended over the operator measure $E_T(F)$ on the σ–algebra $\mathcal{C}$ of Borel sets of the complex plane C. For this purpose we notice that for any nonintersecting sets $B_1, C_1 \in \mathcal{B}$ and $B_2, C_2 \in \mathcal{B}$, $E_1(B_1)E_2(B_2)$ and $E_1(C_1)E_2(C_2)$ represent the projectors on the orthogonal subspaces. Consequently, on the basis of Theorem 7.20 about the sum of the projectors we may extend the measure E_T over the algebra $\mathcal{B}_0$ of the finite unions of the rectangular sets of the complex plane C. After this using Theorem 7.23 about the monotone sequences of the projectors we may extend the measure E_T sequentially on the class $\mathcal{B}_1$ of the countable unions and intersections of the sets of the algebra $\mathcal{B}_0$, and after that on the class $\mathcal{B}_2$ of the countable unions and intersections of the sets of the algebra $\mathcal{B}_1$ and so on (Subsection 2.1.5). So, as a result the measure E_T will be extended over the σ–algebra $\mathcal{C}$ of Borel sets of the complex plane C, and formula (8.129) will take the form

$$T = \int\limits_{-\infty}^{\infty} \int\limits_{-\infty}^{\infty} (t + is) E_T(dt\, ds) = \int z E_T(dz), \qquad (8.131)$$

where dz is the square element of the complex plane C.

But there is no need the measure E_T will be determined on the σ–algebra $\mathcal{C}$ of Borel sets of the complex plane C. It is sufficient to consider integral (8.131) as an integral over the additive operator measure E_T determined on the algebra of the sets $\mathcal{B}_0$. Thus the following theorem is proved.

Theorem 8.20. *To each normal operator T corresponds the decomposition of the identity E_T in terms of which this operator is expressed by spectral decomposition (8.131).*

Decomposition of the identity $E_T(F)$ may be naturally called a *spectral measure* of the normal operator T.

General theory of functions of operators is given in (Dowson 1978, Dunford and Schwarz 1963, 1971). Spectral theory of linear operators is the basis of the modern spectral theory of stationary stochastic processes. For detailed treatment refer to (Cramer and Leadbetter 1967, Pugachev 1965). For nonstationary stochastic processes corresponding theory is based on canonical expansions and integral canonical representations. For details address to (Adomian 1983, Pugachev 1965, Pugachev and Sinitsyn 1987).

Problems

8.4.1. Write in an explicit form the decomposition of the identity for a compact normal operator of Problem 8.1.2.

8.4.2. Consider a self–adjoint integral operator in the space $L_2(X)$, $X \subset R^n$:

$$Ax = \int K(s,u)x(u)du, \quad K(u,s) = \overline{K(s,u)}.$$

After writing the decomposition of the identity for this operator in the form

$$E_A(B)x = \int e_A(B; s, u)x(u)du, \qquad (*)$$

where $e_A(B; s, u)$ is a measure with the values in the space of the functions of two variables s, u find the conditions to which the measure $e_A(B; s, u)$ must satisfy in order the formula $(*)$ should determine the decomposition of the identity. Show that to these conditions satisfies the measure

$$e_Q(B; s, u) = 1_B(s)\delta(u - s)$$

of Example 8.9 and the measure

$$e_{D_i}(B; s, u) = \frac{1}{2\pi} \int_B e^{i\sigma(s-u)}\, d\sigma$$

of Example 8.11. The first one of these Examples shows that the measure e_A in $(*)$ may be a measure with the values in the space of the generalized functions.

8.4.3. Under the conditions of Problem 8.4.2 find a spectral decomposition of the kernel $K(s, u)$ of the self–adjoint integral operator A corresponding to spectral decomposition (8.111) of this operator.

8.4.4. Generalize the results of Problems 8.4.2 and 8.4.3 on the normal integral operators.

8.4.5. Find the measures and the corresponding spectral decompositions of the kernels of the compact self–adjoint operators.

CHAPTER 9
NONLINEAR PROBLEMS OF FUNCTIONAL ANALYSIS

Chapter 9 is devoted to some nonlinear problems of functional analysis. In Section 9.1 a strong (Frechét) differential and derivative and a weak (Gâteaux) differential and derivative of an operator or a functional are defined. The relations between these two kinds of differentials and derivatives are studied. The strong and weak differentials and derivatives of higher orders are defined. The rule for the differentiation of a composite function, the finite increments formula, the Taylor formula and Taylor series are derived for the functionals and the operators. Section 9.2 is devoted to general necessary and sufficient conditions of the extremum of a functional. In the last Section 9.3 the elements of the general theory of differential equations in B-spaces are outlined.

9.1. Differentiation of Operators and Functionals

9.1.1. Strong Differentials and Derivatives

Let $f(x)$ be a function (operator) which maps the normed linear space X into the normed linear space Y. If the difference $f(x+\delta x) - f(x)$ admits in some vicinity of the point x the representation

$$f(x + \delta x) - f(x) = T\delta x + o(\|\delta x\|),$$

where T is the bounded linear operator, $T \in \mathcal{B}(X, Y)$ then the variable $df(x) = T\delta x$ is called *a strong differential (Frechét) differential* of the function f, and the operator T — *a strong derivative (Frechét) derivative* of the function f at the point x and is denoted by $f'(x)$. Here the function f is called *strongly differentiable (Frechét) differentiable* at the point x. The function f is called *strongly differentiable on the set A* if it is strongly differentiable at each point of the set A. In the special case when the function $f(x)$ represents a bounded linear operator $f(x) = Tx$ its strong derivative coincides with the operator T, $f'(x) = T$.

E x a m p l e 9.1. A strong differential of the function $y = f(x)$ which maps n–dimensional space R^n into m–dimensional space R^m is determined by formula

$$d f(x) = f'(x)\, \delta x,$$

where $f'(x)$ is a matrix of the derivatives of the coordinates of the vector function $f(x)$ with respect to the coordinates of the vector x. The function $f(x)$ is strongly differentiable (at the point x, in the region A) if there exist the first derivatives of all its coordinates with respect to all coordinates of the vector x.

E x a m p l e 9.2. Let A be a bounded linear operator mapping the normed linear space X into the space $\mathcal{B}(X, Y)$ of the linear operators which map X into the normed linear space Y. In this case at any $x \in X$ the variable Ax represents an operator mapping X into Y, and the variable $(Ax_1)x_2$ at any $x_1, x_2 \in X$ represents an element of the space Y. The operator A may be considered as a bounded *bilinear operator* which maps the product of the spaces $X^2 = X \times X$ in Y. Consider now the function $f(x) = (Ax)x$. Without loss of the generality we may consider the operator A as *a symmetric* one in the sense that

$$(Ax_2)x_1 = (Ax_1)x_2 \quad \forall x_1, x_2 \in X.$$

Really, the variable $(Ax_2)x_1$ at fixed x_1 represents the result of the action of some linear (bounded) operator on the vector $x_2 \in X$. This operator certainly depends on x_1 and this dependence is linear. Therefore it may be denoted by Bx_1. Then we receive

$$(Ax_2)x_1 = (Bx_1)x_2.$$

Putting $C = (A + B)/2$ we shall have at any x_1 and x_2

$$(Cx_1)x_2 = [(Ax_1)x_2 + (Bx_1)x_2]\,/2 = [(Ax_1)x_2 + (Ax_2)x_1]\,/2,$$

$$(Cx_2)x_1 = [(Ax_2)x_1 + (Bx_2)x_1]\,/2 = [(Ax_2)x_1 + (Ax_1)x_2]\,/2.$$

Thus the operator C is symmetric and

$$(Cx)x = (Ax)x.$$

This result is the generalization of the statement known from linear algebra about the possibility of reducing the matrix of square form in the space R^n to a symmetric matrix.

Considering the operator A as a symmetric one due to its boundedness we get

$$(A(x + \delta x))(x + \delta x) = (Ax)x + 2(Ax)\delta x + o(\|\,\delta x\,\|).$$

Hence it is clear that the function $f(x) = (Ax)x$ is differentiable and its derivative is equal to $2Ax \in \mathcal{B}(X, Y)$. In the special case of the bounded self-adjoint operator A in the H–space the derivative of the square functional (Ax, x) is determined by formula $(Ax, x)' = 2Ax$.

E x a m p l e 9.3. Consider a nonlinear integral operator

$$F(x) = \int f(t, \tau, x(\tau))\,\mu(d\tau),$$

where $f(t, \tau, x)$ is a continuous function which has the continuous derivative $f_x(t, \tau, x)$ with respect to x at any t, τ and μ is a finite measure. As

$$F(x + \delta x) - F(x) = \int [f(t, \tau, x(\tau) + \delta x(\tau)) - f(t, \tau, x(\tau))]\, \mu(d\tau)$$

$$= \int f_x(t, \tau, x(\tau))\, \delta x(\tau)\, \mu(d\tau) + o(|\delta x(\tau)|)\,,$$

then the operator $F(x)$ is differentiable at the point $x = x(t)$ if the function $f_x(t, \tau, x(\tau))\, \delta x(\tau)$ of the variable τ is μ–differentiable at any t. In the case when X is the space $C(T)$ of the continuous functions of the compact T it is sufficient for this purpose that the function $f_x(t, \tau, x)$ of the variable τ at any t and x will be bounded almost everywhere on T relatively to the measure μ. In the case when X is the Lebesgue space $L_p(\Delta, \mathcal{A}, \mu)$, $p > 1$ it is sufficient that the function $f_x(t, \tau, x(\tau))$ at any t will be an element of the space $L_q(\Delta, \mathcal{A}, \mu)$ where $q = p/(p-1)$ (Subsection 3.7.2). It goes without saying that in both cases it is necessary that the function $f(t, \tau, x)$ will be differentiable with respect to x. Here in both cases the derivative of the operator (mapping) F represents a linear integral operator

$$F'(x) = \int f_x(t, \tau, x(\tau)) \cdot \mu(d\tau) \tag{9.1}$$

with the kernel $K(t, \tau) = f_x(t, \tau, x(\tau))$ (the place of function–argument of the operator — in a given case $\delta x(\tau)$ is indicated by a point).

When we have the vector functions $x(\tau)$ and $f(t, \tau, x)$ the derivative $f_x(t, \tau, x)$ represents a matrix of the first derivatives of the coordinates of the vector function f with respect to the coordinates of the vector x.

E x a m p l e 9.4. If the functional $f(x)$ on the space $C(T)$ of the continuous on the compact T scalar functions is differentiable at the point $x = x(t)$ then on the basis of Riesz theorem 6.15 its derivative $f'(x)$ as a continuous linear functional is determined by formula

$$f'(x) = \int_T \cdot\, \mu(dt; x)\,, \tag{9.2}$$

where $\mu(B; x)$ is a finite regular measure determined on the σ–algebra of Borel sets $\mathcal{B}$ of the compact T (certainly which depends on x in the case of the nonlinear functional f).

In the special case when T represents a compact of the space R^n and the measure μ is absolutely continuous relatively to the Lebesgue measure in R^n at the point $x = x(t)$ formula (9.2) takes the form

$$f'(x) = \int_T f'(t; x) \cdot dt\,. \tag{9.3}$$

Here $f'(t; x)$ is the Radon–Nikodym derivative of the measure $\mu(B; x)$, $B \in \mathcal{B}$, with respect to the Lebesgue measure in R^n (Subsection 3.6.3). So, the differential (strong) of the functional $f(x)$ is determined by formula

$$df(x) = f'(x)\,\delta x = \int_T f'(t; x)\,\delta x(t)\,dt\,. \tag{9.4}$$

Comparing this formula with the known formula in the case of m–dimensional vector x,

$$df(x) = \sum_{k=1}^{m} f'_{x_k}\,\delta x_k\,,$$

we see that the function $f'(t; x)$ at each fixed t plays the role of the partial derivative of the functional $f(x)$ with respect to the value of the function $x(t)$ at a given point t. In this case t plays the role of the index k (the number of the coordinate of the vector x).

The function $f'(t; x)$ in formulae (9.3) and (9.4) is called *a functional derivative* of the functional $f(x)$. This notion was introduced by Volterra as an analog of the notion of a partial derivative of the function of a finite–dimensional vector argument with respect to the coordinate of this vector argument.

9.1.2. Differentiation of a Composite Function

The known formula of the differentiation of a composite function is generalized on strong derivatives of the functionals and operators.

Theorem 9.1. *If the function $y = f(x)$ which maps the normed linear space X into the normed linear space Y is strongly differentiable at the point x_0, and the function $z = g(y)$ which maps Y into the normed linear space Z is strongly differentiable at the point $y_0 = f(x_0)$ then the function $gf(x) = g(f(x))$ (the composition of the mappings f and g) is strongly differentiable at the point x_0 and*

$$(gf)'(x_0) = g'(f(x_0))\,f'(x_0)\,. \tag{9.5}$$

▷ From the strong differentiability of the function f follows

$$y_0 + \delta y = f(x_0 + \delta x) = f(x_0) + f'(x_0)\,\delta x + o(\|\,\delta x\,\|)\,,$$

and from the strong differentiability of the function g we get

$$g(y_0 + \delta y) = g(y_0) + g'(y_0)\,\delta y + o(\|\,\delta y\,\|)\,.$$

Substituting here the previous expression $y_0 + \delta y$ and accounting the boundedness of the operators $f'(x_0)$ and $g'(y_0)$ we obtain

$$g(f(x_0 + \delta x)) = g(f(x_0)) + g'(f(x_0))[f'(x_0)\,\delta x + o(\|\,\delta x\,\|)]$$

$$+ o(\|\,f'(x_0)\,\delta x + o(\|\,\delta x\,\|)\,\|)$$

$$= g(f(x_0)) + g'(f(x_0))\,f'(x_0)\,\delta x + o(\|\,\delta x\,\|)\,.$$

Hence follows the existence of the derivative $(gf)'(x)$ at the point x_0 and formula (9.5). ◁

E x a m p l e 9.5. If the space Y in Example 9.2 represents a field of the scalars K then for any differentiable function $g(s)$ of the scalar variable s the derivative of the function $h(x) = g(f(x)) = g(Ax^n)$ is determined by the formula

$$h'(x) = ng'(Ax^n)\,Ax^{n-1}\,.$$

This formula is also valid in the case of an arbitrary normed linear space Y and an arbitrary differentiable mapping g of the space Y into the normed linear space X.

In the special case of square functional (Ax, x) in the real H–space X the previous formula gives

$$h'(x) = 2g'((Ax, x))\,Ax\,.$$

At $A = I$ and $g(s) = s^{p/2}$ we obtain $h(x) = \|\,x\,\|^p = (x, x)^{p/2}$ and

$$h'(x) = p(x, x)^{p/2-1}x = p\,\|\,x\,\|^{p-2}\,x\,.$$

9.1.3. Finite Increments Formula

The following theorem gives the generalization of the known Lagrange finite increments formula.

Theorem 9.2. *If the function $f(x)$ has a continuous derivative $f'(x)$ at all points of some convex set S then for any x_0, $x \in S$ the formula is valid*

$$f(x) - f(x_0) = \int\limits_0^1 f'(x_0 + t(x - x_0))(x - x_0)\,dt\,. \qquad (9.6)$$

▷ For proving consider the function $\varphi(t) = f(x_0 + t(x - x_0))$ of the real variable $t \in [0, 1]$. In consequence of the convexity of the set S

$x_0 + t(x - x_0) \in S$ at all $t \in [0, 1]$. Therefore the function $f(z)$ has a continuous derivative at $z = x_0 + t(x - x_0)$ at all $t \in [0, 1]$. In consequence of this fact the function $\varphi(t)$ has also a continuous derivative at $t \in [0, 1]$. On the basis of formula (9.5)

$$\varphi'(t) = f'(x_0 + t(x - x_0))(x - x_0).$$

Integrating this formula with respect to t we get

$$f(x_0 + 1 \cdot (x - x_0)) - f(x_0 + 0 \cdot (x - x_0)) = f(x) - f(x_0) =$$

$$= \int_0^1 f'(x_0 + t(x - x_0))(x - x_0)\, dt,$$

i.e. formula (9.6). ◁

In the special case when $f(x)$ represents the functional the derivative $f'(x_0 + t(x - x_0))(x - x_0)$ is the numeric function. Therefore we may apply to the integral in (9.6) the theorem about the mean. As a result formula (9.6) will take the form

$$f(x) - f(x_0) = f'(x_0 + \theta(x - x_0))(x - x_0), \tag{9.7}$$

where θ is some number from the interval $[0, 1]$. Thus for the functionals on the normed linear spaces an ordinary formula of the Lagrange finite increments is valid.

R e m a r k. In the general case of the operator $f(x)$ the mean theorem is inapplicable to the integral in (9.6).

9.1.4. Weak Differentials and Derivatives

Let X and Y be the topological linear spaces, $y = f(x)$ be a function which maps X into Y. If there exists the limit

$$\left[\frac{d}{dt} f(x + t\delta x) \right]_{t=0} = \lim_{t \to 0} \frac{f(x + t\delta x) - f(x)}{t}$$

in the topological space Y then it is called *a weak differential (Gâteaux) differential* of the function $f(x)$ at the point x with respect to the direction δx and is denoted by $\delta f(x, \delta x)$:

$$\delta f(x, \delta x) = [df(x + t\delta x)/dt]_{t=0}. \tag{9.8}$$

In the special case when at all δx there exists such continuous linear operator T that $\delta f(x, \delta x) = T\delta x$ this operator T is called *a weak derivative (Gâteaux) derivative* of the function $f(x)$ at the point x and is denoted by $f'(x)$. The function $f(x)$ in such a case is called *weakly differentiable (Gâteaux) differentiable one*.

R e m a r k. In the general case the weak differential $\delta f(x, \delta x)$ may be not a linear function δx. Thus from the existence of a weak differential of the function does not follow its weak differentiability.

9.1.5. Relation between Weak Differentiability and Strong One

In the special case of the normed spaces X and Y we may speak about weak and strong differentiability of the function. Naturally, the following question arises: what is the relation between these two types of the differentiability?

Theorem 9.3. *Any strongly differentiable function is weakly differentiable and its strong and weak derivatives coincide.*

▷ If the function $f(x)$ is strongly differentiable at the point x then

$$f(x + \delta x) = f(x) + f'(x)\,\delta x + o(\|\delta x\|),$$

where $f'(x)$ is a strong derivative of the function $f(x)$. Hence at any fixed δx we find

$$[f(x + t\delta x) - f(x)]/t = f'(x)\,\delta x + o(t)$$

and consequently, there exists a weak differential

$$\delta f(x, \delta x) = f'(x)\,\delta x.$$

This formula shows that the function $f(x)$ is weakly differentiable and its weak derivative is equal to $f'(x)$. ◁

R e m a r k. The inverse conclusion is not true: a weakly differentiable function may be not strongly differentiable function.

E x a m p l e 9.6. Consider an integral operator of Example 9.3 for two–dimensional vector function $x(t) = [x_1(t)\,x_2(t)]^T$ and the Lebesgue measure μ:

$$F(x) = \int_0^1 f(s, \tau, x_1(\tau), x_2(\tau))\,d\tau,$$

where

$$f(s,\tau,x_1,x_2) = \begin{cases} a_1 x_1 + a_2 x_2 & \text{at} \quad x_2 \neq x_1^2 \quad \text{and at} \quad x_1 = x_2 = 0, \\ a_1 x_1 + a_2 x_2 + c & \text{at} \quad x_2 = x_1^2, \quad x_1 \to 0 \end{cases}$$

and $c \neq 0$. In this case the functional $F(x)$ is weakly differentiable at the point $x = 0$, as

$$\frac{F(t\delta x) - F(0)}{t} = \int\limits_0^1 \frac{f(s,\tau,t\delta x_1(\tau),\, t\delta x_2(\tau)) - f(s,\tau,0,0)}{t}\, d\tau$$

$$= \int\limits_0^1 [a_1 \delta x_1(\tau) + a_2 \delta x_2(\tau)]\, d\tau$$

at any $\delta x_1(\tau)$, $\delta x_2(\tau)$ and all sufficiently small t. The weak derivative of the functional $F(x)$ represents a linear integral operator whose kernel serves the constant matrix–row $[a_1 a_2]$. But the functional $F(x)$ has no strong derivative as at infinite small $\delta x_1(\tau)$ and $\delta x_2(\tau) = [\delta x_1(\tau)]^2$

$$F(\delta x) - F(0) = \int\limits_0^1 [a_1 \delta x_1(\tau) + a_2 \delta x_2(\tau) + c]\, d\tau$$

$$= \int\limits_0^1 [a_1 \delta x_1(\tau) + a_2 \delta x_2(\tau)]\, d\tau + c\,.$$

The cited example shows that for a strongly differentiable function at a given point its weak differentiability at this point is not sufficient. But it is valid the following statement.

Theorem 9.4. *If the function $f(x)$ is weakly differentiable in some convex vicinity S of the point x and its weak derivative is continuous at the point x then there exists a strong derivative of the function $f(x)$ at the point x which coincides with its weak derivative.*

▷ Let $f'(x)$ be a weak derivative of the function $f(x)$. Consider the difference

$$\alpha(x,\delta x) = f(x + \delta x) - f(x) - f'(x)\,\delta x\,.$$

We shall choose δx so small that the point $x + t\delta x$ will belong to the vicinity S of the point x at all $t \in [0,1]$. In this case the numeric function $\varphi(t) = gf(x + t\delta x)$ of the variable t is differentiable at all $t \in [0,1]$ at any functionall $g \in Y^*$ as

$$\lim_{\Delta t \to 0} \frac{\varphi(t + \Delta t) - \varphi(t)}{\Delta t} = \lim_{\Delta t \to 0} \frac{g[f(x + (t + \Delta t)\delta x) - f(x + t\delta x)]}{\Delta t}$$

$$= gf'(x + t\delta x)\delta x$$

owing to the weak differentiability of the function $f(x)$ in the vicinity S of the point x and the boundedness of the functional g. Thus $\varphi'(t) = gf'(x + t\delta x)\delta x$. After applying to the function $\varphi(t)$ finite increments formula (9.6) we shall have

$$\varphi(1) - \varphi(0) = \varphi'(t_0)$$

at some $t_0 \in [0, 1]$ or

$$g[f(x + \delta x) - f(x)] = gf'(x + t_0\delta x)\delta x$$

and

$$g\alpha(x, \delta x) = g[f'(x + t_0\delta x) - f'(x)]\,\delta x\,.$$

Hence it follows

$$|g\alpha(x, \delta x)| \le \|g\|\,\|f'(x + t_0\delta x) - f'(x)\|\,\|\delta x\|\,. \qquad (9.9)$$

Now we shall consider $g\alpha(x, \delta x)$ at fixed x and δx as a linear functional of g. By definition (4.19) the norm of this functional is equal

$$\|\alpha(x, \delta x)\| = \sup_g \frac{|g\alpha(x, \delta x)|}{\|g\|}\,.$$

From this formula follows that there exists the functional $g \in Y^*$ for which $|g\alpha(x, \delta x)|\,/\,\|g\|$ as close as possible to $\|\alpha(x, \delta x)\|$, and consequently,

$$\|\alpha(x, \delta x)\| < c\,|g\alpha(x, \delta x)|\,/\,\|g\|$$

at some $c > 1$. Hence and from inequality (9.9) we find

$$\|\alpha(x, \delta x)\| < c\,\|f'(x + t_0\delta x) - f'(x)\|\,\|\delta x\| = o(\|\delta x\|)$$

in consequence of the continuity of $f'(x)$ at the point x. Thus

$$f(x + \delta x) = f(x) + f'(x)\delta x + \alpha(x, \delta x) = f(x) + f'(x)\delta x + o(\|\delta x\|)\,,$$

what proves the strong differentiability of the function $f(x)$ at the point x and the coincidence of its strong derivative with the weak one. ◁

R e m a r k. Notice that formula (9.5) of the differentiation of a composite function in the general case is not true for the weak derivatives. But the following theorem is valid.

Theorem 9.5. *If* $f(x)$ *is a continuous linear function* $f(x) = Ax$ *and the function* $g(y)$ *is weakly differentiable at the point* $y_0 = Ax_0$ *then the function* $gf(x) = g(Ax)$ *is weakly differentiable at the point* x_0 *and its weak derivative at this point is determined by formula (9.5):*

$$(gf)'(x_0) = g'(Ax_0)A. \qquad (9.10)$$

▷ Really, in this case the weak and the strong differentials of the function $f(x) = Ax$ coincide and $\delta y = \delta f(x) = A\delta x$. Therefore

$$[g(y_0 + t\delta y) - g(y_0)]/t = [g(Ax_0 + tA\delta x) - g(Ax_0)]/t.$$

Passing to the limit at $t \to 0$ we get

$$\delta g(y_0, \delta y) = g'(Ax_0)A\delta x.$$

Hence it is clear that the function $g(Ax)$ is weakly differentiable at the point x_0 and its weak derivative is determined by formula (9.10). ◁

9.1.6. Differentials and Derivatives of Higher Orders

A *differential of the second order* of the function $y = f(x)$ is called a differential of its differential,

$$d^2 f(x) = d[df(x)].$$

A *derivative of the second order* or *the second derivative* of the function $f(x)$ is called a derivative of its derivative,

$$f''(x) = [f'(x)]'.$$

A *differential of the order* p of the function $f(x)$ is called a differential of its differential of the $(p-1)$–th order,

$$d^p f(x) = d[d^{p-1} f(x)].$$

A *derivative of the order* p or *the p–th derivative* of the function $f(x)$ is called a derivative of its $(p-1)$–derivative,

$$f^{(p)}(x) = [f^{(p-1)}(x)]'.$$

It is clear that the p–th derivative represents the p–th linear operator whose all arguments coincide with δx, i.e. the p–th degree operator so that

$$d^p f(x) = f^{(p)}(x)\delta x^p \,. \tag{9.11}$$

R e m a r k. Here we use the same denotion for the result of the action of the p–th linear operator on the coinciding values of all its p arguments as in Example 9.3.

All stated denotions refer to both strong differentials and derivatives and to weak ones. In particular, a weak differential of the p–th order of the function $f(x)$ is determined by the formula

$$\delta^p f(x,\delta x) = \left[\frac{d^p}{dt^p} f(x+t\delta x)\right]_{t=0} \,.$$

E x a m p l e 9.7. Consider the functional $f(x)$ of Example 9 .4 in the case when its derivative is determined by formula (9.3). Supposing that the derivative of the functional $f'(t;x)$ is also expressed by formula of the form (9.3) we find

$$f'(t,x+\delta x) - f'(t,x) = \int_T f''(t,s;x)\,\delta x(s)\,ds + o(|\delta x|)$$

and consequently,

$$[f'(x+\delta x) - f'(x)]\delta x = \int_T [\int_T f''(t,s;x)\,\delta x(s)\,ds + o(|\delta x|)]\,\delta x(t)\,dt \,.$$

Hence we find

$$d^2 f(x) = f''(x)\delta x^2 = \int_T \int_T f''(t,s;x)\,\delta x(t)\,\delta x(s)\,dt\,ds \,.$$

Thus the second derivative of the functional $f(x)$ represents a square functional of the form

$$f''(x) = \int_T \int_T f''(t,s;x) \cdots dt\,ds \,,$$

where the points indicate the places where the values of the function δx at the values t and s of the arguments should be substituted. The functional $f''(t,s;x)$ of x which depends on t and s as the parameters is called *a second functional derivative* of the functional $f(x)$.

By the induction we find the p–th differential of the functional $f(x)$

$$d^p f(x) = f^{(p)}(x)\,\delta x^p$$

$$= \int_T \cdots \int_T f^{(p)}(t_1, \ldots, t_p; x)\delta x(t_1) \ldots \delta x(t_p)\, dt_1 \ldots dt_p\,, \qquad (9.12)$$

where the functional $f^{(p)}(t_1, \ldots, t_p; x)$ of x is called *the p-th functional derivative* of the functional $f(x)$.

It is easy to prove that in the case when a functional derivative of the p-th order at any function $x(t)$ is a continuous function of the parameters $t_1, \ldots, t_p$, it is symmetrical relative to $t_1, \ldots, t_p$. It is sufficient to prove this for $p = 2$. In this case, evidently, we have

$$I = \int_T \int_T [f''(t, s; x) - f''(s, t; x)]\, \delta x(t)\, \delta x(s)\, dt\, ds = 0$$

for any continuous function $\delta x(t)$. Supposing that $f''(t_0, s_0; x) - f''(s_0, t_0; x) > 0$ at some t_0, s_0 we shall have $f''(t, s; x) - f''(s, t; x) > 0$ in some vicinity of the point t_0, s_0, $|t - t_0| < \delta$, $|s - s_0| < \delta$. After taking $\delta x(t) > 0$ at $|t - t_0| < \delta$ and at $|t - s_0| < \delta$ and $\delta x(t) = 0$ outside the intervals $|t - t_0| < \delta$, $|t - s_0| < \delta$, we shall have $I > 0$ what is impossible. Analogously it is proved that $f''(t, s, x) - f''(s, t, x)$ cannot be anywhere negative on the set (compact) $T^2 = T \times T$. This property of the functional derivatives is quite similar to the independence of the mixed derivatives of the function of the finite–dimensional vector argument of the order of the differentiation.

9.1.7. Taylor Formula and Series

The generalization of Taylor formula on the functions (mappings) in the linear spaces is given by the following theorem.

Theorem 9.6. *If the function $y = f(x)$ which maps the normed linear space X into the normed linear space Y has at all points of some convex set S the continuous derivatives till the order $n + 1$ inclusively then at all x, $x + \delta x \in S$ Taylor formula is valid*

$$f(x + \delta x) = f(x) + f'(x)\delta x + \cdots + \frac{1}{n!}f^{(n)}(x)\,\delta x^n + R_n\,, \qquad (9.13)$$

where the residual term R_n is determined by formula

$$R_n = R_n(x, \delta x)$$

$$= \frac{1}{n!}\int_0^1 (1 - t)^n f^{(n+1)}(x + t\delta x)\delta x^{n+1}\, dt\,. \qquad (9.14)$$

▷ At $n = 0$ formulae (9.13) and (9.14) give the formula of finite increments (9.6) which was derived earlier. Therefore for proving the theorem we may use the induction method. Supposing that it is true for $n - 1$ we shall have

$$f'(x+\delta x) = f'(x)+f''(x)\delta x+\cdots+\frac{1}{(n-1)!}f^{(n)}(x)\,\delta x^{n-1}+R_{n-1}\,, \quad (9.15)$$

where

$$R_{n-1} = R_{n-1}(x,\delta x) = \frac{1}{(n-1)!}\int_0^1 (1-t)^{n-1}f^{(n+1)}(x+t\delta x)\delta x^n\,dt\,.$$

$$(9.16)$$

Substituting (9.15) into formula of finite increments (9.6), we find

$$f(x+\delta x) - f(x) = \int_0^1 f'(x+t\delta x)\,\delta x\,dt$$

$$= f'(x)\delta x + \frac{1}{2}f''(x)\,\delta x^2 + \cdots + \frac{1}{n!}f^{(n)}(x)\,\delta x^n + \int_0^1 R_{n-1}(x,t\delta x)\,\delta x\,dt\,.$$

Thus formula (9.13) is proved with

$$R_n = R_n(x,\delta x) = \int_0^1 R_{n-1}(x,t\delta x)\,\delta x\,dt\,. \quad (9.17)$$

In order to prove formula (9.14) we change in (9.16) the denotion of the integration variable and substitute δx by $t\delta x$. Then we obtain

$$R_{n-1}(x,t\delta x) = \frac{1}{(n-1)!}\int_0^1 (1-s)^{n-1}f^{(n+1)}(x+ts\,\delta x)\,t^n\,\delta x^n\,ds$$

$$= \frac{1}{(n-1)!}\int_0^t (t-u)^{n-1}f^{(n+1)}(x+u\delta x)\,\delta x^n\,du\,.$$

Substituting this expression into (9.17) we shall have

$$R_n = \int_0^1 R_{n-1}(x,t\delta x)\,\delta x\,dt$$

$$= \frac{1}{(n-1)!} \int_0^1 dt \int_0^t (t-u)^{n-1} f^{(n+1)}(x+u\delta x)\,\delta x^{n+1} du$$

$$= \frac{1}{(n-1)!} \int_0^1 f^{(n+1)}(x+u\delta x)\,\delta x^{n+1} du \int_u^1 (t-u)^{n-1} dt$$

$$= \frac{1}{n!} \int_0^1 (1-u)^n f^{(n+1)}(x+u\delta x)\,\delta x^{n+1} du \, .$$

This formula differs from (9.14) only by the denotion of the integration variable. ◁

Corollary. *In the special case when $f(x)$ represents a functional formula (9.14) for the residual term may be rewritten in the form*

$$R_n = R_n(x, \delta x) = \frac{1}{(n+1)!} f^{(n+1)}(x+\theta\delta x)\,\delta x^{n+1}, \qquad (9.18)$$

where $\theta \in (1,0)$.

▷ As in this case $f^{(n+1)}(x+t\delta x)\delta x^{n+1}$ is the numeric function and $1-t \geq 0$ at $t \in [0,1]$ then we may apply to the integral in (9.14) the theorem about the mean. Then we receive

$$\int_0^1 (1-t)^n f^{(n+1)}(x+t\delta x)\delta x^{n+1} dt$$

$$= f^{(n+1)}(x+\theta\delta x)\delta x^{n+1} \int_0^1 (1-t)^n \, dt$$

$$= \frac{1}{n+1} f^{(n+1)}(x+\theta\delta x)\delta x^{n+1} \, .$$

Substituting this expression into (9.14) we get (9.18). ◁

If the function $f(x)$ is infinitely differentiable and $R_n \to 0$ at $n \to \infty$ then formula (9.13) gives Taylor series expansion of the function $f(x)$

$$f(x+\delta x) = \sum_{k=0}^{\infty} \frac{1}{k!} f^{(k)}(x)\,\delta x^k \, . \qquad (9.19)$$

R e m a r k. Under the conditions of Theorems 9.2 and 9.6 in consequence of Theorems 9.3 and 9.4 the requirement of the existence of strong derivatives and the requirement of the existence of weak derivatives are completely equivalent.

On the basis of formula (9.11) for the differentials of various orders Taylor formula (9.13) may be rewritten in the form

$$f(x + \delta x) = f(x) + \sum_{k=1}^{n} \frac{1}{k!} d^k f(x) + R_n , \qquad (9.20)$$

where

$$R_n = R_n(x, \delta x) = \frac{1}{n!} \int_0^1 (1 - t)^n d^{n+1} f(x + t\delta x) \, dt .$$

E x a m p l e 9.8. In the case of the finite–dimensional space $X = R^m$ any n–linear operator A at the same values of its n arguments is determined by the formula

$$A x^n = \sum_{k_1 + \ldots + k_m = n} a_{k_1, \ldots, k_m} x_1^{k_1} \ldots x_m^{k_m} ,$$

where $x_1, \ldots, x_m$ are the coordinates of the vector x and the summation is taken over all nonnegative integers $k_1, \ldots, k_m$ whose sum is equal to n. So, the common item of the formula and Taylor series is determined by the formula

$$\frac{1}{p!} f^{(p)}(x) \delta x^p = \frac{1}{p!} \sum_{k_1 + \ldots + k_m = p} \frac{\partial^{k_1 + \ldots + k_m} f(x)}{\partial x_1^{k_1} \ldots \partial x_m^{k_m}} \delta x_1^{k_1} \ldots \delta x_m^{k_m} ,$$

and formulae (9.13) and (9.19) are transformed into known formulae of mathematical analysis for the functions of the finite–dimensional arguments.

E x a m p l e 9.9. In the case of the functional $f(x)$ on the space $C(T)$ of the continuous scalar function on the compact T considered in Example 9.7 Taylor formula (9.13) with the residual term in the form of (9.18) on the basis of (9.12) takes the form

$$f(x + \delta x) = f(x) + \int_T f'(t; x) \delta x(t) \, dt +$$

$$\cdots + \frac{1}{n!} \int_T \cdots \int_T f^{(n)}(t_1, \ldots, t_n; x) \delta x(t_1) \ldots \delta x(t_n) \, dt_1 \ldots dt_n$$

$$+ \frac{1}{(n+1)!} \int_T \cdots \int_T f^{(n+1)}(t_1, \ldots, t_{n+1}; x + \theta \delta x) \delta x(t_1)$$

$$\ldots \delta x(t_{n+1}) \, dt_1 \ldots dt_{n+1} .$$

This generalization of Taylor formula and the corresponding generalization of Taylor series on the functionals on $C(T)$ in the case where $T = [a, b]$ was given by Volterra. Therefore the power series in the space $C(T)$ of the form

$$k_0 + \int_T k_1(t)\, x(t)\, dt +$$

$$\cdots + \int_T \cdots \int_T k_n(t_1, \ldots, t_n)\, x(t_1) \ldots x(t_n)\, dt_1 \ldots dt_n + \cdots$$

are usually called *Volterra series*. These series and their finite segments are often used in the control theory for the description of the nonlinear systems.

If we get rid of the explicit expression of the residual term R_n in (9.13) then Theorem 9.6 may be intensified.

Theorem 9.7. *If the function $y = f(x)$ has the continuous derivatives till the order n at all the points of some convex set S then at all x, $x + \delta x \in S$ Taylor formula (9.13) in which $R_n = o(\|\delta x\|^n)$ is valid.*

▷ The proof in a literal sense repeats the proof of Theorem 9.6 and from $R_{n-1} = o(\|\delta x\|^{n-1})$ and formula (9.17) follows

$$\| R_n \| \le \int_0^1 \| R_{n-1}(x, t\delta x) \|\, \| \delta x \|\ dt \le o(\|\delta x\|^{n-1})\, \| \delta x \|,$$

i.e. $R_n = o(\|\delta x\|^n)$. ◁

For detailed theory of differention of functionals and operators refer to (Schwartz 1969, Struwe 1990).

P r o b l e m s

9.1.1. Let A be an operator which maps the space X into the space of the bounded linear operators $\mathcal{B}(X, Z)$ where $Z = \mathcal{B}(X, \mathcal{B}(X, Y))$. Similarly as in Example 9.2 we may consider A as a bilinear operator which maps X^2 into $\mathcal{B}(X, Y)$, $(Ax_1)x_2 = Ax_1x_2 \in \mathcal{B}(X, Y)$. Here $(Ax_1, x_2)x_3 = Ax_1x_2x_3 \in Y$. Consequently, A may be considered as *a trilinear operator* (linear on each from three variables x_1, x_2, x_3 which maps X^3 into Y). Generalizing this construction let us consider a bounded linear operator A which maps X into the space of the bounded linear operators $\mathcal{B}(X, Z_1)$ where $Z_k = \mathcal{B}(X, Z_{k+1})$ $(k = 1, \ldots, n - 2)$, $Z_{n-1} = \mathcal{B}(X, Y)$. The operator A may be considered as n-*linear operator* which maps the product of the spaces $X^n = X \times \cdots \times X$ into Y as $Ax_1 \ldots x_n \in Y$. Apparently, this operator is linear on each from the variables $x_1, \ldots, x_n$ at the fixed values of the others. Prove that without loss of the generality we may consider the operator A as symmetric in that sense that

$$Ax_{i_1} \ldots x_{i_n} = Ax_1 \ldots x_n$$

at all permutations $i_1, \ldots, i_n$ of the numbers $1, \ldots, n$.

I n s t r u c t i o n. Use the induction method. At $x_1 = \cdots = x_n = x$ the operator A will represent *a power operator* of the n–th degree which maps X into Y what may be shortly written in the form Ax^n.

9.1.2. Prove that the function $f(x) = Ax^n$ is differentiable and its derivative is equal

$$f'(x) = nAx^{n-1}.$$

This formula generalizes the known formula of the differentiation of power function in the normed linear spaces.

I n s t r u c t i o n. Using the symmetry of the operator A prove that

$$A(x + \delta x)^n = Ax^n + nAx^{n-1}\delta x + o(\|\delta x\|).$$

9.1.3. Find strong derivatives of the following mappings:
1) $F : R^3 \to R$, $y = x_1 x_2^2 + x_3^3$ at the point $(1, 1, 1)$;
2) $F : R \to R^3$, $y_1 = \sin \pi t$, $y_2 = \cos \pi t$, $y_3 = t$ at the point $t = 1$;
3) $F : R^2 \to R^3$, $y_1 = e^{x_1} \sin \pi x_2$, $y_2 = e^{x_1} \cos \pi x_2$, $y_3 = e^{x_1}$ at the point $(1, 1, 1)$;
4) $F : R^3 \to R^2$, $y_1 = \left[4 - x_1^2 + x_2^2 - x_3^2\right]^{1/2}$, $y_2 = x_1 x_2 x_3$ at the point $(1, 1, 1)$.

9.1.4. Prove that the function

$$f(x_1, x_2) = \begin{cases} \dfrac{x_1^3 x_2}{x_1^4 + x_2^2}, & (x_1, x_2) \neq (0, 0) \\ 0, & (x_1, x_2) = (0, 0) \end{cases}$$

is weakly differentiable at the point $(0, 0)$ and is not strongly differentiable.

9.1.5. Find the first strong derivative of the functional $F(x) = (x, x)$ in the real H–space.

S o l u t i o n. Let fix $x_0 \in H$ and take increment $h \in H$. From formulae

$$F(x_0 + th) - F(x_0) = (x_0 + th) - (x_0, x_0) = 2t(x_0, h) + t^2(h, h),$$

$$\lim_{t \to 0} \frac{F(x_0 + th) - F(x_0)}{t} = 2(x_0, h)$$

we find $F'(x) = 2x$.

9.1.6. Find the first and second strong derivatives of the functional $F(x)$ $= \| x \|$.

S o l u t i o n. Mind that $\| x \| = \sqrt{(x, x)}$ and use the result of Problem 9.1.5. So, we get

$$F'(x) = \frac{1}{2\sqrt{(x, x)}} \cdot 2x = \frac{x}{\| x \|} \quad \text{at } x \neq 0.$$

At $x \neq 0$ the strong derivative does not exist as the function $|t|$ is nondifferentiable at $t = 0$. Using the known property of bilinear operator we get

$$F''(x)(h_1, h_2) = \frac{(h_1, h_2)}{\| x \|} - \frac{(x, h_1)(x, h_2)}{\| x \|^3},$$

$$d^2 F(x) = \frac{\| h \|^2 \| x \|^2 - (x, h)^2}{\| x \|^3}.$$

9.1.7. Find the strong derivatives at the point u_0 of the following functionals:

1) $F(u) = \sin u(x)$ in the space $C[0, \pi]$, $u_0(x) = \cos x$;
2) $F(u) = \cos u(x)$ in the space $C[0, \pi]$, $u_0(x) = \sin x$;
3) $F(u) = u(x) - e^{xu(x)}$ in the space $C[0, 1]$, $u_0(x) \equiv 0$;
4) $F(u) = x^3 u(x) + \operatorname{sh} u(x)$ in the space $C[0, 1]$, $u_0(x) \equiv 0$.

9.1.8. Find the strong derivative relative to u at the point $u_0 \equiv 0$ of the integral operator $F : C[0, \pi] \to C[0, \pi]$ with the parameter λ

$$F(u) = u(x) - \lambda \int_0^\pi \cos[x + u(s)] ds.$$

Prove that all the solutions of equation $F'(0)h = \cos x$ have the form $h(x) = \cos x + C \sin x$, where at $\lambda \neq -1/2$, $C = 0$ and at $\lambda = -1/2$ C is an arbitrary one.

9.1.9. Prove that the derivative $F'(x)$ of the nonlinear integral operator

$$F(x) = \int \cdots \int f(t, \tau_1, \ldots, \tau_n, x(\tau_1), \ldots, x(\tau_n)) \mu(d\tau_1) \ldots \mu(d\tau_n) \quad (9.21)$$

represents a linear integral operator

$$F'(x) = \int K(t, \tau) \cdot \mu(d\tau), \tag{9.22}$$

whose kernel is determined by the formula

$$K(t, \tau) = \int \cdots \int [f_1(t, \tau, \sigma_1, \ldots, \sigma_{n-1}, x(\tau), x(\sigma_1), \ldots, x(\sigma_{n-1})) +$$

$$\cdots + f_k(t, \sigma_1, \ldots, \sigma_{k-1}, \tau, \sigma_k, \ldots, \sigma_{n-1}, x(\sigma_1), \ldots, x(\sigma_{k-1}), x(\tau),$$

$$x(\sigma_k), \ldots, x(\sigma_{n-1})) + \cdots + f_n(t, \sigma_1, \ldots, \sigma_{n-1}, \tau, x(\sigma_1),$$

$$\ldots, x(\sigma_{n-1}), x(\tau))] \, \mu(d\sigma_1) \ldots \mu(d\sigma_{n-1}),$$

where $f_k(t, \tau_1, \ldots, \tau_n, x_1, \ldots, x_n)$ is the first derivative of the function $f(t, \tau_1, \ldots, \tau_n, x_1, \ldots, x_n)$ with respect to x_k (a matrix of the first derivatives of the coordinates of the vector function f with respect to the coordinates of the vector x_k). Formulate the sufficient condition of the differentiability of the operator $F(x)$ in the cases $X = C(T)$ and $X = L_p(\Delta, A, \mu)$, $p > 1$.

9.1.10. Prove that in the case when each of the functions $f(x) = F(x)$ and $g(y) = G(y)$ represents a linear integral operator of Example 9.3 the derivative of their composition $GF(x)$ represents a linear integral operator of the form (9.22) whose kernel $H(u, t)$ represents a composition of the kernels $K(s, t)$ and $L(u, s)$ which correspond to the operators $F(x)$ and $G(y)$:

$$H(u, t) = \int L(u, s) \, K(s, t) \, \nu(ds),$$

where ν is a measure with respect to which the integration is performed in expression (9.1) or (9.21) of the operator $G(y)$.

9.1.11. Consider the square integral operator

$$[F(x, x)](t) = x(t) \int_0^1 K(t, s) x(s) ds,$$

$K(t, s)$ being continuous at $0 \leq t, s \leq 1$. Prove that the corresponding bilinear operator is equal to

$$[F(x, y)](t) = \tfrac{1}{2} x(t) \int_0^1 K(t, s) y(s) ds + \tfrac{1}{2} y(t) \int_0^1 K(t, s) x(s) ds.$$

Find the following formulae for the first and second derivatives:

$$\left[\tfrac{d}{dx} F(x, x) \right] h = 2h(x, h),$$

$$\left[\tfrac{d^2}{dx^2} F(x, x) \right] (h_1, h_2) = 2F(h_1, h_2).$$

9.1.12. Consider the nonlinear integral operator

$$F(x) = x(t) + \int_a^b K(t, s) e^{x(s)} ds,$$

$K(t, s)$ being continuous at $a \leq t, s \leq b$. Prove that $F(x)$ has the following Taylor series:

$$F(x_0 + h) = F(x_0) + h(t) + \int\limits_a^b K(t, s)e^{x_0(s)}h(s)ds$$

$$+ \sum_{n=2}^\infty \frac{1}{n!} \int\limits_a^b K(t, s)e^{x_0(s)}h^n(s)ds .$$

9.1.13. Show that the solution of the Cauchy problem

$$y' + ay + ky^3 = x(t), \quad y(0) = 0 ,$$

in the space $C^1([0, 1])$ is determined by the series

$$y(t) = \int\limits_0^t g_1(t, \tau) x(\tau)$$

$$+ \int\limits_0^t \int\limits_0^t \int\limits_0^t g_3(t, \tau_1, \tau_2, \tau_3) x(\tau_1) x(\tau_2) x(\tau_3) \, d\tau_1 d\tau_2 d\tau_3 +$$

$$\cdots + \int\limits_0^t \cdots \int\limits_0^t g_{2n-1}(t, \tau_1, \ldots, \tau_{2n-1}) x(\tau_1) \ldots x(\tau_{2n-1}) \, d\tau_1 \ldots d\tau_{2n-1} + \cdots ,$$

where

$$g_1(t, \tau) = \exp\{-a(t - \tau)\} ,$$

$$g_3(t, \tau_1, \tau_2, \tau_3) = (k/2a) \exp\{-a(t - \tau_1 - \tau_2 - \tau_3)\} [\exp\{-2at\} - \exp\{-2 \max(\tau_1, \tau_2, \tau_3)\}], \ldots .$$

I n s t r u c t i o n. Determine the successive approximations to the solution of $y(t)$ by the linear equations $y_1' + ay_1 = x$, $y_n' + ay_n = x - ky_{n-1}$ $(n = 2, 3, \ldots)$ which are easily analytically solved.

9.1.14. Consider the nonlinear differential operator

$$F(x) = \frac{d^2 x(t)}{dt^2} + \sin x(t)$$

acting $C^2[0, 1] \rightarrow C[0, 1]$. Prove the following formulae:

1) $d^n F(x_0, h) = \sin \left(t + \frac{n\pi}{2}\right) h^n$, $n \geq 2$, $x_0(t) = t$;

2) $F(x_0 + h) = \sin t + h'' + h \cos t + \sum_{n+2}^\infty \frac{\sin(t + \pi n/2)}{n!} h^n$, $x_0(t) = t$.

9.1.15. Prove that the norms in the spaces l_p and L_p are strongly differentiable if $1 < p < \infty$.

9.2. Extrema of Functionals

9.2.1. Necessary Condition of an Extremum

In many problems of practice arise the problems of searching the values of the arguments of the real function at which it achieves the extremal values. In the case of the function of a scalar or a finite–dimenstional vector argument the methods of the solution of such problems form the basis of the calculus of variations. It is natural in the functional analysis to consider the most general extremal problems for the real functional on the linear spaces. Here we restrict ourselves to the functionals on the normed linear spaces.

Let $f(x)$ be a real functional on the normed linear space X. The point x is called *the maximum point (minimum)* of the functional $f(x)$ if there exists such vicinity V of the point x that $f(x + \delta x) - f(x) < 0$ (> 0) at all $x + \delta x \in V$ (i.e. at all sufficiently small on the norm δx). The maximum and the minimum points of a functional form a set of the *extremum points* of the functional $f(x)$.

Theorem 9.8. *If the functional $f(x)$ achieves the extremum at the point x and has at this point continuous weak derivative $f'(x)$ then at this point*

$$f'(x) = 0. \tag{9.23}$$

▷ On the basis of the known necessary condition of the function extremum $f(x + t\delta x)$ of the scalar variable t at the point $t = 0$ is the equality to zero of the derivative of this function:

$$\left[\frac{d}{dt} f(x + t\delta x) \right]_{t=0} = 0 .$$

But this derivative is a weak differential of the function $f(x)$ (Subsection 9.1.4). Therefore because of weak differentiability of $f(x)$ the previous equality takes the form

$$f'(x)\, \delta x = 0 \quad \forall \delta x ,$$

whence follows condition (9.23). ◁

R e m a r k. Eq. (9.23) is a necessary condition of the extremum of the functional $f(x)$.

E x a m p l e 9.10. Consider the simplest functional which is studied in calculus of variations

$$F(x) = \int_a^b f(t, x(t), x'(t))\, dt \tag{9.24}$$

in the space $C^1([a, b])$ of the functions continuous together with their first derivatives. The differential of this functional on the basis of the results of Example 9.3 is determined by the formula [a]

$$F'(x)\,\delta x = \int_a^b [f_x(t, x(t), x'(t))\,\delta x(t) + f_{x'}(t, x(t), x'(t))\,\delta x'(t)]\, dt \,.$$

Here $x(t)$ and $x'(t)$ may be considered as the coordinates of two–dimensional vector function $[\,x(t)\, x'(t)]^T$ where $f_x(t, x, x')$ and $f_{x'}(t, x, x')$ are partial derivatives of the functions $f(t, x, x')$ with respect to x and x'. After performing the integration by parts in the second item we get

$$F'(x)\delta x = f_{x'}(b, x(b), x'(b))\,\delta x(b) - f_{x'}(a, x(a), x'(a))\,\delta x(a)$$

$$+ \int_a^b [f_x(t, x(t), x'(t)) - \frac{d}{dt} f_{x'}(t, x(t), x'(t))]\,\delta x(t)\, dt \,. \tag{9.25}$$

In order that the function $x(t)$ will realize the extremum of the functional $F(x)$ (will be its point of the extremum) it is necessary that its right–hand side of (9.25) will be equal to zero for all continuous functions $\delta x(t)$ together with its first derivative. In particular, at $\delta x(a) = \delta x(b) = 0$ the integral in right–hand side of (9.25) must be equal to zero at all $\delta x(t)$. But for this purpose it is necessary and sufficient that the expression in square brackets under the sign of the integral will be equal to zero [b] :

$$f_x(t, x(t), x'(t)) - \frac{d}{dt} f_{x'}(t, x(t), x'(t)) = 0 \tag{9.26}$$

or, shortly,

$$f_x - \frac{d}{dt} f_{x'} = 0 \,.$$

---

[a] A weak differential of the functional $F(x)$ in the calculus of variations is called *a variation* of this functional. But in practice in all problems of calculus of variations there exist strong differentials of considered functional which coincide with the weak ones on the basis of Theorem 9.3.

[b] For proving it is sufficient to take $\delta x(t) > 0$ in the region in which the expression in square brackets is positive (or negative) and $\delta x(t) = 0$ outside of this region. For such function $\delta x(t)$ the integral in (9.25) will not be equal to zero.

This differential equation is called in calculus of variations the *Euler equation*.

As regards to the items in the right–hand side of (9.25) which correspond to the ends of the intervals of the integration a, b in calculus of variations two classes of the problems are recognized: the problems with fixed ends when a, $x(a)$, b, $x(b)$ have given values and the problems with moving ends when a, $x(a)$, b, $x(b)$ which are subject to some conditions. In the first case it is always $\delta x(a) = \delta x(b) = 0$ and the differential equation of the second order (9.26) with given $x(a)$ and $x(b)$ form a boundary–value problem. In the second case the conditions to which must satisfy a, $x(a)$, b, $x(b)$ should be given. As a rule it is required that the ends $(a, x(a))$ and $(b, x(b))$ of the segment of the curve $y = x(t)$ with respect to which the integration is performed (9.24) should lie on the given curves $\varphi(t, x) = 0$, $\psi(t, x) = 0$. In this case at given a and b (the points $(a, x(a))$ and $(b, x(b))$ must lie on the lines $t = a$ and $t = b$) the coefficients at $\delta x(a)$ and $\delta x(b)$ in (9.25) must be equal to zero

$$f_{x'}(a, x(a), x'(a)) = f_{x'}(b, x(b), x'(b)) = 0 \,. \tag{9.27}$$

These equalities represent boundary conditions for Eq. (9.26). If a and b are not given then in the right–hand side of (9.25) an item will be added

$$-f(a, x(a), x'(a)) \, \delta a + f(b, x(b), x'(b)) \, \delta b \,,$$

which represents a complete differential of functional (9.24) at fixed function $x(t)$ and conditions (9.27) are substituted by the conditions

$$f(a, x(a), x'(a))\delta a + f_{x'}(a, x(a), x'(a))\delta x(a) = 0 \,,$$
$$f(b, x(b), x'(b)) \, \delta b + f_{x'}(b, x(b), x'(b)) \, \delta x(b) = 0 \,. \tag{9.28}$$

The conditions at the ends of $\varphi(a, x(a)) = 0$, $\psi(b, x(b)) = 0$ give here

$$\varphi_t(a, x(a))\delta a + \varphi_x(a, x(a))\Delta x(a) = 0 \,,$$

$$\psi_t(b, x(b))\delta b + \psi_x(b, x(b)) \, \Delta x(b) = 0 \,,$$

where by indexes are indicated the derivatives with respect to the corresponding arguments, and $\Delta x(a)$ and $\Delta x(b)$ are complete increments of the ordinates of the ends caused by both the changes of the abscissae of the ends a, b and the values of the increments $\delta x(a)$, $\delta x(b)$ of the function $x(t)$ on the ends

$$\Delta x(a) = x'(a) \, \delta a + \delta x(a) \,,$$

$$\Delta x(b) = x'(b) \, \delta b + \delta x(b) \,.$$

Thus the conditions on the ends give the connection equation to which the increments of the coordinates of the ends are subject. After expressing $\delta x(a)$ in terms of δa and $\delta x(b)$ in terms of δb and substituting them in Eqs. (9.28) we obtain instead of Eqs. (9.27) the following boundary conditions:

$$f(a, x(a), x'(a))\varphi_x(a, x(a))$$

$$-f_{x'}(a, x(a), x'(a))[\varphi_t(a, x(a)) + \varphi_x(a, x(a))x'(a)] = 0\,,$$

$$f(b, x(b), x'(b))\psi_x(b, x(b))$$

$$-f_{x'}(b, x(b), x'(b))[\psi_t(b, x(b)) + \psi_x(b, x(b))x'(b)] = 0\,,$$

$$\varphi(a, x(a)) = 0\,, \quad \psi(b, x(b)) = 0\,. \tag{9.29}$$

The mixed problems when one of the ends of curve segment $y = x(t)$ is fixed are also possible.

All mentioned refers to the case of p–dimensional vector function $x(t)$. In this case $f_x(t, x, x')$ and $f_{x'}(t, x, x')$ represent the matrices–rows of the derivatives of the function $f(t, x, x')$ with respect to the coordinates of the vectors x and x'. Here in the case of the m–dimensional vector function φ, $m \leq p$ the differentiation of the connection equation $\varphi(a, x(a)) = 0$ gives m scalar equations for p coordinates of the vector $\delta x(a)$ and δa. After expressing from them m in terms of the residual $p - m$ and δa, substituting the obtained expressions into Eqs. (9.28) and equalizing to zero the coefficients at δa and at coordinates of the vector $\delta x(a)$ which remained arbitrary we shall receive together with the equations $\varphi(a, x(a)) = 0$ $p+1$ scalar boundary conditions for (9.26). The other $p+1$ boundary conditions will be received from the connection equation $\psi(b, x(b)) = 0$ on the other end.

E x a m p l e 9.11. In more general case of the functional

$$F(x) = \int\limits_a^b f(t, x(t), \ldots, x^{(n)}(t))\, dt \tag{9.30}$$

on the space $C^n([a, b])$ of the functions which are continuous together with their derivatives till n order inclusively we obtain in exactly the same way

$$F'(x)\,\delta x = \sum_{l=1}^n (-1)^l \sum_{k=l}^n (-1)^k \frac{d^{k-l}}{dt^{k-l}} f_{x^{(k)}}(t, x(t), \ldots, x^{(n)}(t))$$

$$\times \delta x^{(l-1)}(t)\Big|_a^b + \int\limits_a^b \sum_{k=0}^n (-1)^k \frac{d^k}{dt^k} f_{x^{(k)}}(t, x(t), \ldots, x^{(n)}(t))\,\delta x(t)\, dt\,.$$

$$\tag{9.31}$$

Hence follows at $\delta x(a) = \cdots = \delta x^{(n-1)}(a) = \delta x(b) = \cdots = \delta x^{(n-1)}(b)$ $= 0$ the following Euler equation:

$$\sum_{k=0}^{n} (-1)^k \frac{d^k}{dt^k} f_{x^{(k)}}(t, x, \ldots, x^{(n)}) = 0. \qquad (9.32)$$

This differential equation of the $2n$-th order together with given values $x(a)$, $\ldots$, $x^{(n-1)}(a)$, $x(b)$, $\ldots$, $x^{(n-1)}(b)$ (boundary conditions) is a necessary condition of the extremum of functional (9.30) in the problem with fixed ends a, $x(a)$, $\ldots$, $x^{(n-1)}(a)$, b, $x(b)$, $\ldots$, $x^{(n-1)}(b)$. If the variables $x(a)$, $\ldots$, $x^{(n-1)}(a)$, $x(b)$, $\ldots$, $x^{(n-1)}(b)$ are not given then at known a and b the coefficients at $\delta x(a)$, $\ldots$, $\delta x^{(n-1)}(a)$, $\delta x(b)$, $\ldots$, $\delta x^{(n-1)}(b)$ in (9.31) must be equal to zero:

$$\sum_{k=l}^{n} (-1)^k \frac{d^{k-l}}{dt^{k-l}} f_{x^{(k)}}(t, x(t), \ldots, x^{(n)}(t)) = 0$$

$$\text{at} \quad t = a \quad \text{and at} \quad t = b \quad (l = 1, \ldots, n). \qquad (9.33)$$

These equalities represent the boundary conditions for Eq. (9.32). If a and b are not given then similarly as in Example 9.10 we receive instead of conditions (9.33) at $l = 1$ the conditions

$$f(a, x(a), \ldots, x^{(n)}(a)) \varphi_x(a, x(a))$$

$$- \left[\sum_{k=1}^{n} (-1)^k \frac{d^{k-1}}{dt^{k-1}} f_{x^{(k)}}(t, x(t), \ldots, x^{(n)}(t)) \right]_{t=a}$$

$$\times [\varphi_t(a, x(a)) + \varphi_x(a, x(a)) x'(a)] = 0,$$

$$f(b, x(b), \ldots, x^{(n)}(b)) \psi_x(b, x(b))$$

$$- \left[\sum_{k=1}^{n} (-1)^k \frac{d^{k-1}}{dt^{k-1}} f_{x^{(k)}}(t, x(t), \ldots, x^{(n)}(t)) \right]_{t=b}$$

$$\times [\psi_t(b, x(b)) + \psi_x(b, x(b)) x'(b)] = 0,$$

$$\varphi(a, x(a)) = 0, \quad \psi(b, x(b)) = 0. \qquad (9.34)$$

The other conditions (9.33) $(l = 2, \ldots, n)$ remain unchanged.

The all obtained formulae and equations remain valid in the case of the vector function $x(t)$. In this case $f_{x^{(k)}}(t, x, \ldots, x^{(n)})$ represents a matrix–row of the derivatives of the function $f(t, x, \ldots, x^{(n)})$ with respect to the coordinates of the vector $x^{(k)}$. The boundary conditions for Eq. (9.32) in the problem with moving ends which substitute conditions (9.33) at $l = 1$ are obtained analogously as in Example 9.10.

E x a m p l e 9.12. In applications the problem of a function interpolation given by its values in a discrete series of the points (i.e. by given table) plays an important role. Here it is desirable that the interpolating curve will be more smooth especially in those cases when it is used for finding the derivatives of a given function. If we assume as a measure of the deviation of the curve from the direct line the second derivative of the interpolating function $x(t)$ it is natural to require that the mean square value of the second derivative will be minimal. Thus the problem of finding the most smooth interpolating function for the function which takes the values x_0, x_1, ..., x_n at the given points $a = t_0, t_1, \ldots, t_n = b$ is reduced to finding the function $x(t)$ which minimizes the functional

$$F(x) = \int\limits_a^b x''^2(t)\, dt$$

and satisfying the conditions $x(t_k) = x_k\ (k = 0, 1, \ldots, n)$. Here it is natural to find the function $x(t)$ in the class of the functions which are continuous together with their derivatives of the first and the second order (i.e. in the space $C^2([a, b])$) in the correspondence with general statement of the problem in Example 9.11. The Euler equation in this case has the form $x'''' = 0$ and its general solution represents a cubic polynomial. Boundary conditions (9.33) have in given case the form $x''(a) = x''(b) = 0$. The other two conditions (9.33) are fulfilled automatically at any function $x(t)$. Thus the necessary condition of the minimum of the functional $F(x)$ subjects the function $x(t)$ to $n + 3$ conditions $x(t_k) = x_k$ $(k = 0, 1, , \ldots, n)$, $x''(a) = x''(b) = 0$. But a cubic polynomial may satisfy only four arbitrary taken conditions. Consequently, it is the solution of the problem only at $n = 1$ and in this case represents a linear function. For solving this problem at any n pay attention to the fact that the solution of the equation $x'''' = 0$ is any function composed of the cubic polynomials. Therefore we take

$$x(t) = a_{k0} + a_{k1}t + a_{k2}t^2 + a_{k3}t^3 \quad \text{at} \quad t \in [t_{k-1}, t_k] \quad (k = 1, \ldots, n).$$

Then we receive for determining $4n$ unknown constants $a_{k0}, a_{k1}, a_{k2}, a_{k3}$ $(k = 1, \ldots, n)$ $4n$ conditions: $n + 1$ conditions $x(t_k) = x_k$ $(k = 0, 1, \ldots, n)$, $3n - 3$ conditions of the continuity $x(t), x'(t), x''(t)$ at the points $t_1, \ldots, t_{n-1}$ and two conditions $x''(t_0) = x''(t_n) = 0$. Thus we receive the unique solution of the stated problem.

A piecewise polynomial function which represents a polynomial of the p-th degree on each interval and which is continuous together with its derivatives till $p - q$ order inclusively is called *a spline of the p-th degree of the defect q*. It is clear that the defect of the spline cannot be smaller than 1 as the spline of the p-th degree has the piecewise constant derivative of the p-th order. Thus the solution of the stated problem represents a cubic spline of the defect 1. The splines (spline–functions) are widely used for approximate representation of the functions in modern numerical analysis.

9.2.2. Second Necessary Condition of an Extremum

As it is known from mathematical analysis in order the function of the finite–dimensional vector has an extremum at a given point besides the equality to zero of the first derivatives at this point it is also necessary the nonnegativity (in the case of the minimum) or the nonpositivity (in the case of the maximum) of the second differential in the vicinity of this point. An analogous theorem is valid for the functionals.

Theorem 9.9. *If the functional $f(x)$ has the minimum (the maximum) at the point x and has in some convex vicinity of this point the continuous different from zero second derivative then $f''(x)\delta x^2 \geq 0$ ($f''(x)\delta x^2 \leq 0$) at all sufficiently small on the norm δx.*

▷ For proving it is sufficient to use Taylor formula (9.13) and Theorem 9.7 for the residual term. Accounting that at the point x of the extremum $f'(x) = 0$ we get

$$f(x + \delta x) - f(x) = \frac{1}{2} f''(x)\delta x^2 + o(\|\delta x\|^2).\qquad(9.35)$$

Supposing that the condition $f''(x)\delta x^2 \geq 0$ is not fulfilled for some infinitesimals on the norm δx and accounting that the sign of the difference $f(x+\delta x) - f(x)$ coincides with the sign $f''(x)\delta x^2$ at all sufficiently small on the norm δx we shall have at some infinitesimals on the norm δx $f(x + \delta x) - f(x) < 0$ what contradicts the condition that the functional $f(x)$ achieves at the point x the minimum. ◁

9.2.3. Sufficient Condition of an Extremum

For the existence of the minimum (maximum) of the function of the finite–dimensional vector at the point at which its first differential is equal to zero the strong positivity (negativity) of its second differential at this point is sufficient. In the general case for the extremum of the functional this condition is not sufficient. The following theorem gives more strong sufficient condition of the extremum of the functional.

Theorem 9.10. *If at the point x we have $f'(x)\delta x = 0$ and $f''(x)\delta x^2 > c \parallel \delta x \parallel^2$ at all δx at some $c > 0$ then the functional $f(x)$ has at the point x the minimum.*

▷ If $f'(x)\delta x = 0$, $f''(x)\delta x^2 > c \parallel \delta x \parallel^2$ then at all δx at which $|o(\parallel \delta x \parallel^2)| < \varepsilon \parallel \delta x \parallel^2$ in (9.35) $0 < \varepsilon < c$ formula (9.35) gives

$$f(x + \delta x) - f(x) > (c - \varepsilon) \parallel \delta x \parallel^2 > 0.$$

As this inequality is valid at all sufficiently small δx then the functional $f(x)$ has at the point x the minimum. ◁

Corollary. *If at the point x we have $f'(x)\delta x = 0$ and $f''(x)\delta x^2 < -c \parallel \delta x \parallel^2$ at all δx at some $c > 0$ then the functional $f(x)$ has at the point x the maximum.*

▷ For proving it is sufficient to consider the functional $-f(x)$. As it satisfies the conditions of the theorem it has at the point x the minimum. Consequently, $f(x)$ has at the point x the maximum. ◁

R e m a r k. Theorem 9.10 gives the general sufficient condition of the extremum.

9.2.4. Conditional Extrema

In applications the problems of finding the extremum of the functional $f(x)$ at the given values of some other functionals $f_1(x)$, ..., $f_m(x)$ are widespread.

The point x is called *a point of the conditional maximum (minimum)* of the functional $f(x)$ at the conditions

$$f_k(x) = 0 \quad (k = 1, \ldots, m), \tag{9.36}$$

if $f(x + \delta x) - f(x) < 0 \, (> 0)$ at all x satisfying conditions (9.36) and at all δx satisfying conditions

$$df_k(x) = f'_k(x)\delta x = 0 \quad (k = 1, \ldots, m). \tag{9.37}$$

The problems of the conditional extremum of the functionals are solved similarly as in mathematical analysis the problems of the conditional extremum of the functions of the finite–dimensional vector variable are solved.

Taking into account that in accordance with the necessary condition of the extremum (Theorem 9.8) $f'(x)\delta x = 0$ at all δx satisfying conditions (9.37) we get at all such δx

$$\left[f'(x) + \sum_{k=1}^{m} \lambda_k f'_k(x) \right] \delta x = 0$$

and at all real $\lambda_1, \ldots, \lambda_m$. This suggest that the problem of finding the conditional extremum of the functional at conditions (9.36) may be reduced to the problem of finding the extremum (unconditional) of the functional

$$\varphi(x, \lambda) = f(x) + \sum_{k=1}^{m} \lambda_k f_k(x), \qquad (9.38)$$

which depends on m–dimensional real vector $\lambda = [\lambda_1, \ldots, \lambda_m]^T$.

Theorem 9.11. *If the functional $\varphi(x, \lambda)$ achieves the minimum (maximum) at the point x_λ at some values λ and λ may be chosen in such a way that the point x_λ will belong to the set S of the values x satisfying conditions (9.36) then at this point x_λ the functional $f(x)$ achieves the conditional extremum at conditons (9.36).*

▷ If $\varphi(x, \lambda)$ achieves the extremum at the point x_λ at given λ then

$$\varphi'(x_\lambda, \lambda) = f'(x_\lambda) + \sum_{k=1}^{m} \lambda_k f_k'(x_\lambda) = 0. \qquad (9.39)$$

This equation determines the point of the extremum x_λ as the function of λ. If $x_\lambda \in S$ at some λ then from (9.38) follows $\varphi(x_\lambda, \lambda) = f(x_\lambda)$ at these λ and

$$\varphi(x_\lambda + \delta x, \lambda) - \varphi(x_\lambda, \lambda) = f(x_\lambda + \delta x) - f(x_\lambda)$$

$$+ \sum_{k=1}^{m} \lambda_k [f_k(x_\lambda + \delta x) - f_k(x_\lambda)] = f(x_\lambda + \delta x) - f(x_\lambda) + o(\| \delta x \|)$$

at all δx satisfying conditions (9.37). Hence it is clear that at all sufficiently small on the norm δx satisfying conditions (9.37) the signs of the differences $f(x_\lambda + \delta x) - f(x_\lambda)$ and $\varphi(x_\lambda + \delta x, \lambda) - \varphi(x_\lambda, \lambda)$ coincide what proves the theorem. ◁

R e m a r k. Theorem 9.11 generalizes on the functionals the known method of Lagrange indefinite multipliers for finding the conditional extremum. For finding the conditional extremum of the functional $f(x)$ at conditions (9.36) the unconditional extremum of the functional $\varphi(x, \lambda)$ which is determined by formula (9.38) where $\lambda_1, \ldots, \lambda_m$ are indefinite Lagrange multipliers should be found. Further we need to determine $\lambda_1, \ldots, \lambda_m$ in such a way that the founded point of the extremum will satisfy conditions (9.36).

E x a m p l e 9.13. Let A be a compact self–adjoint operator in the H–space (Subsection 8.1.1). We pose the problem of finding the extremum of square functional $f(x) = (Ax, x)$ at the conditions

$$\|x\|^2 = (x, x) = 1, \quad (x_k, x) = 0 \quad (k = 1, \ldots, n - 1), \qquad (9.40)$$

where $x_1, \ldots, x_{n-1}$ are orthonormal eigenvectors of the operator A which correspond to the first $n - 1$ eigenvalues of $\lambda_1, \ldots, \lambda_{n-1}$ displaces in the nonincreasing order of their moduluses. By Theorem 9.11 it is necessary for this purpose to find the unconditional maximum of the functional

$$\varphi(x, \mu) = (Ax, x) + \sum_{k=1}^{n-1} \mu_k(x_k, x) + \mu_n[(x, x) - 1],$$

where $\mu_1, \ldots, \mu_n$ are the Lagrange multipliers. The necessary condition of the extremum in the correspondence with the results of Example 9.2 gives the equation

$$2Ax + \sum_{k=1}^{n-1} \mu_k x_k + 2\mu_n x = 0. \qquad (9.41)$$

The solution of this equation will satisfy conditions (9.40) if

$$2(Ax, x_p) + \mu_p = 0 \quad (p = 1, \ldots, n - 1).$$

But by virtue of the self–adjointness of the operator A and conditions (9.40) $(Ax, x_p) = (x, Ax_p) = \lambda_p(x, x_p) = 0$. Consequently, the solution x of Eq. (9.41) will satisfy conditions (9.40) at $\mu_1 = \cdots = \mu_{n-1} = 0$. Then Eq. (9.41) will take the form

$$Ax = -\mu_n x.$$

Hence it is clear that $-\mu_n$ should be assumed equal to one of the eigenvalues λ_n, $\lambda_{n+1}, \ldots$ of the operator A. Then x will be the eigenvector orthogonal to the eigenvectors $x_1, \ldots, x_{n-1}$. Here $(Ax, x) = \lambda_k$ where k is equal to one of the numbers $n, n + 1, \ldots$. As λ_n is the modulus–largest from the eigenvalues λ_n, $\lambda_{n+1}, \ldots$ then the problem will be solved if we assume $\mu_n = -\lambda_n$. Then $x = x_n$ will be the eigenvector of the operator A corresponding to the eigenvalue λ_n. Thus we came in another way to the result received in Subsection 8.1.1: the eigenvalue of the operator λ_n in the modulus nonincreasing order represents the extremal value of the square functional (Ax, x) and the corresponding eigenvector x_n realizes this extremal at the condition of the orthogonality of x to all eigenvectors $x_1, \ldots, x_{n-1}$ corresponding to the previous eigenvalues $\lambda_1, \ldots, \lambda_{n-1}$.

For detailed treatment of functionals extrema and its applications for convex and nonsmooth analysis, variational and optimization problems refer to (Clarke 1983, Ekeland and Temam 1976, Schwartz 1969, Struwe 1990).

Problems

9.2.1. The problem of finding the shortest distance between the points (a, c) and (b, d) of Euclidean plane is reduced to finding the minimum of the functional

$$F(x) = \int_a^b \sqrt{1 + x'^2(t)}\, dt \,.$$

Find the function $x(t)$ which realizes the minimum $F(x)$.

9.2.2. In exactly the same way the shortest distance between two points (a, c_1, c_2) and (b, d_1, d_2) of three–dimensional Euclidean space is determined by the minimization of the functional

$$F(x) = \int_a^b \sqrt{1 + x_1'^2(t) + x_2'^2(t)}\, dt \,.$$

Prove that Euler equation (9.26) in this case represents a set of two differential equations:

$$\frac{d}{dt}\left(\frac{x_1'}{\sqrt{1 + x_1'^2 + x_2'^2}}\right) = 0, \quad \frac{d}{dt}\left(\frac{x_2'}{\sqrt{1 + x_1'^2 + x_2'^2}}\right) = 0,$$

which together with the boundary conditions determine the segment of the line connecting the given points.

9.2.3. In the Poincare model (Example 1.21) the shortest distance between the points (a, c) and (b, d), $c, d > 0$ is determined by the minimization of the functional

$$F(x) = \int_a^b \frac{\sqrt{1 + x'^2(t)}}{x(t)}\, dt \,.$$

Prove that Euler equation (9.26) is reduced in this case to the form

$$\frac{1}{x} + \frac{x''}{1 + x'^2} = 0 \qquad (*)$$

and the solution of this equation is given by the formula

$$(t - c_2)^2 + x^2 = c_1, \qquad (**)$$

where the integration constants c_1 and c_2 are expressed in terms of the coordinates a, c, b, d of the given points by the formulae

$$c_1 = \frac{(b - a)^2}{4} + \frac{c^2 + d^2}{2} + \frac{(d^2 - c^2)^2}{4(b - a)^2},$$

$$c_2 = \frac{a + b}{2} + \frac{d^2 - c^2}{2(b - a)}.$$

The equation $(**)$ represents an equation of the circumference on the plane tx of the radius $\sqrt{c_1}$ with the center at the point $(c_2, 0)$ on the abscissa axis. Thus as it was stated in Example 1.21 the semicircles with the centers on the abscissa axis t serve as the direct lines in the Poincare model.

I n s t r u c t i o n. The equation $(*)$ does not contain the explicit independent variable t. Therefore it is reduced by a standard substitution $x' = y$, $x'' = (dy/dx)y$ to the equation of the first order

$$\frac{dx}{x} + \frac{y \, dy}{1 + y^2} = 0,$$

whose solution is determined by the formula

$$x' = y = \sqrt{c_1 - x^2}/x,$$

where c_1 is an arbitrary constant (certainly, $c_1 > 0$). This formula gives a differential equation for the function $x(t)$:

$$\frac{x \, dx}{\sqrt{c_1 - x^2}} = dt.$$

The solution of this equation is given by the formula $\sqrt{c_1 - x^2} = -t + c_2$, where c_2 is the second arbitrary constant. This formula is equivalent to $(**)$.

9.2.4. Find the shortest and the largest distances from given point (a, c) till the curve

$$y = \varphi(t),$$

where $\varphi(t)$ is a differentiable function.

I n s t r u c t i o n. In this case to the Euler equation $x'' = 0$ and its solution $x(t) = c_1 + c_2 t$ the following condition will be added:

$$\sqrt{1 + x'^2(b)} \, \delta b + \frac{x'(b)}{\sqrt{1 + x'^2(b)}} \delta x(b) = 0,$$

and the admissible values $x'(b)\delta b + \delta x(b)$ will be determined from the condition

$$x'(b)\delta b + \delta x(b) = \varphi'(b)\,\delta b\,.$$

Substituting into the obtained equations $x'(b) = c_2$ we find $\delta x(b) = [\varphi'(b) - c_2]\delta b$ and

$$(1 + c_2^2)\,\delta b + c_2[\varphi'(b) - c_2]\,\delta b = 0\,,$$

whence we find

$$c_2 = -1/\varphi'(b)\,.$$

This is the condition of the normality of the line $y = c_1 + c_2 t$ to the curve $y = \varphi(t)$ at the point $t = b$. After finding c_2 the constant c_1 will be determined from the condition on the other end

$$c_1 = c - c_2 a = c + a/\varphi'(b)\,.$$

As far as the variable b is concerned then it is found from the connection equation

$$c_1 + c_2 b = \varphi(b)\,.$$

Thus the shortest and the largest distances from the given point till the given curve are equal to the lengths of the segments of the corresponding normals to the curve which pass through the given point.

9.2.5. Find the curve in the vertical plane while moving on which under the action of the gravity force the particle will pass from one given point to another in the shortest possible time (*brachistochrone problem*).

I n s t r u c t i o n. After placing into the upper point from given ones the origin and directing the ordinate axis vertically downwards we express the displacement time of the particle from the origin into the point (a, b) by the line integral

$$T = \int\limits_{(0,0)}^{(a,b)} \frac{ds}{v}\,, \qquad (*)$$

where $v = v(x, y)$ is the velocity of the particle along the curve, and ds is an element of an arc of the curve. The velocity of the particle v is easily found from the law of the conservation of energy: $mgy = mv^2/2$ (it is supposed that the initial velocity of the particle is equal to zero). Hence we find $v = \sqrt{2gy}$ and the integral $(*)$ takes the form

$$T = \int\limits_0^a \frac{\sqrt{1 + y'^2}}{\sqrt{2gy}}\,dx\,.$$

9.2.6. Find the curve completely placed above the abscissa axis at whose rotation around abscissa axis a body with the smallest surface is obtained. The functional which should be minimized has in a given case the form

$$S = \int\limits_0^a 2\pi y\sqrt{1 + y'^2}dx \,.$$

9.2.7. Find the shortest distances between the given point on the plane and the circumference which does not pass through this point.

9.2.8. Find the shortest distances between the given point R^3 and the sphere which does not pass through this point.

9.2.9. Find the shortest distance between two nonintersecting circumferences on the plane.

9.2.10. Find the shortest distance between two nonintersecting spheres in R^3.

9.3. Differential Equations in Banach Spaces

9.3.1. Linear Differential Equations

In Subsection 6.2.4 the definition of a differential equation in a B–space was given and some examples of equations were cited. In Subsection 6.2.9 Theorems 6.22 and 6.23 were proved which establish the sufficient conditions of the existence of the unique solution of a differential equation in a B–space. But as it was mentioned in Subsection 6.2.9 these conditions impose very strong restrictions on the function in the right–hand side of the equation in consequence of which Theorems 6.22 and 6.23 are inapplicable to a wide important class of the linear and nonlinear equations. Therefore it is necessary to study in detail the differential equations in a B–space. We shall begin with the linear equations.

Let us consider the Cauchy problem for a linear equation

$$\frac{dx}{dt} = A(t)x + y(t), \quad x(t_0) = x_0 \,. \qquad (9.42)$$

Here $A(t)$ is a closed linear operator which acts in the B–space X with the domain D_A dense in X and independent of $t \in [0, T]$ and which represents in the general case a function of the independent variable t, and $y(t)$ is the known function with the values in X. Theorem 6.25 establishes the existence of the unique solution of Eq. (9.42) and the corresponding homogeneous equation

$$\frac{dx}{dt} = A(t)\,x \,, \quad x(t_0) = x_0 \qquad (9.43)$$

only in the case of the bounded operator $A(t)$. Notice that only in this case the right–hand side of Eq. (9.42) satisfies the Lipschitz condition of Subsection 6.2.9. But in practice the equations with the unbounded operator (in particular, with a differential operators of mathematical physics in Subsection 6.2.4) are of a great importance. For finding an approach to the equations with an unbounded operator let us consider a special case of Eq. (9.42) with a constant operator:

$$\frac{dx}{dt} = Ax + y(t), \quad x(0) = x_0 . \tag{9.44}$$

Supposing that the constant operator A is bounded we shall find a solution of Eq. (9.44) by the successive approximations method taking as zero approximation the initial value x_0.

So, at first we solve a homogeneous equation

$$\frac{dx}{dt} = Ax , \quad x(0) = x_0 . \tag{9.45}$$

After determining the successive approximations by the equation

$$\frac{dx_n}{dt} = Ax_{n-1} \quad (n = 1, 2, \ldots) ,$$

we shall have

$$x_1(t) = x_0 + \int_0^t Ax_0 \, d\tau = (I + At) \, x_0 ,$$

$$x_2(t) = x_0 + \int_0^t Ax_1(\tau) \, d\tau = (I + At + A^2 t^2/2) \, x_0$$

and generally

$$x_n(t) = \sum_{k=0}^n \frac{1}{k!} A^k t^k x_0 \quad (n = 1, 2, \ldots) .$$

In the limit at $n \to \infty$ we shall obtain a solution of a homogeneous equation in the form of a series by virtue of which $\| A^k \| \leq \| A \|^k$ the inequalities converges in the uniform topology of the space $\mathcal{B}(X)$ (Subsection 5.1.4) at all t. The sum of this series on the basis of the definition of a function of a bounded operator in Subsection 8.3.1 represents the exponential function e^{At}. Therefore the solution of Eq. (9.45) is determined by the following formula:

$$x(t) = e^{At} x_0.$$

By direct differention of this series it is easy to make sure in the fact that the operator e^{At} satisfies Eq. (9.45).

Let us pass to the solution of nonhomogeneous Eq. (9.44). After determining the successive approximations to the solution at the initial condition $x(0) = 0$ by the equation

$$\frac{dx_n}{dt} = Ax_{n-1} + y(t), \quad (n = 1, 2, \ldots),$$

we get

$$x_1(t) = \int_0^t y(\tau)\, d\tau,$$

$$x_2(t) = \int_0^t Ax_1(\sigma)\, d\sigma + \int_0^t y(\tau)\, d\tau = \int_0^t d\sigma \int_0^\sigma Ay(\tau)\, d\tau + \int_0^t y(\tau)\, d\tau$$

$$= \int_0^t d\tau \int_\tau^t Ay(\tau)\, d\sigma + \int_0^t y(\tau)\, d\tau = \int_0^t [I + A(t - \tau)]\, y(\tau)\, d\tau$$

and generally

$$x_n(t) = \int_0^t \sum_{k=0}^{n-1} \frac{1}{k!} A^k (t - \tau)^k y(\tau)\, d\tau \quad (n = 1, 2, \ldots).$$

Passing to the limit we find the required solution of Eq. (9.44):

$$x(t) = \int_0^t e^{A(t-\tau)} y(\tau)\, d\tau.$$

R e m a r k. Analyzing the obtained results we see that the main role in constructing the solution of Eqs. (9.43) and (9.42) with constant bounded operator A play the operators e^{At} and $e^{A(t-\tau)}$. It is natural to consider that also in the general case of Eq. (9.42) with unbounded variable operator there exist similar operators which are analogous to a fundamental matrix of the solutions of a linear equation in the finite–dimensional space.

9.3.2. Evolution Operators

Let us consider the function $U(t, \tau)$ of two variables with the values in the space of the bounded linear operators $\mathcal{B}(X)$. The values of this function at different values t and τ are called *evolution operators* of Eq. (9.43) if the function $U(t, \tau)$ possesses the following properties:

(i) $U(t, \tau)$ is a continuous function of t and τ in a strong topology of the space $\mathcal{B}(X)$ (i.e. the function $U(t, \tau)x$ is continuous at each x, Subsection 5.1.4) in the triangular $0 \leq \tau < t \leq T$;

(ii) $U(t, t) = I \quad \forall t, \quad U(t, \tau) = U(t, s) U(s, \tau) \quad \forall \tau < s < t;$ (9.46)

(iii) the domain D_A of the operator $A(t)$ serves as an invariant subspace (Subsection 6.3.1) of the operators $U(t, \tau)$ corresponding to all t and $\tau \leq t$;

(iv) the function $U(t, \tau)$ satisfies the differential equations

$$\frac{\partial U(t, \tau)}{\partial t} = A(t) U(t, \tau), \tag{9.47}$$

$$\frac{\partial U(t, \tau)}{\partial \tau} = -U(t, \tau) A(\tau). \tag{9.48}$$

Here the derivatives are recognized in the sense of the equalities

$$\frac{\partial U(t, \tau)x}{\partial t} = \lim_{h \to 0} \frac{U(t + h, \tau)x - U(t, \tau)x}{h} \tag{9.49}$$

at all x at which $U(t, \tau)x \in D_A$ and analogously

$$\frac{\partial U(t, \tau)x}{\partial \tau} = \lim_{h \to 0} \frac{U(t, \tau + h)x - U(t, \tau)x}{h} \tag{9.50}$$

at all $x \in D_A$. This property of the function $U(t, \tau)$ means that the operator function $U(t, \tau)$ represents a solution of two Cauchy problems obtained by the addition of the initial condition $U(\tau, \tau) = I$ to Eq. (9.47) and the finite condition $U(t, t) = I$ to Eq. (9.48).

It follows from Eq. (9.47) that at any x for which $U(\sigma, \tau)x \in D_A$ at all $\sigma \in [s, t]$ the following equality is valid

$$U(t, \tau)x - U(s, \tau)x = \int_s^t A(\sigma)U(\sigma, \tau)x \, d\sigma \,.$$ (9.51)

Similarly from Eq. (9.48) follows that at all $x \in D_A$ the equality is valid

$$U(t, \tau)x - U(t, s)x = \int_\tau^s U(t, \sigma)A(\sigma)x \, d\sigma \,.$$ (9.52)

Theorem 9.12. *The function $U(t, \tau)$ which possesses properties (i)–(iii) forms a set of evolutionary operators of Eq. (9.43) if and only if at all $x \in D_A$ exists the limit which coincides with the operator $A(t)$*

$$\lim_{h \to 0} \frac{U(t+h, t)x - x}{h} = A(t)x \,.$$ (9.53)

▷ On the basis of equality (9.49) Eq. (9.47) may be rewritten in the form:
$$\lim_{h \to 0} \frac{U(t+h, \tau)x - U(t, \tau)x}{h} = A(t)U(t, \tau)x \,.$$

Putting $\tau = t$ by virtue of Eq. (9.46) we get (9.53). If equality (9.53) is valid then after substituting in it x by $U(t, \tau)x$ we receive on the basis of equalities (9.46) and (9.49) Eq. (9.47). Substituting in (9.53) t by τ we obtain the equality

$$\lim_{h \to 0} \frac{U(\tau + h, \tau)x - x}{h} = A(\tau)x \,.$$

Multiplying the right–hand side of this equality on the left by $U(t, \tau)$ and the left–hand side – by the same variable $\lim_{h \to 0} U(t, \tau + h) = U(t, \tau)$ we receive by (9.46)

$$\lim_{h \to 0} \frac{U(t, \tau)x - U(t, \tau + h)x}{h} = U(t, \tau)A(\tau)x$$

or

$$-\frac{\partial U(t,\tau)}{\partial \tau}x = U(t,\tau)\,A(\tau)\,x\,,$$

i.e. Eq. (9.48). ◁

If all the evolutionary operators $U(t,\tau)$ have the bounded inverse operators $U(t,\tau)^{-1}$ then the function $U(t,\tau)^{-1}$ also satisfy some linear differential equations. For obtaining these equations it is sufficient to differentiate the function $U(t,\tau)^{-1}$ with respect to t and to τ and use Eqs. (9.47) and (9.48):

$$\frac{\partial U(t,\tau)^{-1}}{\partial t} = -U(t,\tau)^{-1}\frac{\partial U(t,\tau)}{\partial t}U(t,\tau)^{-1} = -U(t,\tau)^{-1}A(t)\,,$$

$$\frac{\partial U(t,\tau)^{-1}}{\partial \tau} = -U(t,\tau)^{-1}\frac{\partial U(t,\tau)}{\partial \tau}U(t,\tau)^{-1} = A(\tau)U(t,\tau)^{-1}\,.$$

Thus the operator function $U(t,\tau)^{-1}$ satisfies the equations

$$\frac{\partial U(t,\tau)^{-1}}{\partial t} = -U(t,\tau)^{-1}A(t)\,, \qquad \frac{\partial U(t,\tau)^{-1}}{\partial \tau} = A(\tau)U(t,\tau)^{-1}\,. \quad (9.54)$$

Introducing the adjoint operators $V(t,\tau) = [U(t,\tau)^{-1}]^*$ we obtain for them the equations which are quite analogous to the corresponding equations for the differential equations in the finite–dimensional spaces:

$$\frac{\partial V(t,\tau)}{\partial t} = -A(t)^*V(t,\tau)\,, \qquad \frac{\partial V(t,\tau)}{\partial \tau} = V(t,\tau)A(\tau)^* \qquad (9.55)$$

The first equation of Eqs. (9.55) similarly as in the case of the differential equations in the finite–dimensional spaces is called *adjoint* equation with Eq. (9.43).

Theorem 9.13. *The evolutionary operators of* Eq. (9.43) *have bounded inverse operators* $U(t,\tau)^{-1}$ *for which the domain* D_A *of the operator* $A(t)$ *serves as an invariant subspace if and only if*

(i) *at each t there exists a bounded linear operator* $\Phi(t)$ *with the bounded inverse operator* $\Phi(t)^{-1}$;

(ii) *the operators* $\Phi(t)$ *and* $\Phi(t)^{-1}$ *represent the continuous functions in a strong topology of the space* $\mathcal{B}(X)$;

(iii) *there exists the derivative* $\Phi'(t)$ *of the operator* $\Phi(t)$ *which represents a closed operator with constant dense domain D in X serving as an invariant subspace of all the operators* $\Phi(t)$ *and* $\Phi(t)^{-1}$;

(iv) *the coefficient $A(t)$ of* Eq. (9.43) *is expressed by formula*

$$A(t) = \Phi'(t)\Phi(t)^{-1}. \qquad (9.56)$$

▷ If there exist the bounded operators $U(t, \tau)^{-1}$ for which the domain D_A serves as an invariant subspace then putting in Eq. (9.47) $\tau = 0$ we receive

$$A(t) = \frac{\partial U(t, 0)}{\partial t} U(t, 0)^{-1}. \qquad (9.57)$$

Putting $\Phi(t) = U(t, 0)$ we come to equality (9.56). Here the domain D of the operator $\Phi'(t)$ coincides with the domain D_A of the operator $A(t)$ as by virtue of property (iii) of the operators $U(t, \tau)$ and $U(t, \tau)^{-1}$ $x \in D_A$ if and only if $\Phi(t)x \in D_A$. And vice versa if the operator $A(t)$ is expressed by formula (9.56) and the operators $\Phi(t)$ and $\Phi'(t)$ satisfy conditions (i)–(iii) then the domain of the operators $A(t)$ and $\Phi(t)$ coincide and the operators

$$U(t, \tau) = \Phi(t)\Phi(\tau)^{-1} \qquad (9.58)$$

possess properties (i)–(iii) of the evolutionary operators and satisfy Eq. (9.47). Differentiating formula (9.58) with respect to τ we get

$$\frac{\partial U(t, \tau)}{\partial \tau} = -\Phi(t)\Phi(\tau)^{-1}\Phi'(\tau)\Phi(\tau)^{-1}$$

By virtue of formulae (9.58) and (9.56) this equality coincides with Eq. (9.48). ◁

In the case of Eq. (9.43) with the constant operator $A(t) = A$ its solutions are invariant relative to the shifts of the independent variable t: i.e. the solution of Eq. (9.43) at a given t at the initial condition $x(t_0) = x_0$ coincides with its solution at $t + a$ at the initial condition $x(t_0 + a) = x_0$. Therefore $U(t, \tau) = U(t + a, \tau + a)$ at any a. Putting $a = -\tau$ we get $U(t, \tau) = U(t - \tau, 0)$. Thus in the case of the constant operator A the evolutionary operators depend only on the difference of the arguments and we may put $U(t, \tau) = V(t - \tau)$. In the case of the bounded operator A as we have already seen in Subsection 9.3.1 $U(t, \tau) = e^{A(t-\tau)}$ and all the conditions of Theorem 9.13 are satisfied. To ensure in the existence of the inverse operators we may by the direct multiplication of the series for the functions $e^{A(t-\tau)}$ and $e^{-A(t-\tau)}$. The derivative of the operator $e^{A(t-\tau)}$ may be calculated by the differentiation of the corresponding series.

E x a m p l e 9.14. Consider the equation of the heat conduction of Example 6.18

$$\frac{\partial v}{\partial t} = a^2 \frac{\partial^2 v}{\partial x^2}, \quad v(t_0, x) = v_0(x). \tag{9.59}$$

This equation may be considered as an ordinary differential equation in the B–space of the continuous bounded functions $C(R)$ of the variable x.

It is easy to ensure by direct substitution that at any function $v(x)$ the operator

$$U(t, \tau)v = \frac{1}{2a\sqrt{\pi(t-\tau)}} \int_{-\infty}^{\infty} \exp\left\{-\frac{(\xi-x)^2}{4a^2(t-\tau)}\right\} v(\xi) \, d\xi \tag{9.60}$$

satisfies Eq. (9.47) which in a given case has the form

$$\frac{\partial U(t, \tau)v}{\partial t} = a^2 \frac{\partial^2 U(t, \tau)v}{\partial x^2}.$$

It is also clear that it possesses properties (i) and (iii) of the evolutionary operators. In order to make sure in the fact that the first equality of (9.46) is valid for this operator it is sufficient to recall that according to the result of Problem 5.3.4 the exponential function at $\tau \to t$ turns into the δ–function $\delta(\xi - x)$. To prove that for the operator (9.60) the second equality of (9.46) is also valid we put in (9.60) $t = s$ and substitute the obtained function $w(x) = U(s, \tau)v$ into Eq. (9.60) at $\tau = s$ instead of the function $v(x)$. As a result we obtain

$$U(t, s)\, U(s, \tau)v$$

$$= \frac{1}{4a^2\pi\sqrt{(t-s)(s-\tau)}} \int_{-\infty}^{\infty} \exp\left\{-\frac{(\xi-x)^2}{4a^2(t-s)}\right\} d\xi$$

$$\times \int_{-\infty}^{\infty} \exp\left\{-\frac{(\eta-\xi)^2}{4a^2(s-\tau)}\right\} v(\eta) \, d\eta.$$

Performing here the change of variables $\zeta = \xi - \eta$, remaining here η invariable and after integrating with respect to ζ by formula $(*)$ of Example 3.4 we shall obtain

$$U(t, s)U(s, \tau)v$$

$$= \frac{1}{2a\sqrt{\pi(t-\tau)}} \int_{-\infty}^{\infty} \exp\left\{-\frac{(\eta-x)^2}{4a^2(t-\tau)}\right\} v(\eta) \, d\eta = U(t, \tau)v,$$

i.e. the second equality of (9.46).

9.3.3. Solution of Linear Differential Equations

It follows from Eq. (9.47) that formula

$$x(t) = U(t, t_0)x_0 \qquad (9.61)$$

at any x_0, $U(t, t_0)x_0 \in D_A$ $\forall t$ gives the solution of Cauchy problem for homogeneous Eq. (9.43). Prove that this solution is the unique. Let $z(t)$ be some other solution of Cauchy problem (9.43) $z'(t) = A(t)z(t)$, $z(t_0) = x_0$. We determine the function $f(s) = U(t, s)z(s)$. Differentiating this function we shall have by virtue of (9.48)

$$f'(s) = \frac{\partial U(t, s)}{\partial s} z(s) + U(t, s)z'(s)$$
$$= -U(t, s)A(s)z(s) + U(t, s)A(s)z(s) = 0.$$

Thus $f(s) = $ const and $f(t) = f(t_0)$ at any t. But $f(t) = z(t)$, $f(t_0) = U(t, t_0)z(t_0) = U(t, t_0)x_0 = x(t)$. Consequently, $z(t) = x(t)$ what proves the uniqueness of solution (9.61). Thus the following theorem is valid.

Theorem 9.14. *If $U(t, t_0)x_0 \in D_A$ then formula (9.61) determines the unique solution of the Cauchy problem for homogeneous Eq. (9.43).*

R e m a r k. It is easy to see that the inverse theorem is valid in some sense: *if Cauchy problem (9.43) at any $x_0 \in D_A$ has the unique solution $x(t)$ which represents a continuous function of t and continuously depends on x_0 then there exists a set of the evolutionary operators $\{U(t, \tau)\}$ of Eq. (9.43) possessing properties (i)–(iv) of Subsection 9.3.2.*

E x a m p l e 9.15. On the basis of the result of Example 9.14 the unique solution of Eq. (9.59) at the initial condition $v(t_0) = v_0(x)$ is determined by the formula

$$v(t, x) = \frac{1}{2a\sqrt{\pi(t - t_0)}} \int_{-\infty}^{\infty} \exp\left\{-\frac{(\xi - x)^2}{4a^2(t - t_0)}\right\} v_0(\xi)\, d\xi. \qquad (9.62)$$

Let us pass to nonhomogeneous Eq. (9.42). At first we prove one auxiliary supposition.

Lemma. *For any continuous function $f(t)$ with the values in X*

$$\lim_{h \to 0} \frac{1}{h} \int_t^{t+h} f(s)\,ds = f(t). \tag{9.63}$$

▷ We determine the function of the variable v by formula

$$F(v) = \int_t^{t+v} f(s)\,ds.$$

In the consequence of the integrand this function is differentiable and its derivative by Theorem 6.20 is equal

$$F'(v) = f(t+v).$$

Putting $v = 0$ we find $F'(0) = f(t)$. On the other hand, $F'(0)$ represents the left–hand side of formula (9.63). ◁

Theorem 9.15. *If $U(t,t_0)x_0 \in D_A$, $A(t)x$ represents a continuous function of t at any $x \in D_A$, the function $y(t)$ in the right–hand side of Eq. (9.42) is continuous and at any t and τ its range R_y is mapped by the operator $U(t,\tau)$ into the domain D_A of the operator $A(t)$ then formula*

$$x(t) = U(t,t_0)x_0 + \int_{t_0}^t U(t,\tau)\,y(\tau)\,d\tau \tag{9.64}$$

determines the unique solution of the Cauchy problem for Eq. (9.42).

▷ The uniqueness of the solution follows from Theorem 9.14 as the difference of two solutions satisfies the homogeneous Eq. (9.43) at $x_0 = 0$. Thus we must prove only the fact that the second item in the right–hand side of formula (9.64) satisfies Eq. (9.42). For this purpose we find

$$\frac{1}{h} \left[\int_{t_0}^{t+h} U(t+h,\tau)\,y(\tau)\,d\tau - \int_{t_0}^t U(t,\tau)\,y(\tau)\,d\tau \right]$$

$$= \frac{1}{h} \int_{t_0}^t [U(t+h,\tau) - U(t,\tau)]\,y(\tau)\,d\tau + \frac{1}{h} \int_t^{t+h} U(t+h,\tau)\,y(\tau)\,d\tau. \tag{9.65}$$

The limit of the second item is easily found by formula (9.63):

$$\lim_{h \to 0} \frac{1}{h} \int_{t}^{t+h} U(t+h, \tau)\, y(\tau)\, d\tau = U(t,t)\, y(t) = y(t) \,. \tag{9.66}$$

For evaluating the limit of the first item we introduce the operator

$$A_h(t)x = [U(t+h,t)x - x]/h$$

(whose limit on the basis of definition (9.53) serves the operator $A(t)$). Then by virtue of (9.46) we obtain

$$\frac{1}{h} \int_{t_0}^{t} [U(t+h,\tau) - U(t,\tau)]\, y(\tau)\, d\tau = A_h(t) \int_{t_0}^{t} U(t,\tau)y(\tau)\, d\tau \,. \tag{9.67}$$

On the other hand, formula (9.51) gives

$$\frac{1}{h} \int_{t_0}^{t} [U(t+h,\tau) - U(t,\tau)]y(\tau)\, d\tau$$

$$= \frac{1}{h} \int_{t_0}^{t} d\tau \int_{t}^{t+h} A(\sigma)U(\sigma,\tau)\, y(\tau)d\sigma = \frac{1}{h} \int_{t}^{t+h} d\sigma \int_{t_0}^{t} A(\sigma)U(\sigma,\tau)\, y(\tau)\, d\tau \,.$$

The integral with respect to τ represents a continuous function σ. Therefore for evaluating the limit at $h \to 0$ we may use formula (9.63). As a result accounting (9.67) we get

$$\lim_{h \to 0} A_h(t) \int_{t_0}^{t} U(t,\tau)\, y(\tau)\, d\tau = \int_{t_0}^{t} A(t)U(t,\tau)y(\tau)d\tau \,. \tag{9.68}$$

Hence it follows

$$\int_{t_0}^{t} U(t,\tau)y(\tau)d\tau \in D_A \,,$$

$$A(t) \int_{t_0}^{t} U(t,\tau)y(\tau)d\tau = \int_{t_0}^{t} A(t)U(t,\tau)y(\tau)d\tau \,. \tag{9.69}$$

On the basis of formulae (9.66)–(9.69) the passage to the limit at $h \to 0$ in (9.65) gives

$$\frac{d}{dt} \int_{t_0}^{t} U(t,\tau)y(\tau)d\tau = A(t) \int_{t_0}^{t} U(t,\tau)y(\tau)d\tau + y(t) . \quad \triangleleft$$

R e m a r k. Notice that contrary to the equations in the finite–dimensional spaces formula (9.64) is valid not for any x_0 and not for any continuous function $y(t)$.

E x a m p l e 9.16. The unique solution of the nonhomogeneous equation

$$\frac{\partial v}{\partial t} = a^2 \frac{\partial^2 v}{\partial x^2} + f(t,x), \quad v(t_0, x) = v_0(x)\ldots, \tag{9.70}$$

with the continuous function $f(t,x)$ is determined on the basis of (9.64) by formula

$$v(t,x) = \frac{1}{2a\sqrt{\pi(t-t_0)}} \int_{-\infty}^{\infty} \exp\left\{-\frac{(\xi-x)^2}{4a^2(t-t_0)}\right\} v_0(\xi)\, d\xi$$

$$+ \frac{1}{2a\sqrt{\pi}} \int_{t_0}^{t} \frac{d\tau}{\sqrt{t-\tau}} \int_{-\infty}^{\infty} \exp\left\{-\frac{(\xi-x)^2}{4a^2(t-\tau)}\right\} f(\tau,\xi)\, d\xi . \tag{9.71}$$

The conditions of Theorem 9.15 in a given case are fulfilled as according to the result of Example 9.14 the operator $U(t,\tau)$ maps the range of the function $f(\tau,x)$ – the space of the bounded continuous functions $C(R)$ of the variable x – into the space of doubly differentiable functions $C^2(R)$ which represents the domain of the operator $A = \partial^2/\partial x^2$.

9.3.4. Nonlinear Differential Equations

Consider the Cauchy problem for the following nonlinear differential equation:

$$\frac{dx}{dt} = f(t,x), \quad x(t_0) = x_0 . \tag{9.72}$$

We shall suppose that the function $f(t,x)$ is expressed by formula

$$f(t,x) = A(t)x + g(t,x), \tag{9.73}$$

where $A(t)$ is a closed linear operator with dense domain D_A in X independent of $t \in [0,T]$ which represents a continuous function of t in the sense that the function $A(t)x$ is continuous at any $x \in D_A$ and possesses the set of the evolutionary operators $\{U(t,\tau)\}$. After writing Eq. (9.72) in the form

$$\frac{dx}{dt} = A(t)x + g(t,x), \quad x(t_0) = x_0,\qquad (9.74)$$

we may formally solve it as a linear equation not paying attention to the fact that the function $g(t,x)$ depends on the unknown function $x(t)$. As a result formula (9.64) gives the following integral equation:

$$x(t) = U(t,t_0)x_0 + \int\limits_{t_0}^{t} U(t,\tau)\,g(\tau,x(\tau))\,d\tau.\qquad (9.75)$$

The equivalence of differential Eq. (9.74) and integral Eq. (9.75) the following theorem establishes.

Theorem 9.16. *If the range R_g of the function $g(t,x)$ is mapped by the operator $U(t,\tau)$ into the domain D_A of the operator $A(t)$ at all t, τ and the function $U(t,\tau)g(\tau,x)$ is continuous on τ and x at any t then any solution of differential equation (9.74) satisfies integral equation (9.75) and vice versa, any solution of integral equation (9.75) satisfies differential equation (9.74).*

▷ Taking into account that in the correspondence with Theorem 9.14 $dU(t,t_0)x_0/dt = A(t)U(t,t_0)x_0$ and repeating in exactly the same way the evaluations at proving Theorem 9.15 we ensure that any solution of Eq. (9.75) satisfies Eq. (9.74). For proving the inverse statement we substitute into the right–hand side of Eq. (9.75) some solution $x(t)$ of Eq. (9.74) and put

$$y(t) = U(t,t_0)x_0 + \int\limits_{t_0}^{t} U(t,\tau)\,g(\tau,x(\tau))\,d\tau.$$

Differentiating this equation and repeating the evaluatious of the proof of the first part of the theorem we come to the equation

$$\frac{dy}{dt} = A(t)y + g(t,x).$$

We subtract from this equation term by term Eq. (9.74). As a result we obtain for the difference $z(t) = y(t) - x(t)$ a homogeneous linear equation

$dz/dt = A(t)z$ with zero initial condition $z_0 = 0$. By Theorem 9.14 this equation has the unique solution $z(t) = 0$. Consequently, $y(t) = x(t)$. This proves that any solution of Eq. (9.74) satisfies Eq. (9.75). ◁

Eq. (9.75) represents an equation of the form (6.61). Therefore the theorems of the existence of the unique solution proved in Subsection 6.2.8 are applicable to Eq. (9.75). From Theorem 6.20 follows.

Theorem 9.17. *If there exist such numbers $M > 0$, $b > TM$, $c > 0$ that*

(i) $\| U(t, \tau)g(\tau, x) \| < M$ *at all* t_0, τ, $t \in [0, T]$, $t_0 < \tau < t$, $\| x - U(t, t_0)x_0 \| < b$;

(ii) $\| U(t, \tau)g(\tau, x_1) - U(t, \tau)g(\tau, x_2) \| < c \| x_1 - x_2 \|$ *(the Lipschitz condition) at* t, $\tau \in [0, T]$, $\tau < t$, $\| x_i - U(t, t_0)x_0 \| < b$ *$(i = 1, 2)$, then Eq. (9.75), and consequently, Cauchy problem (9.72) have the unique solution $x(t)$ which represents a continuous function satisfying the condition $\| x(t) - U(t, t_0)x_0 \| < b$.*

From Theorem 6.23 follows.

Theorem 9.18. *If Lipschitz condition (ii) of Theorem 9.17 is fulfilled at all x_1, $x_2 \in X$ then Eq. (9.75), and consequently, also Cauchy problem (9.72) have the unique solution $x(t)$ which represents a continuous function.*

E x a m p l e 9.17. A nonlinear equation which is received by the change in Eq. (9.70) of the function $f(t, x)$ by the function $g(t, x, v)$ depends on the unknown function $v(t, x)$ is reduced to integral equation which is obtained from formula (9.71) by the change of the function $f(\tau, \xi)$ by the function $g(\tau, \xi, v(\tau, \xi))$ under the sign of the integral.

R e m a r k. Theorems 9.17 and 9.18 do not impose the rigorous limitations on the function $g(t, x)$. It must not only satisfy any condition of the Lipschitz type but may even contain the items in the form of linear combination of the unbounded linear operators.

E x a m p l e 9.18. Consider a nonlinear equation of the heat conduction

$$\frac{\partial v}{\partial t} = a^2 \frac{\partial^2 v}{\partial x^2} + kv \frac{\partial v}{\partial x}. \qquad (9.76)$$

In this case the function $g(t, x, v) = kv \, \partial v/\partial x$ contains the unbounded linear operator $\partial/\partial x$. Let us evaluate by formula (9.60) the result of the action of this function of the operator $U(t, \tau)$. Intergrating by parts we obtain

$$U(t, \tau)g(\tau, x, v)$$

$$= \frac{k}{2a\sqrt{\pi(t-\tau)}} \int_{-\infty}^{\infty} \exp\left\{-\frac{(\xi-x)^2}{4a^2(t-\tau)}\right\} v(\tau,\xi)\frac{\partial v(\tau,\xi)}{\partial\xi} \, d\xi$$

$$= \frac{k}{8a^3\sqrt{\pi(t-\tau)^3}} \int_{-\infty}^{\infty} (\xi-x)\exp\left\{-\frac{(\xi-x)^2}{4a^2(t-\tau)}\right\} v^2(\tau,\xi) \, d\xi \, .$$

Considering Eq. (9.76) as an equation in the B–space of the bounded continuous functions $C(R)$ of the variable x we ensure that the function $U(t,\tau)g(\tau,x,v)$ satisfies the Lipschitz condition as the function v^2 satisfies it on any finite interval of v. Therefore Eq. (9.76) satisfies the conditions of Theorem 9.17.

Thus the stated theory covers a wide class of different equations which may be presented as an ordinary differential equation in a B–space, in particular, all the equations of mathematical physics. For detailed treatment of differential equations in abstract spaces refer to (Balakrishnan 1976, Curtain and Pritchard 1978, Da Prato and Zabczyk 1992, Gajewski et al. 1974, Ladas ans Lakshmikantam 1972).

P r o b l e m s

9.3.1. The equation considered in Examples 9.14–9.18 describes the extension of the heat in an infinite rod. Now we shall consider the heat conduction equation for semi–infinite rod with the boundary condition at one end which without loss of the generality may be assumed as the point $x = 0$:

$$\frac{\partial v}{\partial t} = a^2\frac{\partial^2 v}{\partial x^2} \, , \quad v_x'(t,0) = hv(t,0) \, , \quad v(0,x) = v_0(x) \, . \tag{9.77}$$

This boundary condition means that to the end of the rod the heat is delivered (is removed at $h < 0$), here the velocity of the temperature change of the end is proportional to its temperature. Eq. (9.77) with the boundary condition represents a differential equation in the B–space $C_1^1([0,\infty))$ of differentiable functions which satisfy to boundary condition.

Formula (9.61) shows that for finding the evolutionary operators it is necessary to solve a homogeneous equation at the initial condition $x(\tau) = x$ in a given case $v(\tau,x) = v(x)$. Formula (9.60) determines the solution of Eq. (9.77) on the whole real axis R. We need the solution for the function $v(x)$ determined only at $x > 0$ and satisfying the boundary condition $v'(0) = hv(0)$. The solution itself must also satisfy this boundary condition. We shall show here standard method whereby we may often pass from the solution of a given equation in partial derivatives considered as a differential equation in another B–space to the solution of the same equation considered as differential equation in another B–space. We write formula (9.60) in the form

$$U(t,\tau)v = \frac{1}{2a\sqrt{\pi(t-\tau)}}\left[\int_0^\infty \exp\left\{-\frac{(\xi-x)^2}{4a^2(t-\tau)}\right\} v(\xi)\,d\xi\right.$$

$$\left.+ \int_0^\infty \exp\left\{-\frac{(\xi+x)^2}{4a^2(t-\tau)}\right\} v(-\xi)\,d\xi\right].$$

In order to use this formula for finding the solution of Eq. (9.77) at the interval $[0,\infty)$ the function $v(x)$ given only for positive x should be extended over the region of negative x in such a way that the function of the variable x which is determined by the previous formula will satisfy the boundary condition. Putting $v(-\xi) = \varphi(\xi)$ we rewrite the previous formula in the form

$$U(t,\tau)v = \frac{1}{2a\sqrt{\pi(t-\tau)}}\left[\int_0^\infty \exp\left\{-\frac{(\xi-x)^2}{4a^2(t-\tau)}\right\} v(\xi)\,d\xi\right.$$

$$\left.+ \int_0^\infty \exp\left\{-\frac{(\xi+x)^2}{4a^2(t-\tau)}\right\} \varphi(\xi)\,d\xi\right].$$

The boundary condition for this function of the variable x gives the equation for the function $\varphi(\xi)$. After solving this equation we express the function $\varphi(\xi)$ in terms of the function $v(\xi)$. As a result we determine the evolutionary operators for Eq. (9.77) with boundary condition:

$$U(t,\tau)v = \frac{1}{2a\sqrt{\pi(t-\tau)}}\int_0^\infty \left\{\exp\left\{-\frac{(\xi-x)^2}{4a^2(t-\tau)}\right\} + \exp\left\{-\frac{(\xi+x)^2}{4a^2(t-\tau)}\right\}\right.$$

$$\times \left.\left[1 - ah\sqrt{2\pi(t-\tau)} + 2ah\Phi\left(\frac{\xi+x}{a\sqrt{2(t-\tau)}} + ah\sqrt{2(t-\tau)}\right)\right]\right\} v(\xi)\,d\xi\,.$$

9.3.2. The heat conduction equation for the bounded rod with the boundary conditions on two ends $x = 0$ and $x = l$,

$$\frac{\partial v}{\partial t} = a^2\frac{\partial^2 v}{\partial x^2}\,,\quad v'_x(t,0) = h_1 v(t,0)\,,$$

$$v'_x(t,l) = h_2 v(t,l)\,,\quad v(0,x) = v_0(x)\,,\tag{9.78}$$

may be considered as a differential equation in the B–space $C_2^1([0,l])$ of the differentiable functions satisfying on the ends of the interval the boundary conditions.

Show that the evolutionary operators for this equation are expressed by the following formula:

$$U(t, \tau)v = \int_0^l \sum_{n=-\infty}^{\infty} B_n \exp\{-\mu_n^2 a^2(t - \tau)\} \, \varphi_n(x)\overline{\varphi_n(\xi)} \, v(\xi) \, d\xi \,,$$

where the functions $\varphi_n(x)$ are determined by

$$\varphi_n(x) = \exp\{i\mu_n x\} + \frac{i\mu_n - h_1}{i\,\mu_n + h_1} \exp\{-i\mu_n x\} \,.$$

They satisfy the differential equations and boundary conditions

$$\varphi_n'' = -\mu_n^2 \varphi_n(x) \,, \quad \varphi_n'(0) = h_1\varphi_n(0) \,, \quad \varphi_n'(l) = h_2\varphi_n(l) \,,$$

from which follows that they are orthogonal:

$$\int_0^l \varphi_n(x)\overline{\varphi_m(x)} \, dx = 0 \quad \text{at} \quad n \neq m \,, \quad B_n^{-1} = \int_0^l |\varphi_n(x)|^2 \, dx \,,$$

and μ_n are the values of the parameter μ at which the fulfillment of the boundary conditions is possible. They are determined by the following transcendental equation:

$$(\mu^2 + h_1 h_2)\text{tg}\,\mu l = (h_2 - h_1)\mu \,.$$

Find the solution of the Cauchy problem for the homogeneous Eq. (9.78) and for the corresponding nonhomogeneuous equation.

I n s t r u c t i o n. While testing the first equality of (9.46) use formula (5.46).

9.3.3.   The heat conduction equation in three–dimensional space R^3 of Example 6.19

$$\frac{\partial v}{\partial t} = a^2 \left(\frac{\partial^2 v}{\partial x^2} + \frac{\partial^2 v}{\partial y^2} + \frac{\partial^2 v}{\partial z^2} \right) \,, \quad v(0, x, y, z) = v_0(x, y, z) \qquad (9.79)$$

may be considered as the differential equation in the B–space of the continuous functions $C(R^3)$. Prove that similarly as in Example 9.14 formula

$$U(t, \tau)v = \int_{-\infty}^{\infty} \int_{-\infty}^{\infty} \int_{-\infty}^{\infty} \exp\left\{ -\frac{(\xi - x)^2 + (\eta - x)^2 + (\xi - z)^2}{4a^2(t - \tau)} \right\}$$

$$\times \frac{v(\xi, \eta, \zeta) \, d\xi \, d\eta \, d\zeta}{8a^3 \pi^{3/2}(t - \tau)^{3/2}}$$

determines the evolutionary operators of Eq. (9.79). Find the solution of the Cauchy problem for Eq. (9.79) and the solution of the Cauchy problem for the corresponding nonhomogeneous equation.

9.3.4. Analogously as in Problem 9.3.1 find the evolutionary operators and the solutions of the Cauchy problems for homogeneous Eq. (9.79) with the corresponding boundary conditions for the problems of the heat extension in the semi–space $x > 0$, in the quadrant x, $y > 0$ and in the octaint x, y, $z > 0$.

9.3.5. The same question (Problem 9.3.4) for two–sided boundary conditions, i.e. for the problems of the heat extension in the layer, in the bim and in the parallelepiped.

9.3.6. Let us consider now a wave equation of Example 6.20 which describes the oscillations of stretched string with fixed ends $x = 0$ and $x = l$:

$$\frac{\partial^2 v}{\partial t^2} = a^2 \frac{\partial^2 v}{\partial x^2}, \quad v(t, 0) = v(t, l) = 0,$$

$$v(0, x) = v_0(x), \quad v_t'(0, x) = w_0(x). \tag{9.80}$$

Putting $w = \partial v / \partial t$ we reduce Eq. (9.80) to the set of two equations

$$\frac{\partial v}{\partial t} = w, \quad \frac{\partial w}{\partial t} = a^2 \frac{\partial^2 v}{\partial x^2}. \tag{9.81}$$

This set of the equations may be considered as the equation of the form (9.43) in the B–space of the continuous two–dimensional vector functions $C([0, l])$ with the constant operator

$$A = \begin{bmatrix} 0 & 1 \\ a^2 \partial^2 / \partial x^2 & 0 \end{bmatrix}.$$

The evolutionary operators in this case represent the matrices

$$U(t, \tau) = \begin{bmatrix} U_{11}(t, \tau) & U_{12}(t, \tau) \\ U_{21}(t, \tau) & U_{22}(t, \tau) \end{bmatrix}.$$

Prove that these evolutionary operators are determined by the formulae

$$U_{11}(t, \tau)v = \frac{2}{l} \int_0^l \sum_{n=1}^{\infty} \cos \frac{n\pi a(t - \tau)}{l} \sin \frac{n\pi x}{l} \sin \frac{n\pi \xi}{l} v(\xi)\, d\xi,$$

$$U_{12}(t, \tau)w = \frac{2}{a\pi} \int_0^l \sum_{n=1}^{\infty} \frac{1}{n} \sin \frac{n\pi a(t - \tau)}{l} \sin \frac{n\pi x}{l} \sin \frac{n\pi \xi}{l} w(\xi)\, d\xi,$$

$$U_{21}(t, \tau)v = -\frac{2a\pi}{l^2} \int_0^l \sum_{n=1}^{\infty} n \sin \frac{n\pi a(t - \tau)}{l} \sin \frac{n\pi x}{l} \sin \frac{n\pi \xi}{l} v(\xi)\, d\xi,$$

$$U_{22}(t, \tau)w = \frac{2}{l} \int_0^l \sum_{n=1}^{\infty} \cos \frac{n\pi a(t - \tau)}{l} \sin \frac{n\pi x}{l} \sin \frac{n\pi \xi}{l} w(\xi)\, d\xi.$$

Find the solution of the Cauchy problem for the homogeneous Eq. (9.80) and for the corresponding nonhomogeneous equation.

9.3.7. The wave equation which describes the oscillations of the thin rectangular membrane fixed along the perimeter has the form

$$\frac{\partial^2 v}{\partial t^2} = a^2 \left(\frac{\partial^2 v}{\partial x^2} + \frac{\partial^2 v}{\partial y^2} \right),$$

$$v(t, 0, y) = v(t, l, y) = v(t, x, 0) = v(t, x, k) = 0,$$

$$v(0, x, y) = v_0(x, y), \quad v_t'(0, x, y) = w_0(x, y). \tag{9.82}$$

In exactly the same way as the wave equation in Problem 9.3.6 this equation is reduced to the set of two equations of the form (9.81). This set of the equations may be considered as the equation of the form (9.43) with the constant operator

$$A = \begin{bmatrix} 0 & 1 \\ a^2(\partial^2/\partial x^2 + \partial^2/\partial y^2) & 0 \end{bmatrix}$$

in the B–space of the continuous functions $C([0, l] \times [0, k])$ satisfying the boundary condition. Prove that the evolutionary operators for Eqs. (9.82) are determined by the following formulae:

$$U_{11}(t, \tau)v = \frac{4}{kl} \int_0^l \int_0^k \sum_{n,m=1}^\infty \cos \mu_{nm} a(t - \tau) \sin \frac{n\pi x}{l} \sin \frac{m\pi y}{k}$$

$$\times \sin \frac{n\pi \xi}{l} \sin \frac{m\pi \eta}{k} \, v(\xi, \eta) \, d\xi \, d\eta,$$

$$U_{12}(t, \tau)w = \frac{4}{kla} \int_0^l \int_0^k \sum_{n,m=1}^\infty \frac{1}{\mu_{nm}} \sin \mu_{nm} a(t - \tau) \sin \frac{n\pi x}{l} \sin \frac{m\pi y}{k}$$

$$\times \sin \frac{n\pi \xi}{l} \sin \frac{m\pi \eta}{k} \, w(\xi, \eta) \, d\xi \, d\eta,$$

$$U_{21}(t, \tau)v = -\frac{4a}{lk} \int_0^l \int_0^k \sum_{n,m=1}^\infty \mu_{nm} \sin \mu_{nm} a(t - \tau) \sin \frac{n\pi x}{l} \sin \frac{m\pi y}{k}$$

$$\times \sin \frac{n\pi \xi}{l} \sin \frac{m\pi \eta}{k} \, v(\xi, \eta) \, d\xi \, d\eta,$$

$$U_{22}(t, \tau)w = \frac{4}{lk} \int_0^l \int_0^k \sum_{n,m=1}^\infty \cos \mu_{nm} a(t - \tau) \sin \frac{n\pi x}{l} \sin \frac{m\pi y}{k}$$

$$\times \sin \frac{n\pi \xi}{l} \sin \frac{m\pi \eta}{k} \, w(\xi, \eta) \, d\xi \, d\eta,$$

where $\mu_{nm} = \pi\sqrt{(n/l)^2 + (m/k)^2}$. Find the solution of the Cauchy problem for Eq. (9.82) and for the corresponding nonhomogneous equation.

9.3.8. Find the evolutionary operators for linear systems with distributed parameters presented in Table A.1.2 (Appendix 1).

9.3.9. Find the evolutionary operator for the Fokker–Planck–Kolmogorov equation

$$\frac{\partial f(z;t)}{\partial t} = -\frac{\partial^T}{\partial z}[a(z,t)f(z;t)]$$

$$+\frac{1}{2}\operatorname{tr}\left\{\frac{\partial}{\partial z}\frac{\partial^T}{\partial z}[b(z,t)\nu(t)b(z,t)^T f(z;t)]\right\},$$

$$f(z,0) = f_0(z) \tag{9.83}$$

at $a(z,t) = a_0 + a_1 z$, $\nu_0 = \nu(t)$, $b(z,t) = b_0$. Here $f(z;t)$ is one–dimensional probability density, z is p–dimensional vector, ν_0 is $(p \times p)$–dimensional intensity matrix of the Wiener process.

9.3.10. Find the evolutionary operator for the Pugachev equation

$$\frac{\partial g(\lambda;t)}{\partial t} = \frac{1}{(2\pi)^p} \int\limits_{-\infty}^{\infty} \int\limits_{-\infty}^{\infty} [i\lambda^T a(z,t)$$

$$+\chi(b(z,t)^T \lambda;t)]e^{i(\lambda^T - \mu^T)z}g(\mu;t)d\mu dz,$$

$$g(\lambda;0) = g_0 \tag{9.84}$$

at $a(z,t) = a_0 + a_1 z$, $b(z,t) = b_0$, $\chi(u;t) = -(u^T \nu_0 u)/2$. Here $g(\lambda;t)$ is one–dimensional characteristic function corresponding to the probability density $f(z;t)$ (Problem 9.3.10),

$$f(z;t) = \frac{1}{(2\pi)^p} \int\limits_{-\infty}^{\infty} e^{-i\mu^T z}g(\mu;t)d\mu, \tag{9.85}$$

z, λ, μ are p–dimensional vectors, ν_0 is $(p \times p)$–dimensional intensity matrix of the Wiener process.

9.3.11. Prove the theorem about the continuous dependence of the solution of the Cauchy problem for Eq. (9.72) on the initial condition and the right–hand side. For this purpose take two equations of the form (9.72) with the functions $f_1(t,x)$ and $f_2(t,x)$ which are determined by formula (9.73) with one and the same operator $A(t)$ and with different functions $g_1(t,x)$ and $g_2(t,x)$ and prove that at any $\varepsilon > 0$ there exists such a number $\delta = \delta_\varepsilon > 0$ that the norm of the difference of the solutions $x_1(t) - x_2(t)$ of these two equations will be smaller than ε at all

$t \in [t, T]$ if $\| x_1(t_0) - x_2(t_0) \| < \delta$, $\| U(t, \tau) g_1(\tau, x) - U(t, \tau) g_2(\tau, x) \| < \delta$ at all t, τ, x.

9.3.12. Consider as the special cases Eqs. (9.42), (9.43) and (9.72) in the separable H-spaces. Formulae (7.33)–(7.35) of Subsection 7.5.4 and the Riesz–Fisher theorem (Problem 7.5.14) show that any separable H-space X is isometrically isomorphic to the space of the sequences l_2: to each $x \in X$ corresponds the sequence of the coefficients $c_k = (x, x_k)$ of the expansion (7.33) and vice versa, to each sequence $\{c_k\}$ corresponds the vector $x \in X$ for which $(x, x_k) = c_k$ and the expansion (7.33) is valid, and according to (7.34) and (7.35) the norm and the scalar product while passing from the space X to the space l_2 remain. Prove that on the basis of this isometric isomorphism of any separable H-space to space l_2 any differenial equation in the separable H-space may be reduced to the differential equation in the space l_2 (i.e. to the infinite set of the scalar differential equations).

9.3.13. The equations of Examples 9.14–9.18 and Problems 9.3.1–9.3.8 may be considered as the differential equations in the corresponding H-spaces: $L_2(R)$ in Examples 9.14–9.18; $L_2([0, \infty))$ with the corresponding boundary condition in Problem 9.3.1; $L_2([0, l])$ with the corresponding boundary conditions in Problem 9.3.2; $L_2(R^3)$ in Problem 9.3.3; the corresponding H-spaces for the equations of Problems 9.3.4–9.3.8. Derive the infinite sets of scalar differential equations which are equivalent to the equations of Examples 9.14–9.18 and Problems 9.3.1–9.3.8.

9.3.14. Prove that if Cauchy problem (9.43) has the unique solution $x(t)$ at any $x_0 \in D_A$ which represents a continuous function of t and continuously depends on x_0 then there exists a set of the evolutionary operators $\{U(t, \tau)\}$.

9.3.15. Prove that if the range $R_{U(t,\tau)}$ of any operator $U(t, \tau)$ is contained in the domain D_A of the operator $A(t)$, $R_{U(t,\tau)} \subset D_A$ then formula (9.64) is valid for any x_0 and for any continuous function $y(t)$.

CHAPTER 10
ELEMENTS OF APPROXIMATE METHODS
IN ABSTRACT SPACES

Chapter 10 is devoted to the elements of approximate methods of functional analysis. In Section 10.1 one of the most efficient methods for approximate solving operator equations, namely the Newton method is presented as well as modified Newton method. Some basic methods of numerical analysis in abstract spaces, namely, the Rayleigh–Ritz method, the Galerkin methods and the finite elements method are outlined in Section 10.2. Some remarks concerning real time numerical analysis are given. The last Section 10.3 is devoted to the methods for solving improper problems.

10.1. Methods of Successive Approximations

10.1.1. General Comments about Approximate Methods

In applications different *operator equations* of the following form are widely encountered:

$$f(x) = y, \tag{10.1}$$

where f is the function (operator) which maps some space X into the space Y (in the special case it may be $Y = X$). Some examples of such equations were given in Section 6.2. Any element x belonging to the domain D_f of the function f for which equality (10.1) is valid is called *a solution* of operator Eq. (10.1) at a given y. It goes without saying that the solution of Eq. (10.1) may exist only in that case when y belongs to the range R_f of the function f.

For solving the simplest equations the different *methods of the successive approximations* or *iterations* are used. We have already acquainted in Subsection 6.2.6 with one of such methods which became the general method of proving the existence of the solutions of the operator equations – the method of contractive mappings.

One of the most important application of the contractive mappings method is the proof of the existence of the function which is given implicitly. Let X be an arbitrary space Y, Z be the B–space, $z = F(x, y)$ be a continuous function of y mapping $X \times Y$ into Z, D_F and R_F be its domain and range correspondingly. Directly from Theorem 6.19 we get the following statement.

Theorem 10.1. *If $0 \in R_F$ and there exists such a natural number p and such a bounded linear operator $B : Z \to Y$ that the operator A^p where*

$$Ay = y + BF(x, y),$$

performs the contractive mapping Y into Y at all $x \in S \subset \{x : \{x, y\} \in D_F, F(x, y) = 0\}$ then the equation

$$F(x, y) = 0$$

determines the unique function $y = f(x)$ with the domain $D_f = S$ and the range $R_f = \{y : \{x, y\} \in D_F, F(x, y) = 0\}$.

Besides the contractive mapping method there are some other methods of the successive approximations, for instance, the tangents methods and the Newton secants method which were initially intended for searching the roots of Eq. (10.1) in the case of the scalar x, y.

There also exist the other types of approximate methods based on the simplification of the problem, on its reduction to the finite–dimensional problem etc. It goes without saying that while simplifying the problem we should try that the solution of the simplified problem will be close to the solution of the initial problem as much as possible.

For the approximate methods theory the question about the possibility to receive a solution of the initial problem with any degree of the accuracy is of great importance. In order to answer this question we usually consider not one problem but a sequence of the simplified problems each of them gives a more precise approximation to the solution of the initial problem then the previous one.

But in practice the question of the estimate of the approximation accuracy to the solution of the initial problem is very important. In most practical problems this question cannot be investigated theoretically and may be solved only experimentally. In this case the use of divergent series and sequences often gives sufficient accuracy for practice and sufficiently more simple algorithms than the use of the convergent procedures.

But it remains unclear what kind of accuracy is sufficient for practice. It is evident that it is impossible to solve this question by means of mathematics. In practice the required accuracy is determined on one hand, by the accuracy of the measurement devices which are used at experimental research, and on the other hand, by the accuracy of the initial data, first of all by the accepted mathematical model of the studied phenomenon. Firstly, all mathematical relations used for searching different phenomena of the environment represent only mathematical model of real phenomena – abstractions and none of the models can completely correspond to the phenomenon which it describes. Secondly, the initial data for searching any phenomenon by means of the accepted model are known only approximately, as they are obtained as a rule by the statistical treatment of the experimental results.

So, the requirement of excessively high accuracy of the calculations is only meaningless but deleterious as it may lead to sufficient complication of the algorithms and calculations, and consequently, to the increase of time consumption and the expenditure of the researches without real increase of the accuracy. Therefore in practice they say about *the rational* accuracy agreed upon available information about the studied phenomenon. And the question about the convergence of approximate procedures is of secondary importance. But it is very important for theoretical base of approximate methods of functional analysis.

10.1.2. Newton Method

As an efficient method of proving the existence of the solutions of different equations the contractive mappings method is suitable for practice only in the simplest special cases. In more complex problems, for instance, in the solution problem of the differential equations of the first order this method may give the sufficient accuracy only after a great number of the steps.

The most feasible method from the successive appoximations methods is that one which was suggested by Newton for finding the roots of the equation $f(x) = y$ where $f(x)$ is a real function of a real variable. This method is based on the fact that while taking in zero approximation some value x_0 (sufficiently by close to the sought root) we substitute the curve $y = f(x)$ by the tangent to it at the the the point x_0, $y = f(x_0) + f'(x_0)(x - x_0)$. After solving the obtained linear equation we receive the first approximation x_1 to the sought root. After this we substitute the curve $y = f(x)$ by the tangent to it at the point x_1. After solving the obtained linear equation we find the second approximation x_2. It is clear from Fig. 10.1 that at appropriate choice of x_0 the successive approximations $x_1, x_2, \ldots, x_n, \ldots$ necessarily converge to the root of the equation $f(x) = 0$. But at unsuccessful choice of x_0 the process of the successive approximations may be "cyclied". On Fig. 10.2 it is shown the case when $x_n = x_{n+2} = x_{n+4} = \ldots, x_{n+1} = x_{n+3} = \ldots$. Therefore the question about the appropriate choice of zero approximation of x_0 is a prime consideration. The high degree of convergence of the successive approximations at the appropriate choice of x_0 led to the case when the Newton method was spread over the sets of the equations and over more general equations and became one of the most efficient general approximate methods of solving the nonlinear operator equations.

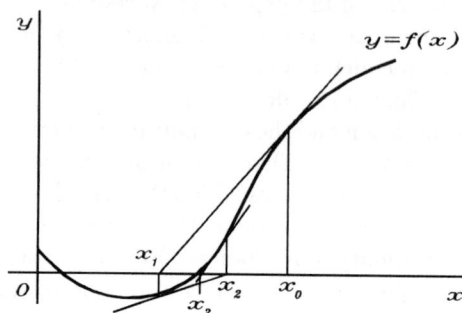

Fig. 10.1

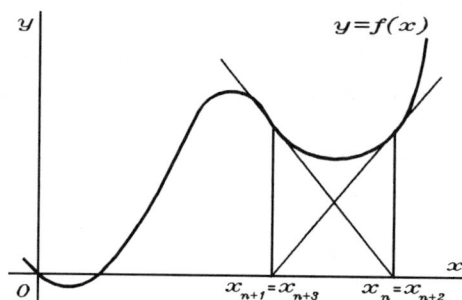

Fig. 10.2

Further we shall study Eq. (10.1) supposing that the spaces X and Y are the B–spaces. After supposing that the function $f(x)$ has a strong derivative in some vicinity of the sought solution of Eq. (10.1) we rewrite it in the form

$$f(a) + f'(a)(x - a) + o(\| x - a \|) = y. \tag{10.2}$$

If the derivative $f'(x)$ represents an invertible operator at the point a then after rejecting in Eq. (10.2) the item $o(\| x - a \|)$ and solving the obtained linear equation we shall have

$$x = a + f'(a)^{-1}[y - f(a)].$$

This formula gives an approximate solution of Eq. (10.1) whose accuracy will be the higher the closer be a to an accurate solution. This gives the opportunity to determine the successive approximations $\{x_n\}$ to the accurate solution x of Eq. (10.1) by formula

$$x_{n+1} = x_n + f'(x_n)^{-1}[y - f(x_n)] \quad (n = 0, 1, 2, \ldots), \tag{10.3}$$

where x_0 is the initial arbitrary sufficiently close to the sought solution the value x. So, the Newton method reduces the solution of the nonlinear Eq. (10.1) to the solution of the sequence of the linear equations.

The answer to the question about the convergence of the sequence of the Newton approximations $\{x_n\}$ gives the following statement.

Theorem 10.2. *If the function $f(x)$ is differentiable in the ball $S_r(x_0)$, its derivative $f'(x)$ satisfies in this ball the Lipschitz condition*

$$\| f'(x') - f'(x) \| < L \, \| x' - x \| \tag{10.4}$$

and represents an operator with a bounded inverse operator whose norm does not exceed $c > 0$, $q = c^2 L \, \| \, y - f(x_0) \, \| \, /2 = c^2 Lk/2 < 1$ and the sum of the series

$$\rho = ck \sum_{\nu=0}^{\infty} q^{2^\nu - 1}$$

does not exceed r then Eq. (10.1) has the solution x in the closed ball $[S_\rho(x_0)]$ and the sequence $\{x_n\}$ converges to x, and

$$\| x_n - x \| < \frac{ckq^{2^n - 1}}{1 - q^{2^n}}. \tag{10.5}$$

▷ At first we establish one auxiliary inequality. Using formula of the finite increments (9.6) we find at any $x, x' \in S_r(x_0)$

$$f(x') - f(x) - f'(x)(x' - x) = \int_0^1 [f'(x + t(x' - x)) - f'(x)](x' - x) \, dt.$$

Hence and from condition (10.4) follows

$$\| f(x') - f(x) - f'(x)(x' - x) \| < L \, \| x' - x \|^2 \int_0^1 t \, dt$$

or

$$\| f(x') - f(x) - f'(x)(x' - x) \| < L \, \| x' - x \|^2 \, /2. \tag{10.6}$$

Now we estimate the norm of the vector $y - f(x_{n+1})$. Bearing in mind that in consequence of formula (10.3) $y = f(x_n) + f'(x_n)(x_{n+1} - x_n)$ and using inequality (10.6) we shall have

$$\| y - f(x_{n+1}) \| = \| f(x_n) - f(x_{n+1}) + f'(x_n)(x_{n+1} - x_n) \|$$

$$< L \, \| x_{n+1} - x_n \|^2 \, /2. \tag{10.7}$$

Applying this inequality at $n = 0, 1, 2, \ldots$ and accounting that $\| y - f(x_0) \| = k$ we find sequentially

$$\| x_1 - x_0 \| \leq \| f'(x_0)^{-1} \| \, \| y - f(x_0) \| \leq ck,$$

$$\| x_2 - x_1 \| \leq \| f'(x_1)^{-1} \| \, \| y - f(x_1) \| < cL \, \| x_1 - x_0 \|^2 \, /2$$

$$< c^3 k^2 L/2 = ckq,$$

$$\| x_3 - x_2 \| < cL \, \| x_2 - x_1 \|^2 \, /2 < c^3 k^2 q^2 L/2 = ckq^3$$

and generally

$$\| x_{n+1} - x_n \| < ckq^{2^n - 1} \quad (n = 0, 1, 2, \ldots). \tag{10.8}$$

For proving this formula at any n we may use the induction method. Supposing that inequality (10.8) is valid at $n = p$ and from (10.3) and (10.7) we find at $n = p + 1$

$$\| x_{p+2} - x_{p+1} \| \leq \| [f'(x_{p+1})]^{-1} \| \, \| y - f(x_{p+1}) \|$$

$$< cL \, \| x_{p+1} - x_p \|^2 \, /2 < c^3 k^2 L q^{2^{p+1} - 2}/2 = ckq^{2^{p+1} - 1}.$$

From inequality (10.8) follows the fundamental property of the sequence $\{x_n\}$ as at any natural p

$$\| x_{n+p} - x_n \| \leq \sum_{\nu=n}^{n+p-1} \| x_{\nu+1} - x_\nu \| < ck \sum_{\nu=n}^{\infty} q^{2^\nu - 1}. \tag{10.9}$$

Consequently, $x_{n+p} - x_n \to 0$ at $n \to \infty$ is uniformly relative to p. Due to the completeness of the B–space X there exists the limit $x = \lim x_n$. Passing in (10.7) to the limit at $n \to \infty$ and accounting the continuity of $f(x)$ we get $y = f(x)$, i.e. $x = \lim x_n$ is the solution of Eq. (10.1).

Further putting in (10.9) $n = 0$ we find at all p

$$\| x_p - x_0 \| < ck \sum_{\nu=0}^{\infty} q^{2^\nu - 1} = \rho,$$

i.e. all the points x_n belong to the ball $S_\rho(x_0)$. Consequently, $x = \lim x_n \in [S_\rho(x_0)]$.

Finally, passing in (10.9) to the limit at $p \to \infty$ we get

$$\| x_n - x \| < ck \sum_{\nu=n}^{\infty} q^{2^\nu - 1} = ckq^{2^n - 1} \sum_{\mu=0}^{\infty} q^{2^{n+\mu} - 2^n}$$

$$= ckq^{2^n - 1} \sum_{\mu=0}^{\infty} q^{2^n(2^\mu - 1)}.$$

Hence accounting that $2^\mu - 1 \geq \mu$ at all integer μ we find

$$\| x_n - x \| < ckq^{2^n - 1} \sum_{\mu=0}^{\infty} q^{2^n \mu} = \frac{ckq^{2^n - 1}}{1 - q^{2^n}}. \quad \triangleleft$$

10.1.3. Modified Newton Method

The largest difficulty while using the Newton method is the inversion of the operators $f'(x_n)$ or what is the same the solution of the linear equations

$$f(x_n) + f'(x_n)(x_{n+1} - x_n) = y. \tag{10.10}$$

Therefore in practice the following modified Newton method is often used. This method is based on the change of all the operators $f'(x_n)$ in Eq. (10.3) by one operator $f'(x_0)$. Here the convergence of the approximations to the solution of Eq. (10.1) certainly will be retarded. But the increase of the number of the iterations has proved itself in the fact that instead of the inversion of the sequence of the operators $\{f'(x_n)\}$ it is sufficient to inverse one operator $f'(x_0)$. The answer to the question about the convergence of the Newton modified iterations process gives the following theorem.

Theorem 10.3. *If the function $f(x)$ is differentiable in the ball $S_r(x_0)$, its derivative satisfies in this ball Lipschitz condition (10.4), $f'(x_0)$ represents an operator with the bounded operator $f'(x_0)^{-1}$, the least root of the quadratic equation*

$$cL\rho^2 - 2\rho + 2h = 0, \tag{10.11}$$

where $c = \| f'(x_0)^{-1} \|$, $h = \| f'(x_0)^{-1}[y - f(x_0)] \|$ does not exceed r and $2cLh < 1$ then Eq. (10.1) has the unique solution in the closed ball $[S_\rho(x_0)]$ and the sequence $\{x_n\}$,

$$x_{n+1} = x_n + f'(x_0)^{-1}[y - f(x_n)], \tag{10.12}$$

converges to x and at this

$$\| x_n - x \| < \frac{(1 - \sqrt{1 - 2cLh})^n}{\sqrt{1 - 2cLh}} h. \tag{10.13}$$

▷ At first we prove that the operator $Ax = x + f'(x_0)^{-1}[y - f(x)]$ maps the ball $S_\rho(x_0)$ into itself. Then if follows that all the points of the sequence $\{x_n\}$ belong to $S_\rho(x_0)$. Using inequality (10.6) we find

$$\| Ax - x_0 \| = \| x - x_0 + f'(x_0)^{-1}[y - f(x)] \|$$

$$= \| [f'(x_0)]^{-1}[f'(x_0)(x - x_0) + y - f(x_0) + f(x_0) - f(x)] \|$$

$$< cL \| x - x_0 \|^2 / 2 + h.$$

At $\| x - x_0 \| < \rho$ by virtue of Eq. (10.11) we have $\| Ax - x_0 \| < cL\rho^2/2 + h = \rho$. Now we prove that Ax is a contraction mapping. As it is differentiable it is sufficient for this purpose to show that the norm of its derivative is smaller than 1 in the closed ball $[S_\rho(x_0)]$. Using inequality (10.4) we obtain at any $x \in S_\rho(x_0)$

$$\| (Ax)' \| = \| I - f'(x_0)^{-1}f'(x) \| = \| f'(x_0)^{-1}[f'(x_0) - f'(x)] \|$$

$$< cL \| x - x_0 \| \leq cL\rho = 1 - \sqrt{1 - 2cLh} < 1.$$

As from (10.12) follows that $x_{n+1} = Ax_n$ then by Theorem 6.19 about contraction mappings $Ax = x$ has the unique solution $x \in [S_\rho(x_0)]$ and the sequence $\{x_n\}$ converges to this solution. But $Ax = x + f'(x_0)^{-1}[y - f(x)]$. Therefore $Ax = x$ if and only if $f(x) = y$, i.e. x is the solution of Eq. (10.1). As the sequence $\{x_n\}$ determined by the contraction mappings method by Theorem 6.21 converges not slower than geometric progression whose denominator α in a given case is equal to $q = cL\rho$ and $\|x_1 - x_0\| = \|f'(x_0)^{-1}[y - f(x_0)]\| = h$ then

$$\|x_n - x\| < \frac{h(cL\rho)^n}{1 - cL\rho} = \frac{(1 - \sqrt{1 - 2cLh})^n}{\sqrt{1 - 2cLh}} h.$$

This proves inequality (10.13). ◁

E x a m p l e 10.1. Consider the nonlinear integral equation

$$\int_a^b f(t, \tau, x(\tau))\, d\tau + g(t, x(t)) = y(t), \qquad (10.14)$$

where $f(t, \tau, x)$ and $g(t, x)$ are the continuous functions which have the continuous derivatives with respect to x. The derivative of the operator

$$F(x) = \int_a^b f(t, \tau, x(\tau))\, d\tau + g(t, x(t))$$

is determined by formula

$$F'(x) = \int_a^b f_x(t, \tau, x(\tau)) \cdot d\tau + g_x(t, x(t)).$$

Using for the solution of the integral equation the Newton method on the basis of Eq. (10.10) we get the sequence of the linear integral equations:

$$\int_a^b f_x(t, \tau, x_n(\tau))[x_{n+1}(\tau) - x_n(\tau)]\, d\tau + g_x(t, x_n(t))[x_{n+1}(t) - x_n(t)]$$

$$= y(t) - \int_a^b f(t, \tau, x_n(\tau))\, d\tau - g(t, x_n(t)). \qquad (10.15)$$

The modified Newton method gives on the basis of Eq. (10.12) the same Eqs. (10.15) with the change of the functions $f_x(t, \tau, x_n(\tau))$ and $g_x(t, x_n(t))$ by the functions $f_x(t, \tau, x_0(\tau))$ and $g_x(t, x_0(t))$ respectively.

R e m a r k.   All obtained equations are also valid in the case of n–dimensional vector functions $x(t)$, $f(t, \tau, x)$, $g(t, x)$ and $y(t)$ in Eq. (10.14). In this case $f_x(t, \tau, x)$ and $g_x(t, x)$ represent the Jacobi matrices of the derivative functions $f(t, \tau, x)$ and $g(t, x)$ with respect to the vector x. Moreover all written formulae and equations are valid in more general case when x is an element of the B–space X, y is an element of the B–space Y, and $f(t, \tau, x)$ and $g(t, x)$ are the operators mapping X into Y (here f depends on t, τ, and g depends on t as the parameters). In this case $f_x(t, \tau, x)$ and $g_x(t, x)$ represent strong derivatives of the operators f and g correspondingly (Subsection 9.1.1).

E x a m p l e 10.2. For solving the vector differential equation

$$\frac{dx}{dt} = f(t, x), \quad x(t_0) = a \tag{10.16}$$

by the Newton method we write Eq. (10.16) in the integral form

$$x(t) = a + \int_{t_0}^{t} f(\tau, x(\tau)) \, d\tau. \tag{10.17}$$

This equation is a special case of Eq. (10.14) at $f(t, \tau, x) = f(\tau, x)1(t - \tau)$, $g(t, x) = -x$, $y(t) = -a$. Therefore the equations which determine the successive approximations to the sought solution have the form

$$\int_{t_0}^{t} f_x(\tau, x_n(\tau))[x_{n+1}(\tau) - x_n(\tau)] \, d\tau - x_{n+1}(t) + x_n(t)$$

$$= -a - \int_{t_0}^{t} f(\tau, x_n(\tau)) \, d\tau + x_n(t).$$

Differentiating this equation we shall obtain the sequence of linear differential equations:

$$\frac{dx_{n+1}}{dt} = f_x(t, x_n)(x_{n+1} - x_n) + f(t, x_n). \tag{10.18}$$

In this case the modified Newton method gives the sequence of the equations:

$$\frac{dx_{n+1}}{dt} = f_x(t, x_0)(x_{n+1} - x_n) + f(t, x_n). \tag{10.19}$$

These equations have the advantages over Eqs. (10.18) as they all have one and the same matrix of the coefficients. It gives the opportunity to find a fundamental matrix of the solutions $u(t, \tau)$, $u(\tau, \tau) = I$ of the corresponding homogenous equation at once for all the equations. Then for determining the approximations $x_n(t)$ to the sought solution $x(t)$ of the initial equation it is sufficient to use the method of constants variation. As a result we obtain

$$x_{n+1}(t) = u(t, t_0)a + \int_{t_0}^{t} u(t, \tau)[f(\tau, x_n(\tau)) - f_x(\tau, x_0(\tau))x_n(\tau)] \, d\tau. \tag{10.20}$$

In the case of the ordinary Newton method we have to find a fundamental matrix of the solution of a homogeneous equation for each equation of Eqs. (10.18) (in this case instead of $u(t, \tau)$ we have to write $u(t, \tau, x_n(\tau))$ indicating by the third argument the dependence of the fundamental matrix of the solutions on $x_n(t)$).

Detailed treatment of the implicitly given contractive mapping is given in (Edwards 1965, Schwartz 1969). Newton methods and similar methods are considered in (Blum 1972, Kahaner et al. 1989).

Problems

10.1.1. Use the Newton method and the modified Newton method for an approximate solution of the differential equation (9.72) in the B–space:

$$\frac{dx}{dt} = f(t, x), \quad x(t_0) = x_0,$$

supposing that the function $f(t, x)$ has the strong derivative $f_x(t, x)$ with respect to x (by the definition it represents a bounded linear operator).

10.1.2. Solve the same problem (Problem 10.1.1) do not supposing that the function $f(t, x)$ is strongly differentiable over x.

I n s t r u c t i o n. Suppose that the function $f(t, x)$ is expressed by formula (9.73) where $g(t, x)$ is strongly differentiable over x and account the equivalence of the differential equation (9.74) and the integral equation (9.75) in this case.

10.2. Some Basic Methods of Numerical Analysis in Abstract Spaces

10.2.1. General Principles of Approximate Methods

For the description of a general scheme of the approximate methods it is convenient to write the operator equation (10.1) in the form:

$$Ax = y, \tag{10.21}$$

where A in the general case is a nonlinear operator, $Ax = f(x)$. Here we shall assume analogously as in Subsection 10.1.2 that the operator A maps the B–space X into the B–space Y (in the special case may be $Y = X$). For approximate solution of Eq. (10.21) it is substituted by the following more simple one:

$$A_n x_n = y_n, \tag{10.22}$$

which yields the immediate solution. For obtaining such simplified equation it is necessary, firstly, to approximate the known right–hand side of Eq. (10.21) by some variable y_n, and secondly, to approximate the operator A by the operator A_n. And this should be made in such a way that on one hand, Eq. (10.22) may be solved with relative ease, and on the other hand, the solution x_n of Eq. (10.22) would approximate with sufficient accuracy the solution x of Eq. (10.21). It is clear that in this case the variables x_n, y_n and the operator A_n must be to some extent more simple than x, y and A. In other words, x_n, y_n must belong to more simple spaces than X and Y. Therefore the question arises about the approximation of the elements of one space by the elements of another.

Let X_n be the B–space whose vectors approximate the vectors of the space X, and Y_n be the B–space whose vectors approximate the vectors of the space Y. The operator A_n will map in this case X_n into Y_n. Usually the finite–dimensional spaces are taken. But in some cases it is expedient to take the infinite–dimensional spaces more often the functional ones. In both cases X_n and Y_n may or may be not the subspaces of the spaces X and Y correspondingly. For establishing the necessary correspondences between x_n and x, y_n and y we introduce the projection operator P_n from X into X_n and the projection operator Q_n from Y into Y_n. It is clear that if X_n is the subspace of the space X then P_n will be an idempotent operator. And also Q_n will be an idempotent operator if Y_n is the subspace of the space Y.

Let us denote by $\| \cdot \|_n$ the norms in the spaces X_n and Y_n. We denote them identically because it will be clear from the formulae to what space one or another norm belongs. In the special case when X_n and Y_n are the subspaces of the spaces X and Y correspondingly the norms $\| \cdot \|_n$ coincide with the norms in the spaces X and Y. As the accuracy measure of the approximation of the vector $x_n \in X_n$ to the vector $x \in X$ it is expedient to take the norm of the difference between the vector x_n and the projection $P_n x$ of the vector x on X_n, $\| x_n - P_n x \|_n$. As the accuracy measure of the approximation of the operator A_n to the operator A at a given $x \in X$ we may take $\| A_n P_n x - Q_n A x \|_n$. We indicate all the variables and the spaces which take part in the approximation of x, y and A by the index n taking into consideration that later on while searching the convergence of approximate solutions to an accurate one we shall consider the sequence of the simplified problems.

Numerical analysis in abstract spaces nowdays presents one of the most rapidly evaluating branch of modern numerical analysis. Tradi-

tionally numerical analysis in abstract spaces contains the following divisions:
- general theory in abstract spaces;
- equations with linear operators;
- equations with nonlinear operators;
- improperly posed problems.

At practice the following basic approximate methods are used:
- the Rayleigh–Ritz and Galerkin methods (Subsections 10.2.3 and 10.2.4) and finite elements and multigrid methods (Subsection 10.2.5);
- the finite elements and multigrid methods (Subsection 10.2.5).

Typical real time problems in dynamics and control, signal and image processing require corresponding approximate methods of numerical analysis in abstract spaces. As usual the following linear methods are used: superposition, convolution, spectral and unitary transforms (Appendix 1, Table A.1.3). Usually the fast versions are implemented. The solution of some specific real time nonlinear problems is based on fast versions of methods described in Subsections 10.2.1–10.2.5.

E x a m p l e 10.3. Consider a linear integral equation in the space of the continuous real scalar functions $C([a, b])$

$$\int\limits_a^b K(t, \tau)x(\tau)\, d\tau + \varphi(t)x(t) = y(t), \quad t \in [a, b], \tag{10.23}$$

where $K(t, \tau)$ is a continuous function. In this case $X = Y = C([a, b])$. We shall approximate the functions from $C([a, b])$ by piecewise linear functions with the points (nodes) of the conjugacy $t_0 = a, t_1, \ldots, t_{N-1}, t_N = b$. As the space $X_n = Y_n$ in this case we may assume the $(N + 1)$–dimensional space R^{N+1}. But we may also assume as $X_n = Y_n$ $N + 1$–dimensional subspace of the space $C([a, b])$ which consists of all piecewise linear functions with the points $t_0 = a, t_1, \ldots, t_{N-1}$, $t_N = b$. In the first case as the operator $P_n = Q_n$ will be the operator which establishes the correspondence between each function $x(t) \in C([a, b])$ and the $(N + 1)$–dimensional vector $[x_0\ x_1\ \ldots\ x_N]^T$, $x_k = x(t_k)$. In the second case the operator $P_n = Q_n$ will establish the correspondence between each function $x(t) \in C([a, b])$ and the piecewise linear function with the same values in the points $t_0 = a, t_1, \ldots, t_{N-1}, t_N = b$. In both cases as the norm in $X_n = Y_n$ will serve $\| x_n \|_n = \max\limits_k | x_k |$ and in the second case this norm coincides with the ordinary norm in $C([a, b])$.

For determining the operator A_n it is natural to substitute in the integral equation the function $x(t)$ by the piecewise linear function and restrict ourselves by the requirement that the equation be satisfied at $t = t_0, t_1, \ldots, t_N$. Then the integral equation will be replaced by the set of linear algebraic equations:

$$\sum_{k=0}^{N} a_{hk} x_k = y_h \quad (h = 0, 1, \ldots, N), \qquad (10.24)$$

and as the operator A_n will serve the linear operator R^{N+1} with the matrix whose elements are determined by the formulae

$$a_{h0} = \int\limits_a^{t_1} \frac{t_1 - \tau}{t_1 - a} K(t_h, \tau) \, d\tau + \varphi(t_h)\delta_{h0} \quad (h = 0, 1, \ldots, N),$$

$$a_{hk} = \int\limits_{t_k}^{t_{k+1}} \frac{t_{k+1} - \tau}{t_{k+1} - t_k} K(t_h, \tau) \, d\tau + \int\limits_{t_{k-1}}^{t_k} \frac{\tau - t_{k-1}}{t_k - t_{k-1}} K(t_h, \tau) \, d\tau$$

$$+\varphi(t_h)\delta_{hk} \quad (h = 0, 1, \ldots, N; \; k = 1, \ldots, N - 1),$$

$$a_{hN} = \int\limits_{t_{N-1}}^{b} \frac{\tau - t_{N-1}}{b - t_{N-1}} K(t_h, \tau) \, d\tau + \varphi(t_h)\delta_{hN} \quad (h = 0, 1, \ldots, N). \quad (10.25)$$

If as $X_n = Y_n$ assume the subspace of piecewise linear functions with the points $t_0, t_1, \ldots, t_N$ then as the operator A_n will serve the linear integral operator with the kernel $K(t, \tau) + \delta(\tau - t)\varphi(\tau)$ at $t \in \{t_0, t_1, \ldots, t_N\}$. Here the integral equation is transformed into a set of the algebraic equations (10.24) which determines the values of the piecewise linear function $x_n(t)$ in the nodes by the given values of the piecewise linear function $y_n(t)$.

E x a m p l e 10.4. For an approximate solution of the linear differential equation in the B–space S

$$\frac{dx}{dt} = Bx + y, \quad x(t_0) = x_0, \qquad (10.26)$$

we shall take in the space S a set of linear independent vectors $\varphi_1, \ldots, \varphi_N$ which belong to the domain D_B of the operator B at all $t \in [a, b]$. We shall approximate the vectors x and y by the linear combinations of the vector $\varphi_1, \ldots, \varphi_N$ which will be certainly dependent on t:

$$x \approx \sum_{k=1}^{N} \alpha_k(t)\varphi_k, \quad y \approx \sum_{k=1}^{N} \beta_k(t)\varphi_k. \qquad (10.27)$$

In this case the space $X = Y$ represents the B–space of the functions continuous on some interval $[a, b]$, $t_0 \in [a, b]$ with the values in the B–space S. This space we shall denote by $C([a, b], S)$ indicating not only the domain of the functions but the space of their values. In a given case the space $X_n = Y_n$ represents the space of N–dimensional vector continuous functions of t which we shall denote by $C([a, b], C^N)$ taking into consideration that the coordinates of the vectors $\alpha = [\alpha_1 \ldots \alpha_N]^T$ and $\beta = [\beta_1 \ldots \beta_N]^T$ in Eq. (10.27) may be complex in the general case. Thus in this case the space $X_n = Y_n$ is infinite–dimensional. As the projection operator $P_n = Q_n$ serves the operator which establishes the correspondence between each function $x(t)$ and N–dimensional vector function $x_n(t) = \alpha(t) = [\alpha_1(t) \ldots \alpha_N(t)]^T$. Here $y_n(t) = \beta(t) = [\beta_1(t) \ldots \beta_N(t)]^T$.

To determine the operator $A_n = I_n d/dt - B_n$ which approximates the operator $A = I d/dt - B$ where I_n is an unity operator in the space X_n, and I is an unity operator in the space X we substitute expressions (10.27) into Eq. (10.26):

$$\sum_{k=1}^{N} \frac{d\alpha_k}{dt} \varphi_k \approx \sum_{k=1}^{N} (\alpha_k B \varphi_k + \beta_k \varphi_k).$$

The vectors $B\varphi_1, \ldots, B\varphi_N$ in the general case do not belong to the subspace formed by the vectors $\varphi_1, \ldots, \varphi_N$. Therefore it is natural to approximate them by the linear combinations of the vectors $\varphi_1, \ldots, \varphi_N$:

$$B\varphi_k \approx \sum_{l=1}^{N} b_{lk} \varphi_l \quad (k = 1, \ldots, N). \tag{10.28}$$

After substituting this expression into the previous equation we obtain

$$\sum_{k=1}^{N} \frac{d\alpha_k}{dt} \varphi_k \approx \sum_{l=1}^{N} \left(\sum_{k=1}^{N} b_{lk} \alpha_k \right) \varphi_l + \sum_{k=1}^{N} \beta_k \varphi_k.$$

After changing the denotion of the indexes of the summation in the double sum of l by k, and k by l we shall have

$$\sum_{k=1}^{N} \frac{d\alpha_k}{dt} \varphi_k = \sum_{k=1}^{N} \left(\sum_{l=1}^{N} b_{kl} \alpha_l + \beta_k \right) \varphi_k. \tag{10.29}$$

Hence follow the equations for the coefficients $\alpha_1, \ldots, \alpha_N$:

$$\frac{d\alpha_k}{dt} = \sum_{l=1}^{N} b_{kl} \alpha_l + \beta_k \quad (k = 1, \ldots, N). \tag{10.30}$$

Thus after determining the coefficients $\alpha_1, \ldots, \alpha_N$ by Eq. (10.30) we may provide an accurate equality in Eq. (10.29). Consequently, as the operator B_n which approximates the operator B it is expedient to assume the linear operator in space

C^N with the matrix whose elements serve the variables b_{kl} $(k, l = 1, \ldots, N)$ which depend in the general case on t. In order to receive the initial conditions for Eq. (10.30) the vector x_0 should be approximated in Eq. (10.26) by linear combination of the vectors $\varphi_1, \ldots, \varphi_N$:

$$x_0 \approx \sum_{k=1}^{N} \alpha_{k0} \varphi_k. \tag{10.31}$$

Then as the initial condition for Eq. (10.30) will serve

$$\alpha_k(t_0) = \alpha_{k0} \quad (k = 1, \ldots, N). \tag{10.32}$$

For finding β_k, b_{kl} and α_{k0} we may introduce in the space S^* which is dual with S such a set of linear independent functionals $\psi_1, \ldots, \psi_N$ that none of them will be orthogonal to all the vectors $\varphi_1, \ldots, \varphi_N$, i.e. that for each ψ_p will be found the vector φ_q for which $\psi_p \varphi_q \neq 0$. Then from (10.27), (10.28) and (10.32) we shall obtain the sets of linear algebraic equations for β_k, b_{kl}, α_{k0}:

$$\sum_{k=1}^{N} (\psi_h \varphi_k) \beta_k = \psi_h y \quad (h = 1, \ldots, N),$$

$$\sum_{k=1}^{N} (\psi_h \varphi_k) b_{kl} = \psi_h B \varphi_l \quad (h = 1, \ldots, N; \ l = 1, \ldots, N),$$

$$\sum_{k=1}^{N} (\psi_h \varphi_k) \alpha_{k0} = \psi_h x_0 \quad (h = 1, \ldots, N).$$

These equations are instantaneously solved if we take a biorthogonal set of the pairs of the vectors from S and S^*, $\psi_h \varphi_k = \delta_{hk}$ (Subsection 7.5.7). Then we get

$$\beta_k = \psi_k y, \quad b_{kl} = \psi_k B \varphi_l, \quad \alpha_{k0} = \psi_k x_0.$$

Eqs. (10.30) are easily solved by any methods of the numerical analysis of the ordinary differential equations. Thus the solution of the Cauchy problem for the linear differential equation in the B-space is reduced by the stated method to the solution of the Cauchy problem for a set of the ordinary differential equations.

For an approximate solution of Eq. (10.26) we may also assume as $X_n = Y_n$ the subspace of the space $X = Y = C([a, b], S)$ which consists of all linear combinations of the vectors $\varphi_1, \ldots, \varphi_N$ whose coefficients serve as the continuous scalar functions of t. In other words, $X_n = Y_n$ in this case represents the subspace of the functions from $C([a, b], S)$ with the values in N-dimensional subspace of the space S in which the vectors $\varphi_1, \ldots, \varphi_N$ serve as the basis. So, as the operator

$P_n = Q_n$ will serve the operator which establishes the correspondence between each function $x(t)$ and its projection

$$x_n(t) = P_n x(t) = \sum_{k=1}^{N} \alpha_k(t)\varphi_k$$

on the subspaces of linear combinations of the vectors $\varphi_1, \ldots, \varphi_N$ with the coefficients which depend on t. Here we have

$$y_n(t) = \sum_{k=1}^{N} \beta_k(t)\varphi_k.$$

As the operator A_n approximating A will serve in this case $P_n A$, $A_n = P_n A$ and Eq. (10.22) will have the form

$$\frac{dx_n}{dt} = P_n B x_n + y_n, \quad x_n(t_0) = P_n x_0. \tag{10.33}$$

Combining the Newton method (or the modified Newton method) with the stated one we may solve approximately the nonlinear differential equations in the B–spaces.

E x a m p l e 10.5. In some cases for reducing Eq. (10.26) to a set of the ordinary differential equations it is expedient to approximate the vectors of the space S by the nonlinear functions of the finite set of the parameters, for instance, by the function $\varphi(\alpha)$ where α is N–dimensional vector of the parameters, $x \approx \varphi(\alpha)$. In this case the operator $P_n = Q_n$ which compares the vector $x \in S$ with N–dimensional vector $x_n = \alpha$ is nonlinear one. Transforming Eq. (10.26) by the operator P_n and taking into consideration that $P_n x = \alpha$ we obtain the differential equation for the vector α

$$\frac{d\alpha}{dt} = P_n B\varphi(\alpha) + \beta, \quad \alpha(t_0) = \alpha_0 = P_n x_0, \tag{10.34}$$

where $\beta = P_n y$ is the value of the argument of the function φ at which it approximates y, $y \approx \varphi(\beta)$. In this case we have $x_n = \alpha$, $y_n = \beta$, $B_n = P_n B\varphi(\cdot)$.

Eq. (10.34) represents a set of N nonlinear scalar differential equations. The function $\varphi(\alpha)$ is not obligatory nonlinear relative to all the coordinates of the vector α. It may be a linear function of ones coordinates of the vector α with the coefficients which depend on the other coordinates of the vector α. In other words, the function $\varphi(\alpha)$ may have the form

$$\varphi(\alpha) = \varphi(\alpha_1, \ldots, \alpha_N) = \sum_{k=1}^{m} \alpha_k \varphi_k(\alpha_{m+1}, \ldots, \alpha_N), \tag{10.35}$$

where $\varphi_1(\alpha_{m+1}, \ldots, \alpha_N), \ldots, \varphi_m(\alpha_{m+1}, \ldots, \alpha_N)$ are linear independent vectors $\varphi_1, \ldots, \varphi_m \in S$ at any values of the parameters $\alpha_{m+1}, \ldots, \alpha_N$. Such approximation often allows sufficiently decrease the number N of the parameters.

The stated in this Example method of solving the differential equations in the B-space S at the approximation of the vectors of the space S by the functions of the form (10.35) is used in modern statistical control theory. In particular, numerical analysis of linear and nonlinear stochastic systems with lumped parameters is based on the numerical solution of the deterministic partial differential equation for probability density (Problem 9.3.10) and integro–differential for characteristic function (Problem 9.3.11).

E x a m p l e 10.6. Use the method of Example 10.4 for solving a wave equation

$$\frac{\partial^2 z}{\partial t^2} = a^2 \frac{\partial^2 z}{\partial x^2} . \tag{10.36}$$

Eq. (10.36) is reduced by the change of variables $z_1 = z$, $z_2 = \partial z/\partial t$ to the form of Eq. (10.26):

$$\frac{d}{dt}\begin{bmatrix} z_1 \\ z_2 \end{bmatrix} = \begin{bmatrix} 0 & 1 \\ a^2 \partial^2/\partial x^2 & 0 \end{bmatrix} \begin{bmatrix} z_1 \\ z_2 \end{bmatrix} .$$

In this case as the space S serves the space of two–dimensional vector functions of the scalar variable x whose domain serves $[0, l]$. It is necessary to add to the obtained equation the boundary conditions $z_1(0, t) = z_1(l, t) = z_2(0, t) = z_2(l, t) = 0$ which are received from the condition of the immovability of the ends of the string and the initial conditions

$$z_1(x, 0) = f_0(x), \quad z_2(x, 0) = \frac{\partial z(x, 0)}{\partial t} = f_1(x), \tag{10.37}$$

where $f_0(x)$ and $f_1(x)$ are arbitrary continuous functions satisfying the boundary conditions $f_0(0) = f_0(l) = f_1(0) = f_1(l) = 0$.

As according to the known from matematical analysis Weierstrass theorem any function which is continuous on the finite closed interval may be uniformly approximated by the trigonometric polynomial then as the basis functions in the space S it is expedient to take (two–dimensional vector) functions which satisfy the boundary conditions

$$\varphi_{kh}(x) = [\sin(k\pi x/l) \ \sin(h\pi x/l)]^T \quad (k, h = 1, \ldots, n).$$

In accordance with Example 10.4 the solution of Eq. (10.36) should be found in the form

$$\begin{bmatrix} z_1(x, t) \\ z_2(x, t) \end{bmatrix} = \sum_{k,h=1}^{n} \alpha_{kh}(t) \begin{bmatrix} \sin(k\pi x/l) \\ \sin(h\pi x/l) \end{bmatrix}$$

or

$$z(x,t) = \sum_{k=1}^{n} \alpha_k(t) \sin \frac{k\pi x}{l}.$$

Substituting this expression into Eq. (10.36) we shall get for the coefficients of $\alpha_k(t)$ the following equations:

$$\frac{d^2\alpha_k}{dt^2} = -k^2\omega^2\alpha_k \quad (k = 1, \ldots, n), \quad \omega = \frac{a\pi}{l}.$$

For receiving the initial conditions for these equations the functions $f_0(x)$ and $f_1(x)$ should be presented by the corresponding segments of the Fourier series extending them on the interval $[-l, 0]$ as the odd function (in order to avoid the occurance of the cosines which may lead to the violation of the boundary conditions). Then we get

$$\alpha_{k0} = \frac{2}{l} \int_0^l f_0(x) \sin \frac{k\pi x}{l} \, dx,$$

$$\alpha'_{k0} = \frac{2}{l} \int_0^l f_1(x) \sin \frac{k\pi x}{l} \, dx \quad (k = 1, \ldots, n).$$

Integrating the equations for α_k at these initial conditions we obtain

$$\alpha_k(t) = \alpha_{k0} \cos k\omega t + \frac{\alpha'_{k0}}{k\omega} \sin k\omega t \quad (k = 1, \ldots, n).$$

As a result the approximate solution of Eq. (10.36) at the initial condition (10.37) and the boundary conditions $z(0, t) = z(l, t) = 0$ will be determined by the following formula:

$$z(x,t) = \sum_{k=1}^{n} \left(\alpha_{k0} \cos k\omega t + \frac{\alpha'_{k0}}{k\omega} \sin k\omega t \right) \sin \frac{k\pi x}{l}.$$

10.2.2. Convergence of Approximate Solutions to Exact Solution

Let us consider a sequence of the B–space $\{X_n\}$ with the corresponding sequence of the projection operators $\{P_n\}$ of the B–space X into X_n, the sequence of the B–space $\{Y_n\}$ with the corresponding sequence of the projection operators $\{Q_n\}$ of the B–space Y into Y_n and the sequence of the solutions $\{x_n\}$ of Eqs. (10.22) at $n = 1, 2, \ldots$. Let x be an exact solution of Eq. (10.21). Before speaking about the convergence of the sequence of approximate solutions to exact one we notice that all the points x, x_n belong to the different spaces. Therefore at first it is necessary to generalize the notion of the convergence over the sequences of the points which belong to the different B–spaces. For approaching to such generalized notion of the convergence recall that as the measure of the proximity of the point $x_n \in X_n$ to $x \in X$ we understood the norm of the difference $x_n - P_n x$ in the space X_n. Therefore it is natural to consider the sequence $\{x_n\}$ *convergent to* x if $\| x_n - P_n x \|_n \to 0$ at $n \to \infty$. For providing the iniqueness of the limit it is necessary to determine the norms in the spaces X_n in such a way that from $\| P_n x - P_n x' \|_n \to 0$ will follow $x = x'$. The set of the norms in the spaces X_n which satisfies this condition is called *nonsingular*. The nonsingularity of the set of the norms in the spaces X_n provides the uniqueness of the limit of the sequence. Correspondingly we shall consider the sequence of the points $\{y_n\}$, $y_n \in Y_n$ convergent to the point $y \in Y$ if $\| y_n - Q_n y \|_n \to 0$ at $n \to \infty$.

For determining the convergence of the sequence of the operators $\{A_n\}$ to the operator A we notice that at any x the point $A_n P_n x$ belongs to the space Y_n, and the point Ax belongs to the space Y. Therefore for determining the convergence of the sequence of the operators $\{A_n\}$ it is natural to use the convergence of the sequence of the points of the space Y_n to the point of the space Y. In the correspondence with this fact we shall call the sequence of the operators $\{A_n\}$, $A_n : X_n \to Y_n$, *convergent* to the operator $A : X \to Y$ at the point $x \in X$ if the sequence $\{A_n P_n x\}$ converges to Ax, i.e. if $\| A_n P_n x - Q_n Ax \|_n \to 0$ at $n \to \infty$.

Theorem 10.4. *If Eqs. (10.21) and (10.22) have the solutions x and x_n, a set of the norms in the spaces X_n is nonsingular, the sequence $\{y_n\}$ converges to y, the sequence of the operators $\{A_n\}$ converges to A on each solution x of Eq. (10.21) and there exists such continuous strictly increasing function $\omega(s)$, $\omega(0) = 0$, $\omega(\infty) = \infty$ that*

$$\| x_n - x_n' \|_n < \omega(\| A_n x_n - A_n x_n' \|_n) \qquad (10.38)$$

at all n, x_n, $x_n' \in X_n$ then the solutions x and x_n of Eqs. (10.21) and (10.22) are unique and the sequence $\{x_n\}$ converges to the solution x of Eq. (10.21).

▷ The uniqueness of the solution of Eq. (10.22) follows immediately from inequality (10.38) at $A_n x_n = A_n x_n' = y_n$. For proving the uniqueness of the solution of Eq. (10.21) we notice that if $Ax = Ax' = y$, then $Ax - Ax' = 0$. So, inequality (10.38) gives

$$\| P_n x - P_n x' \|_n < \omega(\| A_n P_n x - A_n P_n x' \|_n)$$
$$= \omega(\| A_n P_n x - Q_n Ax + Q_n Ax' - A_n P_n x' \|_n)$$
$$\leq \omega(\| A_n P_n x - Q_n Ax \|_n + \| Q_n Ax' - A_n P_n x' \|_n) \to \omega(0) = 0$$

at $n \to \infty$ as the sequence of the operators $\{A_n\}$ converges to the operator A and the function ω strictly increases and is continuous. As the set of the norms in the space X_n is nonsingular then $x = x'$.

Finally, at any n from inequality (10.38) follows

$$\begin{aligned}
\| x_n - P_n x \|_n &< \omega(\| A_n x_n - A_n P_n x \|_n) \\
&= \omega(\| y_n - A_n P_n x \|_n) \leq \omega(\| y_n - Q_n y \|_n \\
&+ \| Q_n A x - A_n P_n x \|_n) \to 0 \text{ at } n \to \infty,
\end{aligned}$$

as the sequence $\{y_n\}$ converges to y and the sequence of the operators A_n converges to the operator A. ◁

In the case of the linear operators A_n condition (10.38) which is often called *the stability condition* of the approximations will be fulfilled if the set of the norms of the inverse operators A_n^{-1} is bounded, i.e. if there exists such $c > 0$ that $\| A_n^{-1} \|_n < c$ at all n. Here $\omega(t) = ct$.

E x a m p l e 10.7. In Example 10.3 taking such sequence of the points $\{t_0^{(n)} = a < t_1^{(n)} < \cdots < t_{N_n}^{(n)} = b\}$, $\Delta_n = \max_k (t_{k+1}^{(n)} - t_k^{(n)}) \to 0$ at $n \to \infty$ we shall have $y_n = [y_0^{(n)} y_1^{(n)} \ldots y_{N_n}^{(n)}]^T \to y$ if at any $\varepsilon > 0$ $y_k^{(n)} - y(t_k^{(n)}) \to 0$ at $n \to \infty$. The sequence of the operators $\{A_n\}$ in this case converges to the operator A. Really, as for any function $x(t) \in C([a,b])$ $x(t_k^{(n)}) = x_k^{(n)}$ then denoting in terms of $x_n(t)$ a piecewise linear function with the same values in the points we shall have

$$\| A_n x_n - Q_n A x \|_n = \max_h \int_a^b K(t_h^{(n)}, \tau)[x_n(\tau) - x(\tau)]\, d\tau \to 0$$

at $n \to \infty$ as $\sup_{\tau \in [a,b]} | x_n(\tau) - x(\tau) | \to 0$ at $n \to \infty$. The condition of the boundedness of the norms of the inverse operators A_n^{-1} will be fulfilled if there exists such constant $c > 0$ that

$$\max_h \left| \int_a^b K(t_h^{(n)}, \tau) x_n(\tau)\, d\tau + \varphi(t_h^{(n)}) x_h^{(n)} \right| > c^{-1} \max_h | x_h^{(n)} |.$$

E x a m p l e 10.8. In Example 10.4 we have $y_n \to y$ if

$$\sum_{k=1}^{N_n} \beta_k^{(n)} \varphi_k^{(n)} \to y \text{ at } n \to \infty.$$

This condition will be, in particular, fulfilled in the case of the separable H–space S if $\varphi_k^{(n)} = \varphi_k$ at all n and the sequence $\{\varphi_k\}$ is orthonormal and compelete. In this case $\beta_k^{(n)} = (y, \varphi_k)$ (Subsection 7.5.4). This condition is also fulfilled in the case of the basis $\{\varphi_k\}$ generated by the biorthonormal system of the pairs of the vectors $\{\varphi_k, \psi_k\}$. So, in this case we have $\beta_k^{(n)} = (y, \psi_k)$ (Subsection 7.5.8).

The sequence of the operators $\{A_n\}$, $A_n = I_n d/dt - B_n$ will converge to the operator $A = Id/dt - B$ on the solution x of Eq. (10.26) if the sequence of the operators $\{B_n\}$ converges to the operator B. Really, as any bounded linear operator is commutative with the operator of the differentiation with respect to scalar variable then

$$A_n P_n x - P_n A x = \frac{d}{dt}(P_n x) - B_n P_n x$$

$$-P_n \frac{dx}{dt} + P_n B x = -(B_n P_n x - P_n B x).$$

We denote by $\gamma_k^{(n)}$ the coefficients at $\varphi_k^{(n)}$ $(k = 1, \ldots, N_n)$ in the approximate formula

$$B x \approx \sum_{k=1}^{N_n} \gamma_k^{(n)} \varphi_k^{(n)}.$$

Then taking into account that $P_n x = \alpha_n = [\alpha_1^{(n)} \ldots \alpha_{N_n}^{(n)}]^T$ we get

$$B_n P_n x = \left[\sum_{l=1}^{N_n} b_{11}^{(n)} \alpha_l^{(n)} \ldots \sum_{l=1}^{N_n} b_{N_n l}^{(n)} \alpha_l^{(n)} \right]^T,$$

$$P_n B x = [\gamma_1^{(n)} \ldots \gamma_{N_n}^{(n)}]^T$$

and

$$B_n P_n x - P_n B x = \left[\sum_{l=1}^{N_n} b_{11}^{(n)} \alpha_l^{(n)} - \gamma_1^{(n)} \ldots \sum_{l=1}^{N_n} b_{N_n l}^{(n)} \alpha_l^{(n)} - \gamma_{N_n}^{(n)} \right]^T.$$

Hence it is clear that at any determination of the norm in the space C^{N_n} the sequence of the operators $\{B_n\}$ will converge to the operator B if

$$\sup_{t \in [a,b]} \max_{1 \leq k \leq N_n} \left| \sum_{l=1}^{N_n} b_{kl}^{(n)} \alpha_l^{(n)} - \gamma_k^{(n)} \right| \to 0 \text{ at } n \to \infty.$$

In particular, if S is the H–space, and $\{\varphi_k\}$ is the basis in it generated by the biorthonormal system $\{\varphi_k, \psi_k\}$ this condition takes the form

$$\sup_{t \in [a,b]} \max_{1 \leq k \leq N_n} \left| \sum_{l=1}^{N_n} (B\varphi_l, \psi_k)(x, \psi_l) - (Bx, \psi_k) \right| \to 0.$$

It is always fulfilled as in this case (Subsection 7.5.8) we have

$$x = \sum_{l=1}^{\infty}(x, \psi_l)\varphi_l, \quad Bx = \sum_{l=1}^{\infty}(x, \psi_l)B\varphi_l.$$

If $\{\varphi_k\}$ is the orthonormal basis then in all previous formulae we get $\psi_k = \varphi_k$.

If we assume as the space $X_n = Y_n$ the subspace of the space $X = Y$ formed by the linear combinations of the vectors $\varphi_1^{(n)} \dots \varphi_{N_n}^{(n)}$ then $B_n = P_n B$ and the condition of the convergence of the sequence of the operators $\{B_n\}$ to the operator B on the solution x of Eq. (10.26) will take the form

$$\| B_n P_n x - P_n Bx \|_n = \| P_n B(P_n x - x) \| \to 0 \text{ at } n \to \infty.$$

This condition is fulfilled if $P_n x \to x$ for the solution x of Eq. (10.26). In the case of the separable H–space S and the basis $\{\varphi_k\}$ in it this condition is always fulfilled.

It is rather difficult to verify condition (10.38) of Theorem 10.4 which in the case of the linear problems is reduced practically to the requirement of the boundedness of the norms set of the inverse operators A_n^{-1}. In our case for verifying this fact use for the solution of Eq. (10.33) formula (9.64). As a result we get

$$x_n(t) = U_n(t, t_0)x_n + \int_{t_0}^{t} U_n(t, \tau)y_n(\tau)\, d\tau, \qquad (10.39)$$

where $U_n(t, \tau)$ are the evolutionary operators of Eq. (10.33) (Subsection 9.3.2). If all the operators $U_n(t, \tau)$ are bounded and the set of their norms on the interval $[a, b]$ is bounded, $\| U_n(t, \tau) \| < c$ at all n and $t, \tau \in [a, b]$ then from Eq. (10.39) follows

$$\sup_{t\in[a,b]} \| x_n(t) \| \leq \sup_{t,\tau\in[a,b]} \| U_n(t, \tau) \| \{\| x_{n0} \| + (b - a) \sup_{t\in[a,b]} \| y_n(t) \|\}$$

$$< c\{\| x_{n0} \| + (b - a) \sup_{t\in[a,b]} \| y_n(t) \|\}. \qquad (10.40)$$

Hence it is clear that for the boundedness of the norms set of the inverse operators A_n^{-1} it is sufficient in our case that the norms set of the operators $U_n(t, \tau)$ be bounded on the interval $[a, b]$. If this condition is fulfilled then according to Theorem 10.4 the sequence of the found approximate solutions converges to the accurate solution of Eq. (10.26).

Condition (10.40) is always fulfilled in the case of the bounded operator B in Eq. (10.26) as in this case we have

$$U_n(t, \tau) = \exp\left\{ P_n \int_{\tau}^{t} B(\tau)\, d\tau \right\}. \qquad (10.41)$$

We make sure in this fact by direct substitution formula (10.41) in Eq. (10.33). Really, from formula (10.41) we get

$$\| U_n(t,\tau) \| \le \exp\{\| P_n \| \sup_{t \in [a,b]} \| B(t) \| \, (b-a)\}. \qquad (10.42)$$

E x a m p l e   10.9.   In Example 10.6 the functions $\sin(k\pi x/l)$ ($k = 1, 2, \ldots$) form the orthogonal basis in the space of the continuous functions which satisfy the boundary conditions. This space may be considered as the subspace of the space $L_2([0, l])$. Therefore from the general results of Example 10.8 follows that the convergence condition of the sequence of the operators $\{B_n\}$ to the operator B is fulfilled.

The stated method of solving the wave equation gives an accurate solution in the form of the Fourier series. Thus we came to the known Fourier method of solving the linear equations in the partial derivatives.

10.2.3. Rayleigh–Ritz Method

This method was at first applied by Rayleigh for solving some boundary value problems and later on was developed by Ritz as a general method of finding the extrema of the functionals which initiated the so–called direct methods of calculus of variations. As it was shown in Subsection 8.1.1 the eigenvectors of the compact self–adjoint operator A realize the extremum of the functional (Ax, x) in the corresponding subspaces, and the extrema values are equal to the corresponding eigenvalues. This extremal property of the eigenvalues was used by Rayleigh. But the searching of the eigenvectors is connected with the solution of some equations of the form of Eq. (10.21). As we have already seen in Examples of Section 9.2 the problems solution of calculus of variations are also reduced to the solution of the operator (differential) equations. This connection between the problems on the extremum and the operator equations gives the opportunity to apply the Rayleigh–Ritz method for solving operator equations.

Let $F(x)$ be a lower bounded real functional on the B–space X. Any sequence of the vectors $\{x_n\}$ convergent to $\inf F(x)$ is called a *minimizing* for the functional $F(x)$. Notice that the problem of finding the maximum of the upper bounded functional $F(x)$ is equivalent to the problem of the minimization of the functional $-F(x)$. The idea of finding the minimizing sequence by the Rayleigh–Ritz method is as follows. In the space B the sequence of the finite–dimensional subspaces $\{X_n\}$ is taken. In each subspace the basis $\{\varphi_1^{(n)}, \ldots, \varphi_{N_n}^{(n)}\}$ is taken, and in each subspace X_n the vector x_n is found which realizes the minimum of the functional $F(x)$ on this subspace. Naturally, the last–named problem is reduced to finding the minimum of the function of the finite number N_n of variables as any vector $x \in X_n$ represents a linear combination of the basis vectors $\varphi_1^{(n)}, \ldots, \varphi_{N_n}^{(n)}$. Thus putting

$$x_n = \sum_{k=1}^{N_n} \alpha_k^{(n)} \varphi_k^{(n)}, \tag{10.43}$$

we find the coefficients $\alpha_1^{(n)}, \ldots, \alpha_{N_n}^{(n)}$ from the condition of the function minimum

$$f(\alpha_1^{(n)}, \ldots, \alpha_{N_n}^{(n)}) = F\left(\sum_{k=1}^{N_n} \alpha_k^{(n)} \varphi_k^{(n)}\right). \tag{10.44}$$

In order to reduce the solution of Eq. (10.21) to the extremal problem it is sufficient to take as the functional $F(x)$ the norm $\| Ax - y \|$. Then the Rayleigh–Ritz method will determine the solution of Eq. (10.21) by formula (10.43) in which the coefficients $\alpha_1^{(n)}, \ldots, \alpha_{N_n}^{(n)}$ are found by the minimization of the function

$$f(\alpha_1^{(n)}, \ldots, \alpha_{N_n}^{(n)}) = \left\| A \sum_{k=1}^{N_n} \alpha_k^{(n)} \varphi_k^{(n)} - y \right\|. \tag{10.45}$$

In the case of the H–space X the Rayleigh–Ritz method is reduced to the known method of least squares. The Rayleigh–Ritz method allows to solve effectively the linear problems in the H–spaces which are reduced to the minimization of the square functionals.

Theorem 10.5. *The solution of* Eq. (10.21) *realizes the minimum of the square functional*

$$F(x) = (Ax, x) - (x, y) - (y, x). \tag{10.46}$$

▷ As A is a bounded operator, then $D_A = X$ (Subsection 7.3.1). If x^* is the solution of Eq. (10.21) then $y = Ax^*$ and for any $x \in X$ we have

$$F(x) = (Ax, x) - (x, Ax^*) - (Ax^*, x) + (Ax^*, x^*)$$

$$-(Ax^*, x^*) = (Ax - Ax^*, x - x^*) - (Ax^*, x^*).$$

Hence taking into account that A is a positive operator (Subsection 6.1.5) we conclude that the functional $F(x)$ achieves the minimal value $-(Ax^*, x^*)$ at $x = x^*$. ◁

The condition of the minimum of the function obtained by the substitution in (10.46) of expression (10.43) gives the equations for the coefficients $\alpha_1^{(n)}, \ldots, \alpha_{N_n}^{(n)}$:

$$\sum_{k=1}^{N_n} (A\varphi_k^{(n)}, \varphi_l^{(n)})\alpha_k^{(n)} = (y, \varphi_l^{(n)}) \quad (l = 1, \ldots, N_n). \ ^a \qquad (10.47)$$

It is very easy to solve these equations if as the basis $\varphi_1^{(n)}, \ldots, \varphi_{N_n}^{(n)}$ in X_n we take the orthonormal vectors relative to the scalar product $(x, y)_A = (Ax, y)$ (Subsection 7.5.7). In this case $(A\varphi_k^{(n)}, \varphi_l^{(n)}) = \delta_{kl}$ and Eq. (10.47) give $\alpha_k^{(n)} = (y, \varphi_k^{(n)})$ $(k = 1, \ldots, N_n)$.

R e m a r k. Theorem 10.5 is easily generalized on the case of the unbounded positive self–adjoint operator A. In this case introducing into its domains D_A the scalar product $(x, y)_A = (Ax, y)$ and supplementing D_A by the limits of all the fundamental sequences relative to the norm $\|x\|_A$ we obtain the H–space H_A. Then Theorem 10.5 will be true if we consider $F(x)$ as the functional on H_A.

In the general case the Rayleigh–Ritz method is not reduced to the general scheme of the approximate methods considered in Subsection 10.2.1 as it is based on the minimization of the functionals but not on the change of Eq. (10.21) by approximate Eqs. (10.22) at different n. But in the linear case of Eq. (10.21) by means of the minimization of the functional (10.46) the Rayleigh–Ritz method may be connected with the general scheme of Subsection 10.2.1. Really, Eqs. (9.47) represent the condition of the orthogonality of the discrepancy $Ax_n - y$ to the subspace X_n. Consequently, the projection of the discrepancy on X_n represents a zero element and Eqs. (10.47) are equivalent to equation

$$P_n A x_n = P_n y,$$

where P_n is an operator of the orthogonal projection on X_n. Putting $A_n = P_n A$, $y_n = P_n y$ we reduce this equation to Eq. (10.22). Here $Y_n = X_n$, $Q_n = P_n$.

---

^{a} For deriving Eqs. (10.47) in the case of the complex H–space X it should be put $\alpha_k^{(n)} = \beta_k + i\gamma_k$ and consider functional (10.46) as the function of real variables β_k, γ_k.

10.2.4. Galerkin Method

This method is based on the fact that the discrepancy, i.e. the difference $Ax - y$ between the left–hand and the right–hand side of operator Eq. (10.21) in the H–space X is equal to zero if and only if it is orthogonal to all the vectors of any complete system. Therefore for searching an approximate solution of Eq. (10.21) in a given subspace X_n of the H–space X it is sufficient to write the condition of the orthogonality of the discrepancy for all the basis vectors of the subspace X_n. In the case of the linear operator A this leads to the same Eq. (10.47) which gives the Rayleigh–Ritz method. But in this case the operator A must be not obligatory self–adjoint and positive but it may be any linear operator. Here Eq. (10.21) may not be connected with the problem of finding the extremum of some functional. This method was firstly applied by Russian scientist Galerkin for the mechanical problems solution. The Galerkin method is more general than the Rayleigh–Ritz method and is applicable for a wide class of linear and nonlinear problems.

In modern general form the Galerkin method implies that for an approximate solution of Eq. (10.21) two systems of the basis vectors are taken. One of them $\{\varphi_k^{(n)}\}$ in the finite–dimensional subspace $X_n \subset D_A \subset X$ (*coordinate system*), and the other $\{\psi_k^{(n)}\}$ in the finite–dimensional subspace Y_n^* of the space Y^* dual with Y (*projection system*) and the coefficients $\alpha_1^{(n)}, \ldots, \alpha_{N_n}^{(n)}$ in expansion (10.43) are determined from the orthogonality condition of the discrepancy $Ax_n - y$ to all the basis vectors of the space Y_n^*:

$$\psi_m^{(n)}\left(A\sum_{k=1}^{N_n}\alpha_k^{(n)}\varphi_k^{(n)} - y\right) = 0 \quad (m = 1, \ldots, N_n). \tag{10.48}$$

For linear operator A Eqs. (10.48) have the form

$$\sum_{k=1}^{N_n}(\psi_m^{(n)}A\varphi_k^{(n)})\alpha_k^{(n)} = \psi_m^{(n)}y \quad (m = 1, \ldots, N_n). \tag{10.49}$$

In the special case of the H–space Y we get $Y^* = Y$ and

$$\psi_m^{(n)}A\varphi_k^{(n)} = (A\varphi_k^{(n)}, \psi_m^{(n)}), \quad \psi_m^{(n)}y = (y, \psi_m^{(n)}).$$

If X is the H–space and $Y = X$ then we may take $\psi_m^{(n)} = \varphi_m^{(n)}$ and Eqs. (10.49) will coincide with Eqs. (10.47) of the Rayleigh–Ritz method.

In the general case Eqs. (10.48) may have no solutions and even in those cases when they have the unique solution the sequence of approximate solutions (10.43) may not converge to the exact solution of Eq. (10.21). In spite of this fact the Galerkin method similarly as the Rayleigh–Ritz method gives effective solution of many practical problems. For successful application of these methods the rational choice of the systems of basis vectors is of great importance. At appropriate choice of these systems the first of the second approximation is found sufficiently accurate. But it is impossible to solve the problem of the choice of the systems of the vectors by pure mathematics. Therefore it is not sufficient to have formal knowledge of mathematics for successful practical application of the Rayleigh–Ritz and Galerkin methods. It is also necessary to be mathematically sophisticated in the field of applications and engineering intuition.

The Galerkin method is reduced to the general scheme of the approximate methods of Subsection 10.2.2. For proving this we determine in the space X_n^* the vectors $f_1^{(n)}, \ldots, f_{N_n}^{(n)}$ which form together with the vectors $\varphi_1^{(n)}, \ldots, \varphi_{N_n}^{(n)}$ the biorthonormal system $f_p^{(n)} \varphi_q^{(n)} = \delta_{pq}$, and in the space Y_n we determine the vectors $g_1^{(n)}, \ldots, g_{N_n}^{(n)}$ which form together with the vectors $\psi_1^{(n)}, \ldots, \psi_{N_n}^{(n)}$ the biorthonormal system $\psi_p^{(n)} g_q^{(n)} = \delta_{pq}$. After that we determine the projection operators P_n and Q_n on X_n and Y_n correspondingly:

$$P_n x = \sum_{k=1}^{N_n} (f_k^{(n)} x) \varphi_k^{(n)}, \quad Q_n y = \sum_{k=1}^{N_n} (\psi_k^{(n)} y) g_k^{(n)}. \tag{10.50}$$

Then the orthogonality condition of the discrepancy $A x_n - y$ to the vectors $\psi_1^{(n)}, \ldots, \psi_{N_n}^{(n)}$ with account of condition (10.48) gives

$$Q_n A x_n = \sum_{k=1}^{N_n} (\psi_k^{(n)} A x_n) g_k^{(n)} = \sum_{k=1}^{N_n} (\psi_k^{(n)} y) g_k^{(n)} = Q_n y,$$

i.e.

$$Q_n A x_n = Q_n y.$$

Putting $A_n = Q_n A$, $y_n = Q_n y$ we reduce this equation to Eq. (10.22).

Now we clarify the sufficient conditions for performing the conditions of Theorem 10.4 about the convergence of the sequence of the solutions $\{x_n\}$ of Eqs. (10.22) to the solution x of Eq. (10.21)

The convergence conditions of the sequence $\{y_n\}$ to y and the sequence of the operators $\{A_n\}$, $A_n = Q_n A$ to the operator A in the case of the continuous operator A are reduced to the requirement that the bases in the subspaces X_n and Y_n will satisfy the following conditions:

$$P_n x \to x \quad \forall x \in X, \qquad Q_n y \to y \quad \forall y \in Y. \tag{10.51}$$

To ensure in this fact it is sufficient to notice that in a given case we have

$$A_n P_n x - Q_n A x = Q_n A P_n x - Q_n A x = Q_n (A P_n x - A x)$$

and the convergence condition of the sequence of the operators $\{A_n\}$ to A takes the form

$$\| Q_n (A P_n x - A x) \| \to 0. \tag{10.52}$$

For the continuous operator A this condition is fulfilled if $P_n x \to x$.

In the case of the separable H–spaces X and Y conditions (10.51) will be fulfilled if as the basis in each of the subspaces X_n and Y_n we assume the first N_n vectors of the basis in X (correspondingly in Y) formed by the biorthonormal system which satisfies the conditions of Theorem 7.31, in particular, by the orthonormal system.

As far as condition (10.38) is concerned then to verify its fulfillment in concrete problems is very difficult. If the operator A possesses such property that at any $x, x' \in D_A$

$$\| x - x' \| < \chi(\| A x - A x' \|), \tag{10.53}$$

where $\chi(s)$ is a continuous strictly increasing function, $\chi(0) = 0$, $\chi(\infty) = \infty$, and the projectors Q_n satisfy the condition

$$\| Q_n y \| \geq c \, \| y \| \tag{10.54}$$

at all $y \in A X_n$ at some $c > 0$ then condition (10.38) is fulfilled as

$$\| x_n - x'_n \| < \chi(\| A x_n - A x'_n \|)$$

$$\leq \chi(c^{-1} \, \| Q_n A x_n - Q_n A x'_n \|) = \chi(c^{-1} \, \| A_n x_n - A_n x'_n \|).$$

Thus for the convergence of the approximations of the Galerkin method, in particular, the Rayleigh–Ritz method of the approximations to the solution of Eq. (10.21) the fulfillment of conditions (10.51) is sufficient. In the special case of the continuous operator A condition (10.52) follows from the first condition (10.51).

For the linear operator A for the fulfillment of condition (10.53) the existence of the bounded inverse operator A^{-1} is sufficient. It goes without saying that instead of conditions (10.53) and (10.54) we may use condition (10.38) which in the case of the linear operator A is fulfilled if the sequence of the norms of the inverse operators A_n^{-1} is bounded. In the case of the linear operator A, the H–spaces X and Y and the the biorthonormal bases $\varphi_1^{(n)}, \ldots, \varphi_{N_n}^{(n)}$ and $\psi_1^{(n)}, \ldots, \psi_{N_n}^{(n)}$ we have the following expansions:

$$Q_n y = \sum_{k=1}^{N_n} (y, \psi_k^{(n)}) \psi_k^{(n)},$$

$$x_n = P_n x = \sum_{k=1}^{N_n} \alpha_k^{(n)} \varphi_k^{(n)},$$

$$A_n x_n = Q_n A x_n = \sum_{k,l=1}^{N_n} \alpha_l^{(n)} (A\varphi_l^{(n)}, \psi_k^{(n)}) \psi_k^{(n)}.$$

The boundedness conditions of the set of the norms of the inverse operators A_n^{-1} are reduced to the following inequalities:

$$\sum_{k=1}^{N_n} \left| \sum_{l=1}^{N_n} \alpha_l^{(n)} (A\varphi_l^{(n)}, \psi_k^{(n)}) \right|^2$$

$$= \sum_{l,m=1}^{N_n} \left(\sum_{k=1}^{N_n} (A\varphi_l^{(n)}, \psi_k^{(n)})(\psi_k^{(n)}, A\varphi_m^{(n)}) \right) \alpha_l^{(n)} \overline{\alpha_m^{(n)}}$$

$$> a \sum_{l=1}^{N_n} |\alpha_l^{(n)}|^2 \quad (n = 1, 2, \ldots) \tag{10.55}$$

at some $a > 0$. This condition replaces in a given case conditions (10.53) and (10.54). Thus the checking of the boundedness of the set of the norms of the inverse operators A_n^{-1} is reduced in a given case to establishing the positive definiteness of several quadratic forms.

The boundedness condition of the sequence of the norms of the inverse operators A_n^{-1} is always fulfilled in the case of positively determined (Subsection 6.1.5) linear operator A (not obligatory self–adjoint) in the

H–space X and the orthoprojectors P_n. Really, in this case in consequence of the self–adjointness of the operator $Q_n = P_n$ for any $x_n \in X_n$ we get

$$(A_n x_n, x_n) = (P_n A x_n, x_n) = (A x_n, P_n x_n) = (A x_n, x_n) > a \, \| x_n \|^2$$

at some $a > 0$. On the other hand,

$$(A_n x_n, x_n) = |(A_n x_n, x_n)| \le \| A_n x_n \| \, \| x_n \| .$$

From the comparison of two obtained inequalities follows

$$\| A_n x_n \| > a \, \| x_n \| .$$

But this is the boundedness condition of the set of the norms of the inverse operators A_n^{-1}. The condition of positive definiteness of the operator A replaces in this case conditions (10.53) and (10.54).

R e m a r k. Notice that the method used in Examples 10.4 and 10.8 represents the extension of the Rayleigh–Ritz and Galerkin methods over the case when the coefficients $\alpha_1^{(n)}, \dots, \alpha_{N_n}^{(n)}$ in (10.43) represent not the numbers but the functions of the variable t.

E x a m p l e 10.10. As the classic example of the problem which is easily solved by the Rayleigh–Ritz method may serve the boundary–value problem for the following linear differential equation:

$$\frac{d}{dt}\left(g(t)\frac{du}{dt} \right) + h(t)u = f(t), \quad u(a) = u(b) = 0, \qquad (10.56)$$

where f, g and h are the continuous functions on the interval $[a, b]$ and here

$$\inf_{t \in [a,b]} g(t) > 0.$$

Obviously it is sufficient to consider the case $g(t) \equiv 1$ as the general case is reduced to this by the change of variables

$$s = \int_a^t \frac{d\tau}{g(\tau)}.$$

Therefore we restrict ourselves by the solution of the boundary–value problem

$$u'' + h(t)u = f(t), \quad u(a) = u(b) = 0. \qquad (10.57)$$

In a given case $A = (d^2/dt^2) + h(t)$. The domain D_A of this operator in the H–space $L_2([a, b])$ represents a set of double differentiable functions satisfying the boundary conditions. Thus $X = Y = L_2([a, b])$.

We prove that at $h(t) \leq c < 2/(b - a)^2$ the operator $-A$ represents a positively determined self–adjoint operator. By the integration by parts we obtain

$$-(Au, v) = -\int\limits_a^b (u'' + hu)v\, dt = \int\limits_a^b (u'v' - huv)\, dt$$

$$= -\int\limits_a^b (v'' + hv)u\, dt = -(u, Av).$$

Hence accounting that $h(t) \leq c$ we find

$$-(Au, u) = \int\limits_a^b (u'^2 - hu^2)\, dt \geq \int\limits_a^b u'^2 dt - c\int\limits_a^b u^2 dt.$$

But it is evident that

$$u(t) = \int\limits_a^t u'(\tau)\, d\tau$$

and in consequence of inequality (1.17)

$$u^2(t) \leq \int\limits_a^t 1 \cdot d\tau \int\limits_a^t u'^2(\tau)\, d\tau \leq (t - a)\int\limits_a^b u'^2(\tau)\, d\tau,$$

$$\int\limits_a^b u^2(t)\, dt \leq \tfrac{1}{2}(b - a)^2 \int\limits_a^b u'^2(t)\, dt,$$

what gives

$$-(Au, u) \geq [2/(b - a)^2 - c]\, \|u\|^2 .$$

From the positive definiteness of the operator $-A$ on the basis of Theorem 10.5 follows that the problem under consideration is equivalent to the problem of finding the maximum of the functional

$$F(u) = \int\limits_a^b (u'' + hu - 2f)u\, dt. \tag{10.58}$$

Notice that Eq. (10.57) is the Euler equation (Example 9.11) for the problem of finding the extremum of functional (10.58).

After taking any sequence of the subspaces $\{X_n\}$ in $X = L_2([a, b])$ and any basis functions $\varphi_1^{(n)}(t), \ldots, \varphi_{N_n}^{(n)}(t)$ satisfying the boundary conditions in each of them we obtain Eqs. (10.47) with

$$(A\varphi_k^{(n)}, \varphi_l^{(n)}) = \int_a^b (\varphi_k^{(n)\prime\prime} + h\varphi_k^{(n)})\varphi_l^{(n)}\, dt,$$

$$(y, \varphi_l^{(n)}) = (f, \varphi_l^{(n)}) = \int_a^b f\varphi_l^{(n)}\, dt.$$

As in a given case the functions

$$\varphi_k(t) = \sin k\omega(t - a), \quad \omega = \pi/(b - a) \quad (k = 1, 2, \ldots)$$

form the orthogonal basis in D_A then it is expedient to assume at all n, $N_n = n$, $\varphi_k^{(n)} = \varphi_k(t) = \sin k\omega(t - a)$. As a result we receive

$$(A\varphi_k^{(n)}, \varphi_l^{(n)}) = \int_a^b [h(t) - k^2\omega^2] \sin k\omega(t - a) \sin l\omega(t - a)\, dt$$

$$= \int_a^b h(t) \sin k\omega(t - a) \sin l\omega(t - a)\, dt - k\omega[1 - (-1)^k]\delta_{kl},$$

$$(f, \varphi_l^{(n)}) = \int_a^b f(t) \sin l\omega(t - a)\, dt.$$

As $-A$ is a positively determined operator then for proving the convergence of the approximations $\{u_n(t)\}$ to the solution of problem (10.57) it remains to show that the operator A and the projectors P_n satisfy condition (10.52). Here due to the boundedness of the operator multiplication by the function h it is sufficient to show that

$$P_n(d^2/dt^2)P_n u - P_n(d^2/dt^2)u \to 0$$

for any function $u \in D_A$. For this purpose we notice that the functions

$$\psi_k(t) = \sqrt{2/(b - a)}\varphi_k(t) = \sqrt{2/(b - a)} \sin k\omega(t - a)$$

represent the orthonormal eigenvectors of the operator d^2/dt^2, corresponding to the eigenvalues $\lambda_k = -k^2\omega^2$. Therefore for any function $u \in D_A$ we have

$$u = \sum_{k=1}^{\infty} (u, \psi_k)\psi_k, \quad P_n u = \sum_{k=1}^{n} (u, \psi_k)\psi_k,$$

$$\frac{d^2}{dt^2}P_n u = \sum_{k=1}^{n}(u, \psi_k)\psi_k'' = -\omega^2 \sum_{k=1}^{n} k^2(u, \psi_k)\psi_k,$$

$$\frac{d^2}{dt^2}u = \sum_{k=1}^{\infty}(u'', \psi_k)\psi_k = \sum_{k=1}^{\infty}(u, \psi_k)\psi_k'' = -\omega^2 \sum_{k=1}^{\infty} k^2(u, \psi_k)\psi_k$$

and

$$P_n \frac{d^2}{dt^2}P_n u = P_n \frac{d^2}{dt^2}u = -\omega^2 \sum_{k=1}^{n} k^2(u, \psi_k)\psi_k.$$

Consequently, $P_n(d^2/dt^2)P_n u - P_n(d^2/dt^2)u = 0$ for any function $u \in D_A$. Thus the sufficient conditions of the convergence of the approximations are fulfilled.

E x a m p l e 10.11. As the second example we consider the known problem of mathematical physics – the Dirichlet problem. This is the problem of solving partial diffrential equation

$$\Delta^2 u + h(x, y)u = f(x, y), \quad \Delta^2 = \frac{\partial^2}{\partial x^2} + \frac{\partial^2}{\partial y^2}, \qquad (10.59)$$

in some region Γ with given boundary condition, usually $u = 0$ at $\{x, y\} \in \partial\Gamma$ (by $\partial\Gamma$ the boundary of the region Γ is denoted). We shall solve this problem for the square region Γ, $\Gamma = [0, l]^2$. Noticing that a set of the functions:

$$\varphi_{km}(x, y) = \sin k\nu \sin m\nu y, \quad \nu = \pi/l \quad (k, m = 1, 2, \ldots)$$

form the orthogonal basis in D_A (D_A represents a set of double diffrentiable functions satisfying the boundary condition in the H–space $X = L_2([0, 1]^2)$). We assume at each n $N_n = n^2$,

$$\varphi_{km}^{(n)}(x, y) = \varphi_{km}(x, y) \quad (k, m = 1, 2, \ldots).$$

Then we obtain Eq. (10.47) with

$$(A\varphi_{km}^{(n)}, \varphi_{pq}^{(n)}) = \int_0^l \int_0^l [h(x, y) - (k^2 + m^2)\nu^2]\varphi_{km}(x, y)\varphi_{pq}(x, y)\, dx\, dy,$$

$$(y, \varphi_{pq}^{(n)}) = (f, \varphi_{pq}^{(n)}) = \int_0^l \int_0^l f(x, y)\varphi_{pq}(x, y)\, dx\, dy.$$

As the operator

$$-A = -\frac{\partial^2}{\partial x^2} - \frac{\partial^2}{\partial y^2} - h(x, y)$$

is a positively determined self–adjoint operator at $h(x, y) < l^2/4$, x, $y \in [0, l]$ then the considered Dirichlet problem on the basis of Theorem 10.5 is equivalent to the problem of finding the maximum of the functional

$$F(u) = \int\limits_0^l \int\limits_0^l \left(\frac{\partial^2 u}{\partial x^2} + \frac{\partial^2 u}{\partial y^2} + hu - 2f \right) u\, dx\, dy,$$

for which Eq. (10.59) serves as the Euler equation.

Accounting that the function $\varphi_{km}(x, y)$ represents the eigenvector of the Laplace operator Δ^2 corresponding to the eigenvalue $-(k^2 + m^2)\nu^2$ similarly as in Example 10.10 we shall prove that the sequence of the operators $A_n = P_n A$ converges to the operator A. Thus the sufficient conditions of the convergence are fulfilled.

10.2.5. Finite Elements Method

Eqs. (10.47) of the Rayleigh–Ritz method similarly as Eqs. (10.49) of the Galerkin method usually require cumbersome computations at their numerical analysis as the matrix of the coefficients is usually complete (i.e. does not contain zero elements) and sometimes it is an ill-posed (i.e. has the determinant close to zero). The orthonormalization of the vectors $\varphi_k^{(n)}$ relative to the scalar product $(x, y)_A$ in the case of the positive self–adjoint operator A (Subsection 7.5.7) also requires cumbersome computations. Therefore for solving the practical problems in the case of functional spaces X and Y it is expedient to take as the basis functions $\varphi_k^{(n)}$ the finite functions with minimal carriers. More often as the basis functions the splines (Example 9.12) with the minimal carrier are taken.

While using the splines the range of the functions argument is partitioned into finite number of pairwise nonintersecting parts of sufficiently simple form (intervals, rectangles, triangles, parallelepipeds and so on) which are called *the finite elements*. After determining on the obtained grid of the finite elements the splines with the minimal carriers we assume these splines as the basis function. Then through the equality to zero of each of the basis functions outside moderate region the matrix of the coefficients of Eq. (10.47) of the Rayleigh–Ritz method contains many zero elements. Such a method of solving the linear operator equations is called *the method of finite elements*.

E x a m p l e 10.12. In the problem of Example 10.10 we partition the interval $[a, b]$ into n equal parts with the length $\Delta = (b - a)/n$ and assume as the basis function piecewise linear splines of the defect 1 with the minimal carriers (Fig. 10.3). These splines are expressed by formulae

$$\varphi_k(t) = \psi\left(\frac{t - t_k}{\Delta}\right), \quad (k = 1, \ldots, n - 1),$$

where

$$\psi(s) = (1 - |s|)\mathbf{1}_{[-1,1]}(s).$$

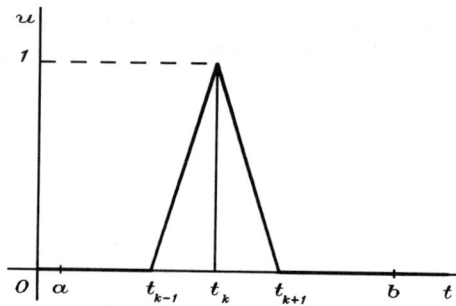

Fig. 10.3

Then the coefficients of Eqs. (10.47) will be determined by the following formulae:

$$(A\varphi_k, \varphi_k) = -\int\limits_a^b \varphi_k^{2\prime} dt + \int\limits_a^b h\varphi_k^2 dt$$

$$= -\frac{2}{\Delta} + \int\limits_{-\Delta}^{\Delta} \left(1 - \frac{|\tau|}{\Delta}\right)^2 h(t_k + \tau)\, d\tau,$$

$$(A\varphi_k, \varphi_{k+1}) = \frac{1}{\Delta} + \frac{1}{\Delta}\int\limits_0^{\Delta} \tau\left(1 - \frac{\tau}{\Delta}\right) h(t_k + \tau)\, d\tau,$$

$$(A\varphi_k, \varphi_{k-1}) = \frac{1}{\Delta} + \frac{1}{\Delta}\int\limits_0^{\Delta} \tau\left(1 - \frac{\tau}{\Delta}\right) h(t_k - \tau)\, d\tau,$$

$$(A\varphi_k, \varphi_l) = 0 \text{ at } |k - l| > 1,$$

$$(f, \varphi_l) = \int\limits_{-\Delta}^{\Delta} \left(1 - \frac{|\tau|}{\Delta}\right) f(t_k + \tau)\, d\tau.$$

Thus we came to Eqs. (10.47) with the tridiagonal matrix of the coefficients $(n - 2)(n - 3)$ whose $(n - 1)^2$ elements are equal to zero.

E x a m p l e 10.13. In Example 10.11 we partition the square $[0, l]^2$ into n^2 equal squares with the lengths of the sides $\Delta = l/n$ and assume as the basis functions the linear splines of the defect 1 (Fig. 10.4). These splines are determined by the formula

$$\varphi_{km}(x,y) = \psi\left(\frac{x - x_k}{\Delta}, \frac{y - y_m}{\Delta}\right), \quad (k, m = 1, \ldots, n - 1),$$

where

$$\psi(\xi, \eta) = [1 - \max(|\xi|, |\eta|)]\mathbf{1}_{[-1,1]^2}(\xi, \eta).$$

Then only $9n^2 - 30n + 29$ from $(n - 1)^4$ of the coefficients of Eqs. (10.47) are different from zero.

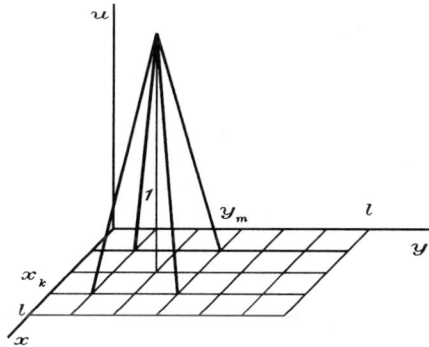

Fig. 10.4

E x a m p l e 10.14. If we divide all the rectangles of Example 10.13 in two by the lines which are parallel to someone bisectrix of the coordinate angle and assume as the finite elements the obtained rectangular triangles then the linear splines of the defect 1 with the minimal carriers will be determined by formula (Fig. 10.5)

$$\varphi_{km}(x, y) = \omega\left(\frac{x - x_k}{\Delta}, \frac{y - y_m}{\Delta}\right) \quad (k, m = 1, \ldots, n - 1),$$

where

$$\omega(\xi, \eta) = \{[1 - \max(|\xi|, |\eta|)]\mathbf{1}(\xi\eta) + (1 - |\xi| - |\eta|)\mathbf{1}(-\xi\eta)\}\mathbf{1}_B(\xi, \eta).$$

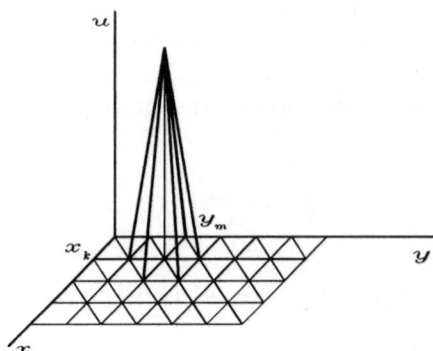

Fig. 10.5

Here as usual $\mathbf{1}(z)$ is the unit step function equal to 1 at $z > 0$ and 0 at $z < 0$, and $\mathbf{1}_B(\xi, \eta)$ is an indicator of the hexagon with the tops at the points $\{0, 1\}$, $\{-1, 0\}$, $\{-1, -1\}$, $\{0, -1\}$, $\{1, 0\}$, $\{1, 1\}$. In this case only $7n^2 - 24n + 21$ from $(n-1)^4$ coefficients of Eqs. (10.47) are different from zero.

Basic methods of numerical analysis in abstract spaces are given in (Aubin 1984, Blum 1972, Kahaner et al. 1989). For some typical applications refer to (Brigham 1988, Burrus and Parks 1985, Pratt 1978).

Problems

10.2.1. It was shown in Problem 9.3.13 that any differential equation in the separable H–space is equivalent to some infinite system of the scalar equations. Truncating this system, i.e. after rejecting all the equations of the system beginning with $n + 1$–th one, and simultaneously excluding from the number of the arguments of the functions in the right–hand sides all the members of the unknown sequence beginning with $n + 1$–th one we shall obtain a finite set of ordinary differential equations which may be integrated by the oridnary methods of the ordinary differential equations theory. On this fact the method of the finite–dimensional approximation of the differential equations in the space l_2 is based. Using the result of Problem 9.3.13 prove that the sequence of equations solutions obtained by means of the finite–dimensional approximation converges at $n \to \infty$ to the exact solution of the given differential equation in the space l_2. In such a way it will be proved that any differential equaton in any separable H–space may be solved with any degree of the accuracy by reducing to the equation in the space l_2 and the subsequent finite–dimensional approximation. Find the estimates of the accuracy of an approximate solution at different n.

10.2.2. At practical use of the finite–dimensional approximation method the choice of the basis in the initial H–space plays an important role. Solve by this method Eqs. (9.76) of Example 9.18 choosing as the basis functions the finite elements.

10.2.3. Let $\{x_n\}$ be the basis in the B–space X, T be the linear operator which acts from the space X into B–space Y. Find an approximate solution of the operator equation of the first kind $Tx = y$ in the form of a finite segment of the expansion over the basis $\{x_n\}$.

10.3. Improper Problems

10.3.1. Two Definitions of Correctness

While solving practical problems it is very important to know how to find such an approximate solution of Eq. (10.21) which will be sufficiently close to exact one at the inexact initial data. Let us consider the operator equation

$$Ax = y \qquad (10.60)$$

at the inexact given y. If here the inverse operator A^{-1} is continuous then in this case an approximate solution x may be sufficiently divergent from exact one at small errors in y. And even at known y the approximate solution x may occur as far from the exact through the inevitable roundoff errors. Not without reason in all theorems about the convergence of the approximate solutions to exact one appears the stability condition. The continuity of the inverse operator A^{-1} is necessary for the fulfillment of this condition. In this connection the necessity of the subdivision of the problems into two classes: *proper* (*correct*) and *improper* (*incorrect*) problems arises.

Let A in Eq. (10.60) be an operator which maps the complete metric space (X, d) into the complete metric space (Y, r). In particular, X and Y may be the B–spaces. The problem of solving Eq. (10.60) is called *properly* (or *properly posed*) *by Hadamard* if

(i) at any $y \in Y$ Eq. (10.60) has the unique solution $x \in D_A$ (in this case we suppose that $R_A = Y$);

(ii) the inverse operator A^{-1} is continuous.

Here the problem is called *an uniformly proper* if the operator A^{-1} is uniformly continuous on Y. If the operator A^{-1} is not continuous then the problem is called *improper* (or *improperly posed*) *by Hadamard*.

This definition of the correctness at first given by Hadamard was generalized by Russian scientist Tychonoff. The point is that according

to Hadamard definition the problem of solving Eq. (10.60) is proper only
in that case when the operator A^{-1} is continuous on the whole space Y.
Meanwhile the region of the continuity of the operator A^{-1} may not
coincide with the whole space Y. Besides, the condition of the existence
and the uniqueness of the solution has the same value both for proper
and improper problems. Therefore it is expedient do not include it into
the definition of the correctness. This is the reason of the necessity of
the modification of the correctness definition.

The problem of solving Eq. (10.60) which has the unique solution
at $y \in C \subset R_A$ is called *proper by Tychonoff* if the inverse operator A^{-1}
is continuous on the set C. Here the inverse image $B = A^{-1}C$ of the set
C in the space X is called *a set of the correctness*. If the sets $C \subset Y$ on
which the operator A^{-1} is continuous does not exist then the problem
is *an improper by Tychonoff*.

A typical improper problem by Hadamard is the problem of solving Eq.(10.60) with the compact operator A (Subsection 7.6.1). These
equations are called *the operator equations of the first kind* if none of the
balls in the space X is not compact [a].

Theorem 10.6.   *The operator inverse to the compact one cannot
be continuous on the whole space if none of the balls in the space X is
not compact* .

▷ If the operator A is compact then the image of any ball which
is determined by it is precompact in consequence of which any sequence
of the points $\{x_n\}$ of any ball in the space X contains such subsequence
$\{x_{n_p}\}$ that the sequence $\{Ax_{n_p}\}$ of the points of the space Y converges.
We take such sequence $\{x_n\}$ in the ball $S_r(0)$ that for any n and m
$d(x_n, x_m) > \delta > 0$. It is possible as the ball $S_r(0)$ is not compact.
Let $\{x_{n_p}\}$ be the subsequence for which the sequence $\{Ax_{n_p}\}$ converges.
Putting $y_n = Ax_n$, $y = \lim y_{n_p}$ we get the convergent sequence $\{y_{n_p}\}$
for which the corresponding sequence $\{x_{n_p}\}$, $x_{n_p} = A^{-1}y_{n_p}$ does not
converge to any limit. Hence by Theorem 4.20 follows that the operator
A^{-1} is not continuous at the point y. ◁

R e m a r k.   This theorem shows that the problem of solving
Eq. (10.60) with the compact operator A is improper by Hadamard.

E x a m p l e 10.15. Consider the Fredholm integral equation

$$\int_T K(s,t)\, x(t)\, dt = y(s), \quad s \in S, \qquad (10.61)$$

----

[a]   Notice that this definition generalizes the definition of a linear operator equation
of the first order given in Subsection 8.1.2.

whose kernel $K(s,t)$ belongs to the space $L_2(T \times S)$. In Subsection 7.6.6 it was shown that the integral operator in this equation which represents the Hilbert – Schmidt operator has the finite absolute norm, and consequently, is compact. And as none of the balls in the H–space $L_2(T)$ is noncompact then Eq. (10.61) represents the *Fredholm integral equation of the first kind*. By Theorem 10.6 the problem of sol ving Eq. (10.61) is improper by Hadamard.

E x a m p l e 10.16. As the differention operator on the space of the conti- nuous functions is unbounded (Example 6.7) the differention problem of a continuous function is improper by Hadamard. It may be considered as the special case of the problem of solving integral Eq. (10.61) in the case of scalar arguments t, s, $S = [a, b]$, $K(s, t) = 1(s - t)$.

As the differentiation operator is continuous in the space of continuous functions with continuous first derivatives (Example 6.6) then the problem of differentiation of such a function is proper.

It should be noticed that one and the same problem may be proper at ones metrics d, r of the spaces X, Y and may be improper at the others. Thus the notion of the correctness of a problem is closely con- nected with the choice of the metrics in the spaces X and Y.

The correctness criterion of the problem of solving Eq.(10.60) by Tychonoff gives the following theorem.

Theorem 10.7. *If the operator A in Eq. (10.60) is continuous, Eq. (10.60) has the unique solution and $y \in C$ and the set $B = A^{-1}C$ is compact then the operator A^{-1} is uniformly continuous on C and consequently, the problem of solving Eq. (10.60) is proper by Tychonoff* (Tychonoff theorem).

▷ If the operator A^{-1} is nonuniformly continuous on C then at any $\delta > 0$ in $B = A^{-1}C$ will occur such pairs of the points $x_n^{(1)}$, $x_n^{(2)}$ that

$$d(x_n^{(1)}, x_n^{(2)}) > \delta, \quad r(Ax_n^{(1)}, Ax_n^{(1)}) < \varepsilon_n \qquad (10.62)$$

for any sequence of the positive numbers $\{\varepsilon_n\}$ which converges to ze- ro. Due to the compactness of B we may assume without loss of the generality the sequences $\{x_n^{(1)}\}$ and $\{x_n^{(2)}\}$ convergent. If they will not convergent then we choose from $\{x_n^{(1)}\}$ the convergent subsequence $\{x_{n_k}^{(1)}\}$ and after that choose from $\{x_{n_k}^{(2)}\}$ the convergent subsequence $\{x_{n_{k_p}}^{(2)}\}$. Here the subsequence $\{x_{n_{k_p}}^{(1)}\}$ will converge to the limit as $\{x_{n_k}^{(1)}\}$. Let $x_1 = \lim x_n^{(1)}$, $x_2 = \lim x_n^{(2)}$. Passing in (10.62) to the limit owing to the continuity of the operator A we shall have

$$d(x_1, x_2) \geq \delta, \quad r(Ax_1, Ax_2) = 0.$$

Consequently, $Ax_1 = Ax_2 = y$, i.e. Eq. (10.60) has two different solutions what contradicts the condition of the theorem. ◁

10.3.2. Regularization of Improper Problems

For the approximate solution of improper problems Tychonoff suggested the *regularization method*. The main idea of this method consists of the substitution of the operator A in Eq.(10.60) by such operator $A(\alpha)$, $A(0) = A$, which depends on nonnegative parameter α that at all $\alpha > 0$ this operator has the continuous inverse operator $A(\alpha)^{-1}$ and that the solution x_α of equation

$$A(\alpha)x = y \tag{10.63}$$

will be as close as possible to the solution x of Eq. (10.60) at sufficiently small α. Thus we come to the definition of regularizing operator.

R e m a r k. As it was noticed by Tychonoff the idea of the regularization of improper problem of function differentiation goes back to Newton; the replacement of the derivative by the ratio of the finite increments essentially represents the regularization of the improper problem.

The operator $R(y, \alpha)$ which depends on the scalar parameter α is called *a regularizing for Eq.(10.60) in the vicinity of the point* $y_0 = Ax_0$ if

(i) it is determined at all α, y, $\alpha \in (0, \alpha_0]$, $r(y, y_0) < \delta_0$ at some α_0, $\delta_0 > 0$;

(ii) at any $\varepsilon > 0$ there will appear such numbers $\delta_\varepsilon \in (0, \delta_0)$, $\alpha_\varepsilon \in (0, \alpha_0)$ that

$$d(x_\varepsilon, x_0) < \varepsilon$$

at all y, $r(y, y_0) < \delta_\varepsilon$ where

$$x_\varepsilon = R(y, \alpha_\varepsilon). \tag{10.64}$$

The parameter α is called *the regularization parameter*.

The operator $R(y, \alpha)$ is called *regularizing* for Eq.(10.60) *on the set* $C \subset R_A$ if it is regularizing in the vicinity of each point $y_0 \in C$.

The operator $R(y, \alpha)$ is called *uniformly regularizing* for Eq.(10.60) *on the set* $C \subset R_A$ if the numbers δ_ε and α_ε at a given $\varepsilon > 0$ do not depend on $y_0 \in C$.

The following theorem gives the criterion for the fact that a given operator will be regularizing for Eq.(10.60).

Theorem 10.8.

(i) *If the operator $R(y, \alpha)$ is determined at all $\alpha \in (0, \alpha_0]$ and $y \in \{y' : r(y', y_0) < \delta_0\}$ at some α_0, δ_0, is continuous at the point $y_0 = Ax_0$ and*

$$\lim_{\alpha \to 0} R(y_0, \alpha) = x_0, \tag{10.65}$$

then it is regularizing for Eq. (10.60) in the vicinity of the point y_0.

(ii) *If the conditions of the theorem are fulfilled at all $y_0 \in C$ then the operator $R(y, \alpha)$ is regularizing for Eq. (10.60) on the set C.*

(iii) *If the operator $R(y, \alpha)$ at all $\alpha \in (0, \alpha_0]$ is uniformly continuous on the set C and the convergence in (10.65) is uniform relative to y_0 on C then the operator $R(y, \alpha)$ is an uniformly regularizing for Eq. (10.60) on the set C.*

▷ From the continuity of the operator $R(y, \alpha)$ at the point y_0 follows that for any $\varepsilon > 0$ at any $\alpha \in (0, \alpha_0]$ there exists such number $\delta_\varepsilon(\alpha)$ that

$$d(R(y, \alpha), R(y_0, \alpha)) < \varepsilon/2 \tag{10.66}$$

at all y, $r(y, y_0) < \delta_\varepsilon(\alpha)$. From (10.65) follows that at the same $\varepsilon > 0$ will occur such α_ε that

$$d(R(y_0, \alpha), x_0) < \varepsilon/2 \tag{10.67}$$

at all $\alpha \leq \alpha_\varepsilon$. From inequalities (10.66) and (10.67), putting $\delta_\varepsilon = \delta_\varepsilon(\alpha_\varepsilon)$ we obtain

$$d(R(y, \alpha_\varepsilon), x_0) \leq d(R(y, \alpha_\varepsilon), \quad R(y_0, \alpha_\varepsilon)) + d(R(y_0, \alpha_\varepsilon), x_0) < \varepsilon$$

at all y, $r(y, y_0) < \delta_\varepsilon$. Hence it follows that the operator $R(y, \alpha)$ is regularizing for Eq.(10.60) in the vicinity of the point y_0. Statements (ii) and (iii) of the theorem follow from (i). ◁

10.3.3. Methods for Finding Regularizing Operator

It was stated in Subsection 10.2.3 that for finding an approximate operator equations in a given set we may minimize the discrepancy, i.e. the distance between the left– and right–hand sides of the equation. Here in the special case when a given set coincides with the domain of the operator, this way gives an exact solution. But in the case of the improper problem the process of the calculations will be unstable. Therefore in order to smooth the process of the calculations and make it stable Tychonoff suggested to add to the variable which serves as the measure of the discrepancy some functional, namely, instead of two measure of the discrepancy to minimize the functional

$$\psi(x, y, \alpha) = \gamma(r(Ax, y)) + \alpha\varphi(x),\qquad(10.68)$$

where $\gamma(s)$ is a strictly increasing continuous function equal to zero at $s = 0$ and α is a positive parameter of the regularization, $\varphi(x)$ is a non-negative functional with the domain $\mathcal{D}_\varphi$ dense in X. Tychonoff showed that at the corresponding choice of the functional $\varphi(x)$ and the parameter of the regularization α we may obtain an approximate solution of Eq.(10.60) which is close to the exact one.

Later on we restrict ourselves to the case of the B–spaces X and Y. We shall say that the nonnegative functional $\varphi(x)$, $\varphi(0) = 0$ determined on the subspace $X_1 \subset X$ generates on X_1 a space with majorant metric if $\sqrt{\varphi(x)}$ satisfies the norm axioms (Subsection 1.3.6) and

$$\| x_1 - x_2 \|_1 = \sqrt{\varphi(x_1 - x_2)} \geq \| x_1 - x_2 \| .\qquad(10.69)$$

Lemma. *If the nonnegative functional $\varphi(x)$ generates on the subspace $X_1 \subset X$ the real H–space $\tilde{X}_1$ with the majorant metric and the set $K_c = \{x : x \in X_1, \varphi(x) \leq c\}$ is precompact at any $c > 0$ then for any continuous nonnegative functional $f(x)$ determined on the whole space X at any $\alpha > 0$ there exists an element $x_\alpha \subset [X_1]$ which realizes the minimum of the functional $g(x) = f(x) + \alpha\varphi(x)$,*

$$g(x_\alpha) = \inf_{x \in [X_1]} g(x).$$

▷ Let $\{x_n\} \subset X_1$ be the sequence for which

$$\lim_{n \to \infty} g(x_n) = \inf_{x \in [X_1]} g(x) = a .\qquad(10.70)$$

Without loss of the generality we may assume that $g(x_{n+1}) \leq g(x_n)$ at all n. Then $g(x_n) \leq g(x_1)$ and consequently, $\alpha\varphi(x_n) \leq g(x_1)$ at all n, i.e. all the points x_n belong to the precompact set K_c at $c = g(x_1)/\alpha$. Hence it follows that the sequence $\{x_n\}$ contains the subsequence $\{x_{n_p}\}$ convergent to some point $x_\alpha \in [X_1]$. Due to the continuity of the functional $f(x)$ $f(x_{n_p}) \to f(x_\alpha)$. It remains to prove that $\varphi(x_{n_p}) \to \varphi(x_\alpha)$. It is sufficient for this purpose to show that $x_{n_p} \to x_\alpha$ and also in the metric of the H–space $\tilde{X}_1$, i.e. that $\| x_{n_p} - x_\alpha \|_1 \to 0$ at $p \to \infty$. At first we prove that the sequence $\{x_{n_p}\}$ is fundamental in the metric of the H–space $\tilde{X}_1$. Putting

$$z_{pq} = (x_{n_p} - x_{n_q})/2, \quad u_{pq} = (x_{n_p} + x_{n_q})/2,$$

we shall have

$$u_{pq} = x_{n_p} - z_{pq} = x_{n_q} + z_{pq} \tag{10.71}$$

and

$$g(u_{pq}) = f(u_{pq}) + \alpha \, \|u_{pq}\|_1^2 = f(x_{n_p}) + \alpha \, \|x_{n_p}\|_1^2$$
$$+ f(u_{pq}) - f(x_{n_p}) + \alpha[-2(x_{n_p}, z_{pq})_1 + \|z_{pq}\|_1^2]$$
$$= g(x_{n_p}) + f(u_{pq}) - f(x_{n_p}) + \alpha[-2(x_{n_p}, z_{pq})_1 + \|z_{pq}\|_1^2].$$

As $g(u_{pq}) > \inf\limits_{x \in X_1} g(x) = a$, then

$$\alpha[2(x_{n_p}, z_{pq})_1 - \|z_{pq}\|_1^2] \leq g(x_{n_p}) + f(u_{pq}) - f(x_{n_p}) - a.$$

Analogously, using the second expression u_{pq} in equality (10.71) we get

$$\alpha[-2(x_{n_q}, z_{pq})_1 - \|z_{pq}\|_1^2] \leq g(x_{n_q}) + f(u_{pq}) - f(x_{n_q}) - a.$$

After summing the obtained inequalities and taking into consideration that $x_{n_p} - x_{n_q} = 2z_{pq}$ we find

$$2\alpha \, \|z_{pq}\|_1^2 \leq g(x_{n_p}) + g(x_{n_q}) - 2a + 2f(u_{pq}) - f(x_{n_p}) - f(x_{n_q}).$$

Now we notice that $x_{n_p}, x_{n_q}, u_{pq} \to x_\alpha$ and by virtue of condition (10.70) $g(x_{n_p}), g(x_{n_q}) \to a$ at $p, q \to \infty$. As the functional $f(x)$ is continuous then $f(x_{n_p}), f(x_{n_q}), f(u_{pq}) \to f(x_\alpha)$ at $p, q \to \infty$. Consequently,

$$\|z_{pq}\|_1 = \|x_{n_p} - x_{n_q}\|_1/2 \to 0 \quad \text{at} \quad p, q \to \infty.$$

This proves the fundamental property of the sequence $\{x_{n_p}\}$ in the metric of the H–space $\tilde{X}_1$. Hence by virtue of the completeness of the H–space follows the existence of the limit $\tilde{x}_\alpha$ to which converges the sequence $\{x_{n_p}\}$ in the metric of the H–space $\tilde{X}_1$. It remains to notice that by virtue of inequality (10.69) the sequence $\{x_{n_p}\}$ also converges to $\tilde{x}_\alpha$ in the metric of the space X. As $x_{n_p} \to x_\alpha$, then $\tilde{x}_\alpha = x_\alpha$. Consequently,

$$g(x_\alpha) = \lim_{p \to \infty} [f(x_{n_p}) + \alpha \, \|x_{n_p}\|_1^2] = f(x_\alpha) + \alpha\varphi(x_\alpha) = a. \; \triangleleft$$

Theorem 10.9. *It is sufficient for the existence at any $\alpha > 0$ of the element x_α in the subspace $[X_1] \subset X$ which realizes the minimum of functional (10.68) that the set $K_c = \{x : x \in X_1, \varphi(x) \leq c\}$ will be precompact at any $c > 0$ and that the functional $\varphi(x)$ will generate on X_1 the H-space $\tilde{X}_1$ with the majorant metrix* [a].

▷ For proving it is sufficient to apply the previous Lemma at $f(x) = \gamma(r(Ax, y)) = \gamma(\|Ax - y\|)$. ◁

Theorem 10.10. *Let $x_0 \in X_1$ be an exact solution of Eq. (10.60) at $y = y_0$, $Ax_0 = y_0$, $x_\alpha \in [X_1]$ be the element which realizes the minimum of functional (10.68) at the fulfillment of the conditions of Theorem 10.9. Then if the operator A is continuous then for any $\varepsilon > 0$ and any strictly increasing functions $\beta_1(\delta)$ and $\beta_2(\delta)$ which satisfy the conditions*

$$\beta_1(0) \neq 0, \quad \beta_2(0) = 0, \quad \beta_1(\delta)\beta_1(\delta) \geq \gamma(\delta) \quad at \quad \delta > 0,$$

there exists such $\delta_\varepsilon > 0$ (dependent on the functions β_1, β_2) that for any $y \in R_A$, δ, α which satisfy the conditions

$$r(y, y_0) = \|y - y_0\| \leq \delta \leq \delta_\varepsilon, \quad \gamma(\delta)/\beta_1(\delta) \leq \alpha \leq \beta_2(\delta), \quad (10.72)$$

the inequality $d(x_\alpha, x_0) = \|x_\alpha - x_0\| < \varepsilon$ is valid.

▷ As functional (10.68) satisfies the inequality $\psi(x_\alpha, y, \alpha) \leq \psi(x_0, y, \alpha)$ then

$$\alpha\varphi(x_\alpha) \leq \psi(x_\alpha, y, \alpha) \leq \gamma(\|y_0 - y\|) + \alpha\varphi(x_0) \leq \gamma(\delta) + \alpha\varphi(x_0), \quad (10.73)$$

$$\gamma(\|Ax_\alpha - y\|) \leq \psi(x_\alpha, y, \alpha) \leq \gamma(\delta) + \alpha\varphi(x_0). \quad (10.74)$$

From inequalities (10.72) and (10.73) we obtain

$$\alpha\varphi(x_\alpha) \leq \alpha[\beta_1(\delta) + \varphi(x_0)], \quad \varphi(x_\alpha) \leq \beta_1(\delta) + \varphi(x_0).$$

Putting $\beta_1(\delta) + \varphi(x_0) = c_0$ we come to the conclusion that the points x_0 and x_α belong to the precompact set K_c at any $c \geq c_0$. By Theorem 10.7 the continuous operator A_c which represents the contraction of the operator A on the compact $[K_c]$ has the uniformly continuous

---

[a] It is clear that the element $x_\alpha \in [X_1]$ which realizes the minimum of functional (10.68) may exist only at $[X_1] \subset D_A$.

inverse operator on the image $A[K_c]$ of the compact $[K_c]$. Consequently, for any $\varepsilon > 0$ there exists such $\eta_\varepsilon > 0$ that

$$\|x_1 - x_2\| < \varepsilon \quad \forall x_1, x_2 \in [K_c], \quad \|Ax_1 - Ax_2\| < \eta_\varepsilon. \tag{10.75}$$

From inequalities (10.74) and (10.72) we receive

$$\gamma(\|Ax_\alpha - y\|) \le [\beta_1(\delta) + \varphi(x_0)]\beta_2(\delta) = c_0\beta_2(\delta),$$

$$\|Ax_\alpha - y\| \le \gamma^{-1}(c_0\beta_2(\delta)).$$

Hence using the inequality of the triangle we find by means of the first equality (10.72)

$$\|Ax_\alpha - Ax_0\| \le \|Ax_\alpha - y\| + \|y - Ax_0\|$$

$$= \|Ax_\alpha - y\| + \|y - y_0\| \le \gamma^{-1}(c_0\beta_2(\delta)) + \delta.$$

Now we choose such a small δ_ε that at all $\delta < \delta_\varepsilon$ will be $\gamma^{-1}(c_0\beta_2(\delta)) + \delta < \eta_\varepsilon$. It is possible as $\gamma(0) = \beta_2(0) = 0$ and the functions γ and β_2 are strictly increasing . Then by virtue of inequality (10.75) we shall have $\|x_\alpha - x_0\| < \varepsilon$. ◁

Corollary. *The operator $R(y, \alpha)$ which establishes the correspondence between any y, α and element $x_\alpha \in [X_1]$ realizing the minimum of functional (10.68) is regularizing for Eq. (10.60) on the image $A[K_c]$ of the closure of the precompact set $K_c = \{x : x \in X_1, \varphi(x) \le c\}$ at any $c > 0$.*

In the correspondence with its role of the stabiliser of the calculations process a nonnegative functional $\varphi(x)$ determined on the set X_1 for which the set $\{x : \varphi(x) \le c\}$ is precompact at any $c > 0$ is called *a stabilizing functional*. A functional $\psi(x, y, \alpha)$ which is determined by formula (10.68) at any stabilizing functional $\varphi(x)$ is called *a smoothing functional* as it smoothes the variations of the calculations process at approximate solution of Eq.(10.60). For the minimization of the smoothing functional we may use any of the approximation methods of finding the extremum of the functionals, for instance, the Rayleigh — Ritz method or the finite elements method.

E x a m p l e 10.17. Consider the integral equations of the first kind (10.61) in the case of the scalar variables t, s, $T = [a, b]$, $S = [c, d]$. In this case as the spaces X and Y serve the real H–spaces $L_2([a, b])$ and $L_2([c, d])$ correspondingly, and as the subspace X_1 we assume the space of the real continuous functions

$C^N([a, b])$ which have continuous derivatives till the order N inclusively. Show that the functional

$$\varphi(x) = \int\limits_a^b \sum_{k=0}^N p_k(t)[x^{(k)}(t)]^2 \, dt \,, \tag{10.76}$$

where $p_0(t), p_1(t), \ldots, p_N(t)$ are the positive functions, $p_0(t) \ge q_0 > 0$, $p_1(t) \ge q_1 > 0$ is a stabilizing, and consequently, the functional

$$\psi(x, y, \alpha) = \int\limits_a^b \left[\int\limits_a^b K(s, t)\, x(t) \, dt - y(s) \right]^2 ds + \alpha\varphi(x) \quad^a \tag{10.77}$$

is a smoothing one. It is sufficient for this purpose to show that the set of the functions $\{x(t) \ : \ x(t) \in C^N([a, b]), \ \varphi(x) \le c\}$ is precompact at any $c > 0$. From the inequality $\varphi(x) \le c$ follows

$$\int\limits_a^b p_0(t)\, x^2(t) \, dt \le c, \quad \int\limits_a^b p_1(t)\, [x'(t)]^2 \, dt \le c. \tag{10.78}$$

The first from these inequalities shows that at the interval $[a, b]$ there exists such point t_0 that

$$|x(t_0)| < \sqrt{c/q_0(b - a)}\,, \tag{10.79}$$

where $q_0 > 0$ is the lower bound of the function $p_0(t)$ on $[a, b]$. As at any t_1, $t_2 \in [a, b]$

$$x(t_2) = x(t_1) + \int\limits_{t_1}^{t_2} x'(t) \, dt \,,$$

then inequality (1.17) and second inequality (10.78) give

$$|x(t_2) - x(t_1)|^2 = \left| \int\limits_{t_1}^{t_2} x'(t) \, dt \right|^2$$

$$\le \int\limits_{t_1}^{t_2} 1 \cdot dt \cdot \int\limits_{t_1}^{t_2} [x'(t)]^2 \, dt \le \frac{t_2 - t_1}{q_1} \int\limits_{t_1}^{t_2} p_1(t)\, [x'(t)]^2 \, dt \le \frac{c}{q_1}(t_2 - t_1), \tag{10.80}$$

where $q_1 > 0$ is the lower bound of the function $p_1(t)$ on $[a, b]$. This inequality shows that all the continuous functions $x(t)$ which satisfy inequalities (10.78) are equicontinuous (See the footnote to Theorem 4.45). From inequalities (10.79) and (10.80) follows

$$|x(t)| \le |x(t_0)| + |x(t) - x(t_0)| < \sqrt{c/q_0(b - a)} + \sqrt{(c/q_1)(b - a)}\,.$$

---

^{a} In the metric of the space $L_2([c, d])$ $\psi(x, y, \alpha) = \|Ax - y\|^2 + \alpha\varphi(x)$.

This inequality shows that the set of all the functions $x(t)$ which satisfy inequalities (10.78) is uniformly bounded. By Arzela – Ascoli theorem 4.46 the set of the continuous functions $x(t)$ for which $\varphi(x) \leq c$ is precompact what proves our statement. Verify that the functional $\sqrt{\varphi(x)}$ which is determined by formula (10.76) possesses all the properties of the norm and that this norm satisfies the identity of the parallelogram and is majorant relative to the norm of the H–space $X = L_2([a, b])$. Consequently, to the given problem Theorem 10.9 is applicable.

The subspace of the space $L_2([a, b])$ with the norm generated by functional (10.76) at $p_0(t) = p_1(t) = \cdots = p_N(t) = 1$ represents the real H–space, namely, the Sobolev space $H^N([a, b])$ (Subsection 3.7.8).

Thus applying the regularization method we reduce the problem of solving the integral equation of the first kind (10.61) to the minimization of smoothing functional

$$\psi(x, y, \alpha) = \int\limits_c^d \left[\int\limits_a^b K(s, t)\, x(t)\, dt - y(s) \right]^2 ds + \alpha \int\limits_a^b \sum_{k=0}^{N} p_k(t)\, [x^{(k)}(t)]^2\, dt \ .$$

In exactly the same way the problem of solving integral Eq.(10.61) in the case of the vector variables t, s and arbitrary regions T and S is regularized. Here the squares of all the derivatives till the order N inclusively should be taken in functional (10.76) with the corresponding positive coefficients.

R e m a r k. Notice that the problem of the minimization of functional (10.68) is related to the problem of finding the conditional minimum of the stabilizing functional $\varphi(x)$ at the condition $\| Ax - y \| \leq \alpha$. According to the Lagrange method this problem is reduced to the minimization of the functional

$$\varphi(x) + \lambda\gamma(\| Ax - y \|) \tag{10.81}$$

with the successive definition of λ from the condition $\| Ax - y \| = \alpha$. In this case by Lemma there exists the vector x_α which realizes the minimum of functional (10.81). Further we show that in this case the operator $R(y, \alpha)$ which establishes the correspondence between x_α and the given y, α is a regularizing for Eq.(10.60).

The stated methods of finding the regularizing operators are easily generalized on the case of any complete metric spaces X, Y and any stabilizing functional $\varphi(x)$. But in this case the element x_α which realizes the minimum of functional (10.68) or (10.81) may not exist. Then the problem of the minimization of functional (10.68) may be substituted by the problem of finding such element $x_\alpha \in [X_1]$ for which

$$\psi(x_\alpha, y, \alpha) \leq q \inf_{x \in [X_1]} \psi(x, y, \alpha) \tag{10.82}$$

at some $q \geq 1$. At $q > 1$ but as close to 1 as possible such element $x_\alpha \in X$ always exists according to the definition of the infimum. In this case the following theorem is valid.

Theorem 10.11 *Let $x_0 \in X_1$ be the exact solution of Eq. (10.60) at $y = y_0$, $Ax_0 = y_0$, $x_\alpha \in X_1$ be an element which satisfies inequality (10.82) (such element evidently, is not uniquely determined). Then if the operator A is continuous then for any $\varepsilon > 0$ and any strictly increasing functions $\beta_1(\delta)$ and $\beta_2(\delta)$ which satisfy the conditions*

$$\beta_1(0) \neq 0, \quad \beta_2(0) = 0, \quad \beta_1(\delta)\beta_2(\delta) \geq \gamma(\delta) \quad at \quad \delta > 0,$$

there exists such $\delta_\varepsilon > 0$ (dependent on the functions β_1, β_2) that for any $y \in R_A$, δ, α which satisfy the conditions

$$r(y, y_0) \leq \delta \leq \delta_\varepsilon, \quad \gamma(\delta)/\beta_1(\delta) \leq \alpha \leq \beta_2(\delta), \tag{10.83}$$

the inequality $d(x_\alpha, x_0) < \varepsilon$ is valid. The operator $R(y, \alpha)$ which establishes the correspondence between the element x_α and the given y, α is regularizing for Eq. (10.60) on the image of the precompact set $K_c = \{x : x \in X_1, \varphi(x) \leq c\}$ at any $c > 0$.

▷ The proof of this theorem is similar to the proof of Theorem 10.10 with the change of the norm by the corresponding distances. ◁

Analogously the problem of the conditional minimization of the functional $\varphi(x)$ at $r(Ax, y) \leq \alpha$ is substituted by the problem of finding such element $x_\alpha \in X$ at some $q \geq 1$ for which

$$\varphi(x_\alpha) \leq q \inf_{r(Ax, y) < \alpha} \varphi(x). \tag{10.84}$$

At $q > 1$ but as close as possible to 1 such an element always exists.

Theorem 10.12 *If Eq. (10.60) has the unique solution at $y \in AX_1$ and the operator A is continuous then the operator $R(y, \alpha)$ which establishes the correspondence of the element x_α satisfying inequality (10.84) to the given y, α (such element, evidently is not uniquely determined), is a regularizing for Eq. (10.60) on the set AX_1.*

▷ Let y_0 be an arbitrary element of the set AX_1, $x_0 \in X_1$ be the solution of the equation $Ax_0 = y_0$. Suppose that the operator $R(y, \alpha)$ is not regularizing. Then there exist such sequences $\{y_k\}$, $\{\alpha_k\}$, $\alpha_k \to 0$ at $k \to \infty$ for which $r(y_k, y_0) < \alpha_k$, $d(x_{\alpha k}, x_0) \geq \varepsilon$ at some $\varepsilon > 0$, $x_{\alpha k} = R(y_k, \alpha_k)$. From inequality (10.84) follows $\varphi(x_{\alpha k}) \leq q\varphi(x_0)$

whence it follows that the sequence $\{x_{\alpha k}\}$ belongs to the precompact set K_c at $c = q\varphi(x_0)$. Denoting by $\bar{x}$ the limit of the convergent subsequence of the sequence $\{x_{\alpha k}\}$ we shall have due to the continuity of the operator A and the inequalities $r(Ax_{\alpha k}, y_k) < \alpha_k$, $r(y_k, y_0) < \alpha_k$ at all k $r(A\bar{x}, y_0) = 0$ [a]. Hence by virtue of the uniqueness of the solution of Eq. (10.60) we obtain $\bar{x} = x_0$ what contradicts the inequalities $d(x_{\alpha k}, x_0) \geq \varepsilon$ at all k. ◁

10.3.4. Choice of Regularization Parameter

It follows from the theory of Subsection 10.3.3 that the regularization parameter α must be sufficiently small that the approximate solution will be close to the exact one, and on the other hand, it must be sufficiently large that the calculations process will be stable and smooth. The bounds of the admissible values α are established by conditions (10.72) and (10.83) of Theorems 10.10 and 10.11. Thus while choosing α the compromise is necessary between the requirements of the closeness of an approximate solution to the exact one and between the stability and the smoothness of the calculations process. In practice this question is solved by inspection. After choosing some α we find for this α an approximate solution. If the calculations process occurs stable and smooth then we try to decrease this value of α, usually it is double decreased . It is continued till the calculations process starts some vibration. After obtaining such vibration effect we take the previous value of α at which the calculations process is sufficiently stable and smooth. If the calculations process which was chosen earlier is unstable this value of α is increased (usually double). This process is continued till the sufficient stability and smoothness of the calculations process is achieved.

10.3.5. Generalization of Regularization Method

The stated theory of the improper problems is extended over the case when the incorrectness is obtained by means of small variations of the operator A in Eq.(10.60). After taking the corresponding measure of the variation of the operator A, for instance, the norm of its increment in the case of the B–spaces X and Y and continuous linear operators A we may include into the definition of the problem correctness the requirement of the continuous dependence of solution of Eq.(10.60) on

---

[a] The inequality $r(Ax_{\alpha k}, y_k) < \alpha_k$ follows from definition (10.84) of the element x_α at $y = y_k$, $\alpha = \alpha_k$.

the operator A and the right–hand side of y. The corresponding changes should be also made in the definition of the regularizing operator. In this case Lemma and Theorems 10.9 and 10.5 with the corresponding changes will be valid.

Usually the regularization of improper problems is connected with the incorrectness of the initial data assignment. Notice that the regularization is also necessary at the accurate assignment of the initial data as without the regularization the computing process at solving the improper problems will be divergent. The regularization reduces the improper problems to the correct ones and thereby extends over them all the methods of approximate solution of the proper problems including all the methods stated in Sections 10.1 and 10.2.

For detailed consideration of the solution of improper problems refer to (Tychonoff and Arsenin 1997).

P r o b l e m s

10.3.1. Consider a general operator equation of the first kind $Tx = y$, where T is a linear operator mapping the real B–space X into the real B–space Y with unbounded inverse operator T^{-1}. Let $\{x_n\}$ be the basis in X, $\{f_n\}$ be an adjoint basis in X^* (Problems 7.5.16 — 7.5.18). Show that the functional $\sqrt{\varphi_N(x)}$ where

$$\varphi_N(x) = k \sum_{n=1}^{N} (f_n x)^2 , \quad k > 0 ,$$

satisfies all the axioms of the norm and the identity of the parallelogram. Therefore at any N the functional $\varphi_N(x)$ generates on the subspace X_N formed by the vectors $x_1, \ldots, x_N$ the H–space with the majorant metric. Prove that the set $K_c = \{x : x \in X_N, \varphi_N(x) \le c\}$ is precompact at any N in consequence of which the functional $\varphi_N(x)$ is stabilizing, and the minimization of the functional $\psi_N(x, y, \alpha) = \|Tx - y\|^2 + \alpha \varphi_N(x)$ at sufficiently large N gives an approximate solution as close as possible to the exact one.

10.3.2. Apply the stabilizing functional of Problem 10.3.1 for solving the integral equation of Example 10.17.

10.3.3. Show that the stabilizing functional of Example 10.17 may be also used for solving the nonlinear equation with the Hammerstein operator

$$\int_a^b K(s, t) f(t, x(t)) \, dt = y(s) , \quad s \in [c, d] ,$$

where $f(t, x)$ is a continuous function of t and x.

APPENDICES

1. Basic Characteristics of Some Typical Linear Systems

Table A.1.1

Operators, Weighting and Transfer Functions for Some Typical Linear Systems with the Lumped Parameters

System	Weighting Function $g(t,\tau), t > \tau$	Transfer Function $\Phi(s)$
1. $y(t) = Kx(t), K > 0$	$K\delta(t-\tau)$	K
2. $y(t) = \int_{t_0}^{t} x(\xi)d\xi, \frac{dy}{dt} = x$	$1(t-\tau)$	$1/s$
3. $y = \frac{dx}{dt}$	$\delta'(t-\tau)$	s
4. $a\frac{dy}{dt} \pm y = bx, a > 0$	$\frac{b}{a}e^{\mp\frac{(t-\tau)}{a}}$	$\frac{b}{as \pm 1}$
5. $\frac{dy}{dx} + a_1(t)y = x$	$1(t-\tau)e^{-\varphi(t,\tau)},$ $\varphi(t,\tau) = \int_{\tau}^{t} a_1(\tau)d\tau$	
6. $a^2\frac{d^2y}{dt^2} + y = bx, a > 0$	$\frac{b}{a}\sin\frac{(t-\tau)}{a}$	$\frac{b}{a^2s^2+1}$
7. $a^2\frac{d^2y}{dt^2} + 2\varepsilon a\frac{dy}{dt} + y = bx,$ $a > 0, 1 > \varepsilon > 0$	$\frac{b}{a\sqrt{1-\varepsilon^2}}e^{-\frac{\varepsilon}{a}(t-\tau)}$ $\times\sin\frac{\sqrt{1-\varepsilon^2}}{a}(t-\tau)$	$\frac{b}{a^2s^2+2\varepsilon as+1}$
8. $a^2\frac{d^2y}{dt^2} + 2\varepsilon a\frac{dy}{dt} + y = bx,$ $\varepsilon > 1$	$\frac{b}{a\sqrt{\varepsilon^2-1}}\big[e^{-\frac{2\varepsilon_1}{a}(t-\tau)}$ $-e^{-\frac{2\varepsilon_2}{a}(t-\tau)}\big],$ $2\varepsilon_{1,2} = 1 \pm \sqrt{\varepsilon^2-1}$	$\frac{b}{a^2s^2+2\varepsilon as+1}$
9. $\frac{d^ny}{dt^n} = x, n \geq 2$	$\frac{(t-\tau)^{n-1}}{(n-1)!}$	$1/s^n$

Table A.1.1 (continued)

System	Weighting Function $g(t,\tau), t > \tau$	Transfer Function $\Phi(s)$
10. $y = b\dfrac{dx}{dt} + x$	$b\delta'(t-\tau) + \delta(t-\tau)$	$bs + 1$
11. $y = b^2\dfrac{d^2x}{dt^2}$ $+2\varepsilon b\dfrac{dx}{dt} + x, b, \varepsilon > 0$	$b^2\delta''(t-\tau)$ $+2\varepsilon b\delta'(t-\tau) + \delta(t-\tau)$	$b^2 s^2$ $+2\varepsilon bs + 1$
12. $a\dfrac{dy}{dt} + y = b\dfrac{dx}{dt} + x$	$\dfrac{b}{a}\delta(t-\tau)$ $+\left(1-\dfrac{b}{a}\right) e^{-\frac{(t-\tau)}{a}}$	$\dfrac{1+bs}{1+as}$

Table A.1.2

Operators, Weighting (Green) and Transfer Functions for Some Typical Linear One-Dimensional Systems with the Distributed Parameters

System	Weighting (Green) Function $g(\xi,\xi_1,t,\tau)$, $t > \tau$	Transfer Function $\Phi(\xi,\xi_1,s)$
1. $y(t) = x(t-\Delta)$ $\forall t > \tau, y(t) = 0, \forall t < \tau,$ $\Delta > 0$	$\delta(t-\tau-\Delta)$	$\exp\{-\Delta s\}$
2. $x(t) = q(0,t),$ $y(t) = q(l,t),$ $\dfrac{\partial^2 q(\xi,t)}{\partial \xi^2} - \dfrac{1}{\gamma^2}\dfrac{\partial q(\xi,t)}{\partial t} = 0,$ $q(\infty,t) = 0, \Delta_1 = l^2/\gamma^2$	$\dfrac{1}{2}\sqrt{\dfrac{\Delta_1}{\pi(t-\tau)^3}}\, e^{-\frac{\Delta_1}{4(t-\tau)}}$	$\exp\{-\sqrt{\Delta_1 s}\}$

System	Weighting (Green) Function $g(\xi, \xi_1, t, \tau)$, $t > \tau$	Transfer Function $\Phi(\xi, \xi_1, s)$
3. $x(t) = -\lambda \dfrac{\partial q(0,t)}{\partial \xi}$, $\quad y(t) = q(0,t)$, $\dfrac{\partial^2 q(\xi,t)}{\partial \xi^2} - \dfrac{1}{\gamma^2}\dfrac{\partial q(\xi,t)}{\partial t} = 0$, $q(\infty,t) = 0, \Delta_2 = \lambda^2/\gamma^2$	$\sqrt{\dfrac{\Delta_2}{\pi(t-\tau)}}$	$\dfrac{1}{\sqrt{\Delta_2}\,s}$
4. $x(t) = -\lambda \dfrac{\partial q(0,t)}{\partial \xi}$ $+\alpha q(0,t),\ y(t) = q(0,t)$, $\dfrac{\partial^2 q(\xi,t)}{\partial \xi^2} - \dfrac{1}{\gamma^2}\dfrac{\partial q(\xi,t)}{\partial t} = 0$, $q(\infty,t) = 0, \beta_1 = 1/\alpha$, $\alpha_1 = \lambda^2/\alpha^2\gamma^2$	$\dfrac{\beta_1}{\alpha_1}\left[\sqrt{\dfrac{\alpha_1}{\pi(t-\tau)^3}}\right.$ $-e^{-\frac{(t-\tau)}{\alpha_1}}$ $\left.\times \mathrm{erf}\sqrt{\dfrac{(t-\tau)}{\alpha_1}}\right]$	$\dfrac{\beta_1}{\sqrt{\alpha_1}\,s+1}$
5. $x(t) = q(0,t)$, $\quad \dot{y}(t) = q(l,t)$, $\dfrac{\partial^2 q(\xi,t)}{\partial \xi^2} - \dfrac{1}{\gamma^2}\dfrac{\partial q(\xi,t)}{\partial t} = 0$, $q(\infty,t) = 0, \Delta_1 = l^2/\gamma^2$	$\mathrm{erfc}\sqrt{\dfrac{a_1}{4(t-\tau)}}$	$\dfrac{\exp\{-\sqrt{\Delta_1 s}\}}{s}$
6. $x(t) = -\lambda \dfrac{\partial q(0,t)}{\partial \xi}$, $\quad \dot{y}(t) = q(0,t)$, $\dfrac{\partial^2 q(\xi,t)}{\partial \xi^2} - \dfrac{1}{\gamma^2}\dfrac{\partial q(\xi,t)}{\partial t} = 0$, $q(\infty,t) = 0, \Delta_2 = \lambda^2/\gamma^2$	$2\sqrt{\dfrac{(t-\tau)}{\pi\Delta_2}}$	$\dfrac{1}{s\sqrt{\Delta_2}\,s}$

System	Weighting (Green) Function $g(\xi, \xi_1, t, \tau)$, $t > \tau$	Transfer Function $\Phi(\xi, \xi_1, s)$
7. $x(t) = -\lambda \dfrac{\partial q(0,t)}{\partial \xi}$ $+\alpha q(0,t), \; \dot{y}(t) = q(0,t),$ $\dfrac{\partial^2 q(\xi,t)}{\partial \xi^2} - \dfrac{1}{\gamma^2} \dfrac{\partial q(\xi,t)}{\partial t} = 0,$ $q(\infty, t) = 0, \; \beta_1 = 1/\alpha,$ $\alpha_1 = \lambda^2/\alpha^2 \gamma^2$	$1 - e^{(t-\tau)/\alpha_1}$ $\times \operatorname{erfc} \sqrt{\dfrac{(t-\tau)}{\alpha_1}}$	$\dfrac{\beta_1}{s\left(\sqrt{\alpha_1 s}+1\right)}$
8. $\dfrac{\partial y}{\partial t} - a^2 \dfrac{\partial^2 y}{\partial \xi^2} = x,$ $y(\xi, 0) = y_0(\xi),$ $\displaystyle\int\limits_{-\infty}^{\infty} |y(\xi,t)| \, d\xi < \infty,$ $t \geq 0, -\infty \leq \xi \leq \infty, a \neq 0$	$\dfrac{1}{2a\sqrt{\pi(t-\tau)}}$ $\times \exp\left[-\dfrac{(\xi-\xi_1)^2}{4a^2(t-\tau)}\right]$	$\dfrac{1}{2a^2 s_1} e^{-s_1 \lvert \xi - \xi_1 \rvert},$ $s_1 = \sqrt{s}/a$
9. $\dfrac{\partial y}{\partial t} - a^2 \dfrac{\partial^2 y}{\partial \xi^2} = x,$ $y(\xi, 0) = y_0(\xi),$ $\dfrac{\partial y}{\partial \xi}(0,t) = y^0(t),$ $t \geq 0, 0 \leq \xi \leq \infty, a \neq 0$	$\dfrac{1}{2a\sqrt{\pi(t-\tau)}}$ $\times \left[e^{-\frac{(\xi-\xi_1)^2}{4a^2(t-\tau)}} \right.$ $\left. + e^{-\frac{(\xi+\xi_1)^2}{4a^2(t-\tau)}} \right]$	$\dfrac{1}{2a^2 s_1} \left[e^{-s_1(\xi-\xi_1)} \right.$ $\left. + e^{-s_1(\xi+\xi_1)} \right],$ $s_1 = \sqrt{s}/a$
10. $\dfrac{\partial y}{\partial t} - a^2 \dfrac{\partial^2 y}{\partial \xi^2} = x,$ $y(\xi, 0) = y_0(\xi),$ $y(0,t) = y^0(t),$ $y(\infty, t) = 0,$ $t \geq 0, 0 \leq \xi \leq \infty, a \neq 0$	$\dfrac{1}{2a\sqrt{\pi(t-\tau)}}$ $\times \left[e^{-\frac{(\xi-\xi_1)^2}{4a^2(t-\tau)}} \right.$ $\left. - e^{-\frac{(\xi+\xi_1)^2}{4a^2(t-\tau)}} \right]$	$\dfrac{1}{2a^2 s_1} \left[e^{-s_1 \lvert \xi - \xi_1 \rvert} \right.$ $\left. - e^{-s_1 \lvert \xi + \xi_1 \rvert} \right],$ $s_1 = \sqrt{s}/a$

Table A.1.2 (continued)

System	Weighting (Green) Function $g(\xi, \xi_1, t, \tau)$, $t > \tau$	Transfer Function $\Phi(\xi, \xi_1, s)$
11. $\dfrac{\partial^2 y}{\partial t^2} - a^2 \dfrac{\partial^2 y}{\partial \xi^2} = x$, $y(\xi, 0) = y_0(\xi)$, $\dfrac{\partial y(\xi, 0)}{\partial t} = y_1(\xi)$, $t \geq 0, -\infty < \xi < \infty, a \neq 0$	$\dfrac{1}{2a}\mathbf{1}[a(t-\tau)-\mid \xi - \xi_1 \mid]$	$\dfrac{1}{2as}e^{-\frac{s}{2}|\xi - \xi_1|}$
12. $\dfrac{\partial^2 y}{\partial t^2} - a^2 \dfrac{\partial^2 y}{\partial \xi^2} = x$, $y(\xi, 0) = y_0(\xi)$, $\dfrac{\partial y(\xi, 0)}{\partial t} = y_1(\xi)$, $y(0, t) = y^0(t)$, $t \geq 0, \xi \geq 0, a \neq 0$	$\dfrac{1}{2a}\mathbf{1}\{[\xi - \xi_1 + a(t-\tau)]$ $-\mathbf{1}[\xi - \xi_1 - a(t-\tau)]$ $-\mathbf{1}[\xi + \xi_1 + a(t-\tau)]$ $+\mathbf{1}[\xi + \xi_1 - a(t-\tau)]\}$	$\dfrac{1}{2as}\left[e^{-\frac{s}{a}|\xi - \xi_1|}\right.$ $\left.-e^{-\frac{s}{a}|\xi + \xi_1|}\right]$

Table A.1.3

Typical Linear Discrete Two–Dimensional Signal and Image Processing Operators

Name	Operator
1. Input and Output Vector and Matrix Representation	$\mathbf{x} = \displaystyle\sum_{n=1}^{N_2} \mathbf{N}_{y,n}\mathbf{X}\mathbf{v}_{y,n}, \ y = \displaystyle\sum_{m=1}^{M_2} \mathbf{N}_{y,n}\mathbf{Y}\mathbf{v}_{y,n},$ $\mathbf{X} = \displaystyle\sum_{n=1}^{N_2} \mathbf{N}_{y,n}^T\mathbf{x}\mathbf{v}_{y,n}^T, \ \mathbf{Y} = \displaystyle\sum_{m=1}^{M_2} \mathbf{N}_{y,n}^T\mathbf{y}\mathbf{v}_{y,n}^T,$ where $\mathbf{X} = [X(n_1, n_2)], \ \mathbf{Y} = [Y(m_1, m_2)]$

Name	Operator
	$$\mathbf{v}_{x,n} = \begin{bmatrix} 0 \\ \vdots \\ 0 \\ 1 \\ 0 \\ \vdots \\ 0 \end{bmatrix} \begin{matrix} 1 \\ \vdots \\ n-1 \\ n \\ n+1 \\ \vdots \\ N_2 \end{matrix} ,\ \mathbf{N}_{x,n} = \begin{bmatrix} \mathbf{0} \\ \vdots \\ \mathbf{0} \\ \mathbf{I} \\ \mathbf{0} \\ \vdots \\ \mathbf{0} \end{bmatrix} \begin{matrix} 1 \\ \vdots \\ n-1 \\ n \\ n+1 \\ \vdots \\ N_2 \end{matrix} ,$$ $$\mathbf{v}_{y,n} = \begin{bmatrix} 0 \\ \vdots \\ 0 \\ 1 \\ 0 \\ \vdots \\ 0 \end{bmatrix} \begin{matrix} 1 \\ \vdots \\ n-1 \\ n \\ n+1 \\ \vdots \\ M_2 \end{matrix} ,\ \mathbf{N}_{y,n} = \begin{bmatrix} \mathbf{0} \\ \vdots \\ \mathbf{0} \\ \mathbf{I} \\ \mathbf{0} \\ \vdots \\ \mathbf{0} \end{bmatrix} \begin{matrix} 1 \\ \vdots \\ n-1 \\ n \\ n+1 \\ \vdots \\ M_2 \end{matrix} ,$$ $\dim \mathbf{v}_{x,n} = N_2 \times 1,\ \dim \mathbf{v}_{y,n} = M_2 \times 1,$ $\dim \mathbf{N}_{x,n} = N_1 N_2 \times N_1,\ \dim \mathbf{N}_{y,n} = M_1 M_2 \times M_1,$ $\dim \mathbf{x} = N_1 N_2 \times 1,\ \dim \mathbf{y} = M_1 M_2 \times 1,$ $\dim \mathbf{X} = N_1 \times N_2,\ \dim \mathbf{Y}_{y,n} = M_1 \times M_2.$
2. Generalized Linear Operator	$$\mathbf{y} = \mathbf{A}\mathbf{x} = \sum_{n=1}^{N} \mathbf{A}\mathbf{N}_{x,n}\mathbf{X}\mathbf{v}_{x,n},$$ $$\mathbf{Y} = \sum_{m=1}^{M}\sum_{n=1}^{N} \mathbf{A}_{mn}\mathbf{X}\mathbf{v}_{x,n}\mathbf{u}_{x,m}^{T},$$ where $$\mathbf{A} = \begin{bmatrix} \mathbf{A}_{1,1} & \mathbf{A}_{1,2} & \cdots & \mathbf{A}_{1,N} \\ \mathbf{A}_{2,1} & \mathbf{A}_{2,2} & \cdots & \mathbf{A}_{2,N} \\ \vdots & \vdots & \cdots & \vdots \\ \mathbf{A}_{M,1} & \mathbf{A}_{M,2} & \cdots & \mathbf{A}_{M,N} \end{bmatrix},$$ $\dim \mathbf{A} = M^2 \times N^2,\ \dim \mathbf{A}_{m,n} = M \times N,$ $\dim \mathbf{x} = N^2 \times 1,\ \dim \mathbf{y}_{m,n} = M^2 \times 1,$

Table A.1.3 (continued)

Name	Operator
	$\dim \mathbf{v}_{x,n} = N \times 1$, $\dim \mathbf{v}_{y,n} = M \times 1$, $\dim \mathbf{N}_{x,n} = N^2 \times N$, $\dim \mathbf{N}_{y,n} = M^2 \times M$, $\mathbf{u}_{x,m} - \mathbf{xx}^T$ eigenvectors, $\mathbf{u}_{y,m} - \mathbf{yy}^T$ eigenvectors, $\mathbf{v}_{x,n} - \mathbf{x}^T\mathbf{x}$ eigenvectors, $\mathbf{v}_{y,n} - \mathbf{y}^T\mathbf{y}$ eigenvectors. For divisible transform $\mathbf{A}$, $\mathbf{A} = \mathbf{A}_C \otimes \mathbf{A}_R$, $$\mathbf{Y} = \mathbf{A}_C \mathbf{X} \mathbf{A}_R^T$$ $\mathbf{A}_R$, $\mathbf{A}_C$ being rows and columns transform operators, $\otimes$ – symbol of direct product of matrices
3. Superposition	$$\mathbf{y} = \mathbf{Dx},$$ $$\mathbf{Y} = \sum_{m=1}^{M} \sum_{n=1}^{N} \mathbf{D}_{m,n} \mathbf{X} \mathbf{v}_{x,n} \mathbf{u}_{x,m}^T,$$ where $$\mathbf{D} = \begin{bmatrix} \mathbf{D}_{11} & \mathbf{D}_{12} & \cdots & \mathbf{D}_{1N} \\ \mathbf{D}_{21} & \mathbf{D}_{22} & \cdots & \mathbf{D}_{2N} \\ \vdots & \vdots & \cdots & \vdots \\ \mathbf{D}_{M1} & \mathbf{D}_{M2} & \cdots & \mathbf{D}_{MN} \end{bmatrix}$$ $$= \begin{bmatrix} \mathbf{D}_{1,1} & 0 & \cdots & & 0 \\ \mathbf{D}_{2,1} & \mathbf{D}_{2,2} & & & \vdots \\ \vdots & \vdots & & & 0 \\ \mathbf{D}_{L,1} & \mathbf{D}_{L,2} & & \mathbf{D}_{M-L+1,N} \\ 0 & \mathbf{D}_{L+1,2} & & & \\ 0 & 0 & & & \vdots \\ \vdots & \vdots & & & \\ 0 & 0 & \cdots & 0 & \mathbf{D}_{M,N} \end{bmatrix},$$ $\dim \mathbf{D} = M^2 \times N^2$, $\dim \mathbf{D}_{m,n} = M \times N$, $\dim \mathbf{x} = N^2 \times 1$, $\dim \mathbf{y} = M^2 \times 1$, $\dim \mathbf{X} = N \times N$, $\dim \mathbf{Y} = M \times M$,

Name	Operator
	$\dim \mathbf{v}_{x,n} = N \times 1,\ \dim \mathbf{v}_{y,n} = M \times 1,$ $\dim \mathbf{N}_{x,n} = N^2 \times N,\ \dim \mathbf{N}_{y,n} = M^2 \times M.$
4. Convolution Operator	See *superposition operator* at $\mathbf{D}_{m_2,n_2}$ $= \mathbf{D}_{m_2+1,n_2+1}.$ So, all columns of the matrix $\mathbf{D}$ are the shifts of the first column. For matrix $\mathbf{D}$ being invariant relative to the shift and divisible, $\mathbf{A} = \mathbf{A}_C \mathbf{A}_R^T,\ \mathbf{D} = \mathbf{D}_C \otimes \mathbf{D}_R$ $$\mathbf{Y} = \mathbf{D}_C \otimes \mathbf{D}_R,$$ where $$\mathbf{D}_{R(C)} = \begin{bmatrix} \mathbf{A}_R(1) & 0 & \ldots & & 0 \\ \mathbf{A}_R(2) & \mathbf{A}_R(1) & \ldots & & 0 \\ \mathbf{A}_R(3) & \mathbf{A}_R(2) & \ldots & & 0 \\ & & & & \mathbf{A}_R(1) \\ \vdots & & & & \vdots \\ \mathbf{A}_R(L) & & & & \\ 0 & & & & \\ \vdots & & & & \vdots \\ 0 & & & \ldots & 0 \quad \mathbf{A}_R(L) \end{bmatrix},$$ $\dim \mathbf{D}_R = \dim \mathbf{D}_C = M \times N.$
5. Operators of General Unitary Transforms	See *generalized linear operator* for the unitary matrix $\left(\mathbf{A} = \mathbf{U},\ \mathbf{U}^{-1} = \mathbf{U}^{*T} \right)$ $$\mathbf{y} = \mathbf{U}\mathbf{x},$$ $$\mathbf{x} = \mathbf{U}^{-1}\mathbf{y}.$$ For divisible unitary transform $\mathbf{U} = \mathbf{U}_C \otimes \mathbf{U}_R$ $$\mathbf{Y} = \mathbf{U}_C \mathbf{X} \mathbf{U}_R^T,$$ $$\mathbf{X} = \mathbf{U}_C^{-1} \mathbf{Y} (\mathbf{U}_R^{-1})^T.$$

Name	Operator
6. Two–dimensional Forward $\mathcal{F}_x$ and Inverse Fourier X Transforms	$\mathcal{F}_x(u,v) = \frac{1}{N} \sum\limits_{j=0}^{N-1} \sum\limits_{k=0}^{N-1} X(j,k)A(j,k;u,v) = A_C X A_R,$ $X(j,k) = \frac{1}{N} \sum\limits_{u=0}^{N-1} \sum\limits_{v=0}^{N-1} \mathcal{F}_x(u,v)B(j,k;u,v) = A_C^* \mathcal{F}_x A_R^*,$ where $$A_R = A_C = \frac{1}{\sqrt{N}} \begin{bmatrix} \varepsilon^0 & \varepsilon^0 & \varepsilon^0 & \cdots & \varepsilon^0 \\ \varepsilon^0 & \varepsilon^1 & \varepsilon^2 & \cdots & \varepsilon^{N-1} \\ \varepsilon^0 & \varepsilon^2 & \varepsilon^4 & \cdots & \varepsilon^{2(N-1)} \\ \vdots & \vdots & \vdots & \vdots & \vdots \\ \varepsilon^0 & \cdot & \cdot & \cdots & \varepsilon^{(N-1)^2} \end{bmatrix},$$ $\varepsilon = \exp\left\{-\frac{2\pi i}{N}\right\},\ \dim \mathcal{F}_x(u) = \dim X(j,k) = N \times N,$ $A = A(j,k;u,v) = \exp\left\{-\frac{2\pi i}{N}(uj+vk)\right\},$ $B = B(j,k;u,v) = \exp\left\{\frac{2\pi i}{N}(uj+vk)\right\} = A^*.$
7. Hadamard Transform Operator	$\mathcal{H}_x(u,v) = \frac{1}{N} \sum\limits_{j=0}^{N-1} \sum\limits_{k=0}^{N-1} X(j,k)(-1)^{p(j,k;u,v)},$ where for symmetric Hadamard matrices $N = 2^n$ $$p(i,k;u,v) = \sum_{i=0}^{n-1}(u_i j_i + v_i k_i),$$ u_i, v_i, j_i, k_i are equal to u, v, j, k in the binary representation. For ordered Hadamard matrices $\mathcal{H}_x(u,v) = \frac{1}{N} \sum\limits_{j=0}^{N-1} \sum\limits_{k=0}^{N-1} X(j,k)(-1)^{q(j,k;u,v)},$ where $$q(j,k;u,v) = \sum_{i=0}^{n-1}[g_i(u)j_i + g_i(u)k_i],$$ $g_0(u) = u_{n-1}, g_1(u) = u_{n-1} + u_{n-2},$ $g_2(u) = u_{n-2} + u_{n-3}, \ldots, g_{n-1}(u) = u_1 + u_0.$

For additional material concerning basic characteristics of typical linear systems refer to (Brigham 1988, Burrus and Parks 1985, Butkovskiy 1982, Elliot 1987, Pratt 1978).

2. Some Definite Integrals and Special Functions

Table A.2.1

Definite Integrals

1.	$$\int_{-\infty}^{\infty} e^{\eta t - ct^2/2}\,dt = \sqrt{\frac{2\pi}{c}}\,e^{\eta^2/2c}$$
2.	$$\int_{-\infty}^{\infty} e^{-\eta^T t - t^T Ct/2}\,dt = \sqrt{\frac{(2\pi)^n}{|C|}}\,e^{\eta^T C^{-1}\eta/2}$$
3.	$$\int_{0}^{\infty} t^{2n} e^{-ct^2/2}\,dt = \frac{(2n-1)!!}{2c^n}\sqrt{\frac{2\pi}{c}},\ \ n = 1, 2, \ldots, c > 0$$
4.	$$\int_{0}^{\infty} t^{2n+1} e^{-ct^2/2}\,dt = \frac{n!}{2^n c^{n+1}},\ \ n = 1, 2, \ldots, c > 0$$
5.	$$\int_{0}^{\infty} \exp\left(\eta t - \frac{c}{2}t^2\right)dt = \sqrt{\frac{\pi}{2c}}e^{\eta^2/2c}\,\mathrm{erfc}\left(-\frac{\eta}{\sqrt{2c}}\right),\ (\mathrm{Re}\,c > 0)$$
6.	$$\int_{-\infty}^{\infty} t\exp\left(\eta t - \frac{c}{2}t^2\right)dt = \frac{\eta}{c}\sqrt{\frac{2\pi}{c}}e^{\eta^2/2c},\ (\mathrm{Re}\,c > 0)$$
7.	$$\int_{0}^{\infty} t\exp\left(\eta t - \frac{c}{2}t^2\right)dt = \frac{1}{c} + \frac{2\eta}{c}\sqrt{\frac{2\pi}{c}}e^{\eta^2/2c}\,\mathrm{erfc}\left(-\frac{\eta}{\sqrt{2c}}\right)$$ $$(\mathrm{Re}\,c > 0,\ |\arg\eta| < \pi)$$
8.	$$\int_{-\infty}^{\infty} t^2\exp\left(\eta t - \frac{c}{2}t^2\right)dt = \frac{1}{c}\sqrt{\frac{2\pi}{c}}\left(1 + \frac{\eta^2}{c}\right)e^{\eta^2/2c}$$ $$(\mathrm{Re}\,c > 0,\ |\arg\eta| < \pi)$$
9.	$$\int_{0}^{\infty} t^2\exp\left(\eta t - \frac{c}{2}t^2\right)dt = \frac{\eta}{c^2} + \sqrt{\frac{2\pi}{c^5}}\frac{c+\eta^2}{4}e^{\eta^2/2c}\,\mathrm{erfc}\left(-\frac{\eta}{\sqrt{2c}}\right)$$ $$(\mathrm{Re}\,c > 0,\ |\arg q| < \pi)$$
10.	$$\int_{-\infty}^{\infty} e^{-q^2 t^2}\sin[p(t + \lambda)]dt = \frac{\sqrt{\pi}}{q}e^{-p^2/4q^2}\sin p\lambda$$
11.	$$\int_{-\infty}^{\infty} e^{-q^2 t^2}\cos[p(t + \lambda)]dt = \frac{\sqrt{\pi}}{q}e^{-p^2/4q^2}\cos p\lambda$$
12.	$$\int_{-\infty}^{\infty} e^{i\mu t + t^2/2}\frac{d^n}{dt^n}e^{-t^2/2} = \sqrt{2\pi}(i)^n e^{\mu^2/2}\frac{d^n}{d\mu^n}e^{-\mu^2}$$

Special Functions and Polynomials

1.	$\operatorname{erf} z = \frac{2}{\sqrt{\pi}} \int\limits_0^z e^{-t^2} dt, \quad \operatorname{erf} z = -\operatorname{erf}(-z),$ $$\operatorname{erf}(0) = 0, \quad \operatorname{erf}(\infty) = 1$$
2.	$\operatorname{erfc} z = \frac{2}{\pi} \int\limits_z^\infty e^{-t^2} dt = 1 - \operatorname{erf} z$ $$\operatorname{erfc}(-z) = 2 - \operatorname{erfc} z, \quad \operatorname{erfc}(0) = 1, \operatorname{erfc}(\infty) = 0$$
3.	$\Phi(z) = \frac{1}{\sqrt{2\pi}} \int\limits_0^z e^{-t^2/2} dt = \frac{1}{2} - \operatorname{erf}(z/\sqrt{2})$ $$\Phi(0) = 0, \quad \Phi(\infty) = \frac{1}{2}, \quad \Phi(-z) = -\Phi(z)$$
4.	$\Gamma(z) = \int\limits_-^\infty e^{-t} t^{z-1} dt \quad (\operatorname{Re} z > 0)$ $$\Gamma(z+1) = z\Gamma(z), \quad \Gamma(z)\Gamma(-z) = -\frac{\pi}{z \sin \pi z}, \quad \Gamma(z)\Gamma(1-z) = \frac{\pi}{\sin \pi z}$$ $$\Gamma(nz) = \sqrt{\frac{n^{2nz-1}}{(2\pi)^{n-1}}} \Gamma(z)\Gamma\left(z + \frac{1}{n}\right) \cdots \Gamma\left(z + \frac{n-1}{n}\right)$$
5.	$J_m(z) = \left(\frac{z}{2}\right)^m \sum\limits_{k=0}^\infty \frac{(-1)^k}{k!\Gamma(m+k+1)} \left(\frac{z}{2}\right)^{2k}$ $$(|\arg z| < \pi, \quad m = 0, \pm 1 \pm 2, \ldots)$$ $$J_m(z) = \frac{1}{\pi} \int\limits_0^\pi \cos(mt - z \sin t) dt = \frac{(-i)^m}{\pi} \int\limits_0^\pi e^{iz \cos t} \cos mt \, dt$$ $$(m = 0, 1, 2, \ldots)$$
6.	$N_m(z) = \frac{1}{\sin m\pi}[J_m(z) \cos m\pi - J_{-m}(z)]$ $$(m \neq 0, \pm 1, \pm 2, \ldots)$$
7.	$J_{m+1/2}(z) = \sqrt{\frac{2}{\pi}} z^{m+1/2} \left(-\frac{1}{z}\frac{d}{dz}\right)^m \frac{\sin z}{z} \quad (m = 1, 2, \ldots)$
8.	$T_n(z) = \cos(n \arccos z) = \frac{n}{2} \sum\limits_{m=0}^{[n/2]} (-1)^m \frac{(n-m-1)!}{m!(n-2m)!} (2z)^{n-2m}$ $$= \frac{(-2)^n n!}{(2n)!} \sqrt{1-z^2} \frac{d^n}{dz^n}(1-z^2)^{n-1/2}$$ $$T_{n+1}(z) = 2zT_n(z) - T_{n-1}(z) \quad (n = 1, 2, \ldots)$$
9.	$U_n(z) = \sin(n \arccos z) = \frac{\sqrt{1-z^2}}{n} \frac{d}{dz} T_n(z), \quad (n = 1, 2, \ldots)$

10.	$L_n(z) = (n!)^2 \sum\limits_{m=0}^{n} (-1)^m \frac{z^m}{(m!)^2(n-m)!} = e^z \frac{d^n}{dz^n}(z^n e^{-z})$, $(n = 1, 2, \ldots)$ $L_{n+1}(z) = (2n+1-z)L_n(z) - n^2 L_{n-1}(z)$ $\frac{dL_{n+1}(z)}{dz} = (n+1)\left[\frac{dL_n(z)}{dz} - L_n(z)\right]$
11.	$H_n(z) = n! \sum\limits_{m=0}^{[n/2]} (-1)^m \frac{(2z)^{n-2m}}{m!(n-2m)!} = (-1)^n e^{z^2} \frac{d^n}{dz^n}(e^{-z^2})$ $H_{n+1}(z) = 2zH_n(z) - 2nH_{n-1}(z)$ $\frac{dH_n(z)}{dz} = 2nH_{n-1}(z), \ (n = 1, 2, \ldots)$
12.	$P_n(z) = 2^{-n} \sum\limits_{m=0}^{[n/2]} (-1)^m \frac{(2n-2m)!}{m!(n-m)!(n-2m)!} z^{n-2m}$ $= \frac{1}{2^n n!} \frac{d^n}{dz^n}(z^2-1)^n, \ (n = 1, 2, \ldots)$ $P_{n+1}(z) = \frac{2n+1}{n+1} z P_n(z) - \frac{n}{n+1} P_{n-1}(z) = z P_n(z) + \frac{z^2-1}{n+1} \frac{dP_n}{dz}$

For additional information concerning definite integrals and special functions refer to (Bateman and Erdelyi 1953, 1955, Korn and Korn 1968).

3. Laplace, Fourier, Sine and Cosine Transforms

The *Laplace transform* connects the one–dimensional single–valued function $F(s)$ of the complex variable s (*transform*) with the corresponding function $f(t)$ of the real variable t (*original*):

$$F(s) = \mathcal{L}[f(t)] = \int_0^\infty e^{-s\tau} f(\tau)d\tau \, . \qquad (A.3.1)$$

Notice that for the generalized functions $\varphi(t)$ the Laplace transform is usually taken in the form of the Stieltjes integral:

$$\mathcal{L}[\varphi(t)] = \int_0^\infty e^{-s\tau} d\varphi(\tau) \, . \qquad (A.3.2)$$

The *inverse Laplace transform* is defined by *inversion formula*

$$f(t) = \mathcal{L}^{-1}[F(s)] = \frac{1}{2\pi i} \int_{c-i\infty}^{c+i\infty} F(s)e^{st}ds \, , \ (i = \sqrt{-1}), \qquad (A.3.3)$$

where c is the abscissa of absolute convergence.

Along with the *one-sided Laplace transforms* (A.3.1), (A.3.3) we introduce the following *two-sided Laplace transforms*:

$$F(s) = \int\limits_{-\infty}^{\infty} e^{-s\tau} f(\tau) d\tau ,\qquad (A.3.4)$$

$$f(t) = \frac{1}{2\pi i} \int\limits_{c-i\infty}^{c+i\infty} e^{-s\tau} F(s) e^{st} ds . \qquad (A.3.5)$$

At practice the relations between the transforms are more simple than relations between the originals. The corresponding typical pairs $F(s)$ and $f(t)$ are usually given in special tables (see Table A.3.1).

Table A.3.1

Laplace Transforms for Some Typical Functions

	$f(t),\ t > 0$	$F(s)$
1.	1	$1/s$
2.	t	$1/s^2$
3.	$\dfrac{t^{n-1}}{(n-1)!},\ n = 1, 2, \ldots$	$1/s^n$
4.	e^{at}	$1/(s-a)$
5.	te^{at}	$1/(s-a)^2$
6.	$\dfrac{t^n}{n!} e^{at},\ n = 1, 2, \ldots$	$1/(s-a)^{n+1}$

	$f(t)$, $t > 0$	$F(s)$
7.	$\left(e^{at} - e^{bt}\right)/(a - b)$	$1/(s - a)(s - b)$
8.	$\left(ae^{at} - be^{bt}\right)/(a - b)$	$s/(s - a)(s - b)$
9.	$(1/a)\sin at$, $a > 0$	$1/(s^2 + a^2)$
10.	$\cos at$, $a > 0$	$s/(s^2 + a^2)$
11.	$\sin(at - \varphi)$, $a > 0$	$\dfrac{a\cos\varphi - s\sin\varphi}{s^2 + a^2}$
12.	$(1/a)\mathrm{sh}\, at$	$1/(s^2 - a^2)$
13.	$\mathrm{ch}\, at$, $a > 0$	$s/(s^2 - a^2)$
14.	$e^{-\alpha t}\sin(at - \varphi)$, $a > 0$	$\dfrac{a\cos\varphi - (s - \alpha)\sin\varphi}{(s - \alpha)^2 + a^2}$
15.	$\dfrac{1}{\sqrt{\pi t}}$	$\dfrac{1}{\sqrt{s}}$
16.	$2\sqrt{\dfrac{t}{\pi}}$	$\dfrac{1}{s\sqrt{s}}$
17.	$\dfrac{2^n t^{n-1/2}}{1 \cdot 3 \cdot 5 \cdots (2n - 1)\sqrt{\pi}}$	$\dfrac{1}{s^{n+1/2}}$
18.	$\dfrac{1}{\sqrt{\pi t}} - ae^{a^2 t}\,\mathrm{erfc}\,(a\sqrt{t})$	$\dfrac{1}{a + \sqrt{s}}$
19.	$\dfrac{1}{\sqrt{\pi t}} + ae^{a^2 t}\,\mathrm{erf}\,(a\sqrt{t})$	$\dfrac{\sqrt{s}}{s - a^2}$
20.	$\dfrac{1}{\sqrt{\pi t}} - \dfrac{2a}{\sqrt{\pi}}e^{-a^2 t}\displaystyle\int_0^{a\sqrt{t}} e^{\lambda^2}\,d\lambda$	$\dfrac{\sqrt{s}}{s + a^2}$
21.	$\dfrac{1}{a}e^{a^2 t}\,\mathrm{erf}\,(a\sqrt{t})$	$\dfrac{1}{\sqrt{s}(s - a^2)}$
22.	$\dfrac{2}{a\sqrt{\pi}}e^{-a^2 t}\displaystyle\int_0^{a\sqrt{t}} e^{\lambda^2}\,d\lambda$	$\dfrac{1}{\sqrt{s}(s + a^2)}$
23.	$e^{a^2 t}\,\mathrm{erfc}\,(a\sqrt{t})$	$\dfrac{1}{\sqrt{s}(\sqrt{s} + a)}$
24.	$\dfrac{1}{\sqrt{b - a}}e^{-at}\,\mathrm{erf}\,\left(\sqrt{(b - a)t}\right)$	$\dfrac{1}{(s + a)\sqrt{s + b}}$

	$f(t)$, $t > 0$	$F(s)$
25.	$J_0(at)$	$\dfrac{1}{\sqrt{s^2 + a^2}}$
26.	$\begin{cases} 0 & at \quad 0 < t < k \\ 1 & at \qquad t > k \end{cases}$	e^{-ks}/s
27.	$\begin{cases} 0 & at \quad 0 < t < k \\ t - k & at \qquad t > k \end{cases}$	e^{-ks}/s^2
28.	$J_0(2\sqrt{kt})$	$\dfrac{1}{s}e^{-k/s}$
29.	$\dfrac{1}{\sqrt{\pi t}}\cos 2\sqrt{kt}$	$\dfrac{1}{\sqrt{s}}e^{-k/s}$
30.	$\dfrac{1}{\sqrt{\pi t}}\,\mathrm{ch}\,2\sqrt{kt}$	$\dfrac{1}{\sqrt{s}}e^{k/s}$
31.	$\dfrac{k}{2\sqrt{\pi t^3}}\exp\left(-\dfrac{k^2}{4t}\right)$	$e^{-k\sqrt{s}}$, $k > 0$
32.	$\mathrm{erfc}\left(\dfrac{k}{2\sqrt{t}}\right)$	$\dfrac{1}{s}e^{-k\sqrt{s}}$, $k \geq 0$
33.	$\dfrac{1}{\sqrt{\pi t}}\exp\left(-\dfrac{k^2}{4t}\right)$	$\dfrac{1}{\sqrt{s}}e^{-k\sqrt{s}}$, $k \geq 0$
34.	$\Gamma'(1) - \ln t$, $\Gamma'(1) = -0.5772$	$\dfrac{1}{s}\ln s$
35.	$\dfrac{1}{t}\left(e^{bt} - e^{at}\right)$	$\ln\dfrac{s - a}{s - b}$
36.	$\dfrac{1}{\sqrt{\pi t}}\sin(2k\sqrt{t})$	$\mathrm{erf}\left(\dfrac{k}{\sqrt{s}}\right)$
37.	$\dfrac{1}{t}\sin kt$	$\mathrm{arctg}\left(\dfrac{k}{s}\right)$
38.	$\delta(t)$	$1/2$
39.	$\delta(t - a)$	e^{-as}, $a > 0$
40.	$\delta^{(n)}(t - a)$, $a > 0$, $n = 1, 2, \ldots$	$s^n e^{-as}$

Putting in two–sided Laplace transforms (A.3.4) and (A.3.5) $s = i\omega$ we get complex one–dimensional *Fourier transforms*:

$$F(i\omega) = \mathcal{F}[f(t)] = \int\limits_{-\infty}^{\infty} e^{-i\omega\tau} f(\tau)d\tau . \qquad (A.3.6)$$

$$f(t) = \mathcal{F}^{-1}[F(\omega)] = \frac{1}{2\pi} \int\limits_{-\infty}^{\infty} F(i\omega)e^{i\omega t} d\omega . \qquad (A.3.7)$$

Formulae (A.3.6) and (A.3.7) may be presented in the following real form:

$$\mathcal{F}[f(t)] = \sqrt{2\pi}C(\omega) , \qquad (A.3.8)$$

$$f(t) = \begin{cases} \frac{1}{\sqrt{2\pi}} \int\limits_{-\infty}^{\infty} C(\omega)e^{i\omega t} d\omega, \\[2mm] \sqrt{\frac{2}{\pi}} \int\limits_{-\infty}^{\infty} C_c(\omega) \cos \omega t d\omega, \\[2mm] \sqrt{\frac{2}{\pi}} \int\limits_{-\infty}^{\infty} C_s(\omega) \sin \omega t d\omega . \end{cases} \qquad (A.3.9)$$

The correspondences between some typical pairs $f(t)$ and $C(\omega)$, $f(t)$ and the cosine transform $C_c(\omega)$, $f(t)$ and the sine transform $C_s(\omega)$ are usually given in special tables (see Tables A.3.2 – A.3.4).

Table A.3.2

Fourier Transforms for Some Typical Functions

	$f(t), t > 0$	$C(\omega)$
1.	$\dfrac{\sin at}{t}$ $(a > 0)$	$\left(\dfrac{\pi}{2}\right)^{1/2}$, $|\omega| < n,$ $0, |\omega| > a$
2.	$e^{i\alpha t}, p < t < q,$ $0, t < p, t > q$	$\dfrac{i}{(2\pi)^{1/2}} \dfrac{e^{iq(\alpha-\omega)} - e^{ip(\alpha-\omega)}}{\omega - a}$, $\omega \neq a,$ $\dfrac{q - p}{(2\pi)^{1/2}}, \omega = a$

	$f(t), t > 0$	$C(\omega)$
3.	$e^{-ct+i\alpha t}, \quad t > 0,$ $\quad (c > 0)$ $0, t < 0$	$\dfrac{i}{(2\pi)^{1/2}(\alpha - \omega + ic)}$
4.	$e^{-pt^2} \ (\operatorname{Re} p > 0)$	$(2p)^{-1/2}e^{-\omega^2/4p}$
5.	$e^{i\alpha t^2}$	$(2\alpha)^{-1/2}e^{-i(\omega^2/4a-\pi/4)}$
6.	$\cos\alpha t^2 \ (a > 0)$	$(2\alpha)^{-1/2}\cos(\omega^2/4a - \pi/4)$
7.	$\sin\alpha t^2$	$(2\alpha)^{-1/2}\sin(\pi/4 - \omega^2/4a)$
8.	$|t|^{-s} \ (0 < \operatorname{Re} s < 1)$	$(2/\pi)^{1/2}\Gamma(1 - s)\sin(s\pi/2)|\omega|^{s-1}$
9.	$\dfrac{e^{-a|t|}}{|t|^{1/2}} \ (a > 0)$	$\dfrac{[(a^2 + \omega^2)^{1/2} + a]^{1/2}}{(a^2 + \omega^2)^{1/2}}$
10.	$\dfrac{\operatorname{ch} at}{\operatorname{ch} \pi t} \ (-\pi < a < \pi)$	$\left(\dfrac{2}{\pi}\right)^{1/2}\dfrac{\cos(a/2)\operatorname{ch}(\omega/2)}{\operatorname{ch}\omega + \cos\alpha}$
11.	$\dfrac{\operatorname{sh} at}{\operatorname{sh} \pi t} \ (-\pi < a < \pi)$	$\left(\dfrac{2}{\pi}\right)^{1/2}\dfrac{\sin a}{\operatorname{ch}\omega + \cos a}$
12.	$\dfrac{\operatorname{ch} at}{\operatorname{sh} \pi t} \ (-\pi < a < \pi)$	$\left(\dfrac{2}{\pi}\right)^{1/2}\dfrac{\operatorname{sh}\omega}{i(\operatorname{ch}\omega + \cos a)}$
13.	$\dfrac{\operatorname{sh} at}{\operatorname{ch} \pi t} \ (-\pi < a < \pi)$	$\left(\dfrac{2}{\pi}\right)^{1/2}\dfrac{\sin(a/2)\operatorname{ch}(\omega/2)}{i(\operatorname{ch}\omega + \cos a)}$
14.	$(a^2 - t^2)^{-1/2}, \ |t| < a,$ $0, \qquad\quad |t| > a$	$(\pi/2)^{1/2}J_0(a\omega)$
15.	$\dfrac{\sin[b(a^2 + t^2)^{1/2}]}{(a^2 + t^2)^{1/2}}$	$0, \qquad\qquad\qquad |\omega| > b,$ $\pi/2)^{1/2}J_0(a\sqrt{b^2 - \omega^2}), \quad |\omega| < b$
16.	$\dfrac{\cos(b\sqrt{a^2 - t^2})}{(a^2 - t^2)^{1/2}}, \ |t| < a,$ $0, \ |t| > a$	$(\pi/2)^{1/2}J_0(\sqrt{\omega^2 + b^2})$
17.	$\dfrac{\operatorname{ch}(b\sqrt{a^2 - t^2})}{(a^2 - t^2)^{1/2}}, \ |t| < a,$ $0, \ |t| > a$	$(\pi/2)^{1/2}J_0(\sqrt{\omega^2 - b^2})$
18.	$P_n(t), \ |t| < 1,$ $0, \ |t| > 1$	$i^n\pi^{1/2}J_{n+1/2}(\omega)$

Cosine Transforms for Some Typical Functions

	$f(t), t > 0$	$C_c(\omega)$
1.	$1, \quad 0 < t < a,$ $0, \qquad t > a$	$\left(\dfrac{2}{\pi}\right)^{1/2} \dfrac{\sin \omega a}{\omega}$
2.	$t^{-\alpha}(0 < \operatorname{Re}\alpha < 1)$	$\left(\dfrac{2}{\pi}\right)^{1/2} \Gamma(1-\alpha)\sin\left(\tfrac{1}{2}\alpha\pi\right)\omega^{s-1}$
3.	$\cos t, \quad 0 < t < a,$ $0, \qquad t > a$	$\left(\dfrac{1}{2\pi}\right)^{1/2}\left[\dfrac{\sin a(1-\omega)}{1-\omega}\right.$ $\left. +\dfrac{\sin a(1+\omega)}{1+\omega}\right]$
4.	e^{-t}	$\left(\dfrac{2}{\pi}\right)^{1/2} \dfrac{1}{1+\omega^2}$
5.	e^{-t^2}	$\tfrac{1}{2}e^{-\omega^2/4}$
6.	$\cos\left(\tfrac{1}{2}t^2\right)$	$\cos\left(\dfrac{\omega^2}{2}-\dfrac{\pi}{4}\right)$
7.	$\sin\left(\tfrac{1}{2}t^2\right)$	$\sin\left(\dfrac{\pi}{4}-\dfrac{\omega^2}{2}\right)$
8.	$(1-t^2)^{\nu}, \quad 0 < t < 1,$ $0, \qquad\quad t > 1$ $(\operatorname{Re}\nu > -1)$	$2^{\nu}\Gamma(\nu+1)\omega^{-\nu-1/2}J_{\nu+1/2}(\omega)$

Sine Transforms for Some Typical Functions

	$f(t), t > 0$	$C_s(\omega)$
1.	$1, \quad 0 < t < a,$ $0, \qquad t > a$	$\left(\dfrac{2}{\pi}\right)^{1/2} \dfrac{1-\cos a\omega}{\omega}$
2.	$t^{-\alpha}, 0 < \operatorname{Re}\alpha < 2$	$\left(\dfrac{2}{\pi}\right)^{1/2} \Gamma(1-\alpha)\cos\left(\tfrac{1}{2}\alpha\pi\right)\omega^{\alpha-1}$
3.	e^{-t}	$\left(\dfrac{2}{\pi}\right)^{1/2} \dfrac{\omega}{1+\omega^2}$

	$f(t),\ t > 0$	$C_s(\omega)$
4.	$t^{n-1}e^{at},\ \ a > 0$	$\left(\dfrac{2}{\pi}\right)^{1/2}(n-1)!(a^2+\omega^2)^{-n/2}$ $\times \sin(n\,\mathrm{arctg}\,\omega/a)$
5.	$te^{-t^2/2}$	$\omega e^{-\omega^2/2}$
6.	$\dfrac{\sin t}{t}$	$(2\pi)^{-1/2}\ln\left\|\dfrac{1+\omega}{1-\omega}\right\|$
7.	$\begin{array}{ll}0, & 0 < t < a,\\ (t^2-a^2)^{1/2}, & t > a\end{array}$	$\sin\left(\dfrac{\pi}{2}\right)^{1/2}J_0(a\omega)$
8.	$\begin{array}{ll}t(1-t^2)^\nu, & 0 < t < 1,\\ 0, & t > 1\end{array}$ $(\mathrm{Re}\,\nu > -1)$	$2^\nu\Gamma(\nu+1)\omega^{-\nu-1/2}J_{\nu+3/2}(\omega)$

Discrete Laplace transforms are defined by formulae

$$F^*(s) = \mathcal{D}[f(mT)] = \sum_{m=0}^{\infty} e^{-smT}f(mT)\,, \qquad (A.3.10)$$

$$f(mT) = \frac{1}{i\omega_0}\int_{c-i\omega_0/2}^{c+i\omega_0/2} F^*(s)e^{-smT}\,ds\,,\quad (\omega_0 = 2\pi/T)\,. \qquad (A.3.11)$$

Here the lattice function $f(mT)$ of the discrete time mT $(m = 0, 1, 2, \ldots, T$ is sampling time) is called an *orinigal*. As a rule the Laplace transforms of lattice functions $f(mT)$ depend upon e^{sT}. So, introducing the new variable $z = e^{sT}$ we get instead of formula (A.3.10)

$$F^*\!\left(s = \frac{1}{T}\ln z\right) = F(z) = \sum_{m=0}^{\infty} z^{-m}f(mT)$$

or shortly

$$F(z) = Z[f(mT)]\,. \qquad (A.3.12)$$

This transform is called *Z–transform*.

Formulae (A.3.10) – (A.3.12) are valid for *one–sided discrete Laplace transforms*. *For two–sided discrete Laplace transforms* we have the following formulae:

$$F^*(s) = \sum_{m=-\infty}^{\infty} e^{-smT}f(mT)\,, \qquad (A.3.13)$$

$$f(mT) = \frac{1}{i\omega_0} \int_{c-i\omega_0/2}^{c+i\omega_0/2} F^*(s)e^{smT}\,ds\,. \qquad (A.3.14)$$

Some typical correspondences between discrete Laplace transforms for some typical functions are given in Table A.3.5.

Putting in formulae (A.3.13) and (A.3.14) $s = i\omega$ we get corresponding formulae for one–dimensional *discrete Fourier transforms*:

$$F^*(i\omega) = \mathcal{D}[f(mT)] = \sum_{m=-\infty}^{\infty} e^{-i\omega mT} f(mT)\,, \qquad (A.3.15)$$

$$f(mT) = \mathcal{D}^{-1}[F^*(i\omega)] = \frac{1}{\omega_0} \int_{-\omega_0/2}^{\omega_0/2} F^*(i\omega)e^{i\omega mT}\,d\omega\,, \quad (\omega_0 = 2\pi/T)\,.$$
$$(A.3.16)$$

Table A.3.5

Discrete Laplace Transforms for Some Typical Functions

	$f(n)\,,\ \ n = mT$	$F^*(s)\,,\ \ q = sT$
1.	$1[n]$	$\dfrac{e^q}{e^q - 1}$
2.	$\Delta 1[n-1] = 1[n] - 1[n-1]$	1
3.	n	$\dfrac{e^q}{(e^q - 1)^2}$
4.	n^2	$\dfrac{e^q}{(e^q - 1)^3}(e^q + 1)$
5.	n^3	$\dfrac{e^q}{(e^q - 1)^4}(e^{2q} + 4e^q + 1)$
6.	$n^{(k)} = n(n-1)\ldots(n-k+1)$	$k!\dfrac{e^q}{(e^q - 1)^{k+1}}$
7.	$(n \pm k)^{(k)}$	$k!e^{\pm kq}\dfrac{e^q}{(e^q - 1)^{k+1}}$

	$f(n)$, $n = mT$	$F^*(s)$, $q = sT$
8.	a^n	$\dfrac{e^q}{e^q - a}$
9.	$e^{\alpha n}$	$\dfrac{e^q}{e^q - e^\alpha}$
10.	$1 - e^\alpha(n - k)$, $n > k$	$e^{-kq}\dfrac{e^q(1 - e^\alpha)}{(e^q - e^\alpha)(e^q - 1)}$
11.	$Ae^{\alpha n} - Be^{\alpha_1 n}$	$\dfrac{e^{2q}(A - B) - e^q(Ae^{\alpha_1} - Be^\alpha)}{(e^q - e^\alpha)(e^q - e^{\alpha_1})}$
12.	$e^{\alpha n} - e^{\alpha(n-1)}$	$\dfrac{e^q - 1}{e^q - e^\alpha}$
13.	$ne^{\alpha n}$	$\dfrac{e^q e^\alpha}{(e^q - e^\alpha)^2}$
14.	$n^2 e^{\alpha n}$	$\dfrac{e^q e^\alpha}{(e^q - e^\alpha)^3}(e^q + e^\alpha)$
15.	$n^3 e^{\alpha n}$	$\dfrac{e^q e^\alpha}{(e^q - e^\alpha)^4}(e^{2q} + 4e^q e^\alpha + e^{2\alpha})$
16.	$\dfrac{(n \mp k)^m}{m!}e^{\alpha(n \mp k)}$	$\dfrac{e^{q(1 \pm k)}e^{m\alpha}}{(e^q - e^\alpha)^{m+1}}$
17.	e^{ian}	$\dfrac{e^q}{e^q - e^{ia}}$
18.	$\cos an$	$\dfrac{(e^q - \cos a)e^q}{e^{2q} - 2e^q \cos a + 1}$
19.	$\sin an$	$\dfrac{e^q \sin a}{e^{2q} - 2e^q \cos a + 1}$
20.	$\sin(an + \psi)$	$\dfrac{e^{2q}\sin\varphi + e^q \sin(a - \psi)}{e^{2q} - 2e^q \cos a + 1}$
21.	$(-1)^n$	$\dfrac{e^q}{e^q + 1}$
22.	$(-1)^n e^{\alpha n}$	$\dfrac{e^q}{e^q + e^\alpha}$
23.	$(-1)^n \cos an$	$\dfrac{(e^q + \cos a)e^q}{e^{2q} + 2e^q \cos a + 1}$
24.	$(-1)^n \sin an$	$-\dfrac{e^q \sin a}{e^{2q} + 2e^q \cos a + 1}$
25.	$(-1)^n n$	$-\dfrac{e^q}{(e^q + 1)^2}$

	$f(n),\ n = mT$	$F^*(s),\ q = sT$
26.	$\operatorname{ch} \alpha_1 n$	$\dfrac{(e^q - \operatorname{ch}\alpha_1)e^q}{e^{2q} - 2e^q \operatorname{ch}\alpha_1 + 1}$
27.	$\operatorname{sh} \alpha_1 n$	$\dfrac{e^q \operatorname{sh}\alpha_1}{e^{2q} - 2e^q \operatorname{ch}\alpha_1 + 1}$
28.	$\operatorname{sh}(\alpha_1 n + \varphi)$	$\dfrac{e^{2q}\operatorname{sh}\psi + e^q \operatorname{sh}(\alpha_1 - \psi)}{e^{2q} - 2e^q \operatorname{ch}\alpha_1 + 1}$
29.	$\dfrac{\cos an}{n!}$	$e^{e^{\frac{\cos a}{q}}}\cos\left(\dfrac{\sin a}{e^q}\right)$
30.	$\dfrac{\sin an}{n!}$	$e^{e^{\frac{\cos a}{q}}}\sin\left(\dfrac{\sin a}{e^q}\right)$
31.	$\dbinom{k}{n} e^{\alpha n}$	$\left(1 + \dfrac{e^\alpha}{e^q}\right)^k$
32.	$\displaystyle\sum_{m=1}^{n} \dfrac{1}{m}$	$\dfrac{e^q}{e^q - 1}\ln\dfrac{e^q}{e^q - 1}$
33.	$\displaystyle\sum_{m=1}^{n-1} \dfrac{1}{m!}$	$\dfrac{e^{e^{\frac{1}{q}}}}{e^q - 1}$
34.	$\dfrac{1}{n!}$	$e^{e^{\frac{1}{q}}}$
35.	$\dbinom{k}{n} = \dfrac{k!}{(k-n)!n!},\ n \le k$	$(1 + e^{-q})^k$
36.	$(-1)^n e^{\alpha^2}\dfrac{d^n e^{-\alpha^2}}{d\alpha^n} = \dfrac{H_n(\alpha)}{n!}$	$\exp\left\{-\dfrac{1 + 2\alpha e^q}{e^{2q}}\right\}$
37.	$e^\alpha \dfrac{d^n}{d\alpha^n} = \dfrac{L_n(\alpha)}{n!}$	$\dfrac{e^q}{e^q - 1}\exp\left\{-\dfrac{\alpha}{e^q - 1}\right\}$
38.	$\dfrac{1}{2^n n!}\dfrac{d^n(\alpha^2 - 1)^n}{d\alpha^n} = P_n(\alpha)$	$\dfrac{e^q}{\sqrt{e^{2q} - 2\alpha e^q + 1}}$
39.	$\sin(n\arccos\alpha) = U_n(\alpha)$	$\dfrac{e^q}{e^{2q} - 2\alpha e^q + 1}$
40.	$\cos(n\arccos\alpha) = T_n(\alpha)$	$\dfrac{e^q(e^q - \alpha)}{e^{2q} - 2\alpha e^q + 1}$

For detailed theory of Laplace and Fourier transforms, mathematical software and applications refer to (Blum 1972, Brigham 1988, D'Azzo and Houpis 1975, Elliot 1987, Kahaner et al. 1989, Korn and Korn 1968).

REFERENCES

Adams, R.A. (1975) *Sobolev Spaces*. Academic Press, New York.

Adomian, G. (1983) *Stochastic Systems*. Academic Press, New York.

Antosik, P., Mikusinski J. and R.Sikorski (1973) *Theory of Distributions*. Elsevier, Publ., Co., Amsterdam.

Aubin, J. – P. (1984) *Applied Abstract Analysis*. Wiley, New York.

Balakrishnan, A.V. (1976) *Applied Functional Analysis*. Springer – Verlag, New York.

Bateman, H. and A.Erdelyi (1953) *Higher Transcendental Functions*. V. 1, V. 2. McGraw Hill, New York.

Bateman, H. and A.Erdelyi (1955) *Higher Transcendental Functions*. V. 3. McGraw Hill, New York.

Bose, N.K. (1988) *Digital Filters: Theory and Application*. Elsevier, New York.

Brignam, E.O. (1988) *The Fast Fourier Transform and its Applications*. Prentice Hall, Englewood Cliffs, N.J.

Burkill, J.C. and H.Burkill (1970) *A Second Course in Mathematical Analysis*. Cambridge University Press, Cambridge.

Burrus, C.S. and T.W.Parks (1985) *DFT/FFT and Convolution Algorithms*. Wiley – Interscience, New York.

Butkovskiy, A.G. (1982) *Green Functions and Transfer Functions Handbook*. Ellis Hordwood, New York.

Butkovskiy, A.G. (1983) *Structural Theory of Distributed Systems*. Ellis Hordwood: John Wiley, Chichester.

Blum, E.K. (1972) *Numerical Analysis and Computation. Theory and Practice*. Addison Wesley.

Clarke, F.H. (1983) *Optimization and Nonsmooth Analysis*. Wiley – Interscience, New York.

Cramer, H. and M.R.Leadbetter (1967) *Stationary and Related Stochastic Processes. Sample Function Properties and their Applications*. John Wiley, N.J.

Curtain, R.F. and A.J.Pritchard (1978) *Infinite Dimensional Linear System Theory*. Springer – Verlag, New York.

Da Prato, G. and J.Zabczyk (1992) *Stochastic Equations in Infinite Dimensions*. Cambridge University Press, Cambridge.

D'Azzo, J.J. and C.H.Houpis (1975) *Linear Control System Analysis and Design: Conventional and Modern*. McGraw–Hill, New York.

Dowson, H.R. (1978) *Spectral Theory of Linear Operators*. Academic Press, New York.

Dunford, N. and J.Schwarz (1958) *Linear Operators. Part I: General Theory*. Wiley – Interscience, New York.

Dunford, N. and J.Schwarz (1963) *Linear Operators. Part II: Spectral Theory. Self Adjoint Operators in Hilbert Space*. Wiley – Interscience, Publishers, New York.

Dunford, N. and J.Schwarz (1971) *Linear Operator. Part III: Spectral Operators*. Wiley – Interscience, New York.

Edwards, R.E. (1965) *Functional Analysis. Theory and Application*. Holt, Rinehart and Winston, New York.

Ekeland, I. and R.Temam (1976) *Convex Analysis and Variational Problems*. Elsevier, North – Holland, Amsterdam.

Elliot, D.F. (1987) *Handbook of Digital Signal Processing, Engineering Applications*. Academic Press, San Diego.

Gajewski, H., Groger K. and K. Zacharias (1974) *Nichtlineare Operatorgleichungen und Operatordifferentialgleichungen*. Akademie – Verlag, Berlin.

Halmos, P.R. (1950) *Measure Theory*. D.Van Nostrand Co., Princeton, N.J.

Halmos, P.R. (1957) *Introduction to Hilbert Space*. Chelsa, Bronx, New York.

Halmos, P.R. (1960) *Naive Set Theory*. D.Van Nostrand Co., Princeton, N.J.

Il'in, V.A. and E.G.Posnajk (1982) *Fundamentals of Mathematical Analysis*. MIR, Moscow.

Jazwinski, A.H. (1990) *Stochastic Processes and Filtering Theory*. Academic Press, San Diego.

Kahaner, D., Moler C. and S.Nash (1989) *Numerical Methods and Software*. Prentice Hall, Englewood Cliffs, N.J.

Kelly, J. L. (1957) *General Topology*. D.Van Nostrand Co., Princeton, N.J.

Kolmogorov, A.N. (1956) *Foundations of the Theory of Probability*. (2nd English ed.) Chelsea, New York.

Korn, G.A. and T.M.Korn (1968) *Mathematical Handbook for Scientists and Engineers*. McGraw–Hill, London.

Ladas, G. and V.Lakshmikantam (1972) *Differential Equations in Abstract Spaces*. Academic Press, New York.

Lancaster, P. (1969) *Theory of Matrices*. Academic Press, New York.

Laning, J.H. (Jr.) and R.H.Battin (1956) *Random Processes in Automatic Control*. McGraw – Hill, New York.

McShane, E.J. (1974) *Stochastic Calculus and Stochastic Models*. Academic Press, New York.

Noble, B. and J.Daniel (1977) *Applied Linear Algebra*. Prentice Hall, Englewood Cliffs, N.J.

Oppenheim, A.V. and R.W.Schafer (1975) *Digital Signal Processing*. Prentice Hall, Englewood Cliffs, N.J.

Papoulis, A. (1968) *Systems and Transforms with Applications in Optics*. McGraw–Hill, New York.

Parthasarathy, K.R. (1980) *Introduction to Probability and Measure*. Academic Press, New York.

Pratt, W.K. (1978) *Digital Image Processing*. Wiley – Interscience, New York.

Pugachev, V.S. (1965) *Theory of Random Functions and its Application to Control Problems*. Pergamon Press, Oxford.

Pugachev, V.S. and I.N.Sinitsyn (1987) *Stochastic Differential Systems. Analysis and Filtering*. John Wiley, Chichester.

Rabiner, L. and B.Gold (1975) *Theory and Application of Digital Signal Processing*. Prentice Hall, Englewood Cliffs, N.J.

Reed, M. and B.Simon (1972) *Methods of Modern Mathematical Physics. 1. Functional Analysis*. Academic Press, New York.

Ringrose, I.R. (1971) *Compact Non Self Adjoint Operators*. Van Nostrand Reinhold Co., New York.

Rudin, W. (1964) *Principles of Mathematical Analysis*. McGraw – Hill, New York.

Rudin, W. (1973) *Functional Analysis*. McGraw – Hill, New York.

Schaefer, H.H. (1966) *Topological Vector Spaces*. Macmillian, New York.

Schwartz, Y.T. (1969) *Nonlinear Functional Analysis*. Gordon and Breach, New York.

Struwe, M. (1990) *Variational Methods*. Springer, New York.

Suppes, S. (1960) *Axiomatic Set Theory*. D.Van Nostrand Co., Princeton, N.J.

Tychonoff, A.N. and A.A.Samarskii (1963) *Equations of Mathematical Physics*. Pergamon Press, Oxford.

Tychonoff, A.N. and U.Ya. Arsenin (1997) *Solution of Ill–Posed Problems*. John Wiley, New York.

Volterra, V. (1913) *Leçons sur les Fonctions de Lignes*. Gauthier – Villars, Paris.

Volterra, V. and J.Pérés (1935) *Théorie General des Fonctionelles*. Gauthier – Villars, Paris.

Wilkinson, J.H. (1965) *The Algebraic Eigenvalue Problem*.   Oxford University Press, Oxford.

Xia Dao–Xing (1972) *Measure and Integration Theory on Infinite-Dimensional Spaces*. Academic Press, New York.

Yosida, K. (1965) *Functional Analysis*. Springer – Verlag, Berlin.

Zadeh, A. and C.Desoer C. (1963) *Linear System Theory. The State Space Approach*. McGraw–Hill, New York.

SUBJECT INDEX